Cyclostationarity in Communications and Signal Processing

IEEE Press
445 Hoes Lane, PO Box 1331
Piscataway, NJ 08855-1331

IEEE Communications Society, *Sponsor*

IEEE Communications Society Liaison
Jack Holzman

Technical Reviewers

Stephan P. Bruzzone
Radix Technologies, Inc., Mountainview, CA
Douglas Cochran
Arizona State University, Tempe, AZ
Bart Rice
Lockheed Missiles and Space Company, Inc., Sunnyvale, CA

Cyclostationarity in Communications and Signal Processing

Edited by

William A. Gardner

IEEE PRESS

IEEE Communications Society, *Sponsor*

The Institute of Electrical and Electronics Engineers, Inc., New York

This book may be purchased at a discount from the publisher when ordered in bulk quantities. For more information contact:

IEEE PRESS Marketing
Attn: Special Sales
P.O. Box 1331
445 Hoes Lane
Piscataway, NJ 08855-1331

Printed in the United States of America

10 9 8 7 6 5 4 3 2 1

ISBN 0-7803-1023-3
IEEE Order Number: PC0373-1

Library of Congress Cataloging-in-Publication Data

Cyclostationarity in communications and signal processing / edited by William A. Gardner.
p. cm.
Includes bibliographical references and index.
ISBN 0-7803-1023-3
1. Signal processing—Mathematics. 2. Cyclostationary signals--Mathematical models. 3. Spectral theory (Mathematics) I. Gardner, William A.
TK5102.9.C94 1994
621.382′2—dc20 93-21059
CIP

Contents

PART I

Chapter 1 **An Introduction to Cyclostationary Signals** **1**
William A. Gardner

1 Background 2
2 Fundamental Concepts, Philosophy, and Definitions 10
3 Introduction to the Principles of Second-Order (Wide-Sense) Cyclostationarity 32
4 Exploitation of Cyclostationarity 65
References 81

Chapter 2 **Higher-Order Statistics for Nonlinear Processing of Cyclostationary Signals** **91**
Chad M. Spooner

1 Introduction to Higher-Order Cyclostationarity 91
2 The Temporal Parameters of HOCS 104
3 The Spectral Parameters of HOCS 119
4 Discussion 125
5 Development of the Theory 128
6 Digital Quadrature-Amplitude Modulation 135
7 Measurement of the Parameters 140
8 Applications 151
9 Conclusions 157
10 Acknowledgments 157
Appendix: Pure Sine Waves and Temporal Cumulants 157
References 159

Chapter 3 **An Overview of Sensor Array Processing for Cyclostationary Signals** **168**
Stephan V. Schell

1 Introduction 169
2 Modeling Data from Sensor Arrays 170
3 Cyclostationarity 174
4 Spatial Filtering 178
5 Spatial and Fractionally Spaced Temporal Equalization 204
6 Direction Finding 212
7 Estimating the Number of Signals 222
8 Conclusions 227
9 Acknowledgments 227
Appendix: Moments of Cyclic Correlation Matrices 227
References 231

Chapter 4 **Polyperiodic Linear Filtering** **240**
L. E. Franks

1 Introduction 240
2 Periodic and Polyperiodic Linear Systems 241
3 Finite-Order PTV Systems 244
4 Periodic Filtering of Cyclostationary Signals 247
5 Polyperiodic Filtering of Polycyclostationary Signals 258
6 Conclusions 264
References 265

Chapter 5 **Representation, Prediction, and Identification of Cyclostationary Processes—A State-Space Approach** **267**
Sergio Bittanti, Paolo Bolzern, Giuseppe De Nicolao, and Luigi Piroddi

1 Introduction 267
2 Basics of Linear Periodic Systems 270
3 State-Space Representation of Cyclostationary Processes 274
4 Parma Representation of Cyclostationary Processes 284
5 Conclusions 291
6 Acknowledgments 292
References 292

Chapter 6 **Representation and Estimation for Periodically and Almost Periodically Correlated Random Processes** **295**

Dominique Dehay and Harry L. Hurd

1 Introduction 295
2 Spectral Theory for the Covariance 297
3 Spectral Theory for the Harmonizable Case 301
4 Process Representations 303
5 Estimation of the Coefficient Functions and their Spectral Densities 311
6 Conclusions and New Directions of Research 320
References 321

PART II

Article 1 **Joint Transmitter/Receiver Optimization for Multiuser Communications** **329**

Sumit Roy, Jian Yang, and P. Sarath Kumar

1 Introduction 329
2 Mimo System Optimization: Coordinated Users 332
3 Mimo System Optimization: Uncoordinated Users 342
4 Numerical Examples and Discussions 354
5 Conclusions 357
6 Acknowledgments 357
References 359

Article 2 **Beneficial Effects of Spectral Correlation on Synchronization** **362**

Nelson M. Blachman

1 Introduction 362
2 Spectral Analysis of the PAM Signal Itself 365
3 Spectral Analysis of $u(t)^2$ 373
4 Conclusions 387
5 Acknowledgments 388
References 388

Article 3 **Cyclostationarity-Exploiting Methods for Multipath-Channel Identification** **391**
Luciano Izzo, Antonio Napolitano, and Luigi Paura

1 Introduction 391
2 Cyclostationary Model for MCI 393
3 Cyclostationarity-Exploiting Methods for MCI 394
4 Implementation of the Methods 397
5 Simulation Results 400
6 Conclusions 413
7 Appendix 413
References 415

Article 4 **Blind Channel Identification and Equalization Using Spectral Correlation Measurements, Part I: Frequency-Domain Analysis** **417**
Zhi Ding

1 Introduction 418
2 Fundamentals of Blind Channel Equalization and Identification 419
3 Cyclostationarity of the Channel Output Signal 424
4 Parametric Channel Identification 427
5 SCD Estimation and Simulation 430
6 Conclusions 432
References 433

Article 5 **Blind Channel Identification and Equalization Using Spectral Correlation Measurements, Part II: A Time-Domain Approach** **437**
Lang Tong, Guanghan Xu, and Thomas Kailath

1 Introduction 438
2 Exploitation of Cyclostationarity in the Time Domain 439
3 Channel Identification 441
4 Blind Equalization with Antenna Arrays 450
5 Conclusions 453
6 Acknowledgments 453
References 453

Article 6 **A Review of Digital Spectral Correlation Analysis: Theory and Implementation** **455**
Randy S. Roberts, William A. Brown, and Herschel H. Loomis, Jr.

1 Introduction 455
2 Time Smoothing Algorithms 456
3 Frequency Smoothing Algorithms 470
4 On the Design of Digital Spectral Correlation Analyzers 474
5 Conclusions 477
References 478

Article 7 **On Recent Developments in Prediction Theory for Cyclostationary Processes** **480**
Abol G. Miamee

1 Introduction 478
2 Preliminaries 479
3 Discrete-Time Cyclostationary Processes 480
4 Continuous-Time Cyclostationary Processes 486
References 491

Editor's Introduction

Many conventional statistical signal-processing methods treat random signals as if they were statistically stationary, that is, as if the parameters of the underlying physical mechanisms that generate the signals do not vary with time. But for most man-made signals encountered in communication, telemetry, radar, and sonar systems, some parameters do vary periodically with time. In some cases even multiple incommensurate (not harmonically related) periodicities are involved. Examples include sinusoidal carriers in amplitude, phase, and frequency modulation systems, periodic keying of the amplitude, phase, or frequency in digital modulation systems, periodic scanning in television, facsimile, and some radar systems, and periodic motion in rotating machinery. Although in some cases these periodicities can be ignored by signal processors, such as receivers that must detect the presence of signals of interest, estimate their parameters, and/or extract their messages, in many cases there can be much to gain in terms of improvements in performance of these signal processors by recognizing and exploiting underlying periodicity. This typically requires that the random signal be modeled as *cyclostationary* or, for multiple periodicities, *polycyclostationary*, in which case the statistical parameters vary in time with single or multiple periods. Cyclostationarity also arises in signals of natural origins, because of the presence of rhythmic, seasonal, or other cyclic behavior. Examples include time-series data encountered in meteorology, climatology, atmospheric science, oceanology, astronomy, hydrology, biomedicine, and economics.

Important work on cyclostationary processes and time-series dates back over three decades, but only recently has the number of published papers in this area grown exponentially. Fueled by recent advances in applications to communications, signal processing, and time-series analysis that demonstrate substantial advantages of exploiting cyclostationarity in both design and analysis, the appetite for learning about cyclostationarity exhibited by research and development communities in areas such as wireless and cable communications, signals intelligence and covert communications, and modeling and prediction for natural systems (hydrology, climatology, meteorology, oceanology, biology/medicine, economics, etc.) has outgrown the available tutorial literature. This edited book is intended to help fill this void by presenting individual tutorial treatments of the major subtopics of cyclostationarity and by featuring selected articles that survey the latest developments in various specific areas.

The book is composed of two parts. Part I consists of six chapters, the first four of which are adapted from the four plenary lectures at the Workshop on Cyclostationary Signals, which was held August 16–18, 1992, at the Napa Valley Lodge in Yountville, California. Part II consists of seven articles. Part I is strongly tutorial and provides in-depth surveys of major areas of work. Similarly, Part II, which focuses on more specific topics, also has a tutorial survey flavor. Each of these two parts treats both theory and application.

Chapter 1 provides a historical perspective on cyclostationarity and discusses in detail both the practical and mathematical motives for studying cyclostationarity. It also treats the philosophy of aesthetics and utility that underlies alternative conceptual/mathematical frameworks within which theory and method can be developed. The latter half of the chapter surveys the theory and application of wide-sense cyclostationarity, touching on the problems of detection, modulation recognition, source-location, and extraction of highly corrupted signals, and the roles that the spectral-line generation and spectral-redundancy properties of cyclostationarity play in tackling these and other problems. This chapter provides an introduction to cyclostationary signals that serves as a foundation for the remainder of the book.

Chapter 2 provides an overview of the recently formulated theory of higher-order temporal and spectral moments and cumulants of cyclostationary time-series. It is shown that the nth-order polyperiodic cumulant of a polycyclostationary time-series is the solution to the problem of characterizing the strengths of all sine waves that are produced by multiplying n delayed versions of the time-series together, with the parts of those sine waves that result from products of sine waves that are present in lower-order factors of the nth-order product removed. Thus, the study of higher-order cumulants is motivated by a practical problem that arises in signal processing. The chapter also discusses other motivations for studying the moments and cumulants and provides a historical account of cumulants and their uses. The properties of these statistical functions that render them useful in signal processing are discussed and compared to the properties of similar statistical functions for stationary time-series. Applications of the unique signal-selectivity property of the polyperiodic cumulants to the tasks of weak-signal detection and source location are briefly described.

Chapter 3 provides an overview of sensor array processing for cyclostationary signals, focusing on adaptive spatial filtering and direction-of-arrival estimation, and presenting some new results on blind equalization and channel identification. It briefly describes many recently introduced methods and highlights their advantages and disadvantages relative to each other and to more conventional techniques that ignore cyclostationarity. Applications of cyclostationarity-exploiting methods to existing problems in array processing and to the design of new wireless communication systems are suggested.

Chapter 4 supplements the material on cyclostationary processes by reviewing the basic theory of periodically and polyperiodically time-varying linear systems. Such systems are extensively employed as filters for processing and modeling cyclostationary signals. Various input-output and state-variable descriptions together with filter structures that are appropriate for implementing the desired response characteristics in both continuous- and discrete-time are discussed. The chapter concludes with a brief discussion and some examples of polyperiodic filtering for waveform extraction.

Chapter 5 provides an overview of the state-space theory of cyclostationary processes in discrete time. The three alternative descriptions, (i) jointly periodic autocovariance functions, (ii) state-space stochastic models (Markovian representations), and (iii) autoregressive moving average models with periodic coefficients, are investigated, and connections among them are explained. Innovations representations, linear prediction, spectral factorization, and model identification are all studied and the current state of knowledge on these topics is summarized.

Chapter 6 provides a review of the spectral theory of cyclostationary (periodically and almost periodically correlated) random processes and of existing results on the consistent estimation of the Fourier coefficients of the autocorrelation function and their Fourier transforms, the spectral correlation densities. The representation of these processes in terms of sets of jointly stationary processes and in terms of unitary operators also is reviewed.

Article 1 in Part II addresses the joint transmitter/receiver optimization problem for multiuser communications and presents a coherent view of system design approaches that include different but related multiinput/multioutput models on the basis of analytical optimization. The present state of knowledge in this area is summarized, and the potential for suppression of cochannel interference that is afforded by the cyclostationarity of the signals is emphasized. The results demonstrate analytically that greatly improved cross-talk rejection is achievable when the spectral correlation property of the cyclostationary signals is properly exploited.

In Article 2 the objective is to provide insight into the nature of the self-noise that is present in the timing wave produced by a square-law synchronizer acting on a cyclostationary pulse-amplitude modulated signal and to provide a quantitative analysis of the mean square phase jitter in the timing wave. The results obtained show explicitly how the design and performance analysis of the square-law synchronizer is characterized by the spectral correlation function and the fourth-order spectral-moment function of the signal.

Article 3 provides a tutorial review of recent methods for multipath channel identification using known test signals. By exploiting the signal-selectivity properties of the cyclic autocorrelation function or the associated spectral correlation function, these methods can perform well in severely corruptive noise and interference environments. Several such identification methods are compared in terms of their performance characteristics by analysis and simulation.

Article 4 provides a brief overview of the various approaches to blind channel equalization and identification that have been reported in the literature and then explains the potential advantages to be gained by exploiting the cyclostationarity of digital-quadrature-amplitude-modulated signals. The theoretical possibility of accomplishing blind identification with the use of only second-order statistics is explained, and a frequency-domain approach is described.

The fifth article presents a time-domain approach to the blind channel equalization and identification problem. The results of simulations presented therein suggest that exploitation of second-order cyclostationarity can be an effective alternative to methods that ignore it in favor of higher-than-second-order statistics. A connection between the frequency-domain and time-domain approaches also is explained.

Article 6 reviews the theory and implementation of digital spectral correlation analysis. The performance characteristics and computational requirements of various algorithms based on either time smoothing or frequency smoothing are compared analytically, and two specific implementation studies are briefly presented.

Article 7 briefly reviews recent developments in the theory of prediction for cyclostationary processes. The fundamental role in the theory played by multivariate stationary representations of univariate cyclostationary processes is explained, and both discrete-time and continuous-time processes are considered.

The chapters in Part I and articles in Part II collectively cover a wide range of topics in the theory and application of cyclostationarity. We hope that the tutorial style of these contributions coupled with the broad survey and comprehensive reference lists they provide will make this volume instrumental in furthering progress in understanding and using cyclostationarity not only in the fields of communications and signal processing, but in all fields where cyclostationary data arises.

ACKNOWLEDGMENTS

The editor gratefully acknowledges the sponsorship of the 1992 Workshop on Cyclostationary Signals by the National Science Foundation (Grant # MIP-91-12800), the Army Research Office (Grant # DAAL03-92-G-0297), the Air Force Office of Scientific Research (Grant # F49620-92-J-0303), and the Office of Naval Research (Grant # N0001492-J-1218). This workshop motivated this volume and most of the contributions contained herein grew out of presentations made at this workshop.

PART I

Chapter 1

An Introduction to Cyclostationary Signals

William A. Gardner
Department of Electrical and Computer Engineering
University of California, Davis
Davis, CA 95616

and

Statistical Signal Processing, Inc.
6950 Yount St.
Yountville, CA 94599

This introductory chapter has five objectives. The first is to give a broad but thorough view of the status of the development of the theory and application of cyclostationary signals. This entails discussing and answering the following questions: What is cyclostationarity? How is it useful? Why publish a book on cyclostationarity? What are some of the seminal contributions to the study of cyclostationarity? and What are some of the specific motivations—both practical and mathematical—for studying cyclostationarity?

The second objective is to explain the philosophies behind the two alternative approaches to the subject: the orthodox approach based on stochastic processes and ensemble averaging and the more recently developed approach based on nonstochastic time-series and time averaging. Since some controversy regarding these two approaches is said to exist (it is more misunderstanding than it is controversy), the

---

This chapter is adapted from the opening plenary lecture at the Workshop on Cyclostationary Signals, held August 16-18, 1992 in Yountville, CA. The reference style (author(s), date(s)) is used in this chapter to help the reader put the contributions surveyed into historical perspective. In the remainder of the book, references are identified by number according to the order listed at the end of each chapter and article.

discussion here is intended to be particularly thorough, including both pragmatic and mathematical arguments and illuminating both strengths and weaknesses of each approach. The goal is to provide a sound basis for choice for everyone interested in studying cyclostationarity.

The third objective is to provide a comprehensive introduction to the principles of second-order cyclostationarity, which involve only second-order statistical moments of signals in the time and frequency domains. This treatment considers primarily discrete-time signals, and in this way it complements previous treatments by this author, which focus on continuous-time signals.

The fourth objective of this chapter is to survey applications of second-order cyclostationarity in the areas of communications and signal processing. The focus here is on exploiting the spectral redundancy and sine-wave generation properties of cyclostationary signals to perform difficult signal-processing tasks.

The fifth objective is to provide a reasonably comprehensive bibliography of work on the theory and application of cyclostationarity (which is complemented by the more focused bibliographies in subsequent chapters and articles).

Altogether, this chapter provides a foundation for the rest of the book that will help the reader to put each individual contribution into perspective and to integrate the parts into a whole reference source that not only will chart the past, but also will serve as a primary vehicle for taking us into the future.

1 BACKGROUND

1.1 What Is Cyclostationarity?

Let us begin with the most obvious question: "What is a *cyclostationary* signal?"[1] One answer is that a signal is cyclostationary of order n (in the wide sense) if and only if we can find some nth-order nonlinear transformation of the signal that will generate finite-strength additive sine-wave components, which result in spectral lines. For example, for $n = 2$, a quadratic transformation (like the squared signal or the product of the signal with a delayed version of itself, or the weighted sum of such products) will generate spectral lines. For $n = 3$ or $n = 4$, cubic or quartic transformations (i.e., sums of weighted products of 3 or 4 delayed versions of the signal) will generate spectral lines. In contrast, for stationary signals, only a spectral line at frequency zero can be generated.

Another answer to this question, which is completely equivalent to the first answer but does not appear to be so upon first encounter, is that a signal is cyclostationary of order n (in the wide sense) if and only if the time fluctuations in n spectral bands with center frequencies that sum to certain discrete nonzero values are statistically dependent in the sense that their joint nth-order moment (the infinite time average of their product in which each factor is shifted in frequency to have a center frequency of

[1] For the moment, it is not important to be specific about whether or not we conceive of a signal as a member of the ensemble of some stochastic process. This issue is addressed later. Similarly the modifier *wide sense* is also explained later, in footnote 12.

zero) is nonzero. In contrast, for stationary signals, only those bands whose center frequencies sum to zero can exhibit statistical dependence.

In fact, for a cyclostationary signal, each sum of center frequencies for which the nth-order spectral moment is nonzero is identical to the frequency of a sine wave that can be generated by putting the signal through an appropriate nth-order nonlinear transformation.

For the simplest nontrivial case, which is $n = 2$, this means that a signal $x(t)$ is cyclostationary with *cycle frequency* α if and only if at least some of its delay-product waveforms, $y(t) = x(t-\tau)x(t)$ or $z(t) = x(t-\tau)x^*(t)$ (where $(\cdot)^*$ denotes conjugation) for some delays τ, exhibit a spectral line at frequency α, and if and only if the time fluctuations in at least some pairs of spectral bands of $x(t)$, whose two center frequencies sum (for the case of $y(t)$) or difference (for the case of $z(t)$) to α, are correlated.

If not all cycle frequencies α for which a signal is cyclostationary are multiples of a single fundamental frequency (equal to the reciprocal of a fundamental period), then the signal is said to be *polycyclostationary* (although the term cyclostationary also can be used in this more general case when the distinction is not important). This means that there is more than one statistical periodicity present in the signal.

1.2 Is Cyclostationarity Useful?

Perhaps the second most obvious question an engineer would ask is, "Is the property of cyclostationarity useful?" The answer is emphatically "Yes!" Cyclostationarity can generally be exploited to enhance the accuracy and reliability of information gleaned from data sets such as measurements of corrupted signals. This enhancement is relative to the accuracy and reliability of information that can be gleaned from stationary data sets or from cyclostationary data sets that are treated as if they were stationary. Such information includes the following:

1. A decision as to the presence or absence of a random signal, or about the number of random signals present, with a particular modulation type in a data set that also contains background noise and other modulated signals,
2. A classification of multiple received signals present in a noisy data set according to their modulation types,
3. An estimate of a signal parameter, such as carrier phase, pulse timing, or direction of arrival, based on a noise-and-interference-corrupted data set,
4. An estimate of an analog or digital message being communicated by a signal over a channel corrupted by noise, interference, and distortion,
5. A prediction of a future value of a random signal,
6. An estimate of the input-output relation of a linear or nonlinear system based on measurements of the system's response to random excitation,
7. An estimate of the degree of causality between two data sets, and
8. An estimate of the parameters of a model for a data set.

1.3 Why Publish a Book on Cyclostationarity?

The next question we should consider is "Why publish a book on cyclostationarity?" Some of the primary reasons are

1. There is a growing awareness in signal processing and communications communities that the cyclostationarity inherent in many man-made random signals and some signals of natural origins (that were previously modeled as stationary) must be properly recognized and modeled if analyses of systems involving such signals are to properly reflect actual behavior;
2. There is a growing awareness of the potential for considerable enhancement of performance of signal-processing algorithms by recognizing and exploiting cyclostationarity in the design process rather than ignoring it by treating signals as if they were stationary;
3. There is a growing awareness by theoreticians that cyclostationary processes are, in many ways, much more than a trivial variation on stationary processes and do, therefore, merit their attention to further develop and refine the theory of these processes;
4. There is a perception by engineers and scientists that cyclostationary processes are much more than a trivial variation on stationary processes and do, therefore, merit their effort to retrain—to expand their theoretical background (their analytical/conceptual "tool boxes") from stationary to cyclostationary processes; and
5. Technological advances, which enable the implementation of increasingly sophisticated signal-processing algorithms, have made the exploitation of cyclostationarity more viable in practice.

We have important work on cyclostationary processes dating back twenty to thirty years (Bennett, 1958; Gladyshev, 1961; Brelsford, 1967; Franks, 1969; Hurd, 1969; Gardner, 1972) and the author's research group at the University of California, Davis, has contributed for the last twenty years. Also, there have been relatively isolated contributions from many others to the development of this subject over the last twenty years. However, the growth in the number of research papers has recently accelerated, and it is only in the last five years that research groups, journal editors, and program directors at funding agencies have shown real interest. The accelerated growth in research activity is illustrated by the histogram of the number of papers on cyclostationarity published per two-year period that is shown in Fig. 1.[2]

[2]The statistics in this graph were compiled by the author using a comprehensive bibliography that he has created over the last five years using his personal files, computerized literature searches, and the assistance of colleagues and students, most notably L. Paura, C. M. Spooner, and K. Vokurka.

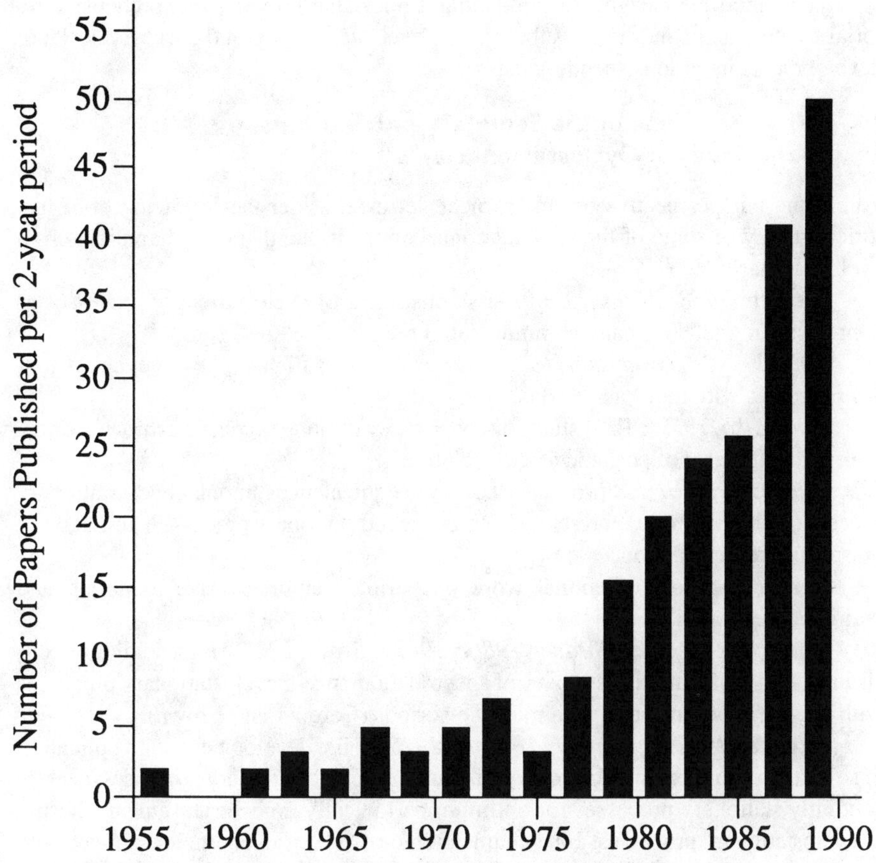

Figure 1: Histogram of papers on cyclostationarity.

Considering the following indicators, it appears that a "critical mass" of interest has been reached and, as a result, that research activity will undergo explosive growth:

1. Acceleration in production of research papers on cyclostationarity;
2. Interest of the National Science Foundation, Army Research Office, Air Force Office of Scientific Research, and Office of Naval Research in supporting the recent workshop on cyclostationarity;
3. Interest demonstrated by the participants of the recent workshop on cyclostationarity;
4. Recent increases in both industrial and government funding of research on cyclostationarity.

Thus, the time is right for publishing a book that provides comprehensive tutorial treatments of the major subtopics of cyclostationarity and surveys the latest developments in various specific areas.

1.4 What Are Some of the Seminal Contributions to the Study of Cyclostationarity?

To expand our perspective on this subject, let us consider the following brief historical survey of some of the seminal contributions to the theory and application of cyclostationarity:[3]

(*Bennett, 1958; Franks, 1969*): Establishment of cyclostationary processes as appropriate models for many communications signals.

(*Jacobs, 1958; Gladyshev, 1963; Gardner, 1978*): First studies of cyclostationary processes with multiple periods.

(*Gudzenko, 1959*): First study of consistency of nonparametric estimates of the Fourier coefficients of periodic autocorrelations.

(*Gladyshev, 1961 and 1963*): Discovery of equivalences among a cyclostationary process (with one period) and several vector-valued stationary processes. Initial work on spectral representation.

(*Brelsford, 1967*): Seminal work on periodic autoregressive modeling and periodic linear prediction.

(*Hurd, 1969, 1989a; Gardner, 1986c, 1987a; Brown 1987*): First studies of consistency of nonparametric estimates of spectral moments of cyclostationary processes with one period (Hurd) and with multiple periods (Gardner and Brown).

(*Gardner, 1972; Gardner and Franks, 1975*): First development and application of several series representations of continuous-time cyclostationary processes in terms of jointly stationary processes for optimum periodically time-variant linear filtering of cyclostationary processes. First characterization of Fourier coefficients of periodic autocorrelations and periodic spectra (the cyclic autocorrelations and cyclic spectra) as crosscorrelations and cross-spectra of frequency-shifted versions of the process.

(*Rootenberg and Ghozati, 1977, 1978; Bittanti 1987; Bittanti and DeNicolao, 1993*): First efforts to develop the Gauss-Markov theory of cyclostationary processes; formulation and partial solution of the cyclospectral factorization problem.

(*Pagano, 1978*): Development of equivalence between univariate periodic AR modeling and multivariate constant AR modeling.

(*Miamee and Salehi, 1980*): Extension—from stationary to cyclostationary processes—of the Wold-Cramér decomposition of a process (and its spectrum) into regular (continuous) and singular (discrete) components.

(*Nedoma, 1963; Boyles and Gardner, 1983*): First formulation and development of cycloergodicity for cyclostationary processes with single (Nedoma) and multiple (Boyles and Gardner) periods.

[3]Contributions from the untranslated Russian literature are not included here, but it is mentioned that several Russian authors, most notably Ya. P. Dragan, have published a substantial amount on cyclostationarity.

(*Gardner, 1985*): First general treatise on cyclostationary processes and their applications to signal processing and communications (1 book chapter).

(*Gardner, 1986b, 1987a*): First formulation and development of the nonstochastic statistical theory of cyclostationary time-series and its applications to signal processing and communications (6 book chapters).

(*Gardner, 1987a; Brown, 1987; Chen, 1989; Agee et al., 1990; Schell, 1990; Spooner, 1992*): First studies of the exploitability of the separability of individual-signal contributions to cyclic temporal and spectral moments (of second order) of multiple interfering signals for the problems of detection, modulation recognition, time-delay estimation, blind-adaptive spatial filtering, and high-resolution direction finding. Discovery that spectrally overlapping signals can be separated with linear temporal processing by exploiting spectral redundancy.

(*Gardner and Spooner, 1992b; Spooner and Gardner, 1992a, b; Spooner, 1992*): First formulation and development of the temporal and spectral moment and cumulant theory of cyclostationary time of order series $n > 2$.

1.5 What about Terminology?

A few words about terminology are in order. The first term given to this class of processes is the term *cyclostationary*, which was introduced by Bennett (1958), who also introduced the term *cycloergodic*. Other terms used include *periodically stationary*, *periodically nonstationary*, and *periodically correlated*. This last term is appropriate only for second-order (wide-sense) cyclostationarity, whereas the preceding three terms admit the modifiers wide-sense, nth-order, and strict-sense, and are, therefore, more general. The most commonly used term is *cyclostationary*. When multiple periodicities exist, this term is modified to *polycyclostationary*, although the terms *almost cyclostationary* and *almost periodically correlated* are used also.

1.6 What Are Some of the Specific Motivations for Studying Cyclostationarity?

There is a great deal of motivation for studying cyclostationarity. Let us consider first some of the practical motives and then some of the mathematical motives and, while we are at it, we can recognize many of the existing contributions to the study of cyclostationarity. The practical motives cited here are specified in terms of a series of facts.

Fact 1: Cyclostationary models, such as PAR (periodic autoregressive), PMA (periodic moving average), and PARMA (periodic autoregressive moving average), can be more parsimonious—better fit with fewer parameters—than stationary models (AR, MA, and ARMA) are. This has been illustrated with real data from

- climatology/meteorology (Brelsford, 1967; Hasselmann and Barnett, 1981; Barnett, 1983; Barnett et al., 1984; Johnson et al., 1985)
- hydrology (Salas, 1972; Salas and Smith, 1980; Vecchia, 1983, 1985; Thompstone et al., 1985; Obeysekera and Salas, 1986; McLeod et al., 1987; Bartolini et al., 1988)

- medicine/biology (Newton, 1982)
- oceanology (Dragan and Yavorskii, 1982; Dragan et al., 1984, 1987)
- economics (Parzen and Pagano, 1979).

Fact 2: Periodic prediction of cyclostationary processes can be done (and periodic causality between cyclostationary processes can be found) when time-invariant prediction is not possible or is inferior (and time-invariant causality is not found or is weaker). Examples are given in Section 2.

Fact 3: Spectrally overlapping cyclostationary signals can never be separated using time-invariant linear filters (e.g., optimum filters of the Wiener and Kalman type for stationary models of the cyclostationary signals). But they can possibly be separated using periodic filters that exploit spectral redundancy. This has been demonstrated for PAM (pulse-amplitude modulation), digital QAM (quadrature-amplitude modulation), AM (amplitude modulation), ASK (amplitude-shift-keying), and PSK (phase-shift-keying) signals (Brown, 1987; Gardner, 1987a; Gardner and Brown, 1989; Gardner and Venkataraman, 1990; Reed and Hsia, 1990; Petersen, 1992; Gardner, 1993).

Fact 4: The biases and variances of parameter estimators (e.g., for TDOA (time-difference-of-arrival), FDOA (frequency-difference-of-arrival), and AOA (angle-of-arrival) of propagating waves) can be much lower, especially for multiple interfering signals, when algorithms that exploit the signal selectivity associated with cyclostationarity (rather than ignore it by treating the signals as if they were stationary) are used. This has been demonstrated for various types of communications signals (Gardner, 1987a, 1988a, 1990a; Gardner and Chen, 1988, 1992; Chen 1989; Chen and Gardner, 1992; Schell and Gardner 1989, 1990a,b,c, 1991, 1992, 1993a; Schell, 1990; Gardner and Spooner, 1993; Izzo et al., 1989, 1990, 1992; Xu and Kailath, 1992).

Fact 5: For the design and analysis of systems that synchronize local digital clocks and sine-wave generators to the frequencies and phases of periodicities embedded in received communications and telemetry signals, the property of cyclostationarity is crucial (Franks, 1980; Franks and Bubrouski, 1974; Moeneclaey, 1982, 1983, 1984; Gardner 1986a).

Fact 6: For the design of algorithms that blindly adapt sensor arrays to perform spatial filtering (for beam/null steering and/or mitigation of multipath fading effects), exploitation of signal selectivity associated with cyclostationarity has proven to be extremely powerful (Agee et al., 1987, 1988, 1990; Schell and Gardner, 1990a; Gardner 1990a) and application to multiuser wireless communications appears to be promising (Gardner et al., 1992; Schell et al., 1993).

Fact 7: For the design of algorithms that adapt channel equalizers to remove intersymbol interference in digital communication systems, exploitation of the phase information contained in second-order cyclostationary statistics of the channel output enables blind adaptation without the use of higher-order statistics (cf. Chapter 3 and Articles 4 and 5 in this volume).

Fact 8: For radio-signal analysis, including detection, classification, modulation recognition, source location, etc., the cyclic spectrum analyzer and related algorithms that exploit cyclostationarity have proven to be ideally suited (Gardner, 1985, 1986b,c, 1987a,b, 1988b,c, 1990a,c, 1991a; Gardner et al., 1987; Brown, 1987; Roberts, 1989; Roberts et al., 1991; Brown and Loomis, 1992; Spooner and Gardner, 1991, 1992a,b; Gardner and Spooner, 1990, 1992a; Spooner, 1992).

Fact 9: For the design and analysis of communications systems that accommodate unintentional nonlinearities that inadvertently generate spectral lines from modulated message signals, the property of cyclostationarity is crucial (Campbell et al., 1983; Albuquerque et al., 1984).

Fact 10: For acoustic-noise analysis for rotating machinery, the cyclic spectrum analyzer holds promise for improved diagnosis of machine wear (e.g., in ground, air, and water vehicles, and hydroelectric plants) and for detection, classification, and location of cyclostationary noise sources (e.g., submarines) (Sherman, 1992).

Fact 11: Many statistical inference and decision problems involving multiple interfering cyclostationary signals in noise can exploit the cyclostationarity to great advantage because of the inherent noise-tolerance and separability of the cyclic features in the signals (Gardner, 1987a, 1990a, 1991a, 1992).

Let us now consider some of the mathematical motives for studying cyclostationarity. Cyclostationary processes (including one or more periods), as a subclass of nonstationary processes, have more in common with stationary processes than do other subclasses of nonstationary processes. The common structure shared by cyclostationary processes suggests (and in some ways this has already been proven) that important theorems and special theories for stationary processes can be extended and/or generalized, and that important theorems for generally nonstationary processes can be specialized, to cyclostationary processes. This potential for mathematical progress, coupled with the increasingly recognized importance of cyclostationarity to practical problems, provides strong motivation for mathematicians to study these processes.

A few examples of important theorems/theories for stationary (or nonstationary) processes that should be—or have been—extended/generalized (or specialized) are given here (consult the key given below[4]).

Topic 1: †† *Wiener-Khinchin and Shiryaev-Kolmogorov theorems relating temporal and spectral moments and cumulants* (Gardner 1986b, 1987a, 1990c; Gardner and Spooner, 1990; Spooner and Gardner, 1992a; Spooner, 1992; Chapter 2 in this volume)

Topic 2: † *Spectral representation theory* (e.g., for harmonizable processes) (Gladyshev, 1963; Hurd, 1974a, 1989b, 1991; Honda, 1982; Rao and Chang, 1988; Chapter 6 in this volume)

---

[4]* Some progress has been made for cyclostationary processes with one period.
† Substantial progress has been made for cyclostationary processes with one period.
** Some progress has been made for polycyclostationary processes with multiple periods.
†† Substantial progress has been made for polycyclostationary processes with multiple periods.

Topic 3: † *Wold-Cramér theorem on decomposition of a process into singular and regular components and decomposition of its spectrum into discrete and continuous components* (Miamee and Salehi, 1980; article 7 in this volume)

Topic 4: ** *Wiener and Kalman smoothing, filtering, and prediction theory* (Gardner, 1972; Gardner and Franks, 1975; Gardner 1985, 1987a, 1993; Brown, 1987; Gardner and Brown, 1989; Chapter 5 and Article 1 in this volume)

Topic 5: * *Theory of AR, MA, and ARMA models, linear prediction, and parametric spectral estimation* (Brelsford, 1967; Pagano, 1978; Miamee and Salehi, 1980; Tiao and Grupe, 1980; Sakai, 1982, 1983, 1990, 1991; Pourahmadi and Salehi, 1983; Vecchia, 1985; Obeysekera and Salas, 1986; Li and Hui, 1988; Anderson and Vecchia, 1992; Chapter 5 and Article 7 in this volume)

Topic 6: * *Theory of fast algorithms for linear prediction and filtering* (Sakai, 1982, 1983)

Topic 7: * *Markov theory of state-space representations* (Rootenberg and Ghozati, 1977, 1978; Bittanti, 1987; Bittanti and DeNicolao, 1993; Chapter 5 in this volume)

Topic 8: * *Birkhoff Ergodic Theorem and associated ergodic theory* (Nedoma, 1963; Blum and Hansen, 1966; Boyles and Gardner, 1983; Honda, 1990)

Topic 9: ** *Theory of consistent nonparametric estimation of temporal and spectral moments and cumulants* (Gudzenko, 1959; Hurd, 1969, 1989a; Alekseev, 1988, 1991; Gardner, 1985, 1986c, 1987a, 1991b; Dehay, 1991; Spooner, 1992; Spooner and Gardner, 1991, 1992c; Genossar et al., 1993; Hurd and Leskow, 1992a, 1992b; Chapter 2 in this volume)

Topic 10: †† *Theory of higher-order statistics* (*temporal and spectral moments and cumulants*) (Gardner, 1990c; Gardner and Spooner, 1990, 1992b; Spooner and Gardner, 1992a,b; Spooner, 1992; Chapter 2 in this volume)

2 FUNDAMENTAL CONCEPTS, PHILOSOPHY, AND DEFINITIONS

What do we need to accomplish here? We need a general description of the types of signals that motivate the work being done under the name of cyclostationarity; we need a generally useful definition of the signal property called cyclostationarity, and we need to understand what mathematical/conceptual frameworks are particularly useful for formulating and solving practical problems involving cyclostationary signals, particularly those arising in communication system design and analysis and more general signal processing.

We shall see that an empirically motivated approach to accomplishing these things leads naturally to a probabilistic conceptual framework. However, this framework is distinct from that of stochastic processes in that it does not involve the concept of an ensemble of random samples.

2.1 Signal Types

The types of signals of primary interest here are those normally encountered in communication systems. These signals are typically unpredictable and occur over long periods of time. That is, they are in some sense random (this does not necessarily mean stochastic) and they are persistent rather than transient. These signals also typically originate from physical sources with parameters that are either time-invariant, periodic, or polyperiodic. Thus, the characteristics of the physical signal-generating mechanism vary polyperiodically with time (this includes as special cases periodic variation and time invariance). In some cases the signal-generating mechanism can be decomposed into more elementary signal generators whose outputs are mixed together to form the signal of interest. Some of these more elementary signal generators can have characteristics all of which are time-invariant, thereby giving rise to stationary random signals. Other elementary signals can be simply periodic or polyperiodic functions of time. Thus, the signals of interest often consist of combinations—additive, multiplicative, and other types—of stationary and polyperiodic signals, and are called *polycyclostationary* signals.

2.2 Operational Definition of Polycyclostationarity

What physical evidence in a signal reveals that there is polyperiodic time variation present in its generating mechanism? Fortunately, there is a unique unambiguous answer to this question that appears to be adequate for the general purpose of designing and analyzing signal-processing algorithms that exploit or in some way involve the underlying polyperiodic time variation: We shall say that polyperiodic time variation exists in the generating mechanism of a signal if and only[5] if it is possible to generate finite-amplitude additive polyperiodic components from the signal by passing it through some appropriate nonlinear transformation that is time-invariant and stable. We can take this as an operational definition of polycyclostationarity.

2.3 Operational Origin of Probabilistic Models

There are two particularly interesting ways to characterize an adequately large class of all nonlinear transformations that could potentially generate polyperiodic components from polycyclostationary signals. One way is to require that all transformations of interest be representable in a generalized[6] Volterra series (which is a multivariate Taylor series with a continuum of variables indexed by time). Thus, for a signal x

[5]One can conceive of polyperiodic variation in a signal generator that is unobservable in the signal. This is analogous to the concept of unobservability in system theory. Since the focus here is on modeling signals in which there is physical evidence of underlying polyperiodic variation, we are not interested in unobservable polyperiodic variation.

[6]"Generalized" in the sense that the transformations are not constrained to be causal.

and a transformation $g(\cdot)$, we have

$$g(x) = \sum_n \int_{\tau_n} k_n(\tau_n) L_x(t, \tau_n)_n \, d\tau,$$

where $L_x(t, \tau_n)_n$ is the nth-order delay product

$$L_x(t, \tau_n)_n = \prod_{j=1}^{n} x(t + \tau_j),$$

and $k_n(\tau_n)$ is the nth-order Volterra kernel. For example, all transformations $g(\cdot)$ that are continuous and have finite memory admit a convergent Volterra series representation.

Another way is to require that all transformations be representable as a convolution with a finite product of Dirac deltas:

$$\begin{aligned} g(x) &= \int g(y_1, y_2, \ldots, y_n) \prod_{j=1}^{n} \delta\left[y_j - x(t + \tau_j)\right] dy_j \\ &= g[x(t + \tau_1), x(t + \tau_2), \ldots, x(t + \tau_n)], \end{aligned}$$

which can be reexpressed as a Riemann-Stieltjes integral

$$g(x) = \int g(\mathbf{y}_n) d^n I_x(t, \tau_n, \mathbf{y}_n)_n$$

where

$$I_x(t, \tau_n, \mathbf{y}_n)_n \triangleq \prod_{j=1}^{n} I\left[y_j - x_j(t + \tau_j)\right]$$

and $I(\cdot)$ is the indicator function

$$I(z) = \begin{cases} 1, & z > 0 \\ 0, & z \leq 0, \end{cases}$$

for which $dI(z) = \delta(z)dz$. For example, all continuous transformations with finite discrete memory admit this representation. Although this representation of $g(\cdot)$ in terms of itself appears to accomplish nothing, we shall see that it is very useful for our purpose here.

The operation, denoted by $P\{\cdot\}$, for extracting the additive polyperiodic component of a signal is linear (as explained subsequently). Therefore, the polyperiodic component of a weighted sum of signals is the weighted sum of polyperiodic components. Consequently, the polyperiodic component of the first type of transformed signal is given by

$$P\{g(x)\} = \sum_n \int_{\tau_n} k_n(\tau_n) P\{L_x(t, \tau_n)_n\} \, d\tau_n$$

and that of the second type of transformed signal is given by

$$P\{g(x)\} = \int g(y_n) d^n P\{I_x(t, \tau_n, y_n)_n\}.$$

As explained later on, the function $P\{L_x(t, \tau_n)_n\}$ is mathematically equivalent to the joint nth-order moment of n random variables $X_j = x(t + \tau_j)$, $j = 1, 2, \ldots, n$, and the function $P\{I_x(t, \tau_n, y_n)_n\}$ is mathematically equivalent to the joint nth-order probability distribution for the same N random variables, and these equivalences reveal that the polyperiodic component extraction operation $P\{\cdot\}$ is mathematically equivalent to the probabilistic expectation operation. In fact, choosing

$$g(x) = L_x(t, \tau_n)_n$$

in the preceding equation yields

$$P\{L_x(t, \tau_n)_n\} = \int \left[\prod_{j=1}^{n} y_j \right] d^n P\{I_x(t, \tau_n, y_n)_n\}$$

which is the standard formula from probability theory for the nth-order moment in terms of the nth-order probability distribution.

We see, then, that the physical evidence in a signal of polyperiodic time-variation in the generating mechanism of the signal is completely characterized by the signal's temporal moment functions or its temporal probability distribution functions. That is, the polyperiodic component of the delay product of the signal is a temporal moment function and the polyperiodic component of the indicator product is a temporal probability distribution. Hence, we are led naturally by a practically motivated inquiry into the problem of mathematically characterizing physical evidence of polyperiodic time-variation in an unpredictable signal, to a probabilistic description of the signal. Moreover, as explained later on, these moments and distributions are identical to those corresponding to a polycyclostationary stochastic process with appropriate ergodic properties (called *cycloergodicity*), in which case the signal $x(t)$ can be *interpreted* as a sample path (one ensemble member) of the stochastic process. However, in spite of this equivalence between the mathematical model of polyperiodic time variation underlying a signal and a corresponding stochastic process model, the conceptual framework of a stochastic process and its associated ensemble is fundamentally different from the conceptual framework of a single signal that is characterized by all the polyperiodic components that can be generated from it using nonlinear transformations. It is the latter conceptual framework, not the former, that is motivated by the desire to design and analyze signal processors that exploit the generatable polyperiodic components.

2.4 Stochastic vs. Nonstochastic Operational Models

We need to understand the similarities and differences between the stochastic-process approach and the nonstochastic signal (or time-series) approach to conceptualizing, defining, and modeling stationary (S), cyclostationary (CS), and polycyclostationary (PCS) signals, and to developing theory—like the classical theory of statistical inference and decision—to guide the practice of designing and analyzing signal-processing algorithms.

The nonstochastic time-series approach to this subject has not gained the wide level of acceptance that the stochastic-process approach enjoys, particularly for stationary processes. This is believed to be primarily a result of the limited exposure that the time-series approach has received. The aim in recent work (Gardner, 1987a) on developing the time-series approach has been to bring the aesthetics of mathematics and the utility of engineering pragmatism together to produce elegant problem solving.[7] The treatment presented here aims at the same target: the focusing of attention on important concepts for mathematicians who care about the applicability of the mathematics of polycyclostationary signals and for engineers who seek more than a superficial understanding of not only the "how" but also the "why" of polycyclostationary signal processing.

However, before embarking on a discussion of specific mathematical definitions and properties, questions of mathematical existence, and unsolved mathematical problems, a brief summary of the essence of the differences and similarities, from an operational standpoint, of the two alternative approaches is presented.

When properly restricted to appropriate domains of definition (i.e., requiring stochastic process models to exhibit certain ergodic properties and requiring time-series models to exhibit certain regularity properties that guarantee the existence of infinitely long time averages), either approach can be used to obtain the same results in deriving signal-processing algorithms and analyzing their performances (Gardner, 1990a). However, it is not guaranteed that any particular user will in fact obtain the same results regardless of the approach used, because each approach has its own unique conceptual attributes. Thus, it is argued here that the most proficient problem solvers need to understand how to use both approaches.[8] Some problems may naturally fit one approach or the other, and some other problems may benefit from application of both approaches. For example, sometimes it is easier to see how to carry out a particular mathematical calculation using one or the other approach (the stochastic-process approach seems to be favored here), and sometimes it is easier

[7] The title of the book (Gardner, 1987a), which seems to have sparked some controversy, *Statistical Spectral Analysis: A Nonprobabilistic Theory*, can be misleading since it is shown in this book that an empirically motivated inquiry into the problem of quantifying the average behavior of spectral measurements leads naturally to a probabilistic theory. Since this probabilistic theory is nonstochastic (it involves only time averages, not ensemble averages), the title could have been *Statistical Spectral Analysis: A Nonstochastic Theory*. Nevertheless, the majority of the concepts and methods developed in the book are not only nonstochastic, they are indeed nonprobabilistic, and a primary goal of the book is to show that in an empirically motivated development of the fundamental concepts and methods of statistical spectral analysis, probability does not play a seminal role. It does play an important role in the mechanics of quantifying average behavior, but it plays no role in conceptualizing the objectives and methods (parametric and nonparametric) of statistical spectral analysis of single time-series.

[8] It is curious that some followers of the stochastic-process approach insist that the alternative approach is of no value or, worse yet, has negative value. Perhaps the stochastic process faith should be formally recognized as a religion.

to relate the mathematics to the real-world problem at hand using one or the other approach (the nonstochastic time-series approach seems to be favored here with regard to many of the applications discussed in this book[9]).

Mathematicians have, for the most part, chosen the stochastic-process framework for their work because it is apparently more amenable to deep mathematical treatment. Statisticians have, for the most part, chosen the approach to statistical inference and decision that is based on stochastic processes because it does naturally fit the problem of making inferences about a total population on the basis of limited "random" samples from the population, which is the statistician's classical problem. The concept of a population, or ensemble, also naturally fits a number of situations in communications engineering and signal processing or time-series analysis for engineering purposes; however, there are many other engineering (and science) problems involving time-series data where the ensemble concept is fictitious, irrelevant, or otherwise inappropriate. In these cases, users often force an application of the theory of stochastic processes onto their real-world problem because they have not learned that there is a viable alternative for statistical inference and decision. This can lead to substantial confusion and less effective engineering.

2.5 Nonstochastic Statistical Inference and Decision

Let us briefly consider how a theory of statistical inference and decision can be based on the concept of a single time-series without reference to an ensemble. Many—but by no means all—real-world problems in engineering and science involve time-series data for which no population exists; that is, for which replication of the "experiment" is impossible or impractical. However, many of these time-series arise from physical phenomena that can be considered to be unchanging in their basic nature for a very long time. In such cases, conceptually idealizing this time-invariance by extending the length of time without bound enables us to conceive of a model that is derivable from the data in the limit as the amount of data used for measuring the parameters of the model approaches infinity. This leads us to the concept of a fraction-of-time (FOT) probability model that is free from the abstract concept of a population. For example, the FOT probability that a time-series exceeds some specified level is defined to be the fraction of time that this event occurs over the life of the time-series.

Once we have accepted the idea of an infinitely long time-series with an FOT probability model, we can develop a theory of statistical inference and decision that is isomorphic to the theory for stationary stochastic processes. This was briefly pointed out in (Wold, 1948), developed in (Hofstetter, 1964), and extended from stationary to cyclostationary and polycyclostationary time-series in (Gardner, 1987a), (Gardner and Brown, 1991). But one might ask what it is that motivates the development of such a FOT probability theory. One of the answers to this question is analogous to that which motivates the theory that is based on the concept of a population: We want to make inferences about the physical phenomenon that gave rise to the observed

[9]A compelling example of this is the novel derivation of the cumulant in the study of higher-order cyclostationarity presented in Chapter 2.

time-series. To the extent that this phenomenon is characterized by the FOT model for the time-series $x(t)$, (i.e., the set of joint FOT probability distributions for all finite sets of time translates $\{x(t+t_i) : i = 1, \ldots, n\}$ for all n translations τ_i and all natural numbers n), we can interpret our objective as that of making inferences about the infinitely long time-series or its generating mechanism on the basis of finite-length observations. We can use the FOT probability model to calculate bias, variance, and confidence intervals for parameter estimates, and we can calculate probabilities of correct and incorrect decisions. We also can formulate and solve optimization problems.

2.6 A Historical Perspective

The stochastic-process approach (to the exclusion of the nonstochastic time-series approach) is currently the orthodox approach because this is the approach that dominated for sixty years in mathematics and statistics and it is, therefore, the approach in terms of which the theory of statistical inference and decision has been formulated and is taught. It does not follow that the stochastic-process approach is orthodox because it is always the superior approach. This last point can be illustrated with a brief history of statistical inference and decision in communications engineering.

Why have communications engineers focused on using theoretical measures of performance that average over an ensemble of signals and/or noises? Because they have wanted to design systems that would perform well on the average over the ensemble, and because mathematicians and statisticians had developed a powerful theory of statistical inference and decision that was based on ensemble averages, and because probability theory itself is an immensely powerful conceptual tool, and few engineers have realized that probability theory can be based entirely on time-averages. But why then have communications engineers focused almost exclusively on measuring system performance in practice by averaging over time for a single system? Because of economics (the high cost of making measurements on many systems) and because they also want each system to perform well on the average over time.

In order to match the theory based on ensembles of data to the practice based on a single record of data, they invoked the concept of ergodicity. That is, they agreed to use stationary stochastic-process models that were ergodic so that the mathematically calculated expected values (ensemble averages) would equal the measured time averages (in the limit as averaging time approaches infinity).

Unfortunately, however, the logic seems to have stopped at this point. It apparently was not recognized (except by too few to make a difference) that once consideration was restricted to ergodic stationary models, the stochastic process and its associated ensemble could be dispensed with because a completely equivalent theory of statistical inference and decision that was based entirely on time-averages over a single record of data could be used (Hofstetter, 1964). Any calculations made using a model based on the time-average theory could be applied to any one member of an ensemble if one so desired because the arguments that justify the ergodic stochastic-process model also guarantee that the time-average for one ensemble member will be the same (with probability one) as the time-average for any other ensemble member.

Because the time-average framework is more conceptually straightforward for application to problems where time-average performance is of primary concern, it is a more natural choice; but because of history and inertia, it may never gain its rightful place in engineering. This is even more likely to be the case when the utility of nonergodic stochastic processes is taken into account. For example, whenever transient behavior is of interest, ergodic models are ruled out, because all transient behavior is lost in an infinitely long time-average. Thus, to counter the conceptual simplicity and realism offered by the time-average approach, the stochastic-process approach offers the advantage of more general applicability.

Nevertheless, there is a special class of signals that includes more than just those that can be modeled as stationary ergodic processes, for which there is a compelling argument, which has only recently surfaced, to adopt an alternative nonstochastic approach. And these are the signals that are appropriately modeled as polycyclostationary time-series. As explained earlier here, the use of time-averages to extract additive polyperiodic components from nonlinear transformations of these signals leads naturally to a probability theory based entirely on time averaging. Let us now consider in some detail these two alternative approaches to conceptualizing, defining, and modeling signals.

2.7 Dual Theoretical Frameworks

The concepts and definitions presented here apply equally well to continuous-time and discrete-time signals. We need only choose either the continuous-time-averaging operation

$$\langle\cdot\rangle \triangleq \lim_{Z\to\infty} \frac{1}{2Z}\int_{-Z}^{Z} (\cdot)\, dt$$

or its discrete-time counterpart

$$\langle\cdot\rangle \triangleq \lim_{Z\to\infty} \frac{1}{2Z+1}\sum_{t=-Z}^{Z} (\cdot).$$

We consider first stochastic processes, and then we consider nonstochastic time-series. Let $X(t)$ be a real-valued stochastic process on the real line $-\infty < t < \infty$, with measure μ on the probability space Ω. Consider the event indicator

$$I[x - X(t)] \triangleq \begin{cases} 1, & X(t) < x \\ 0, & X(t) \geq x. \end{cases}$$

The expected value of this event indicator is the probability distribution (PD) function for the random variable $X(t)$,

$$F_{X(t)}(x) \triangleq Prob\{X(t) < x\} = E\{I[x - X(t)]\}$$

where the expectation operation $E\{\cdot\}$ is defined by

$$E\{H\} \triangleq \int_{\Omega} H(\omega)\, d\mu(\omega)$$

for any random variable H defined on Ω. Therefore, the joint PD function for the set of random variables

$$\boldsymbol{X}(t) \triangleq \{X(t+t_1), \ldots, X(t+t_n)\},$$

where t is the time-translation parameter, is given by the expectation

$$F_{\boldsymbol{X}(t)}(\boldsymbol{x}) = E\left\{\prod_{j=1}^{n} I\left[x - X(t+t_j)\right]\right\},$$

and the joint probability density (Pd) function is

$$f_{\boldsymbol{X}(t)}(\boldsymbol{x}) = \frac{\partial^n}{\partial x_1 \ldots \partial x_n} F_{\boldsymbol{X}(t)}(\boldsymbol{x}),$$

which contains Dirac deltas when the PD contains step discontinuities. We have the following theorem from probability theory.

Fundamental Theorem of Expectation

For any nonrandom function $g(\cdot)$ for which $E\{g[\boldsymbol{X}(t)]\}$ exists, we have

$$E\{g[\boldsymbol{X}(t)]\} \triangleq \int y f_{g[\boldsymbol{X}(t)]}(y)\, dy$$

$$= \int g(\boldsymbol{x}) F_{\boldsymbol{X}(t)}(\boldsymbol{x})\, d\boldsymbol{x}$$

That is, the Pd for $g[\boldsymbol{X}(t)]$ need not be found from the Pd of $\boldsymbol{X}(t)$ in order to evaluate the expected value of this random variable. This theorem can be used to verify that the PD is indeed equal to the expected value of the event indicator by letting

$$g[\boldsymbol{X}(t)] = \prod_{j=1}^{n} I\left[x_j - X(t+t_j)\right]$$

to obtain

$$E\{g[\boldsymbol{X}(t)]\} = \int \prod_{j=1}^{n} I\left[x_j - z_j\right] f_{\boldsymbol{X}(t)}(\boldsymbol{z})\, d\boldsymbol{z}$$

$$= \int_{-\infty}^{\boldsymbol{x}} f_{\boldsymbol{X}(t)}(\boldsymbol{z})\, d\boldsymbol{z}$$

$$= F_{\boldsymbol{X}(t)}(\boldsymbol{x}).$$

Let us now consider time-series. Let $x(t)$ be a well-behaved[10] real-valued time-series (a nonstochastic real-valued function) on the real line $-\infty < t < \infty$. Consider the event indicator

$$I[x - x(t)] \triangleq \begin{cases} 1, & x(t) < x \\ 0, & x(t) \geq x. \end{cases}$$

The time-average of this event indicator is the *fraction-of-time* (FOT) PD

$$\hat{F}^0_{x(t)}(x) \triangleq \hat{P}rob\{x(t) < x\} = \hat{E}^0\{I[x - x(t)]\}$$

where the time-average operation $\hat{E}^0\{\cdot\}$ is defined by

$$\hat{E}^0\{h(t)\} \triangleq \lim_{Z\to\infty} \frac{1}{2Z} \int_{-Z}^{Z} h(t + t')\, dt'$$

for any time function h. (The superscript 0 will be explained subsequently.) Therefore the joint FOT PD for the set of variables $\boldsymbol{x}(t) \triangleq \{x(t + t_1), \ldots, x(t + t_n)\}$ is given by

$$\hat{F}^0_{\boldsymbol{x}(t)}(\boldsymbol{x}) = \hat{E}^0 \left\{ \prod_{j=1}^{n} I\left[x_j - x(t + t_j)\right] \right\}$$

and the joint FOT Pd is

$$\hat{f}^0_{\boldsymbol{x}(t)}(\boldsymbol{x}) = \frac{\partial^n}{\partial x_1 \ldots \partial x_n} \hat{F}^0_{\boldsymbol{x}(t)}(\boldsymbol{x}).$$

We have the following theorem.

Fundamental Theorem of Time-Averaging

For every time-invariant function $g(\cdot)$ for which $\hat{E}^0\{g[\boldsymbol{x}(t)]\}$ exists, we have

$$\begin{aligned} \hat{E}^0\{g[\boldsymbol{x}(t)]\} &\triangleq \lim_{Z\to\infty} \frac{1}{2Z} \int_{-Z}^{Z} g[\boldsymbol{x}(t + t')]\, dt' \\ &= \int g(\boldsymbol{x}) \hat{f}^0_{\boldsymbol{x}(t)}(\boldsymbol{x})\, d\boldsymbol{x}. \end{aligned}$$

This theorem motivates us to call the time-average operation $\hat{E}^0\{\cdot\}$ the *temporal-expectation operation*. To illustrate the validity of this theorem, we can substitute the definition of the FOT PD into the definition of the FOT Pd, which can be substituted into the result of this theorem to obtain

---

[10]We mean "well-behaved" in the sense that $x(t)$ exhibits the regularity required for all time averages of interest to exist.

$$
\begin{aligned}
\int g(\boldsymbol{x}) f_{x(t)}(\boldsymbol{x})\, d\boldsymbol{x} &= \int g(\boldsymbol{x}) \frac{\partial^n}{\partial x_1 \ldots \partial x_n} F_{x(t)}(\boldsymbol{x})\, d\boldsymbol{x} \\
&= \int g(\boldsymbol{x}) \frac{\partial^n}{\partial x_1 \ldots \partial x_n} \hat{E}^0 \left\{ \prod_{j=1}^{n} I\left[x_j - x(t+t_j)\right] \right\} d\boldsymbol{x} \\
&= \int g(\boldsymbol{x}) \hat{E}^0 \left\{ \prod_{j=1}^{n} \delta\left[x_j - x(t+t_j)\right] \right\} d\boldsymbol{x} \\
&= \hat{E}^0 \left\{ \int g(\boldsymbol{x}) \prod_{j=1}^{n} \delta\left[x_j - x(t+t_j)\right] \right\} d\boldsymbol{x} \\
&= \hat{E}^0\{g[\boldsymbol{x}(t)]\},
\end{aligned}
$$

where we have used the sampling property of the Dirac delta δ, which is the derivative of the unit-step function $I(\cdot)$.

For any function $h(t)$ for which $\hat{E}^0\{h(t)\}$ exists, we have

$$h(t) = c + r(t)$$

where $c = \hat{E}^0\{h(t)\}$ is a constant (independent of t) and $r(t) = h(t) - c$ is the residual for which $\hat{E}^0\{r(t)\} = 0$. Consequently, the temporal expectation operation can also be called the *constant-component extractor*.

We can see from these two theorems that there is a *duality* between the *probability-space* theory of stochastic processes based on the operation $E\{\cdot\}$ and what we shall call the *time-space* theory of time-series based on $\hat{E}^0\{\cdot\}$. Wold (Wold, 1948) tried to formalize this duality in terms of an isomorphism based on the mapping

$$x(t+\sigma) \to X(t, \omega(\sigma))$$

where $X(t, \omega)$ is a sample path of the stochastic process $X(t)$ corresponding to the sample point $\omega = \omega(\sigma)$, indexed by σ, in Ω. That is, the ensemble members of $X(t)$ correspond to translates of $x(t)$ in this isomorphism. While this isomorphism is conceptually useful, a mathematically rigorous study of it has not (to my knowledge) been performed. (For example, does such a stochastic process corresponding to a given time-series actually exist?)

We can justifiably ask, "Just how viable is the time-space theory?—do 'well-behaved' time-series models exist?" The answer to the latter question is "Yes"; examples are provided by typical sample paths of ergodic stochastic processes. But, "Can we construct useful time-series models?" The answer, again, is "Yes": We can construct models in the same way we do for stochastic processes, except we specify $\hat{F}_{x(t)}(\boldsymbol{x})$ instead of $F_{X(t)}(\boldsymbol{x})$.

Regarding the answer to the first question, we might ask "Does this apparent reliance, of existence of time-series, on stochastic processes detract from the conceptual simplicity of working with time-series rather than stochastic processes?" The answer, in my opinion, is "No."

Let us trace the conceptual paths for both stochastic processes and time-series so that we can see specifically where they are parallel and where they diverge. As before, we begin with stochastic processes by giving the definitions of the classes of processes of interest in the study of cyclostationarity, namely, processes that are S, CS, or PCS of order n (in the strict sense).

2.8 Stochastic-Process Definitions

Definition 1: $X(t)$ is a S process if and only if $F_{X(t)}(\boldsymbol{x})$ is independent of the time-translation parameter t.

Definition 2: $X(t)$ is a CS process with period T if and only if $F_{X(t)}(\boldsymbol{x})$ is periodic in t with period T.

Definition 3: $X(t)$ is a PCS process with periods $\{T\} = T_1, T_2, T_3, \ldots$ if and only if $F_{X(t)}(\boldsymbol{x})$ is polyperiodic in t with periods $\{T\}$ (which is a sum of periodic functions with single periods $T_1, T_2, T_3, \ldots$).

The relationships among the class of generally nonstationary (NS) processes and the three classes S, CS, and PCS can be described with the Venn diagram shown in Fig. 2.

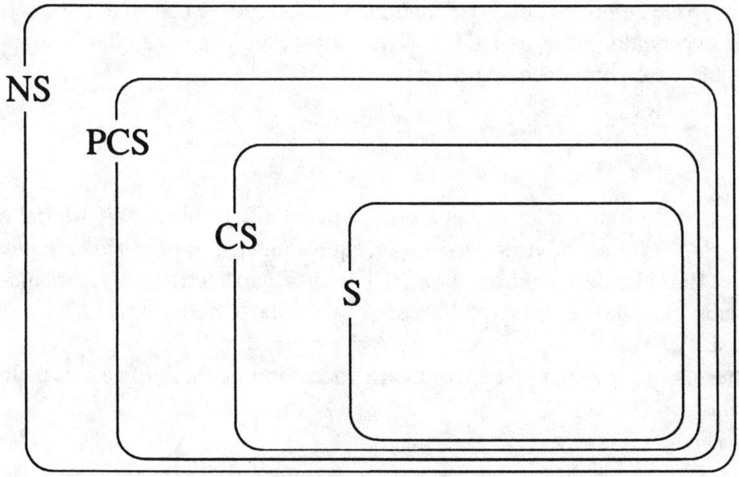

Figure 2: Venn diagram of classes of stochastic processes. With the class NS omitted, also the Venn diagram of classes of time-series.

It is useful to expand the polyperiodic PD function in a Fourier series:

$$F_{X(t)}(\boldsymbol{x}) = \sum_{\alpha} F^{\alpha}_{X(0)}(\boldsymbol{x})\, e^{i2\pi\alpha t} = \sum_{\alpha} F^{\alpha}_{X(t)}(\boldsymbol{x}),$$

where $F^{\alpha}_{X(0)}(\boldsymbol{x})$ are the Fourier-coefficients, and where the sinusoidal-component functions are given by

$$F^{\alpha}_{X(t)}(\boldsymbol{x}) \triangleq \lim_{Z\to\infty} \frac{1}{2Z} \int_{-Z}^{Z} F_{X(t+t')}(\boldsymbol{x})\, e^{-i2\pi\alpha t'}\, dt'$$
$$= \hat{E}^{0}\left\{F_{X(t)}(\boldsymbol{x})\, e^{-i2\pi\alpha t}\right\} e^{i2\pi\alpha t}$$
$$\triangleq \hat{E}^{\alpha}\left\{F_{X(t)}(\boldsymbol{x})\right\}.$$

For any function $h(t)$ for which $\hat{E}^{\alpha}\{h(t)\}$ exists, we have

$$h(t) = c\, e^{i2\pi\alpha t} + r(t)$$

where c is a constant, $ce^{i2\pi\alpha t} = \hat{E}^{\alpha}\{h(t)\}$, and $r(t) = h(t) - ce^{i2\pi\alpha t}$ is the residual for which $\hat{E}^{\alpha}\{r(t)\} = 0$. Consequently, we call the operation

$$\hat{E}^{\alpha}\{\cdot\} \triangleq \hat{E}^{0}\left\{(\cdot)e^{-i2\pi\alpha t}\right\} e^{i2\pi\alpha t}$$

the *sine-wave-component extractor*. It can be thought of as the limit, as bandwidth goes to zero, of a bandpass filter with center frequency α and unity gain at α. For $\alpha = 0$, it reduces to the constant-component extractor $\hat{E}^{0}\{\cdot\}$.

2.9 Time-Series Definitions

Now let us turn from stochastic processes to time-series. Before we can give the dual time-space definitions of S, CS, and PCS time-series, we need to generalize the temporal expectation operation $\hat{E}^{0}\{\cdot\}$. The appropriate generalization is simply the sum of sine-wave-component extractors

$$\hat{E}^{\{\alpha\}}\{\cdot\} \triangleq \sum_{\alpha\in\{\alpha\}} \hat{E}^{\alpha}\{\cdot\}$$

for all sine-wave frequencies in some set $\{\alpha\}$ of interest. Thus, $\hat{E}^{\{\alpha\}}\{\cdot\}$ is called the *multiple-sine-wave-component extractor* or, equivalently, the *polyperiodic-component extractor*. (It is identical to the operator $P\{\cdot\}$ discussed in Section 2.3.) The sine-wave frequencies α are the harmonics of the reciprocals of the periods $1/T_1$, $1/T_2$, $1/T_3, \ldots$ of interest.

In terms of the generalized temporal expectation operation, we can define the *polyperiodic FOT PD*:

$$\hat{F}^{\{\alpha\}}_{x(t)}(\boldsymbol{x}) = \hat{E}^{\{\alpha\}}\left\{\prod_{j=1}^{n} I\left[x_j - x(t+t_j)\right]\right\}$$

and the *polyperiodic FOT Pd*:

$$\hat{f}^{\{\alpha\}}_{x(t)}(\boldsymbol{x}) = \frac{\partial^n}{\partial x_1 \ldots \partial x_n} \hat{F}^{\{\alpha\}}_{x(t)}(\boldsymbol{x}).$$

It is not obvious that $\hat{F}^{\{\alpha\}}_{x(t)}(\boldsymbol{x})$ is indeed a valid probability distribution function, but this proposition is proved in (Gardner and Brown, 1991). We have the following theorem.

Fundamental Theorem of Polyperiodic-Component Extraction

For every time-invariant function $g(\cdot)$ for which $\hat{E}^{\{\alpha\}}\{g[\boldsymbol{x}(t)]\}$ exists, we have

$$\hat{E}^{\{\alpha\}}\{g[\boldsymbol{x}(t)]\} \triangleq \sum_{\alpha} \lim_{Z\to\infty} \frac{1}{2Z} \int_{-Z}^{Z} g[\boldsymbol{x}(t+t')]\, e^{-i2\pi\alpha t'}\, dt'$$
$$= \int g(\boldsymbol{x})\, \hat{f}^{\{\alpha\}}_{\boldsymbol{x}(t)}(\boldsymbol{x})\, d\boldsymbol{x}$$

The validity of this theorem can be illustrated in the same way the validity of the fundamental theorem of constant-component extraction is illustrated. Also, this theorem is valid more generally if $g(\cdot) = g(t, \cdot)$ is polyperiodic in time t.

We are now in a position to define the classes of S, CS, and PCS of order n (in the strict-sense) time-series. Let $\{\alpha\}$ be the set of all α for which $\hat{F}^{\{\alpha\}}_{\boldsymbol{x}(t)} \not\equiv 0$.

Definition 4: $x(t)$ is a S time-series if and only if $\hat{F}^{\{\alpha\}}_{\boldsymbol{x}(t)}(\boldsymbol{x})$ exists and $\not\equiv 0$ and is independent of the time-translation parameter t (that is, $\{\alpha\} = \{0\}$).

Definition 5: $x(t)$ is a CS time-series with period T if and only if $\hat{F}^{\{\alpha\}}_{\boldsymbol{x}(t)}(\boldsymbol{x})$ exists and $\not\equiv 0$ and is periodic in t with period T (that is, $\{\alpha\}$ = {harmonics of $1/T$}).

Definition 6: $x(t)$ is a PCS time-series with periods $\{T\} = T_1, T_2, T_3, \ldots$ if and only if $\hat{F}^{\{\alpha\}}_{\boldsymbol{x}(t)}(\boldsymbol{x})$ exists and $\not\equiv 0$ and is polyperiodic in t with periods $\{T\}$).

The relationships among these three classes of time-series can be described with the Venn diagram shown in Fig. 2 except that for time-series, unlike stochastic processes, the superclass NS does not exist. Generally nonstationary FOT PDs cannot be defined (although locally S FOT PDs, which are NS, can be defined by limiting the time-averaging interval used in $\hat{E}^0\{\cdot\}$ to one of finite length Z).

These definitions of S, CS, and PCS time-series represent a modification of previous terminology. Wold (Wold, 1948) defined a stationary (S′) time-series to be one for which $\hat{F}^0_{\boldsymbol{x}(t)}(\boldsymbol{x})$ exists and $\not\equiv 0$. To refine this definition, (Gardner, 1987a) defined a purely S′ time-series to be a S′ time-series for which $\hat{F}^{\alpha}_{\boldsymbol{x}(t)}(\boldsymbol{x}) \equiv 0$ for all $\alpha \neq 0$. Following Wold, (Gardner, 1987a) also defined a cyclostationary (CS′) time-series with period T to be one for which $\hat{F}^{\{\alpha\}}_{\boldsymbol{x}(t)}(\boldsymbol{x})$ exists and $\not\equiv 0$ for some $\{\alpha\} \in$ {harmonics of $1/T$}, and to refine this a purely CS′ time-series was defined to be a CS′ time-series for which $\hat{F}^{\{\alpha\}}_{\boldsymbol{x}(t)}(\boldsymbol{x}) \equiv 0$ for all $\alpha \notin$ {harmonics of $1/T$}.

The relationships among these previously defined classes of time-series can be described with the Venn diagram shown in Fig. 3. Observe that the nesting of the classes S′, CS′, and PCS′ ≡ PCS is inverted from that in Fig. 2 for the classes S, CS, and PCS. The definitions of S, CS, and PCS form the basis for a theory of time-series that in some ways has a stronger duality with the theory of stochastic processes than does the theory that could be based on the previous definitions of S′, CS′, and PCS′.

On the other hand, the previous concepts of pure stationarity and pure cyclostationarity arise also within the framework of stochastic processes. Since these concepts depend on notions of ergodicity, let us now consider the relevant types of ergodicity.

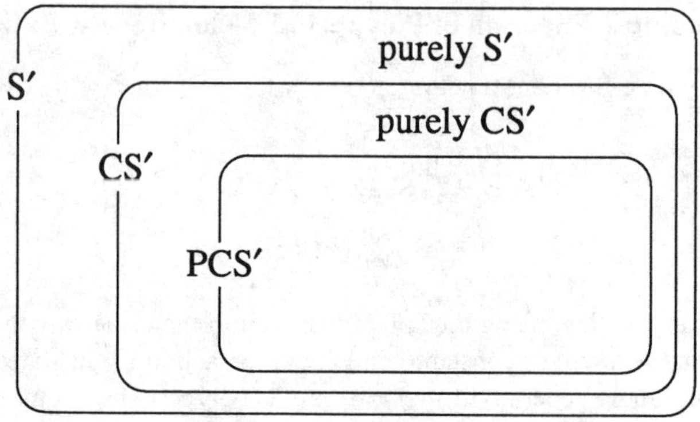

Figure 3: Venn diagram for previously defined classes of time-series.

2.10 Cycloergodicity and Refined Stochastic-Process Definitions

Definition 7: $X(t)$ is an *ergodic* (E) process if and only if for every natural number n and every nonrandom function $g(\cdot)$ of n variables for which $E\{g[\boldsymbol{X}(t)]\}$ exists, we have

$$\hat{E}^0\{E\{g[\boldsymbol{X}(t)]\}\} = \hat{E}^0\{g[\boldsymbol{X}(t)]\} \ w.p.\ 1$$

(where w.p. 1 means with probability equal to 1). For a S process, $\hat{E}^0\{E\{\cdot\}\} = E\{\cdot\}$ (since the constant component of a constant is that constant) and the outer operation on the left side of this defining equation can be deleted.

Definition 8: $X(t)$ is a *cycloergodic* (CE) process with period T if and only if for every natural number n and every nonrandom function $g(\cdot)$ of n variables for which $E\{g[\boldsymbol{X}(t)]\}$ exists, we have (with $\{\alpha\} \in \{\text{harmonics of } 1/T\}$)

$$\hat{E}^{\{\alpha\}}\{E\{g[\boldsymbol{X}(t)]\}\} = \hat{E}^{\{\alpha\}}\{g[\boldsymbol{X}(t)]\} \ w.p.\ 1.$$

For a CS process, $\hat{E}^{\{\alpha\}}\{E\{\cdot\}\} = E\{\cdot\}$ (since the periodic component of a periodic function is that periodic function) and the outer operation on the left side of this defining equation can be deleted.

Definition 9: $X(t)$ is a *polycycloergodic* (PCE) process with periods $\{T\}$ if and only if it is cycloergodic with period T_k for $k = 1, 2, 3, \ldots$.

Stochastic processes that are not CE or PCE can exhibit hidden cyclostationarity. For example, if $X(t)$ is S and PCE with all periods, then its sample paths are stationary time-series (w.p. 1); however, if $X(t)$ is S (and possibly E) but not CE, its sample paths can be CS (w.p. 1). Similarly, if $X(t)$ is CS and PCE with all periods, then its sample paths are CS time-series (w.p. 1); however, if $X(t)$ is CS (and possibly CE), but not PCE, its sample paths can be PCS (w.p. 1). Such non-CE and non-PCE models typically result from the (explicit or implicit) inclusion of random-phase variables in the stochastic-process model. This hidden cyclostationarity motivates the following refined probability-space definitions.

Definition 10: If $X(t)$ is S and PCE with all periods, then it is defined to be *purely stationary*, and its sample paths are purely stationary (S or purely S′) time-series (w.p. 1): there is no hidden CS.

Definition 11: If $X(t)$ is CS and PCE with all periods, then it is defined to be *purely CS*, and its sample paths are purely cyclostationarity (CS or purely CS′) time-series (w.p. 1): there is no hidden PCS.

Definition 12: If $X(t)$ is PCS and PCE with all periods, then it is defined to be *purely PCS*: there is no hidden PCS.

The relationships among all the classes of stochastic processes defined so far are illustrated with the Venn diagram shown in Fig. 4.

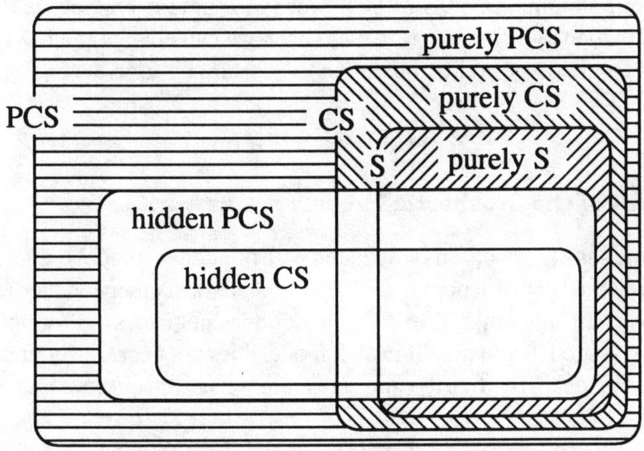

Figure 4: Venn diagram of classes of stochastic processes.

Let us now consider an example that illustrates the various classes of stochastic processes and time-series that have been introduced. Let the *stochastic process* $X(t)$ be specified by

$$X(t) = A(t) + B(t)\cos(\omega_1 t + \theta_1) + C(t)\cos(\omega_2 t + \theta_2),$$

where $A(t)$, $B(t)$, and $C(t)$ are purely stationary ergodic processes. If θ_1 and θ_2 are nonrandom, then the *stochastic process* $X(t)$ is PCS and PCE. On the other hand, if θ_1 and/or θ_2 is random, then (depending on their PDs) the *stochastic process* $X(t)$ can be PCS (with periods T_1 and T_2), or it can be CS (with period T_1 or T_2), or it can be S, and $X(t)$ is not PCE. Furthermore, with probability one, the sample paths of $X(t)$ are PCS *time-series*, the sample paths of $A(t)$, $B(t)$, and $C(t)$ are S *time-series*, and the sample paths of the components $B(t)\cos(\omega_1 t + \theta_1)$ and $C(t)\cos(\omega_2 t + \theta_2)$ are CS *time-series*.

2.11 Phase Randomization

To pursue the concept of phase randomization a little further, it is noted that even if θ_1 and θ_2 are both nonrandom, we can introduce a random phase Θ into $X(t)$ to obtain

$$Y(t) = X(t + \Theta)$$

which can be changed from PCS to CS or to S by choosing the distribution for Θ (Gardner, 1978; Hurd, 1974b). Thus, we see that there is a nonuniqueness of models for stochastic processes. We can change the stochastic process from PCS to CS to S by phase-randomizing with a single phase variable: $X(t) \to X(t + \Theta)$. Or, equivalently, we can change the PD function from polyperiodic to periodic to constant by time-averaging; e.g.,

$$\hat{E}^0 \left\{ F_{X(t)}^{\{\alpha\}}(\boldsymbol{x}) \right\} = F_{X(t)}^{0}(\boldsymbol{x}).$$

That is, phase-randomizing a CS or PCS process or time-averaging its PD function can result in hidden cyclostationarity. Similarly, we can change the PD function for a time-series from polyperiodic to periodic to constant by time-averaging; e.g.,

$$\hat{E}^0 \left\{ \hat{F}_{x(t)}^{\{\alpha\}}(\boldsymbol{x}) \right\} = \hat{F}_{x(t)}^{0}(\boldsymbol{x}).$$

2.12 Pitfalls of the Stochastic-Process Framework

There are some significant pitfalls associated with nonunique models. One such pitfall is "hidden statistical dependence." Let SI denote statistical independence (e.g., of two variables). We can show that SI in a CS model does not necessarily imply SI in the corresponding S model, and that SI in an S model does not necessarily imply SI in the associated CS model. To prove the first statement we simply observe that the equality

$$f_{X_1(t),X_2(t)} = f_{X_1(t)} f_{X_2(t)}$$

that results from the SI of jointly CS processes $X_1(t)$ and $X_2(t)$ does not necessarily imply the equality

$$\hat{E}^0 \left\{ f_{X_1(t),X_2(t)} \right\} = \hat{E}^0 \left\{ f_{X_1(t)} \right\} \hat{E}^0 \left\{ f_{X_2(t)} \right\},$$

which would hold if, in the associated S model, $X_1(t)$ and $X_2(t)$ were SI. To prove the second statement, we consider the example of discrete-time processes

$$X_1(t) = Z(t) = i.i.d. \pm 1$$

$$X_2(t) = Z(t)\cos(t).$$

We can easily show that

$$\hat{E}^0 E \left\{ X_1^n(t) X_2^m(t) \right\} = \hat{E}^0 E \left\{ X_1^n(t) \right\} \hat{E}^0 E \left\{ X_2^m(t) \right\}$$

for all n and m and, therefore, $X_1(t)$ and $X_2(t)$ are SI in the S model. However,

$$E\left\{X_1^n(t)X_2^m(t)\right\} \neq E\left\{X_1^n(t)\right\}E\left\{X_2^m(t)\right\}$$

for n and m odd and, therefore, $X_1(t)$ and $X_2(t)$ are not SI in the CS model.

We are now ready to take stock. One conclusion we can draw is that when a process is not PCE, the hidden CS or hidden PCS can result in single-sample-path behavior (w.p. 1) that cannot be predicted from probabilistic analysis (unless the hidden CS can be revealed by conditioning on certain random phase variables).

An important fact concerning this conclusion is that the theory of PCE is mostly nonexistent and appears to require nontrivial extensions/generalizations of the theory of E and the incomplete theory of CE.

Another important fact is that the commonplace approach to deriving ad hoc signal-processing algorithms, of replacing expectation operations $E\{\cdot\}$ in analytical expressions with time-average operations $\hat{E}^0\{\cdot\}$, or (when a composition of both operations are present in the analytical expression) of deleting the expectation operation, cannot be justified (and will often fail to produce the desired results) when the stochastic-process model used is not PCE.

A related important fact is that the "optimum" solutions to inference and decision problems (e.g., for signal estimation and detection) that are based on S and E, but not CE (or based on CS and CE, but not PCE), process models can be highly inferior to inference and decision rules that exploit the hidden CS (or hidden PCS).

Let us consider some examples that illustrate the ramifications of the preceding conclusion and associated facts.

Example 1: Let $Y(t)$ be the output of a time-invariant nonlinear transformation with input $X(t)$. Let $X(t)$ be S (for all n) and E, but not CE, with no spectral lines. Then $Y(t)$ is S (for all n) and E but, because of the hidden CS in $X(t)$, contains spectral lines. The presence of these spectral lines cannot be explained except by virtue of the hidden CS in $X(t)$. As a specific example, let $X(t) = A(t)\cos(\omega_1 t + \Theta_1)$, where $A(t)$ is S and E, and Θ_1 is independent of $A(t)$ and uniformly distributed on $[0, 2\pi]$, and let $Y(t) = X^2(t)$. Then $Y(t)$ has spectral lines at frequencies 0 and $\pm\omega_1/\pi$ Hz.

Example 2: Let $X(t)$ be S (for $n = 2$) and E, but not CE, and let $X(t)$ admit an exact AR model (with white residuals). The sample paths of the white residuals can be partially predictable (w.p. 1) using linear periodic predictors derived from the sample-path statistics. For example, let $U(t)$ be any nonwhite CS process and let $V(t)$ be an independent S process with complementary spectrum; that is, the sum of the spectra of $V(t)$ and the stationarized version $U(t+\Theta)$ of $U(t)$ equals a constant over all frequency. Then the S process $W(t+\Theta) = U(t+\Theta) + V(t+\Theta)$ is white (the spectra of $V(t)$ and $V(t+\Theta)$ are identical) but the CS process $W(t)$ is nonwhite. That is, the autocorrelation of $W(t+\Theta)$ is proportional to a delta function in the lag variable but the autocorrelation of $W(t)$ is not. Thus, $W(t)$ is predictable using linear periodic predictors and so too are its sample paths. Since the sample paths of $W(t+\Theta)$ and $W(t)$ are the same except for a time-shift, the sample paths of $W(t+\Theta)$

are also predictable with a periodic predictor. Furthermore, the sample paths of a S and E, but not CE, process $X(t)$ can admit periodic AR (PAR) models (w.p. 1) with finite periodic order even though the stochastic process $X(t)$ admits no finite-order AR model. For example, let the CS and CE process $Z(t)$ be a first-order PAR process. The stationary process $X(t) = Z(t + \Theta)$ will not, in general, admit any finite-order AR model. But its sample paths, being the same as the sample paths of $Z(t)$ except for a time shift, admit first-order PAR models that can be identified from the sample path statistics.

Example 3: Let $X(t)$ and $Y(t)$ be jointly S (for all n) and E, but not CE, and, according to the usual definition of *causality*, let there be no causal relationship of $X(t)$ to $Y(t)$. That is, no linear or nonlinear *time-invariant* operation on $X(t)$ and its past has any prediction capability for $Y(t)$ and its future. Yet, each sample path of $Y(t)$ can possibly be perfectly *cyclically caused* by the corresponding sample path of $X(t)$. That is, a periodic operation on $X(t)$ can possibly perfectly predict $Y(t)$. As a specific example, consider the two continuous-time processes

$$X(t) = Z(t) = \pm 1$$

$$Y(t) = Z(t - \tau)\cos(t + \Theta)$$

where the transition times between $+1$ and -1 are arbitrary, and Θ is independent of $Z(t)$ and uniformly distributed on $[0, 2\pi]$. It can easily be shown that

$$E\left\{X^n(t - v)Y^m(t)\right\} = E\left\{X^n(t - v)\right\} E\left\{Y^m(t)\right\}$$

for all n and m, and all v; that is, $Y(t)$ is statistically independent of the past of $X(t)$. Nevertheless, each sample path of $Y(t)$ can be perfectly predicted from the past of the corresponding sample path of $X(t)$:

$$Y(t) = X(t - \tau)\cos(t + \Theta).$$

Example 4: Let $Z(t) = X(t) + Y(t)$, where $X(t)$ and $Y(t)$ are statistically independent, S (for $n = 2$), and E, but not CE, processes that have identical spectral densities. The Wiener filter for extracting $X(t)$ from $Z(t)$ (separating $X(t)$ and $Y(t)$) is essentially useless. Its transfer function is a constant. Yet the sample paths of $X(t)$ and $Y(t)$ can possibly be perfectly separated with a periodic filter. Examples include communication signals such as digital QAM, AM, PSK, ASK, and PAM. As a specific example, it can be shown (Gardner, 1993) that up to N spectrally coincident digital QAM signals with excess bandwidth $\geq (N - 1)100\%$ can be perfectly separated.

Example 5: Let $\boldsymbol{X}(t) = \{X_1(t), X_2(t)\}$ be purely S (for all n) with a probabilistic model that is very similar to that of $\boldsymbol{Y}(t) = \{Y_1(t), Y_2(t)\}$, which is CS (e.g., $\boldsymbol{X}(t)$ and $\boldsymbol{Y}(t)$ are both Gaussian processes and the PSDs of $\boldsymbol{X}(t)$ equal those of the stationarized $\boldsymbol{Y}(t)$). The Cramér-Rao bounds of the same parameters in each of $\boldsymbol{X}(t)$ and $\boldsymbol{Y}(t)$ (e.g., the relative time delay (TD) of an additive signal component common to both $X_1(t)$ and $X_2(t)$) can be drastically different. This has been demonstrated for TD at two reception platforms and for angle-of-arrival at a sensor array (Chen and

Gardner, 1992; Schell and Gardner, 1992b). Moreover, even the Cramér-Rao bound of the stationarized $\boldsymbol{Y}(t)$ can be drastically different from that of the purely S $\boldsymbol{X}(t)$.

Example 6: Let $X(t)$ and $Y(t)$ be independent, S (for all n), and E, but not CE, and let $Z(t)$ be specified under two hypotheses—

under hypothesis 1: $Z(t) = X(t) + Y(t)$

under hypothesis 2: $Z(t) = Y(t)$

The "optimum" (e.g., maximum-posterior-probability) detector for the presence of $X(t)$ in $Z(t)$ can be greatly outperformed by detectors that exploit the hidden CS in $X(t)$ and/or $Y(t)$, e.g., the joint maximum-posterior-probability detector and phase estimator (Gardner, 1988b; Gardner and Spooner, 1992a).

2.13 Two Paths into the Future

One approach to this unsettling situation, which is illustrated by the preceding examples, that should appeal to mathematicians is to take what shall be called *Path 1:* Develop the needed theory of PCE. The current status of the theory of CE and PCE is that substantial progress has been made for (1) CE w.p. 1 for discrete-time CS processes and Gaussian continuous-time CS processes, and (2) PCE in mean-square for finite-order moments of discrete- and continuous-time PCS processes. Little or no progress has been made for (1) PCE w.p. 1 for discrete-time PCS processes, (2) CE w.p. 1 for non-Gaussian continuous-time CS and PCS processes, and (3) PCE w.p. 1 for continuous-time CS and PCS processes.

The only paper to address PCE w.p. 1 (Boyles and Gardner, 1983) suggests that a substantial breakthrough will be required (even for the much less technical case of discrete time): conventional approaches and ideas apparently lead to dead ends. This suggests a challenge not unlike that Birkhoff faced around 1930 when he formulated and proved the fundamental ergodic theorem to replace the very unsatisfying "ergodic hypothesis." We need a fundamental polycycloergodic theorem that elegantly formalizes our informal notion of a PCE process in terms of a necessary and sufficient condition on the associated probability measure.

The most useful concept regarding PCE that we have for applications is the following unproved proposition.

Proposition *PCS processes constructed from stable (decaying-memory) nonrandom polyperiodic transformations of purely stationary ergodic processes are PCE.*

Stochastic-process models for many, if not most, communications signals can be constructed in this way.

In view of the difficulties before us, we should ask what some of the advantages of the stochastic-process approach are. The most apparent advantages are listed here:

1. It is the orthodox approach to modeling and studying evolutionary random phenomena and it is, therefore, attractive to those already familiar with it.

2. Mathematicians do know how, in principle, to construct stochastic-process models from elementary mathematical constructs (Borel fields, sigma algebras, probability measures, etc.). Therefore, there is a greater likelihood of success (compared with time-series) in constructing a mathematical theory of PCS and PCE processes from a few basic axioms.
3. It is possible, in principle, to exploit the hidden CS (or PCS) in a non-CE (or non-PCE) process within the conventional framework of stochastic processes. But, this requires that one have a model of the hidden CS (or PCS) that is *explicitly dependent* on one or more random phase variables Θ that are responsible for the lack of CE so that one can calculate probability densities and expectations *conditioned on* Θ.
4. Development of the theory of PCE will help clear the way for making the time-space theory of time-series mathematically rigorous.

In spite of these advantages, there is an alternative approach that should appeal to pragmatic engineers and scientists. Let us begin with the following perspective: The probability-space approach based on expectation introduces abstractions that, in many applications (e.g., many problems for which single-sample-path signal processing is of interest), have no redeeming *practical* value. Some of these abstractions can be properly dealt with only with a theory of PCE that is presently nonexistent. Regressing back to pre-1930 and adopting a "PCE hypothesis" is very unappealing (because the hypothesis can be false).

So, let us consider taking what shall be called *Path 2*: Adopt the time-space approach whose theory is in many ways dual to that of the probability-space approach, but without the practical drawbacks associated with cycloergodicity and the distracting abstraction associated with expectation over ensembles.

The essence of cyclostationarity from an operational standpoint is the fact that sine waves making up additive polyperiodic components can be generated from random data by applying certain nonlinear transformations. And, the time-space theory of cyclostationarity arises *naturally* out of the fundamental theorem of polyperiodic-component extraction using the generalized temporal expectation operation $\hat{E}^{\{\alpha\}}$; whereas the expectation E that gives rise to the probability-space theory has little to do with the essence of cyclostationarity.

But, can we construct time-series models? The answer is yes. Time-series models for many, if not most, communication signals can be constructed by subjecting one or more elementary time-series (e.g., purely stationary and white) to elementary transformations such as filters, periodic modulators, multiplexors, etc. (Gardner, 1987a). The defining properties of a discrete-time purely stationary white time-series are:

1. Whiteness: $\hat{f}^{0}_{\boldsymbol{x}(t)}(\boldsymbol{x}) = \hat{f}^{0}_{x(t+t_1)}(x_1)\hat{f}^{0}_{x(t+t_2)}(x_2)\ldots\hat{f}^{0}_{x(t+t_n)}(x_n)$ for unequal $t_1, t_2, \ldots, t_n$.
2. Pure stationarity: $\hat{f}^{\{\alpha\}}_{\boldsymbol{x}(t)}(\boldsymbol{x}) \equiv \hat{f}^{0}_{\boldsymbol{x}(t)}(\boldsymbol{x})$ for all $\{\alpha\}$.
3. Existence: any sample path of any i.i.d. stochastic process will do (w.p. 1).

Okay. But, can we do probabilistic analysis using time-space theory? The answer, again, is yes. Performance measures such as bias, variance, Cramér-Rao bounds, confidence intervals, probabilities of decision-errors, etc., can be calculated using time-space theory just as well as they can using probability-space theory (Gardner, 1987a). But can we use the theory of statistical inference and decision? Yes, indeed.

The author's current assessment of progress along Path 2 can be summarized as follows: The considerable progress in the development and application of the time-space (or *temporal-probability*, or *fraction-of-time probability*) theory of CS and PCS time-series that has been made since its adoption by the UCD and SSPI groups in 1985, cf. (Gardner, 1987a, 1991a) includes:

1. Temporal and spectral second-order-moment theory (cyclic autocorrelation and cyclic spectra, or spectral correlation functions) (Gardner, 1987a; Section 3 in this chapter).
2. Temporal and spectral higher-order-moment and cumulant theory (cyclic cumulants and cyclic polyspectra, or spectral cumulants) (Gardner and Spooner, 1992b; Spooner and Gardner, 1992a,b; Spooner, 1992; Chapter 2 in this volume).
3. The rudiments of fraction-of-time probability distribution theory (Gardner, 1987a; Gardner and Brown, 1991).
4. A wide variety of applications of the theory to signal-processing and communications problems involving signal detection, signal classification, signal-parameter estimation, and signal-waveform estimation (Agee, et al., 1987, 1988, 1990; Brown, 1987; Chen, 1989; Chen and Gardner, 1992; Gardner, 1987a,b, 1988a,b,c, 1990a,b, 1991a,c, 1992, 1993; Gardner and Archer, 1993; Gardner and Brown, 1989; Gardner and Chen, 1988, 1992; Gardner and Paura, 1992; Gardner and Spooner, 1992a, 1993; Gardner and Venkataraman, 1990; Gardner et al., 1987, 1992; Schell, 1990; Schell and Agee, 1988; Schell and Gardner, 1989, 1990a,b,c, 1991, 1992, 1993a,b; Schell et al., 1989, 1993; Spooner, 1992; Spooner and Gardner, 1992b; Section 4 in this chapter; Chapter 2; Chapter 3).

Also, the conceptual gap between the existing time-space theory and its application to many signal-processing problems in communications is perceived by its current users to be much narrower than it is for the dual probability-space theory.

Further support for taking Path 2 includes the fact that the temporal-probability approach, which is centered on the concrete sine-wave extraction operation, has led naturally to a *derivation* of the cumulant as the solution to a fundamental problem in characterizing higher-order CS and PCS. It is doubtful that this derivation would have been discovered within the stochastic-process framework, which is centered on the abstract expectation operation. This derivation is discussed in Chapter 2 in this volume.

But, in the final analysis, the duality between the time-space and probability-space theories will likely result in either path taking the sufficiently persistent practical problem solver to the same places, although not necessarily in the same elapsed time or with the same energy. This duality can be formalized with the following loosely stated conjecture.

Conjecture *For every theorem that can be proved for a PCE PCS process, a dual theorem can be proved for a PCS time-series—and vice versa.*

This can be viewed as a generalization of Wold's isomorphism from S to PCS processes.

Nevertheless, there does remain a fundamental question that is not yet always answerable: Given a self-consistent set of probability distributions $\hat{F}^{\{\alpha\}}_{x(t)}$ for all orders n, does there exist a corresponding time-series $x(t)$? We have sufficient conditions on $\hat{F}^{\{\alpha\}}_{x(t)}$ that guarantee existence of $x(t)$: They are identical to the conditions that guarantee that $F_{X(t)}$ is PCE and they are called *mixing conditions* in the theory of stochastic processes. But we do not yet have a *necessary* and sufficient condition. (This presents another challenge for mathematicians.)

With regard to the taking of Paths 1 and 2, we can draw three conclusions:

1. The more abstract theory of PCS stochastic processes will undoubtedly be found to be of considerable value as it is developed, and those who are sufficiently mathematically inclined are encouraged to pursue this approach.
2. The less abstract theory of cyclostationary time-series is more accessible to engineers and scientists interested in theory as a *conceptual aid for solving practical problems*. It should be the preferred approach for the practically oriented whenever ensembles are not, in and of themselves, of primary concern.[11]
3. Both theories present important challenges to mathematicians.

In the remainder of this chapter, Path 2 is taken, and the theory and application of second-order (wide-sense) cyclostationarity is pursued in some detail.

3 INTRODUCTION TO THE PRINCIPLES OF SECOND-ORDER (WIDE-SENSE) CYCLOSTATIONARITY

The second-order (wide-sense[12]) theory of discrete-time stochastic processes deals with the probability-space autocorrelation function

$$R_X(t, t-\tau) = E\{X(t)X^*(t-\tau)\}.$$

---

[11]The practical value of this approach is amply demonstrated for parametric and nonparametric spectral analysis of S as well as CS and PCS time-series in (Gardner, 1987a).

[12]Wide-sense theory deals with moments, whereas strict-sense theory deals with probability distributions.

For a PCS process $X(t)$, this function is polyperiodic in t for each τ. The associated Fourier series for this function is

$$R_X(t, t-\tau) = \sum_{\{\alpha\}} R_X^\alpha(\tau)\, e^{i2\pi\alpha(t-\tau/2)}$$

where $\{\alpha\}$ includes all values of α in the principal domain $(-\frac{1}{2}, \frac{1}{2}]$ for which the corresponding Fourier coefficient is not identically zero as a function of τ:

$$R_X^\alpha(\tau) \triangleq \left\langle R_X(t, t-\tau)\, e^{-i2\pi\alpha(t-\tau/2)} \right\rangle \not\equiv 0.$$

If this PCS process is PCE, then (with probability equal to one)

$$R_X^\alpha(\tau) = \hat{R}_X^\alpha(\tau) \triangleq \left\langle X(t)X^*(t-\tau)\, e^{-i2\pi\alpha(t-\tau/2)} \right\rangle.$$

The sine waves $\exp[i2\pi\alpha(t-\tau/2)]$ in the Fourier series introduced here contain the time shift $-\tau/2$ so that the discrete-time theory presented here will match the continuous-time theory (cf. Gardner, 1985) in which the function $R_X(t+\tau/2, t-\tau/2)$ is expanded in a Fourier series with unshifted sine waves $\exp(i2\pi\alpha t)$.

The second-order theory of PCS discrete time-series $x(t)$ deals with the time-space autocorrelation function

$$\begin{aligned}\hat{E}^{\{\alpha\}}\left\{x(t)x^*(t-\tau)\right\} &= \sum_{\{\alpha\}} \hat{E}^0\left\{x(t)x^*(t-\tau)\, e^{-i2\pi\alpha t}\right\} e^{i2\pi\alpha t} \\ &= \sum_{\{\alpha\}} \hat{R}_x^\alpha(\tau)\, e^{i2\pi\alpha(t-\tau/2)},\end{aligned}$$

where

$$\hat{R}_x^\alpha(\tau) \triangleq \hat{E}^0\{x(t)x^*(t-\tau)\, e^{-i2\pi\alpha(t-\tau/2)}\}.$$

That is, this theory deals with the sine-wave components in the delay product $x(t)x^*(t-\tau)$, whereas in the stochastic-process framework, we deal with an ensemble average that happens to be made up entirely of a sum of sine waves.

When our primary concern is the sine-wave components generated from $x(t)$ by the quadratic transformation $x(t)x^*(t-\tau)$, then the expectation operation $E\{\cdot\}$ and the associated ensemble are irrelevant. This being the case here, we proceed with the time-space theory. However, it is mentioned that the time-space theory presented can be translated to a probability-space theory (cf. Section 2) simply by following the rule:

> For all sinusoidally-weighted time averages $\langle z(t)e^{-i2\pi\alpha t}\rangle$ of time-series $z(t)$, replace $z(t)$ by the expected value $E\{Z(t)\}$ of the corresponding stochastic process $Z(t)$ to obtain $\langle E\{Z(t)\}e^{-i2\pi\alpha t}\rangle$ (when $\alpha = 0$, the operation $\langle\cdot\rangle$ can be omitted to obtain $E\{Z(t)\}$ *only* if $Z(t)$ is purely stationary).
>
> Common examples of $z(t)$ appearing in this presentation include delay products $z(t) = x(t)x^*(t-\tau)$ and cross products $z(t) = u(t)v^*(t)$.

In the remainder of this chapter, the circumflex notation (that was introduced in Section 2) on all time-average quantities is omitted for simplicity.

In the first part of this section, the possibility of generating spectral lines by simply squaring the signal is illustrated for two types of signals: the random-amplitude modulated sine wave and the random-amplitude modulated periodic pulse train. Then in the second part, it is explained that the property that enables spectral-line generation with some type of quadratic time-invariant transformation is called *cyclostationarity of order 2* (in the wide sense) and is characterized by the *cyclic autocorrelation function*, which is a generalization of the conventional autocorrelation function. Following this, it is shown that a signal exhibits cyclostationarity if and only if the signal is correlated with certain frequency-shifted versions of itself.

In the third and last part of this section, the correlation of frequency-shifted versions of a signal is localized in the frequency domain and this leads to the definition of a *spectral correlation density function*. It is then explained that this function is the Fourier transform of the cyclic autocorrelation function. This Fourier-transform relation between these two functions includes as a special case the well-known *Wiener relation* between the power spectral density function and the autocorrelation function. A normalization of the spectral correlation density function that converts it into a spectral correlation coefficient, whose magnitude is between zero and unity, is then introduced as a convenient measure of the degree of spectral redundancy in a signal.

Continuing in the final part of this section, the effects on the spectral correlation density function of several signal-processing operations are described. These include filtering and waveform multiplication, which in turn include the special cases of time delay and multipath propagation, bandlimiting, frequency conversion, and time sampling. These results are used to derive the spectral correlation density function for the random-amplitude modulated sine wave, the random-amplitude modulated pulse train, and the binary phase-shift keyed sine wave. The spectral correlation density functions for some other types of phase-shift keyed signals are also described graphically.

To conclude this section, the measurement of the (estimation of the ideal) spectral correlation density function is discussed and a particular algorithm for this purpose is illustrated with a simulation of a phase-shift keyed signal.

To complement similar treatments of this material (Gardner, 1987a; Gardner, 1991a), attention is focused in this section primarily on discrete-time signals rather than continuous-time signals.[13]

3.1 Spectral Line Generation

A discrete-time signal $x(t)$, for $t = 0, \pm 1, \pm 2, \pm 3, \ldots$, contains a *finite-strength additive sine-wave component* (an ac component) with frequency α, say

$$a \cos(2\pi\alpha t + \theta) \qquad \text{with } \alpha \neq 0 \tag{1}$$

[13]For convenience, the notation herein is modified from that in (Gardner, 1987a; Gardner 1991a): here, $\tilde{R}_x^\alpha$ and $\tilde{S}_x^\alpha$ are used for continuous time and R_x^α and S_x^α are used for discrete time.

if the Fourier coefficient

$$M_x^\alpha = \langle x(t)\, e^{-i2\pi\alpha t} \rangle \tag{2}$$

is not zero, in which case (1) gives

$$M_x^\alpha = \frac{1}{2} a\, e^{i\theta}.$$

In (2), the operation $\langle\cdot\rangle$ is the time-averaging operation

$$\langle\cdot\rangle \triangleq \lim_{Z\to\infty} \frac{1}{2Z+1} \sum_{t=-Z}^{Z} (\cdot).$$

In this case, the power spectral density (PSD) of $x(t)$ includes a spectral line at $f = \alpha$ and its image $f = -\alpha$. (The PSD is defined later in this section.) That is, the PSD in the principal domain $(-1/2, 1/2]$ contains the additive term[14]

$$\left|M_x^\alpha\right|^2 [\delta(f-\alpha) + \delta(f+\alpha)], \tag{3}$$

where $\delta(\cdot)$ is the Dirac delta, or impulse, function. For convenience in the sequel, it is said that such a signal exhibits *first-order periodicity*, with frequency α.

Let $x(t)$ be decomposed into the sum of its finite-strength sine-wave component, with frequency α, and its residual, say $n(t)$,

$$x(t) = a\cos(2\pi\alpha t + \theta) + n(t), \tag{4}$$

where $n(t)$ is defined to be that which is left after subtraction of (1) from $x(t)$. It is assumed that $n(t)$ is random. Here, the term *random* is used to denote nothing more than the vague notion of erratic or unpredictable behavior. If the sine wave is weak relative to the random residual, it might not be evident from visual inspection of $x(t)$ that it contains a periodic component. Hence, it is said to contain *hidden periodicity*. However, because of the associated spectral lines, hidden periodicity can be detected and in some applications exploited through techniques of spectral analysis.

This presentation is concerned with signals that contain more subtle types of hidden periodicity that, unlike first-order periodicity, do not give rise to spectral lines in the PSD, but that can be converted into first-order periodicity by a nonlinear time-invariant transformation of the signal. In particular, we shall focus on the type of hidden periodicity that can be converted by a quadratic transformation to yield spectral lines in the PSD.

The discussion begins with two motivating examples. In the convention used here, the PSD for $x(t)$ is denoted by $S_x(f)$ and is periodic with unity period. $\tilde{S}_x(f)$

---

[14]The strength of the spectral line is $\left|M_x^\alpha\right|^2$ as indicated in (3) if and only if the limit (2) exists in the temporal mean square sense with respect to the time parameter u obtained by replacing t with $t+u$ in (2) (Gardner, 1987a, Chapter 15, exc. 6).

denotes the PSD restricted to the principal domain $(-1/2, 1/2]$; therefore,

$$S_x(f) = \sum_{n=-\infty}^{\infty} \tilde{S}_x(f+n).$$

On occasion, continuous-time signals also are discussed herein. In such cases it is assumed that the signal is time-scaled and bandlimited so that the PSD is restricted to the band $(-1/2, 1/2]$. Consequently, the PSD of the discrete-time sampled version, restricted to the principal domain, will be identical to the PSD of the continuous-time signal. Consequently, the same notation, $\tilde{S}_x(f)$, is used for both.

Example 1: AM Let $a(t)$ be a real random lowpass signal (say lowpass filtered thermal noise) with the PSD $S_a(f)$ shown in Fig. 5a, which contains no spectral lines. If $a(t)$ is used to modulate the amplitude of a sine wave, we obtain the amplitude-modulated (AM) signal

$$x(t) = a(t)\cos(2\pi f_o t), \tag{5}$$

whose PSD $S_x(f)$ is given by (Gardner, 1987a, Chapter 3, Sec. D)

$$S_x(f) = \frac{1}{4}S_a(f+f_o) + \frac{1}{4}S_a(f-f_o) \tag{6}$$

as shown in Fig. 5b.

Although the PSD is centered about $f = f_o$ and $f = -f_o$, there is no spectral line at f_o or $-f_o$. The reason for this is that, as shown in Fig. 5a, there is no spectral line in $S_a(f)$ at $f = 0$. This means that the dc component

$$M_a^0 \triangleq \langle a(t) \rangle \tag{7}$$

is zero, since the strength of any spectral line at $f = 0$ is $|M_a^0|^2$.

Let us now square $x(t)$ to obtain

$$\begin{aligned} y(t) &= x^2(t) = a^2(t)\cos^2(2\pi f_o t) \\ &= \tfrac{1}{2}\left[b(t) + b(t)\cos(4\pi f_o t)\right] \end{aligned} \tag{8}$$

where

$$b(t) = a^2(t). \tag{9}$$

Since $b(t)$ is nonnegative, its dc value must be positive: $M_b^0 > 0$. Consequently, the PSD of $b(t)$ contains a spectral line at $f = 0$, as shown in Fig. 5c. The PSD for $y(t)$ is given by

$$S_y(f) = \frac{1}{4}\left[S_b(f) + \frac{1}{4}S_b(f+2f_o) + \frac{1}{4}S_b(f-2f_o)\right] \tag{10}$$

and, as shown in Fig. 5d, it contains spectral lines at $f = \pm 2f_o$ as well as at $f = 0$. Thus, by putting $x(t)$ through a quadratic transformation (a squarer in this case) we

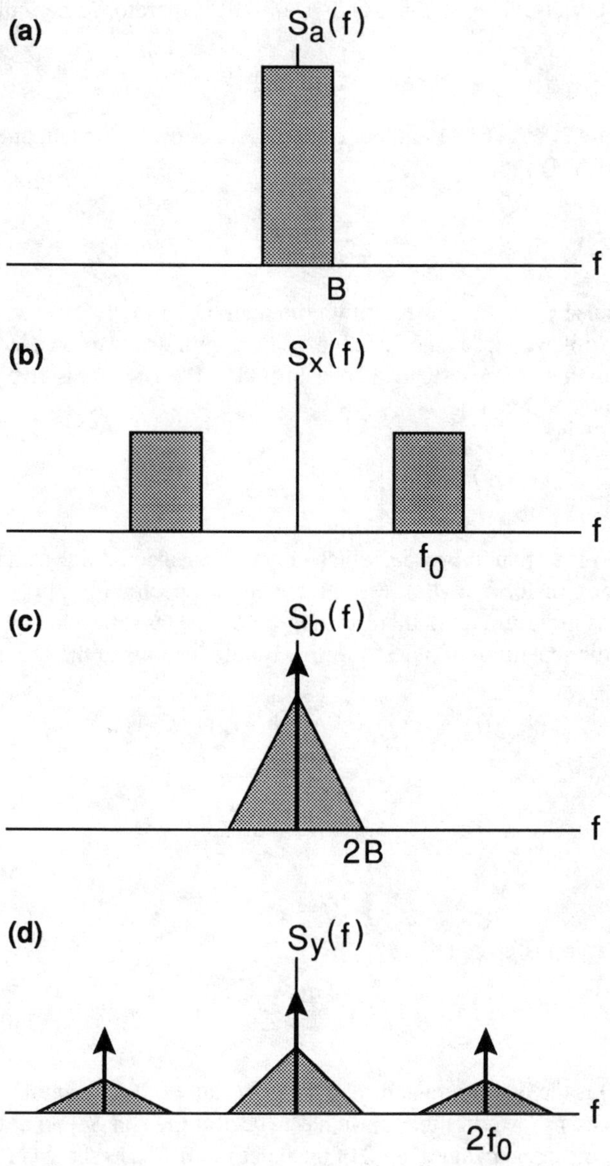

Figure 5: (a) Power spectral density (PSD) of a lowpass signal. (b) PSD of an amplitude-modulated (AM) signal. (c) PSD of a squared lowpass signal. (d) PSD of a squared AM signal.

have converted the hidden periodicity resulting from the sine-wave factor $\cos(2\pi f_o t)$ in (5) into first-order periodicity with associated spectral lines. This is particularly easy to see if $a(t)$ is a random binary sequence that switches back and forth between

$+1$ and -1 because then $b(t) \equiv 1$ and $y(t)$ in (8) is therefore a periodic signal:

$$y(t) = \frac{1}{2} + \frac{1}{2}\cos(4\pi f_o t).$$

Example 2: PAM As another example, we consider the real pulse-amplitude-modulated (PAM) signal

$$x(t) = \sum_{n=-\infty}^{\infty} a(nT_o)\, p(t - nT_o), \tag{11}$$

where the pulse $p(t)$ is confined within the interval $(-T_o/2, T_o/2)$ so that the pulse translates do not overlap, as shown in Fig. 6. For simplicity, we consider a continuous-time signal in this example (to avoid aliasing). The PSD of $x(t)$ is given by (Gardner, 1987a, Chapter 3, Sec. D)

$$\tilde{S}_x(f) = \frac{1}{T_o}\left|\tilde{P}(f)\right|^2 \sum_{m=-\infty}^{\infty} \tilde{S}_a(f - m/T_o), \tag{12}$$

where $\tilde{S}_a(f)$ is shown in Fig. 5a, which contains no spectral lines, and where $\tilde{P}(f)$ is the Fourier transform of $p(t)$. Since there are no spectral lines in $\tilde{S}_a(f)$ (or $\tilde{P}(f)$ since $p(t)$ has finite duration), there are none in $\tilde{S}_x(f)$, as shown in Fig. 7a, regardless of the periodic repetition of pulses in $x(t)$. But, let us look at the square of $x(t)$:

$$y(t) = x^2(t) = \sum_{n=-\infty}^{\infty} b(nT_o)\, q(t - nT_o), \tag{13}$$

where

$$b(nT_o) = a^2(nT_o) \tag{14a}$$

and

$$q(t) = p^2(t). \tag{14b}$$

The PSD for $y(t)$ is given by

$$\tilde{S}_y(f) = \frac{1}{T_o}\left|\tilde{Q}(f)\right|^2 \sum_{m=-\infty}^{\infty} \tilde{S}_b(f - m/T_o), \tag{15}$$

where $\tilde{Q}(f)$ is the Fourier transform of $q(t)$. Because of the spectral line at $f = 0$ in $\tilde{S}_b(f)$, which is shown in Fig. 5c, we have spectral lines in $\tilde{S}_y(f)$ at the harmonics m/T_o (for some integer values of m) of the pulse rate $1/T_o$, as shown in Fig. 7b. Thus, again, we have converted the hidden periodicity in $x(t)$ into first-order periodicity with associated spectral lines by using a quadratic transformation. This is particularly easy to see if $a(nT_o)$ is a random binary sequence with values ± 1, because then $b(nT_o) \equiv 1$ and $y(t)$ in (13) is therefore a periodic signal

$$y(t) = \sum_{n=-\infty}^{\infty} q(t - nT_o). \tag{16}$$

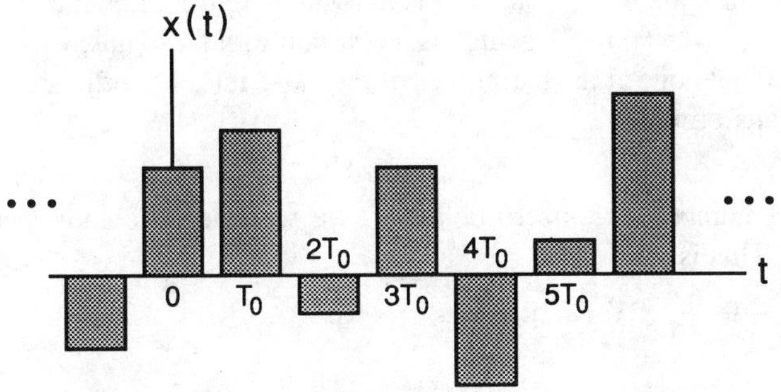

Figure 6: A pulse-amplitude-modulated (PAM) signal with pulse width less than interpulse time.

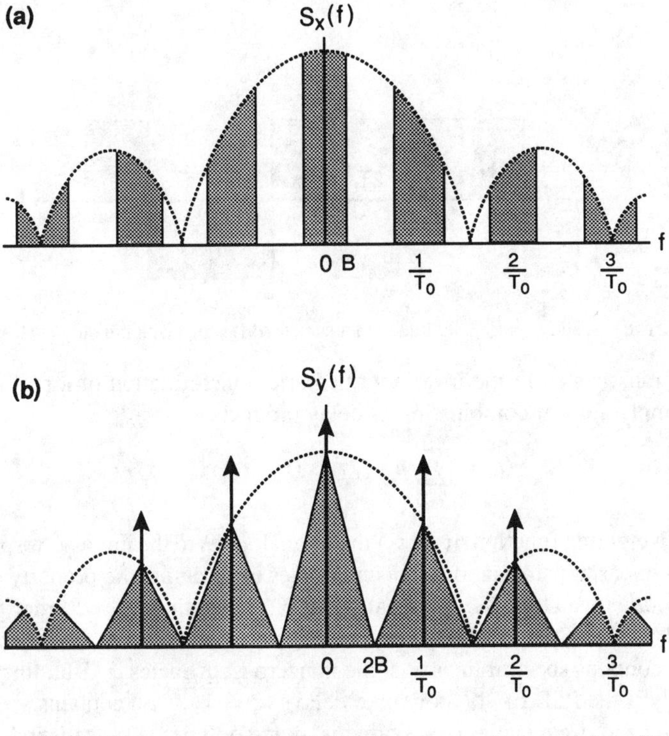

Figure 7: (a) Power spectral density (PSD) of a pulse-amplitude-modulated (PAM) signal with 67% duty-cycle pulses. (b) PSD of the squared PAM signal.

3.2 The Cyclic Autocorrelation Function

Although the squaring transformation works in these examples, a different quadratic transformation involving delays can be required in some cases. For example, if $a(nT_o)$

in Example 2 is again binary, but $p(t)$ is flat with height 1 and width T_o, as shown in Fig. 8, then $y(t) = x^2(t) = 1$, which is a constant for all t. Thus, we have a spectral line at $f = 0$ but none at the harmonics of the pulse rate. Nevertheless, if we use the quadratic transformation

$$y(t) = x(t)x(t - \tau) \tag{17}$$

for any of a number of nonzero delays τ, we will indeed obtain spectral lines at $f = m/T_o$. That is,

$$\begin{aligned} M_y^\alpha &= \langle y(t)\, e^{-i2\pi\alpha t} \rangle \\ &= \langle x(t)x(t - \tau)\, e^{-i2\pi\alpha t} \rangle \neq 0 \end{aligned} \tag{18}$$

for $\alpha = m/T_o$ for some m.

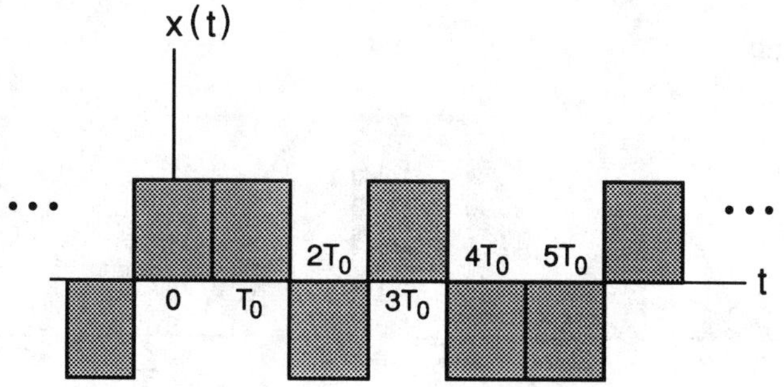

Figure 8: A binary pulse-amplitude-modulated (PAM) signal with full duty-cycle pulses.

The most general time-invariant quadratic transformation of a real time-series $x(t)$ is simply a linear combination of delay products

$$y(t) = \sum_{\tau_1, \tau_2} h(\tau_1, \tau_2)x(t - \tau_1)x(t - \tau_2)$$

for some weighting function $h(\tau_1, \tau_2)$ that is analogous to the impulse-response function for a linear transformation. This motivates us to define the property of *second-order periodicity* as follows: The real signal $x(t)$ contains second-order periodicity if and only if the PSD of the delay-product signal $x(t - \tau_1)x(t - \tau_2)$ for some delays τ_1 and τ_2 contains spectral lines at some nonzero frequencies α. But, this will be so if and only if the PSD of (17) for some delays ($\tau = \tau_2 - \tau_1$) contains spectral lines at some nonzero frequencies $\alpha \neq 0$; that is, if and only if (18) is satisfied.

In developing the continuous-time theory of second-order periodicity it has been found to be more convenient to work with the symmetric delay product

$$y_\tau(t) = x(t + \tau/2)x^*(t - \tau/2). \tag{19}$$

The complex conjugate $*$ is introduced here for generality to accommodate complex-valued signals, but it is mentioned that for some complex-valued signals, the quadratic transformation without the conjugate can also be useful (Gardner, 1987a, Chapter 10, Sec. C). From (19), the fundamental parameter (18) of second-order periodicity for continuous time becomes

$$\tilde{R}_x^\alpha(\tau) \triangleq \langle x(t+\tau/2)x^*(t-\tau/2)\, e^{-i2\pi\alpha t}\rangle, \tag{20a}$$

which is the Fourier coefficient $M_{y_\tau}^\alpha$ of the additive sine-wave component with frequency α contained in the delay-product signal $y_\tau(t)$. However, for discrete-time signals, delays equal to half the sampling increment are not allowed. Nevertheless, since

$$\langle x(t)x^*(t-\tau)e^{-i2\pi\alpha t}\rangle = \tilde{R}_x^\alpha(\tau)\, e^{-i\pi\alpha\tau}$$

for continuous time, then we can define the fundamental parameter of second-order periodicity for discrete time as follows

$$R_x^\alpha(\tau) \triangleq \langle x(t)x^*(t-\tau)\, e^{-i2\pi\alpha t}\rangle\, e^{i\pi\alpha\tau} \tag{20b}$$

in order to maintain the strongest analogy between the continuous- and discrete-time theories. Observe that since t and τ take on only integer values, then $R_x^\alpha(\tau)$ is periodic in α with period two, and also $R_x^{\alpha+1}(\tau) = R_x^\alpha(\tau)e^{i\pi\tau}$.

The notation $R_x^\alpha(\tau)$ is introduced for this Fourier coefficient because, for $\alpha = 0$, (20) reduces to the conventional autocorrelation function

$$R_x^0(\tau) = \langle x(t)x^*(t-\tau)\rangle,$$

for which the notation $R_x(\tau)$ is commonly used. Furthermore, since $R_x^\alpha(\tau)$ is a generalization of the autocorrelation function, in which a cyclic (sinusoidal) weighting factor $e^{-i2\pi\alpha t}$ is included before the time-averaging is carried out, $R_x^\alpha(\tau)$ is called the *cyclic autocorrelation function.* Also, the *conjugate cyclic autocorrelation* for complex-valued signals obtained from (20) by deleting the conjugate,

$$R_{xx^*}^\alpha(\tau) = \langle x(t)x(t-\tau)\, e^{-i2\pi\alpha t}\rangle\, e^{i\pi\alpha\tau}, \tag{21}$$

is a further modification of the conventional autocorrelation.[15]

Thus, we have two distinct interpretations of $R_x^\alpha(\tau) = M_{y_\tau}^\alpha$. In fact, we have yet a third distinct interpretation, which can be obtained by simply factoring $\exp(-i2\pi\alpha t)$ in order to reexpress (20) as

$$R_x^\alpha(\tau) = \Big\langle \big[x(t)\, e^{-i\pi\alpha t}\big]\big[x(t-\tau)\, e^{+i\pi\alpha(t-\tau)}\big]^* \Big\rangle. \tag{22}$$

[15] Although some readers will recognize the similarity between the cyclic autocorrelation function and the radar ambiguity function, the relationship between these two functions is only superficial. The concepts and theory underlying the cyclic autocorrelation function, as summarized in this article, have little in common with the concepts and theory of radar ambiguity (cf. (Gardner, 1987a, Chapter 10, Sec. C)). For example, the radar ambiguity function has no meaning relevant to ambiguity (in Doppler) when applied to a real signal, or when applied to a complex signal without the conjugate.

That is, $R_x^\alpha(\tau)$ is actually a conventional crosscorrelation function

$$R_{uv}(\tau) \triangleq \langle u(t)v^*(t-\tau)\rangle = R_x^\alpha(\tau), \tag{23}$$

where

$$u(t) = x(t)\, e^{-i\pi\alpha t} \tag{24a}$$

and

$$v(t) = x(t)\, e^{+i\pi\alpha t} \tag{24b}$$

are frequency translates of $x(t)$. Recall that multiplying a signal by $\exp\{\pm i\pi\alpha t\}$ shifts the spectral content of the signal by $\pm\alpha/2$. For example, the PSDs of $u(t)$ and $v(t)$ are

$$S_u(f) = S_x(f+\alpha/2) \tag{25}$$

and

$$S_v(f) = S_x(f-\alpha/2). \tag{26}$$

It follows from (23) and (24) that $x(t)$ exhibits second-order periodicity ((20) is not identically zero as a function of τ for some $\alpha \neq 0$) if and only if *frequency translates* of $x(t)$ are correlated with each other in the sense that (23) is not identically zero as a function of τ for some $\alpha \neq 0$ in (24). This third interpretation of $R_x^\alpha(\tau)$ suggests an appropriate way to normalize $R_x^\alpha(\tau)$ as explained next.

As long as the mean values of the frequency translates $u(t)$ and $v(t)$ are zero (which means that $x(t)$ does not contain finite-strength[16] additive sine-wave components at frequencies $\pm\alpha/2$ and, therefore, that $S_x(f)$ has no spectral lines at $f = \pm\alpha/2$), the crosscorrelation $R_{uv}(\tau) \equiv R_x^\alpha(\tau)$ is actually a temporal crosscovariance $K_{uv}(\tau)$. That is,

$$\begin{aligned} K_{uv}(\tau) &\triangleq \langle [u(t)-\langle u(t)\rangle]\,[v(t-\tau)-\langle v(t-\tau)\rangle]^*\rangle \\ &= \langle u(t)v^*(t-\tau)\rangle = R_{uv}(\tau) \end{aligned} \tag{27}$$

An appropriate normalization for the temporal crosscovariance is the geometric mean of the two corresponding temporal variances. This yields a temporal correlation coefficient, the magnitude of which is upper bounded by unity. It follows from (24) that the two variances are given by

$$K_u(0) = R_u(0) = \langle |u(t)|^2\rangle = R_x(0) \tag{28a}$$

and

$$K_v(0) = R_v(0) = \langle |v(t)|^2\rangle = R_x(0). \tag{28b}$$

Therefore, the temporal correlation coefficient for frequency translates is given by

$$\frac{K_{uv}(\tau)}{[K_u(0)K_v(0)]^{1/2}} = \frac{R_x^\alpha(\tau)}{R_x(0)} \triangleq \gamma_x^\alpha(\tau). \tag{29}$$

[16]It does contain infinitesimal sine-wave components.

Hence, the appropriate normalization factor for the cyclic autocovariance $R_x^\alpha(\tau)$ is simply $1/R_x(0)$ (and it is the same for the conjugate cyclic autocovariance).

This is a good point at which to introduce some more terminology. A signal $x(t)$ for which the autocorrelation $R_x(\tau)$ exists (e.g., remains finite as the averaging time goes to infinity) and is not identically zero (as it is for transient signals) is commonly said to be *stationary* of second order (in the wide sense). But we need to refine the terminology to distinguish between those stationary signals that exhibit second-order periodicity ($R_x^\alpha(\tau) \not\equiv 0$ for some $\alpha \neq 0$) and those stationary signals that do not ($R_x^\alpha(\tau) \equiv 0$ for all $\alpha \neq 0$). Consequently, we shall call the latter for which $R_x^\alpha(\tau) \equiv 0$ *stationary* of second order (in the wide sense) and the former for which $R_x^\alpha(\tau) \not\equiv 0$ for some values of α that are integer multiples of a single fundamental frequency $1/T$ (corresponding to the period T) *cyclostationary* of second order (in the wide sense). If there is more than one fundamental frequency, then we call the signal *polycyclostationary* of second order (in the wide sense). We shall also call any nonzero value of the frequency parameter α in the principal domain $(-\frac{1}{2}, \frac{1}{2}]$ for which $R_x^\alpha(\tau) \not\equiv 0$ a *cycle frequency*. The discrete set of cycle frequencies is called the *cycle spectrum*. For example, if a signal is cyclostationary, the cycle spectrum contains only harmonics (integer multiples) of the fundamental cycle frequency, which is the reciprocal of the fundamental period. But if the signal is polycyclostationary, then the cycle spectrum contains harmonics of each of the incommensurate fundamental cycle frequencies.

We conclude this section by reconsidering the AM example and determining the cyclic autocorrelation function for the AM signal.

Example 1 continued: AM Let $a(t)$ be a real random stationary signal with zero mean:

$$\langle a(t) \rangle = 0, \tag{30a}$$

$$\langle a(t)a^*(t-\tau) \rangle \not\equiv 0, \tag{30b}$$

$$\langle a(t)a^*(t-\tau)\, e^{-i2\pi\alpha t} \rangle \equiv 0 \qquad \text{for all } \alpha \neq 0. \tag{30c}$$

Equation (30c) guarantees that

$$\langle a(t)\, e^{-i2\pi\alpha t} \rangle \equiv 0 \qquad \text{for all } \alpha \neq 0. \tag{30d}$$

We consider the amplitude-modulated sine wave

$$\begin{aligned} x(t) &= a(t)\cos(2\pi f_o t + \theta) \\ &= \frac{1}{2} a(t) \left[e^{i(2\pi f_o t + \theta)} + e^{-i(2\pi f_o t + \theta)} \right]. \end{aligned} \tag{31}$$

Because of (30d), $a(t)$ contains no finite-strength additive sine-wave components and, therefore (together with (30a)), $x(t)$ contains no finite-strength additive sine-wave components. This means that its power spectral density contains no spectral lines. However, the quadratic transformation

$$y_\tau(t) = x(t)x^*(t-\tau)$$
$$= a(t)a^*(t-\tau)\frac{1}{4}\left[e^{i2\pi f_o\tau} + e^{-i2\pi f_o\tau} + e^{i(4\pi f_o t + 2\theta)}e^{-i2\pi f_o\tau} + e^{-i(4\pi f_o t+2\theta)}e^{i2\pi f_o\tau}\right] \tag{32}$$

does contain finite-strength additive sine-wave components with frequencies $\alpha = \pm 2f_o$, since (30b) renders one or the other of the last two terms in the quantity

$$\begin{aligned}\langle y_\tau(t)\,e^{-i2\pi\alpha t}\rangle &= \frac{1}{4}\,e^{i2\pi f_o\tau}\langle a(t)a^*(t-\tau)e^{-i2\pi\alpha t}\rangle \\ &+\frac{1}{4}\,e^{-i2\pi f_o\tau}\langle a(t)a^*(t-\tau)\,e^{-i2\pi\alpha t}\rangle \\ &+\frac{1}{4}\,e^{i2\theta}\,e^{-i2\pi f_o\tau}\langle a(t)a^*(t-\tau)\,e^{-i2\pi(\alpha-2f_o)t}\rangle \\ &+\frac{1}{4}\,e^{-i2\theta}\,e^{i2\pi f_o\tau}\langle a(t)a^*(t-\tau)\,e^{-i2\pi(\alpha+2f_o)t}\rangle\end{aligned} \tag{33}$$

nonzero for $\alpha = \pm 2f_o$. That these are the only two nonzero cycle frequencies follows from the fact that (30c) renders (33) equal to zero for all α except $\alpha = 0$ and $\alpha = \pm 2f_o$. Thus, the cycle spectrum consists of only the two cycle frequencies $\alpha = \pm 2f_o$ and the degenerate cycle frequency $\alpha = 0$.

Hence, the versions $u(t)$ and $v(t)$ of $x(t)$ obtained by frequency shifting $x(t)$ up and down by $\alpha/2 = f_o$ are correlated. This is not surprising since (31) reveals that $x(t)$ is obtained from $a(t)$ by frequency shifting up and down by f_o and then adding. In conclusion, we have the cyclic autocorrelation function (in the principal domain of α)

$$R_x^\alpha(\tau) = \begin{cases} \frac{1}{4}e^{\pm i2\theta}R_a(\tau) & \text{for } \alpha = \pm 2f_o \\ \frac{1}{2}R_a(\tau)\cos(2\pi f_o\tau) & \text{for } \alpha = 0 \\ 0 & \text{otherwise,}\end{cases} \tag{34}$$

the magnitude of which is graphed in Fig. 9 for a typical autocorrelation $R_a(\tau)$. It follows from (34) that the temporal correlation coefficient is given by

$$\gamma_x^\alpha(\tau) = \begin{cases} \frac{1}{2}e^{\pm i2\theta}\gamma_a^0(\tau) & \text{for } \alpha = \pm 2f_o \\ \gamma_a^0(\tau)\cos(2\pi f_o\tau) & \text{for } \alpha = 0 \\ 0 & \text{otherwise.}\end{cases} \tag{35a}$$

Thus, the strength of correlation between $x(t)\exp(-i\pi\alpha t)$ and $x(t-\tau)\exp(i\pi\alpha[t-\tau])$, which is given by

$$\left|\gamma_x^\alpha(\tau)\right| = \frac{1}{2}\left|\gamma_a^0(\tau)\right|, \tag{35b}$$

can be substantial (as large as $1/2$ for this amplitude-modulated signal.

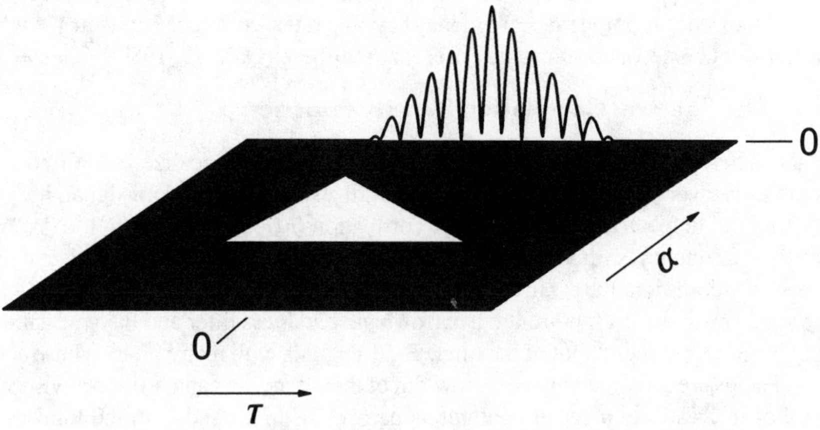

Figure 9: Magnitude of the cyclic autocorrelation function for an AM signal graphed as the height of a surface above the time-frequency plane with coordinates τ and α.

As an especially simple specific example of $a(t)$, we consider as before a random binary sequence, which switches back and forth between $+1$ and -1. If we set $\tau = 0$ in (32), we obtain

$$y_0(t) = |x(t)|^2 = |a(t)|^2 \cos^2(2\pi f_o t + \theta)$$
$$= \frac{1}{2} + \frac{1}{2}\cos(4\pi f_o t + 2\theta),$$

which clearly contains finite-strength additive sine-wave components with frequencies $\alpha = \pm 2 f_o$. In fact, in this very special case, there is no random component in $y_0(t)$. On the other hand, for $\tau \neq 0$, $y_\tau(t)$ contains both a sine-wave component and a random component.

To illustrate the conjugate cyclic autocorrelation (21), let us consider the analytic signal for AM,

$$z(t) = \frac{1}{2} a(t)\, e^{i(2\pi f_o t + \theta)}.$$

For this signal, we have

$$R^\alpha_{zz^*}(\tau) \triangleq \langle z(t) z(t-\tau)\, e^{-i2\pi\alpha t}\rangle\, e^{i\pi\alpha\tau}$$
$$= \frac{1}{4}\langle a(t)a(t-\tau)\, e^{i2\pi(2f_o - \alpha)t}\rangle\, e^{-i[2\pi(f_o - \alpha/2)\tau - 2\theta]}$$
$$= \begin{cases} \frac{1}{4} R_a(\tau)\, e^{i2\theta} & \text{for } \alpha = 2 f_o \\ 0 & \text{otherwise.} \end{cases}$$

Other examples of cyclostationary and polycyclostationary signals can be similarly viewed as mixtures of stationarity and periodicity. Examples are cited in

Section 1. Typical cycle spectra include harmonics of pulse rates, keying rates, spreading-code chipping rates, frequency hopping rates, code repetition rates, doubled carrier frequencies, and sums and differences of these (Gardner, 1987a, Chapter 12).

3.3 The Spectral Correlation Density Function

In the same way that it is beneficial for some purposes to localize in the frequency domain the average power $\langle |x(t)|^2 \rangle = R_x(0)$ in a stationary random signal, it can be very helpful to localize in frequency the correlation $\langle u(t)v^*(t) \rangle = \langle |x(t)|^2 e^{-i2\pi\alpha t} \rangle \equiv R_x^\alpha(0)$ of frequency-shifted signals $u(t)$ and $v(t)$ for a cyclostationary or polycyclostationary random signal $x(t)$. In the former case of localizing the power, we simply pass the signal of interest $x(t)$ through a narrowband bandpass filter and then measure the average power at the output of the filter. By doing this with many filters whose center frequencies are separated by the bandwidth of the filters, we can partition any spectral band of interest into a set of contiguous narrow disjoint bands. In the limit as the bandwidths approach zero, the corresponding set of measurements of average power, normalized by the bandwidth, constitute the power spectral density (PSD) function. That is, at any particular frequency f (in the principal domain $(-1/2, 1/2]$), the PSD for $x(t)$ is given by

$$S_x(f) \triangleq \lim_{B\to 0} \frac{1}{B} \left\langle \left| h_B^f(t) \otimes x(t) \right|^2 \right\rangle, \tag{36}$$

where $\otimes$ denotes convolution and $h_B^f(t)$ is the discrete-impulse response of a one-sided bandpass filter with center frequency f, bandwidth B, and unity gain at the band center (see Fig. 10).

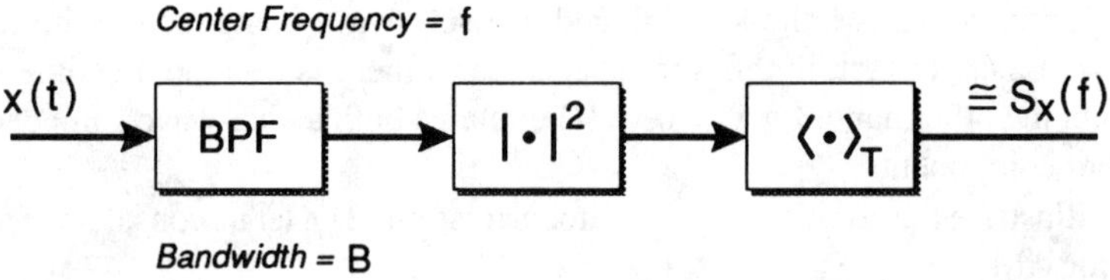

Figure 10: One channel of a spectrum analyzer for measuring the power spectral density (PSD). (The symbol $\cong$ indicates that the output only approximates the ideal function $S_x(f)$ for finite T and B.)

In the latter case of localizing the correlation, we simply pass both of the two frequency translates $u(t)$ and $v(t)$ of $x(t)$ through the same set of bandpass filters that are used for the PSD and then measure the temporal correlation of the filtered signals (see Fig. 11) to obtain

$$S_x^\alpha(f) \triangleq \lim_{B\to 0} \frac{1}{B} \left\langle \left[h_B^f(t) \otimes u(t) \right] \left[h_B^f(t) \otimes v(t) \right]^* \right\rangle, \tag{37}$$

which is called the *spectral correlation density (SCD) function.* This yields the spectral density of correlation in $u(t)$ and $v(t)$ at frequency f, which is identical to the spectral density of correlation in $x(t)$ at frequencies $f+\alpha/2$ and $f-\alpha/2$ (see Fig. 12). That is, $S_x^\alpha(f)$ is the bandwidth-normalized (i.e., divided by B) correlation of the amplitude and phase fluctuations of the narrowband spectral components in $x(t)$ centered at frequencies $f+\alpha/2$ and $f-\alpha/2$, in the limit as the bandwidth B of these narrowband components approaches zero. For complex-valued signals, the conjugate SCD obtained from (37) by deleting the complex conjugate is also of interest for some signals (Gardner, 1987a, Chapter 10, Sec. C).

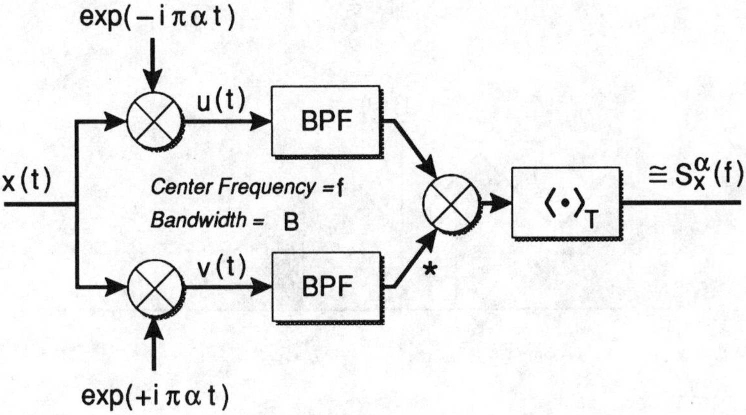

Figure 11: One channel-pair of a spectral correlation analyzer (or a cyclic spectrum analyzer) for measuring the spectral correlation density (or cyclic spectral density).

Strictly speaking, the SCD is not a valid density function in the usual sense, since it is not nonnegative and, in fact, not even real-valued. However, its integral over all frequencies does equal the correlation of $u(t)$ and $v(t)$ and, when $u(t)$ and $v(t)$ are decomposed into narrowband spectral components, the correlation of the components centered at f is indeed the SCD evaluated at f. Because of the lack of the nonnegativity property of the SCD, the correlation of $u(t)$ and $v(t)$ can equal zero without the SCD being identically zero because the integral of the SCD over all f can be zero even though the SCD is not identically zero. Nevertheless, because of the properties that the SCD does share with densities like the PSD, the term *density* is retained.

It is well known (see, for example, (Gardner, 1987a, Chapter 3, Sec. C) for a proof for continuous time) that the PSD obtained from (36) is equal to the Fourier transform of the autocorrelation function,

$$S_x(f) = \sum_{\tau=-\infty}^{\infty} R_x(\tau)\, e^{-i2\pi f\tau}. \tag{38}$$

Similarly, it can be shown (cf. (Gardner, 1987a, Chapter 11, Sec. C) for continuous time) that the SCD (or conjugate SCD) obtained from (37) is the Fourier transform of the cyclic autocorrelation function (or conjugate cyclic autocorrelation),

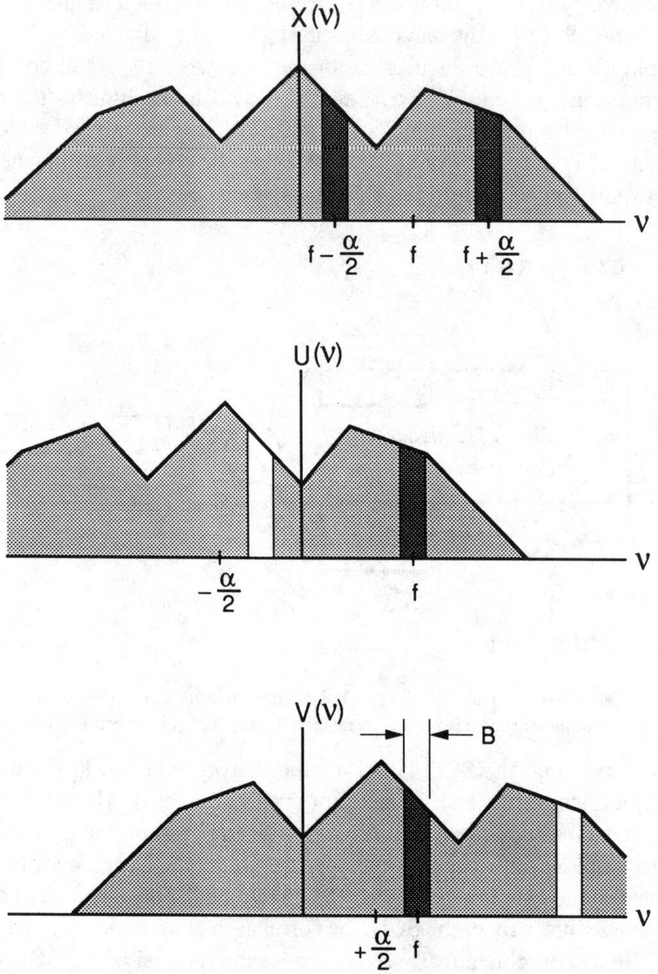

Figure 12: Illustration of spectral bands used in the measurement of the spectral correlation density $S_x^\alpha(f)$. (ν is a dummy frequency variable; the shaded bands are the bands selected by the BPFs.)

$$S_x^\alpha(f) = \sum_{\tau=-\infty}^{\infty} R_x^\alpha(\tau)\, e^{-i2\pi f\tau} \tag{39a}$$

and, therefore, $R_x^\alpha(\tau)$ is given by the inverse transform

$$R_x^\alpha(\tau) = \int_{-1/2}^{1/2} S_x^\alpha(f)\, e^{i2\pi f\tau}\, df. \tag{39b}$$

Since $R_x^\alpha(\tau)$ is periodic in α with period two, so too is $S_x^\alpha(f)$. Also, since τ takes on only integer values, then $S_x^\alpha(f)$ is periodic in f with period one. Furthermore, since increasing $f \pm \alpha/2$ by ± 1 has no effect on the spectral components at these frequencies, then it follows that $S_x^\alpha(f)$ also exhibits the periodicity $S_x^{\alpha+1}(f+\frac{1}{2}) = S_x^\alpha(f)$. Consequently, the principal domain for $S_x^\alpha(f)$ can be taken to be either the square with vertices $(f, \alpha) = (\pm\frac{1}{2}, \pm\frac{1}{2})$ or the diamond with vertices $(f, \alpha) = (0, \pm 1)$ and $\pm\frac{1}{2}, 0)$. Relation (38) is known as the *Wiener relation* (see, for example, (Gardner, 1987a, Chapter 3, Sec. C)), and (39) is therefore called the *cyclic Wiener relation* (Gardner, 1987a, Chapter 11, Sec. C). The cyclic Wiener relation includes the Wiener relation as the special case of $\alpha = 0$. (In the probabilistic framework of stochastic processes, which is based on expected values [ensemble averages] instead of time averages, the probabilistic counterpart of (38) is known as the *Wiener-Khinchin relation* and, therefore, the probabilistic counterpart of (39) is called the *cyclic Wiener-Khinchin relation* (Gardner, 1990a, Chapter 12, Sec. 12.2).) Because of the relation (39), the SCD is also called the *cyclic spectral density* function (Gardner, 1987a, Chapter 10, Sec. B).

It follows from (39) and the interpretation (23) of $R_x^\alpha(\tau)$ as $R_{uv}(\tau)$ that the SCD is the Fourier transform of the crosscorrelation function $R_{uv}(\tau)$ and is therefore identical to the cross-spectral density function for the frequency translates $u(t)$ and $v(t)$:

$$S_x^\alpha(f) \equiv S_{uv}(f), \tag{40}$$

where $S_{uv}(f)$ is defined by the right-hand side of (37) for arbitrary $u(t)$ and $v(t)$. This is to be expected since the cross-spectral density $S_{uv}(f)$ is known (cf. (Gardner, 1987a, Chapter 7, Sec. A)) to be the spectral correlation density for spectral components in $u(t)$ and $v(t)$ at frequency f, and $u(t)$ and $v(t)$ are frequency-shifted versions of $x(t)$. The identity (40) suggests an appropriate normalization for $S_x^\alpha(f)$: As long as the PSDs of $u(t)$ and $v(t)$ contain no spectral lines at frequency f, which means that the PSD of $x(t)$ contains no spectral lines at either of the frequencies $f \pm \alpha/2$, then the correlation of the spectral components (40) is actually a covariance since the means of the spectral components are zero (Gardner, 1987a, Chapter 11, Sec. C). When normalized by the geometric mean of the corresponding variances, which are given by

$$S_u(f) = S_x(f + \alpha/2) \tag{41a}$$

and

$$S_v(f) = S_x(f - \alpha/2), \tag{41b}$$

the covariance becomes a correlation coefficient:

$$\frac{S_{uv}(f)}{[S_u(f)S_v(f)]^{1/2}} = \frac{S_x^\alpha(f)}{[S_x(f+\alpha/2)S_x(f-\alpha/2)]^{1/2}} \triangleq \rho_x^\alpha(f). \tag{42}$$

Since $|\rho_x^\alpha(f)|$ is bounded to the interval [0, 1], it is a convenient measure of the degree of local spectral redundancy that results from spectral correlation. For example, for $|\rho_x^\alpha(f)| = 1$, we have complete spectral redundancy at $f + \alpha/2$ and $f - \alpha/2$. For conjugate spectral redundancy of complex-valued signals, (42) is modified by replacing the numerator with the conjugate SCD.

Let us now return to the AM example considered previously.

Example 1 continued: AM By Fourier transforming (34) and invoking the cyclic Wiener relation (39), we obtain the following SCD function on the principal domain for the amplitude-modulated signal (31):

$$S_x^\alpha(f) = \begin{cases} \frac{1}{4}e^{\pm i2\theta}S_a(f) & \text{for } \alpha = \pm 2f_o \\ \frac{1}{4}S_a(f+f_o) + \frac{1}{4}S_a(f-f_o) & \text{for } \alpha = 0 \\ 0 & \text{otherwise,} \end{cases} \tag{43}$$

where it has been assumed that $S_a(f \pm f_o) = 0$ for $|f| > 1/2$ to avoid aliasing effects in the principal domain. The magnitude of this SCD is graphed in Fig. 13 as the height of a surface above the bifrequency plane with coordinates f and α. For purposes of illustration, $a(t)$ is assumed to have an arbitrary low pass PSD for this graph. Observe that although the argument f of the SCD is continuous, as it always will be for a random signal, the argument α is discrete, as it always will be since it represents the harmonic frequencies of periodicities underlying the random time-series (the sine-wave carrier in this example).

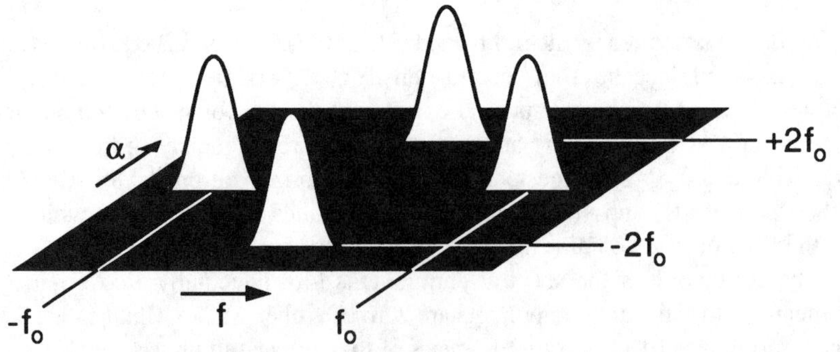

Figure 13: Magnitude of the spectral correlation density function for an AM signal graphed as a height above the bifrequency plane with coordinates f and α.

It follows from (43) that the spectral correlation coefficient is given by

$$\rho_x^\alpha(f) = \frac{S_a(f)\, e^{\pm i2\theta}}{\{[S_a(f+2f_o) + S_a(f)]\,[S_a(f) + S_a(f-2f_o)]\}^{1/2}} \quad \text{for } \alpha = \pm 2f_o. \tag{44a}$$

Thus, the strength of correlation between spectral component in $x(t)$ at frequencies $f+\alpha/2$ and $f-\alpha/2$ is unity:

$$\left|\rho_x^\alpha(f)\right| = 1 \quad \text{for } |f| < f_o \text{ and } \alpha = \pm 2f_o, \tag{44b}$$

provided that $a(t)$ is bandlimited to $|f| \le f_o$,

$$S_a(f) = 0 \quad \text{for } |f| \ge f_o. \tag{45}$$

This is not surprising since the two spectral components in $x(t)$ at frequencies $f \pm \alpha/2 = f \pm f_o$ are obtained from the single spectral component in $a(t)$ at frequency f simply by shifting and scaling. Thus, they are perfectly correlated. That is, the upper (lower) sideband for $f > 0$ carries exactly the same information as the lower (upper) sideband for $f < 0$. Techniques for exploiting this *spectral redundancy* are described in Section 4.

To illustrate the conjugate SCD, we consider the analytic signal $z(t)$ for AM.

$$S^{\alpha}_{zz^*}(f) = \sum_{\tau=-\infty}^{\infty} R^{\alpha}_{zz^*}(\tau)\, e^{-i2\pi f\tau}$$
$$= \begin{cases} \frac{1}{4} S_a(f)\, e^{i2\theta} & \text{for } \alpha = 2f_o \\ 0 & \text{otherwise.} \end{cases}$$

Before considering other examples of the SCD, let us first gain an understanding of the effects of some basic signal-processing operations on the SCD. This greatly facilitates the determination of the SCD for commonly encountered man-made signals.

3.4 Filtering

When a signal $x(t)$ undergoes a linear time-invariant (LTI) transformation (i.e., a convolution or a filtering operation),

$$\begin{aligned} z(t) &= h(t) \otimes x(t) \\ &\triangleq \sum_{u=-\infty}^{\infty} h(u)x(t-u), \end{aligned} \tag{46}$$

the spectral components in $x(t)$ are simply scaled by the complex-valued transfer function $H(f)$, which is the Fourier transform

$$H(f) = \sum_{t=-\infty}^{\infty} h(t)\, e^{-i2\pi ft} \tag{47}$$

of the discrete-impulse-response function $h(t)$ of the transformation. As a result, the PSD gets scaled by the squared magnitude of $H(f)$ (see, for example, (Gardner, 1987a, Chapter 3, Sec. C) or (Gardner, 1990a, Chapter 10, Sec. 10.1) for continuous time)

$$S_z(f) = |H(f)|^2\, S_x(f). \tag{48}$$

Equation (48) can be derived from the definition (36) of the PSD. Similarly, because the spectral components of $x(t)$ at frequencies $f \pm \alpha/2$ are scaled by $H(f \pm \alpha/2)$, the SCD gets scaled by the product $H(f + \alpha/2)H^*(f - \alpha/2)$:

$$S^{\alpha}_z(f) = H(f + \alpha/2)H^*(f - \alpha/2)S^{\alpha}_x(f). \tag{49}$$

This result, called the *input-output SCD relation for filtering*, which can be derived from the definition (37) of the SCD, includes (48) as the special case of $\alpha = 0$. Observe that it follows from (49) and the definition (42) that

$$\left|\rho_z^\alpha(f)\right| \equiv \left|\rho_x^\alpha(f)\right|. \tag{50}$$

That is, the magnitude of the spectral correlation coefficient is unaffected by filtering (if $H(f \pm \alpha/2) \neq 0$).

Example 3: Time Delay As our first example of (49), we consider a filter that simply delays the input by some integer t_o; then $h(t) = \delta(t - t_o)$, where δ is the Kronecker delta, and $H(f) = e^{-i2\pi f t_o}$. Therefore, for $z(t) = x(t - t_o)$, we obtain from the input-output SCD relation (49)

$$S_z^\alpha(f) = S_x^\alpha(f)\, e^{-i2\pi\alpha t_o}, \tag{51}$$

which indicates that, unlike the PSD, the SCD of a cyclostationary signal is sensitive to the timing or phase of the signal.

Example 4: Multipath Propagation As a second example of (49), if $x(t)$ undergoes multipath propagation during transmission to yield a received signal

$$z(t) = \sum_n a_n x(t - t_n),$$

where a_n and the integer t_n are the attenuation factor and delay of the nth propagation path, we have

$$H(f) = \sum_n a_n e^{-i2\pi f t_n} \tag{52}$$

and therefore (49) yields

$$S_z^\alpha(f) = S_x^\alpha(f) \sum_{n,m} a_n a_m^* \exp(-i2\pi[f(t_n - t_m) + \alpha(t_n + t_m)/2]). \tag{53}$$

Example 5: Bandpass Signals As a third example of the utility of the relation (49), let us determine the support region in the (f, α) plane for a bandpass signal with lowest frequency b and highest frequency B. To enforce such a spectrum, we can simply put any signal $x(t)$ through an ideal bandpass filter with transfer function (on the principal domain $(-1/2, 1/2]$)

$$H(f) = \begin{cases} 1 & \text{for } b < |f| < B \\ 0 & \text{otherwise.} \end{cases}$$

It then follows directly from the input-output SCD relation (49) that the SCD for the output of this filter can be nonzero only for $||f| - |\alpha|/2| > b$ and $|f| + |\alpha|/2 < B$:

$$S_z^\alpha(f) = \begin{cases} 0 & \text{for } ||f| - |\alpha|/2| \leq b \text{ or } |f| + |\alpha|/2 \geq B, \\ S_x^\alpha(f) & \text{otherwise.} \end{cases} \tag{54}$$

This shows that the support region in the (f, α) plane for a bandpass signal is the four diamonds located at the vertices of a larger diamond, depicted in Fig. 14a. By letting $b \to 0$, we obtain the support region for a lowpass signal, and by letting $B \to 1/2$, we obtain the support region for a highpass signal. This is shown in Figs. 14b and 14c.

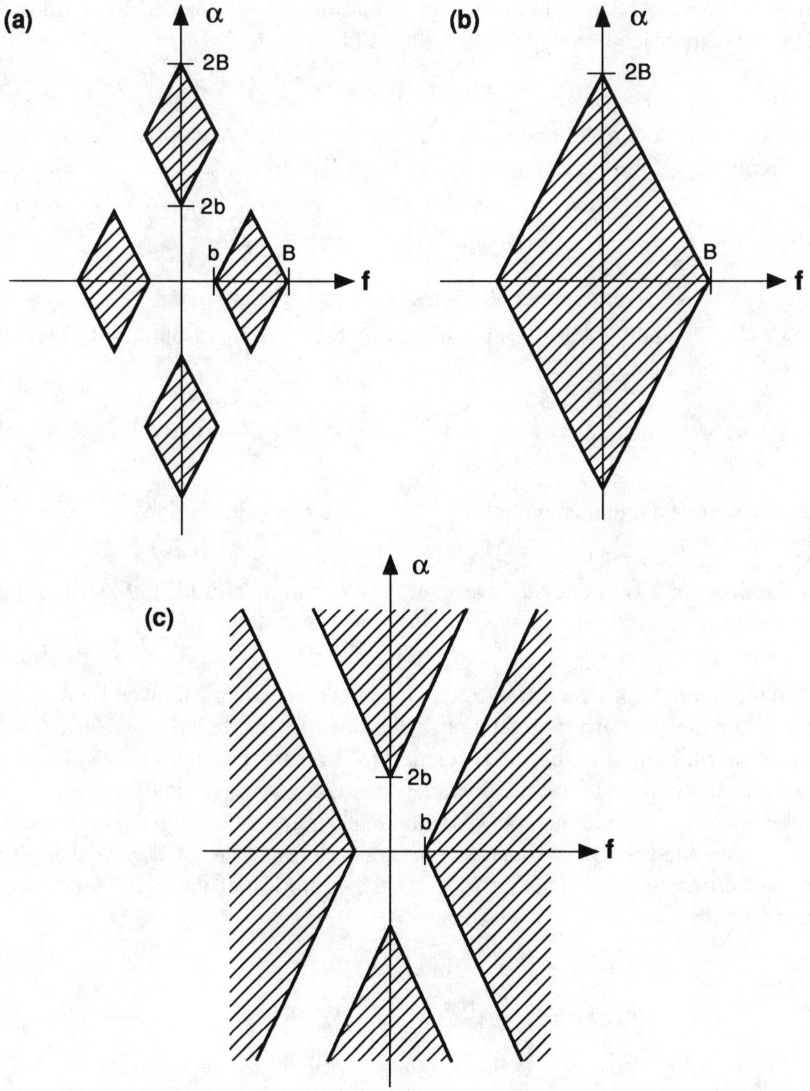

Figure 14: (a) Support region in the bifrequency plane for the spectral correlation density function of a bandpass signal. (b) Support region for a lowpass signal. (c) Support region for a highpass signal (shown over a small fraction of the diamond-shaped principal domain).

3.5 Signal Multiplication and Time Sampling

When two signals are multiplied together, we know from the convolution theorem that their Fourier transforms get convolved. From this, we expect some sort of convolution relation to hold for the SCDs of signals passing through a product modulator. In fact, it can be shown (cf. (Gardner, 1987a, Chapter 11, Sec. C) or (Gardner, 1990a, Chapter 12, exc. 41) for continuous time) that if $x(t)$ is obtained by multiplying together two statistically independent[17] time-series $r(t)$ and $s(t)$,

$$x(t) = r(t)s(t), \tag{55}$$

then the cyclic autocorrelation of $x(t)$ is given by the discrete circular convolution in cycle frequency of the cyclic autocorrelations of $r(t)$ and $s(t)$:

$$R_x^\alpha(\tau) = \sum_{\beta \in (-\frac{1}{2}, \frac{1}{2}]} R_r^\beta(\tau) \, R_s^{\alpha-\beta}(\tau) \, , \tag{56}$$

where, for each α, β ranges over all values in the principal domain $(-\frac{1}{2}, \frac{1}{2}]$ for which $R_r^\beta(\tau) \not\equiv 0$. By Fourier transforming (56), we obtain the input-output SCD relation for signal multiplication:

$$S_x^\alpha(f) = \int_{-1/2}^{1/2} \sum_{\beta \in (-\frac{1}{2}, \frac{1}{2}]} S_r^\beta(\nu) S_s^{\alpha-\beta}(f-\nu) d\nu \tag{57}$$

which is a double circular convolution that is continuous in the variable f and discrete in the variable α.

Example 6: Frequency Conversion As an example of (57), if $s(t)$ is simply a sinusoid,

$$s(t) = \cos(2\pi f_o t + \theta),$$

the product modulator becomes a frequency converter when followed by a filter to select either the up-converted version or the down-converted version of $r(t)$. By applying first the input-output SCD relation (57) for the product modulator (which applies since a sinusoid is statistically independent of all time-series (Gardner, 1987a, Chapter 15, Sec. A)), and then (49) for the filter, we can determine the up-converted or down-converted SCD. To illustrate, we first determine the SCD for the sinusoid $s(t)$. By substituting the sinusoid $s(t)$ into the definition of the cyclic autocorrelation, we obtain

$$R_s^\alpha(\tau) = \begin{cases} \frac{1}{2}\cos(2\pi f_o \tau) & \text{for } \alpha = 0 \\ \frac{1}{4} e^{\pm i 2\theta} & \text{for } \alpha = \pm 2 f_o \\ 0 & \text{otherwise} \end{cases} \tag{58}$$

on the principal domain of α.

---

[17] Time-series are statistically independent if their joint fraction-of-time probability densities factor into products of individual fraction-of-time probability densities, as explained in (Gardner, 1987a, Chapter 15, Sec. A).

Fourier transforming then yields the SCD

$$S_s^\alpha(f) = \begin{cases} \frac{1}{4}\delta(f - f_o) + \frac{1}{4}\delta(f + f_o) & \text{for } \alpha = 0 \\ \frac{1}{4}e^{\pm i2\theta}\delta(f) & \text{for } \alpha = \pm 2f_o \\ 0 & \text{otherwise} \end{cases} \tag{59}$$

on the principal domain, which is illustrated in Fig. 15a. Using (57), we circularly convolve this SCD with that of a stationary signal $r(t)$, for which

$$S_r^\alpha(f) = \begin{cases} S_r(f) & \text{for } \alpha = 0 \\ 0 & \text{for } \alpha \neq 0 \end{cases} \tag{60}$$

on the principal domain (see Fig. 15b). The result is that the SCD of the stationary signal simply gets replicated and scaled at the four locations of the impulses in the SCD of the sinusoid, as illustrated in Fig. 15c (provided that $S_r(f \pm f_o) = 0$ for $|f| \geq 1/2$ to avoid aliasing effects in the principal domain).

Example 7: Time Sampling Another important signal-processing operation is periodic time sampling. It is known that for a stationary signal $x(t)$, the PSD $S_x(f)$ of the sequence of samples $\{x(nT_s) : n = 0, \pm 1, \pm 2, \ldots\}$ is related to the PSD $\tilde{S}_x(f)$ of the continuous-time waveform by the aliasing formula (cf. (Gardner, 1987a, Chapter 3, Sec. E) or (Gardner, 1990a, Chapter 11, Sec. 11.1))

$$S_x(f) = \frac{1}{T_s} \sum_{n=-\infty}^{\infty} \tilde{S}_x\left(f - \frac{n}{T_s}\right). \tag{61}$$

It is shown in (Gardner, 1987a, Chapter 11, Sec. C), (Gardner, 1990a, Chapter 12, Sec. 12.4) that this aliasing formula generalizes for the SCD to

$$S_x^\alpha(f) = \frac{1}{T_s} \sum_{m,n=-\infty}^{\infty} \tilde{S}_x^{\alpha+m/T_s}\left(f - \frac{m}{2T_s} - \frac{n}{T_s}\right). \tag{62}$$

Observe that, when $x(t)$ is not stationary (i.e., when $\tilde{S}_x^\alpha(f) \not\equiv 0$ for $\alpha = m/T$ for some nonzero integers m), the conventional PSD aliasing formula (61) must be corrected according to (62) evaluated at $\alpha = 0$:

$$S_x(f) = \frac{1}{T_s} \sum_{m,n=-\infty}^{\infty} \tilde{S}_x^{m/T_s}\left(f - \frac{m}{2T_s} - \frac{n}{T_s}\right). \tag{63}$$

This reflects the fact that, when aliased overlapping spectral components add together, their PSD values add only if they are uncorrelated. When they are correlated, as in a cyclostationary signal, the PSD value of the sum of overlapping aliased components depends on the particular magnitudes and phases of their correlations. The SCD aliasing formula (62) is illustrated graphically in Fig. 16, where the support regions for

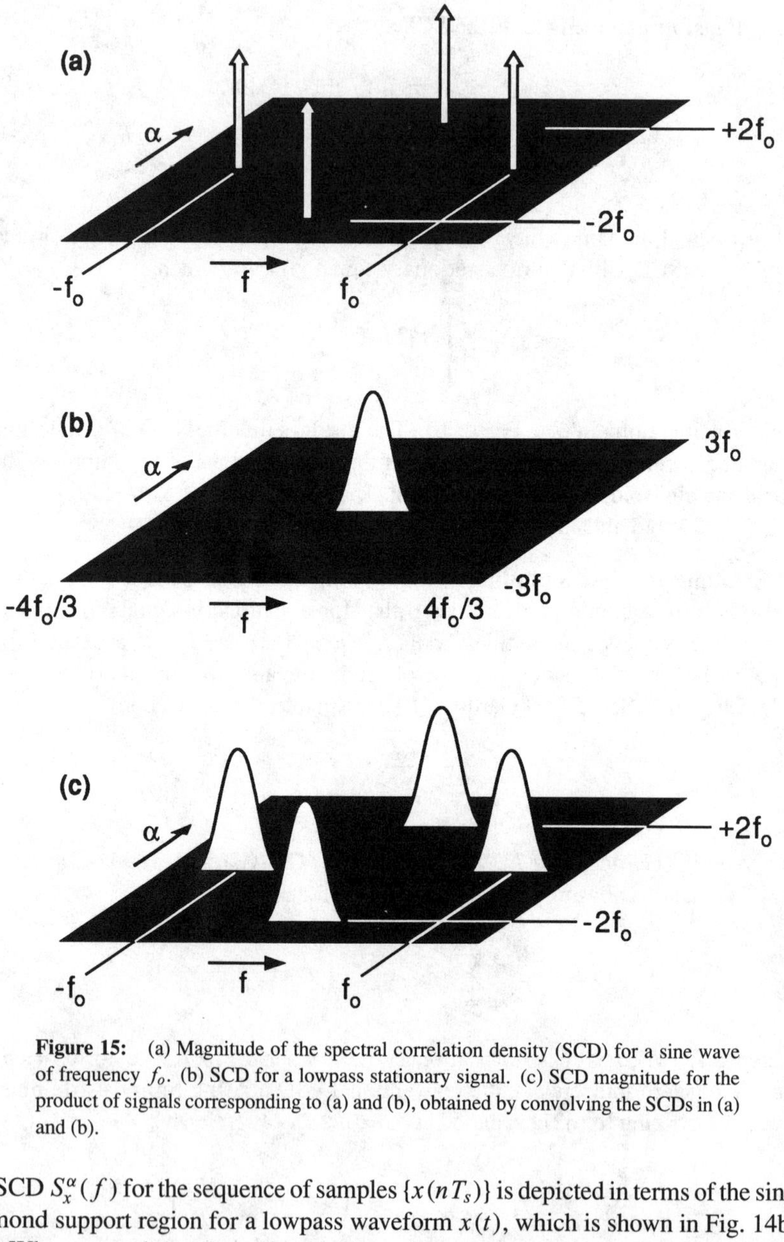

Figure 15: (a) Magnitude of the spectral correlation density (SCD) for a sine wave of frequency f_o. (b) SCD for a lowpass stationary signal. (c) SCD magnitude for the product of signals corresponding to (a) and (b), obtained by convolving the SCDs in (a) and (b).

the SCD $S_x^\alpha(f)$ for the sequence of samples $\{x(nT_s)\}$ is depicted in terms of the single diamond support region for a lowpass waveform $x(t)$, which is shown in Fig. 14b.

When we subsample a discrete-time signal $x(t)$ with sampling rate $1/T_s$ for some integer T_s to obtain the signal $z(t) = x(tT_s)$, we obtain the discrete-time analog of (62),

$$S_z^\alpha(f) = \frac{1}{T_s} \sum_{q \in P_\alpha} S_x^{(\alpha+q)/T_s}\left(\frac{f+q/2}{T_s}\right), \tag{64}$$

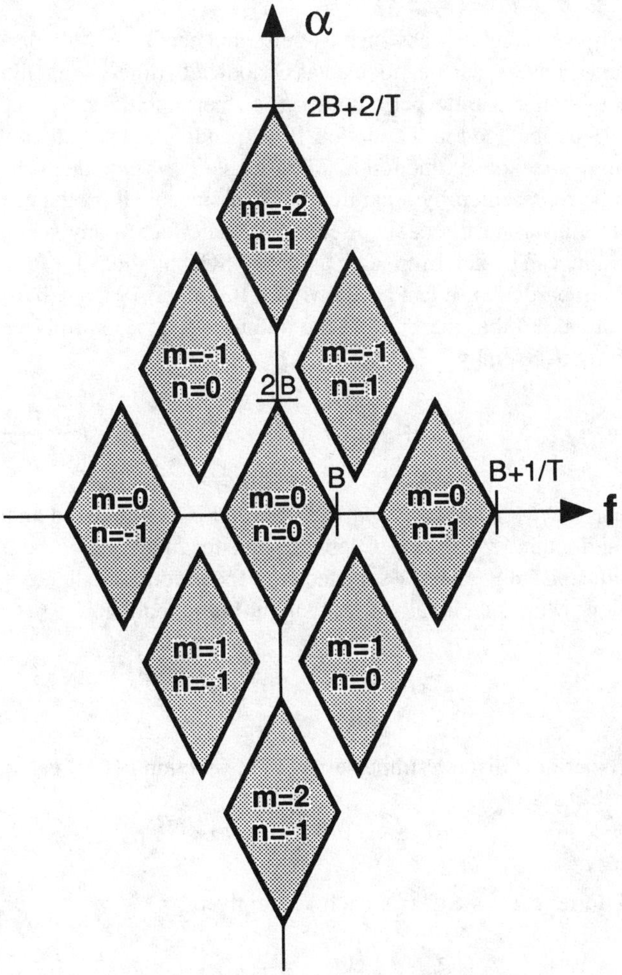

Figure 16: Illustration of support regions in the bifrequency plane for the spectral correlation densities that are aliased by periodic time sampling.

where P_α is the set of all integers $q = \beta T_s - \alpha$ for which $\alpha \in (-\frac{1}{2}, \frac{1}{2}]$. Similarly, when we resample a discrete-time signal $x(t)$, by (effectively) interpolating back to a continuous-time waveform and then time sampling at the new rate $1/T_s$ to obtain $z(t)$, the SCD is given by

$$S_z^\alpha(f) = \frac{1}{T_s} \sum_{m,n=-\infty}^{\infty} \tilde{S}_x^{\alpha+m/T_s}\left(f - \frac{m}{2T_s} - \frac{n}{T_s}\right), \tag{65}$$

where $\tilde{S}_x^\alpha(f)$ is the SCD $S_x^\alpha(f)$ restricted to its principal domain.

3.6 Periodically Time-Variant Filtering

Many signal-processing devices such as pulse and carrier modulators, multiplexors, samplers, and scanners, can be modeled as periodically time-variant filters, especially if multiple incommensurate periodicities (i.e., periodicities that are not harmonically related) are included in the model. By expanding the periodically time-variant discrete-impulse-response function in a Fourier series as explained shortly, any such system can be represented by a parallel bank of sinusoidal product modulators followed by time-invariant filters. Consequently, the effect of any such system on the SCD of its input can be determined by using the SCD relations for filters and product modulators. In particular, it can be shown (cf. (Gardner, 1987a, Chapter 11, Sec. D) for continuous time) that the SCD of the output $z(t)$ of a multiply-periodic system with input $x(t)$ is given by

$$S_z^\alpha(f) = \sum_{\beta,\gamma\in A} G_\beta(f+\alpha/2)G_\gamma^*(f-\alpha/2)S_x^{\alpha-\beta+\gamma}\left(f-\frac{\beta+\gamma}{2}\right), \tag{66}$$

provided that $S_x(f+\beta)=0$ for $|f|\geq 1/2$ for all $\beta\in A$ to avoid aliasing effects in the principal domain, where $G_\beta(f)$ are the transfer functions of the filters and A is the set of sinusoid frequencies associated with the product modulators in the system representation. More specifically, for the input-output equation

$$z(t) = \sum_{u=-\infty}^{\infty} h(t,u)x(u), \tag{67}$$

the multiply-periodic discrete-impulse-response function $h(t,u)$ can be expanded in the Fourier series

$$h(t+\tau,t) = \sum_{\beta\in A} g_\beta(\tau)\, e^{i2\pi\beta t}, \tag{68}$$

where the Fourier coefficients (for each τ) are given by

$$g_\beta(\tau) = \langle h(t+\tau,t)\, e^{-i2\pi\beta t}\rangle. \tag{69}$$

It follows from (67) and (68) that the filter output can be expressed as

$$z(t) = \sum_{\beta\in A}\left[x(t)\, e^{i2\pi\beta t}\right]\otimes g_\beta(t), \tag{70}$$

where $g_\beta(t)$ are the discrete-impulse-response functions of the filters with corresponding transfer functions $G_\beta(f)$. Thus, periodically time-variant filters perform time-invariant filtering on frequency-shifted versions $x(t)e^{i2\pi\beta t}$ of the input. This results in summing scaled, frequency-shifted, cycle-frequency-shifted versions of the SCD for the input $x(t)$ to obtain the SCD for the output $z(t)$, as indicated in (66).

Let us now consider some additional examples of modulation types, making use of the results obtained in the preceding paragraphs to determine SCDs. However, in the interest of realism and for the sake of analytical simplicity, continuous-time signal models are used.

Example 2 continued: PAM Let $\{a_n\}$ be a stationary random sequence, and let us interpret these random variables as the time samples of a continuous-time random waveform, $a_n = a(nT_o)$, with PSD $\tilde{S}_a(f)$. We consider the continuous-time PAM signal

$$x(t) = \sum_{n=-\infty}^{\infty} a_n\, p(t - nT_o + \varepsilon), \tag{71}$$

where $p(t)$ is a deterministic finite-energy pulse and ε is a fixed pulse-timing phase parameter. To determine the SCD of $x(t)$, we can recognize that $x(t)$ is the output of a periodically time-variant linear system with input $a(t)$, and impulse response

$$h(t, u) = \sum_{n=-\infty}^{\infty} p(t - nT_o + \varepsilon)\, \delta(u - nT_o),$$

where δ is the Dirac delta. We can then use the continuous-time counterpart of the input-output SCD relation (66), which is identical in form except that continuous-time Fourier transforms are used (cf. (Gardner, 1987a, Chapter 11, Sec. D)). Or we can recognize that this particular periodically time-variant system is composed of a product modulator that implements an impulse sampler, followed by a linear time-invariant pulse-shaping filter with impulse-response function $h(t) = p(t)$, as shown in Fig. 17. We can then use the continuous-time counterpart of the input-output SCD relation (57), which is identical in form except the convolutions are linear (cf. (Gardner, 1987a, Chapter 11, Sec. C)), as it applies to impulse sampling, together with the relation (49) for filtering. The result is

$$\tilde{S}_x^\alpha(f) = \frac{1}{T_o^2} \tilde{P}(f + \alpha/2)\tilde{P}^*(f - \alpha/2) \sum_{m,n=-\infty}^{\infty} \tilde{S}_x^{\alpha+m/T_o}\left(f - \frac{m}{2T_o} - \frac{n}{T_o}\right) e^{i2\pi\alpha\varepsilon}. \tag{72}$$

Using the SCD aliasing formula (62) for $a(t)$ we can reexpress (72) as

$$\tilde{S}_x^\alpha(f) = \frac{1}{T_o} \tilde{P}(f + \alpha/2)\tilde{P}^*(f - \alpha/2) S_a^\alpha(f)\, e^{i2\pi\alpha\varepsilon}, \tag{73}$$

where $S_a^\alpha(f)$ is the SCD for the pulse-amplitude sequence $\{a_n\}$. Having assumed that $\{a_n\}$ is stationary, and using the periodicity property

$$S_a^\alpha(f) = \begin{cases} S_a(f + \alpha/2) & \text{for } \alpha = k/T_o \\ 0 & \text{otherwise} \end{cases} \tag{74}$$

for $k = 0, \pm 1, \pm 2, \ldots$, we can express (73) as

$$\tilde{S}_x^\alpha(f) = \begin{cases} \dfrac{1}{T_o} \tilde{P}(f+\alpha/2)\tilde{P}^*(f - \alpha/2) S_a(f + \alpha/2)\, e^{i2\pi\alpha\varepsilon} & \text{for } \alpha = k/T_o \\ 0 & \text{otherwise.} \end{cases} \tag{75}$$

A graph of the magnitude of this SCD for the full-duty-cycle rectangular pulse

$$p(t) = \begin{cases} 1 & \text{for } |t| \le T_o/2 \\ 0 & \text{otherwise} \end{cases} \tag{76}$$

and a white-noise amplitude sequence with PSD

$$S_a(f) = 1 \tag{77}$$

is shown in Fig. 18.

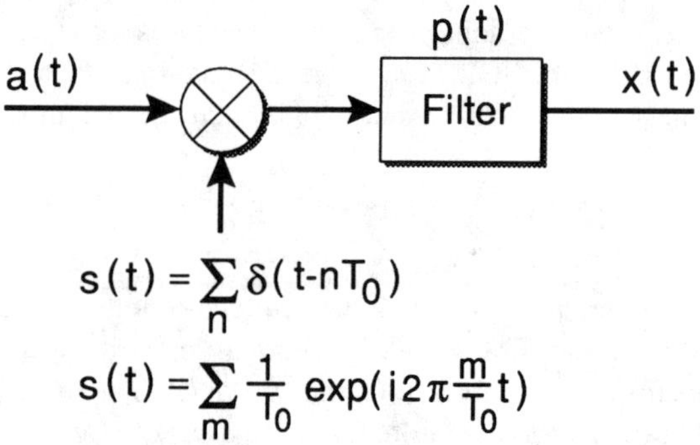

Figure 17: Interpretation of PAM signal generator as the cascade of an impulse sampler and a pulse-shaping filter.

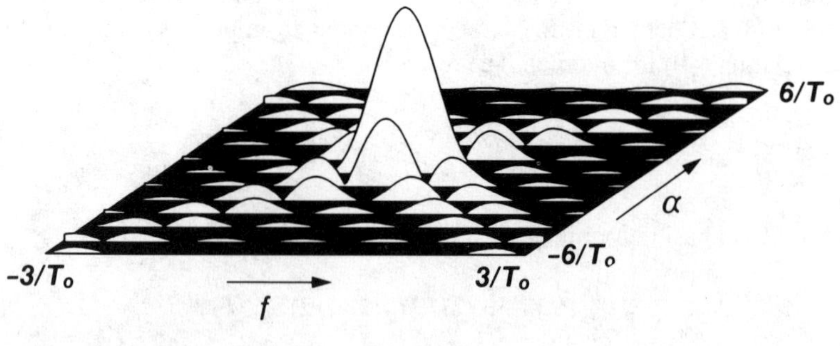

Figure 18: Magnitude of the spectral correlation density for a PAM signal with full-duty-cycle rectangular pulses.

It follows from (77) that for all $\alpha = k/T_o$ for which $S_a(f \pm \alpha/2) \neq 0$ and $\tilde{P}(f+\alpha/2)\tilde{P}^*(f-\alpha/2) \neq 0$, the spectral correlation coefficient $\rho_x^\alpha(f)$ is unity in magnitude:

$$\left|\rho_x^\alpha(f)\right| = 1. \tag{78}$$

Thus, all spectral components outside the band $|f| < 1/2T_o$ are completely redundant with respect to those inside this band. Techniques for exploiting this spectral redundancy are described in Section 4.

The conjugate SCD for the PAM signal (71) is given by (73) with $\tilde{P}^*(f-\alpha/2)$ replaced by $\tilde{P}^*(\alpha/2-f)$ and $S_a^\alpha(f)$ replaced by $S_{aa^*}^\alpha(f)$. For a real PAM signal, the conjugate SCD is identical to the SCD; however, for complex PAM the conjugate SCD is, in general, different and is, in fact, zero for the complex PAM that models the complex envelopes of most digital QAM signals, including QPSK. This follows from the fact that $\langle a_n a_{n+m}\rangle = 0$ for all m for such signals; consequently, $S_{aa^*}^\alpha(f) = 0$ for all α.

By inverse Fourier transforming the SCD (75), we obtain the cyclic autocorrelation function

$$\tilde{R}_x^\alpha(\tau) = \begin{cases} \dfrac{1}{T_o}\displaystyle\sum_{n=-\infty}^{\infty} \tilde{R}_a(nT_o) r_p^\alpha(\tau - nT_o)\, e^{i2\pi\alpha\varepsilon} & \text{for } \alpha = k/T_o \\ 0 & \text{otherwise,} \end{cases} \tag{79}$$

where

$$r_p^\alpha(\tau) \triangleq \int_{-\infty}^{\infty} p(t+\tau/2)p^*(t-\tau/2)e^{-i2\pi\alpha t}\,dt. \tag{80}$$

For a white-noise amplitude-sequence as in (77), (79) reduces to

$$\tilde{R}_x^\alpha(\tau) = \frac{1}{T_o} r_p^\alpha(\tau)\, e^{i2\pi\alpha\varepsilon} \qquad \text{for } \alpha = k/T_o, \tag{81}$$

and, for a rectangular pulse as in (76), this yields the temporal correlation coefficient

$$\tilde{\gamma}_x^\alpha(\tau) = \frac{\sin[\pi\alpha(T_o - |\tau|)]}{\pi\alpha T_o} e^{i2\pi\alpha\varepsilon} \qquad \text{for } |\tau| \le T_o, \tag{82}$$

the magnitude of which is shown in Fig. 19. This correlation coefficient peaks for $\alpha = 1/T_o$ at $\tau = T_o/2$, where it takes on the value

$$\left|\tilde{\gamma}_x^\alpha(T_o/2)\right| = 1/\pi \qquad \text{for } \alpha = 1/T_o. \tag{83}$$

That is, the strongest possible spectral line that can be regenerated in a delay-product signal for this particular PAM signal occurs when the delay equals half the pulse period. In contrast to this, when the more bandwidth-efficient pulse whose transform is a raised cosine is used, the optimal delay for sine-wave regeneration is zero.

An especially simple example of a sequence of pulse amplitudes $\{a_n\}$ is a binary sequence with values ± 1. If we consider $\tau = 0$ in the delay-product signal, then we obtain

$$y_0(t) = |x(t)|^2 = \sum_{m,n=-\infty}^{\infty} a_n a_m\, p(t - nT_o + \varepsilon)p(t - mT_o + \varepsilon).$$

If the pulses do not overlap (i.e., if $p(t) = 0$ for $|t| \ge T_o/2$), this reduces to

$$\begin{aligned} y_0(t) &= \sum_{n=-\infty}^{\infty} a_n^2\, p^2(t - nT_o + \varepsilon) \\ &= \sum_{n=-\infty}^{\infty} p^2(t - nT_o + \varepsilon), \end{aligned}$$

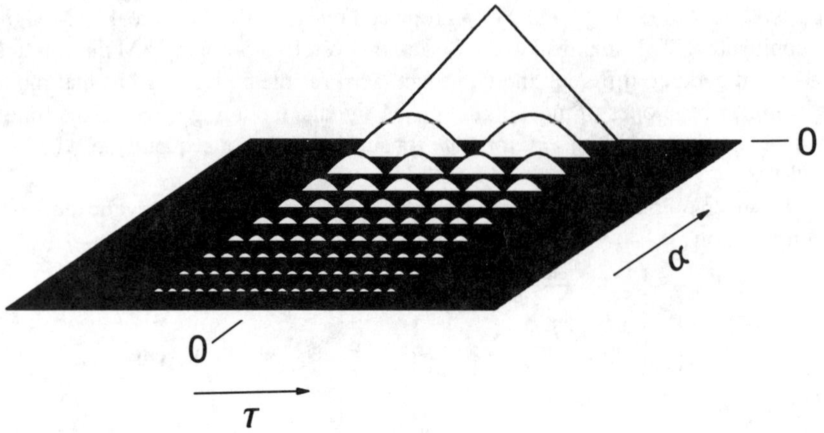

Figure 19: Magnitude of the cyclic autocorrelation function (normalized to form a correlation coefficient) for a PAM signal with full duty-cycle rectangular pulses.

which is periodic with period T_o and therefore contains finite-strength additive sine-wave components with frequencies k/T_o (except when $p(t)$ is flat as in (76)). In this very special case where $\{a_n\}$ is binary and the pulses do not overlap, there is no random component in $y_0(t)$; but, for $\tau \neq 0$, $y_\tau(t)$ contains both sine-wave components and random components (even when $p(t)$ is flat).

Example 8: ASK and PSK By combining the amplitude-modulated sine wave and the digital amplitude-modulated pulse train, we obtain the amplitude-shift-keyed (ASK) signal

$$x(t) = a(t)\cos(2\pi f_o t + \theta), \tag{84}$$

where

$$a(t) = \sum_{n=-\infty}^{\infty} a_n\, p(t - nT_o + \varepsilon), \tag{85}$$

and $\{a_n\}$ are digital amplitudes. By using the continuous-time counterpart of the SCD relation (57) for signal multiplication and the result (75) for the SCD of $a(t)$, we can obtain the SCD for the signal (84) by simply convolving the SCD functions shown in Figs. 15a and 18. The result is shown in Fig. 20a, where the cycle frequencies shown are $\alpha = \pm 2f_o + m/T_o$ and $\alpha = m/T_o$ for integers m, and where $f_o = 3.3/T_o$. When $f_o T_o$ is irrational, the ASK signal is polycyclostationary with fundamental periods T_o and $1/2f_o$.

For a binary sequence with each $a_n = \pm 1$, this amplitude-shift-keyed signal, with the pulse (76), is identical to the binary phase-shift-keyed (BPSK) signal

$$x(t) = \cos\left[2\pi f_o t + \theta + \sum_{n=-\infty}^{\infty} \phi_n\, p(t - nT_o)\right], \tag{86}$$

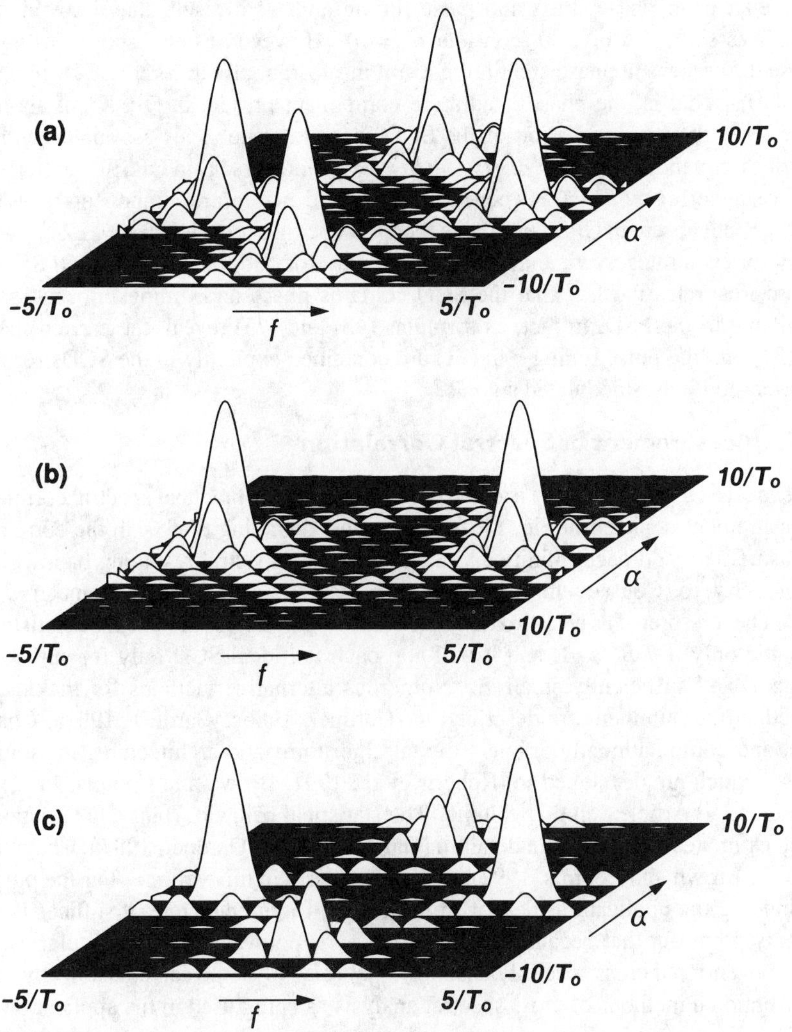

Figure 20: Magnitude of spectral correlation densities. (a) BPSK, (b) QPSK, and (c) SQPSK. (Each signal has a rectangular keying envelope.)

where $\phi_n \triangleq (a_n - 1)/2$, since shifting the phase of a sine wave by 0 or π is the same as multiplying its amplitude by 1 or -1. Other commonly used types of phase-shift-keyed signals include quaternary phase-shift keying (QPSK) and staggered QPSK (SQPSK). The details of these signal types are available in the literature (see, for example (Gardner, 1987a, Chapter 12, Sec. E) or (Gardner, 1990a, Chapter 12, Sec. 12.5)). Only their SCD-magnitude surfaces are shown here in Fig. 20b, c, where again $f_o = 3.3/T_o$.

It is emphasized that the three signals BPSK, QPSK, and SQPSK differ only in their carrier phase shifts and pulse timing and, as a result, they have identical PSDs, as shown in Fig. 20 (consider $\alpha = 0$). However, as also shown in Fig. 20, these differences in phase and timing result in substantially different SCDs (consider $\alpha \neq 0$). That is, the phase-quadrature component present in QPSK but absent in BPSK results in cancellation of the SCD at cycle frequencies associated with the carrier frequency (viz., $\alpha = \pm 2f_o + m/T_o$ for all integers m) in QPSK. Similarly, the pulse staggering by $T_o/2$ (between the in-phase and quadrature components) present in SQPSK but absent in QPSK results in the SCDs being cancelled at $\alpha = \pm 2f_o + m/T_o$ only for even integers m, and at $\alpha = m/T_o$ only for odd integers m in SQPSK. This again illustrates the fact that the SCD contains phase and timing information not available in the PSD. In fact, as formulas (43) and (75) reveal, the carrier phase θ in (31) and the pulse timing ε in (71) are contained explicitly in the SCDs for these carrier- and pulse-modulated signals.

3.7 Measurement of Spectral Correlation

The ideal SCD function (37) is derived by idealizing the practical spectral correlation measurement depicted in Fig. 11, by letting the averaging time T in the correlation measurement approach infinity and then letting the spectral resolving bandwidth B approach zero. Consequently, the practical measurement with finite parameters T and B can be interpreted as an estimate of the ideal SCD. This estimate will be statistically reliable only if $TB \gg 1$, and it will approach the ideal SCD only for sufficiently large T and sufficiently small B. Numerous alternative methods for making this practical measurement are described in (Gardner, 1986c; Gardner, 1987a, Chapter 13), and computationally efficient digital algorithms and architectures for some of these, which are developed in (Roberts et al., 1991; Brown and Loomis, 1992), are presented in Article 6 in this volume. The statistical behavior (bias and variance) of such estimates is analyzed in detail in (Gardner, 1986c; Gardner, 1987a, Chapter 15, Sec. B; Brown and Loomis, 1992), and in Chapter 6 in this volume. For the purpose of making the applications described in Section 4 more concrete, it suffices here to simply point out that because the SCD $S_x^\alpha(f)$ is equivalent to a particular case of the conventional cross spectral density $S_{uv}(f)$ (cf. (40)), one can envision any of the conventional methods of cross spectral analysis as being used in the applications.

Example 9: QPSK As an example, the result of using the Wiener-Daniell method (Gardner, 1987a, Chapter 7, Sec. D), based on frequency smoothing of the cross-periodogram of $u(t)$ and $v(t)$ (the conjugate product of their FFTs), is illustrated in Fig. 21 for a QPSK signal with carrier frequency $f_o = 1/4T_s$ and keying rate $1/T_o = 1/8T_s$, where $1/T_s$ is the sampling rate. An FFT of length 128 ($T = 128T_s$) was used in Fig. 21a, and only four frequency bins were averaged together ($B = 4/T$) to produce each output point, whereas, in Fig. 21b, the FFT length used was 32,768 ($T = 32{,}768T_s$) and 1,024 bins were averaged together ($B = 1{,}024/T$). It is easily seen by comparing with the ideal SCD in Fig. 20b that without adequate spectral smoothing the variability of the SCD estimate can be very large.

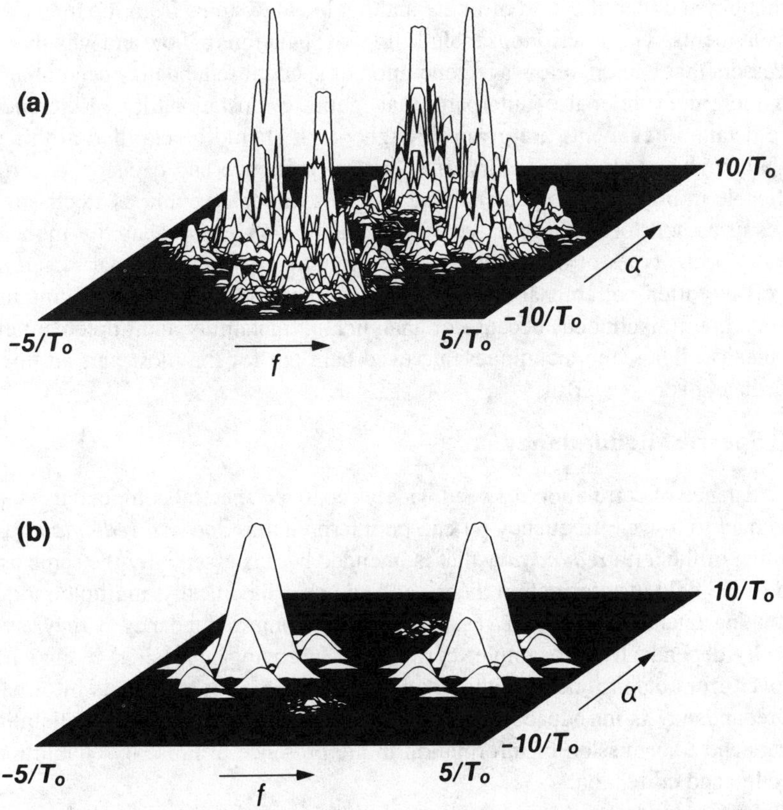

Figure 21: Magnitude of a spectral correlation density (SCD) estimate obtained from a finite-length data record for the QPSK signal whose ideal SCD is shown in Fig. 20b. (a) Record length is 128 time samples, and four adjacent frequency (f) bins are averaged together. (b) Record length is 32,768 and 1,024 adjacent frequency (f) bins are averaged together. (The sampling rate in both (a) and (b) is $10/T_o$, where $1/T_o$ is the keying rate of the QPSK signal.)

4 EXPLOITATION OF CYCLOSTATIONARITY

This section describes some ways of exploiting sine-wave generation and the inherent spectral redundancy associated with the spectral correlation in cyclostationary signals to perform various signal-processing tasks. These include detecting the presence of signals buried in noise and/or severely masked by interference; recognizing such corrupted signals according to modulation type; estimating parameters such as time-difference-of-arrival at two reception platforms and direction of arrival at a reception array on a single platform; blind-adaptive spatial filtering of signals impinging on a reception array; reduction of signal corruption due to cochannel interference and/or channel fading for single-receiver systems; linear periodically time-variant prediction; and identification of linear and nonlinear systems from input and output measurements. The descriptions include brief explanations of how and why the signal processors that exploit sine-wave generation or spectral redundancy can outperform their more conventional counterparts that ignore cyclostationarity. References to more detailed treatments are given throughout. It should be clarified at this point that although the classical theory of statistical inference and decision is certainly applicable in principle to cyclostationary signals (modeled either as stochastic processes or as nonstochastic time-series with fraction-of-time probability models, cf. Section 2), the types of problems where exploitation of cyclostationarity can really pay off are often not amenable to the classical theories (e.g., Bayes minimum risk and maximum likelihood) because of analytical intractability and implementational complexity. Thus, the techniques surveyed here are for the most part ad hoc, but nevertheless very powerful.

4.1 Spectral Redundancy

The existence of correlation between widely separated spectral components (separation equal to a cycle frequency α) can be interpreted as *spectral redundancy*. The meaning of the term *redundancy* that is intended here is essentially the same as that used in the field of information theory and coding. Specifically, multiple randomly fluctuating quantities (random variables) exhibit some redundancy if they are statistically dependent, for example, correlated. In coding, undesired redundancy is removed from data to increase the efficiency with which it represents information, and redundancy is introduced in a controlled manner to increase the reliability of storage and transmission of information in the presence of noise by enabling error detection and correction.

Here, redundancy that is inadvertently introduced into signals by the modulation process is to be exploited to enhance the accuracy and reliability of information gleaned from the measurements of corrupted signals, but the term *information* is interpreted in a broad sense. For instance, it includes the eight examples outlined in Section 1.2. In all of these examples, the performance of the signal processors that make the decisions and/or produce the estimates can be substantially improved by suitably exploiting spectral redundancy. The degree of improvement relative to the performance of more commonly used signal processors that ignore spectral redundancy depends on both the severity of the signal corruption (noise, interference, distortion) and the degree of redundancy in the signal $x(t)$, as measured by the magnitude of the spectral correlation coefficient $\left|\rho_x^\alpha(f)\right|$ (or its conjugate counterpart) defined in Section 3. The degree of improvement also depends on the amount of data available for processing (the collection time). The utility of exploiting spectral redundancy can also be enhanced by intentionally designing the signal to exhibit a sufficient amount of spectral redundancy.

The primary feature of spectral redundancy that enables it to be readily exploited is its distinctive character. That is, most man-made signals exhibit spectral redundancy, but most noise (all noise that is not cyclostationary) does not and, more importantly, when multiple signals of interest and signals not of interest (interference) overlap in both time and frequency, their spectral redundancy functions are nonoverlapping because their cycle frequencies α are distinct. This is a result of signals having distinct carrier frequencies and/or pulse rates or keying rates, even when occupying the same spectral band.

The distinctive character of spectral redundancy makes signal-selective measurements possible. Specifically, for the received composite signal

$$x(t) = \sum_{\ell=1}^{L} s_\ell(t) + n(t), \tag{87}$$

where the set $\{s_\ell(t)\}_1^L$ includes both signals of interest and interference—all of which are statistically independent of each other—and where $n(t)$ is background noise, we have the SCD (for measurement time $T \to \infty$)

$$S_x^\alpha(f) = \sum_{\ell=1}^{L} S_{s_\ell}^\alpha(f) + S_n^\alpha(f). \tag{88}$$

But if the only signal with the particular cycle frequency $\alpha = \alpha_k$ is $s_k(t)$, then (for $T \to \infty$) we have

$$S_x^\alpha(f) = S_{s_k}^\alpha(f), \tag{89}$$

regardless of the temporal or spectral overlap among $\{s_\ell(t)\}_1^L$ and also $n(t)$. This perfect signal selectivity of ideal SCDs implies that practical measurements of SCDs or their parameters can be made signal selective for measurement times τ that are long enough.

Example 1: BPSK Signal in Multiple AM Interference and Noise To illustrate the concept of signal selectivity, let us consider the situation in which a broadband BPSK signal of interest is received in the presence of white noise and five interfering AM signals with narrower bandwidths that together cover the entire band of the BPSK signal. The noise and each of the five interfering signals have equal average power. Therefore, the total signal-to-interference-and-noise ratio (SINR) is approximately −8 dB. The BPSK signal has carrier frequency $f_o = 0.25/T_s$ and keying rate $\alpha_o = 0.0625/T_s$. It has full-duty-cycle half-cosine envelope, which results in an approximate bandwidth of $B_o = 0.1875/T_s$. The five AM signals have carrier frequencies $f_1 = 0.156/T_s$, $f_2 = 0.203/T_s$, $f_3 = 0.266/T_s$, $f_4 = 0.313/T_s$, $f_5 = 0.375/T_s$, and bandwidths $B_1 = 0.04/T_s$, $B_2 = 0.05/T_s$, $B_3 = 0.045/T_s$, $B_4 = 0.04/T_s$, $B_5 = 0.08/T_s$. With the use of the same measurement parameters (FFT length = 32,768) as in Example 9 in Section 3 for the measurement of the SCD of QPSK, the SCD for these six signals in noise was measured. The resultant SCD magnitude is shown in Fig. 22a. Also shown in Fig. 22b, c are the SCD magnitudes for the BPSK signal alone and for the five AM interferences plus noise alone. Although all six signals exhibit strong spectral redundancy $\left(\left|\rho_{s_\ell}^\alpha(f)\right| = 1\right)$, the cycle frequencies α at which this redundancy exists are distinct because the carrier frequen-

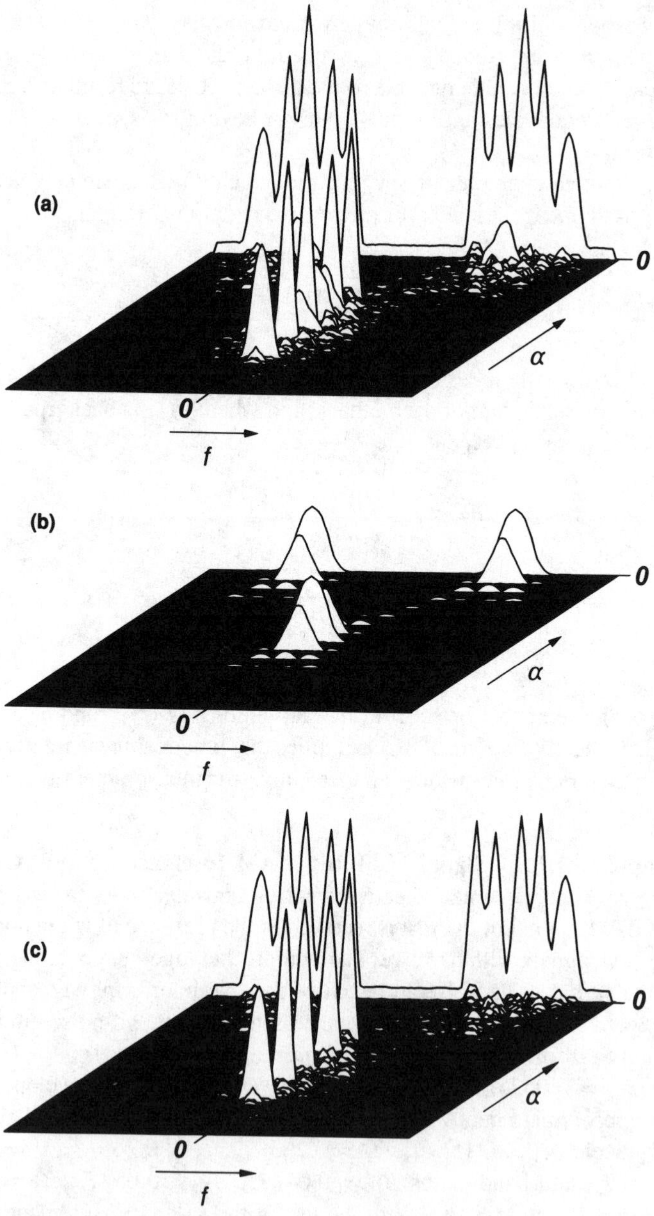

Figure 22: Magnitudes of estimated spectral correlation densities (SCDs). (a) SCD magnitude for a BPSK signal corrupted by white noise and five AM interferences. (b) SCD magnitude for the BPSK signal alone. (c) SCD magnitude for the white noise and five AM interferences. (The power levels, center frequencies, and bandwidths for the signals and noise are specified in the text; the record length used is 32,768 time samples, and 1,024 adjacent frequency (f) bins are averaged together.)

cies are all distinct. Thus, an accurate estimate of the SCD for the BPSK signal is easily extracted from the SCD for the corrupted measurements. Similarly, accurate estimates of the SCDs for each of the five AM signals can be extracted. Consequently, any information contained in these SCDs can be reliably extracted.

In connection with this example, let us briefly consider some of the signal-processing tasks outlined in Section 1.2.

4.2 Detection and Classification

We can see from Fig. 22 that knowing the particular pattern of the SCDs for BPSK and AM signals (see Figs. 13 and 20) enables us to detect the presence of six signals and to classify them according to modulation type. This would be impossible if only PSD (SCD at $\alpha = 0$) measurements were used. One approach to exploiting the spectral redundancy of a signal to detect its presence is to generate a spectral line at one of its cycle frequencies and then detect the presence of the spectral line (cf. Section 3). It has been shown that the maximum-SNR spectral-line generator for a signal $s(t)$ in additive Gaussian noise and interference with PSD $S_n(f)$ produces the detection statistic (cf. (Gardner, 1987a, Chapter 14, Sec. E) for continuous time)

$$z = \int_{-1/2}^{1/2} \hat{S}_x^\alpha(f) \frac{S_s^\alpha(f)^*}{S_n(f+\alpha/2)S_n(f-\alpha/2)} \, df \tag{90}$$

for comparison to a threshold. In (90), $\hat{S}_x^\alpha(f)$ is a crude estimate of $S_x^\alpha(f)$ obtained by deleting the time-averaging operation $\langle\cdot\rangle$ and the limiting operation from (37) and choosing B equal to the reciprocal of the record length of $x(t)$. It can be shown that (90) is equivalent to whitening the noise and interference using a filter with transfer function $1/[S_n(f)]^{1/2}$, and then correlating the measured SCD for the noise-and-interference-whitened data with the ideal SCD of the signal, as transformed by the whitener, to be detected (Gardner, 1987a, Chapter 14, Sec. E). Equivalently, for noise consisting of a white component plus strong narrowband components, (90) corresponds to attenuating the narrowband components well below the white-noise component—i.e., excising the narrowband components—using a filter with transfer function $1/S_n(f)$, and then correlating the measured SCD for the narrowband-excised data with the ideal SCD of the signal (untransformed by the excision filter).

A detailed study of both optimum (e.g., maximum-likelihood and maximum-SNR) and more practical suboptimum detection on the basis of SCD measurement is reported in (Gardner, 1988b), and receiver operating characteristics for these detectors obtained by simulation are presented in (Gardner and Spooner, 1992a, 1993). See also (Zivanovich and Gardner, 1991).

4.3 Parameter Estimation

Once the six signals have been detected and classified, their carrier frequencies and phases and the keying rate and phase of the BPSK signal can—with sufficiently long signal duration—be accurately estimated from the magnitude and phase of the SCD (cf., f_o, θ in (43) and T_o, ε in (75)) (Gardner and Spooner, 1993). It is clear from the theory discussed in Section 3 that SCD measurement is intimately related to the measurement of the amplitudes and phases of sine waves generated by quadratic transformations of the data. Thus, the fact that an SCD feature occurs at $\alpha = 2f_o$ for each carrier frequency f_o is a direct result of the fact that a sine wave (spectral line) with frequency $\alpha = 2f_o$ and phase 2θ can be generated by putting the data through a quadratic transformation. Similarly, for the SCD feature at $\alpha = 1/T_0$, where $1/T_0$ is the keying rate, a spectral line with frequency $\alpha = 1/T_0$ and phase ε can be quadratically generated. Consequently, SCD measurement is useful either directly or indirectly for estimation of synchronization parameters (frequencies and phases) required for the operation of synchronized receivers. The link between synchronization problems and spectral redundancy is pursued in (Gardner, 1986a) and also in Article 2 in this volume.

4.4 Time-difference-of-arrival Estimation

The cross SCD $S_{wx}^{\alpha}(f)$ for two signals $x(t)$ and $w(t)$ is defined in a way that is analogous to the definition (37) and (24) of the auto SCD $S_x^{\alpha}(f)$. That is, $x(t)$ in (24a) is simply replaced with $w(t)$. If we were to compute the cross SCD for two sets of corrupted measurements obtained from two reception platforms, then the cross SCD magnitude would look very similar to that in Fig. 22 (except that the low flat feature at $\alpha = 0$, which represents the PSD of the receiver noise, would be absent), but the phase of the cross SCD would contain a term linear in f at each value of α where the auto SCD of one of the six signals is nonzero. The slope of this linear phase is proportional to the time-difference-of-arrival (TDOA) of the wavefront at the two platforms for the particular signal with that feature. That is, for $x(t)$ from one platform given by (87) and $w(t)$ from the other platform given by

$$w(t) = \sum_{\ell=1}^{L} a_\ell s_\ell(t - t_\ell) + m(t) \tag{91}$$

where $\{t_\ell\}$ are the TDOAs, we have

$$S_{wx}^{\alpha}(f) = S_{s_\ell}^{\alpha}(f) a_\ell e^{-i2\pi(f+\alpha/2)t_\ell} \tag{92}$$

provided that $s_\ell(t)$ is the only signal with cycle frequency α. Consequently, accurate estimates of the TDOAs of each of these signals can be obtained from the cross SCD measurement, regardless of temporal and spectral overlap or of the closeness of the individual TDOAs. In other words, the signal selectivity in the α domain eliminates the problem of resolving TDOAs of overlapping signals.

For example, it follows from (89) and (92) that

$$\frac{S_{wx}^{\alpha}(f)}{S_x^{\alpha}(f)} = a_\ell \, e^{-i2\pi(f+\alpha/2)t_\ell} \tag{93}$$

over the support band of $S_{s_\ell}^{\alpha}(f)$. This suggests doing a weighted least-squares fit, with respect to a_ℓ and t_ℓ, of a measurement of the left side of (93) to the right side:

$$\min_{\hat{a}_\ell, \hat{t}_\ell} \left\{ \int_{-1/2}^{1/2} \left| W_\alpha(f) \left[\frac{\hat{S}_{wx}^\alpha(f)}{\hat{S}_x^\alpha(f)} - \hat{a}_\ell \, e^{-i2\pi(f+\alpha/2)\hat{t}_\ell} \right] \right|^2 df \right\} \tag{94}$$

where $W_\alpha(f)$ is some weighting function. After minimization with respect to $\hat{a}_\ell$, this reduces to

$$\max_{\hat{t}_\ell} \left\{ \int_{-1/2}^{1/2} |W_\alpha(f)|^2 \, \frac{S_{wx}^\alpha(f)}{S_x^\alpha(f)} \, e^{-i2\pi(f+\alpha/2)\hat{t}_\ell} \, df \right\}. \tag{95}$$

The two algorithms corresponding to the two choices $W_\alpha(f) = S_x^\alpha(f)$ (which yields the SPECCOA method) and $W_\alpha(f) = 1$ over some band (which yields the SPECCORR method), along with several other related algorithms are studied in detail in (Gardner and Chen, 1992; Chen and Gardner, 1992; Gardner and Spooner, 1993), where excellent robustness to unknown and/or varying noise and interference is demonstrated. It is also shown in Article 3 in this volume that this approach is easily generalized to the problem of multipath channel identification where multiple t_ℓ and a_ℓ for a single signal are to be estimated using the least-squares criterion (94) with a sum over ℓ included (provided that the multiple t_ℓ are resolvable, i.e., spaced farther apart than the width of the inverse discrete Fourier transform of $|W_\alpha(f)|^2 \, S_{wx}^\alpha(f)/S_x^\alpha(f)$).

4.5 Spatial Filtering

Continuing in the same vein, we consider receiving these same six signals in noise with an antenna array. Then we can use the signal selectivity in α to blindly adapt (without any training information other than knowledge of the cycle frequency α of each signal of interest) adapt a linear combiner of the complex-valued outputs from the elements in the array to perform spatial filtering. Specifically, by directing the linear combiner to enhance or restore spectral redundancy (or conjugate spectral redundancy) in its output at a particular cycle frequency α, the combiner will adapt to null out all other signals (if there are enough elements in the array to make this nulling possible). This behavior of the combiner can be seen from the fact that the spectral correlation coefficient for $x(t)$ in (87) is (from (89))

$$\rho_x^\alpha(f) = \frac{S_{s_\ell}^\alpha(f)}{[S_x(f+\alpha/2)S_x(f-\alpha/2)]^{1/2}}, \tag{96}$$

where

$$S_x(f) = \sum_{k=1}^{L} S_{s_k}(f) + S_n(f), \tag{97}$$

and, similarly, the temporal correlation coefficient for the frequency-shifted versions of $x(t)$ is

$$\gamma_x^\alpha(\tau) = \frac{R_{s_\ell}^\alpha(\tau)}{R_x(0)}, \tag{98}$$

where

$$R_x(0) = \sum_{k=1}^{L} R_{s_k}(0) + R_n(0). \tag{99}$$

Thus, nulling signals other than $s_\ell(t)$ in the output $x(t)$ of the linear combiner and attenuating the noise $n(t)$ in $x(t)$ reduces the denominators in (96) and (98) but not the numerators. Hence, $|\rho_x^\alpha(f)|$ and $|\gamma_x^\alpha(\tau)|$ can be increased by attenuating the noise and nulling any of the signals other than $s_\ell(t)$. Moreover, the linear combiner needs no knowledge of the reception characteristics of the array (no calibration) to accomplish this attenuation and nulling. A thorough study of spectral-coherence-restoral algorithms that perform this blind adaptive spatial filtering is reported in (Schell and Agee, 1988; Agee et al., 1990; Schell and Gardner, 1993b) and a tutorial discussion is given in Chapter 3 in this volume.

4.6 Direction Finding

We can take this approach one step further if we do indeed have calibration data for the reception characteristics of an antenna array because we can then also exploit signal selectivity in α to perform high-resolution direction finding (DF) without some of the drawbacks (described below) of conventional methods for high-resolution DF, such as subspace fitting methods (Schell and Gardner, 1993a), that do not exploit spectral redundancy. In particular, let us consider the narrowband model

$$\boldsymbol{x}(t) = \sum_{\ell=1}^{L} \boldsymbol{a}(\theta_\ell) s_\ell(t) + \boldsymbol{n}(t) \tag{100}$$

for the analytic signal (or complex envelope) $\boldsymbol{x}$ of the received data vector of dimension r, where $\boldsymbol{a}(\theta_\ell)$ is the direction vector associated with the ℓ-th received signal $s_\ell(t)$, and the function $\boldsymbol{a}(\cdot)$ is specified by the calibration data for the array. Then, by working with the magnitude and phase information contained in the $r \times r$ cyclic correlation matrix

$$\boldsymbol{R}_{\boldsymbol{x}}^\alpha(\tau) = \boldsymbol{R}_{s_k}^\alpha(\tau) = \boldsymbol{a}(\theta_k) R_{s_k}^\alpha(\tau) \boldsymbol{a}^\dagger(\theta_k) \tag{101}$$

for some fixed τ (where $\dagger$ denotes conjugate transpose), instead of working with the information contained in the conventional correlation matrix

$$\boldsymbol{R}_{\boldsymbol{x}}(0) = \sum_{\ell=1}^{L} \boldsymbol{R}_{s_\ell}(0) + \boldsymbol{R}_{\boldsymbol{n}}(0) = \sum_{\ell=1}^{L} \boldsymbol{a}(\theta_\ell) \tilde{R}_{s_\ell}(0) \boldsymbol{a}^\dagger(\theta_\ell) + \boldsymbol{R}_{\boldsymbol{n}}(0) \tag{102}$$

we can avoid the need for advance knowledge of the correlation properties of the noise $\boldsymbol{R}_{\boldsymbol{n}}(0)$ and interference $R_{s_\ell}(0)$ for $\ell \neq k$, and we can avoid the constraint imposed by conventional methods that the number of elements in the array exceed the total number L of signals impinging on the array. Also, by resolving signals in α, we need not resolve them in direction of arrival. Consequently, superior effective spatial resolution is another advantage available through the exploitation of spectral redundancy. As an example of a cyclostationarity-exploiting DF method, we can use the fact that the $r \times r$ matrix in (101) has a rank of unity and the $(r-1)$-dimensional null space of this matrix is orthogonal to $\boldsymbol{a}(\theta_k)$. Therefore, we can choose as our estimate of θ_k that value $\hat{\theta}_k$ which renders $\boldsymbol{a}(\hat{\theta}_k)$ most nearly orthogonal to the null space of an estimate of the matrix $\boldsymbol{R}_{\boldsymbol{x}}^{\alpha}(\tau)$ obtained from finite-time averaging. Similar remarks apply to the conjugate cyclic autocorrelation matrix. A thorough study of this approach to signal-selective DF is reported in (Schell et al., 1989; Schell, 1990; Schell and Gardner, 1991), where various algorithms are introduced and their performances are evaluated, and a tutorial discussion is given in Chapter 3 in this volume.

In the preceding paragraphs of this Section 4, the signal-processing tasks (with the exception of spatial filtering) involve decision or parameter estimation, but do not involve estimating (or extracting) an entire signal or an information-bearing message carried by the signal. Nevertheless, for the signal-extraction problem, the utility of spectral redundancy is just as apparent, as explained in the following paragraphs.

4.7 Signal Extraction

Spectrally redundant signals that are corrupted by other interfering signals can be more effectively extracted in some applications by exploiting spectral correlation through the use of periodic or multiply-periodic linear time-variant filters, instead of the more common time-invariant filters. These time-variant filters enable spectral redundancy to be exploited for signal extraction, because such filters perform frequency-shifting operations (cf. (70)) as well as the frequency-dependent magnitude-weighting and phase-shifting operations performed by time-invariant filters. The utility of this is easily seen for the simple example in which interference in some portions of the spectral band of the signal is so strong that it overpowers the signal in those partial bands. In this case, a time-invariant filter can only reject both the signal and the interference in those highly corrupted bands, whereas a time-variant filter can replace the rejected spectral components of the signal of interest with spectral components from other uncorrupted (or less corrupted) bands that are highly correlated with the rejected components from the signal.

AM is an obvious example of this because of the complete redundancy that exists between its upper sideband (above the carrier frequency) and its lower sideband (below the carrier frequency). Although this redundancy is exploited in the conventional double sideband demodulator to obtain a 3-dB gain in SNR performance, it is seldom exploited properly when partial-band interference is present. The proper exploitation in this case is illustrated in Fig. 23. Figure 23a shows the spectral content (Fourier transform magnitude of a finite segment of data) for an AM signal with partial-band interference in the upper sideband. Figure 23b shows the spectral content after the interference has been rejected by time-invariant filtering. The signal distortion caused by rejection of the signal components along with the interference can be completely removed by simply shifting replicas of perfectly correlated components from the lower sideband into the upper sideband, and then properly adjusting their magnitudes and phases, as suggested in Fig. 23c.

A less easily explained example involves two spectrally overlapping linearly modulated signals such as AM, PAM, ASK, PSK, or digital QAM (quadrature AM).

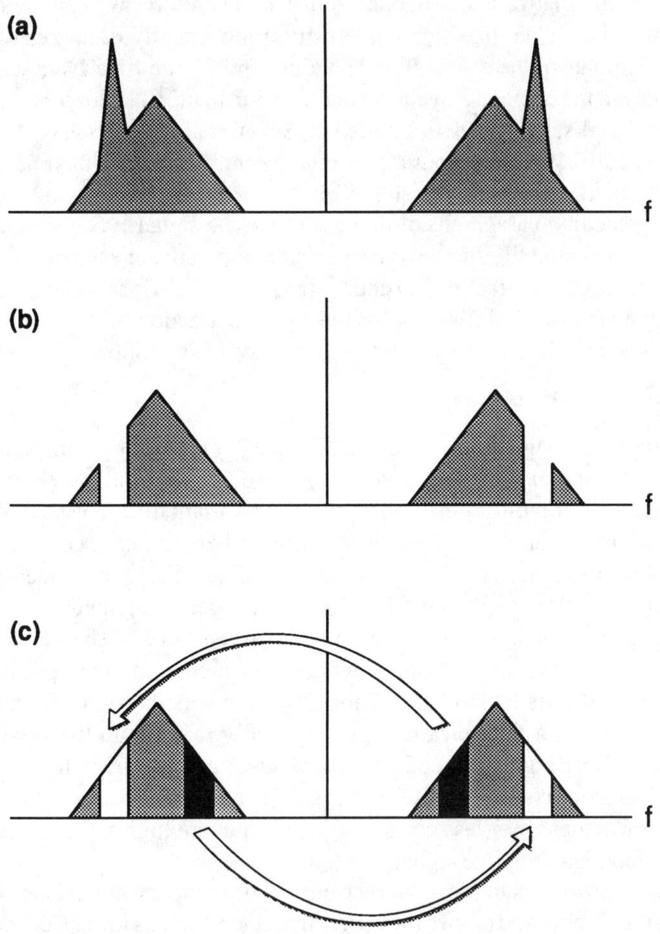

Figure 23: Illustration of power spectral densities (PSDs) for cochannel-interference removal with minimal signal distortion. (a) PSD for AM signal plus interference. (b) PSD after interference removal by time-invariant filtering. (c) PSD after distortion removal by frequency-shifting.

It can be shown that, regardless of the degree of spectral and temporal overlap, each of the two interfering signals can be perfectly extracted by using frequency shifting and complex weighting, provided only that they have either different carrier frequencies or phases (AM, ASK, BPSK) or different keying rates or phases (PAM, ASK, PSK, digital QAM) and at least 100% excess bandwidth (bandwidth in excess of the minimum Nyquist bandwidth for zero intersymbol interference). In addition, when the excess bandwidth is $(L-1)100\%$, L spectrally overlapping signals can be separated if they have the same keying rate but different keying phases or carrier frequencies. Also, when broadband noise is present, extraction of each of the signals can in many cases be accomplished without substantial noise amplification. To illustrate the potential for signal separation in this case, consider L digital QAM signals with $(L-1)100\%$ excess bandwidth, all sharing the same carrier frequency and keying rate, but with distinct keying phases. Then for any particular frequency f in the Nyquist band the received spectral component at that frequency is a weighted sum of the L spectral components of the L individual signals at that same frequency f and the same is true at the $L-1$ additional frequencies separated by the keying rate, except that the sets of L weights in each of these L weighted sums are distinct, although the L sets of L spectral components are all identical (because the spectral correlation coefficients are unity in magnitude). Thus, for each frequency within the Nyquist band, we have L equations with L unknowns. In practice, the $L \times L$ array of weights also will be unknown and will have to be adaptively learned.

This particular problem of separating multiple digital QAM signals sharing the same carrier frequency (or baseband PAM signals) and sharing the same keying rate is explored in Article 1 in this volume.

To gain additional insight into how spectrally overlapping signals can be separated by frequency-shift filtering, we consider the case of two QPSK (quadrature-phase-shift-keyed) signals with unequal carrier frequencies and unequal keying rates and 100% excess bandwidth. The graphs in Fig. 24 show the overlapping spectra for these two signals. Starting from the top of this figure, each pair of graphs illustrates the result of one filtering and frequency-shifting stage. The subband shaded with a single set of parallel lines represents spectral components from one signal that are not corrupted by the other signal. These components are selected and complex-weighted by a filter and then frequency-shifted to cancel the components in another subband, which is identified by crosshatched shading. The result of this cancellation is shown in the second graph (which contains no shading) of each pair. After five such stages, a full sideband of each of the two QPSK signals has been completely separated. In each stage the complete spectral redundancy between components separated by the keying rate is being exploited, and this same spectral redundancy can be used to reconstruct the entire QPSK signal from either one of its sidebands.

The five cascaded stages of filtering, frequency-shifting, and adding operations can be converted into one parallel connection of frequency-shifters, each followed by a filter, simply by using standard system-transformations to move all frequency-shifters to the input.

Further insight into how spectrally overlapping signals can be separated by frequency-shift filtering can be gained by considering the case of two double sideband AM signals with suppressed carrier (or, equivalently, two ASK signals, or one AM and one ASK) with different carrier frequencies and any amount of spectral overlap. For each of these signals the upper sideband is completely redundant with the lower sideband. Consequently, if we were to reflect the complex spectrum about its center—its downconverted carrier frequency—say f_1, by replacing frequency f with

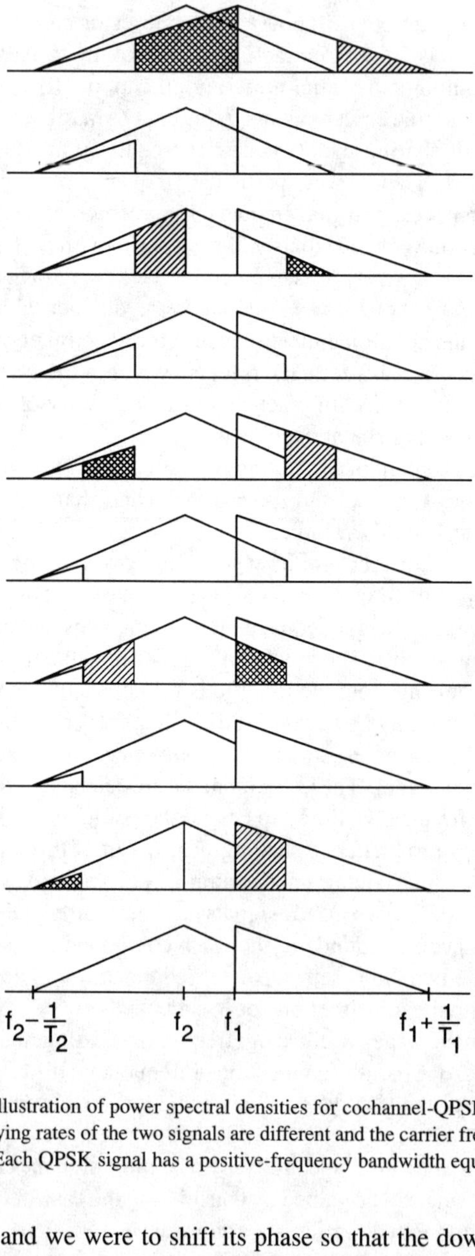

Figure 24: Illustration of power spectral densities for cochannel-QPSK-signal separation. The keying rates of the two signals are different and the carrier frequencies also are different. Each QPSK signal has a positive-frequency bandwidth equal to twice its keying rate.

$2f_1 - f$ for all f, and we were to shift its phase so that the downconverted carrier phase becomes zero, and we were to conjugate this reflected phase-shifted spectrum, then we would obtain precisely the original spectrum. Thus, if we subtracted the conjugated, phase-shifted, reflected spectrum from the original spectrum, we would cancel the signal. This cancellation in the frequency domain is equivalent to simply downconverting the signal to a carrier frequency and phase of zero and then conjugating the time-domain signal and subtracting it from the unconjugated downconverted time-domain signal. However, in the process of cancelling the signal downconverted to zero frequency, we introduce severe distortion to the other signal present, which is downconverted to frequency $f_2 - f_1 \neq 0$: there will be two replicas of the signal present, with carrier frequencies of $f_2 - f_1$ and $f_1 - f_2$. Nevertheless, by downconverting the processed data so that one of the two replicas present (either the original signal or its conjugate that was added in the first stage of processing) has carrier frequency and phase of zero, we can again subtract the conjugate of this processed data to cancel one of the two replicas present. This will introduce new distortion; that is, there will again be two replicas of the signal present, but this time the carrier frequencies will be $\pm 2\,|f_2 - f_1|$. By proceeding through $N = B/2\,|f_2 - f_1| - 1$ stages of downconverting, conjugating, and subtracting (where B is the bandwidth of the signal with original carrier frequency f_2), we end up with two nonoverlapping replicas of the signal, which can be separated with a filter.

When signal distortion due to convolution (e.g., from passage through a channel) is present, this procedure will still work, in principle, provided that a filtered version of the conjugated data is subtracted at each stage. The challenge in practice is to find a way to adapt the filter needed to obtain effective cancellation.

A final example involves the reduction of the signal distortion due to frequency-selective fading caused by multipath propagation. Straightforward amplification in faded portions of the spectrum using a time-invariant filter suffers from the resultant amplification of noise. In contrast to this, a periodically time-variant filter can replace the faded spectral components with stronger highly correlated components from other bands. If these correlated spectral components are weaker than the original components before fading there will be some noise enhancement when they are amplified. But the amount of noise enhancement can be much less than that which would result from the time-invariant filter, which can only amplify the very weak faded components.

Detailed studies of the principles of operation and the mean-squared-error performance of both optimum and adaptive frequency-shift filters are reported in (Gardner, 1987a, Chapter 14, Secs. A, B; Gardner, 1990a, Chapter 12, Sec. 12.8; Gardner and Brown, 1989; Reed and Hsia, 1990; Gardner, 1993). See also (Zivanovich and Gardner, 1991).

4.8 Prediction and Causality

If a signal is correlated with time-shifted versions of itself (i.e., if it is not a white-noise signal), then its past can be used to predict its future. The higher the degree of temporal coherence $\left|\gamma_x^0(\tau)\right|$, the better the prediction. A signal that exhibits cyclostationarity is also correlated with frequency-shifted versions of itself. Consequently, its future can be better predicted if frequency-shifted versions of its past also are used, so that its spectral coherence as well as its temporal coherence can be exploited. For example, if $x(t)$ has cycle frequencies $\{\alpha_1, \ldots, \alpha_{N-1}\}$ then we can estimate the future value

$x(t+\tau)$ for some $\tau > 0$ using a linear combination of the present and past values of the N signals

$$x_q(t) = x(t)\, e^{i2\pi\alpha_q t} \qquad \text{for } q = 0, \ldots, N-1. \tag{103}$$

That is, the predicted value is given by

$$\hat{x}(t+\tau) = \sum_{u=0}^{M-1} \sum_{q=0}^{N-1} h_q(u)\, x_q[t-u], \tag{104}$$

where M is the memory-length of the predictor. The set of MN prediction coefficients that minimize the time-averaged (over t) squared magnitude of the prediction error $\hat{x}(t+\tau) - x(t+\tau)$ can be shown to be fully specified by the cyclic correlation functions for the N cycle frequencies. Specifically, the set of MN coefficients $\{h_q(u)\}$ is the solution to the set of MN simultaneous linear equations

$$\sum_{u=0}^{M-1} \sum_{q=0}^{N-1} h_q(u)\, R_x^{\alpha_p-\alpha_q}(t-u)\, e^{i\pi(\alpha_p-\alpha_q)(t-u)} = R_x^{\alpha_p}(t+\tau)\, e^{i\pi\alpha_p(t+\tau)} \tag{105}$$

for $t = 0, \ldots, M-1$ and $p = 0, \ldots, N-1$. Also, the percent accuracy of prediction is determined solely by the temporal coherence functions (29) for the frequency translates. It can be shown that for each cycle frequency α_q exploited, there is a corresponding increase in the percent accuracy of the prediction.

In the same way that time-invariant autoregressive model-fitting of stationary time-series data is mathematically equivalent to time-invariant linear prediction (Gardner, 1987a, Chapter 9, Sec. B), it can be shown that frequency-shift (or polyperiodic time-variant) autoregressive model-fitting is mathematically equivalent to frequency-shift linear prediction. Studies of this problem are reported in (Brelsford, 1967; Pagano, 1978; Miamee and Salehi, 1980; Tiao and Grupe, 1980; Sakai, 1982, 1983, 1990, 1991; Vecchia, 1985; Obeysekera and Salas, 1986; Li and Hui, 1988; Anderson and Vecchia, 1992). Also, the univariate prediction problem for cyclostationary (not polycyclostationary) time-series is equivalent to the multivariate prediction problem for stationary time-series (Pagano, 1978). This follows from the representation of univariate cyclostationary time-series in terms of multivariate stationary time-series (Gladyshev, 1961; Gardner and Franks, 1975). A survey of recent results in prediction theory for cyclostationary processes is given in Article 7 in this volume.

A measure of the degree to which one time-series causes another time-series is the degree to which the present and past of the former can linearly predict the future of the latter. If the two time-series are jointly cyclostationary, then cyclic as well as constant causality is possible. In fact, by considering only time-invariant predictors, it is possible to conclude for some pairs of time-series that no causality exists when, in fact, one time-series is perfectly cyclically caused by the other. An example of this is $x(t) = z(t)$ and $y(t) = z(t-\tau)\cos(t)$, where $\tau > 0$ and $z(t)$ is an independent identically distributed sequence. The best linear time-invariant predictor of $y(t)$ using the past of $x(t)$ is $\hat{y}(t) = 0$, whereas the best linear periodically time-variant predictor is $\hat{y}(t) = x(t-\tau)\cos(t) = y(t)$, which yields perfect prediction. Moreover, if $z(t)$ takes on values of only ± 1, it is particularly easy to show for this example that

$$\langle x^n(t-s)y^m(t)\rangle = \langle x^n(t-s)\rangle\langle y^m(t)\rangle$$

for all positive integers m, n, and s. Consequently, even the best *nonlinear* time-invariant predictor is $\hat{y}(t) = 0$.

4.9 Linear and Nonlinear System Identification

The cyclostationarity of signals passing through linear time-invariant systems can be exploited in several ways for the purpose of using input/output or output-only measurements to identify the system. In the case where the input/output measurements are corrupted by additive noise and/or interfering signals, the signal selectivity property associated with cyclostationarity can be used to obtain (asymptotically) complete immunity to this corruption. As explained in more detail in (Gardner, 1990b), the transfer function $H(f)$ of a system with corrupted input $w(t)$ and corrupted output $x(t)$ is given by

$$\frac{S^{\alpha}_{xw}(f-\alpha/2)}{S_w(f-\alpha/2)}$$

regardless of the additive corruption in $w(t)$ and $x(t)$, provided only that α is a cycle frequency of the uncorrupted system-input and is not a cycle frequency of the corruption, and that the support of the SCD $S^{\alpha}_{xw}(f-\alpha/2)$ in f covers the whole passband of the system.

Also, in the case where corrupted output-only measurements are available, the fact that the spectral correlation function $S^{\alpha}_{x}(f)$ of the system output contains information about the phase as well as the magnitude of the transfer function $H(f)$ (cf. (49)) means that blind identification of the system using only second-order statistics (SCD and PSD) is possible (Gardner, 1991c). One particularly simple scheme for blind channel equalization for digital QAM signals (or PAM signals) uses the fact that over each and every symbol interval, the channel output is the sum of noise and a linear combination of the same functions (viz., $\{h(t-nT) : n = 0, \pm 1, \pm 2, \ldots\}$ for $0 \le t < T$, where T is the length of the symbol interval, and $h(t)$ is the combined impulse response function of the transmitter's pulse-shaping filter and the channel. Consequently, the first term of an empirical Karhunen-Loeve expansion of the channel output over one symbol interval obtained from an eigendecomposition of the empirical output autocovariance matrix over one symbol interval (measured by performing synchronized averaging over multiple symbol intervals) can be used to equalize the channel. (This is particularly so when the symbol sequence and noise are both white.) That is, the eigenvector φ_1 corresponding to the largest eigenvalue will tend to be colinear with the channel output pulse over one symbol interval and orthogonal to the tails within this interval from the pulses centered in other symbol intervals. Thus, the inner product

$$\hat{a}_n = \sum_{t=(n-1)T}^{nT-1} w(t)\,\varphi_1^*(t)$$

of this eigenvector with the channel output (for the nth symbol interval with length T) can suppress intersymbol interference and provide an accurate estimate of the symbol in each interval.

The aforementioned synchronized average that provides the empirical autocovariance matrix is given by

$$\frac{1}{2N+1}\sum_{n=-N}^{N} x(t+nT)\,x^*(s+nT)$$

for the ts element of the $T \times T$ matrix (assuming the sampling increment is $T_s = 1$).

This straightforward approach is the special case of a more general approach (corresponding to the choice of parameters $K = L = 0$) proposed by Schell in Chapter 3 in this volume, where the results of a simulation of this special case are presented. Schell's more general method is derived from another approach based on least-squares filtering of a cyclostationary signal model to the channel output. Other approaches are described in Articles 4 and 5 in this volume.

A popular approach to the identification of nonlinear dynamical systems from input-output measurements is to model the system in terms of the Volterra series, which is a generalization of the power-series (or polynomial) representation of a memoryless system to systems with memory, and then to identify one-by-one the Volterra kernels, each one of which characterizes one term in the series representation. The first kernel is the impulse response of the linear part of the system. The second kernel is a two-dimensional generalization of the impulse response of the quadratic part of the system, and so on. Common approaches to identifying the kernels are based on crosscorrelation measurements between the unknown-system output and specially designed nonlinear functions of the system input.

Although the fundamental theory of this crosscorrelation approach to nonlinear system identification is built on the foundation of stationary random processes or time-series (Schetzen, 1989), it has recently been shown (Gardner and Archer, 1993) that substantial advantages can be gained by using cyclostationary inputs to the unknown system and cyclic crosscorrelations. In particular, desirable orthogonality (zero-correlation) properties between the system output and nonlinear functions of the input that are not possible for stationary inputs are possible for cyclostationary inputs, and this leads to particularly convenient designs for the inputs and the nonlinear functions. Moreover, this approach of exploiting cyclostationarity to identify time-invariant systems has recently been generalized to identify polyperiodic nonlinear systems (Gardner and Paura, 1992). In (Gardner and Archer, 1993; Gardner and Paura, 1992), the basic theory of this new approach is presented for both a time-domain method, which directly identifies the Volterra kernels or their polyperiodic counterparts, and a frequency-domain method, which directly identifies the multidimensional Fourier transforms of the kernels—the Volterra transfer functions, and several examples of cyclostationary inputs and corresponding nonlinear functions are given. This work exploits higher-than-second-order cyclostationarity, the principles of which are given in Chapter 2 in this volume.

REFERENCES

Agee, B. G., S. V. Schell, and W. A. Gardner (1987) "Self-coherence restoral: A new approach to blind adaptation of antenna arrays," *Proceedings of the Twenty-First Annual Asilomar Conference on Signals, Systems, and Computers*, Pacific Grove, CA, 1987, pp. 589–593.

Agee, B. G., S. V. Schell, and W. A. Gardner (1988) "The SCORE approach to blind adaptive signal extraction: An application of the theory of spectral correlation," *Proceedings of the IEEE/ASSP Fourth Workshop on Spectral Estimation and Modeling*, 3-5 August 1988, Minneapolis, MN, pp. 277–282.

Agee, B. G., S. V. Schell, and W. A. Gardner (1990) "Spectral self-coherence restoral: A new approach to blind adaptive signal extraction," *Proceedings of the IEEE*, Vol. 78, pp. 756–767.

Albuquerque, J. P. A., O. Shimbo, and L. N. Ngugen (1984) "Modulation transfer noise effects from a continuous digital carrier to FDM/FM carriers in memoryless nonlinear devices," *IEEE Transactions on Communications*, Vol. COM-32, pp. 337–353.

Alekseev, V. G. (1988) "Estimating the spectral densities of a Gaussian periodically correlated stochastic process," *Problems of Information Transmission*, Vol. 24, pp. 109–115.

Alekseev, V. G. (1991) "Spectral density estimators of a periodically correlated stochastic process," *Problems of Information Transmission*, Vol. 26, pp. 286–288.

Anderson, P. L., and A. V. Vecchia (1992) "Asymptotic results for periodic autoregressive moving average processes," *Journal of Time Series Analysis*, Vol. 13.

Barnett, T. P. (1983) "Interaction of the monsoon and pacific trade wind system at interannual time scales. I. The equatorial zone," *Monthly Weather Review*, Vol. III, pp. 756–773.

Barnett, T. P., H. D. Heinz, and K. Hasselmann (1984) "Statistical prediction of seasonal air temperature over Eurasia," *Tellus Ser. A* (Sweden), Vol. 36A, pp. 132–146.

Bartolini, P., J. D. Salas, and J. T. B. Obeysekera (1988) "Multivariate periodic ARMA (1, 1) processes," *Water Resources Research*, Vol. 24, pp. 1237–1246.

Bennett, W. R. (1958) "Statistics of regenerative digital transmission," *Bell System Tech. Jour.*, Vol. 37, pp. 1501–1542.

Bittanti, S. (1987) "The periodic prediction problem for cyclostationary processes– An introduction," *Proc. of the NATO Advanced Research Workshop*, Groningen 1986, pp. 239–249.

Bittanti, S., and G. De Nicolao (1993) "Spectral factorization of linear periodic systems with application to the optimal prediction of periodic ARMA models," *Automatica*, Vol. 29, March.

Blum, J. R., and D. L. Hanson (1966) "On quasi-periodic probability measures," *Journal of Math. Mech.*, Vol. 15, pp. 291–303.

Boyles, R. A., and W. A. Gardner (1983) "Cycloergodic properties of discrete-parameter non-stationary stochastic processes," *IEEE Transactions on Information Theory*, Vol. IT-29, pp. 105–114.

Brelsford, W. M. (1967) "Probability predictions and time series with periodic structure," Ph.D. Dissertation, Johns Hopkins University, Baltimore, Maryland.

Brown, W. A. (1987) "On the theory of cyclostationary signals," Ph.D. Dissertation, Department of Electrical Engineering and Computer Science, University of California, Davis, California.

Brown, W. A., and H. H. Loomis, Jr. (1992) "Digital implementation of spectral correlation analyzers," *IEEE Transactions on Signal Processing*, Vol. 40, pp. 703–720.

Campbell, J. C., A. J. Gibbs, and B. M. Smith (1983) "The cyclostationary nature of crosstalk interference from digital signals in multipair cable—Part I: Fundamentals," *IEEE Transactions on Communications*, Vol. COM-31, pp. 629–637.

Chen, C. K. (1989) "Spectral correlation characterization of modulated signals with application to signal detection and source location," Ph.D. Dissertation, Department of Electrical Engineering and Computer Science, University of California, Davis, California.

Chen, C. K., and W. A. Gardner (1992) "Signal-selective time-difference-of-arrival estimation for passive location of manmade signal sources in highly corruptive environments. Part II: Algorithms and performance," *IEEE Transactions on Signal Processing*, Vol. 40, pp. 1185–1197.

Dehay, D. (1991) "Contributions to the spectral analysis of nonstationary processes," Ph.D. Dissertation, Universite des Sciences et Techniques de Lille Flandres Artois, June 1991.

Dragan, Ya. P., and I. N. Yavorskii (1982) *Rhythmics of Sea Waves and Underwater Acoustic Signals*, Kiev: Naukova Dumka (Russian).

Dragan, Ya. P., V. A. Rozhkov, and I. N. Yavorskii (1984) "Applications of the theory of periodically correlated random processes to the probabilistic analysis of oceanological time series," *Probabilistic Analysis and Modeling of Oceanological Processes* (Russian) (V. A. Rozhkov, ed.), pp. 4–23, Leningrad: Gidrometeoizdat.

Dragan, Ya. P., V. A. Rozhkov, and I. N. Yavorskii (1987) *Methods of Probabilistic Analysis of Oceanological Rhythmics* (Russian), Leningrad: Gidrometeoizdat.

Franks, L. E. (1969) *Signal Theory*, Englewood Cliffs, NJ: Prentice-Hall.

Franks, L. E., and J. Bubrouski (1974) "Statistical properties of timing jitter in a PAM timing recovery scheme," *IEEE Transactions on Communications*, Vol. COM-22, pp. 913–930.

Franks, L. E. (1980) "Carrier and bit synchronization in data communication—A tutorial review," *IEEE Transactions on Communications*, Vol. COM-28, pp. 1107–1121.

Gardner, W. A. (1972) "Representation and estimation of cyclostationary processes," Ph.D. Dissertation, Department of Electrical and Computer Engineering, University of Massachusetts, reprinted as *Signal and Image Processing Lab Technical Report No. SIPL-82-1*, Department of Electrical and Computer Engineering, University of California, Davis, California, 95616 (1982).

Gardner, W. A., and L. E. Franks (1975) "Characterization of cyclostationary random signal processes," *IEEE Transactions on Information Theory*, Vol. IT-21, pp. 4–14.

Gardner, W. A. (1978) "Stationarizable random processes," *IEEE Transactions on Information Theory*, Vol. IT-24, pp. 8–22.

Gardner, W. A. (1985) *Introduction to Random Processes with Applications to Signals and Systems*, New York: Macmillan.

Gardner, W. A. (1986a) "The role of spectral correlation in design and performance analysis of synchronizers," *IEEE Transactions on Information Theory*, Vol. COM-34, pp. 1089–1095.

Gardner, W. A. (1986b) "The spectral correlation theory of cyclostationary time-series," *Signal Processing* (EURASIP), Vol. 11, pp. 13–36.

Gardner, W. A. (1986c) "Measurement of spectral correlation," *IEEE Transactions on Acoustics, Speech, and Signal Processing*, Vol. ASSP-34, pp. 1111–1123.

Gardner, W. A. (1987a) *Statistical Spectral Analysis: A Nonprobabilistic Theory*, Englewood Cliffs, NJ: Prentice-Hall.

Gardner, W. A. (1987b) "Spectral correlation of modulated signals, Part I—Analog modulation," *IEEE Transactions on Communications*, Vol. COM-35, pp. 584–594.

Gardner, W. A., W. A. Brown, and C.-K. Chen (1987) "Spectral correlation of modulated signals, Part II—Digital modulation," *IEEE Transactions on Communications*, Vol. COM-35, pp. 595–601.

Gardner, W. A. (1988a) "Simplification of MUSIC and ESPRIT by exploitation of cyclostationarity," *Proceedings of the IEEE*, Vol. 76, No. 7, pp. 845–847.

Gardner, W. A. (1988b) "Signal interception: A unifying theoretical framework for feature detection," *IEEE Transactions on Communications*, Vol. COM-36, pp. 897–906.

Gardner, W. A. (1988c) "Exploitation of spectral correlation in cyclostationary signals," *Proc. Fourth Annual ASSP Workshop on Spectrum Estimation and Modeling*, 3-5 August 1988, Minneapolis, MN, pp. 1–6.

Gardner, W. A., and C-K. Chen (1988) "Interference-tolerant time-difference-of-arrival estimation for modulated signals," *IEEE Transactions on Acoustics, Speech, and Signal Processing*, Vol. ASSP-36, pp. 1385–1395.

Gardner, W. A., and W. A. Brown (1989) "Frequency-shift filtering theory for adaptive co-channel interference removal," *Conference Record, Twenty-Third Asilomar Conference on Signals, Systems and Computers*, 30 Oct.-1 Nov. 1989, Pacific Grove, CA, pp. 562–567.

Gardner, W. A. (1990a) *Introduction to Random Processes with Applications to Signals and Systems*, 2nd ed., New York: McGraw-Hill.

Gardner, W. A. (1990b) "Identification of systems with cyclostationary input and correlated input/output measurement noise," *IEEE Transactions on Automatic Control*, Vol. 35, pp. 449–452.

Gardner, W. A. (1990c) "Spectral characterization of n-th order cyclostationarity," *Proceedings of the Fifth IEEE/ASSP Workshop on Spectrum Estimation and Modeling*, 10-12 Oct. 1990, Rochester, NY, pp. 251–255. (Invited paper)

Gardner, W. A., and S. Venkataraman (1990) "Performance of optimum and adaptive frequency-shift filters for co-channel interference and fading," *Proceedings of the Twenty-Fourth Annual Asilomar Conference on Signals, Systems, and Computers*, 5-7 Nov. 1990, Pacific Grove, CA, pp. 242–247.

Gardner, W. A., and C. M. Spooner (1990) "Higher order cyclostationarity, cyclic cumulants, and cyclic polyspectra," *Proceedings of the 1990 International Symposium on Information Theory and Its Applications*, 27-30 Nov. 1990, Honolulu, HI, pp. 355–358.

Gardner, W. A. (1991a) "Exploitation of spectral redundancy in cyclostationary signals," *IEEE Signal Processing Magazine*, Vol. 8, pp. 14–36.

Gardner, W. A. (1991b) "Two alternative philosophies for estimation of the parameters of time-series," *IEEE Transactions on Information Theory*, Vol. 37, pp. 216–218.

Gardner, W. A. (1991c) "A new method of channel identification," *IEEE Transactions on Communications*, Vol. 39, pp. 813–817.

Gardner, W. A., and W. A. Brown (1991) "Fraction-of-time probability for time-series that exhibit cyclostationarity," *Signal Processing*, Vol. 23, No. 3, pp. 273–292.

Gardner, W. A. (1992) "A unifying view of coherence in signal processing," *Signal Processing* (EURASIP), Vol. 29, pp. 113–140.

Gardner, W. A., and Chen, C. K. (1992) "Signal-selective time-difference-of-arrival estimation for passive location of manmade signal sources in highly corruptive environments. Part I: Theory and method," *IEEE Transactions on Signal Processing*, Vol. 40, pp. 1168–1184.

Gardner, W. A., and L. Paura (1992) "Identification of polyperiodic nonlinear systems," Technical Report, December 1992, Department of Electrical and Computer Engineering, University of California, Davis.

Gardner, W. A., S. V. Schell, and P. Murphy (1992) "Multiplication of cellular radio capacity by blind adaptive spatial filtering," accepted for publication in *1992 Inter-*

national Conference on Selected Topics in Wireless Communications, June 25-26 and *Proceedings of the 22nd Annual IEEE Communication Theory Workshop*, June 28-July 21, 1992.

Gardner, W. A., and C. M. Spooner (1992a) "Signal interception: Performance advantages of cyclic feature detectors," *IEEE Transactions on Communications*, Vol. 40, No. 1, pp. 149–159.

Gardner, W. A., and C. M. Spooner (1992b) "The cumulant theory of cyclostationary time-series, Part I: Foundation," Technical Report, March 1992, Department of Electrical and Computer Engineering, University of California, Davis, and *IEEE Transactions on Signal Processing* (in press).

Gardner, W. A. (1993) "Cyclic Wiener filtering: Theory and method," *IEEE Transactions on Communications*, Vol. 41, pp. 151–163.

Gardner, W. A., and T. L. Archer (1993) "Exploitation of cyclostationarity for identifying the Volterra kernels of nonlinear systems," *IEEE Transactions on Information Theory*, Vol. 39, pp. 535–542.

Gardner, W. A., and C. M. Spooner (1993) "Detection and source location of weak cyclostationary signals: Simplifications of the maximum-likelihood receiver," *IEEE Transactions on Communications*, Vol. 41 (in press).

Genossar, M. J., H. Lev-Ari, and T. Kailath (1993) "Consistent estimation of the cyclic autocorrelation," *IEEE Transactions on Signal Processing*, Vol. 41 (in press).

Gladyshev, E. G. (1961) "Periodically correlated random sequences," *Soviet Math. Dokl.*, Vol. 2, pp. 385–388.

Gladyshev, E. G. (1963) "Periodically and almost periodically correlated random processes with continuous time parameter," *Theory Prob. Appl.*, Vol. 8, pp. 173–177.

Gudzenko, L. I. (1959) "On periodic nonstationary processes," *Radio Eng. Electron. Phys.* (Russian), Vol. 4, pp. 220–224.

Hasselmann, K., and T. P. Barnett (1981) "Techniques of linear prediction for systems with periodic statistics," *Journal of Atmospheric Science*, Vol. 38, pp. 2275–2283.

Hofstetter, E. M. (1964) "Random processes," Chapter 3 in *The Mathematics of Physics and Chemistry*, Vol. II, H. Margenau and G. M. Murphy eds., Princeton, NJ: D. Van Nostrand Co.

Honda, I. (1982) "On the spectral representation and related properties of periodically correlated stochastic process," *The Transactions of the IECE of Japan*, Vol. E65, pp. 723–729.

Honda, I. (1990) "On the ergodicity of Gaussian periodically correlated stochastic processes," *Transactions of the Institute of Electronics, Information and Communication Engineers E*, Vol. E73, pp. 1729–1737.

Hurd, H. L. (1969) "An investigation of periodically correlated stochastic processes," Ph.D. Dissertation, Duke University, Durham, North Carolina.

Hurd, H. L. (1974a) "Periodically correlated processes with discontinuous correlation functions," *Theory Prob. Appls.*, Vol. 19, pp. 804–807.

Hurd, H. L. (1974b) "Stationarizing properties of random shifts," *SIAM Journal of Applied Math.*, Vol. 26, pp. 203–212.

Hurd, H. L. (1989a) "Nonparametric time series analysis for periodically correlated processes," *IEEE Transactions on Information Theory*, March 1989, Vol. 35, pp. 350–359.

Hurd, H. L. (1989b) "Representation of strongly harmonizable periodically correlated processes and their covariances," *Jour. Multivariate Anal.*, Vol. 29, pp. 53–67.

Hurd, H. L. (1991) "Correlation theory of almost periodically correlated processes," *Jour. Multivariate Anal.*, Vol. 37, No. 1, pp. 24–45.

Hurd, H. L., and J. Leskow (1992a) "Estimation of the Fourier coefficient functions and their spectral densities for Φ-mixing almost periodically correlated processes," *Statistics and Probability Letters*, Vol. 14, pp. 299–306.

Hurd, H. L., and J. Leskow (1992b) "Strongly consistent and asymptotically normal estimation of the covariance for almost periodically correlated processes," *Statistics and Decision*, Vol. 10, pp. 201–225.

Izzo, L., L. Paura, and G. Poggi (1989) "Multiple-source localization: A new method exploiting the cyclostationarity property," *Proceedings of Douzième Colloque sur le Traitement du Signal et des Images*, Juan-les-Pins, France, pp. 481–484.

Izzo, L., A. Napolitano, L. Paura (1990) "Interference-tolerant estimation of amplitude and time-delay parameters of a composite signal," *Proceedings of Fifth European Signal Processing Conference* (EUSIPCO '90), Barcelona, 18-21 September 1990, pp. 103–106.

Izzo, L., L. Paura, and M. Tanda (1992) "Signal interception in non-Gaussian noise," *IEEE Transactions on Communications*, Vol. 40, pp. 1030–1037, June 1992.

Izzo, L., L. Paura, and G. Poggi (1992) "An interference-tolerant algorithm for localization of cyclostationary-signal sources," *IEEE Transactions on Signal Processing*, Vol. 40, pp. 1682–1686, July 1992.

Jacobs, K. (1958) "Almost periodic Markov processes," *Mathematics Annual*, Vol. 134, pp. 408–427 (German).

Johnson, C. M., P. Lemke, and T. P. Barnett (1985) "Linear prediction of sea ice anomalies," *Journal of Geophys. Res.*, Vol. 90, pp. 5665–5675.

Li, W. K., and Y. V. Hui (1988) "An algorithm for the exact likelihood of periodic autoregressive moving average models," *Comm. Stat., Simulation Comput.*, Vol. 17, pp. 1483–1494.

McLeod, A. I., D. J. Noakes, K. W. Hipel, and R. M. Thompstone (1987) "Combining hydrologic forecasts," *Journal of Water Resources Planning and Management*, Vol. 113, pp. 29–41, January 1987.

Miamee, A. G., and H. Salehi (1980) "On the prediction of periodically correlated stochastic processes," *Multivariate Analysis-V*, P. R. Krishnaiah (ed.), Amsterdam: North Holland Publishing Company, pp. 167–179.

Moeneclaey, M. (1982) "Linear phase-locked loop theory for cyclostationary input disturbances," *IEEE Transactions on Communications*, Vol. COM-30, pp. 2253–2259.

Moeneclaey, M. (1983) "The optimum closed-loop transfer function of a phase-locked loop used for synchronization purposes," *IEEE Transactions on Communications*, Vol. COM-31, pp. 549–553.

Moeneclaey, M. (1984) "A fundamental lower bound on the performance of practical joint carrier and bit synchronizers," *IEEE Transactions on Communications*, Vol. COM-32, pp. 1007–1012.

Nedoma, J. (1963) "Über die ergodizität and r-ergodizität stationärer wahrscheinlichkeitsmasse," *Z. Wahrscheinlichkeitstheorie*, Vol. 2, pp. 90–97.

Newton, H. J. (1982) "Using periodic autoregressions for multiple spectral estimation," *Technometrics*, Vol. 24, pp. 109–116.

Obeysekera, J. T. B., and J. D. Salas (1986) "Modeling of aggregated hydrologic time series," *Journal of Hydrol.* Vol. 86, pp. 197–219.

Pagano, M. (1978) "On periodic and multiple autoregressions," *Ann. Statist.*, Vol. 6, pp. 1310–1317.

Parzen, E., and M. Pagano (1979) "An approach to modeling seasonally stationary time-series," *Journal of Econometrics*, Vol. 9, North Holland Publishing Company, pp. 137–153.

Petersen, B. R. (1992) "Equalization in Cyclostationary Interference," Ph.D. Dissertation, Department of Systems and Computer Engineering, Carleton University, Ottawa, Ontario, Canada.

Pourahmadi, M., and H. Salehi (1983) "On subordination and linear transformation of harmonizable and periodically correlated processes," *Probability Theory on Vector Spaces, III* (Lublin), Berlin-New York: Springer, pp. 195–213.

Rao, M. M., and D. K. Chang (1988) "Special representations of weakly harmonizable processes," *Stochastic Analysis and Applications*, Vol. 6, pp. 169–190.

Reed, J. H., and T. C. Hsia (1990) "The performance of time-dependent adaptive filters for interference rejection," *IEEE Transactions on Acoustics, Speech and Signal Processing*, Vol. 38, pp. 1373–1385.

Roberts, R. S. (1989) "Architectures for digital cyclic spectral analysis," Ph.D. Dissertation, Department of Electrical Engineering and Computer Science, University of California, Davis, California.

Roberts, R. S., W. A. Brown, and H. H. Loomis, Jr. (1991) "Computationally efficient algorithms for cyclic spectral analysis," *IEEE Signal Processing Magazine*, Vol. 8, pp. 38–49.

Rootenberg, J., and S. A. Ghozati (1977) "Stability properties of periodic filters," *International Journal of Systems and Science*, Vol. 8, pp. 953–959.

Rootenberg, J., and S. A. Ghozati (1978) "Generation of a class of nonstationary random processes," *International Journal of Systems and Science*, Vol. 9, pp. 935–947.

Sakai, H. (1982) "Circular lattice filtering using Pagano's method," *IEEE Transactions on Acoustics, Speech, and Signal Processing*, Vol. ASSP-30, pp. 279–287.

Sakai, H. (1983) "Covariance matrices characterization by a set of scalar partial autocorrelation coefficients," *The Annals of Statistics*, Vol. 11, pp. 337–340.

Sakai, H. (1990) "Spectral analysis and lattice filter for periodic autoregressive processes," *Electronics and Communications in Japan, Part 3*, January 1990, Vol. 73, pp. 9–15.

Sakai, H. (1991) "On the spectral density matrix of a periodic ARMA process," *Journal of Time Series Analysis*, Vol. 12, pp. 73–82.

Salas, J. D. (1972) "Range analysis for storage problems of periodic stochastic processes," Ph.D. Dissertation, Department of Civil Engineering, Colorado State University, Fort Collins, CO.

Salas, J. D., and R. A. Smith (1980) "Correlation properties of periodic AR(ρ) models," *Proceedings of the 3rd International Symposium on Stochastic Hydraulics*, 5-7 August 1980, Tokyo, pp. 107–115.

Schell, S. V., and B. G. Agee (1988) "Application of the SCORE algorithm and SCORE extensions to sorting in the rank-L spectral self-coherence environment," *Conference Record, Twenty-Second Asilomar Conference on Signals, Systems and Computers*, 31 Oct.-2 Nov. 1988, Pacific Grove, CA, Vol. 1, pp. 274–278.

Schell, S. V., R. A. Calabretta, W. A. Gardner, and B. G. Agee (1989) "Cyclic MUSIC algorithms for signal-selective direction estimation," *Proceedings of the International Conference on Acoustics, Speech, and Signal Processing*, Glasgow, 23-26 May 1989, pp. 2278–2281.

Schell, S.V., and W. A. Gardner (1989) "Signal-selective direction finding for fully correlated signals," *Abstracts of Sixth Multidimensional Signal Processing Workshop of the IEEE/ASSP Society*, 6-8 Sept. 1989, Pacific Grove, CA, pp. 139–140.

Schell, S. V. (1990) "Exploitation of spectral correlation for signal-selective direction finding," Ph.D. Dissertation, Department of Electrical Engineering and Computer Science, University of California, Davis, California.

Schell, S. V., and W. A. Gardner (1990a) "Signal-selective high-resolution direction finding in multipath," *Proceedings of the International Conference on Acoustics, Speech, and Signal Processing*, Albuquerque, NM, 3-6 April 1990, pp. 2667–2670.

Schell, S. V., and W. A. Gardner (1990b) "Progress on signal-selective direction finding," *Proceedings of the Fifth IEEE/ASSP Workshop on Spectrum Estimation and Modeling*, 10-12 Oct. 1990, Rochester, NY, pp. 144–148.

Schell, S. V., and W. A. Gardner (1990c) "Detection of the number of cyclostationary signals in unknown interference and noise," *Proceedings of the Twenty-Fourth Annual Asilomar Conference on Signals, Systems, and Computers*, 5-6 Nov. 1990, Pacific Grove, CA, pp. 473–477.

Schell, S. V., and W. A. Gardner (1991) "Estimating the directions of arrival of cyclostationary signals—Part I: Theory and methods," Technical Report, November 1991, Department of Electrical and Computer Engineering, University of California, Davis.

Schell, S. V., and W. A. Gardner (1992) "Cramér-Rao lower bound for parameters of Gaussian cyclostationary signals," *IEEE Transactions on Information Theory*, Vol. 38, pp. 1418–1422, July.1992.

Schell, S. V., and W. A. Gardner (1993a) "High resolution direction finding," Chapter in *Handbook of Statistics*, N. K. Bose and C. R. Rao (Eds.), Amsterdam: Elsevier.

Schell, S. V., and W. A. Gardner (1993b) "Blind adaptive spatio-temporal filtering for wideband cyclostationary signals," *IEEE Transactions on Signal Processing*, Vol. 41, pp. 1961–1964.

Schell, S. V., W. A. Gardner, and P. A. Murphy (1993) "Blind adaptive antenna arrays in cellular communications for increased capacity" in *Proc. of 3rd Virginia Tech. Symp. on Wireless Personal Comm.*, June 1993.

Schetzen, M. (1989) *The Volterra and Wiener Theories of Nonlinear Systems*, Second Edition, Malabar, FL: Krieger Publ. Co.

Sherman, P. J. (1992) "Random processes from rotating machinery," *Proceedings of the Workshop on Nonstationary Stochastic Processes and Their Applications*, Hampton, VA, 1-2 Aug. 1991, Singapore: World Scientific.

Spooner, C. M., and W. A. Gardner (1991) "Estimation of cyclic polyspectra," *Proceedings of Twenty-Fifth Annual Asilomar Conference on Signals, Systems, and Computers*, Pacific Grove, CA, 4-6 Nov. 1991, pp. 370–376.

Spooner, C. M. (1992) "Theory and application of higher-order cyclostationarity," Ph.D. Dissertation, Department of Electrical and Computer Engineering, University of California, Davis.

Spooner, C. M., and W. A. Gardner (1992a) "An overview of higher-order cyclostationarity," *Proceedings of the Workshop on Nonstationary Stochastic Processes and Their Applications*, Hampton, VA, 1-2 Aug. 1991, Singapore: World Scientific.

Spooner, C. M., and W. A. Gardner (1992b) "The cumulant theory of cyclostationary time-series, Part II: Development and applications," Technical Report, March 1992, Department of Electrical and Computer Engineering, University of California, Davis, and *IEEE Transactions on Signal Processing* (in press).

Spooner, C. M., and W. A. Gardner (1992c) "Performance evaluation of cyclic polyspectrum estimators," *Proceedings of Twenty-Sixth Asilomar Conference on Signals, Systems, and Computers*, Pacific Grove, CA, 26-29 Oct. 1992.

Thompstone, R. M., W. Hipel, and A. I. McLeod (1985) "Grouping of periodic autoregressive models," *Time Series Analysis: Theory and Practice 6*, edited by O. D. Anderson, J. K. Ord, and E. A. Robinson, pp. 35–49, Amsterdam: Elsevier, North-Holland.

Tiao, G. C., and M. R. Grupe (1980) "Hidden periodic autoregressive-moving average models in time series data," *Biometrika*, Vol. 67, pp. 365–373.

Vecchia, A. V. (1983) "Aggregation and estimation for periodic autoregressive-moving average models," Ph.D. Dissertation, Colorado State University, Fort Collins.

Vecchia, A. V. (1985) "Maximum likelihood estimation for periodic autoregressive moving average models," *Technometrics*, Vol. 27, pp. 375–384.

Wold, H. O. A. (1948) "On prediction in stationary time-series," *Ann. Math. Stat.*, Vol. 19, pp. 558–567.

Xu, G., and T. Kailath (1992) "Direction-of-arrival estimation via exploitation of cyclostationarity—A combination of temporal and spatial processing," *IEEE Transactions on Signal Processing*, Vol. 40, pp. 1775–1786.

Zivanovich, G. D., and W. A. Gardner (1991) "Degrees of cyclostationarity and their application to signal detection and estimation," *Signal Processing* (EURASIP), Vol. 22, pp. 287–297.

Chapter 2

Higher-Order Statistics for Nonlinear Processing of Cyclostationary Signals

Chad M. Spooner
Department of Electrical and Computer Engineering
University of California
Davis, CA 95616

1 INTRODUCTION TO HIGHER-ORDER CYCLOSTATIONARITY

This chapter presents a tutorial introduction to the theory and application of higher-order cyclostationarity and is an adaptation and extension of the fourth plenary lecture given at the Workshop on Cyclostationary Signals [111] held in August 1992. As the title indicates, this chapter is generally concerned with higher-order statistics (HOS), which means moments and cumulants with orders greater than two. It is specifically concerned with the higher-order statistics of polycyclostationary signals, the study of which is called *higher-order cyclostationarity* (HOCS). This tutorial explains the mathematical structure of the moments and cumulants of polycyclostationary signals in both the time and frequency domains, provides practical interpretations of these moments and cumulants, discusses several applications of the theory, and presents nonparametric estimators of the higher-order statistics, which are generalizations of nonparametric autocorrelation and spectrum estimators for stationary signals and for order two to polycyclostationary signals and to arbitrary orders.

The word *signal* is used in this chapter to mean a single persistent time-series with finite time-averaged power. This is somewhat unusual because the study of

This work was supported jointly by the National Science Foundation under grants MIP–88–12902 and MIP–91–12800, PI: W. A. Gardner, and by the United States Army Research Office under contract DAAL03–91–C–0018, PI: W. A. Gardner.

higher-order statistics is usually carried out in the framework of stochastic process theory. However, it has been demonstrated that the theory of second-order polycyclostationarity (SOCS) is more naturally developed in the time-series framework because it closes the conceptual gap between the abstract probability space quantities of the theory and the empirical quantities of interest in practice [34, 35, 43, 45, 52, 53, 110]. For example, it has the benefit of permitting the sine-wave-generation interpretation of cyclostationarity, which in turn has led to a unique derivation of the cumulant (Section 2.3).

A signal $x(t)$ is second-order cyclostationary (in the wide sense, cf. [52]) if there exists a quadratic transformation, such as a squarer, of $x(t)$ such that the output of this transformation contains at least one finite-strength additive sine-wave component with nonzero frequency. Equivalently—but not obviously—a signal $x(t)$ is second-order cyclostationary if the amplitude and phase fluctuations of some narrowband frequency components with center frequencies that are separated by a nonzero amount are temporally correlated; that is, $x(t)$ is second-order cyclostationary if it exhibits *spectral correlation.* The frequencies of the quadratically generated sine-wave components and the frequency separations of the correlated spectral components are specified by the same set of numbers. These numbers are most commonly called *cycle frequencies*, but are also called *frequency separations*. If the cycle frequencies are not all harmonically related, then the signal is said to be *polycyclostationary* (cf. Chapter 1). The subject of HOCS is the study of the additive sine-wave components in the outputs of higher-than-second-order nonlinear transformations of $x(t)$. Equivalently—but not obviously—it is the study of the temporal statistical dependence between more than two spectral components of $x(t)$. Important examples of modulations that produce polycyclostationary signals are analog and digital amplitude and frequency modulation, and digital quadrature-amplitude modulation (QAM) (which includes some phase-shift-keyed and amplitude-shift-keyed signals as special cases) [37, 38]. Naturally occurring signals can also exhibit polycyclostationarity [35, 52].

The existence of unintentional, unavoidable, and purposely designed nonlinear transformations forms the basic motivation for studying the statistics of arbitrary nonlinear transformations of polycyclostationary signals. Unintentional and unavoidable nonlinearities can be found in the components of receivers and signal processors, such as amplifiers and modulators. When polycyclostationary signals are processed by these components, spectral lines can appear in the output. To characterize the performance of systems containing these components, a statistical characterization of the dependence of the frequencies, amplitudes, and phases of these spectral lines on the inputs is necessary. Purposely designed nonlinearities include quadratic timing-recovery circuits, such as the delay-and-multiply circuit, fourth-power carrier-phase recovery devices for digital quadrature-amplitude-modulated signals with balanced M-ary symbol constellations for $M > 2$, and other similar quadratic, quartic, and higher-order processors for signal detection, modulation recognition, and parameter estimation.

Nonlinear system models are common in many areas of scientific inquiry, such as physics, biology, and engineering. It is often of interest to determine an appropriate model for a particular system, and then to design and carry out experiments to determine the model's parameters. A common model for the relation between the input and output of a nonlinear system is the Volterra series [102]. This model is quite general and is used to motivate the choice of the particular statistical quantities to be dealt with by the theory of HOCS that is developed in this chapter. Before proceeding with this, however, the analytical tools of fraction-of-time probability are defined, the relationship of the theory of HOCS to other theories that deal with the statistical properties of nonlinearly transformed signals is clarified, and two specific signal-processing problems are described in order to properly motivate the subsequent development.

1.1 Fraction-of-Time Probability

In direct analogy with the joint probability distribution function defined for stochastic processes, the nth-order fraction-of-time (FOT) probability distribution function for the time-series $x(t)$, $t \in (-\infty, \infty)$ is defined by

$$F_{\boldsymbol{x}(t)}(\boldsymbol{y}) \triangleq \hat{E}^{\{\alpha\}}\left\{\prod_{j=1}^{n} U[y_j - x(t+t_j)]\right\} \tag{1}$$

where

$$\hat{E}^{\{\alpha\}}\{z(t)\} \triangleq \sum_{\alpha} \left\langle z(t+u)e^{-i2\pi\alpha u}\right\rangle = \sum_{\alpha} \left\langle z(u)e^{-i2\pi\alpha u}\right\rangle e^{i2\pi\alpha t} \tag{2}$$

is the multiple sine-wave extraction operation and

$$\langle w(u)\rangle \triangleq \lim_{Z\to\infty} \frac{1}{Z}\int_{-Z/2}^{Z/2} w(u)\,du$$

is the usual time-averaging operation. In (1), $U[\cdot]$ is the event-indicator function

$$U[y_j - x(t+t_j)] \triangleq \begin{cases} 1, & x(t+t_j) < y_j \\ 0, & \text{otherwise,} \end{cases}$$

and the vectors $\boldsymbol{x}(t)$ and $\boldsymbol{y}$ are specified by

$$\boldsymbol{x}(t) \triangleq [x(t+t_1) \ \cdots \ x(t+t_n)]^{\dagger},$$

$$\boldsymbol{y} \triangleq [y_1 \ \cdots \ y_n]^{\dagger},$$

where $\dagger$ denotes matrix transpose. The sum in (2) is over all real numbers α for which $\left\langle z(u)e^{-i2\pi\alpha u}\right\rangle \neq 0$. The multiple sine-wave extraction operation $\hat{E}^{\{\alpha\}}\{\cdot\}$ is completely analogous to the expectation operation $E\{\cdot\}$. That is, it is shown in [45] that the function (1) is a valid probability distribution function for the time-series for which it exists.

If the only nonzero term in the sum over α in (2) is that corresponding to $\alpha = 0$, then the time-series $x(t)$ is said to be *stationary of order n* (in the strict sense, cf. [52]). On the other hand, if some terms corresponding to $\alpha \neq 0$ are also nonzero, then $x(t)$ is said to *exhibit nth-order cyclostationarity*. If all such values of α are integer multiples of a single fundamental frequency, corresponding to a period T_0, then $x(t)$ is said to be *cyclostationary of order n*; otherwise $x(t)$ is said to be *polycyclostationary* (or *multiply cyclostationary* or *almost cyclostationary*) *of order n*. In the latter case, the FOT distribution function can be expressed as

$$F_{\boldsymbol{x}(t)}(\boldsymbol{y}) = F_{\boldsymbol{x}}^{0}(\boldsymbol{y}) + \sum_{q} \left[F_{\boldsymbol{x}(t);T_q}(\boldsymbol{y}) - F_{\boldsymbol{x}}^{0}(\boldsymbol{y}) \right],$$

where

$$F_{\boldsymbol{x}}^{0}(\boldsymbol{y}) \triangleq \left\langle \prod_{j=1}^{n} U\left[y_j - x(t + t_j)\right] \right\rangle,$$

and $F_{\boldsymbol{x}(t);T_q}(\boldsymbol{y})$ is the FOT distribution obtained from (1) by summing in (2) over only those values of α that are integer multiples of the distinct fundamental frequency $1/T_q$.

The FOT probability density function is given by the n-fold derivative of the FOT distribution function,

$$f_{\boldsymbol{x}(t)}(\boldsymbol{y}) \triangleq \frac{\partial^n}{\partial y_1 \cdots \partial y_n} F_{\boldsymbol{x}(t)}(\boldsymbol{y}),$$

and has the properties normally associated with a probability density function. The FOT expectation operation is defined in terms of the FOT probability distribution in the natural way. Let $g[\boldsymbol{x}(t)]$ be any function of the vector of time-samples $\boldsymbol{x}(t)$. If we redefine the symbol $\hat{E}^{\{\alpha\}}\{\cdot\}$ to mean the expected value with respect to the probability density function (PDF) $f_{\boldsymbol{x}(t)}(\boldsymbol{y})$,

$$\hat{E}^{\{\alpha\}}\{g[\boldsymbol{x}(t)]\} \triangleq \int_{-\infty}^{\infty} \cdots \int_{-\infty}^{\infty} g[\boldsymbol{y}] f_{\boldsymbol{x}(t)}(\boldsymbol{y})\, d\boldsymbol{y}, \tag{3}$$

then it can be shown that this expected value consists exactly of the finite-strength additive sine-wave components that are present in $g[\boldsymbol{x}(t)]$ (cf. [45, 52]). Specifically

$$\hat{E}^{\{\alpha\}}\{g[\boldsymbol{x}(t)]\} = \sum_{\alpha} M_g^{\alpha} e^{i2\pi\alpha t}, \tag{4}$$

where

$$M_g^{\alpha} \triangleq \left\langle g[\boldsymbol{x}(t)] e^{-i2\pi\alpha t} \right\rangle, \tag{5}$$

which is consistent with the meaning given to $\hat{E}^{\{\alpha\}}\{\cdot\}$ in (2).

The FOT distributions and densities are not used explicitly (as in (3)) in this chapter. Instead, the time-averages that characterize the FOT expectations are used directly as in (4) and (5). Thus, the primary purpose of the preceding discussion of FOT probability is to show that the multiple-sine-wave extraction operation $\hat{E}^{\{\alpha\}}\{\cdot\}$ is completely analogous to the familiar probabilistic expectation operation $E\{\cdot\}$ and, therefore, that the subsequent use of terms such as moment, cumulant, and characteristic function is both appropriate and mathematically justified. For a more detailed overview of the theory of FOT probability, see Chapter 1.

It is important to understand that the theory herein can be developed using stochastic processes and $E\{\cdot\}$ instead of time-series and $\hat{E}^{\{\alpha\}}\{\cdot\}$. Doing so, however, results in a more abstract theory in the sense that it is another step removed from practice. Specifically, developing the theory of HOCS using stochastic processes: (i) requires a thorough understanding of ergodicity and cycloergodicity, (ii) does not lend itself to the important interpretation of moments and cumulants as nonlinearly generated sine waves, and (iii) complicates the development of single-data-record estimators of the parameters of the theory, because the parameters are based on an ensemble, whereas the estimators operate on a single sample path. However, analytical calculation of parameters, such as cumulants, can sometimes be easier using stochastic processes rather than FOT probability. Both frameworks have been used to *develop* the theory herein, with the time-average framework heavily favored and the stochastic-process framework used as a tool to facilitate certain calculations. Nevertheless, only the time-average framework is used to *present* the theory herein so as to develop a unified whole in the reader's mind.

1.2 Other Theories for Statistical Analysis of Time-Series

Because the autocorrelation function is the expected value (either stochastic or FOT) of a signal $x(t)$ multiplied by a delayed version of itself—which is a quadratic transformation—it can be viewed as the expected value of a signal $y(t,\tau)$ that is obtained by nonlinearly transforming $x(t)$,

$$y(t,\tau) = x(t+\tau/2)x(t-\tau/2).$$

If the expected value of $y(t,\tau)$ does not depend on t, then the signal $x(t)$ is said to be *second-order stationary in the wide sense*, and the autocorrelation function is said to be *translation invariant* because when t is replaced by $t+t_0$, the autocorrelation remains the same. The theory of wide-sense stationary signals is well developed and the basics of this theory are assumed to be familiar to the reader. The point here is that we can think of the autocorrelation as a second-order statistic of $x(t)$, or as the mean of the output of a quadratic transformation of $x(t)$, and when we use the expectation $\hat{E}^{\{\alpha\}}\{\cdot\}$ this mean is just the dc value or the constant component of the output of the transformation.

For wide-sense polycyclostationary signals $x(t)$, the autocorrelation depends on t in a specific way:

$$\hat{E}^{\{\alpha\}}\{y(t,\tau)\} = R_x(t,\tau) = \sum_{\alpha} R_x^{\alpha}(\tau)e^{i2\pi\alpha t},$$

where

$$R_x^\alpha(\tau) \triangleq \left\langle x(t+\tau/2)x(t-\tau/2)e^{-i2\pi\alpha t}\right\rangle.$$

If we think of this autocorrelation as the expected value of $y(t,\tau)$, then it is clear that the output of the quadratic transformation contains finite-strength additive sine-wave components. Alternatively, if we think of the autocorrelation as a second-order statistic of $x(t)$, then it is clear that this autocorrelation is polyperiodic in t.

We can thus interpret the theories of second-order wide-sense stationarity and polycyclostationarity in two different ways: (i) in terms of translation invariant or polyperiodic second-order probabilistic parameters, or (ii) in terms of the constant or sine-wave components in the outputs of quadratic transformations of the signal.

These two interpretations extend to the case of higher-order statistical parameters such as moments. A signal $x(t)$ is *nth-order stationary in the wide sense* if each kth-order moment is translation invariant for $k = 1, \cdots, n$. Such signals are the subject of the theory of *higher-order statistics* [9, 54, 81, 83, 99] (which is somewhat of a misnomer since the restriction to stationarity is not made explicit).

A signal $x(t)$ is *nth-order polycyclostationary in the wide sense* if the output $y(t,\boldsymbol{\tau}) = \prod_{j=1}^n x(t+\tau_j)$ of some nth-order nonlinear transformation of $x(t)$ contains at least one finite-strength additive sine-wave component with nonzero frequency α:

$$R_x^\alpha(\boldsymbol{\tau})_n = \left\langle \prod_{j=1}^n x(t+\tau_j)e^{-i2\pi\alpha t}\right\rangle = \left\langle y(t,\boldsymbol{\tau})e^{-i2\pi\alpha t}\right\rangle \not\equiv 0.$$

Thus, for a polycyclostationary signal

$$\hat{E}^{\{\alpha\}}\{y(t,\boldsymbol{\tau})\} = \sum_\alpha R_x^\alpha(\boldsymbol{\tau})_n\, e^{i2\pi\alpha t},$$

where the sum is over the frequencies of all of the sine-wave components of $y(t,\boldsymbol{\tau})$. That is, the nth-order nonlinearity generates sine waves from an nth-order polycyclostationary signal.

It is the interpretation of sine-wave generation, rather than periodicity of the moment, that is most important from both conceptual and practical standpoints because it is by examining this interpretation that *cumulants* naturally assume their central role in the theory of HOCS. This is explained in detail in Section 2.

The difference between the theories of SOCS, HOS, and HOCS are pointed out throughout the chapter where appropriate. The major difference between HOS and HOCS is that HOS has been developed and applied within the probabilistic framework of (stationary) stochastic processes, whereas HOCS has been developed and applied primarily within the fraction-of-time probability framework. Whenever the theory of HOS is applied to problems in which only a single data record is available, the question of ergodicity is of primary importance. And when the single data record is itself polycyclostationary—which is quite possible even when the stochastic process from which it arises is strict-sense stationary (cf. Chapter 1)—the theory of HOCS presented in this chapter is required to obtain correct results for single-record processing—it is not optional. Specific ways in which the absence of ergodicity can cause erroneous results in applying the theory of HOS are discussed in Section 7.6.

1.3 Two Motivating Examples

To motivate the development of the theory, two sample problems are presented. These problems are especially difficult to solve by using the theories of SOCS or HOS, and the classical theory of statistical inference and decision is intractable. Consider the signal model

$$x(t) = s(t) + m_x(t) + \sum_{k=1}^{M} i_k(t)$$

where $s(t)$ is a signal of interest, $\{i_k(t)\}$ are non-Gaussian interfering signals that temporally and spectrally overlap $s(t)$, and $m_x(t)$ is broadband noise. All the signals and noise are statistically independent of each other. The signal of interest $s(t)$ is polycyclostationary of order n but is not polycyclostationary of order two. We assume that the power level of the noise is unknown and is varying within the observation interval, and that the interferers can be turning on and off within the observation interval. The first of the sample problems is to detect the presence of $s(t)$.

Methods of detecting $s(t)$ that are based on measuring the power spectrum are ineffective because the power level of the noise and interference is unknown and time-varying: Setting a threshold for energy detection is difficult. Detection methods that are based on measuring the cyclic spectrum are not applicable because the signal of interest is not second-order polycyclostationary: There is no spectral correlation to measure. Detection methods that are based on measuring the (stationary) higher-order statistics of $x(t)$ (e.g., the bispectrum) can fail for the same reason that the energy detection methods do: All non-Gaussian signals contribute to the HOS. Nevertheless, as explained in Section 8.1, the theory of HOCS provides a practical solution to this problem provided that $s(t)$ possesses a unique nth-order cycle frequency.

Next let's assume that we have data available from another receiver:

$$y(t) = a_0 s(t - d_0) + m_y(t) + \sum_{k=1}^{M} a_k i_k(t - d_k),$$

where the a_k and d_k are the attenuation factors and delays, respectively, relative to the corresponding signals in $x(t)$, and all the signals and noise in $y(t)$ are statistically independent of each other. The second of the two sample problems mentioned previously is to estimate a_0 and d_0. This estimation is difficult to accomplish by computing the cross-correlation between filtered versions of $x(t)$ and $y(t)$ (known as *generalized cross-correlation* [GCC] methods) because the signals are spectrally overlapping and cannot be separated by linear time-invariant filtering. Thus, each signal will contribute a peak to the GCC output function. SOCS-based estimators cannot be used because, as in the detection problem, there is no SOCS to exploit, and HOS-based methods suffer from the same problem as the GCC: All signals contribute a peak.

However, as explained in Section 8.2, this problem can be solved by using the theory of HOCS, provided only that $s(t)$ possesses a unique nth-order cycle frequency.

The remainder of this chapter is organized as follows: In Section 1.4, the history of the cumulant is traced, starting with the original contributions from the late nineteenth century and ending with the recent workshops on higher-order statistics sponsored by the Institute of Electrical and Electronics Engineers. The time-domain, or temporal, moments and cumulants of polycyclostationary signals are defined in Section 2, where a complete introduction to cumulants is provided. The frequency-domain, or spectral, moments and cumulants are the subject of Section 3. These spectral moments and cumulants are shown to be related to the temporal moments and cumulants by Fourier transformation. In Section 4, the utility of the theory of moments and cumulants for solving statistical inference and decision problems is discussed. The theory of HOCS is developed further in Section 5 by accommodating complex-valued signals and by determining the effects of signal-processing operations on the higher-order moments and cumulants of polycyclostationary signals. The moments and cumulants of the complex envelopes of digital QAM signals (which includes real and complex pulse-amplitude-modulated (PAM) signals as special cases) are derived in Section 6 for arbitrary orders. Nonparametric estimation of the moments and cumulants of arbitrary polycyclostationary signals is discussed and illustrated with examples in Section 7, and the two motivating examples discussed in this section—weak-signal detection and interference-tolerant time-delay estimation—are studied in Section 8. Concluding remarks are given in Section 9.

1.4 Relevant Previous Work

In this section, relevant work in the areas of higher-order statistics and second-order polycyclostationarity is reviewed. In addition, since cumulants are assumed to be unfamiliar to the reader, a complete history of the development of the concept of the cumulant is included.

1.4.1 Cumulants and Polyspectra

The history of the cumulant is traced in this section. A concise history, to begin with, is that the cumulant was born in mathematical statistics, developed in the probabilistic theory of stochastic processes, and after nearly one hundred years found its way into electrical engineering through the field of higher-order statistics as applied primarily to problems of time-series modeling and system identification.

In 1903, Danish astronomer Thorvald Nicolai Thiele published a book called *The Theory of Observations* [113] in which he tried to quantify the statistical nature of measurement errors. Thiele developed functions that he called *laws of presumptive errors*, which are probability density functions. By expressing these functions in Maclaurin series form, he found that they could be characterized by moments or by cumulants, which he called *half-invariants*, because cumulants are invariant to additive constants in a random variable, but not to multiplicative constants. Thiele recognized that the half-invariants provided an easy way to test for the normal distribution: The higher-order half-invariants are zero for Gaussian random variables. Although Thiele's introduction of cumulants was practically motivated, he did not arrive at them as the solution to a particular problem. Thiele gave no indication that he was aware of any other work on cumulants, and he did not use any term other than half-invariant to describe cumulants. Cramer [16], A. Fisher [28], and Graham [73] each claim that Thiele discovered cumulants.

Cumulants were introduced into the theory of sampling distributions, that is, the theory of the probability distribution of sample statistics, largely through the work of R. A. Fisher, Kendall, and Wishart, [14, 29, 68, 118]. The basic problem here is to determine the distribution of statistics such as the sample moments or sample cumulants by, say, determining the moments or the cumulants of the sample statistic. Fisher knew that the mean of a statistic does not necessarily equal the corresponding population parameter. In the case of the sample variance $\hat{\sigma}^2$, the mean is

$$\begin{aligned} E\{\hat{\sigma}^2\} &= E\left\{\frac{1}{N}\sum_{i=1}^{N} x_i^2 - \frac{1}{N^2}\sum_{i=1}^{N}\left(x_i \sum_{j=1}^{N} x_j\right)\right\} \\ &= \frac{N-1}{N^2} E\left\{\sum_{i=1}^{N} x_i^2\right\} - \frac{1}{N^2} E\left\{\sum_{i,j\ i\neq j} x_i x_j\right\} \\ &= \frac{N-1}{N}\sigma^2, \end{aligned}$$

where σ^2 is the population variance. Fisher proposed a new set of cumulant statistics, called *k-statistics*, for which the expected value is equal to the corresponding population cumulant. This set of statistics greatly facilitated work in sampling distribution theory. A detailed treatment of this topic is given in [68]. In 1937, Cornish [14] attributed the term *cumulant* to Laplace, without referencing a particular work of Laplace's, because Laplace called the logarithm of the characteristic function the *cumulative function* (CF), and cumulants are the coefficients of this function in Maclaurin series form. The cumulative function gets its name from the property that the CF for a sum of independent random variables is the sum of their CFs: The CF is cumulative. However, in 1928, Wishart [118] called the CF the *kappa generating function*, and called the cumulants *cumulative moment functions*. The terms *cumulant* and *log-characteristic function* are used today.

In the early 1950s, cumulants were used in an engineering context for the first time by Kuznetsov, Stratonovich, and Tikhonov in a study of the passage of stochastic processes through linear and nonlinear systems [75, 76]. Apparently without motivation, the authors decided to characterize the output stochastic process by using the logarithm of the joint characteristic function of samples of the output. They were therefore faced with a function that contained cumulants that were parameterized by the specific values of the sampling times. They called these cumulants *generalized correlation functions* because they were equal to the familiar correlation function for order two (the covariance). The main result in these two references is that the generalized correlation functions for the output process are related in a simple way to those for the input process. The authors also used the generalized correlation functions to characterize the degree to which the output process deviated from normality. A more general form of the relation between moments and cumulants than that given by Thiele in [113] is given in [76] apparently for the first time, but without proof, and is not central to the work therein. The authors of [75, 76] used none of the terminology described here in the previous paragraphs.

In the late 1950s and early 1960s the theory of cumulants, which are sometimes called *semi-invariants* (or *seminvariants*), was put on a firm theoretical foundation by Shiryaev and Leonov [78, 103]. In [78], the cumulants of the output of a polynomial nonlinearity are obtained in terms of the cumulants of the input process, thereby formalizing the earlier work in [75, 76]. Also, a combinatorial proof of the relationships between moments and cumulants is given for arbitrary order. This appears to be the only published proof of these relations. The results of the work in [103] are essentially the probabilistic counterpart of the material presented in Sections 2 and 3 herein, restricted (for the most part) to stationary processes; [103] is a measure-theoretic approach to understanding higher-order moments and spectra of stochastic processes. Shiryaev defines the polyspectrum as a cumulant with respect to the logarithm of a spectral characteristic function, which is the characteristic function of the spectral increments of the process, and also shows that the polyspectrum is equal to the Fourier transform of the time-domain cumulant function for generally nonstationary as well as stationary processes. Shiryaev does not specialize his results to the case of cyclostationary or polycyclostationary processes for which cumulants take on a special meaning that leads to special applications, which is explained in subsequent sections of this chapter.

In the 1960s the properties of cumulants, both temporal and spectral, were investigated [9] and measurement techniques were developed [10, 11]. It is here that the term *polyspectrum* is introduced (which Brillinger attributes to J. W. Tukey) for the spectral cumulants or, equivalently, the Fourier transform of the temporal cumulants, and a case for the superiority of cumulants over moments for use in the theoretical development of HOS is made in [9]. The processes involved in [9] are assumed to be nth-order stationary, which means that all moments up to and including order n are translation invariant. In [9–11], polyspectra are defined to be the Fourier transforms of time-domain cumulants, but are also recognized to be spectral cumulants.

The 1970s saw minimal application of polyspectra and cumulants. The focus was on the third-order polyspectrum, called the *bispectrum*; the application was to the area of detection of phase-coupling in sinusoids [61, 69, 70, 84, 96, 97]. Three sinusoids with frequencies $\{f_i\}_{i=1}^3$ and random phases $\{\theta_i\}_{i=1}^3$ are said to be phase-coupled if $f_1 + f_2 = f_3$ and the sum $\theta_1 + \theta_2$ is statistically dependent on θ_3. This can be the case, for example, in the output of a linear-plus-quadratic system with the sum of sine waves with frequencies f_1 and f_2 at the input. Some progress in this area was made, and a corresponding interest in the statistical properties of estimates of the bispectrum was piqued. (Interestingly, this work has nothing to do with statistical inference based on single-sample-path processing since the phase coupling can be measured from a single sample path if and only if the phase is time-varying and the coupling that can be measured in this case is correlation—or more general dependences—over time, which has nothing to do with correlation over the ensemble unless the phases are ergodic stochastic processes. A time-invariant random phase is a nonergodic stochastic process.)

In the 1980s a sector of the electrical engineering community became interested in HOS as a tool that could be used to perform system identification. Since the autocorrelation (and power spectral density) of a second-order stationary process does not contain phase information, it cannot be used to identify nonminimum-phase systems. Researchers were led to higher-order statistics because higher-order moments and cumulants do contain phase information. Many researchers in the area of HOS claim that second-order statistics cannot be used to obtain phase information, but this is incorrect since the cyclic autocorrelation, which is a second-order statistic, is sensitive to phase for cyclostationary and polycyclostationary signals, and this fact is currently being exploited for nonminimum-phase system identification (cf. Chapter 3 and Articles 4 and 5 in this volume). Nevertheless, if the signal is stationary, or does not exhibit SOCS, then third- or higher-order statistics must be used to obtain phase information. System identification is still the dominant application area in HOS, as can be seen by noting that over half of the recent HOS tutorial paper by Mendel [81] is concerned with parametric system identification, that is, determining the coefficients of AR, MA, and ARMA system models (see references in [81]). Other recent applications include synchronization [6, 26], random signal detection [54, 71], image reconstruction [90], tests for the Gaussian property and linearity of stochastic process [59], neural-network based estimation [115], radar signal processing [25], equalization [91], and direction-finding (source location) [12, 31, 60, 92, 121]. In most of these applications, the signals of interest are modeled as stationary stochastic processes (the exception is synchronization), and in many cases the highest order employed is three. There are good reasons for the latter restriction. For example, if the input signal-to-noise ratio (SNR) for a signal in noise is below 0 dB, then the output SNR is approximately proportional to the input SNR raised to the nth power for an nth-order nonlinear transformation, such as a measurement of an nth-order moment or cumulant. Thus, the output SNR decreases as the order n increases for a given data-record length. Also, computational complexity of algorithms that exploit HOS can grow rapidly as the order increases.

The subject of cumulants has been largely neglected by the authors of the classic (or at least popular) texts in probability theory, mathematical statistics, stochastic processes, and time-series analysis. This is largely because of the long-standing emphasis on the correlation theory of processes and time-series wherein only the first and second moments are of interest. This theory is very powerful because it is sufficient for the explanation of the behavior of Gaussian processes, handles linear transformations of data easily and elegantly, and is computationally simpler (more tractable) than higher-order theories. The limited treatment of cumulants and polyspectra in the most well-known texts is discussed next, followed by a brief description of three modern texts that treat the topic of cumulants and polyspectra in some detail.

The texts considered in time-series analysis are [4, 8, 58, 64, 67, 74, 93]. The book [93] by Priestley contains the most material on cumulants. There is a short history that starts with Shiryaev's contribution [103], and then cumulants are defined through the cumulative function (CF). Polyspectra are defined to be the Fourier transforms of these cumulants. Brillinger [8], on the other hand, defines cumulants in terms of their relation to moments and lists several of their elementary properties. The other books in the area of time-series analysis offer little (such as the moment/cumulant relations for $n = 1, 2, 3, 4$) or nothing on cumulants.

The texts considered in the area of stochastic processes are [7, 23, 24, 33, 57, 65, 66, 82, 86, 88, 95, 112, 114, 119, 120]. These texts pay very little attention to cumulants and polyspectra. The three texts [82], [95], and [112] each define the cumulant through the series expansion of the CF, but do little with them. The other references in the area of stochastic processes do not even mention cumulants.

References [15, 27, 28, 79, 80, 87, 89] are the texts considered in probability theory. Parzen [89] and Papoulis [87] both define the cumulant through the series expansion of the CF, although Parzen does it in an exercise. Neither theory nor application of cumulants is developed in either of these books. A. Fisher essentially reproduces Thiele's work in his 1923 book [28]. The other references in the area of probability theory do not mention cumulants.

Finally, the texts considered in mathematical statistics are [16, 30, 68, 77, 98, 117]. Both Fisz [30] and Cramer [16] define the cumulant through the series expansion of the CF in the usual way, but go no further with the theory. The unique book by Kendall and Stuart [68] devotes a great deal of attention to cumulants, mostly in the context of sampling distribution theory (as mentioned previously). The authors are not concerned with stochastic processes nor time-series analysis and do not define polyspectra. The other texts in mathematical statistics do not mention cumulants.

Three modern texts do treat the topics of cumulants and polyspectra for time-series and stochastic processes. The first is by Rosenblatt [99]. The material on cumulants and polyspectra in this text is essentially the same as in the two papers [10, 11], in which the cumulants and polyspectra of stationary stochastic processes are investigated, with emphasis on estimating these parameters from finite-length data records. The book by Priestley [94] contains a chapter devoted to estimation of the polyspectrum of a stationary stochastic process from a finite-length data record. The methods considered therein are the same as in [99]. The material in both of these texts is considered further in Section 7 herein, where the measurement of the parameters of HOCS is studied, and where an error in Priestley's description of the frequency-domain method of estimating the cyclic polyspectrum is brought to light. The third modern text that treats cumulants is a collection of papers edited by Haykin [84]. The chapter by Nikias treats the topic of estimation of the polyspectrum, but the emphasis of the chapter is on the use of such estimates in solving the problem of parametric (ARMA) system identification.

The nine research papers [40], [42], [51], and [104–109], all of which are from the same research group, address the topic of HOCS directly. In [40], the higher-order temporal moments of polycyclostationary time-series are introduced and shown to be related to the higher-order moments of spectral components of the time-series by Fourier transformation, and it is established for the first time that the nth-order lag product of a time-series contains a finite-strength additive sine-wave component with frequency α if and only if the joint moment of n spectral components of the time-series is nonzero for some sets of n frequencies that sum to α. In [42], the concept of a *pure nth-order sine wave* is introduced and is shown to be characterized by the temporal cumulant function, whose Fourier transform is shown to be a spectral cumulant function, and the relationship between these functions and those that arise in the theory of HOS for stationary processes is discussed. Higher-order temporal and spectral moments of cyclostationary time-series are used in [51] to identify Volterra models of nonlinear systems. New techniques for measuring the higher-order statistical functions (cyclic moments, cumulants, and polyspectra) for polycyclostationary time-series are presented in [104], and some results on their performance are presented in [105]. A brief overview of the theory of higher-order cyclostationary time-series is given in [106]. Finally, the application of the theory of HOCS to the problems of weak-signal detection and time-delay estimation is considered in [109].

Three research papers by other researchers also treat HOCS directly. One of these is the recent paper [123], in which the stochastic-process framework is used, and in which no connection is made to the sine-wave generation idea that is central to this chapter, nor to cumulants, which are also central to this chapter. The second is [5], in which a cyclic spectral analysis of the powers of a PAM signal is carried out, that is, the cyclic spectrum of the output of a nonlinear system with a PAM signal at the input is calculated. The results herein are more general than those in [5]. The third is [2], in which the symmetry properties of nth-order polyspectra for $n \leq 6$ for cyclostationary stochastic processes are investigated.

Finally, there are several research papers that deal with HOCS indirectly [17–22, 55, 56]. These papers, which are the work of a single research group, treat the problem of higher-order statistical analysis of generally nonstationary stochastic processes, which includes cyclostationarity as a special case, and present several applications of the theory to cyclostationary signals, such as system identification and signal detection.

The strong recent interest in HOS (cumulants) in a sector of the electrical engineering community is reflected in the recent workshops on higher-order statistics, in the two *IEEE Proceedings* tutorial papers [81, 83], and in the special sections on HOS in the July 1990 *IEEE Transactions on ASSP*, and the January 1990 *IEEE Transactions on Automatic Control.*

1.4.2 Second-Order Polycyclostationarity

The research on second-order polycyclostationarity can be divided into two sets: that which adheres to the time-average analysis (FOT) framework, and that which uses the stochastic-process framework. The former is more relevant to this chapter because the time-average framework is used herein. The basic references for the theory and application of second-order polycyclostationarity for the time-average framework are [34, 35, 44, 45] and the references therein and in Chapter 1 of this volume and for the stochastic-process framework in [33, 63] and the references therein and in Chapter 1. Examples of the application of cyclostationarity include weak-signal detection [39, 46], time-delay estimation [49, 50], high-resolution direction finding [100], blind-adaptive spatial filtering for signal extraction [1, 101], system identification [41, 51], and frequency-shift filtering for signal separation [48]. A bibliography on cyclostationarity and its applications, and a detailed overview of second-order cyclostationarity is given in Chapter 1 of this volume.

2 THE TEMPORAL PARAMETERS OF HOCS

2.1 Moments

Let us consider all nonlinear signal-processing operations that can be represented by a Volterra series. This includes (but not exclusively) all continuous, time-invariant, finite-memory, causal systems [85]. The output $y(t)$ of such an operation is expressed as[1]

$$y(t) = \int_{-\infty}^{\infty} h_1(\tau_1) x(t+\tau_1)\, d\tau_1 + \int_{-\infty}^{\infty}\int_{-\infty}^{\infty} h_2(\tau_1, \tau_2) x(t+\tau_1) x(t+\tau_2)\, d\tau_1 d\tau_2 + \cdots$$

$$= \sum_n \int_{-\infty}^{\infty} \cdots \int_{-\infty}^{\infty} h_n(\boldsymbol{\tau}) L_x(t, \boldsymbol{\tau})_n \, d\boldsymbol{\tau},$$

where $\boldsymbol{\tau} \triangleq [\tau_1 \cdots \tau_n]^\dagger$ and $L_x(t, \boldsymbol{\tau})_n$ is the *nth-order lag product* of the input $x(t)$

$$L_x(t, \boldsymbol{\tau})_n \triangleq \prod_{j=1}^{n} x(t+\tau_j). \tag{6}$$

We are interested in the finite-strength additive sine-wave components present in the output but absent in the input, that is, those sine waves that are generated by the action of the nonlinear operation on the input $x(t)$, which is assumed to exhibit cyclostationarity. For example, the strength (magnitude and phase) of the sine wave with frequency α in $y(t)$ is given by (assuming the order of $\hat{E}^{\{\alpha\}}$, $\sum$, and $\int$ can be interchanged)

$$\left\langle y(t) e^{-i2\pi\alpha t} \right\rangle = \sum_n \int_{-\infty}^{\infty} \cdots \int_{-\infty}^{\infty} h_n(\boldsymbol{\tau}) \left\langle L_x(t, \boldsymbol{\tau})_n e^{-i2\pi\alpha t} \right\rangle d\boldsymbol{\tau}.$$

Thus, we need only study the statistical quantities

$$\left\langle L_x(t, \boldsymbol{\tau})_n e^{-i2\pi\alpha t} \right\rangle,$$

[1] Strictly speaking, the name Volterra is reserved for causal systems, for which $h_n(\boldsymbol{\tau}) = 0$ for any $\tau_j < 0$, but this restriction is lifted here. Also, infinite-memory systems, for which the $h_n(\cdot)$ have infinite support, are included here.

for arbitrary positive integers n, which are the strengths of the sine-wave components contained in the nth-order lag products of $x(t)$.

The nth-order lag product is an elementary nth-order homogeneous polynomial transformation of $x(t)$. This transformed time-series can be decomposed into a periodic or polyperiodic part and an aperiodic residual part,

$$L_x(t, \boldsymbol{\tau})_n = p(t, \boldsymbol{\tau})_n + m(t, \boldsymbol{\tau})_n,$$

where

$$\left\langle m(t, \boldsymbol{\tau})_n \, e^{-i2\pi\alpha t} \right\rangle = 0, \tag{7}$$

for all real α. The polyperiodic portion of $L_x(t, \boldsymbol{\tau})_n$ has associated with it the Fourier series

$$p(t, \boldsymbol{\tau})_n = \sum_{\alpha} R_x^{\alpha}(\boldsymbol{\tau})_n \, e^{i2\pi\alpha t}, \tag{8}$$

where

$$R_x^{\alpha}(\boldsymbol{\tau})_n \triangleq \left\langle p(t, \boldsymbol{\tau})_n \, e^{-i2\pi\alpha t} \right\rangle. \tag{9}$$

It is assumed herein that the partial sums in the Fourier series (8) converge uniformly in t for each $\boldsymbol{\tau}$ to $p(t, \boldsymbol{\tau})_n$. Then $p(\cdot, \boldsymbol{\tau})_n$ is an almost periodic function, the limit (9) exists for each $\boldsymbol{\tau}$, and the set of values of the real variable α for which $R_x^{\alpha}(\boldsymbol{\tau})_n \not\equiv 0$ for each $\boldsymbol{\tau}$ is denumerable [13]. That is, there is at most a denumerable set of incommensurate periods in the polyperiodic function $p(t, \boldsymbol{\tau})_n$ for each $\boldsymbol{\tau}$. It is further assumed that the union over all $\boldsymbol{\tau}$ of the sets of values of α for which $R_x^{\alpha}(\boldsymbol{\tau})_n \not\equiv 0$ is denumerable. For example, it is shown in [63] that this union is denumerable for $n = 2$ if $p(t, \boldsymbol{\tau})_2$ is uniformly continuous in t and $\boldsymbol{\tau}$.

The lag-product time-series can therefore be expressed as

$$L_x(t, \boldsymbol{\tau})_n = \sum_{\alpha} R_x^{\alpha}(\boldsymbol{\tau})_n \, e^{i2\pi\alpha t} + m(t, \boldsymbol{\tau})_n, \tag{10}$$

where the sum is over the denumerable set of real α for which $R_x^{\alpha}(\boldsymbol{\tau})_n \not\equiv 0$. From (7) and (10), we have

$$R_x^{\alpha}(\boldsymbol{\tau})_n = \left\langle L_x(t, \boldsymbol{\tau})_n \, e^{-i2\pi\alpha t} \right\rangle. \tag{11}$$

Each value of α in the representation (10) is called an *impure cycle frequency of order n* (to distinguish it from a pure nth-order cycle frequency, which is defined subsequently), and $R_x^{\alpha}(\boldsymbol{\tau})_n$ in (11) is called the *cyclic temporal moment function* (CTMF) *of order n*. From (11), it is evident that the CTMF arises quite naturally from a consideration of the finite-strength additive sine-wave components in the lag product (6). The sum of all such sine waves in $L_x(t, \boldsymbol{\tau})_n$ is given by the temporal expected value of the lag product (cf. Section 1.1),

$$\hat{E}^{\{\alpha\}} \{L_x(t, \boldsymbol{\tau})_n\} = \sum_{\alpha} R_x^{\alpha}(\boldsymbol{\tau})_n \, e^{i2\pi\alpha t},$$

which is called the *temporal moment function* (TMF), and is denoted by $R_x(t, \boldsymbol{\tau})_n$,

$$R_x(t, \boldsymbol{\tau})_n \triangleq \hat{E}^{\{\alpha\}}\{L_x(t, \boldsymbol{\tau})_n\} = \sum_{\alpha} R_x^{\alpha}(\boldsymbol{\tau})_n \, e^{i2\pi\alpha t}. \tag{12}$$

An individual component of the TMF, such as

$$R_x^{\alpha}(\boldsymbol{\tau})_n \, e^{i2\pi\alpha t},$$

is called an *nth-order moment sine wave* or an *impure nth-order sine wave.* Time-series for which there exists at least one moment sine wave of order n (with $\alpha \neq 0$) are said to exhibit *nth-order cyclostationarity* (CS_n) *in the wide sense* (such a time-series can be nth-order cyclostationary or nth-order polycyclostationary). A potentially confusing property of CS_n time-series is that such time-series are in general CS_{2n}. A simple example illustrates this fact. Consider the time-series given by

$$x(t) = \cos(\omega t) + m(t),$$

where the zero-mean signal $m(t)$ contains no sine-wave components. The signal $x(t)$ is CS_1. Every second-order lag product contains sine waves as well,

$$x(t+\tau_1)x(t+\tau_2) = \frac{1}{2}\cos(2\omega t + \omega[\tau_1 + \tau_2]) + \frac{1}{2}\cos(\omega[\tau_1 - \tau_2]) + \text{residue}.$$

More interesting cases involve random time-series that do not themselves contain additive sine-wave components, because it is still true that, for example, CS_2 implies CS_4. This is illustrated in the next section.

2.2 Impure *n*th-Order Sine Waves

It is often the case that an nth-order moment sine wave is *impure* in the sense that it is made up—in part or wholly—of products of k m_jth-order moment sine waves with $\sum_{j=1}^{k} m_j = n$ and $k \leq n$, for various values of k. For example, consider the following signal, which is an amplitude-modulated (AM) signal with an added carrier tone:

$$x(t) = a(t)\cos(2\pi f_c t + \theta) + B\cos(2\pi f_c t + \theta). \tag{13}$$

The second-order lag product for this signal clearly contains (complex) sine-wave components with the frequencies 0, $2f_c$, and $-2f_c$. Let's examine the sine wave with frequency $2f_c$:

$$\begin{aligned} R_x^{2f_c}(\boldsymbol{\tau})_2 &= \left\langle x(t+\tau_1)x(t+\tau_2)e^{-i2\pi 2f_c t}\right\rangle \\ &= \frac{1}{4}R_a^0(\boldsymbol{\tau})_2 \, e^{i(2\pi f_c[\tau_1+\tau_2]+2\theta)} + \frac{B^2}{4}e^{i(2\pi f_c[\tau_1+\tau_2]+2\theta)}, \end{aligned}$$

where we have assumed that $a(t)$ has zero mean,

$$\hat{E}^{\{\alpha\}}\{a(t)\} = 0.$$

Thus, there are two components that make up the second-order moment sine wave with frequency $2f_c$. The first component

$$\frac{1}{4} R_a^0(\boldsymbol{\tau})_2 \, e^{i(2\pi f_c[\tau_1+\tau_2]+2\theta)} e^{i2\pi 2 f_c t}$$

does not result from the multiplication of any sine-wave components in either of the factors of the lag product, whereas the second component

$$\frac{B^2}{4} e^{i(2\pi f_c[\tau_1+\tau_2]+2\theta)} e^{i2\pi 2 f_c t}$$

is nothing more than the result of a sine-wave term in $x(t+\tau_1)$ multiplying a sine-wave term in $x(t+\tau_2)$.

Identifying the pure and impure components of an nth-order moment sine wave is not always so easy. Let's consider the AM signal (13) with $B = 0$,

$$x(t) = a(t)\cos(2\pi f_c t + \theta), \tag{14}$$

and a fourth-order lag product

$$L_x(t, \boldsymbol{\tau})_4 = \prod_{j=1}^{4} x(t+\tau_j). \tag{15}$$

The sine wave with frequency $4f_c$ in the lag product (15) is given by

$$R_x^{4f_c}(\boldsymbol{\tau})_4 \, e^{i2\pi 4 f_c t} = \frac{1}{16}\left\langle \prod_{j=1}^{4} a(t+\tau_j) \right\rangle e^{i(2\pi f_c[\tau_1+\tau_2+\tau_3+\tau_4]+4\theta)} e^{i2\pi 4 f_c t}. \tag{16}$$

Since

$$R_a^0(\boldsymbol{\tau})_2 = \langle a(t+\tau_1)a(t+\tau_2)\rangle ,$$

then the lag product for $a(t)$ can be represented by

$$a(t+\tau_1)a(t+\tau_2) = R_a^0(\tau_1,\tau_2)_2 + b(t,\tau_1,\tau_2) \tag{17}$$

for which

$$\langle b(t,\tau_1,\tau_2)\rangle \equiv 0$$

$$R_a^0(\tau_1,\tau_2)_2 \equiv R_a^0(\boldsymbol{\tau})_2.$$

By making use of (17) in the fourth-order lag product for $a(t)$, we can begin to see the pure and impure components of the fourth-order sine wave (16):

$$R_x^{4f_c}(\boldsymbol{\tau})_4$$

$$= \frac{1}{16}\left\langle [R_a^0(\tau_1,\tau_2)_2 + b(t,\tau_1,\tau_2)][R_a^0(\tau_3,\tau_4)_2 + b(t,\tau_3,\tau_4)]\right\rangle e^{i(2\pi f_c[\tau_1+\tau_2+\tau_3+\tau_4]+4\theta)}$$

$$= \frac{1}{16}\left[R_a^0(\tau_1,\tau_2)_2 R_a^0(\tau_3,\tau_4)_2 + \langle b(t,\tau_1,\tau_2)b(t,\tau_3,\tau_4)\rangle\right] e^{i(2\pi f_c[\tau_1+\tau_2+\tau_3+\tau_4]+4\theta)}.$$

So, there are components of the fourth-order moment sine wave (16) that consist of products of second-order moment sine waves, and there are—potentially—components that are not. We say potentially because there are other products of lower-order sine waves, namely those obtained by using a different factorization of the fourth-order lag product, and we are unsure at this point if these other impure sine waves are the only components of $\langle b(t, \tau_1, \tau_2) b(t, \tau_3, \tau_4) \rangle$.

In the case of the second-order lag products of (13), we can purify the second-order sine waves by operating directly on the data: We simply remove the sine-wave component $B \cos(2\pi f_c t + \theta)$ from $x(t)$. In the case of the fourth-order lag products of (14) (or (13)), we cannot purify the fourth-order moment sine waves by operating on the data because there are no sine waves in the data. Furthermore, we cannot simply subtract the second-order sine waves from the lag products $x(t + \tau_1)x(t + \tau_2)$ and $x(t + \tau_3)x(t + \tau_4)$ because there are similar sine waves in other factorizations of the fourth-order lag product. In the next section we show how to properly purify the nth-order moment sine waves, thereby obtaining the *pure nth-order sine waves.*

2.3 Pure nth-Order Sine Waves

For low orders n, it is easy to mathematically characterize a pure nth-order sine wave in a way that matches our intuition. For $n = 1$, the moment sine waves are, by definition, pure first-order sine waves. For $n = 2$, all products of first-order moment sine waves can be subtracted from the second-order moment sine waves to obtain the pure second-order sine waves, which are denoted by $\sigma_x(t, \tau_1, \tau_2)_2$,

$$\begin{aligned} \sigma_x(t, \tau_1, \tau_2)_2 &\triangleq \hat{E}^{\{\alpha\}} \{x(t + \tau_1)x(t + \tau_2)\} - \hat{E}^{\{\alpha\}} \{x(t + \tau_1)\} \, \hat{E}^{\{\alpha\}} \{x(t + \tau_2)\} \\ &= R_x(t, \boldsymbol{\tau})_2 - R_x(t, \tau_1)_1 \, R_x(t, \tau_2)_1. \end{aligned}$$

There are several interesting points to be made concerning pure second-order sine waves.

1. Since $R_x(t, \tau_i)_1, i = 1, 2$, and $R_x(t, \boldsymbol{\tau})_2$ are first- and second-order moments, then $\sigma_x(t, \tau_1, \tau_2)_2$ is a temporal covariance function.
2. If $R_x(t, \tau)_1 \equiv 0$, then there are no lower-than-second-order sine waves, and the second-order moment sine waves are equal to the pure second-order sine waves.
3. If the variables $x(t + \tau_1)$ and $x(t + \tau_2)$ are statistically independent (in the temporal sense [35, 45]), then $\hat{E}^{\{\alpha\}} \{x(t + \tau_1)x(t + \tau_2)\} = \hat{E}^{\{\alpha\}} \{x(t + \tau_1)\} \, \hat{E}^{\{\alpha\}} \{x(t + \tau_2)\}$ and therefore $\sigma_x(t, \tau_1, \tau_2)_2 = 0$, that is, there is no pure second-order sine wave for this particular pair of lags τ_1 and τ_2.

A recursion can be used to compute the pure third-order sine waves. Each distinct product of lower-order sine waves must be subtracted from the third-order moment sine waves. Thus, products of *pure* second-order and *pure* first-order sine waves are subtracted from the third-order moment sine waves:

$$\sigma_x(t,\boldsymbol{\tau})_3 = \hat{E}^{\{\alpha\}}\left\{\prod_{j=1}^{3} x(t+\tau_j)\right\} - \sigma_x(t,\tau_1,\tau_2)_2\,\sigma_x(t,\tau_3)_1 - \sigma_x(t,\tau_1,\tau_3)_2\,\sigma_x(t,\tau_2)_1$$
$$-\sigma_x(t,\tau_2,\tau_3)_2\,\sigma_x(t,\tau_1)_1 - \sigma_x(t,\tau_1)_1\,\sigma_x(t,\tau_2)_1\,\sigma_x(t,\tau_3)_1\,. \tag{18}$$

Note that all possible products of pure lower-order sine waves appear in (18). The terms in the sum of products that are subtracted can be enumerated easily by considering the distinct *partitions* of the index set $\{1, 2, 3\}$. A partition of a set G is a collection of p subsets of G, $\{\nu_i\}_{i=1}^{p}$, having the following properties[2] [3]:

$$G = \bigcup_{j=1}^{p} \nu_j,$$

$$\nu_j \bigcap \nu_k = \emptyset \quad \text{for} \quad j \neq k.$$

The set P_3 of distinct partitions of $\{1, 2, 3\}$ is

$$\begin{aligned} &p = 1 : \{1,2,3\} \\ &p = 2 : \{1,2\},\{3\} \quad \{1,3\},\{2\} \quad \{2,3\},\{1\} \\ &p = 3 : \{1\},\{2\},\{3\}\,. \end{aligned}$$

Thus, we can express the pure third-order sine waves $\sigma_x(t, \boldsymbol{\tau})_3$ as a sum over the elements of P_3:

$$\sigma_x(t,\boldsymbol{\tau})_3 = R_x(t,\boldsymbol{\tau})_3 - \sum_{\substack{P_3\\ p\neq 1}} \left[\prod_{j=1}^{p} \sigma_x(t,\boldsymbol{\tau}_{\nu_j})_{n_j}\right],$$

where $\boldsymbol{\tau}_{\nu_j}$ is the vector of elements of $\{\tau_j\}_{j=1}^{3}$ that have indices in ν_j, and n_j is the number of elements in ν_j.

Note that, as in the case of $n = 2$, if the first-order moment sine waves are zero, $\hat{E}^{\{\alpha\}}\{x(t)\} = 0$, then the third-order moment sine waves are equal to the pure third-order sine waves. In this case, there are no products of lower-order sine waves that can be subtracted from the moment.

Because there is a one-to-one correspondence between the set of distinct factorizations of a product of n factors and the set of distinct partitions of the set $\{1, 2, \cdots, n\}$ (as illustrated for $n = 3$ above), the formula for the pure nth-order sine waves can be expressed in terms of these partitions,

$$\sigma_x(t,\boldsymbol{\tau})_n = R_x(t,\boldsymbol{\tau})_n - \sum_{\substack{P_n\\ p\neq 1}} \left[\prod_{j=1}^{p} \sigma_x(t,\boldsymbol{\tau}_{\nu_j})_{n_j}\right], \tag{19}$$

---

[2] The total number of distinct partitions of a set is called Bell's number, which must be computed by a recursion involving Stirling numbers of the second kind [73, 110]. This same recursion can be modified to yield the partitions themselves.

where P_n is the set of distinct partitions of $\{1, 2, \cdots, n\}$. The pure-sine-wave formula[3] (19) gives all the pure nth-order sine waves associated with the lag vector $\boldsymbol{\tau}$. A single pure nth-order sine wave with frequency β can be selected by computing the Fourier coefficient

$$\sigma_x^{\beta}(\boldsymbol{\tau})_n \, e^{i2\pi\beta t} = \left\langle \sigma_x(u, \boldsymbol{\tau})_n \; e^{i2\pi\beta(t-u)} \right\rangle, \tag{20}$$

and can be expressed in terms of pure lower-order sine waves by substituting the Fourier series for each σ_x,

$$\sigma_x(t, \boldsymbol{w})_k = \sum_{\beta_k} \sigma_x^{\beta_k}(\boldsymbol{w})_k \, e^{i2\pi\beta_k t}, \tag{21}$$

where the sum is over all pure cycle frequencies β_k of order k, into (19). Thus,

$$\sigma_x^{\beta}(\boldsymbol{\tau})_n = R_x^{\beta}(\boldsymbol{\tau})_n - \sum_{\substack{P_n \\ p \neq 1}} \left[\sum_{\boldsymbol{\beta}^{\dagger}\mathbf{1}=\beta} \prod_{j=1}^{p} \sigma_x^{\beta_j}(\boldsymbol{\tau}_{\nu_j})_{n_j} \right], \tag{22}$$

where $\boldsymbol{\beta}$ is the p-dimensional vector of pure cycle frequencies $[\beta_1 \;\cdots\; \beta_p]^{\dagger}$ and $\mathbf{1}$ is the p-dimensional vector of ones. Hence, the pure-sine-wave strength $\sigma_x^{\beta}(\boldsymbol{\tau})_n$ is given by the CTMF $R_x^{\beta}(\boldsymbol{\tau})_n$ with all products of pure lower-order sine-wave strengths, for sine waves whose frequencies sum to β, subtracted out.

The next step in the development of the temporal parameters is to introduce the cumulant function for the set $\{x(t + \tau_j)\}_{j=1}^{n}$. Before doing this, a general introduction to cumulants is provided.

2.4 Cumulants

The references cited in this and succeeding sections dealing with cumulants of random variables, stochastic processes, and nonstochastic time-series are believed to be the original sources where these quantities were first introduced.

2.4.1 Cumulants of a Single Random Variable

Let the real-valued random variable X have probability density function (PDF) $f_X(u)$ and characteristic function $\Phi_X(\omega)$:

$$\Phi_X(\omega) = \int_{-\infty}^{\infty} f_X(u) e^{i\omega u} \, du.$$

It is well known that the moments of X can be obtained from the characteristic function by differentiation,

$$E\{X^n\} = (-i)^n \frac{\partial^n}{\partial \omega^n} \Phi_X(\omega) \bigg|_{\omega=0} = m_n.$$

[3] This approach to obtaining pure nth-order sine waves can break down in some special anomalous cases involving degenerate time-series, which are described in the Appendix.

The nth-order moment of X is, therefore, the coefficient of the term corresponding to $(i\omega)^n/n!$ in the Maclaurin series expansion of the characteristic function:

$$\Phi_X(\omega) = \sum_{n=0}^{\infty} m_n \frac{(i\omega)^n}{n!}. \tag{23}$$

The characteristic function is a useful tool in the study of random variables, but it does have a drawback. If Y is the sum of two independent random variables X_1 and X_2

$$Y = X_1 + X_2,$$

then the PDF for Y is the convolution of the PDFs for X_1 and X_2,

$$f_Y(u) = \int_{-\infty}^{\infty} f_{X_1}(u - \lambda) f_{X_2}(\lambda)\, d\lambda,$$

which implies that the characteristic function for Y is the product of characteristic functions for X_1 and X_2,

$$\Phi_Y(\omega) = \Phi_{X_1}(\omega)\Phi_{X_2}(\omega). \tag{24}$$

By using (23) in (24), it can be shown that the nth-order moment of Y is explicitly dependent on the moments of X_1 and X_2 of all orders n and lower.

If we transform the multiplication in (24) to addition by applying the natural logarithm, we obtain the relation

$$\ln \Phi_Y(\omega) = \ln \Phi_{X_1}(\omega) + \ln \Phi_{X_2}(\omega). \tag{25}$$

These new functions are called *cumulative functions* [14], a term that is due to Laplace, and the coefficient of the term corresponding to $(i\omega)^n/n!$ in the Maclaurin series expansion of the left side of (25) (provided that it exists), is called the nth-order *cumulant* of the random variable Y [14]. The nth-order cumulant for Y is, therefore, the sum of the nth-order cumulants for X_1 and X_2. Note that if X is a Gaussian random variable, its cumulative function is a second-order polynomial in ω and, therefore, all higher-order (greater than two) cumulants of this random variable are zero:

$$\frac{\partial^n}{\partial \omega^n} \ln \Phi_X(\omega) \equiv 0, \quad n \geq 3.$$

2.4.2 Cumulants of a Set of Random Variables

The multivariate PDF for the set of r random variables $\{X_j\}_{j=1}^r$ is given by

$$f_{\mathbf{X}}(\boldsymbol{x}) \triangleq \frac{\partial^r}{\partial x_1 \cdots \partial x_r} F_{\mathbf{X}}(\boldsymbol{x}), \tag{26}$$

where $F_{\mathbf{X}}(\boldsymbol{x})$ is the multivariate distribution function defined by

$$F_{\mathbf{X}}(\boldsymbol{x}) \triangleq \text{Prob}\left\{ \bigcap_{j=1}^{r} [X_j < x_j] \right\}.$$

The multivariate characteristic function is the multidimensional conjugate Fourier transform of (26):

$$\Phi_{\mathbf{X}}(\omega) = \int_{-\infty}^{\infty} \cdots \int_{-\infty}^{\infty} f_{\mathbf{X}}(\boldsymbol{x}) e^{i\omega^{\dagger}x}\, dx.$$

The nth-order moment corresponding to the product

$$\prod_{j=1}^{r} X_j^{q_j}$$

where

$$\sum_{j=1}^{r} q_j = n, \quad q_j \geq 1 \text{ for all } j,$$

is given by the coefficient of the term corresponding to

$$\frac{i^n \prod_{j=1}^{r} \omega_j^{q_j}}{\prod_{j=1}^{r} q_j!}$$

in the multidimensional series expansion of the characteristic function. We need only consider the case where $r = n$ and therefore $q_j = 1$ for each j. This is so because if some of the q_j are greater than one, we can simply consider a larger set of variables $\{X_j'\}_{j=1}^n$, where some of the X_j' are identical according to the values of q_j.

The cumulants are given by the coefficients in the series expansion of the cumulative function $\ln \Phi_{\mathbf{X}}(\omega)$. Since we consider only $r = n$, the resulting cumulants are called *simple cumulants*. Thus, the nth-order simple cumulant for the variables $\{X_j\}_{j=1}^n$ is defined by

$$C_{\mathbf{X}} \triangleq (-i)^n \frac{\partial^n}{\partial\omega_1 \cdots \partial\omega_n} \ln \Phi_{\mathbf{X}}(\omega) \bigg|_{\omega=\mathbf{0}}. \tag{27}$$

2.4.3 Multivariate Moment and Cumulant Relations

For the set of random variables $\{X_j\}_{j=1}^n$, the simple moment is given by

$$R_{\mathbf{X}} = E\left\{\prod_{j=1}^{n} X_j\right\}. \tag{28}$$

Let ν_k be some nonempty subset of the set of indices $\{1, 2, \cdots, n\}$. Then the moment of order $n_k = |\nu_k|$ for those variables with subscripts in ν_k is

$$R_{\mathbf{X},\nu_k} = E\left\{\prod_{j\in\nu_k} X_j\right\}.$$

The nth-order simple cumulant for $\{X_j\}_{j=1}^n$ can be expressed in terms of the simple moments $R_{\mathbf{X},\nu_k}$ by using the distinct partitions P_n of the index set $\{1, 2, \cdots, n\}$ [78]:

$$C_{\mathbf{X}} = \sum_{P_n} \left[(-1)^{p-1}(p-1)! \prod_{j=1}^{p} R_{\mathbf{X},\nu_j} \right]. \tag{29}$$

Similarly, the nth-order simple moment $R_{\mathbf{X}}$ can be expressed in terms of lower-order simple cumulants [78]:

$$R_{\mathbf{X}} = \sum_{P_n} \left[\prod_{j=1}^{p} C_{\mathbf{X},\nu_j} \right], \tag{30}$$

where $C_{\mathbf{X},\nu_j}$ is the simple cumulant of the variables $\{X_k\}_{k\in\nu_j}$. The relationships between simple moments and cumulants (29) and (30) are valid for both real- and complex-valued random variables [9, 99, 108].

An important and useful property of multivariate cumulants is the *independence property.* Consider the set of variables

$$\{Z_m\}_{m=1}^n = \{X_j : j = 1, \cdots, r;\ Y_k : k = 1, \cdots, s\}, \quad n = r + s,$$

where the X_j are independent of the Y_k. The nth-order joint PDF for these variables factors

$$f_{\mathbf{XY}}(\mathbf{z}) = f_{\mathbf{X}}(\mathbf{x}) f_{\mathbf{Y}}(\mathbf{y}), \quad \mathbf{z} = [x_1 \cdots x_r\ y_1 \cdots y_s]^\dagger,$$

which implies that the characteristic function is the product of characteristic functions for **X** and **Y**:

$$\Phi_{\mathbf{XY}}(\boldsymbol{\omega}) = \Phi_{\mathbf{X}}(\boldsymbol{\omega}_x)\Phi_{\mathbf{Y}}(\boldsymbol{\omega}_y).$$

Therefore,

$$\ln \Phi_{\mathbf{XY}}(\boldsymbol{\omega}) = \ln \Phi_{\mathbf{X}}(\boldsymbol{\omega}_x) + \ln \Phi_{\mathbf{Y}}(\boldsymbol{\omega}_y)$$

and the n-fold derivative in (27) is zero. Thus, if there is a subset of variables in the set $\{x_j\}_{j=1}^n$ that are independent of the remaining variables, then the simple cumulant for this set is equal to zero.

Another important property of multivariate cumulants is the following *addition rule*. Suppose the set $\{Z_m\}_{m=1}^n$ is given by

$$Z_m = X_m + Y_m, \quad m = 1, 2, \cdots, n, \tag{31}$$

where $\{X_m\}$ is statistically independent of $\{Y_m\}$. Then the cumulant of $\{Z_m\}$ is given by the sum of the cumulants of $\{X_m\}$ and $\{Y_m\}$:

$$C_{\mathbf{Z}} = C_{\mathbf{X}} + C_{\mathbf{Y}}. \tag{32}$$

Finally, if the variables $\{X_j\}_{j=1}^n$ are jointly Gaussian, then the cumulative function $\ln \Phi_{\mathbf{X}}(\boldsymbol{\omega})$ is a quadratic function of $\boldsymbol{\omega}$ and, therefore, the simple cumulants are zero for $n \geq 3$. Thus, multivariate cumulants can be used to measure the dependence between random variables, and to test for the Gaussian property [59].

2.4.4 Cumulants of a Time-Series

The relations (29) and (30) are more accessible than the log-characteristic function expression (27) and are, therefore, used for the cumulants and moments of time-series. Comparing equations (19) and (30), it is apparent that the relationship for $\{x(t+\tau_j)\}_{j=1}^n$ between the pure-sine-wave function $\sigma_x(t, \boldsymbol{\tau})_n$ and the temporal moment function $R_x(t, \boldsymbol{\tau})_n$ is equivalent to that between the simple cumulant and moment for $\{X_j\}_{j=1}^n$. By using the sine-wave-extraction operation, which is an expectation operation, we can reexpress the TMF in terms of lower-order simple cumulants:

$$R_x(t, \boldsymbol{\tau})_n = \sum_{P_n} \left[\prod_{j=1}^{p} C_x(t, \boldsymbol{\tau}_{\nu_j})_{n_j} \right] \tag{33}$$

where $C_x(t, \boldsymbol{\tau}_{\nu_j})_{n_j}$ is the simple cumulant of the n_j variables $\{x(t+\tau_k)\}_{k \in \nu_j}$ that is obtained by using the sine-wave-extraction operation $\hat{E}^{\{\alpha\}}\{\cdot\}$. Moreover, the relation (29) can be used to obtain an expression for the simple cumulant in terms of the lower-order simple moments:

$$C_x(t, \boldsymbol{\tau})_n = \sum_{P_n} \left[(-1)^{p-1}(p-1)! \prod_{j=1}^{p} R_x(t, \boldsymbol{\tau}_{\nu_j})_{n_j} \right]. \tag{34}$$

The function $C_x(t, \boldsymbol{\tau})_n$ in (34) is called the *temporal cumulant function* (TCF). Using the equivalence between (19) and (33), we obtain

$$C_x(t, \boldsymbol{\tau})_n = \sigma_x(t, \boldsymbol{\tau})_n \, . \tag{35}$$

That is, the pure nth-order sine-wave function $\sigma_x(t, \boldsymbol{\tau})_n$ is identical to the nth-order temporal cumulant function! The Fourier coefficient of this polyperiodic function of t is given by

$$\begin{aligned} C_x^{\beta}(\boldsymbol{\tau})_n &\triangleq \left\langle C_x(t, \boldsymbol{\tau})_n \, e^{-i2\pi\beta t} \right\rangle \\ &= \sum_{P_n} \left[(-1)^{p-1}(p-1)! \sum_{\boldsymbol{\alpha}^{\dagger}\mathbf{1}=\beta} \prod_{j=1}^{p} R_x^{\boldsymbol{\alpha}_j}(\boldsymbol{\tau}_{\nu_j})_{n_j} \right] \end{aligned} \tag{36}$$

and is called the *cyclic temporal cumulant function* (CTCF) [42]. An individual component of the TCF, $C_x^{\beta}(\boldsymbol{\tau})_n e^{i2\pi\beta t}$, is called an *nth-order cumulant sine wave* to distinguish it from an nth-order moment sine wave. It can be seen from (20) and (35) that the CTCF is identical to the (complex-valued) strength of the pure nth-order sine wave with frequency β that is contained in the nth-order lag product $L_x(t, \boldsymbol{\tau})_n$.

This instance is the first (to the best of our knowledge) in which cumulants have arisen as the solution to a practically motivated problem, namely the problem of pure nth-order sine-wave generation [42], rather than as a mathematical observation concerning the characteristic function [14, 68, 76, 113].

2.5 Properties of the Temporal Parameters of HOCS

We have seen that the TMF can be constructed from all of the CTMFs (cf. (12)), that the TCF can be constructed from all of the CTCFs (cf. (22)), that the TMF can be constructed from all of the lower-order TCFs (cf. (33)), and that the TCF can be constructed from all of the lower-order TMFs (cf. (34)). Thus, any CTCF can be obtained from all of the appropriate CTMFs, and vice versa. In other words, the set of moment functions for orders 1 through n contains the same information as the set of cumulant functions for orders 1 through n. How then should we determine which set of functions to work with in the study of sine-wave generation? To assist us in making the correct choice, we consider some important properties of these functions.

2.5.1 Signal Selectivity

Suppose our time-series $x(t)$ consists of the sum of M mutually independent time-series,

$$x(t) = \sum_{m=1}^{M} y_m(t). \tag{37}$$

Then, from the addition rule for cumulants, the TCF for $x(t)$ is the sum of TCFs for $\{y_m(t)\}$,

$$C_x(t, \boldsymbol{\tau})_n = \sum_{m=1}^{M} C_{y_m}(t, \boldsymbol{\tau})_n. \tag{38}$$

Thus, the pure nth-order sine waves in the lag products of each of $y_m(t)$ add to form the pure nth-order sine waves in the lag product of $x(t)$. The TMF does not obey this very useful cumulative relation.

To illustrate how (38) can be applied in practice, consider the situation where $\{y_m(t)\}_{m=1}^{M}$ represent M interfering signals that overlap in time and frequency, but which possess some distinct pure nth-order cycle frequencies, say $\{\beta_m\}_{m=1}^{M}$. Then it follows from (38) that

$$C_x^{\beta_m}(\boldsymbol{\tau})_n = C_{y_m}^{\beta_m}(\boldsymbol{\tau})_n, \quad m = 1, 2, \cdots, M.$$

This indicates that the presence or absence of each of the signals $y_m(t)$ can be detected by measuring (estimating) the CTCFs of $x(t)$ for the cycle frequencies $\{\beta_m\}$, and that parameters of each of the signals (on which these CTCFs depend) can be estimated. As illustrated in [1, 35, 39, 41, 46–50, and 100] for second order and in Section 8 for higher order, this signal-selectivity property can be exploited in numerous ways to accomplish noise-and-interference-tolerant signal detection and estimation.

As another application, let $M = 2$, $y_1(t)$ be non-Gaussian, and $y_2(t)$ be Gaussian. Then $C_{y_2}(t, \boldsymbol{\tau})_n \equiv 0$ for $n \geq 3$ and, from (37), we have

$$C_x(t, \boldsymbol{\tau})_n = C_{y_1}(t, \boldsymbol{\tau})_n, \quad n \geq 3,$$

which indicates the detectability of $y_1(t)$ with no knowledge about $y_2(t)$ except that it is Gaussian.

2.5.2 Mathematical Properties

It can be shown that both the CTMF and the CTCF are sinusoidal jointly in the n variables $\boldsymbol{\tau}$:

$$C_x^{\beta}(\boldsymbol{\tau}+\mathbf{1}\Delta)_n = C_x^{\beta}(\boldsymbol{\tau})_n e^{i2\pi\beta\Delta}, \tag{39}$$

$$R_x^{\alpha}(\boldsymbol{\tau}+\mathbf{1}\Delta)_n = R_x^{\alpha}(\boldsymbol{\tau})_n e^{i2\pi\alpha\Delta}. \tag{40}$$

Hence, these functions are not absolutely integrable with respect to $\boldsymbol{\tau}$. The periodicity of these functions suggests that we might reduce the dimension of these functions without loss of information. Reducing the dimension by one yields[4]

$$\bar{C}_x^{\beta}(\boldsymbol{u})_n \triangleq C_x^{\beta}([\boldsymbol{u}^{\dagger}\ 0]^{\dagger})_n,$$

$$\bar{R}_x^{\alpha}(\boldsymbol{u})_n \triangleq R_x^{\alpha}([\boldsymbol{u}^{\dagger}\ 0]^{\dagger})_n,$$

(where $\boldsymbol{u} = [u_1 \cdots u_{n-1}]^{\dagger}$), which are not sinusoidal in general. The value of $C_x^{\beta}(\boldsymbol{\tau})_n$ ($R_x^{\alpha}(\boldsymbol{\tau})_n$) for any $\boldsymbol{\tau}$ can be obtained from the value of $\bar{C}_x^{\beta}(\boldsymbol{u})_n$ ($\bar{R}_x^{\alpha}(\boldsymbol{u})_n$) by using (39) ((40)). This leads us to ask if these reduced-dimension (RD) functions are integrable. We shall show that the function $\bar{R}_x^{\alpha}(\boldsymbol{u})_n$ (RD-CTMF) is not in general, whereas the function $\bar{C}_x^{\alpha}(\boldsymbol{u})_n$ (RD-CTCF) is in general for time-series possessing an asymptotic independence property. That is, consider the arbitrary two-set partition $\boldsymbol{\tau} = [\boldsymbol{\tau}_0\ \boldsymbol{\tau}_1]$ and assume that the FOT density for the set of variables

$$\boldsymbol{x}(t) \triangleq \{x(t+\tau_j)\}_{j=1}^{n} = \{\boldsymbol{x}_0(t)\ \boldsymbol{x}_1(t)\}$$

factors asymptotically:

$$f_{\boldsymbol{x}(t)}(\boldsymbol{y}) \rightarrow f_{\boldsymbol{x}_0(t)}(\boldsymbol{y}_0) f_{\boldsymbol{x}_1(t)}(\boldsymbol{y}_1) \quad \text{as } \boldsymbol{\tau}_0 \rightarrow \infty,$$

where $\boldsymbol{\tau}_0 \rightarrow \infty$ means that all of the elements of $\boldsymbol{\tau}_0$ are tending to infinity. This asymptotic factorization implies that the TMF, which is a moment corresponding to the PDF $f_{\boldsymbol{x}(t)}(\cdot)$, is asymptotically factorable as well,

$$\begin{aligned} \hat{E}^{\{\alpha\}}\{L_x(t,\boldsymbol{\tau})_n\} &= \hat{E}^{\{\alpha\}}\left\{L_x(t,\boldsymbol{\tau}_0)_{n_0} L_x(t,\boldsymbol{\tau}_1)_{n_1}\right\} \\ &\rightarrow \hat{E}^{\{\alpha\}}\left\{L_x(t,\boldsymbol{\tau}_0)_{n_0}\right\} \hat{E}^{\{\alpha\}}\left\{L_x(t,\boldsymbol{\tau}_1)_{n_1}\right\} \quad \text{as } \boldsymbol{\tau}_0 \rightarrow \infty. \end{aligned}$$

Thus the TCF is asymptotically zero

$$C_x(t,\boldsymbol{\tau})_n \rightarrow 0 \quad \text{as } \boldsymbol{\tau}_0 \rightarrow \infty,$$

[4]The reason for this particular choice of dimension reduction is made clear in Section 3.

because of the independence property of cumulants (cf. Section 2.4.3), which implies that each CTCF is asymptotically zero. Generally, then, the CTCF is asymptotically zero as long as at least one of the n lag values—say τ_n—is fixed, since the set of variables associated with $L_x(t, \boldsymbol{\tau}_0)_{n_0}$ is asymptotically independent of the set of variables associated with $L_x(t, \boldsymbol{\tau}_1)_{n_1}$, where $\boldsymbol{\tau}_0 = [\tau_1 \cdots \tau_{n-1}]^\dagger$ and $\boldsymbol{\tau}_1 = [\tau_n]$. On the other hand, the cyclic temporal moment becomes

$$\begin{aligned} R_x^\alpha(\boldsymbol{\tau})_n &\to \left\langle R_x(t, \boldsymbol{\tau}_0)_{n_0} R_x(t, \boldsymbol{\tau}_1)_{n_1} e^{-i2\pi\alpha t}\right\rangle \quad \text{as } \boldsymbol{\tau}_0 \to \infty \\ &= \sum_\gamma R_x^\gamma(\boldsymbol{\tau}_0)_{n_0} R_x^{\alpha-\gamma}(\boldsymbol{\tau}_1)_{n_1}, \end{aligned}$$

which is not necessarily zero. In fact, it is often nonzero.

If the rate of decay of the RD-CTCF is sufficiently large (e.g., $O(||\boldsymbol{u}||^{-2})$), then $\bar{C}_x^\alpha(\boldsymbol{u})_n$ is absolutely integrable and, therefore, Fourier transformable. The RD-CTMF is not, in general, Fourier transformable except in a generalized sense that accommodates (products of) Dirac delta functions, because it does not in general decay as its arguments grow without bound, but rather it oscillates. We shall see in Section 3 that the Fourier transforms of the RD-CTCF and RD-CTMF can be very useful in characterizing a signal's higher-order statistical behavior in the frequency-domain. Before leaving the subject of temporal parameters, however, an example that illustrates the temporal parameters is presented.

2.6 Example: Sine Waves in Noise

Let $x(t)$ be the sum of two sinusoids in broadband noise

$$\begin{aligned} x(t) &= s(t) + m(t) \\ &= \sum_{j=1}^{2} A_j \cos(2\pi g_j t + \theta_j) + m(t). \end{aligned}$$

Since the sinusoids are statistically independent of the noise $m(t)$, the nth-order cumulant for $x(t)$ is

$$C_x(t, \boldsymbol{\tau})_n = C_s(t, \boldsymbol{\tau})_n + C_m(t, \boldsymbol{\tau})_n.$$

For $n = 1$, the TCF is given by

$$C_x(t, \tau)_1 = R_x(t, \tau)_1 = s(t + \tau) + \hat{E}^{\{\alpha\}}\{m(t + \tau)\},$$

whereas for $n > 1$, $C_s(t, \boldsymbol{\tau})_n$ is identically zero. This is proven next.

It is sufficient to compute the nth-order TCF for the time-series that consists of only a single sine wave

$$y(t) = e^{i2\pi f_c t + i\theta},$$

because if $y(t)$ consists of the sum of M sine waves, then the nth-order TCF is the sum of the nth-order TCFs for each of the M sine waves. This results from the fact that a sine wave is statistically independent of every time-series, including itself [35] and, therefore, the addition rule for cumulants can be used. The first-order TCF for $y(t)$ is

$$\begin{aligned} C_y(t, \tau)_1 &= \hat{E}^{\{\alpha\}}\left\{y(t+\tau)\right\} \\ &= y(t+\tau) \\ &= e^{i2\pi f_c(t+\tau)+i\theta}. \end{aligned}$$

The second-order TCF is the second-order TMF with the product of first-order TMFs subtracted off,

$$\begin{aligned} C_y(t, \boldsymbol{\tau})_2 &= \hat{E}^{\{\alpha\}}\left\{y(t+\tau_1)y(t+\tau_2)\right\} - \hat{E}^{\{\alpha\}}\left\{y(t+\tau_1)\right\}\hat{E}^{\{\alpha\}}\left\{y(t+\tau_2)\right\} \\ &= \hat{E}^{\{\alpha\}}\left\{y(t+\tau_1)\right\}\hat{E}^{\{\alpha\}}\left\{y(t+\tau_2)\right\} - \hat{E}^{\{\alpha\}}\left\{y(t+\tau_1)\right\}\hat{E}^{\{\alpha\}}\left\{y(t+\tau_2)\right\} \\ &= 0. \end{aligned}$$

From (19), the nth-order TCF is given by

$$C_y(t, \boldsymbol{\tau})_n = R_y(t, \boldsymbol{\tau})_n - \sum_{\substack{P_n \\ p\neq 1}}\left[\prod_{j=1}^{p} C_y(t, \boldsymbol{\tau}_{\nu_j})_{n_j}\right].$$

For $n = 3$, all the terms in the sum are zero except for the one that corresponds to $p = 3$ because the pure first- and second-order sine waves are equal to zero. Thus,

$$\begin{aligned} C_y(t, \boldsymbol{\tau})_3 &= R_y(t, \boldsymbol{\tau})_3 - \prod_{j=1}^{3} C_y(t, \tau_j)_1 \\ &= \hat{E}^{\{\alpha\}}\left\{\prod_{j=1}^{3} y(t+\tau_j)\right\} - \prod_{j=1}^{3}\hat{E}^{\{\alpha\}}\left\{y(t+\tau_j)\right\} \\ &= 0. \end{aligned}$$

By induction,

$$C_y(t, \boldsymbol{\tau})_n = 0 \quad n \geq 2.$$

We can further conclude that the nth-order cumulant for any polyperiodic function $y(t)$ is zero for $n \geq 2$ because all such functions obey

$$\hat{E}^{\{\alpha\}}\left\{L_y(t, \boldsymbol{\tau})_n\right\} \equiv L_y(t, \boldsymbol{\tau})_n.$$

Now, back to our example. For $n > 1$,

$$C_x(t, \boldsymbol{\tau})_n = C_m(t, \boldsymbol{\tau})_n, \quad n > 1.$$

If $m(t)$ is Gaussian, then $C_m(t, \boldsymbol{\tau})_n \equiv 0$ for $n > 2$. The nth-order moment $R_x(t, \boldsymbol{\tau})_n$ is difficult to compute by straightforward calculation, but it can be more easily computed by using the lower-order cumulants as in (33). Assuming that $m(t)$ is a zero-mean Gaussian signal, the third-order moment is given by

$$
\begin{aligned}
R_x(t, \boldsymbol{\tau})_3 &= \sum_{\substack{P_3 \\ p \neq 1}} \left[\prod_{j=1}^{p} C_x(t, \boldsymbol{\tau}_{\nu_j})_{n_j} \right] \\
&= C_m(t, \tau_1, \tau_2)_2\, s(t+\tau_3) + C_m(t, \tau_1, \tau_3)_2\, s(t+\tau_2) \qquad (41) \\
&\quad + C_m(t, \tau_2, \tau_3)_2\, s(t+\tau_1) + L_s(t, \boldsymbol{\tau})_3.
\end{aligned}
$$

If $m(t)$ is also stationary and white, then

$$
C_m(t, \tau_j, \tau_k)_2 = N_0 \delta(\tau_j - \tau_k),
$$

and

$$
\begin{aligned}
&R_x(t, \boldsymbol{\tau})_3 \\
&\quad = N_0\left[\delta(\tau_1-\tau_2)s(t+\tau_3) + \delta(\tau_1-\tau_3)s(t+\tau_2) + \delta(\tau_2-\tau_3)s(t+\tau_1)\right] + L_s(t, \boldsymbol{\tau})_3.
\end{aligned}
$$

The impure third-order cycle frequencies for $x(t)$ for this $m(t)$ are $\pm g_1, \pm g_2, \pm g_1 \pm g_2$, and $\pm g_i \pm g_j \pm g_k$ for any choices of $i, j, k \in \{1, 2\}$. There are no pure third-order cycle frequencies, which means that the set of impure cycle frequencies $\{\alpha\}$ is quite different from the set of pure cycle frequencies $\{\beta\}$. The CTMF for $\alpha = 3g_1$ is given by

$$
\begin{aligned}
R_x^{3g_1}(\boldsymbol{\tau})_3 &= \left\langle R_x(t, \boldsymbol{\tau})_3 e^{-i2\pi 3 g_1 t} \right\rangle \\
&= \frac{A_1^3}{8} e^{i(2\pi 3 g_1 [\tau_1+\tau_2+\tau_3] + 3\theta_1)}.
\end{aligned} \qquad (42)
$$

Alternatively, if the broadband noise $m(t)$ is a Gaussian polycyclostationary signal [35], then it will contribute its second-order cycle frequencies to the TMF in (41) by mixing with the cycle frequencies of $s(t)$ in the first three terms on the right side.

3 THE SPECTRAL PARAMETERS OF HOCS

The Fourier transform of $\bar{R}_x^0(\boldsymbol{u})_2$ is the power spectral density (PSD) of $x(t)$ and the Fourier transform of $\bar{C}_x^0(\boldsymbol{u})_2$ is the PSD of the centered version[5] of $x(t)$ (this is the Wiener relation [116], cf. Section 4.2.1 and [35]). The Fourier transform of a symmetrized version[6] of $\bar{R}_x^\alpha(\boldsymbol{u})_2$ for nonzero α is the spectral correlation function (cyclic spectral density function), and the Fourier transform of a symmetrized version of $\bar{C}_x^\alpha(\boldsymbol{u})_2$ is the spectral correlation function for the time-series $x(t)$ with its first-order sine waves removed (this is the cyclic Wiener relation [35], cf. Section 4.2.1). Therefore, the spectral parameters of HOCS could be defined to be the Fourier transforms of $\bar{R}_x^\alpha(\boldsymbol{u})_n$ and $\bar{C}_x^\alpha(\boldsymbol{u})_n$, whenever such transforms exist. These transforms are indeed

---

[5] $x(t) - \hat{E}^{\{\alpha\}}\{x(t)\}$

[6] $\left\langle x(t+u/2)x^*(t-u/2)e^{-i2\pi\alpha t}\right\rangle$

the central spectral parameters of the theory of HOCS, but it is more natural to derive them from a consideration of spectral moments and spectral cumulants; that is, from limiting versions (as bandwidth approaches zero) of moments and cumulants of narrowband spectral components of $x(t)$, and then to show that they can be characterized as Fourier transforms of temporal moments and cumulants.

3.1 Moments

It is assumed that $x(t)$ is absolutely integrable on finite intervals. We consider the complex envelope of the spectral component of a segment of $x(u)$ that is centered at t and has width T:

$$X_T(t, f) \triangleq \int_{t-T/2}^{t+T/2} x(v)e^{-i2\pi f v}\, dv. \tag{43}$$

The temporal moment of the set of n variables $\{X_T(t, f_j)\}_{j=1}^n$ is defined by[7]

$$\begin{aligned} S_{x_T}(\boldsymbol{f})_n &\triangleq \left\langle \prod_{j=1}^n X_T(t, f_j) \right\rangle, \quad \boldsymbol{f} \triangleq [f_1 \cdots f_n]^\dagger \\ &= \lim_{Z\to\infty} \frac{1}{Z} \int_{-Z/2}^{Z/2} \prod_{j=1}^n X_T(t, f_j)\, dt, \end{aligned} \tag{44}$$

and is assumed for the time being to exist. If the integration time T in (43) is now allowed to tend to infinity in (45), the *spectral moment function* (SMF)

$$\begin{aligned} S_x(\boldsymbol{f})_n &\triangleq \lim_{T\to\infty} S_{x_T}(\boldsymbol{f})_n \\ &= \lim_{T\to\infty} \lim_{Z\to\infty} \frac{1}{Z} \int_{-Z/2}^{Z/2} \prod_{j=1}^n X_T(t, f_j)\, dt \end{aligned} \tag{45}$$

is obtained. However, this limit exists only in a generalized sense that accommodates products of Dirac deltas (impulse functions). We shall see that Dirac deltas can be avoided by working with the cumulant counterpart of this moment.

To see that the SMF (46) is composed of products of impulse functions, we proceed as follows. The function (45) can be expressed in terms of the CTMF as follows

$$\begin{aligned} S_{x_T}(\boldsymbol{f})_n &= \left\langle \int_{-T/2}^{T/2} \cdots \int_{-T/2}^{T/2} \prod_{j=1}^n x(t + v_j)e^{-i2\pi f_j (t+v_j)}\, d\boldsymbol{v} \right\rangle \\ &= \int_{-\infty}^{\infty} \cdots \int_{-\infty}^{\infty} w_T(\boldsymbol{v})_n\, R_x^{\alpha_0}(\boldsymbol{v})_n\, e^{-i2\pi \boldsymbol{f}^\dagger \boldsymbol{v}}\, d\boldsymbol{v}, \end{aligned} \tag{46}$$

---

[7] It would be more consistent to use $\hat{E}^{\{\alpha\}}\{\cdot\}$ in place of $\langle\cdot\rangle$ in the definition of $S_{x_T}(\boldsymbol{f})_n$, but it is easy to show that these two operations lead to the same function (45). Thus, we start with the time-average operator $\langle\cdot\rangle$.

where

$$\alpha_0 \triangleq \sum_{j=1}^{n} f_j = \boldsymbol{f}^\dagger \mathbf{1},$$

$$w_T(\boldsymbol{v})_n \triangleq \prod_{j=1}^{n} \operatorname{rect}(v_j/T),$$

$$\operatorname{rect}(t) \triangleq \begin{cases} 1, & |t| \le 1/2, \\ 0, & |t| > 1/2. \end{cases}$$

Thus, $S_{x_T}(\boldsymbol{f})_n$ is nonzero only if the sum α_0 of the frequencies is equal to an impure nth-order cycle frequency α of $x(t)$. Now, assuming that $R_x^\alpha(\boldsymbol{v})_n$ is absolutely integrable on the hypercube of size T on a side, we see that the Fourier transform (47) exists and that (45) therefore exists. Assuming for the time being that $R_x^\alpha(\boldsymbol{v})_n$ is Fourier transformable on the entire space

$$S_x^\alpha(\boldsymbol{f})_n \triangleq \int_{-\infty}^{\infty} \cdots \int_{-\infty}^{\infty} R_x^\alpha(\boldsymbol{\tau})_n e^{-i2\pi \boldsymbol{f}^\dagger \boldsymbol{\tau}} \, d\boldsymbol{\tau} \tag{47}$$

(46) yields

$$S_{x_T}(\boldsymbol{f})_n = \int_{-\infty}^{\infty} \cdots \int_{-\infty}^{\infty} S_x^{\alpha_0}(\boldsymbol{f}-\boldsymbol{g})_n \prod_{k=1}^{n} T \operatorname{sinc}(g_k T) \, d\boldsymbol{g},$$

where

$$\operatorname{sinc}(f) \triangleq \frac{\sin(\pi f)}{\pi f}.$$

Thus, the finite-time spectral moment $S_{x_T}(\boldsymbol{f})_n$ converges to

$$\begin{aligned} S_x(\boldsymbol{f})_n \triangleq \lim_{T\to\infty} S_{x_T}(\boldsymbol{f})_n &= \int_{-\infty}^{\infty} \cdots \int_{-\infty}^{\infty} S_x^{\alpha_0}(\boldsymbol{f}-\boldsymbol{g})_n \prod_{k=1}^{n} \delta(g_k) \, d\boldsymbol{g} \\ &= S_x^{\alpha_0}(\boldsymbol{f})_n, \end{aligned} \tag{48}$$

where $\delta(\cdot)$ is the Dirac delta. Let us investigate this hypothetical Fourier transform (47). Using the fact that $R_x^\alpha(\boldsymbol{\tau})_n$ is sinusoidal in the translation variables (cf. (40)), it can be shown, formally, that

$$S_x^\alpha(\boldsymbol{f})_n = \bar{S}_x^\alpha(\boldsymbol{f}')_n \delta(\boldsymbol{f}^\dagger \mathbf{1} - \alpha), \quad \boldsymbol{f}' \triangleq [f_1 \cdots f_{n-1}]^\dagger, \tag{49}$$

where $\bar{S}_x^\alpha(\boldsymbol{f}')_n$ is the Fourier transform of the RD-CTMF

$$\bar{S}_x^\alpha(\boldsymbol{f}')_n \triangleq \int_{-\infty}^{\infty} \cdots \int_{-\infty}^{\infty} \bar{R}_x^\alpha(\boldsymbol{u})_n \, e^{-i2\pi \boldsymbol{u}^\dagger \boldsymbol{f}'} \, d\boldsymbol{u}. \tag{50}$$

Thus, we have the formal result

$$S_x(\boldsymbol{f})_n = \begin{cases} \bar{S}_x^\alpha(\boldsymbol{f}')_n \delta(\boldsymbol{f}^\dagger \mathbf{1} - \alpha), & \boldsymbol{f}^\dagger \mathbf{1} = \alpha, \\ 0, & \boldsymbol{f}^\dagger \mathbf{1} \ne \alpha, \end{cases} \tag{51}$$

for all cycle frequencies α of $x(t)$. The SMF can, therefore, be reexpressed as

$$S_x(\boldsymbol{f})_n = \sum_{\alpha} \bar{S}_x^{\alpha}(\boldsymbol{f}')_n \delta(\boldsymbol{f}^{\dagger}\mathbf{1} - \alpha), \tag{52}$$

which reveals that the SMF is a sum of components with impulsive factors. Moreover, it is shown in the next section that $\bar{S}_x^{\alpha}(\boldsymbol{f}')_n$ can also be a sum of components with impulsive factors and even products of impulses. Thus, neither the SMF nor the reduced-dimension SMF (RD-SMF) (50) are well-behaved functions.

3.2 Cumulants

The *spectral cumulant function* (SCF) is better behaved than the SMF. To establish this fact, we proceed in a manner analogous to that used for the SMF to obtain a characterization of the SCF in terms of the Fourier transform of the RD-CTCF. The simple cumulant of the variables $\{X_T(t, f_j)\}_{j=1}^n$ is given by

$$\begin{aligned} P_{x_T}(\boldsymbol{f})_n &\triangleq \text{Cumulant}\,\{X_T(t, f_j)\}_{j=1}^n \\ &= \sum_{P_n} \left[(-1)^{p-1}(p-1)! \prod_{j=1}^{p} S_{X_T}(\boldsymbol{f}_{\nu_j})_{n_j} \right]. \end{aligned} \tag{53}$$

This function is well defined for finite T since each moment $S_{X_T}(\cdot)$ is finite. The spectral cumulant function is defined to be the limit

$$P_x(\boldsymbol{f})_n \triangleq \lim_{T\to\infty} P_{x_T}(\boldsymbol{f})_n. \tag{54}$$

Equation (47) can be used to reexpress $P_{x_T}(\boldsymbol{f})_n$ in terms of lower-order CTMFs:

$$P_{x_T}(\boldsymbol{f})_n = \sum_{P_n} k(p) \prod_{j=1}^{p} \left[\underbrace{\int_{-\infty}^{\infty} \cdots \int_{-\infty}^{\infty}}_{n_j} w_T(\boldsymbol{v}_{\nu_j})_{n_j}\, R_x^{\alpha_j}(\boldsymbol{v}_{\nu_j})_{n_j}\, e^{-i2\pi \boldsymbol{f}_{\nu_j}^{\dagger} \boldsymbol{v}_{\nu_j}}\, d\boldsymbol{v}_{\nu_j} \right],$$

where $k(p) \triangleq (-1)^{p-1}(p-1)!$ and

$$\alpha_j \triangleq \sum_{k\in\nu_j} f_k.$$

From this expression we see that if for every partition in the set P_n (except that for $p = 1$), there is some α_j that is *not* a cycle frequency of order $n_j = |\nu_j|$, then the function $P_{x_T}(\boldsymbol{f})_n$ is equal to the function $S_{x_T}(\boldsymbol{f})_n$. If there is at least one partition such that all the α_j for that partition are cycle frequencies of order n_j, then the function $P_{x_T}(\boldsymbol{f})_n$ differs from $S_{x_T}(\boldsymbol{f})_n$. This is important when considering methods for measuring $P_{x_T}(\boldsymbol{f})_n$, as in Section 7.

The function $P_{x_T}(\boldsymbol{f})_n$ can be expressed more compactly as

$$P_{x_T}(\boldsymbol{f})_n = \int_{-\infty}^{\infty} \cdots \int_{-\infty}^{\infty} \sum_{P_n} k(p) \left[\prod_{j=1}^{p} w_T(\boldsymbol{v}_{\nu_j})_{n_j} R_x^{\alpha_j}(\boldsymbol{v}_{\nu_j})_{n_j} \right] e^{-i2\pi \boldsymbol{f}^\dagger \boldsymbol{v}} \, d\boldsymbol{v}$$

$$= \int_{-\infty}^{\infty} \cdots \int_{-\infty}^{\infty} w_T(\boldsymbol{v})_n \left[\sum_{P_n} k(p) \prod_{j=1}^{p} R_x^{\alpha_j}(\boldsymbol{v}_{\nu_j})_{n_j} \right] e^{-i2\pi \boldsymbol{f}^\dagger \boldsymbol{v}} \, d\boldsymbol{v}. \tag{55}$$

By using (47), it can be shown that this last expression is equivalent to the Fourier transform of the CTCF on a hypercube of size T on a side:

$$P_{x_T}(\boldsymbol{f})_n = \int_{-\infty}^{\infty} \cdots \int_{-\infty}^{\infty} w_T(\boldsymbol{v})_n C_x^{\alpha_0}(\boldsymbol{v})_n e^{-i2\pi \boldsymbol{f}^\dagger \boldsymbol{v}} \, d\boldsymbol{v}. \tag{56}$$

By analogy with the preceding argument for the SMF, the SCF is given by

$$P_x(\boldsymbol{f})_n = \sum_{\beta} \bar{P}_x^{\beta}(\boldsymbol{f}')_n \delta(\boldsymbol{f}^\dagger \boldsymbol{1} - \beta), \tag{57}$$

where

$$\bar{P}_x^{\beta}(\boldsymbol{f}')_n \triangleq \int_{-\infty}^{\infty} \cdots \int_{-\infty}^{\infty} \bar{C}_x^{\beta}(\boldsymbol{u})_n e^{-i2\pi \boldsymbol{u}^\dagger \boldsymbol{f}'} \, d\boldsymbol{u}, \tag{58}$$

is defined to be the *cyclic polyspectrum* (CP). The transform (58) does exist (in the strict sense that excludes Dirac deltas) in general for time-series with asymptotically independent variables such that the reduced-dimension CTCF decays sufficiently rapidly in all directions so that $\bar{C}_x^{\beta}(\boldsymbol{u})_n$ is absolutely integrable and, hence, Fourier transformable (cf. Section 2.5.2).

It is now easy to show that the RD-SMF $\bar{S}_x^{\alpha}(\boldsymbol{f}')_n$ can itself contain impulses. The decomposition of this function in terms of lower-order CTCFs can be obtained by computing the Fourier coefficient of the TMF given in (33):

$$R_x^{\alpha}(\boldsymbol{\tau})_n = C_x^{\alpha}(\boldsymbol{\tau})_n + \sum_{\substack{P_n \\ p \neq 1}} \left[\sum_{\boldsymbol{\beta}^\dagger \boldsymbol{1} = \alpha} \prod_{j=1}^{p} C_x^{\beta_j}(\boldsymbol{\tau}_{\nu_j})_{n_j} \right]. \tag{59}$$

Setting $\boldsymbol{\tau} = [\boldsymbol{u}\ 0]$ and Fourier transforming in $\boldsymbol{u}$ yields

$$\bar{S}_x^{\alpha}(\boldsymbol{f}')_n = \bar{P}_x^{\alpha}(\boldsymbol{f}')_n + \sum_{\substack{P_n \\ p \neq 1}} \left[\sum_{\boldsymbol{\beta}^\dagger \boldsymbol{1} = \alpha} \bar{P}_x^{\beta_p}(\boldsymbol{f}'_{\nu_p})_{n_p} \prod_{j=1}^{p-1} \bar{P}_x^{\beta_j}(\boldsymbol{f}'_{\nu_j})_{n_j} \delta(\boldsymbol{1}^\dagger \boldsymbol{f}_{\nu_j} - \beta_j) \right] \tag{60}$$

where it is assumed that the partitions in (59) are ordered so that ν_p always contains n. In (60), each coefficient $\bar{P}_x^{\beta_j}(\cdot)$ is well behaved so that there are no hidden impulse functions on the right-hand side. For the case in which there is no cyclostationarity of order less than n associated with $\{x(t+\tau_j)\}_{j=1}^{n}$, the sum over P_n in (60) is zero for nonzero values of α. If $x(t)$ exhibits cyclostationarity of order less than n such that there is at least one $\boldsymbol{\beta}$ such that $\boldsymbol{\beta}^\dagger \mathbf{1} = \alpha$, then the sum cannot be identically zero as a function of $\boldsymbol{f}'$. Since it is known that $\bar{P}_x^\alpha(\boldsymbol{f}')_n$ is well behaved (contains no impulses), then the RD-SMF must contain impulses or products of impulses and, therefore, so must the SMF.

Because of the characterization (57) of the SCF, we see that the SCF is nonzero only on the hyperplanes specified by $\boldsymbol{f}^\dagger \mathbf{1} = \beta$, where β is a pure nth-order cycle frequency of the time-series $x(t)$.

In this section we have seen that the CP is, in general, the only well-behaved spectral function in the theory of HOCS. The SMF and its reduced-dimension version $\bar{S}_x^\alpha(\boldsymbol{f}')_n$ in general contain products of impulses and are, therefore, not well-behaved functions. However, in the special case where the lowest order of cyclostationarity of $x(t)$ is n, the impure nth-order sine waves (with strengths given by the CTMFs) are identical to the pure nth-order sine waves (CTCFs) for nonzero α and, as a result, the nth-order SCF is identical to the nth-order SMF, which results in equality between the CP and the RD-SMF. In addition, there can be many values of the frequency vector $\boldsymbol{f}'$ for which the RD-SMF and the CP are equal even when $x(t)$ exhibits lower-order cyclostationarity. For these $\boldsymbol{f}'$, the CP can be measured by measuring the RD-SMF. This is explained more fully in Section 7.

3.3 Example: Sine Waves in Noise

We reconsider the example of Section 2.6. The third-order RD-SMF for the signal consisting of two sinusoids in noise,

$$\begin{aligned} x(t) &= \sum_{j=1}^{2} A_j \cos(2\pi g_j t + \theta_j) + m(t) \\ &= s(t) + m(t), \end{aligned}$$

for $\alpha = 3g_1$ is given by the Fourier transform of the RD-CTMF

$$\bar{R}_x^\alpha(\boldsymbol{u})_3 = \frac{A_1^3}{8} e^{i(2\pi 3 g_1 [u_1 + u_2] + 3\theta_1)}$$

obtained by setting $\tau_3 = 0$ in (42). This Fourier transform is

$$\bar{S}_x^\alpha(\boldsymbol{f}')_3 = \frac{A_1^3}{8} e^{i3\theta_1} \delta(f_1 - 3g_1)\delta(f_2 - 3g_1)$$

which demonstrates that the RD-SMF can be multiply impulsive. For Gaussian $m(t)$, the nth-order CP for $x(t)$ is identically zero for $n > 2$. If $m(t)$ is not Gaussian, then

$$\bar{P}_x^\beta(\boldsymbol{f}')_n = \bar{P}_m^\beta(\boldsymbol{f}')_n, \quad n \geq 2,$$

which must be nonzero for some β and n.

4 DISCUSSION

4.1 Utility of the Theory

The theory of HOCS and, in particular, the theory of temporal and spectral cumulants, is expected to be useful because it possesses the following attributes:

1. It characterizes the sine-wave components present in the output of nonlinear transformations (both with and without memory) of polycyclostationary signals.
2. It characterizes the statistical dependence between the amplitude and phase fluctuations in distinct spectral bands, which can be useful for polyperiodically time-variant nonlinear filtering.
3. The cyclic cumulants and cyclic polyspectra are signal selective in the same way that the second-order parameters, the cyclic autocorrelation and cyclic spectrum, are signal selective.
4. The cyclic parameters depend (explicitly in some cases) on the phases or timing references of the signal (for example, carrier phase and/or clock timing).
5. The cyclic cumulants and polyspectra are well-behaved mathematical quantities for signal models that are physically appropriate.

The theory is expected to be useful in the following specific ways:

1. Design and analysis of detection and parameter estimation algorithms for polycyclostationary signals with weak or nonexistent second-order polycyclostationarity (cf. Section 8).
2. Performance analysis of any detector, estimator, or other signal processor that operates nonlinearly on a polycyclostationary signal.
3. Study of the effects of unintentional or unavoidable nonlinearities in system elements on polycyclostationary signals.
4. Optimization of sine-wave generation devices for synchronization purposes.
5. Modulation classification and recognition based on frequencies of generatable sine-wave and the orders at which the sine waves occur (cf. Section 8).

4.2 Special Cases of the Parameters

4.2.1 HOCS, the Cyclic Spectrum, and the Power Spectrum

In this section we examine the relationship between the parameters of HOCS for $n = 2$ and the well-established parameters of second-order polycyclostationarity (SOCS), which include the (nonprobabilistic) autocorrelation and power spectrum as special cases.

The autocorrelation function for a real time-series $x(t)$ is defined to be

$$R_x(\tau) \triangleq \langle x(t+\tau/2)x(t-\tau/2)\rangle, \tag{61}$$

which is obtained by using the time-averaging operation $\langle\cdot\rangle$, that is,

$$R_x(\tau) = \hat{E}^{\{0\}}\{x(t+\tau/2)x(t-\tau/2)\}.$$

This function does not in general describe the second-order polycyclostationarity (if any exists) of $x(t)$. To do that we need to use the sine-wave extraction operator to obtain the second-order TMF for $\tau_1 = \tau/2$ and $\tau_2 = -\tau/2$,

$$R_x(t,\tau) \triangleq \hat{E}^{\{\alpha\}}\{x(t+\tau/2)x(t-\tau/2)\}. \tag{62}$$

For a stationary time-series, (61) and (62) are identical, but for a polycyclostationary time-series we have

$$R_x(t,\tau) = \sum_{\alpha} R_x^{\alpha}(\tau)e^{i2\pi\alpha t} \tag{63}$$

where

$$R_x^{\alpha}(\tau) \triangleq \left\langle x(t+\tau/2)x(t-\tau/2)e^{-i2\pi\alpha t}\right\rangle \tag{64}$$

and, therefore, $R_x(\tau) = R_x^0(\tau)$. Equations (63) and (64) define the central time-domain parameters of SOCS for real time-series [35]. The function (64) is called the *cyclic autocorrelation function.* We can easily relate the cyclic autocorrelation function to the RD-CTMF for $n = 2$. Since

$$\begin{aligned}\bar{R}_x^{\alpha}(u)_2 &= \left\langle x(t+u)x(t)e^{-i2\pi\alpha t}\right\rangle \\ &= \left\langle x(t+u/2)x(t-u/2)e^{-i2\pi\alpha t}\right\rangle e^{i\pi\alpha u},\end{aligned}$$

then the RD-CTMF for $n = 2$ is related to the cyclic autocorrelation by a sinusoidal factor:

$$\bar{R}_x^{\alpha}(u)_2 = R_x^{\alpha}(u)e^{i\pi\alpha u}. \tag{65}$$

The *spectral correlation function* or *cyclic spectrum* is the limit as the bandwidth tends to zero ($T \to \infty$) of the time-averaged product of spectral components with approximate bandwidth $1/T$ and frequency separation α:

$$S_x^{\alpha}(f) = \lim_{T\to\infty}\left\langle \frac{1}{T}X_T(t, f+\alpha/2)X_T^*(t, f-\alpha/2)\right\rangle. \tag{66}$$

This function is the Fourier transform of the cyclic autocorrelation function (64),

$$S_x^{\alpha}(f) = \int_{-\infty}^{\infty} R_x^{\alpha}(\tau)e^{-i2\pi f\tau}\, d\tau. \tag{67}$$

The relation (67) is the *cyclic Wiener relation*, and it reduces to the *Wiener relation* between the power spectrum and the autocorrelation for $\alpha = 0$,

$$S_x^0(f) = \int_{-\infty}^{\infty} R_x^0(\tau) e^{-i2\pi f\tau} \, d\tau. \tag{68}$$

Combining (50), (65), and (67) yields

$$\begin{aligned}\bar{S}_x^\alpha(f')_2 &= \int_{-\infty}^{\infty} \bar{R}_x^\alpha(u)_2 \, e^{-i2\pi f' u} \, du \\ &= \int_{-\infty}^{\infty} R_x^\alpha(u) e^{i\pi\alpha u} e^{-i2\pi f' u} \, du \\ &= S_x^\alpha(f' - \alpha/2),\end{aligned} \tag{69}$$

which implies that the Fourier transform of the RD-CTMF, $\bar{S}_x^\alpha(f')_2$, is related to the cyclic spectrum by a frequency shift of $\alpha/2$. Note that the function $\bar{S}_x^\alpha(f')_2$ can contain impulses that are due to the first-order sine-wave components contained in the data $x(t)$. In the development of the theory of SOCS, it is most natural to assume that the data does not contain such finite-strength additive sine waves. In this case, the moments and cumulants are equal, and $\bar{S}_x^\alpha(f')_2$ does not contain impulses,

$$\begin{aligned}\bar{R}_x^\alpha(u)_2 &\equiv \bar{C}_x^\alpha(u)_2, \\ \bar{P}_x^\alpha(f')_2 &\equiv \bar{S}_x^\alpha(f')_2 \\ &= S_x^\alpha(f' - \alpha/2)\end{aligned}$$

Thus, in this special case, the CP is equal to the shifted cyclic spectrum, and the CP for $\alpha = 0$ is equal to the PSD, $\bar{P}_x^0(f')_2 = S_x^0(f')$. We conclude that the parameters of HOCS that are defined in this paper are consistent (to within a frequency shift) with the previously developed second-order parameters for polycyclostationary time-series, and are consistent with the notions of autocorrelation and power spectrum and are therefore properly referred to as generalizations of the second-order parameters. The same can be said of the parameters of HOCS and SOCS generalized from real time-series to complex time-series as done in Section 5.1 for HOCS and in [35] for SOCS.

4.2.2 HOCS and Higher-Order Stationarity

A signal is called *kth-order stationary* (in the wide sense, cf. Chapter 1) if its nth-order moments (stochastic or FOT) for $n = 1, \cdots, k$ are translation invariant. Such signals, within the stochastic-process framework, are the subject of the literature on higher-order statistics [81, 83]. The cyclic polyspectrum for such signals is zero except for $\beta = 0$. Thus, the stochastic (FOT) HOCS parameters of an nth-order stationary signal match those of stochastic (FOT) higher-order statistics. But, there is not, in general, a match between the same types of statistics from the two different frameworks (cf. Chapter 1).

Consider a zero-mean polycyclostationary ergodic stochastic process $X(t)$. If we compute the theoretical FOT power spectrum for a sample path of the process and the theoretical stochastic power spectrum for the stationarized stochastic process obtained by phase randomizing (cf. Chapter 1), then we get the same function provided that the phase randomization is done properly [36]. But the phase-randomized process is not *cycloergodic*. In other words, it is possible for a polycyclostationary time-series to be a sample path of a stationary stochastic process: if the power spectrum is the only quantity of interest, this modelling discrepancy is not apparent. This is so simply because of ergodicity: The $\alpha = 0$ component in the FOT second-order moment (63)

$$R_x(t,\tau) = \sum_{\alpha} R_x^{\alpha}(\tau) e^{i2\pi\alpha t} \tag{70}$$

must be equal (with probability one) to the stochastic second-order moment $\mathcal{R}_X(\tau)$:

$$R_x^0(\tau) = \mathcal{R}_X(\tau) = E\{X(t+\tau/2)X(t-\tau/2)\}.$$

This actually occurs often in the literature on communications theory. Because of this correspondence, the $\alpha = 0$ component in (70) is sometimes referred to as the "stationary component" of the second-order moment.

Can this be generalized to higher-orders? What is the interpretation of the $\alpha = 0$ component of the nth-order TMF (12) or the $\beta = 0$ component of the nth-order TCF (34)? The answer is that neither can be interpreted as a stationary component because the $\alpha = 0$ and $\beta = 0$ components of the moment and cumulant, respectively, are not in general equal to the moments and cumulants of a stationarized process. This is a result of the fact that each of these quantities depends explicitly on lower-order cyclic moments and/or cumulants (cf. (36) with $\beta = 0$). This is demonstrated in Section 7.

The relationships between all the parameters of higher-order cyclostationarity are shown graphically in Fig. 1. In this figure, the labels on the lines represent functional relationships between the quantities at the arrowheads. The portions of the figure that correspond to the special cases of second-order stationarity, second-order cyclostationarity, and higher-order stationarity are as indicated in the figure caption.

5 DEVELOPMENT OF THE THEORY

5.1 Complex Time-Series

To allow for arbitrary conjugations in the lag product of a complex-valued time-series $x(t)$, we use the following notation

$$L_x(t, \boldsymbol{\tau})_n = \prod_{j=1}^{n} x^{(*)_j}(t+\tau_j), \tag{71}$$

where $(*)_j$ is either a conjugation $*$ or nothing, that is, $(*)_j$ is an optional conjugation of the jth factor $x(t+\tau_j)$. For *each* of the 2^n different choices of conjugations in

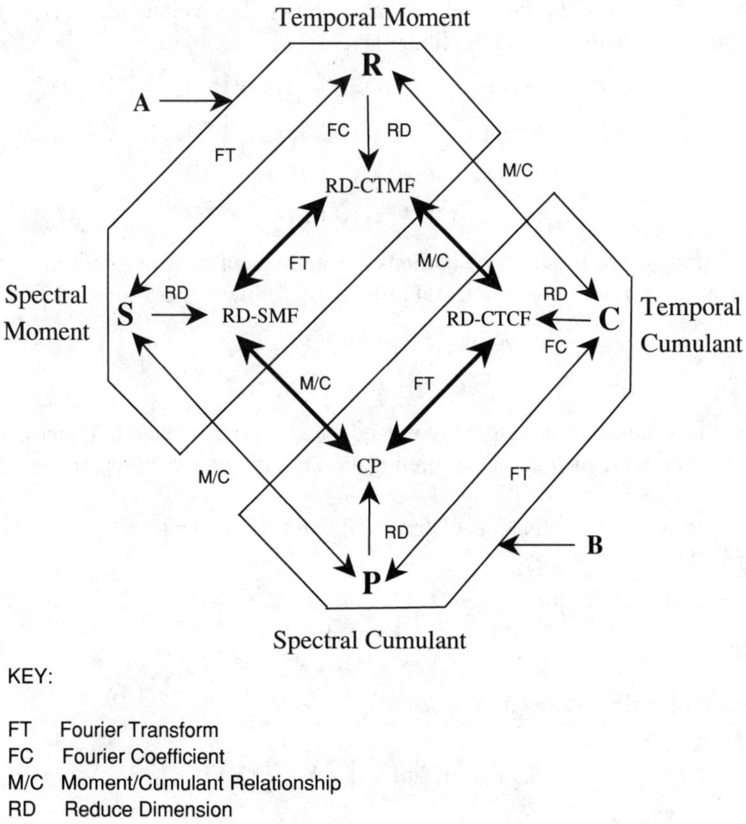

Figure 1: A pictorial representation of the relationships between the parameters of higher-order cyclostationarity. The parameters of higher-order stationarity as they are typically defined correspond to the inner diamond (bold arrows) because in this case the translation invariance of the time-domain quantities suggests the reduced-dimension quantities. The quantities in the boxes **A** and **B** are equivalent for zero-mean signals and $n = 2$ and 3 because in this case moments and cumulants are equal. Thus, box **A** (or **B**) contains the relation between the cyclic spectrum and the cyclic autocorrelation for zero-mean polycyclostationary signals, and as a special case contains the Wiener relation (bold arrow).

(71) we define the CTMF by

$$R_x^\alpha(\boldsymbol{\tau})_n \triangleq \left\langle L_x(t, \boldsymbol{\tau})_n e^{-i2\pi\alpha t} \right\rangle,$$

and its reduced-dimension counterpart by

$$\bar{R}_x^\alpha(\boldsymbol{u})_n \triangleq R_x^\alpha(\boldsymbol{\tau})_n, \quad \boldsymbol{\tau} = [\boldsymbol{u}\ 0],$$

as in Section 2 for real time-series.

In general, the 2^n functions $R_x^\alpha(\boldsymbol{\tau})_n$ are distinct. This is immediately clear in the case of $n = 2$, where we have the functions

$$R_{x_1}^{\alpha}(\boldsymbol{\tau})_2 = \left\langle x(t+\tau_1)x(t+\tau_2)e^{-i2\pi\alpha t}\right\rangle,$$
$$R_{x_2}^{\eta}(\boldsymbol{\tau})_2 = \left\langle x^*(t+\tau_1)x(t+\tau_2)e^{-i2\pi\eta t}\right\rangle,$$
$$R_{x_3}^{\eta}(\boldsymbol{\tau})_2 = \left\langle x(t+\tau_1)x^*(t+\tau_2)e^{-i2\pi\eta t}\right\rangle,$$
$$R_{x_4}^{\alpha}(\boldsymbol{\tau})_2 = \left\langle x^*(t+\tau_1)x^*(t+\tau_2)e^{-i2\pi\alpha t}\right\rangle.$$

For real time-series these four functions are equivalent provided that $\eta = \alpha$. For complex time-series they are not, and instead the following relations hold:

$$R_{x_1}^{\alpha}(\tau_1, \tau_2)_2 = R_{x_4}^{-\alpha}(\tau_1, \tau_2)_2^*$$
$$R_{x_2}^{\eta}(\tau_1, \tau_2)_2 = R_{x_3}^{-\eta}(\tau_1, \tau_2)_2^*.$$

For certain complex-valued signal types and values of n, the cycle frequency sets that are associated with each choice of conjugation are distinct, as illustrated in [35] for the case of $n = 2$.

For an arbitrary collection of time-series translates $\{y_j(t+\tau_j)\}_{j=1}^n$, the cross (joint) CTMF is defined by

$$R_y^{\alpha}(\boldsymbol{\tau})_n = \left\langle \prod_{j=1}^{n} y_j(t+\tau_j)e^{-i2\pi\alpha t}\right\rangle, \tag{72}$$

and the cross SMF is defined by

$$S_y(\boldsymbol{f})_n = \lim_{T\to\infty}\left\langle \prod_{j=1}^{n} Y_{jT}(t, f_j)\right\rangle. \tag{73}$$

We know from Section 3 that (73) can be nonzero only if $\boldsymbol{f}^\dagger\mathbf{1} = \alpha$, where α is an impure nth-order cycle frequency of $\{y_j(t+\tau_j)\}_{j=1}^n$ (i.e., (72) is not identically zero as a function of $\boldsymbol{\tau}$ for this α), in which case (73) is the n-dimensional Fourier transform of (72).

For the choice of $y_j(t+\tau_j) = x^{(*)_j}(t+\tau_j)$, the following is obtained,

$$\begin{aligned} Y_{jT}(t, f_j) &= \int_{t-T/2}^{t+T/2} x^{(*)_j}(u)e^{-i2\pi f_j u}\,du \\ &= \left[\int_{t-T/2}^{t+T/2} x(u)e^{-i2\pi(-)_j f_j u}\,du\right]^{(*)_j} \\ &= X_T^{(*)_j}(t, (-)_j f_j), \end{aligned}$$

where $(-)_j$ is the optional minus sign associated with the optional conjugation $(*)_j$. So, the SMF for $\{y_i(t+\tau_i)\}$ can be expressed in terms of $X_T(\cdot,\cdot)$ by

$$S_y(\boldsymbol{f})_n = \lim_{T\to\infty}\left\langle \prod_{j=1}^{n} X_T^{(*)_j}(t, (-)_j f_j)\right\rangle, \tag{74}$$

which is nonzero for $\boldsymbol{f}^{\dagger}\mathbf{1} = \alpha$. The RD-CTCF can be constructed in the same manner as in Section 2 for real time-series, that is, by combining lower-order CTMFs. The CP is the $(n-1)$-dimensional Fourier transform of the RD-CTCF, and is nonzero only for $\boldsymbol{f}^{\dagger}\mathbf{1} = \beta$ where β is a pure nth-order cycle frequency for $\{x^{(*)_j}(t+\tau_j)\}_{j=1}^{n}$. Just as in the case of the SMF (74), the SCF can be thought of as the limit (as $T \to \infty$) of the joint simple cumulant of the set $\{Y_{jT}(t, f_j)\}_{j=1}^{n}$. Thus, the CP for the set $\{x^{(*)_j}(t+\tau_j)\}_{j=1}^{n}$ is characterized by the limit (as $T \to \infty$) of the cumulant of the set $\{X_T^{(*)_j}(t, (-)_j f_j)\}_{j=1}^{n}$, analogous to the characterization for real time-series in Section 3.

5.2 Signal-Processing Operations

In this section we obtain input-output relations for the higher-order moments and cumulants of time-series subjected to the signal-processing operations of addition, multiplication, convolution, and periodic time-sampling. The derived relations can help in the calculation of higher-order parameters of modulated signals if such signals can be represented as a series of operations on a simpler signal, for which the higher-order parameters are known or are easily determined.

5.2.1 Addition

Let $z(t)$ be equal to the sum of two statistically independent time-series $x(t)$ and $y(t)$,

$$z(t) = x(t) + y(t).$$

In this case, the TCF for $z(t)$ is given by

$$C_z(t, \boldsymbol{\tau})_n = C_x(t, \boldsymbol{\tau})_n + C_y(t, \boldsymbol{\tau})_n, \tag{75}$$

which implies that the CTCF is given by

$$C_z^{\beta}(\boldsymbol{\tau})_n = C_x^{\beta}(\boldsymbol{\tau})_n + C_y^{\beta}(\boldsymbol{\tau})_n, \tag{76}$$

and the RD-CTCF is given by

$$\bar{C}_z^{\beta}(\boldsymbol{u})_n = \bar{C}_x^{\beta}(\boldsymbol{u})_n + \bar{C}_y^{\beta}(\boldsymbol{u})_n. \tag{77}$$

The CP is, therefore, given by

$$\bar{P}_z^{\beta}(\boldsymbol{f}')_n = \bar{P}_x^{\beta}(\boldsymbol{f}')_n + \bar{P}_y^{\beta}(\boldsymbol{f}')_n. \tag{78}$$

No equally simple additive relations hold in general for the CTMF and SMF. The results (75)–(78) can be extended by induction to the case in which $z(t)$ consists of the sum of M statistically independent time-series.

5.2.2 Product Modulation

Let the time-series $z(t)$ be the product of two statistically independent time-series $x(t)$ and $y(t)$

$$z(t) = x(t)y(t).$$

The statistical independence [35, 45] of $x(t)$ and $y(t)$ implies that

$$\hat{E}^{\{\alpha\}}\left\{\prod_{j=1}^{n} x^{(*)_j}(t+\tau_j)\prod_{k=1}^{m} y^{(*)_k}(t+v_k)\right\} = \hat{E}^{\{\alpha\}}\left\{\prod_{j=1}^{n} x^{(*)_j}(t+\tau_j)\right\}\hat{E}^{\{\alpha\}}\left\{\prod_{k=1}^{m} y^{(*)_k}(t+v_k)\right\},$$

for all values of $n, m, \boldsymbol{\tau}$, and $\boldsymbol{v}$. The TMF (12) for $z(t)$ is simply

$$\begin{aligned} R_z(t,\boldsymbol{\tau})_n &= \hat{E}^{\{\alpha\}}\left\{\prod_{j=1}^{n} z^{(*)_j}(t+\tau_j)\right\} \\ &= \hat{E}^{\{\alpha\}}\left\{\prod_{j=1}^{n} x^{(*)_j}(t+\tau_j)y^{(*)_j}(t+\tau_j)\right\} \\ &= R_x(t,\boldsymbol{\tau})_n R_y(t,\boldsymbol{\tau})_n. \end{aligned}$$

Using (12), the TMF can be expressed in terms of CTMFs,

$$\begin{aligned} R_z(t,\boldsymbol{\tau})_n &= \sum_{\alpha} R_z^{\alpha}(\boldsymbol{\tau})_n e^{i2\pi\alpha t} \\ &= \left[\sum_{\eta} R_x^{\eta}(\boldsymbol{\tau})_n e^{i2\pi\eta t}\right]\left[\sum_{\gamma} R_y^{\gamma}(\boldsymbol{\tau})_n e^{i2\pi\gamma t}\right], \end{aligned}$$

which implies that the CTMF for $\{z(t+\tau_j)\}_{j=1}^{n}$ is given by

$$\begin{aligned} R_z^{\alpha}(\boldsymbol{\tau})_n &= \left\langle R_z(t,\boldsymbol{\tau})_n e^{-i2\pi\alpha t}\right\rangle \\ &= \sum_{\eta} R_x^{\eta}(\boldsymbol{\tau})_n R_y^{\alpha-\eta}(\boldsymbol{\tau})_n \qquad (79) \\ &= \sum_{\gamma} R_x^{\alpha-\gamma}(\boldsymbol{\tau})_n R_y^{\gamma}(\boldsymbol{\tau})_n, \qquad (80) \end{aligned}$$

which are discrete convolutions, where $\alpha - \eta$ is equal to an impure nth-order cycle frequency for $y(t)$ in (80), and $\alpha - \gamma$ is equal to an impure nth-order cycle frequency for $x(t)$ in (80). The TCF and CTCF for $\{z(t+\tau_j)\}_{j=1}^{n}$ can be constructed by using (34), (36), and (80) or (80). The SMF can be obtained from (80) using the convolution theorem for the Fourier transform,

$$S_z^{\alpha}(\boldsymbol{f})_n = \sum_{\gamma}\int_{-\infty}^{\infty}\cdots\int_{-\infty}^{\infty} S_x^{\alpha-\gamma}(\boldsymbol{f}-\boldsymbol{g})_n S_y^{\gamma}(\boldsymbol{g})_n d\boldsymbol{g}, \qquad (81)$$

which is a joint continuous and discrete convolution. No equally simple input-output relations hold for the CTCF and CP if both $x(t)$ and $y(t)$ are random. However, if one of these time-series is nonrandom (which in the FOT framework means it is constant, periodic, or polyperiodic [35, 45]), say $x(t)$, then its lag product is identically equal to its TMF:

$$R_x(t, \boldsymbol{\tau})_n = \hat{E}^{\{\alpha\}}\{L_x(t, \boldsymbol{\tau})_n\} \equiv L_x(t, \boldsymbol{\tau})_n.$$

In this case, there is a simple formula for the temporal and spectral moments and cumulants for $z(t)$, which is obtained next. Expressing the TCF for $z(t)$ in terms of the TMFs for $x(t)$ and $y(t)$ yields

$$\begin{aligned} C_z(t, \boldsymbol{\tau})_n &= \sum_{P_n}\left[(-1)^{p-1}(p-1)!\prod_{j=1}^{p} R_z(t, \boldsymbol{\tau}_{\nu_j})_{n_j}\right] \\ &= \sum_{P_n}\left[(-1)^{p-1}(p-1)!\left\{\prod_{j=1}^{p} R_x(t, \boldsymbol{\tau}_{\nu_j})_{n_j}\right\}\left\{\prod_{k=1}^{p} R_y(t, \boldsymbol{\tau}_{\nu_k})_{n_k}\right\}\right]. \end{aligned}$$

Because $x(t)$ is nonrandom, the product of lower-order TMFs for $x(t)$ is equal to the nth-order TMF for $x(t)$ for every partition and can, therefore, be factored out of the sum, which leaves the cumulant of $y(t)$:

$$C_z(t, \boldsymbol{\tau})_n = L_x(t, \boldsymbol{\tau})_n C_y(t, \boldsymbol{\tau})_n = R_x(t, \boldsymbol{\tau})_n C_y(t, \boldsymbol{\tau})_n.$$

Thus, in the special case where $x(t)$ is nonrandom (polyperiodic), the formulas (80)–(81) hold with R_z, $\bar{S}_z$, R_y, $\bar{S}_y$ replaced by C_z, $\bar{P}_z$, C_y, $\bar{P}_y$, respectively.

5.2.3 Linear Time-Invariant Filtering

Let $z(t)$ be equal to a filtered version of $x(t)$,

$$z(t) = \int_{-\infty}^{\infty} h(\lambda)x(t-\lambda)\, d\lambda,$$

where the impulse-response function $h(\cdot)$ is assumed to be absolutely integrable. It is easy to show that the CTMF for $\{z(t+\tau_j)\}_{j=1}^{n}$ is given by

$$R_z^{\alpha}(\boldsymbol{\tau})_n = \int_{-\infty}^{\infty}\cdots\int_{-\infty}^{\infty}\left[\prod_{j=1}^{n} h^{(*)_j}(\lambda_k)\right] R_x^{\alpha}(\boldsymbol{\tau}-\boldsymbol{\lambda})_n\, d\boldsymbol{\lambda}, \quad \boldsymbol{\lambda} \triangleq [\lambda_1 \cdots \lambda_n]^{\dagger}.$$

Assuming that it exists, the spectral moment function can be obtained by using the convolution theorem for the Fourier transform,

$$S_z(\boldsymbol{f})_n = \left[\prod_{j=1}^{n} H^{(*)_j}((-)_j f_j)\right] S_x(\boldsymbol{f})_n, \tag{82}$$

where

$$H(f) \triangleq \int_{-\infty}^{\infty} h(t)e^{-i2\pi f t}\, dt$$

is the transfer function of the filter. The input-output relation (82) is intuitively pleasing since the effect of filtering $x(t)$ is to scale the spectral component in $x(t)$ with frequency ν by the complex number $H(\nu)$. Thus, the individual spectral components that are averaged to form the SMF are each scaled by the appropriate value $H^{(*)_j}_{((-)_j f_j)}$. It follows from (34) and (54) that the effect of filtering on the CTCF and the CP is the same as that for the CTMF and SMF:

$$C_z^\beta(\boldsymbol{\tau})_n = \int_{-\infty}^{\infty} \left[\prod_{j=1}^{n} h^{(*)_j}(\lambda_h) \right] C_x^\beta(\boldsymbol{\tau} - \boldsymbol{\lambda})_n \, d\boldsymbol{\lambda}, \tag{83}$$

and

$$\bar{P}_z^\beta(\boldsymbol{f}')_n = \left[H^{(*)_n}((-)_n[\beta - \mathbf{1}^\dagger \boldsymbol{f}']) \prod_{j=1}^{n-1} H^{(*)_j}((-)_j f_j) \right] \bar{P}_x^\beta(\boldsymbol{f}')_n. \tag{84}$$

Thus, the effect of filtering is conveniently represented in terms of both cumulant and moment functions.

5.2.4 Periodic Time-Sampling

Let $z(t)$ be the product of an impulse train and the time-series $x(t)$:

$$z(t) = y(t)x(t),$$
$$y(t) = \sum_{m=-\infty}^{\infty} \delta(t - mT_s),$$

where T_s is the sampling increment, and $f_s = 1/T_s$ is the sampling rate. Since $y(t)$ is periodic it is statistically independent of $x(t)$, and the results of Section 5.2.2 can be used to find the nth-order statistical parameters for $z(t)$. The nth-order RD-CTMF for $z(t)$ is given by (cf. Section 5.2.2)

$$\bar{R}_z^\alpha(\boldsymbol{u})_n = \sum_\gamma \bar{R}_x^\gamma(\boldsymbol{u})_n \bar{R}_y^{\alpha-\gamma}(\boldsymbol{u})_n.$$

By using the formal identity

$$\sum_{m=-\infty}^{\infty} \delta(t - mT_s) = f_s \sum_{m=-\infty}^{\infty} e^{i2\pi m t f_s},$$

it is straightforward to show that the RD-CTMF for $y(t)$ is given by

$$\bar{R}_y^\eta(\boldsymbol{u})_n = f_s^n \sum_{\boldsymbol{m}} \left[\exp\left\{ i2\pi f_s \boldsymbol{u}^\dagger \boldsymbol{m}' \right\} \kappa(\boldsymbol{m}^\dagger \mathbf{1} f_s - \eta) \right]$$

where

$$\boldsymbol{m} \triangleq [m_1 \cdots m_n]^\dagger \quad \boldsymbol{m}' \triangleq [m_1 \cdots m_{n-1}]^\dagger,$$

and

$$\kappa(x) = \begin{cases} 1, & x = 0, \\ 0, & x \neq 0. \end{cases}$$

Thus, the RD-SMF for $z(t)$ is given by

$$\bar{S}_z^{\alpha}(\boldsymbol{f}')_n = f_s^n \sum_{\gamma} \left[\sum_{\boldsymbol{m}} \kappa(\boldsymbol{m}^{\dagger}\mathbf{1} f_s - \alpha + \gamma) \bar{S}_x^{\gamma}(\boldsymbol{f}' - \boldsymbol{m}' f_s)_n \right]. \tag{85}$$

Because $y(t)$ is periodic, the cumulant for $\{z(t+\tau_j)\}_{j=1}^n$ is given by (cf. Section 5.2.2):

$$C_z(t, \boldsymbol{\tau})_n = R_y(t, \boldsymbol{\tau})_n C_x(t, \boldsymbol{\tau})_n,$$

and therefore the analysis above for the RD-CTMF holds also for the RD-CTCF,

$$\bar{C}_z^{\beta}(\boldsymbol{u})_n = \sum_{\eta} \bar{C}_x^{\eta}(\boldsymbol{u})_n \bar{R}_y^{\beta-\eta}(\boldsymbol{u})_n,$$

and for the CP,

$$\bar{P}_z^{\beta}(\boldsymbol{f}')_n = f_s^n \sum_{\gamma} \left[\sum_{\boldsymbol{m}} \kappa(\boldsymbol{m}^{\dagger}\mathbf{1} f_s - \beta + \gamma) \bar{P}_x^{\gamma}(\boldsymbol{f}' - \boldsymbol{m}' f_s)_n \right]. \tag{86}$$

The formulas (85) and (86) show that there are two kinds of aliasing effects due to sampling: (i) frequency aliasing, which is due to the overlapping of images of the CP (RD-SMF) with the same cycle frequency that occurs when $\gamma = \beta$ ($\gamma = \alpha$) in the sum and, (ii) cycle aliasing, which is due to the overlapping of images of the CP (RD-SMF) with cycle frequencies other than β (α).

6 DIGITAL QUADRATURE-AMPLITUDE MODULATION

6.1 Cumulant Formulas for QAM

In this section, we present the higher-order parameters for real- and complex-valued pulse-amplitude-modulated (PAM) signals, which provide useful models for the classes of digital baseband and quadrature-amplitude-modulated (QAM) signals. The PAM time-series is given by

$$x(t) = \sum_{m=-\infty}^{\infty} a_m p(t + mT_0 + t_0), \tag{87}$$

where $\{a_m\}$ is an independent and identically distributed (IID) symbol sequence, $1/T_0$ is the symbol rate, t_0 is an unknown constant that represents the absolute timing of the waveform, and $p(t)$ is the pulse function with Fourier transform given by $P(f)$,

$$P(f) = \int_{-\infty}^{\infty} p(t) e^{-i2\pi f t}\, dt.$$

It is desired to calculate the CTCF and CP for $\{x^{(*)_j}(t+\tau_j)\}_{j=1}^n$.

The higher-order cumulants for PAM time-series can be derived by using the results of Section 5. This derivation is sketched next. The cumulants for arbitrary

PAM signals can be derived from the cumulants for PAM signals that have rectangular pulses (or any other pulse with duration limited to an interval with length equal to the reciprocal of the symbol rate). These latter cumulants can be shown [110] to be given by

$$\begin{aligned} C_x(t, \boldsymbol{\tau})_n &= \text{Cumulant } \{x^{(*)_j}(t+\tau_j)\}_{j=1}^n \\ &= C_{a,n} \sum_{m=-\infty}^{\infty} \prod_{j=1}^{n} p(t + mT_0 + \tau_j), \end{aligned} \tag{88}$$

where $C_{a,n}$ is the nth-order cumulant of the symbol sequence, and is given by

$$C_{a,n} \triangleq \sum_{P_n} \left[(-1)^{p-1}(p-1)! \prod_{j=1}^{n} R_{a,\nu_j} \right],$$

in which

$$R_{a,\nu_j} \triangleq \lim_{K\to\infty} \frac{1}{2K+1} \sum_{k=-K}^{K} \prod_{q\in\nu_j} a_k^{(*)_q}. \tag{89}$$

The RD-CTCF and CP follow directly from (88):

$$\bar{C}_x^{\beta}(\boldsymbol{u})_n = \frac{C_{a,n}}{T_0} \int_{-\infty}^{\infty} p(t) \prod_{j=1}^{n-1} p(t+u_j) e^{-i2\pi\beta t}\, dt e^{i2\pi\beta t_0}, \tag{90}$$

$$\bar{P}_x^{\beta}(\boldsymbol{f}')_n = \frac{C_{a,n}}{T_0} P(\beta - \mathbf{1}^{\dagger}\boldsymbol{f}') \prod_{j=1}^{n-1} P(f_j) e^{i2\pi\beta t_0}, \tag{91}$$

for $\beta = k/T_0$.

An arbitrary PAM time-series (87) can be represented as a filtered product of an impulse train with a time-series $a(t)$:

$$\begin{aligned} x(t) &= \sum_{m=-\infty}^{\infty} a_m p(t + mT_0 + t_0) \\ &= \left[a(t) \sum_{m=-\infty}^{\infty} \delta(t + mT_0) \right] \otimes p(t + t_0) \\ &= y(t) \otimes p(t + t_0), \end{aligned}$$

where $\otimes$ represents the convolution operation, and $a(t)$ is a rectangular-pulse PAM signal of the type just analyzed. The cyclic cumulants and cyclic polyspectra for $y(t)$ can be determined by using the results of Section 5.2.4, and the results of Section 5.2.3 can then be applied to determine the effect of filtering on these cyclic parameters.

The CP for $y(t)$ follows from (86):

$$\bar{P}_y^{\beta}(\boldsymbol{f}')_n = T_0^{-n} \sum_{\gamma} \left[\sum_{\boldsymbol{m}} \kappa(\boldsymbol{m}^{\dagger}\mathbf{1}/T_0 - \beta + \gamma) \bar{P}_a^{\gamma}(\boldsymbol{f}' - \boldsymbol{m}'/T_0)_n \right]. \tag{92}$$

where $\boldsymbol{m} \triangleq [m_1 \cdots m_n]$, $\boldsymbol{m}' \triangleq [m_1 \cdots m_{n-1}]$, and $\kappa(\cdot)$ is the Kronecker delta function. The form of (92) implies that the pure nth-order cycle frequencies for $y(t)$ are $\{\beta = k/T_0\}$ for all integers k. It can be shown that the function $\bar{P}_y^{\beta}(\boldsymbol{f}')_n$ is a constant function:

$$\bar{P}_y^{\beta}(\boldsymbol{f}')_n = \frac{C_{a,n}}{T_0}.$$

The effect of filtering the time-series $y(t)$ is easily determined by using (84) with the filter transfer function

$$H(f) = \int_{-\infty}^{\infty} p(t+t_0)e^{-i2\pi ft}\, dt = P(f)e^{i2\pi f t_0}.$$

Thus, the nth-order CP for $x(t)$ is given by

$$\bar{P}_x^{\beta}(\boldsymbol{f}')_n = \frac{C_{a,n}}{T_0} P((-)_n[\beta - \mathbf{1}^{\dagger}\boldsymbol{f}'])^{(*)_n} \prod_{j=1}^{n-1} P((-)_j f_j)^{(*)_j} e^{i2\pi\beta t_0}, \tag{93}$$

which reduces to the following simpler form for real-valued time-series (no conjugations):

$$\bar{P}_x^{\beta}(\boldsymbol{f}')_n = \frac{C_{a,n}}{T_0} P(\beta - \mathbf{1}^{\dagger}\boldsymbol{f}') \prod_{j=1}^{n-1} P(f_j) e^{i2\pi\beta t_0}. \tag{94}$$

Inverse Fourier transforming (93) yields the RD-CTCF for PAM:

$$\bar{C}_x^{\beta}(\boldsymbol{u})_n = \frac{C_{a,n}}{T_0} \int_{-\infty}^{\infty} p^{(*)_n}(t) \prod_{j=1}^{n-1} p^{(*)_j}(t+u_j) e^{-i2\pi\beta t}\, dt\, e^{i2\pi\beta t_0}, \quad \beta = k/T_0. \tag{95}$$

6.2 Real-Valued PAM

In the case of binary symmetric real PAM, the symbols take on the values ± 1 with equal FOT probability. For $n = 2$, the cumulant for the symbol variables is

$$C_{a,2} = 1$$

Therefore, the second-order RD-CTCF is given by

$$\bar{C}_x^{\beta}(u)_2 = \frac{1}{T_0} \int_{-\infty}^{\infty} p(t)p(t+u)e^{-i2\pi\beta t}\, dt\, e^{i2\pi\beta t_0}, \quad \beta = k/T_0,$$

and the second-order CP is given by

$$\bar{P}_x^{\beta}(f)_2 = \frac{1}{T_0} P(\beta - f)P(f)e^{i2\pi\beta t_0},$$

which for $\beta = 0$ reduces to the well-known formula for the PSD of a unit-power PAM signal,

$$\bar{P}_x^{0}(f)_2 = \frac{1}{T_0}|P(f)|^2.$$

For $n = 4$, the cumulant for the symbol variables is given by

$$\begin{aligned} C_{a,4} &= R_{a,4} - 3R_{a,2}^2 - 4R_{a,3}R_{a,1} + 12R_{a,2}R_{a,1}^2 - 6R_{a,1}^4 \\ &= R_{a,4} - 3R_{a,2}^2 = -2, \end{aligned}$$

where $R_{a,k}$ is given by

$$R_{a,k} \triangleq \lim_{K\to\infty} \frac{1}{2K+1} \sum_{m=-K}^{K} a_m^k.$$

Thus, the fourth-order RD-CTCF is given by

$$\bar{C}_x^{\beta}(\boldsymbol{u})_4 = \frac{-2}{T_0} \int_{-\infty}^{\infty} p(t) \prod_{j=1}^{3} p(t+u_j) e^{-i2\pi\beta t}\, dt\, e^{i2\pi\beta t_0}.$$

The magnitude of this function for $u_3 = 0$ is shown in Figs. 2 and 3 for $\beta = 0$ and $\beta = 1/T_0$ for two different pulse shapes. The pulse shape for Fig. 2 is rectangular,

$$p(t) = \begin{cases} 1, & |t| \le T_0/2, \\ 0, & \text{otherwise}, \end{cases}$$

and the pulse shape for Fig. 3 is the inverse transform of the bandlimited pulse transform

$$P(f) = \begin{cases} 1, & |f| \le 1/2T_0, \\ 0, & |f| > 1/2T_0, \end{cases} \tag{96}$$

which is a $\sin(x)/x$ shaped pulse.

6.3 Complex-Valued PAM

Consider the complex PAM signal (87) with symbol constellation $\{\pm 1, \pm i\}$ (equally probable), and a pulse with transform given by (96). This is a model for the complex envelope of a bandwidth-efficient quaternary-phase-shift-keyed (QPSK) signal. This signal has no second-order cyclostationarity because for $n = 2$ and the choice of no conjugations (or for two conjugations) $C_{a,2} = 0$, and for the choice of one conjugation, $C_{a,2} = 1$. But because of (96), the RD-CTCF is zero:

$$\begin{aligned} &\int_{-\infty}^{\infty} p(t) p^*(t+u) e^{-i2\pi\beta t}\, dt \\ &\quad = \int_{-\infty}^{\infty} P(f) P^*(f-\beta) e^{-i2\pi f u}\, df e^{i2\pi\beta u} = 0, \quad \beta = q/T_0, \ q \ne 0. \end{aligned}$$

Nevertheless, this PAM signal exhibits fourth-order cyclostationarity for two choices of conjugations. In the first case there are no conjugations and therefore

$$\begin{aligned} C_{a,4} &= R_{a,4} - 3R_{a,2}^2 - 4R_{a,3}R_{a,1} + 12R_{a,2}R_{a,1}^2 - 6R_{a,1}^4 \\ &= R_{a,4} - 3R_{a,2}^2 = R_{a,4} = 1, \end{aligned}$$

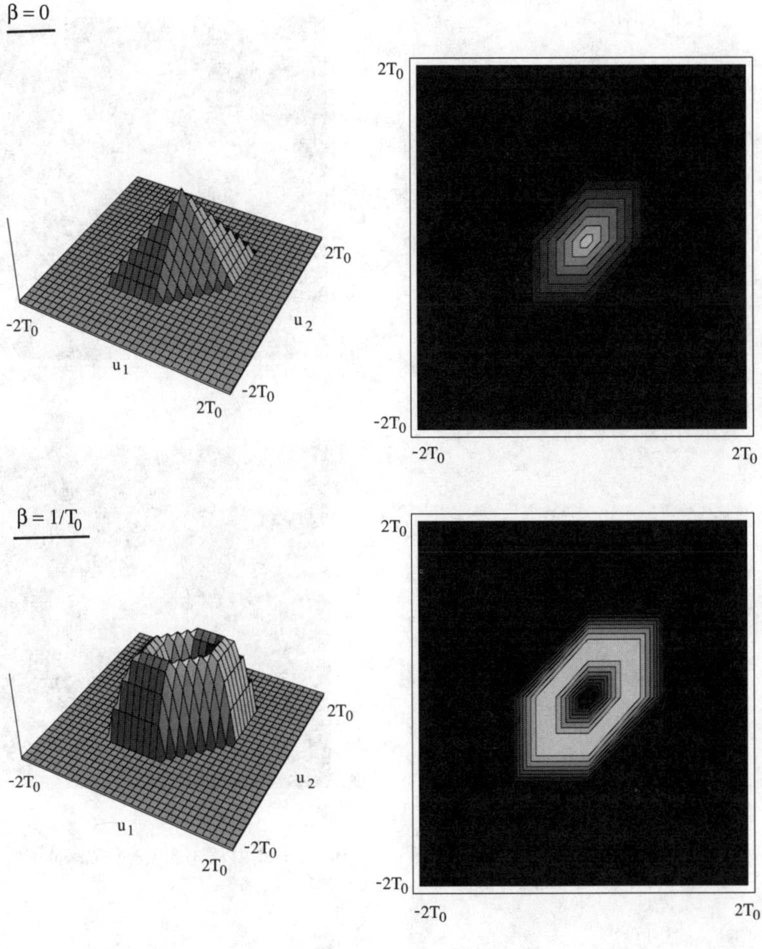

Figure 2: Surface and contour plots of the theoretical RD-CTCF for a real-valued binary PAM signal with a rectangular pulse shape.

which implies that

$$\bar{C}_x^{\beta}(\boldsymbol{u})_4 = \frac{1}{T_0}\int_{-\infty}^{\infty} p(t)\prod_{j=1}^{3} p(t+u_j)e^{-i2\pi\beta t}\,dt e^{i2\pi\beta t_0}, \quad \beta = k/T_0.$$

In the second case, two variables are conjugated, that is, the set of variables under consideration is

$$\{x(t+\tau_1)\,x(t+\tau_2)\,x^*(t+\tau_3)\,x^*(t+\tau_4)\}.$$

In this case, the cumulant for the symbol variables simplifies to $C_{a,4} = -1$, and the cumulant for the PAM time-series is given by

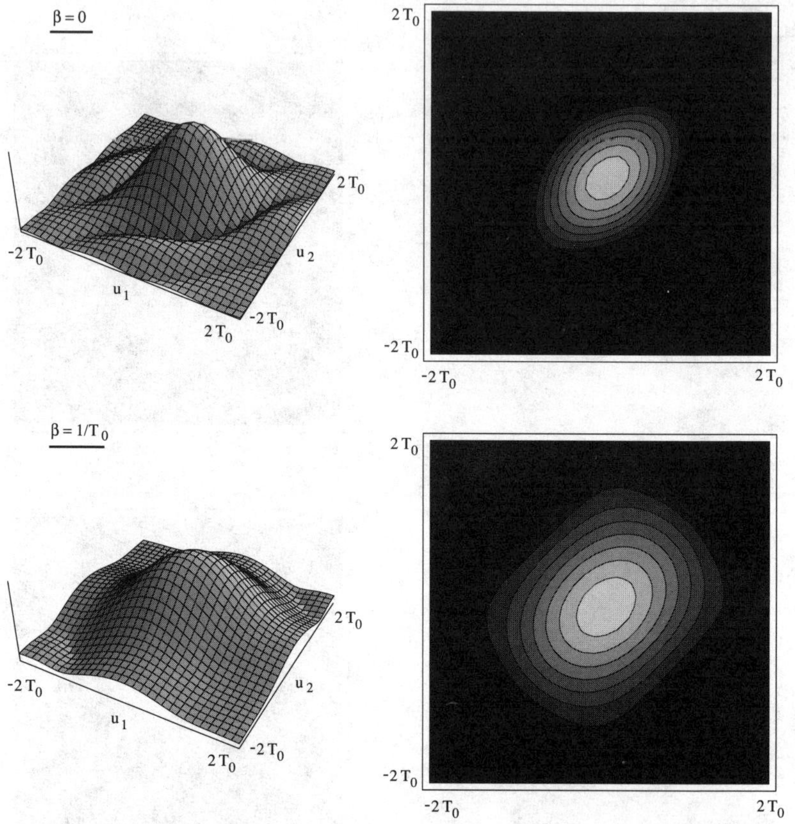

Figure 3: Surface and contour plots of the theoretical RD-CTCF for a real-valued binary PAM signal with a Nyquist-shaped pulse.

$$\bar{C}_x^{\beta}(\boldsymbol{u})_4 = \frac{-1}{T_0}\int_{-\infty}^{\infty} p^*(t)p(t+u_1)p(t+u_2)p^*(t+u_3)e^{-i2\pi\beta t}\,dt\,e^{i2\pi\beta t_0}, \quad \beta = k/T_0.$$

Note, however, that a different symbol distribution (e.g., uniform over the 8th roots of unity) could render $C_{a,4} = 0$ in both of the preceding cases, but some higher-order cumulant (e.g., $n = 8$) would be nonzero.

7 MEASUREMENT OF THE PARAMETERS OF HOCS

7.1 The Flavor of the Measurement Problem

The measurement of HOCS parameters from a single finite-length data record is considered in this section. This study is motivated by the need for such estimates that is demonstrated by the applications of the theory that are considered in Section 8.

The measurement problem as formulated here may be somewhat confusing because the ideal parameters to be measured are all mathematically derived from a single infinite-length time-series, whereas the more familiar estimation problem involves using a single finite-length data record to estimate an ideal parameter (e.g., a stochastic polyspectrum) of a stochastic process of which the single data record is assumed to be a sample path. The estimation methods described here can be viewed as the result of "backing off" from the limits that define the ideal parameters (e.g., FOT cyclic polyspectrum) and then manipulating the resulting expressions to derive various estimators.

Measurement of the polyspectrum for strictly stationary stochastic processes is considered in detail in [11, 94, 99, 122], and to some extent estimation of the cyclic polyspectrum is similar. Because a natural first step in constructing estimators for HOCS parameters is to generalize this work, especially if the stochastic-process framework is used, it is important to understand the estimation methods outlined in these references and to compare them with the methods that are natural in the time-average framework. To do this it is required to understand the conditions under which the RD-SMF is equal to the CP because the methods of [11, 94, 99] use estimates of the stochastic counterpart of the RD-SMF to estimate the stochastic polyspectrum. It should be emphasized that the objective is not to estimate the RD-SMF because it is generally impulsive and can contain products of impulses. However, it is desired to estimate the RD-SMF wherever it is equal to the CP, because it is the well-behaved CP that we are interested in.

The conditions for equality between the RD-SMF and the CP can be determined by expressing the RD-CTMF in terms of the RD-CTCF and lower-order CTMFs (cf. (36)),

$$R_x^{\beta}(\boldsymbol{\tau})_n = C_x^{\beta}(\boldsymbol{\tau})_n - \sum_{\substack{P_n \\ p\neq 1}} \left[(-1)^{p-1}(p-1)! \sum_{\boldsymbol{\alpha}^{\dagger}\mathbf{1}=\beta} \prod_{j=1}^{p} R_x^{\alpha_j}(\boldsymbol{\tau}_{\nu_j})_{n_j} \right], \tag{97}$$

where $\boldsymbol{\tau} = [u_1 \cdots u_{n-1}\ 0]$, and Fourier transforming in the u_j variables to obtain $\bar{S}_x^{\beta}(\boldsymbol{f}')_n$:

$$\bar{S}_x^{\beta}(\boldsymbol{f}')_n = \bar{P}_x^{\beta}(\boldsymbol{f}')_n - \sum_{\substack{P_n \\ p\neq 1}} \left[k(p) \sum_{\boldsymbol{\alpha}^{\dagger}\mathbf{1}=\beta} \bar{S}_{\boldsymbol{x}}^{\alpha_p}(\boldsymbol{f}'_{\nu_p})_{n_p} \prod_{j=1}^{p-1} \bar{S}_{\boldsymbol{x}}^{\alpha_j}(\boldsymbol{f}'_{\nu_j})_{n_j} \delta(\boldsymbol{f}_{\nu_j}^{\dagger}\mathbf{1} - \alpha_j) \right], \tag{98}$$

where $k(p) = (-1)^{p-1}(p-1)!$. To derive (98), it is assumed that the partition elements (members of P_n) are ordered such that ν_p always contains n in (97), and (49) is used to transform each of the CTMFs in the products in the sum over P_n in (97), except for the one with reduced dimension (corresponding to the partition element ν_p), for which (50) is used.

It is clear that $\bar{S}_x^{\beta}(\boldsymbol{f}')_n$ is equal to $\bar{P}_x^{\beta}(\boldsymbol{f}')_n$ only if the sum over P_n in (98) is zero, which will happen if one or more of the impulse functions is zero (cf. (60)). This is the case for all $\boldsymbol{f}'$ except those that lie on a *β-submanifold.* β-Submanifolds are simply the $\boldsymbol{f}'$ vectors for which there is at least one partition with $p > 1$ for which there is at least one $\boldsymbol{\alpha}$ in (98) such that the argument of each associated impulse function is zero, in which case that impulse is nonzero[8]. It is important to note that the function $\bar{S}_x^{\beta}(\boldsymbol{f}')_n$ is *not* impulsive at a value of $\boldsymbol{f}'$ that lies on a β-submanifold unless all the lower-order coefficients $\bar{S}_x^{\alpha_j}(\boldsymbol{f}'_{\nu_j})_{n_j}$ of the impulses are themselves nonzero.

Note that if $x(t)$ is strictly stationary, then the set of impure (and pure) mth-order cycle frequencies is either the null set or $\{0\}$ for each m, and our condition for equality between the RD-SMF and CP (for $\alpha = \beta = 0$) is that there does not exist any proper subset of $\boldsymbol{f}$ such that the frequencies in this subset sum to zero, which is exactly the condition stated in [99] for the equality of the stochastic RD-SMF and polyspectrum for a strictly stationary stochastic process.

The method of estimating the polyspectrum for strictly stationary processes that is proposed in [11, 99] is based on the *higher-order periodogram* defined by

$$I_{X_T}^{0}(t,\boldsymbol{f}')_n \triangleq \frac{1}{T} X_T(t,(-)_n[-\mathbf{1}^{\dagger}\boldsymbol{f}'])^{(*)_n} \prod_{j=1}^{n-1} X_T(t,(-)_j f_j)^{(*)_j}.$$

The idea is to smooth this function over the $n-1$ variables $\boldsymbol{f}'$ with an $(n-1)$-dimensional window $W(\boldsymbol{f}')$ *while avoiding the inclusion of values of $I_{X_T}^{0}(t,\boldsymbol{f}')_n$ for $\boldsymbol{f}'$ that lie on a* 0*-submanifold.* It is shown in [99] that this method is asymptotically unbiased and consistent provided that the window function, which depends on the data-length T and on n, satisfies certain conditions related to its rate of decay. This frequency-smoothing method is also presented in Priestley's book [94]. However, Priestley does not force the smoothing window to be zero on the 0-submanifolds. His single example of the method uses a zero-mean time-series and $n = 3$, in which case there are no 0-submanifolds to avoid.

In [10, 11, 56, 99], the measurement parameter T is coupled with the width of the window $W(\boldsymbol{f}')$. In the approaches for estimating the CP discussed in this paper, the width of the spectral smoothing window is a parameter that is decoupled from the data-length parameter T. That is, in the methods for estimating the frequency-domain parameters of RD-SMF and CP, two independent measurement parameters are explicitly used: the frequency-smoothing window width Δf and the data length T. This is consistent with the theory of second-order polycyclostationarity [35], and is appropriate because in actual measurement situations, the operator should be able to choose T and Δf independently, although $T\Delta f \gg 1$ is required to obtain a reliable estimate.

[8]The β-submanifolds can be defined by using the sum on the right-hand side of (60) as well. However, since cumulants are generally computed and estimated from moments, it is conceptually helpful to assume that the impure cycle frequencies—rather than the pure cycle frequencies—are known.

7.2 Time-Domain Parameters

Since the CTMF is the simplest of the parameters of HOCS and can be used to build the CTCF, which in turn can be used to compute the CP, we start by considering its estimator.

We are given the finite-length portion of a persistent time-series that has length T and center t,

$$x(u) : u \in [t - T/2, t + T/2].$$

This segment of data can be expressed as

$$x(u)\,\text{rect}\left[\frac{u-t}{T}\right],$$

where

$$\text{rect}(t) \triangleq \begin{cases} 1, & |t| \le 1/2, \\ 0, & |t| > 1/2, \end{cases}$$

and the nth-order lag product for this segment is given by

$$L_{x_T}(u, t, \boldsymbol{\tau})_n = \prod_{j=1}^{n} x(u+\tau_j)\,\text{rect}\left[\frac{u+\tau_j - t}{T}\right].$$

The estimator for the CTMF $R_x^\alpha(\boldsymbol{\tau})_n$ is simply

$$R_{x_T}^\alpha(t, \boldsymbol{\tau})_n \triangleq \frac{1}{T}\int_{-\infty}^{\infty} L_{x_T}(u, t, \boldsymbol{\tau})_n e^{-i2\pi\alpha u}\, du,$$

which can be expressed as

$$R_{x_T}^\alpha(t, \boldsymbol{\tau})_n \triangleq \frac{1}{T}\int_{t_l}^{t_u} \prod_{j=1}^{n} x^{(*)_j}(v+\tau_j)\, e^{-i2\pi\alpha v}\, dv, \tag{99}$$

where $t_l = t - T/2 - \min\{\tau_j\}$, $t_u = t + T/2 - \max\{\tau_j\}$, and $t_u \ge t_l$. If $t_u < t_l$ the estimate is defined to be zero because in this case the delays are so widely separated that the shifted data segments do not overlap. The estimate (99) converges to the ideal CTMF,

$$\lim_{T\to\infty} R_{x_T}^\alpha(t, \boldsymbol{\tau})_n = R_x^\alpha(\boldsymbol{\tau})_n$$

since $R_x^\alpha(\boldsymbol{\tau})_n$ is *defined* to be the pointwise limit of $R_{x_T}^\alpha(t, \boldsymbol{\tau})_n$. A detailed analysis of the temporal bias and variance of this estimator for arbitrary $x(t)$ as functions of T is given in [110] (cf. [62] for the case of $n = 2$).

The estimator for the CTCF is given by the combination of lower-order CTMF estimates that is specified by (36),

$$C_{x_T}^\beta(t, \boldsymbol{\tau})_n \triangleq \sum_{P_n}\left[(-1)^{p-1}(p-1)! \sum_{\boldsymbol{\alpha}^\dagger \mathbf{1} = \beta} \prod_{j=1}^{p} R_{x_T}^{\alpha_j}(t, \boldsymbol{\tau}_{\nu_j})_{n_j}\right], \tag{100}$$

where the second sum is over all vectors $\boldsymbol{\alpha} = [\alpha_1 \cdots \alpha_p]^\dagger$ of cycle frequencies with orders $n_1 \cdots n_p$ that sum to β: $\boldsymbol{\alpha}^\dagger \mathbf{1} = \beta$. Since each CTMF estimate converges to its ideal parameter, the following convergence (assuming that the possibly infinite but denumerable sum over $\boldsymbol{\alpha}$ can be interchanged with the limit) holds:

$$\lim_{T\to\infty} C_{x_T}^{\beta}(t, \boldsymbol{\tau})_n = C_x^{\beta}(\boldsymbol{\tau})_n.$$

In order to construct the estimate (100), all impure lower-order cycle frequencies $x(t)$ must be known or estimated; an algorithm for estimating these frequencies is described in Section 8. The CTMF estimator (99) converges pointwise in t to the CTMF (11) and, therefore, the CTCF estimator (100) converges pointwise to the CTCF (36) [104, 105, 110]. To estimate the RD-CTMF and RD-CTCF, we set $\boldsymbol{\tau} = [u_1 \cdots u_{n-1}\ 0]$ in (99) and (100), respectively.

7.3 Frequency-Domain Parameters

As explained in [104, 110], the CP can be estimated by Fourier transforming a windowed estimate of the RD-CTCF:

$$\bar{P}_{x_T}^{\beta}(t, \boldsymbol{f}')_{\Delta f} = \int_{-\infty}^{\infty} \cdots \int_{-\infty}^{\infty} w_{1/\Delta f}(\boldsymbol{u}) \bar{C}_{x_T}^{\beta}(t, \boldsymbol{u})_n e^{-i2\pi \boldsymbol{u}^\dagger \boldsymbol{f}'}\, d\boldsymbol{u}, \tag{101}$$

$$w_{1/\Delta f}(\boldsymbol{u}) = \prod_{j=1}^{n-1} \operatorname{rect}(u_j \Delta f).$$

The multidimensional window $w_{1/\Delta f}(\boldsymbol{u})$ can be replaced by any function with finite support in $(n-1)$-dimensional Euclidean space such that its Fourier transform converges (formally) to a product of Dirac delta functions:

$$\lim_{\Delta f \to 0} \mathcal{F}^{n-1}\{w_{1/\Delta f}(\boldsymbol{u})\} = \prod_{j=1}^{n-1} \delta(f_j).$$

The CP can also be estimated by first constructing the *nth-order cyclic periodogram*

$$\begin{aligned} I_{x_T}^{\beta}(t, \boldsymbol{f}')_n &\triangleq \frac{1}{T} X_T^{(*)_n}(t, (-)_n[\beta - \mathbf{1}^\dagger \boldsymbol{f}']) \prod_{j=1}^{n-1} X_T^{(*)_j}(t, (-)_j f_j) \\ &= \mathcal{F}^{n-1}\{\bar{R}_{x_T}^{\beta}(t, \boldsymbol{u})_n\}, \end{aligned}$$

masking it by a special function $Z_\beta(\boldsymbol{f}')$, and then convolving with a multidimensional smoothing window:

$$\bar{P}_{x_T}^{\beta}(t, \boldsymbol{f}')_{\Delta f} = W_{\Delta f}(\boldsymbol{f}') \otimes \left[I_{x_T}^{\beta}(t, \boldsymbol{f}')_n Z_\beta(\boldsymbol{f}') \right]. \tag{102}$$

In (102), $Z^{\beta}(\boldsymbol{f}')$ is equal to one everywhere except at those $\boldsymbol{f}'$ that lie on *β-submanifolds*, in which case it is equal to zero. The vector $[g_1 \cdots g_n]$ for $g_n = \beta - \sum_{j=1}^{n-1} g_j$, lies on a β-submanifold if there is at least one partition $\{\nu_j\}_{j=1}^p$ in P_n with $p > 1$ such that each sum $\alpha_k = \sum_{k\in\nu_j} g_k$ is an n_jth-order cycle frequency of $x(t)$ (for the conjugations chosen). As mentioned previously, these β-submanifolds must be avoided in the convolution (102) because the smoothed nth-order cyclic periodogram converges (formally) to the function $\bar{S}_x^\beta(\boldsymbol{f}')_n$, which can contain multiple impulsive factors for values of $\boldsymbol{f}'$ that lie on β-submanifolds, but which, for all other $\boldsymbol{f}'$, is equal to the nonimpulsive function $\bar{P}_x^\beta(\boldsymbol{f}')_n$. These impulses are avoided in the method (101) because the additive sine-wave components in the $\boldsymbol{u}$ variables of the RD-CTMF estimate $\bar{R}_{x_T}^\beta(t, \boldsymbol{u})_n$ are removed in forming the RD-CTCF estimate $\bar{C}_{x_T}^\beta(t, \boldsymbol{u})_n$, and it is these additive sine waves that give rise to the (smoothed) spectral lines in the transform $I_{x_T}^\beta(t, \boldsymbol{f}')_n$ of $\bar{R}_{x_T}^\beta(t, \boldsymbol{u})_n$.

There are several difficulties with this frequency-smoothing method. The first is that although the impulsive parts of the RD-SMF are avoided by the smoothing operation, there can be substantial leakage into nearby regions in $\boldsymbol{f}'$, which are exactly the regions used to compute the CP estimate for $\boldsymbol{f}'$ that are on the β-submanifolds. This can be seen by considering the simple case of $n = 2$. If the data contains additive finite-strength sine-wave components, then the spectrum will contain impulses. To estimate the continuous portion of the spectrum at a point where there is an impulse due to the discrete portion of the spectrum is problematic when using frequency smoothing because of leakage. This leakage problem is discussed further in Section 7.4, and is illustrated with numerical examples in Section 7.5. This leakage problem is not mentioned in [10, 11, 94] for the case of stationary stochastic processes, nor in [56] for the case of cyclostationary stochastic processes. In addition to the leakage problem, the frequency-smoothed cyclic periodogram method is computationally costlier than the Fourier transformed RD-CTCF method for $n > 2$ even when there are no β-submanifolds to avoid [110].

Finally, the CP can be estimated for values of $\boldsymbol{f}'$ that do not lie on β-submanifolds by time-averaging the masked higher-order cyclic periodogram

$$\bar{P}_{x_T}^\beta(t, \boldsymbol{f}')_{1/\Delta f} = g_T(t) \otimes \left[I_{x_{1/\Delta f}}^\beta(t, \boldsymbol{f}')_n Z_\beta(\boldsymbol{f}') \right] \tag{103}$$

where

$$g_T(t) = \begin{cases} 1/T, & |t| \le T/2, \\ 0, & \text{otherwise.} \end{cases}$$

Unlike the frequency-smoothing method, the time-averaging method cannot produce estimates of the CP for frequencies $\boldsymbol{f}'$ that lie on β-submanifolds without modification. An example of such a modification to (103) is to compute estimates of the CP for values of $\boldsymbol{f}'$ that are near to the β-submanifolds, and then use interpolation to estimate the value of the CP for the $\boldsymbol{f}'$ that are on the β-submanifolds.

The three methods of estimating the CP are consistent [110] in the sense that

$$\begin{aligned} \bar{P}_x^\beta(\boldsymbol{f}')_n &= \lim_{\Delta f \to 0} \lim_{T \to \infty} \bar{P}_{x_T}^\beta(t, \boldsymbol{f}')_{\Delta f} \\ &= \lim_{\Delta f \to 0} \lim_{T \to \infty} \bar{P}_{x_T}^\beta(t, \boldsymbol{f}')_{1/\Delta f}. \end{aligned} \tag{104}$$

The estimators (101) and (102) are generalizations of estimators of the polyspectrum that were proposed by Brillinger and Rosenblatt [9–11, 99] from stationary signals to polycyclostationary signals. The three estimators of the CP presented in this section each reduce to a well-known estimator of the PSD for the case of $n = 2$ and zero-mean stationary signals. Specifically, (101) reduces to the Blackman-Tukey method (Fourier transformation of a tapered autocorrelation estimate), (102) reduces to the Wiener-Daniell method (frequency smoothing of the periodogram), and (103) reduces to the Bartlett-Welch method (time-averaging of the periodogram) [cf. Chapter 4 in 35]. Similarly, each of the estimators of the CP is a generalization of an estimator of the cyclic spectrum from second-order to higher-orders [Chapter 13 in 35].

7.4 Leakage from β-Submanifolds

As mentioned in the previous section, a leakage effect exists in the method (102) that is due to the smearing of the impulses on the β-submanifolds to neighboring regions off the β-submanifolds. To see this, the temporal mean of (102) is computed. From [110],

$$\left\langle \bar{R}^{\beta}_{x_T}(t, \boldsymbol{u})_n \right\rangle = \bar{R}^{\beta}_{x}(\boldsymbol{u})_n v(\boldsymbol{u}),$$

where

$$v_T(\boldsymbol{u}) \triangleq \begin{cases} 1 + u_0(\boldsymbol{u})/T, & |u_0(\boldsymbol{u})| \leq T \\ 0, & \text{otherwise,} \end{cases}$$

and

$$u_0(\boldsymbol{u}) \triangleq \min\{0, u_1, \cdots, u_{n-1}\} - \max\{0, u_1, \cdots, u_{n-1}\}.$$

Using this result, it is easy to show that the temporal mean of (102) is given by the convolution

$$W_{\Delta f}(\boldsymbol{f}') \otimes \left[\left(\bar{S}^{\beta}_{x}(\boldsymbol{f}')_n \otimes V_T(\boldsymbol{f}') \right) Z_{\beta}(\boldsymbol{f}') \right],$$

where $V_T(\boldsymbol{f}')$ is the $(n-1)$-dimensional Fourier transform of $v_T(\boldsymbol{u})$. The effect of the convolution with $V_T(\boldsymbol{f}')$ is to smear the impulses in $\bar{S}^{\beta}_{x}(\boldsymbol{f}')_n$, thus producing spectral leakage into nearby regions, which cannot be removed by the masking function $Z_{\beta}(\boldsymbol{f}')$.

The method (101) also exhibits leakage when the CTMFs (11) are computed using an FFT algorithm, and the cycle frequencies α_j are not "on bin center." This leakage can be substantially reduced by computing each CTMF by evaluating the FST (the Fourier series transform, which is like the DFT, but is a function of a continuous frequency variable) for every α_j appearing in the sum in (100), but this can greatly increase the computational cost of the method [105, 110].

7.5 Numerical Examples

In this section several fourth-order CP measurements are displayed graphically for the purpose of illustrating the relative performances of the CP estimators (101), (102), and (103).

The signal of interest is a binary pulse-amplitude-modulated signal with keying period $T_0 = 7T_s$ and IID symbols. Both rectangular keying pulses $\text{rect}(t/T_0)$ and Nyquist-shaped pulses (96) are simulated. For the purposes of the simulations, only the second-order cycle frequencies k/T_0 for $|k| \leq 3$ were used to compute the lower-order CTMFs and find the β-submanifolds for the case of rectangular pulses because the higher harmonics ($|k| > 3$) produce relatively weak features and can, therefore, be neglected. For Nyquist pulses, the only lower-order cycle frequency is $\alpha = 0$ for $n = 2$. Neither signal has any cycle frequencies for odd values of the order n.

The fourth-order CP for $\beta = 0$ and $\beta = 1/T_0$ was estimated for an observation interval length of 2048 samples using the following methods:

FS: the frequency-smoothing method (102),

FT: the Fourier-transformed RD-CTCF method (101) using FFTs to estimate the lower-order CTMFs,

FST: the Fourier-transformed RD-CTCF (101) using the FST for estimating the lower-order CTMFs, and

TA: the time-averaging method (103).

For the time-domain methods (FT and FST), the RD-CTCF was estimated on the cubic grid of integers $\boldsymbol{u} = [u_1\ u_2\ u_3] \in [-8, 7]^3$, and then Fourier transformed. In the FS method, the spectral smoothing window width was set equal to 128 samples, and in the TA method, the block size was chosen to be 32 samples. Thus, each method uses approximately the same temporal and spectral resolution parameters, and can therefore be compared fairly.

Because computing the entire CP using (102) is relatively costly and is difficult to display, only two "slices" of the CP were estimated. The slices correspond to $S_1 = \{[0.0\ 1/32\ f]\}$ for the case of rectangular pulses and S_2: $\{[1/32\ 1/32\ f]\}$ for the case of Nyquist pulses for $f \in \{k/32, k = -16, \cdots, 15\}$. Two points in S_1 lie on submanifolds for both values of β, and no points of S_2 lie on submanifolds for both values of β. The value of the TA estimate for $\boldsymbol{f}'$ on β-submanifolds is defined to be zero. The ideal CP was computed by numerically evaluating the formula (94).

The CP estimates for ten independent realizations (with the same timing parameter t_0 in (87)) of the signals were averaged to produce the final set of estimates. The magnitudes of the estimates are shown in Figs. 4–7. Generally speaking, the FST method delivers the best performance because it suffers the least from the aforementioned leakage effect. However, it is not the best method from a computational cost point of view [110]. All of the methods produce results that are reasonable approximations to the theoretical fourth-order CP.

7.6 Cycloergodicity and Measurement of Cumulants

We illustrate here the kind of error that can result from modelling a signal as a stationary stochastic process when the sample paths are actually polycyclostationary time-series, and then measuring higher-order parameters from a single sample path.

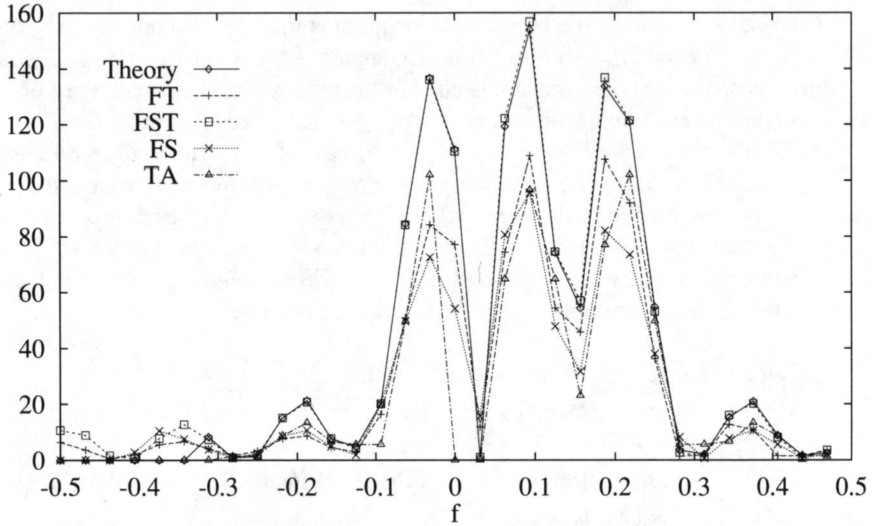

Figure 4: Estimates of the fourth-order CP for binary PAM with rectangular pulses for $\beta = 1/T_0$, a collect time of $292T_0$, and $f \in S_1$.

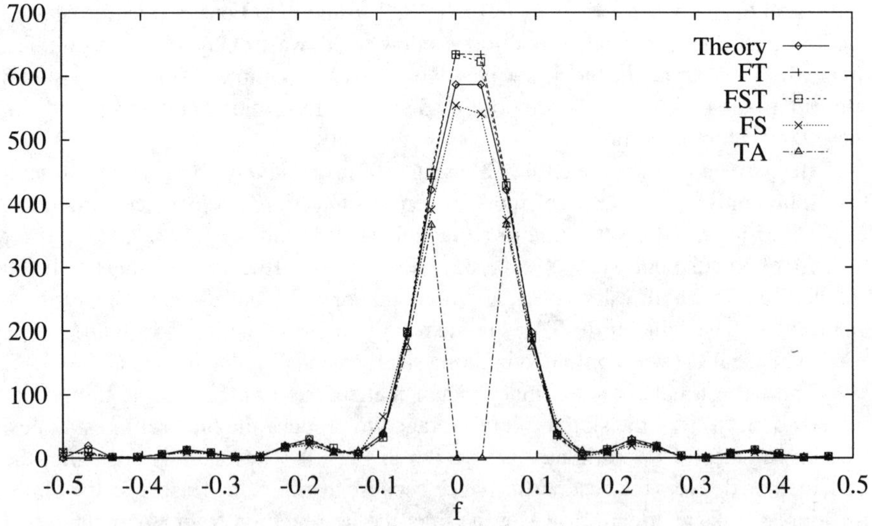

Figure 5: Estimates of the fourth-order CP for binary PAM with rectangular pulses for $\beta = 0$, a collect time of $292T_0$, and $f \in S_1$.

Specifically, we consider a zero-mean, noiseless binary PAM signal with rectangular pulses, $\beta = 0$, $T = 512T_0$, and $n = 4$. If the signal is thought of as stationary (e.g., by making the pulse timing parameter t_0 random and uniformly distributed over $[0, T_0]$ in the stochastic-process model, a stationary process is obtained [32]), and the

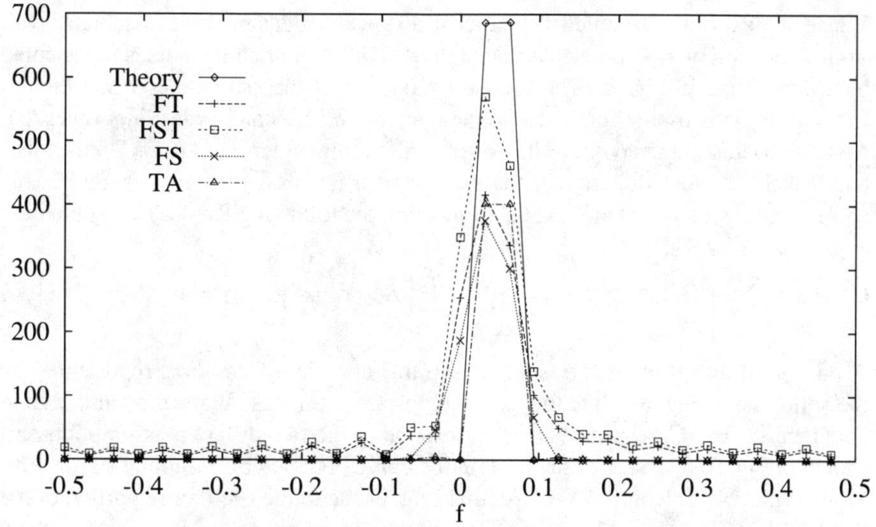

Figure 6: Estimates of the fourth-order CP for binary PAM with Nyquist pulses for $\beta = 1/T_0$, a collect time of $292T_0$, and $f \in S_2$.

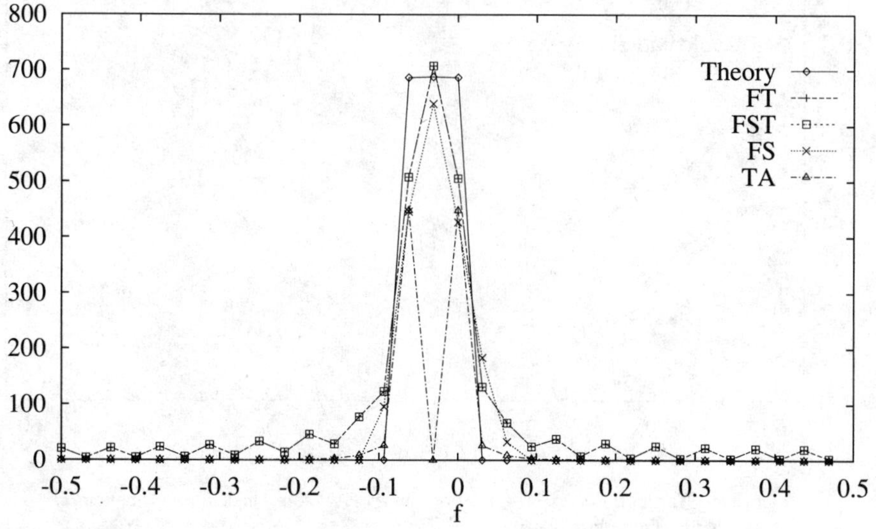

Figure 7: Estimates of the fourth-order CP for binary PAM with Nyquist pulses for $\beta = 0$, a collect time of $292T_0$, and $f \in S_2$.

theoretical RD-CTCF is computed, the following result is obtained:

$$\bar{C}_x^0(\boldsymbol{u})_4 = \sum_{P_n} \left[(-1)^{p-1}(p-1)! \prod_{j=1}^{p} R_x^0(\boldsymbol{\tau}_{\nu_j})_{n_j} \right], \qquad \tau_i = u_i,\, \tau_4 = 0. \quad (105)$$

Hence, it is explicitly assumed that there are no lower-order sine waves associated with $x(t)$. A portion of an estimate corresponding to this cumulant formula is represented by a dotted line in Fig. 8. The oscillatory portion of the estimate does not die out for large u. This oscillation is due to the interaction of second-order sine waves that have not been subtracted out of the fourth-order moment term in (105). On the other hand, if the cyclostationarity of the sample path is recognized and, therefore, the second-order sine waves are taken into account, the following RD-CTCF is obtained:

$$\bar{C}_x^0(\boldsymbol{u})_4 = \sum_{P_n} \left[(-1)^{p-1}(p-1)! \sum_{\boldsymbol{\alpha}^\dagger \mathbf{1}=0} \prod_{j=1}^{p} R_x^{\alpha_j}(\boldsymbol{\tau}_{\nu_j})_{n_j} \right], \quad \tau_i = u_i,\ \tau_4 = 0. \quad (106)$$

A portion of an estimate corresponding to this cumulant formula is represented by the solid line in Fig. 8. Note that both graphs are estimates obtained from the same data record. The CP estimates that correspond to the two different assumptions are shown in Fig. 9. The size of the peak in the curve obtained by assuming stationarity grows with collect length T because this peak is due to the oscillatory portion of the RD-CTCF estimate, which does not die out with increasing u.

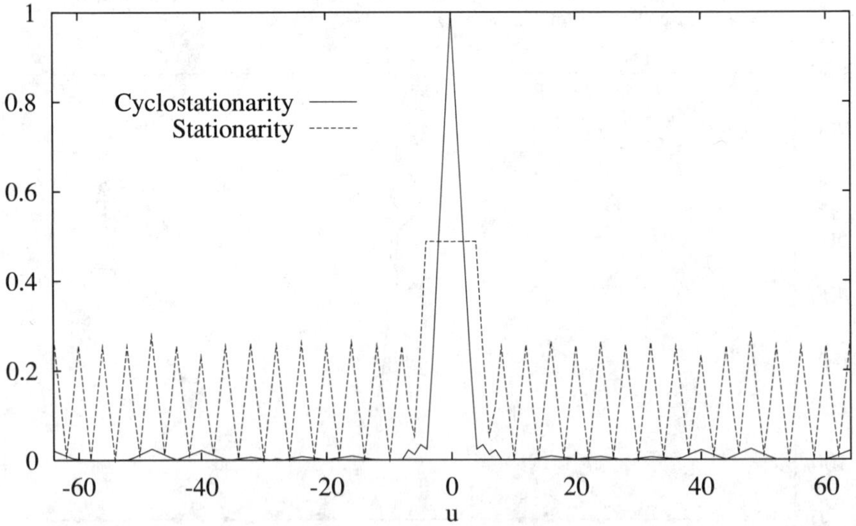

Figure 8: The fourth-order RD-CTCF for a binary PAM signal under the assumptions of stationarity and cyclostationarity.

Note that this phenomenon is not observed in the case of $n = 2$ because we typically consider zero-mean signals that do not contain finite-strength additive sine-wave components. Thus, the PSD estimates agree with the calculated formula based on a stationary stochastic model, even when the sample paths are polycyclostationary, because there are no lower-order sine waves to deal with (cf. Section 4). This example illustrates the fact that the theory of HOCS is important even in the case where cyclostationarity is not of direct interest.

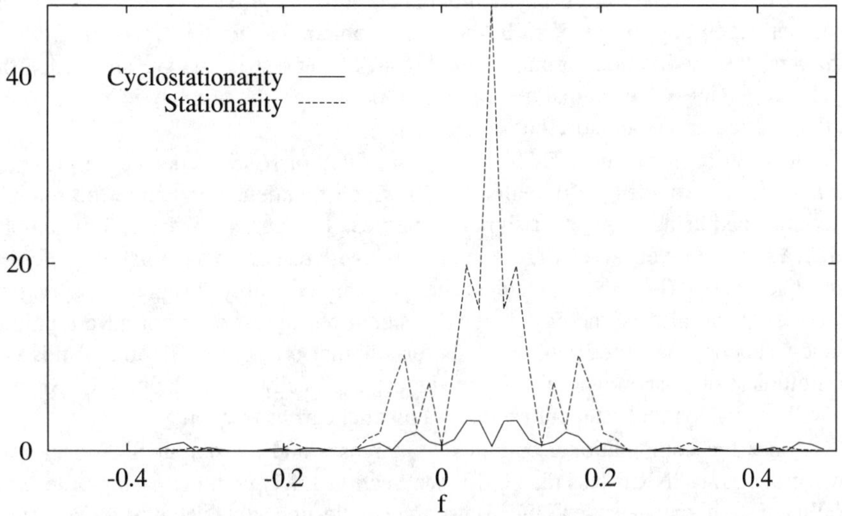

Figure 9: The fourth-order CP for a binary PAM signal under the assumptions stationarity and cyclostationarity.

8 APPLICATIONS

In general, the applications of the theory of HOCS are similar to those for the theory of SOCS and/or the theory of HOS, namely weak-signal detection, signal parameter estimation, signal waveform extraction, system identification and equalization, and array-based direction finding and blind-adaptive spatial filtering. The use of cyclic cumulants or cyclic polyspectra rather than conventional cumulants or polyspectra can be advantageous when the corrupting signals are not Gaussian (man-made signals are rarely Gaussian), and the use of higher-order cyclic parameters rather than second-order cyclic parameters can be advantageous when the signal does not exhibit second-order polycyclostationarity, or when there are no cycle frequencies unique to the signal of interest for order two, but there are unique cycle frequencies for a higher order.

8.1 Weak-Signal Detection

In this section, we consider the problem of detecting the presence of a desired signal (or signals) $s(t)$ in a received data set $x(t)$ with noise and interference $n(t)$,

$$x(t) = s(t) + n(t), \quad -T/2 \leq t \leq T/2.$$

Several versions of the detection problem are of interest. The first is called the *general search problem*, in which a data set is analyzed to determine if there are any polycyclostationary signals present. No information about the received data is assumed in the general search problem. In the second problem, called the *known-cycle-frequency problem*, a specific pure cycle frequency/order pair (β, n_0) is given and the data set is analyzed in an attempt to determine if there is a signal present in the data corresponding to this pair. In the third problem, called the *known-modulation problem*, the modulation format of the signal of interest is known, and hence the cyclic cumulants of the signal are known (in principle); the problem is to determine if that particular signal is present in the data.

Motivation for using HOCS, instead of SOCS or HOS, to detect the presence of a signal is provided by the situation in which this detection cannot be reliably accomplished using these other signal properties. This can be the case, for example, when a signal has very weak SOCS and is weak with respect to the noise and interference background (low SNR), which is changing unpredictably during the observation interval. It can also be the case when the signal of interest does not have a unique cycle frequency for order two, but does for a higher order. An example of this is a communication system wherein all the signals are spectrally overlapping QPSK and have the same symbol rate, but each has a distinct carrier frequency.

Since the computational complexity increases and the output SNR decreases (when the input SNR is less than 0 dB, see Section 1.4.1) with increasing order n of nonlinearity, it is always desirable to use the smallest possible value of n. This typically corresponds to the lowest order of (substantial) cyclostationarity of the signal. Accordingly, it is often the case in the known-modulation problem that the CTCF and CTMF are (approximately) equal for the chosen value of n, $R_s^{\alpha}(\boldsymbol{\tau})_n = C_s^{\alpha}(\boldsymbol{\tau})_n$. In addition, if the noise and interference (hereafter referred to as the environment) is such that $R_n^{\beta}(\boldsymbol{\tau})_n = 0$, where β is the pure cycle frequency of interest in $s(t)$, in which case $R_x^{\beta}(\boldsymbol{\tau})_n = C_s^{\beta}(\boldsymbol{\tau})_n$, then any detection method that requires an estimate of the CTCF for $s(t)$ can be implemented by using an estimate of the CTMF for $x(t)$, thereby reducing the computational complexity of the method. However, if the environment is unknown, it is best to use estimates of the CTCF so that potential lower-order sine-wave interactions can be avoided. Let us now turn to an examination of each of the three detection problems.

8.1.1 The General Search Problem

In the general search problem there is a maximum order N that is to be used for processing. The goal of the processing is to produce a list of pure cycle frequencies $\{\beta_n\}$ for each value of n from 1 to N. The list $\{\beta_n\}$ should characterize the detectable cyclostationarity of order n (and only n) that is associated with $x(t)$. Thus, these lists should not be contaminated by entries that are due to lower-order sine-wave interactions. To accomplish this task, the TCF for $x(t)$ is estimated for each order n. From this estimate, the cycle frequencies $\{\beta_n\}$, which are needed for the estimate of the TCF for order $n+1$, can be found. More explicitly, the general search problem can be tackled using the following algorithm:

0. Let $n = 1$
1. Compute $\hat{C}'_x(t, \boldsymbol{\tau})_n = L_x(t, \boldsymbol{\tau})_n - \sum_{\substack{P_n \\ p \neq 1}}{}' \prod_{j=1}^{p} \hat{C}_x(t, \boldsymbol{\tau}_{\nu_j})_{n_j}$

2. Compute $Y(f) = \text{FFT}_t\ \{\hat{C}'_x(t, \tau)_n\}$
3. Threshold detect the bins of Y to find $\{\beta_n\}$
4. Compute $\hat{C}_x^{\beta_n}(\tau)_n = \left\langle \hat{C}'_x(t, \tau)_n\, e^{-i2\pi\beta_n t} \right\rangle$
5. Compute $\hat{C}_x(t, \tau)_n = \sum_{\beta_n} \hat{C}_x^{\beta_n}(\tau)_n e^{i2\pi\beta_n t}$
6. $n \to n+1$; if $n \leq N$ then go to 1.

In step 4, the interval over which the average $\langle\cdot\rangle$ is performed is determined by the amount of data $x(t)$ available. If any of the detected cycle frequencies are of particular interest, a cyclic polyspectral analysis can be performed from which the modulation type can possibly be determined. Also, a simple analysis of the relationship between the estimated cycle frequencies for the different values of n can be used to advantage for modulation recognition.

8.1.2 The Known-Cycle-Frequency Problem

In this problem, one or more of the signal's modulation frequencies are known, such as a symbol rate or carrier frequency, but the shape of the CTCF is unknown. The environment is still assumed to be unknown and, therefore, the general search algorithm is still of interest. However, it can be improved for the known-cycle-frequency problem by combining it with a least-squares estimation technique. Let (β, n_0) be the cycle frequency/order pair of interest. Use the general search algorithm up to order $n_0 - 1$. Form $\hat{C}'_x(t, \tau)_{n_0}$, and use a least-squares estimator to detect the presence of the signal of interest using the statistic

$$\begin{aligned} Y &= \left\langle \hat{\boldsymbol{w}}^\dagger \hat{\boldsymbol{C}}'_x(t, \boldsymbol{\tau})_{n_0}\, e^{-i2\pi\beta t} \right\rangle \\ &= \hat{\boldsymbol{w}}^\dagger \hat{\boldsymbol{C}}_x^{\beta}(\boldsymbol{\tau})_{n_0}, \end{aligned}$$

where

$$\hat{\boldsymbol{C}}'_x(t, \boldsymbol{\tau})_{n_0} = \left[\hat{C}'_x(t, \boldsymbol{\tau}_1)_{n_0} \cdots \hat{C}'_x(t, \boldsymbol{\tau}_K)_{n_0}\right]^\dagger$$

$$\hat{\boldsymbol{C}}_x^{\beta}(\boldsymbol{\tau})_{n_0} = \left[\hat{C}_x^{\beta}(\boldsymbol{\tau}_1)_{n_0} \cdots \hat{C}_x^{\beta}(\boldsymbol{\tau}_K)_{n_0}\right]^\dagger$$

and where $\hat{\boldsymbol{w}}$ is the unit-norm version of the least-squares weight vector

$$\hat{\boldsymbol{w}} = \arg\min_{\boldsymbol{w}} \left\langle \left| \boldsymbol{w}^\dagger \hat{\boldsymbol{C}}'_x(t, \boldsymbol{\tau})_{n_0} - e^{i2\pi\beta t} \right|^2 \right\rangle. \tag{107}$$

The solution to (107) is

$$\hat{\boldsymbol{w}} = \boldsymbol{R}^{-1} \hat{\boldsymbol{C}}_x^{\beta}(\boldsymbol{\tau})_{n_0}$$

where

$$\boldsymbol{R} = \left\langle \hat{\boldsymbol{C}}'_x(t, \boldsymbol{\tau})_{n_0} \hat{\boldsymbol{C}}'_x(t, \boldsymbol{\tau})_{n_0}^H \right\rangle,$$

in which H denotes conjugate transpose. Thus, the detection statistic is

$$Y = \hat{\boldsymbol{C}}_x^{\beta}(\boldsymbol{\tau})_{n_0}^H \boldsymbol{R}^{-1} \hat{\boldsymbol{C}}_x^{\beta}(\boldsymbol{\tau})_{n_0}.$$

This detection statistic is obtained by forming the particular linear combination of data sets $\hat{C}'(t, \tau_1)_{n_0}, \cdots, \hat{C}'(t, \tau_K)_{n_0}$ that optimally combines the regenerated sine waves with frequency β present in each set, and then correlates this composite regenerated sine wave with the stored sine wave $e^{i2\pi\beta t}$.

8.1.3 The Known-Modulation Problem

In this problem, the data is analyzed to determine if it contains a signal with known modulation type. In particular, we know the CTCF of $s(t)$ for $n = n_0$ and pure cycle frequency β. The general search algorithm can be used to remove all lower-order sine waves up to order $n_0 - 1$. Then, from $\hat{C}'_x(t, \tau)_{n_0}$ the CTCF estimate $\bar{C}^{\beta}_{x_T}(\boldsymbol{u})_{n_0}$ for cycle frequency β can be determined. The proposed detection statistic is the correlation of measured and ideal CTCFs

$$Y = \int_{-\infty}^{\infty} \cdots \int_{-\infty}^{\infty} \bar{C}^{\beta}_{x_T}(\boldsymbol{u})_{n_0} \bar{C}^{\beta}_{s}(\boldsymbol{u})^*_{n_0} \, d\boldsymbol{u}.$$

The primary justification for this particular statistic is that when no signal is present with nth-order pure cycle frequency β, then $\bar{C}^{\beta}_{x_T}(\boldsymbol{u})_{n_0} \to 0$, which implies that $Y \to 0$; when the signal of interest is present, then

$$Y \to \int_{-\infty}^{\infty} \cdots \int_{-\infty}^{\infty} \left|\bar{C}^{\beta}_{s}(\boldsymbol{u})_n\right|^2 \, d\boldsymbol{u}. \tag{108}$$

Thus, Y is an asymptotically noise-free statistic on both the signal-present and signal-absent hypotheses. (Furthermore, the integral (108) is finite [107].) Hence, this statistic is the natural generalization of the single-cycle detector that exploits second-order polycyclostationarity [35]

$$Y_{SCD} = \int_{-\infty}^{\infty} R^{\alpha}_{x_T}(\tau) R^{\alpha}_{s}(\tau)^* \, d\tau = \int_{-\infty}^{\infty} S^{\alpha}_{X_T}(f) S^{\alpha}_{s}(f)^* \, df, \tag{109}$$

which has several optimality properties [39] and has been shown to outperform radiometric (energy) detectors for weak polycyclostationary signals in time-varying environments [46]. In (109), $R^{\alpha}_{s}(\tau)$ is the cyclic autocorrelation for $s(t)$, $R^{\alpha}_{x_T}(\tau)$ is the cyclic autocorrelation estimate (cyclic correlogram), $S^{\alpha}_{X_T}(f)$ is the cyclic periodogram, and $S^{\alpha}_{s}(f)$ is the cyclic spectrum for $s(t)$ [35].

The detection statistic Y can be generalized to include only a portion of $\boldsymbol{u}$-space, denoted by $G \subset \mathbf{R}^{n_0}$,

$$Y = \int_G \bar{C}^{\beta}_{x_T}(\boldsymbol{u})_{n_0} \bar{C}^{\beta}_{s}(\boldsymbol{u})^*_{n_0} \, d\boldsymbol{u}.$$

Choices for G might include those values of $\boldsymbol{u}$ for which the RD-CTCF $\bar{C}^{\beta}_{s}(\boldsymbol{u})_{n_0}$ is particularly large, or for which the coefficient of variation (variance divided by squared mean) of the estimator $\bar{C}^{\beta}_{x_T}(\boldsymbol{u})_{n_0}$ of the RD-CTCF is particularly small.

8.2 Time-Delay Estimation

The received signal model for the time-delay estimation problem is given by

$$x(t) = s(t) + n(t)$$

$$y(t) = As(t + d) + m(t).$$

It is desired to estimate the time-delay parameter d, which is sometimes called the time-difference-of-arrival (TDOA) for $s(t)$. It is assumed that $s(t)$ is nth-order polycyclostationary with cycle frequency β. The time-series $n(t)$ and $m(t)$ consist of arbitrary noise and interfering signals, except that it is assumed that neither $n(t)$ nor $m(t)$ is nth-order polycyclostationary with the pure cycle frequency β.

Conventional approaches to this problem (those that do not exploit cyclostationarity) can be collectively referred to as *generalized cross-correlation* (GCC) methods [72]. In the GCC methods, filtered versions of the sensor outputs $x(t)$ and $y(t)$ are cross correlated, and the estimate of d is taken to be the location of the peak in the cross-correlation estimate. These methods suffer when $n(t)$ and $m(t)$ contain signals common to both, each with its own TDOA, because each such signal contributes a peak of its own to the cross-correlation function. This causes two problems. The first is a resolution problem which, to be solved, requires that the differences in the TDOAs for each of the signals be greater than the widths of the cross-correlation functions so that the peaks can be resolved. The second problem is that it is difficult to correctly associate each peak with its corresponding signal. Both of these problems arise because the GCC methods are not signal selective; they produce TDOA peaks for all the signals in the received data unless they are spectrally disjoint and can, therefore, be separated by filtering. Signal-selective methods that exploit the SOCS of the desired signal, which is assumed to be unique to that signal, are studied in [49, 50]. These methods have been shown to outperform the GCC methods and have been shown to produce unbiased TDOA estimates with variance that is smaller than the Cramer-Rao lower bound on the variance of TDOA estimators that are based on the assumption that the signal and environment are stationary. However, these methods do not apply when there is no SOCS to exploit. In this case, we can turn to HOCS in order to develop signal-selective TDOA estimators.

Following the approach in [49] for SOCS, the general methodology considered here for HOCS is least-squares estimation. The following two examples illustrate the methodology. To keep the notation simple only real-valued signals are considered.

Let us define a cross cumulant as follows

$$C_{xy}(t, \boldsymbol{\tau})_n \triangleq \text{Cumulant}\,\{x(t + \tau_1) \cdots x(t + \tau_{n-1})\; y(t + \tau_n)\}.$$

The cyclic component of this cross cumulant for the signal-specific cycle frequency β is

$$\begin{aligned} C_{xy}^{\beta}(\boldsymbol{\tau})_n &\triangleq \left\langle C_{xy}(t,\boldsymbol{\tau})_n e^{-i2\pi\beta t}\right\rangle \\ &= AC_s^{\beta}(\boldsymbol{\tau}+\boldsymbol{\delta}_n d)_n, \end{aligned}$$

where $\boldsymbol{\delta}_n$ is the unit vector along the nth coordinate. It is easy to show that the following relations involving RD-CTCFs hold:

$$\bar{C}_{xy}^{\beta}(\boldsymbol{u})_n = A\bar{C}_s^{\beta}(\boldsymbol{u}-\mathbf{1}d)_n e^{i2\pi\beta d},$$

$$\bar{C}_x^{\beta}(\boldsymbol{u})_n = \bar{C}_s^{\beta}(\boldsymbol{u})_n.$$

Thus, we can do a least-squares fit of a measurement of $\bar{C}_{xy}^{\beta}$ to a measurement of $\bar{C}_x^{\beta}$ over a region G of $\boldsymbol{u}$-space of interest:

$$\min_{A,\hat{d}} \int_G \left|\bar{C}_{xy_T}^{\beta}(\boldsymbol{u})_n - A\bar{C}_{x_T}^{\beta}(\boldsymbol{u}-\mathbf{1}\hat{d})_n e^{i2\pi\beta\hat{d}}\right|^2 d\boldsymbol{u}.$$

This leads (cf. [49]) to the following estimator of the delay d:

$$\hat{d}_1 = \arg\max_{\hat{d}} \Re\left\{\int_G \bar{C}_{x_T}^{\beta}(\boldsymbol{u})_n \bar{C}_{xy_T}^{\beta}(\boldsymbol{u}+\mathbf{1}\hat{d})_n^* e^{i2\pi\beta\hat{d}}\, d\boldsymbol{u}\right\},$$

where *arg max* means the value at which the maximum occurs, and $\Re\{\cdot\}$ means the real part. This estimator is a higher-order generalization of the spectral coherence alignment algorithm for TDOA estimation [49], which exploits second-order polycyclostationarity and has been shown to possess several optimality properties [47]. If the data is complex, the estimator $\hat{d}_1$ is modified by replacing the operation $\Re\{\cdot\}$ with the magnitude operation.

As an alternative, we can avoid using cross-sensor measurements entirely by noting that

$$\bar{C}_y^{\beta}(\boldsymbol{u})_n = A^n\bar{C}_s^{\beta}(\boldsymbol{u})_n e^{i2\pi\beta d},$$

and

$$\bar{C}_x^{\beta}(\boldsymbol{u})_n = \bar{C}_s^{\beta}(\boldsymbol{u})_n,$$

which suggests the following least-squares approach:

$$\hat{d}_2 = \arg\min_{A,\hat{d}} \int_G \left|\bar{C}_{y_T}^{\beta}(\boldsymbol{u})_n - A^n\bar{C}_{x_T}^{\beta}(\boldsymbol{u})_n e^{i2\pi\beta\hat{d}}\right|^2 d\boldsymbol{u}.$$

The estimator for d is given explicitly by

$$\hat{d}_2 = \frac{-1}{2\pi\beta}\text{angle}\left\{\int_G \bar{C}_{x_T}^{\beta}(\boldsymbol{u})_n \bar{C}_{y_T}^{\beta}(\boldsymbol{u})_n^*\, d\boldsymbol{u}\right\},$$

which is a higher-order generalization of the second-order cyclic phase difference algorithm for TDOA estimation without cross-sensor measurements [49]. This algorithm has the drawback that the TDOA can be properly estimated only if it has magnitude smaller than $\frac{1}{2\beta}$ because the TDOA estimate is a function of an angle estimate.

9 CONCLUSIONS

The theory of the higher-order statistics of polycyclostationary signals, called higher-order cyclostationarity, holds greatest promise for applications involving statistical inference and decision using data from a multiple-signal environment. This is because the higher-order cyclic cumulants possess the property of signal selectivity, which means that these cumulants can be used to extract information about each of the signals in turn, without (that is, asymptotically) interference from the other signals. This can be accomplished if each signal in the environment possesses a unique cycle frequency, for instance, a baud rate, chip rate, or carrier frequency.

The theory also shows promise as a general theoretical framework for performance analyses of signal-processing systems that employ second- or higher-order nonlinear transformations of polycyclostationarity input signals.

Finally, since the theory is fundamentally concerned with sine-wave generation, it holds promise for fruitful application to problems in the area of symbol and carrier synchronization.

The primary difficulties with higher-order statistics are computational in nature. First, for a given data set, higher statistics require more operations to compute than do second-order statistics. Second, the relevant estimators often have larger variance for the same amount of data when compared to estimators of lower-order statistics, and so need more data to achieve acceptable levels of estimation accuracy. The severity of these difficulties may be lessened in the future as hardware speed increases and more efficient estimation methods are found.

ACKNOWLEDGMENTS

The novel research reported in this chapter was carried out jointly with Professor William A. Gardner. The author is also grateful for Professor Gardner's editorial assistance during the preparation of this chapter. This work was supported jointly by the National Science Foundation under grants MIP–88–12902 and MIP–91–12800, PI: W. A. Gardner, and by the United States Army Research Office under contract DAAL03–91–C–0018, PI: W. A. Gardner.

APPENDIX PURE SINE WAVES AND TEMPORAL CUMULANTS APPENDIX

In Section 2.3, the temporal cumulant function is derived by considering the problem of pure sine-wave generation. That is, it is found that the strengths of pure nth-order sine waves are characterized by the CTCF. However, we have found that some time-series possess a degeneracy such that, for certain values of the lag vector $\boldsymbol{\tau}$, the

CTCF is nonzero when there are no pure sine waves present in the lag product. This degeneracy is illustrated and explained here for the case of $n = 4$.

Consider the fourth-order lag product

$$L_x(t, \boldsymbol{\tau})_4 = x(t + \tau_1)x(t + \tau_2)x(t + \tau_3)x(t + \tau_4) \tag{110}$$

for the binary (± 1) PAM signal (87) with rectangular pulses,

$$p(t) = \begin{cases} 1, & |t| \leq T_0/2 \\ 0, & \text{otherwise.} \end{cases}$$

There can be sine waves associated with the factors

$$f_1(t) = x(t + \tau_1)x(t + \tau_2),$$

and

$$f_2(t) = x(t + \tau_3)x(t + \tau_4)$$

that make up $L_x(t, \boldsymbol{\tau})_4$, and if so, the product sine waves

$$\hat{E}^{\{\alpha\}}\{f_1(t)\}\,\hat{E}^{\{\alpha\}}\{f_2(t)\}$$

are subtracted, along with product sine waves corresponding to all other unique factorizations of $L_x(t, \boldsymbol{\tau})_4$, from the moment $\hat{E}^{\{\alpha\}}\{L_x(t, \boldsymbol{\tau})_4\}$ to obtain the pure fourth-order sine waves (cf. (34)). Thus, a pure fourth-order sine wave contains no products of lower-order sine waves, and cannot be equal to such a product; this is simply the intuitive notion of a pure fourth-order sine wave. To make clear the degeneracy that we have alluded to, $\tau_3 = \tau_4$ is chosen in (110),

$$L_x(t, \boldsymbol{\tau})_4 = x(t + \tau_1)x(t + \tau_2)x^2(t + \tau_3). \tag{111}$$

In this case, since $x^2(t + \tau_3) \equiv 1$, we have the equivalence

$$L_x(t, \boldsymbol{\tau})_4 = [x(t + \tau_1)x(t + \tau_2)][1] = L_x(t, \tau_1, \tau_2)_2.$$

Thus, $L_x(t, \boldsymbol{\tau})_4$ contains nothing other than products of second-order sine waves, namely the sine waves in $x(t+\tau_1)x(t+\tau_2)$ multiplied by the second-order sine wave with frequency zero given by $x^2(t + \tau_3) \equiv 1$. Consequently, $L_x(t, \boldsymbol{\tau})_4$ can contain no pure fourth-order sine waves. However, from (95) we have the CTCF

$$C_x^{\beta}(\boldsymbol{\tau})_4 = \frac{C_{a,4}}{T_0}\int_{-\infty}^{\infty} p(t + \tau_1)p(t + \tau_2)p^2(t + \tau_3)e^{-i2\pi\beta t}\,dt\,e^{i2\pi\beta t_0}, \quad \beta = k/T_0,$$

which is not zero for all τ_1, τ_2, τ_3. Another way to see that the cumulant is nonzero is to use (34) to compute the TCF:

$$\begin{aligned} C_x(t, \boldsymbol{\tau})_4 &= \hat{E}^{\{\alpha\}}\{L_x(t, \boldsymbol{\tau})_4\} - \hat{E}^{\{\alpha\}}\{x(t + \tau_1)x(t + \tau_2)\}\,\hat{E}^{\{\alpha\}}\{x(t + \tau_3)x(t + \tau_3)\} \\ &\quad - \hat{E}^{\{\alpha\}}\{x(t + \tau_1)x(t + \tau_3)\}\,\hat{E}^{\{\alpha\}}\{x(t + \tau_2)x(t + \tau_3)\} \\ &\quad - \hat{E}^{\{\alpha\}}\{x(t + \tau_1)x(t + \tau_3)\}\,\hat{E}^{\{\alpha\}}\{x(t + \tau_2)x(t + \tau_3)\} \\ &= -2\hat{E}^{\{\alpha\}}\{x(t + \tau_1)x(t + \tau_3)\}\,\hat{E}^{\{\alpha\}}\{x(t + \tau_2)x(t + \tau_3)\}. \end{aligned}$$

It is, in fact, reasonable to expect a nonzero result for the fourth-order TCF of the set $\{x(t+\tau_1), x(t+\tau_2), x(t+\tau_3), x(t+\tau_3)\}$ since it contains no proper subset that is statistically independent of the remaining elements (cf. Section 2.4.3).

However, since there is a repeated factor in (111), we can treat the lag product as the product of three, rather than four, factors $\{x(t+\tau_1), x(t+\tau_2), x^2(t+\tau_3)\}$. In this case, there *is* an independent subset among the *three* variables and

$$\hat{E}^{\{\alpha\}}\left\{x(t+\tau_1)x(t+\tau_2)x^2(t+\tau_3)\right\} \equiv \hat{E}^{\{\alpha\}}\left\{x(t+\tau_1)x(t+\tau_2)\right\}\hat{E}^{\{\alpha\}}\left\{x^2(t+\tau_3)\right\}.$$

The third-order cumulant for these three variables is zero. However, the presence of a repeated factor in the lag product is not in itself the cause of the failure of the pure-sine-wave interpretation of the TCF to hold. It is the fact that the repeated factor is degenerate:

$$x^2(t+\tau_3) \equiv 1.$$

This fact leads us to conjecture that the only degenerate time-series (those for which the nth-order TCF is not zero, yet there are clearly no pure nth order sine waves) are those piece-wise constant signals with all values equal to the mth roots of unity (to within a single factor, that is, the roots of K will do). In such a case,

$$x^m(t+\tau_0) \equiv 1 \quad \text{for all } \tau_0,$$

and the same fix applies: The $(n-m+1)$th-order cumulant for the variables

$$\{x(t+\tau_1),\ x(t+\tau_2), \cdots, x(t+\tau_{n-m}),\ x^m(t+\tau_{n-m+1})\}$$

is zero, which is in agreement with the absence of any pure sine waves.

REFERENCES

[1] B. G. Agee, S. V. Schell, and W. A. Gardner. "Spectral Self-Coherence Restoral: A New Approach to Blind Adaptive Signal Extraction Using Antenna Arrays," *Proceedings of the IEEE,* Vol. 78, pp. 753–767, 1990.

[2] V. G. Alekseev. "Symmetry Properties of High-Order Spectral Densities of Stationary and Periodic-Nonstationary Stochastic Processes," *Problemy Peredachi Informatsii,* Vol. 23, No. 3, pp. 48–53, 1987.

[3] G. E. Andrews. *The Theory of Partitions,* Reading, MA: Addison-Wesley, 1976.

[4] J. S. Bendat and A. G. Piersol. *Engineering Applications of Correlation and Spectral Analysis,* New York: Wiley, 1980.

[5] G. Bilardi and S. Pupolin. "Spectral Analysis of the Powers of a PAM Digital Signal," *Alta Frequenza,* Vol. LIII, No. 2, pp. 70–76, March 1984.

[6] N. M. Blachman and S. H. Mousavinezhad. "The Spectrum of the Square of a Synchronous Random Pulse Train," *IEEE Transactions on Communications,* Vol. 38, No. 1, pp. 13–17, January 1990.

[7] A. Blanc-Lapierre and R. Fortet. *Theory of Random Functions, Volume II,* New York: Gordon and Breach, 1968.

[8] D. R. Brillinger. *Time Series: Data Analysis and Theory,* New York: Holt, Rinehart, and Winston, 1975.

[9] D. R. Brillinger. "An Introduction to Polyspectra," *Annals of Mathematical Statistics,* Vol. 36, pp. 1351–1374, 1965.

[10] D. R. Brillinger and M. Rosenblatt. "Asymptotic Theory of Estimates of k-Th Order Spectra," in *Spectral Analysis of Time Series,* B. Harris, ed., New York: Wiley, 1967.

[11] D. R. Brillinger and M. Rosenblatt. "Computation and Interpretation of k-Th Order Spectra," in *Spectral Analysis of Time Series,* B. Harris, ed., New York: Wiley, 1967.

[12] H-H. Chiang and C. L. Nikias. "A New Method for Adaptive Time Delay Estimation for Non-Gaussian Signals," *IEEE Transactions on Signal Processing,* Vol. 38, No. 2, pp. 209–219, February 1990.

[13] C. Corduneanu. *Almost Periodic Functions,* 2nd Ed., New York: Chelsea Publishing Co., 1989.

[14] E. A. Cornish and R. A. Fisher. "Moments and Cumulants in the Specification of Distributions," *Revue International Stat. Inst.,* Vol. 5, pp. 307–320, 1937.

[15] H. Cramer. *Random Variables and Probability Distributions,* London: Cambridge University Press, 1937.

[16] H. Cramer. *Mathematical Methods of Statistics,* Princeton, NJ: Princeton University Press, 1946.

[17] A. V. Dandawate and G. B. Giannakis. "Polyspectral Analysis of Non-Stationary Signals: System Identification, Classification and Ambiguity Functions," *Proceedings of the International Signal Processing Workshop on Higher Order Statistics,* Chamrousse, France, pp. 147–150, July 10–12, 1991.

[18] A. V. Dandawate and G. B. Giannakis. "Ergodicity Results for Non-Stationary Signals: Cumulants, Ambiguity Functions and Wavelets," *Proceedings of the 25th Annual Conference on Information Sciences and Systems,* Johns Hopkins University, Baltimore, Maryland, pp. 976–983, March 20–22, 1991.

[19] A. V. Dandawate and G. B. Giannakis. "Nonparametric Polyspectral Analysis of AM Signals and Processes with Missing Observations," in *Proc. of the 26th Conference on Information Sciences and Systems*, Princeton University, NJ, March 18–20, 1992.

[20] A. V. Dandawate and G. B. Giannakis. "Cyclic-cumulant Based Identification of Almost Periodically Time-Varying Systems: Parametric Methods," in *ICASSP 1992*, San Francisco, CA, pp. 229–232, March 23–26, 1992.

[21] A. V. Dandawate and G. B. Giannakis. "Detection and Classification of Cyclostationary Signals using Cyclic-HOS: A Unified Approach," in *Proc. of the Society of Photo-Opt. Instr. Engr., Advanced Signal Processing Algorithms, Architectures, and Implementations,* San Diego, CA, July 1992.

[22] A. V. Dandawate and G. B. Giannakis. "Nonparametric Identification of Linear (Almost) Periodically Time-Varying Systems using Cyclic Polyspectra," in *Proc. Sixth Workshop on Statistical Signal and Array Processing,* Victoria, British Columbia, Canada, pp. 152–155, October 7–9, 1992.

[23] W. B. Davenport and W. L. Root. *Random Signals and Noise,* New York: McGraw-Hill, 1958.

[24] J. L. Doob. *Stochastic Processes,* New York: Wiley, 1953.

[25] R. F. Dwyer. "The Extraction of Range and Doppler from Fourth-Order Spectra," in *Proceedings of the 1990 ICASSP,* pp. 2403–2406, 1990.

[26] T. T. Fang. "Analysis of Self-Noise in a Fourth-Power Clock Regenerator," *IEEE Transactions on Communications,* Vol. 39, No. 1, pp. 133–140, January 1991.

[27] W. Feller. *An Introduction to Probability Theory and Its Applications, Vol. I and II*, 2nd Ed., New York: Wiley, 1957.

[28] A. Fisher. *The Mathematical Theory of Probabilities,* New York: Macmillan, 1923.

[29] R. A. Fisher. "Moments and Product Moments of Sampling Distributions," *Proc. London Math. Soc.* (Series 2), Vol. 30, pp. 199–238, 1930.

[30] M. Fisz. *Probability Theory and Mathematical Statistics*, New York: Wiley, 1963.

[31] P. Forster and C. L. Nikias. "Bearing Estimation in the Bispectrum Domain," *IEEE Transactions on Signal Processing*, Vol. 39, No. 9, pp. 1994–2006, September 1991.

[32] W. A. Gardner. "Stationarizable Random Processes," *IEEE Transactions on Information Theory,* Vol. 24, No. 1, pp. 8–22, 1978.

[33] W. A. Gardner. *Introduction to Random Processes with Applications to Signals and Systems.* New York: Macmillan, 1985 (2nd ed., New York: McGraw-Hill, 1990).

[34] W. A. Gardner. "The Spectral Correlation Theory of Cyclostationary Time-Series," *Signal Processing* [EURASIP], Vol. 11, pp. 13–36, 1986.

[35] W. A. Gardner. *Statistical Spectral Analysis: A Nonprobabilistic Theory*, Englewood Cliffs, NJ: Prentice-Hall, 1987.

[36] W. A. Gardner. "Common Pitfalls in the Application of Stationary Process Theory to Time-Sampled and Modulated Signals," *IEEE Transactions on Communications,* Vol. 35, No. 5, pp. 529–534, May 1987.

[37] W. A. Gardner. "Spectral Correlation of Modulated Signals: Part I – Analog Modulation," *IEEE Transactions on Communications,* Vol. 35, No. 6, pp. 584–594, June 1987.

[38] W. A. Gardner, W. A. Brown, and C-K. Chen. "Spectral Correlation of Modulated Signals: Part II – Digital Modulation," *IEEE Transactions on Communications,* Vol. 35, No. 6, pp. 595–601, June 1987.

[39] W. A. Gardner. "Signal Interception: A Unifying Theoretical Framework for Feature Detection," *IEEE Transactions on Communications,* Vol. 36, No. 8, pp. 897–906, August 1988.

[40] W. A. Gardner. "Spectral Characterization of N-th Order Cyclostationarity," in *Proceedings of the IEEE/ASSP Workshop on Spectrum Estimation,* pp. 251–255, New York, 1990.

[41] W. A. Gardner. "Identification of Systems with Cyclostationary Input and Correlated Input/Output Measurement Noise," *IEEE Transactions on Automatic Control,* Vol. 35, No. 4, pp. 449–452, April 1990.

[42] W. A. Gardner and C. M. Spooner. "Higher-Order Cyclostationarity, Cyclic Cumulants, and Cyclic Polyspectra," in *Proceedings of the International Symposium on Information Theory and Its Applications (ISITA),* pp. 355–358, Honolulu, Hawaii, November 1990.

[43] W. A. Gardner. "Two Alternative Philosophies for Estimation of the Parameters of Time-Series," *IEEE Transactions on Information Theory,* Vol. 37, No. 1, pp. 216–218, January 1991.

[44] W. A. Gardner. "Exploitation of Spectral Redundancy in Cyclostationary Signals," *IEEE Signal Processing Magazine,* Vol. 8, No. 2, pp. 14–36, April 1991.

[45] W. A. Gardner and W. A. Brown. "Fraction-of-Time Probability for Time-Series that Exhibit Cyclostationarity," *Signal Processing* [EURASIP], Vol. 23, No. 3, pp. 273–292, June 1991.

[46] W. A. Gardner and C. M. Spooner. "Signal Interception: Performance Advantages of Cyclic-Feature Detectors," *IEEE Transactions on Communications,* Vol. 40, No. 1, pp. 149–159, January 1992.

[47] W. A. Gardner and C. M. Spooner. "Weak-Signal Detection and Source Location: Simplifications of the Maximum-Likelihood Receiver," *IEEE Transactions on Communications,* Vol. 41, No. 6, pp. 905–916, June 1993.

[48] W. A. Gardner. "Cyclic Wiener Filtering: Theory and Method," *IEEE Transactions on Communications,* Vol. 41, pp. 151–163, January 1993.

[49] W. A. Gardner and C. K. Chen. "Signal-Selective Time-Difference-of-Arrival Estimation for Passive Location of Man-made Signal Sources in Highly Corruptive Environments. Part I: Theory and Method," *IEEE Transactions on Signal Processing,* Vol. 40, pp. 1168–1184, May 1992.

[50] C. K. Chen and W. A. Gardner. "Signal-Selective Time-Difference-of-Arrival Estimation for Passive Location of Man-made Signal Sources in Highly Corruptive Environments. Part II: Algorithms and Performance," *IEEE Transactions on Signal Processing,* Vol. 40, pp. 1185–1197, May 1992.

[51] W. A. Gardner and T. L. Archer. "Exploitation of Cyclostationarity for Identifying the Volterra Kernels of Nonlinear Systems," *IEEE Transactions on Information Theory,* Vol. 39, pp. 535–542, March 1993.

[52] W. A. Gardner. "An Introduction to Cyclostationary Signals," Chapter 1 in this volume, *Cyclostationarity in Communications and Signal Processing*, W. A. Gardner, ed., New York: IEEE Press, 1994.

[53] W. A. Gardner and C. M. Spooner. "Cyclostationary Signal Processing," in *Digital Signal Processing Techniques and Applications,* C. T. Leondes, ed., New York: Academic Press, 1994 (in press).

[54] G. B. Giannakis and M. K. Tsatsanis. "Signal Detection and Classification Using Matched Filtering and Higher Order Statistics," *IEEE Transactions on Acoustics, Speech, and Signal Processing,* Vol. 38, No. 7, pp. 1284–1296, July 1990.

[55] G. B. Giannakis and A. V. Dandawate. "Polyspectral Analysis of Non-Stationary Signals: Bases, Consistency and HOS-WV," *Proceedings of the International Signal Processing Workshop on Higher Order Statistics,* Chamrousse, France, pp. 167–170, July 10–12, 1991.

[56] G. B. Giannakis and A. V. Dandawate. "Polyspectral Analysis of (Almost) Cyclostationary Signals: LPTV System Identification and Related Applications," in *Proceedings of the Twenty-Fifth Asilomar Conference on Signals, Systems and Computers*, Pacific Grove, CA, pp. 377–382, November 4-6, 1991.

[57] I. I. Gikhman and A. V. Skorokhod. *Introduction to the Theory of Random Processes,* Philadelphia: W. B. Saunders, 1969.

[58] E. J. Hannan. *Time-Series Analysis,* New York: Wiley, 1960.

[59] M. J. Hinich. "Testing for Gaussianity and Linearity of a Stationary Time Series," *Journal of Time Series Analysis,* Vol. 3, No. 3, pp. 169–176, 1982.

[60] M. J. Hinich and G. R. Wilson. "Time Delay Estimation Using the Cross Bispectrum," *IEEE Transactions on Signal Processing,* Vol. 40, No. 1, pp. 106–113, January 1992.

[61] P. J. Huber, B. Kleiner, T. Glasser, and G. Dumermuth. "Statistical Methods for Investigating Phase Relations in Stationary Stochastic Processes," *IEEE Transactions on Audio and Electroacoustics,* Vol. 19, No. 1, pp. 78–86, March 1971.

[62] H. L. Hurd. "Nonparametric Time Series Analysis for Periodically Correlated Processes," *IEEE Transactions on Information Theory,* Vol. 35, No. 2, pp. 350–359, March 1989.

[63] H. L. Hurd. "Correlation Theory of Almost Periodically Correlated Processes," *Journal of Multivariate Analysis,* Vol. 37, pp. 24–45, 1991.

[64] G. M. Jenkins and D. G. Watts. *Spectral Analysis and Its Applications,* San Francisco: Holden-Day, 1968.

[65] S. Karlin and H. M. Taylor. *A First Course in Stochastic Processes,* New York: Academic Press, 1975.

[66] S. Karlin and H. M. Taylor. *A Second Course in Stochastic Processes,* New York: Academic Press, 1981.

[67] S. M. Kay. *Modern Spectral Estimation,* Englewood Cliffs, NJ: Prentice-Hall, 1988.

[68] M. G. Kendall and A. Stuart. *The Advanced Theory of Statistics: Volume 1, Distribution Theory,* London: Griffin, 1958.

[69] Y. C. Kim and E. J. Powers. "Digital Bispectral Analysis of Self-excited Fluctuation Spectra," *Physics of Fluids,* Vol. 21, No. 8, pp. 1452–1453, August 1978.

[70] Y. C. Kim, J. M. Beall, and E. J. Powers. "Bispectrum and Nonlinear Wave Coupling," *Physics of Fluids,* Vol. 23, No. 2, pp. 258–263, February 1980.

[71] D. Kletter and H. Messer. "Suboptimal Detection of Non-Gaussian Signals by Third-Order Spectral Analysis," *IEEE Transactions on Acoustics, Speech, and Signal Processing,* Vol. 38, No. 6, pp. 901–909, June 1990.

[72] C. H. Knapp and G. C. Carter. "The Generalized Correlation Method for Estimation of Time Delay," *IEEE Transactions on Acoustics, Speech, and Signal Processing,* Vol. 24, No. 4, August 1976.

[73] R. L. Graham, D. E. Knuth, and O. Patashnik. *Concrete Mathematics,* Reading, MA: Addison-Wesley, 1989.

[74] L. H. Koopmans. *The Spectral Theory of Time Series,* New York: Academic Press, 1974.

[75] P. I. Kuznetsov, R. L. Stratonovich, and V. I. Tikhonov. "The Transmission of Certain Random Functions Through Linear Systems," in *Non-linear Transformations of Stochastic Processes,* P. I. Kuznetsov, ed., Oxford: Pergamon Press, 1965.

[76] P. I. Kuznetsov, R. L. Stratonovich, and V. I. Tikhonov. "The Transmission of Random Functions Through Non-Linear Systems," in *Non-Linear Transformations of Stochastic Processes,* P. I. Kuznetsov, ed., Oxford: Pergamon Press, 1965.

[77] E. L. Lehmann. Testing Statistical Hypothesis, New York: Wiley, 1959.

[78] V. P. Leonov and A. N. Shiryaev. "On a Method of Calculation of Semi-Invariants," *Theory of Probability and Its Applications,* Vol. 4, No. 3, pp. 319–328, 1959.

[79] M. Loeve. *Probability Theory,* New York: Van Nostrand, 1955.

[80] M. Loeve. *Probability Theory II*, 4th Ed., New York: Springer-Verlag, 1963.

[81] J. M. Mendel. "Tutorial on Higher-Order Statistics (Spectra) in Signal Processing and System Theory: Theoretical Results and Some Applications," *Proceedings of the IEEE,* Vol. 79, No. 3, pp. 277–305, March 1991.

[82] D. Middleton. *Statistical Communication Theory,* New York: McGraw-Hill, 1960.

[83] C. L. Nikias and M. R. Raghuveer. "Bispectrum Estimation: A Digital Signal Processing Framework," *Proceedings of the IEEE,* Vol. 75, No. 7, pp. 869–891, July 1987.

[84] C. L. Nikias. "Higher-Order Spectral Analysis," in *Advances in Spectrum Analysis and Array Processing,* S. Haykin, ed., Englewood Cliffs, NJ: Prentice-Hall, 1991.

[85] G. Palm and T. Poggio. "The Volterra Representation and the Wiener Expansion: Validity and Pitfalls," *SIAM Journal of Applied Mathematics,* Vol. 33, No. 2, pp. 195–216, September 1977.

[86] A. Papoulis. *Signal Analysis,* New York: McGraw-Hill, 1977.

[87] A. Papoulis. *Probability, Random Variables, and Stochastic Processes*, 3rd Ed., New York: McGraw-Hill, 1991.

[88] E. Parzen. *Stochastic Processes,* San Francisco: Holden-Day, 1962.

[89] E. Parzen. *Modern Probability Theory and Its Applications,* New York: Wiley, 1960.

[90] A. P. Petropulu and C. L. Nikias. "Signal Reconstruction from the Phase of the Bispectrum," *IEEE Transactions on Signal Processing,* Vol. 40, No. 3, pp. 601–610, March 1992.

[91] B. Porat and B. Friedlander. "Blind Equalization of Digital Communication Channels Using High-Order Moments," *IEEE Transactions on Signal Processing,* Vol. 39, No. 2, pp. 522–526, February 1991.

[92] B. Porat and B. Friedlander. "Direction Finding Algorithms Based on High-Order Statistics," *IEEE Transactions on Signal Processing,* Vol. 39, No. 9, pp. 2016–2024, September 1991.

[93] M. B. Priestley. *Spectral Analysis and Time Series,* London: Academic Press, 1981.

[94] M. B. Priestley. *Non-Linear and Non-Stationary Time Series Analysis,* London: Academic Press, 1988.

[95] V. S. Pugachev. *Theory of Random Functions,* Oxford: Pergamon Press, 1965.

[96] M. R. Raghuveer and C. L. Nikias. "Bispectrum Estimation: A Parametric Approach," *IEEE Transactions on Acoustics, Speech, and Signal Processing,* Vol. 33, No. 4, pp. 1213–1230, October 1985.

[97] M. R. Raghuveer. "Time-Domain Approaches to Quadratic Phase Coupling Estimation," *IEEE Transactions on Automatic Control,* Vol. 35, No. 1, pp. 48–56, January 1990.

[98] H. Raiffa and R. Schlaifer. *Applied Statistical Decision Theory,* Cambridge, MA: Harvard Business School, 1961.

[99] M. Rosenblatt. *Stationary Sequences and Random Fields,* Boston: Birkhauser, 1985.

[100] S. V. Schell and W. A. Gardner. "High Resolution Direction Finding," chapter in *Handbook of Statistics,* Vol. 10, N. K. Bose and C. R. Rao (eds.), Amsterdam: North Holland, 1992.

[101] S. V. Schell. "An Overview of Sensor-Array Processing for Cyclostationary Signals," Chapter 3 in this volume, *Cyclostationarity in Communications and Signal Processing*, W. A. Gardner, ed., New York: IEEE Press, 1994.

[102] M. Schetzen. *The Volterra and Wiener Theories of Nonlinear Systems,* 2nd Ed., Malabar, FL: Krieger Publ. Co., 1989.

[103] A. N. Shiryaev. "Some Problems in the Spectral Theory of Higher-Order Moments—I," *Theory of Probability and Its Applications,* Vol. 5, No. 3, pp. 265–284, 1960.

[104] C. M. Spooner and W. A. Gardner. "Estimation of Cyclic Polyspectra," *Proceedings of the Twenty-Fifth Annual Asilomar Conference on Signals, Systems and Computers,* Pacific Grove, CA, pp. 370–376, November 4-6, 1991.

[105] C. M. Spooner and W. A. Gardner. "Performance Evaluation of Cyclic Polyspectrum Estimators," *Proceedings of the Twenty-Sixth Annual Asilomar Conference on Signals, Systems and Computers,* Pacific Grove, CA, Oct. 26–28, 1992.

[106] C. M. Spooner and W. A. Gardner. "An Overview of the Theory of Higher-Order Cyclostationarity," in *Proceedings of the Workshop on Nonstationary Stochastic Processes and Their Applications,* (A. G. Miamee, ed.), World Scientific, Singapore, 1992.

[107] W. A. Gardner and C. M. Spooner. "The Cumulant Theory of Cyclostationary Time-Series, Part I: Foundation," Technical Report; Dept. of Electrical and Computer Engineering, University of California, CA, Davis, May 1992.

[108] C. M. Spooner and W. A. Gardner. "The Cumulant Theory of Cyclostationary Time-Series, Part II: Development and Applications," Technical Report; Dept. of Electrical and Computer Engineering, University of California, CA, Davis, May 1992.

[109] C. M. Spooner and W. A. Gardner. "Exploitation of Higher-Order Cyclostationarity for Weak-Signal Detection and Time-Delay Estimation," *Proceedings of the Sixth Workshop on Statistical Signal & Array Processing*, Victoria, British Columbia, Canada, pp. 197–201, Oct. 7–9, 1992.

[110] C. M. Spooner. "Theory and Application of Higher-Order Cyclostationarity," Ph.D. Dissertation, Dept. of Electrical and Computer Engineering, University of California, Davis, CA, June 1992.

[111] C. M. Spooner. "Higher-Order Statistics for Nonlinear Processing of Cyclostationary Signals," (invited plenary lecture) in *Proceedings of the Workshop on Cyclostationary Signals*, Yountville, CA, August 16–18, 1992, Davis, CA: University of California.

[112] R. L. Stratonovich. *Topics in the Theory of Random Noise, Volume I,* New York: Gordon and Breach, 1963.

[113] T. N. Thiele. *The Theory of Observations (1903)*, reprinted in *Annals of Math. Stat.,* Vol. 2, pp. 165–308, 1931.

[114] H. Urkowitz. *Signal Theory and Random Processes,* Dedham, MA: Artech House, 1983.

[115] L. Wang and J. M. Mendel. "Cumulant-Based Parameter Estimation Using Neural Networks," in *Proceedings of the 1990 ICASSP,* pp. 2369–2372, 1990.

[116] N. Wiener. "Generalized Harmonic Analysis," *Acta Mathematika,* Vol. 55, pp. 117–258, 1930.

[117] S. S. Wilks. *Mathematical Statistics,* Princeton, NJ: Princeton University Press, 1950.

[118] J. Wishart. "A Problem in Combinatorial Analysis Giving the Distribution of Certain Moment Statistics," *Proc. London Math. Soc.* (Series 2), Vol. 29, pp. 309–321, 1929.

[119] A. M. Yaglom. *Correlation Theory of Stationary and Related Random Functions II,* New York: Springer-Verlag, 1987.

[120] A. M. Yaglom. *An Introduction to the Theory of Stationary Random Functions,* Englewood Cliffs, NJ: Prentice-Hall, 1962.

[121] W. Zhang and M. Raghuveer. "Nonparametric Bispectrum-Based Time-Delay Estimators for Multiple Sensor Data," *IEEE Transactions on Signal Processing,* Vol. 39, No. 3, pp. 770–774, March 1991.

[122] I. G. Zhurbenko. "Statistical Estimation of Higher-Order Spectra," *Theory of Probability and Its Applications,* Vol. 30, pp. 75–86, 1985.

[123] G. D. Zivanovich. "Some Aspects of the Higher-Order Cyclostationary Theory," in *Proceedings of the IEEE International Conference on Acoustics, Speech, and Signal Processing*, 1991.

Chapter 3

An Overview of Sensor Array Processing for Cyclostationary Signals

Stephan V. Schell
Department of Electrical Engineering
The Pennsylvania State University
University Park, PA 16802

Abstract

The three primary goals of this overview are the following: 1) to describe in a brief but tutorial manner the state of the art of processing cyclostationary signals with sensor arrays and the performance of the current methods, 2) to show how the understanding of these methods can be used in the design of communication systems to increase capacity and/or improve signal quality relative to existing systems, and 3) to describe several open research problems in sensor array processing for cyclostationary signals. Whereas the first goal is intended to speed the self-education of colleagues who are knowledgeable about cyclostationary signals or sensor array processing but not necessarily both, the last is intended to speed the involvement of colleagues who are already knowledgeable about both but are simply interested in another perspective and additional ideas on important and challenging research problems. The middle goal provides one possible bridge between the first and last goals: by demonstrating the advantages of understanding sensor array processing for cyclostationary signals, by identifying problems to be solved in the communication system design presented here, and by pointing out that the role of cyclostationarity-exploiting methods in the design of communication systems is itself an open problem.

1 INTRODUCTION

Arrays of sensors such as radio antennas are useful in the process of detecting the presence of propagating signals, estimating their directions of arrival and other parameters, and estimating the signal waveforms themselves. Application areas include radar, sonar, commercial communications monitoring, signals intelligence, biomedical signal processing, geophysical exploration, communication systems, and others. However, in some of these applications the prior knowledge required by conventional methods for performing these tasks is difficult, costly, or simply impossible to obtain. Furthermore, conventional methods that are derived for sine waves or stationary Gaussian noise cannot exploit many of the statistical properties of structured man-made signals, especially those used in communication systems.

More specifically, almost all man-made communication signals exhibit a statistical property called *cyclostationarity*. A key point is that this property can be exploited in signal processors to favor desired signals and to discriminate against undesired signals, interference, and noise. This signal selectivity offers appealing benefits: post-processing that determines which signals and their parameters are of interest can sometimes be reduced or eliminated, and applicability to some difficult environments is enhanced. In addition, many cyclostationarity-exploiting algorithms offer these benefits while using much less prior knowledge of signal characteristics than conventional methods. In particular, some methods of direction estimation and adaptive spatial filtering that exploit cyclostationarity require only knowledge of the baud rate, carrier frequency, or other frequency that characterizes the underlying periodicity exhibited by the desired signals. It is explained in this overview how this knowledge is used to avoid the need for training signals, estimates of directions of arrival, and array calibration data in spatial filtering schemes, and to avoid the need for knowledge of noise characteristics and to soften the strict requirements on the number and angular spacing of signals in direction finding schemes. It is also explained how the cyclostationarity of bauded digital communication signals can be used to blindly identify and equalize a distortive channel.

An application of this understanding to the design of a cellular communication system is also presented. The current demand for telecommunications exceeds current capacity in some markets and may increase sharply as more bandwidth-intensive commercial services such as multimedia-based conferencing, information retrieval, and electronic banking and shopping are introduced. In this overview, it is shown how understanding of sensor array processing methods for cyclostationary signals can be used in the design of a scheme that substantially increases capacity relative to existing systems.

Throughout the overview, open problems are identified by the appearance of $\boxed{\alpha}$. These problems range from highly applied, such as efficient implementation of these signal-processing methods and design of communication systems, to relatively theoretical, such as statistically optimum detection and estimation of cyclostationary signals and analytical performance prediction.

The remainder of the overview is organized as follows. In Section 2, the mathematical models for the sensor array data are described. In Section 3, notation and basic results on the measurement of spatial characteristics of cyclostationary signals are summarized, and their implications for the basic problems of spatial filtering, direction finding, and estimating the number of signals are briefly discussed. In Section 4, algorithms for blindly adapting (i.e., without using a training signal) a spatial or spatio-temporal filter to extract high-quality estimates of signal waveforms are summarized, including the class of SCORE algorithms. Also in this section, two possible applications of blind adaptive spatial filters to problems in commercial communication systems are summarized. Although the material in Section 5 on blind channel identification and equalization is closely related to Section 4, its primary focus is on simultaneously adapting a spatial filter and demodulating a bauded digital communication signal. In Section 6, algorithms for finding the directions of arrival of cyclostationary signals at a sensor array are summarized. In Section 7, the problem of estimating the number of cyclostationary signals arriving at an array, which must often be solved prior to or jointly with estimating the directions of arrival, is addressed. Finally, in Section 8 a few brief conclusions are drawn and the chapter is briefly summarized.

2 MODELING DATA FROM SENSOR ARRAYS

In this section the mathematical model for the signals at the output of a sensor array is developed from basic physical considerations. The general model for wideband data is derived by assuming that a single sinusoid arrives at the array and then applying superposition to build up the expressions for multiple nonsinusoidal signals. An extremely useful simplifying approximation is then justified and applied to yield a description referred to in the sensor array signal-processing literature as the *narrowband model.* Extensive use is made of this model in this overview.

2.1 General Wideband Model

Consider the analytic signal $\exp(j2\pi ft)$ corresponding to a real sine wave having frequency f and arriving at the array from angle θ. For simplicity, assume that the sensors in the array and the signal source are coplanar so that ordered pairs and a single angle suffice to describe the positions of the sensors and the direction of arrival of the signal, respectively, and assume that the wavefronts impinging on the array are planar. If the propagation medium does not significantly affect the signal as it propagates from one end of the array to the other, then the signal received at one sensor differs from the signal received at another sensor only by a delay. As suggested by Fig. 1, the dependence of the delay on the locations of the sensors and on the angle of arrival can be determined by using elementary geometry. Specifically, if we assume that the coordinates of the M sensors are $(q_1, r_1), \ldots, (q_M, r_M)$, then it can be shown that the delay t_m of the signal at the mth sensor relative to the signal at the origin of the coordinate system can be expressed as $t_m = -[q_m \sin(\theta) + r_m \cos(\theta)]/c$, where c is the propagation speed and θ is measured clockwise from the r axis. Since the signal is sinusoidal, the propagation delay t_m is equivalent to a phase shift by the amount $\psi_m = -2\pi f t_m$, which is in turn equivalent to multiplication by $\exp(j\psi_m)$. Thus, the signal received by the array can be expressed in the vector form

$$\boldsymbol{x}(t) \triangleq \begin{bmatrix} x_1(t) \\ \vdots \\ x_M(t) \end{bmatrix} = \begin{bmatrix} \exp(j\psi_1(\theta, f)) \\ \vdots \\ \exp(j\psi_M(\theta, f)) \end{bmatrix} e^{j2\pi f t}, \quad (1)$$

where

$$\psi_m(\theta, f) = [q_m \sin(\theta) + r_m \cos(\theta)]2\pi f/c. \quad (2)$$

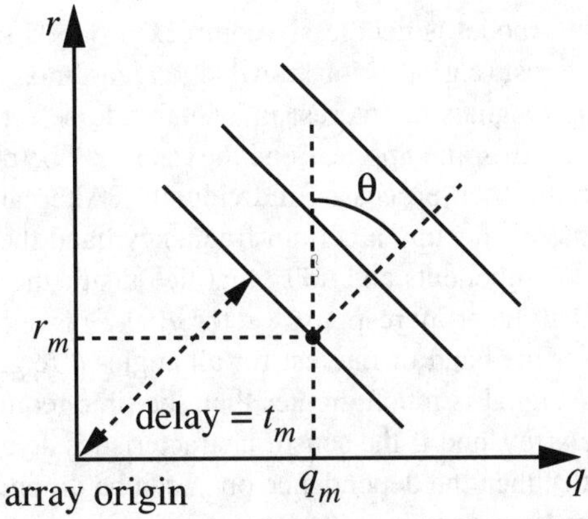

Figure 1: Plane waves propagating from angle θ toward the array origin.

More generally, the sensors can have differing directional and frequency-dependent characteristics, which can be modeled by applying differing gains and phases to the elements of the vector in (1). Denoting the gain and phase of the mth sensor by $g_m(\theta, f)$ and $\phi_m(\theta, f)$, respectively, the analytic signal at the outputs of the sensors can be expressed as

$$\boldsymbol{x}(t) = \boldsymbol{a}(\theta, f)e^{j2\pi f t} \quad (3)$$

where $a_m(\theta, f) = g_m(\theta, f) \exp[j(\psi_m(\theta, f) + \phi_m(\theta, f))]$ is the mth element of the vector $\boldsymbol{a}(\theta, f)$, which is referred to in this overview as the *array response vector*, although the terms *aperture vector*, *array vector*, *array manifold vector*, *DOA vector*, *direction vector*, and *steering vector* also appear in the literature. The collection of array response vectors for all angles θ and all frequencies f of interest is referred to as the *array manifold*.

In the more general (and interesting) case in which multiple nonsinusoidal signals arrive at the array, the data can be modeled by decomposing it in the frequency domain (temporarily assuming that the signals are Fourier-transformable) and using linear superposition:

$$\begin{aligned}
x(f) &= \sum_{l=1}^{L} a(\theta_l, f)s_l(f) + i(f) \\
&= [a(\theta_1, f) \cdots a(\theta_L, f)]\,[s_1(f) \cdots s_L(f)]^T + i(f) \\
&= A(\Theta, f)s(f) + i(f),
\end{aligned} \tag{4}$$

where L signals $s_1(f), \ldots, s_L(f)$ arrive from angles $\theta_1, \ldots, \theta_L$ and $i(f)$ represents interference and noise components (e.g., thermal noise from the sensors and associated electronics, background noise from the environment, and spatially diffuse sources of man-made interference such as cities). That is, the array data is linear with respect to the signals and is linear (in the frequency domain) with respect to $a(\theta, f)$.

2.2 Narrowband Model

The general wideband model is needlessly complex if only a relatively narrow frequency band is of interest (e.g., if prior knowledge regarding the center frequencies and bandwidths of the signals of interest is available to select the narrow band of interest). For example, in some applications the data may be channelized into very narrow bands which are then processed individually. Alternatively, if it is known that the signals of interest occupy a certain frequency band then it is advantageous to reject interference components and noise that lie outside the band. If this band is sufficiently narrow that the array response vector $a(\theta, f)$ is approximately constant with respect to f over the band of interest for all angles θ (e.g., if the reciprocal of the bandwidth of the signal is much greater than the time required for the signal to propagate across the array, and if the sensor characteristics do not vary significantly across this bandwidth), then the dependence on f can be dropped and the array data can be modeled in the time domain as the analytic signal

$$\begin{aligned}
x(t) &= \sum_{l=1}^{L} a(\theta_l)s_l(t) + i(t) \\
&= [a(\theta_1) \cdots a(\theta_L)]\,[s_1(t) \cdots s_L(t)]^T + i(t) \\
&= A(\Theta)s(t) + i(t),
\end{aligned} \tag{5}$$

where $s(t)$ and $i(t)$ are analytic signals. Although the signals $s_l(t)$ are not sinusoids, the spatial characteristics of the array response can be approximately modeled as if they were. This observation is the essence of the narrowband model.

A more detailed discussion of the conditions under which this assumption is valid, as well as a detailed investigation of the representation of wideband array data, can be found in [16]. However, some justification can be offered here without undue complexity. Consider a single signal $s(t)$ having flat power spectral density over the band $[f_0 - B/2, f_0 + B/2]$, and arriving at a uniform linear array (ULA) for which $(q_m, r_m) = (dm, 0)$ where d is the sensor spacing. The spectral density of $x(t)$ is $S_{xx}(f) = a(\theta, f)a^H(\theta, f)$ for $|f - f_0| \leq B/2$, and the autocorrelation at lag $\tau = 0$ can be expressed as

$$[\boldsymbol{R}_{\boldsymbol{xx}}(0)]_{m,n} = B \exp\left(-j2\pi \frac{f_0 d}{c}(m-n)\sin\theta\right) \operatorname{sinc}\left(\pi \frac{Bd}{c}(m-n)\sin\theta\right).$$

The total average power P_{tot} of $\boldsymbol{x}(t)$ is $P_{tot} = \operatorname{tr}\{\boldsymbol{R}_{\boldsymbol{xx}}(0)\} = MB$, and the average power P_{f_0} in $\boldsymbol{x}(t)$ that can be represented by the narrowband model $\boldsymbol{a}(\theta, f_0)\, s(t)$ can be expressed as

$$\begin{aligned} P_{f_0} &= \operatorname{tr}\left\{\boldsymbol{P}_{\boldsymbol{a}(\theta, f_0)} \boldsymbol{R}_{\boldsymbol{xx}}(0)\right\} \\ &= \frac{B}{M} \sum_{k,m=1}^{M} \operatorname{sinc}\left(\frac{\pi\eta}{2}(k-m)\sin\theta\right) \end{aligned}$$

where $\eta \triangleq B/f_0$ is the relative bandwidth, f_0 is the frequency chosen for use in the narrowband model (it can be any frequency in the reception band), and $\boldsymbol{P}_{\boldsymbol{v}}$ denotes the orthogonal projection matrix for vector $\boldsymbol{v}$. Thus, the ratio $\gamma \triangleq P_{f_0}/P_{tot}$ is a measure of the degree to which $\boldsymbol{x}(t)$ admits a narrowband model:

$$\gamma \triangleq \frac{P_{f_0}}{P_{tot}} = \frac{1}{M^2} \sum_{k,m=1}^{M} \operatorname{sinc}\left(\frac{\pi\eta}{2}(k-m)\sin\theta\right).$$

Clearly, the narrowband model is exact ($P_{f_0} = P_{tot}$) for $\eta = 0$ but degrades as the relative bandwidth increases, as shown in Fig. 2. For example, at the worst-case value of θ ($\theta = 90$ degrees), the narrowband model for the 4-element ULA accounts for 99% of the received power for $\eta = 10\%$, and 90% of the received power for $\eta = 30\%$.

$\boxed{\alpha}$ Despite this justification for the narrowband model, a more thorough investigation of the sensitivity of narrowband sensor array processing methods to deviations from the narrowband model is needed.

In this overview and in much of the literature on direction finding, the sampled complex envelope of the array data is used in the description and analysis of the various algorithms because the algorithms are typically implemented on a digital computer and therefore operate on sampled data. Since the complex envelope of a bandlimited analytic signal can be obtained by performing a complex down-conversion (i.e., by multiplying the data by $\exp(-j2\pi f t)$ for some appropriate f), the corresponding model for the sampled complex envelope is essentially the same as in (5), except that $\boldsymbol{x}(t)$, $\boldsymbol{s}(t)$, and $\boldsymbol{i}(t)$ denote the complex envelopes of the array data, the signals, and the noise, respectively, and t is replaced with n:

$$\boldsymbol{x}(n) = \sum_{l=1}^{L} \boldsymbol{a}(\theta_l) s_l(n) + \boldsymbol{i}(n) = \boldsymbol{A}(\Theta)\boldsymbol{s}(n) + \boldsymbol{i}(n). \tag{6}$$

Almost all of the algorithms discussed in this overview are based on this model.

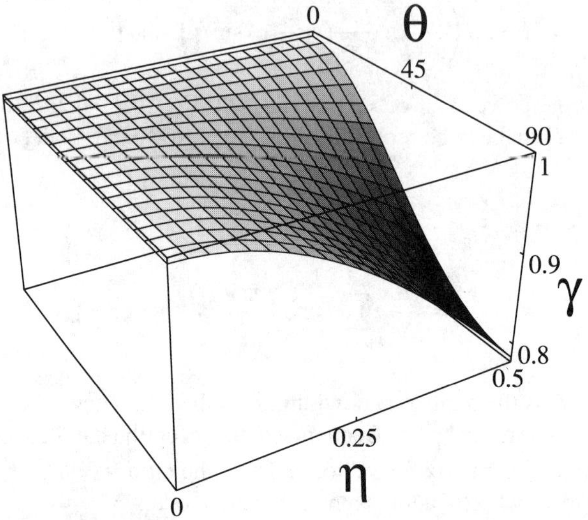

Figure 2: Accuracy of narrowband model versus relative bandwidth η and angle of arrival θ for a 4-element ULA.

3 CYCLOSTATIONARITY

In this section the relevant results from the theory of cyclostationary signals are briefly summarized in the context of the sensor array processing problem. First, notation used in the rest of this overview is summarized. Second, recent results from [69] on the mean and covariance of the cyclic correlogram matrix, which is an estimate of the cyclic autocorrelation matrix, are summarized. These results generalize those in [43]. Finally, the implications of these results are stated for the sensor array signal-processing tasks of waveform estimation, direction finding, and estimating the number of cyclostationary signals arriving at the array.

3.1 Notation

The following abbreviations are used throughout this overview:

SOI: Signal of interest
SNOI: Signal not of interest
STF: Spatio-temporal filter
DF: Direction finding
DOA: Direction of arrival
LCL-PTV: Linear-conjugate-linear polyperiodic time varying
SNR: Signal-to-noise ratio
SINR: Signal-to-interference and noise ratio
MSE: Mean-squared error

RMSE: Root-mean-squared error
BPSK: Binary phase-shift-keyed

With reference to the abbreviation LCL-PTV, it is noted that the term *linear-conjugate-linear multiply-periodic time-varying* (LCL-MPTV) is used in an earlier work [81] to denote the same concept. The new term is chosen for consistency with the terminology used in [32]. The following definitions are used throughout this overview:

D-1 The notation $\langle f(n)\rangle_N$ denotes the time average of $f(n)$:$\langle f(n)\rangle_N \triangleq \frac{1}{N}\sum_{n=0}^{N-1} f(n)$.

D-2 The notation $f(N) = O(g(N))$, where $g(N)$ is a monotonically decreasing positive function of N, means that $\lim_{N\to\infty} f(N)/g(N) = c$ for some constant c.

D-3 The function $Z_N(f,\tau)$ is defined as

$$Z_N(f,\tau) \triangleq \frac{1}{N}\sum_{n=0}^{N-1-\tau} e^{-j2\pi fn} = \left(1-\frac{\tau}{N}\right)\frac{\operatorname{sinc}(\pi f(N-\tau))}{\operatorname{sinc}(\pi f)} e^{-j\pi f(N-\tau-1)}$$

and is $O\left(\frac{1}{N}\right)$ for fixed $f \neq 0$ and τ, where $\operatorname{sinc}(x) \triangleq \sin x/x$.

D-4 The superscripts $*$, T, H, $+$, and $\perp$ denote conjugation, matrix transposition, matrix conjugate transposition, pseudo-inversion, and the orthogonal complement operation, respectively. Furthermore, $\star$ and $\otimes$ denote the convolution and Kronecker (tensor) product (cf. [41]) operators, respectively.

D-5 The expectation operation is denoted by $E\{\}$ and can denote either the ensemble-averaging operation for which the argument is typically a stochastic process, or the sinewave (or polyperiodic component) extraction operation based on time-averaging. In this paper, this operation can be interpreted in either way with complete equivalence of the results, provided that underlying assumptions of cycloergodicity and so forth are properly attended to. Extensive discussion of these two philosophies can be found elsewhere in this volume [32].

D-6 The time-variant cross-correlation $\boldsymbol{R}_{\boldsymbol{xy}}(m,n)$ between $\boldsymbol{x}(m)$ and $\boldsymbol{y}(n)$ is defined as

$$\boldsymbol{R}_{\boldsymbol{xy}}(m,n) \triangleq E\left\{\boldsymbol{x}(m)\boldsymbol{y}^H(n)\right\}.$$

D-7 If $\boldsymbol{R}_{\boldsymbol{xy}}(n+\tau,n)$ has a Fourier series representation in n, it is given by

$$\boldsymbol{R}_{\boldsymbol{xy}}(n+\tau,n) = \sum_{\beta} \boldsymbol{R}_{\boldsymbol{xy}}^{\beta}(\tau)e^{j2\pi\beta n}$$

where the sum is taken over all β for which the cyclic cross-correlation $\boldsymbol{R}_{\boldsymbol{xy}}^{\beta}(\tau)$, defined by

$$\boldsymbol{R}_{\boldsymbol{xy}}^{\beta}(\tau) \triangleq \left\langle \boldsymbol{R}_{\boldsymbol{xy}}(n+\tau,n)e^{-j2\pi\beta n}\right\rangle_{\infty},$$

is not identically zero as a function of τ.

D-8 The symbol $\hat{\boldsymbol{R}}_{\boldsymbol{xy}}^{\alpha}(\tau)$ denotes the cyclic cross-correlogram of $\boldsymbol{x}(n)$ and $\boldsymbol{y}(n)$ at cycle frequency α and lag τ:

$$\hat{\boldsymbol{R}}_{\boldsymbol{xy}}^{\alpha}(\tau) \triangleq \begin{cases} \frac{1}{N}\sum_{n=0}^{N-1-\tau} \boldsymbol{x}(n+\tau)\boldsymbol{y}^{H}(n)\, e^{-j2\pi\alpha n}, & \text{for } \tau \geq 0, \\ \frac{1}{N}\sum_{n=-\tau}^{N-1} \boldsymbol{x}(n+\tau)\boldsymbol{y}^{H}(n)\, e^{-j2\pi\alpha n}, & \text{for } \tau < 0, \end{cases}$$

and it is useful to note that $\hat{\boldsymbol{R}}_{\boldsymbol{xy}}^{\alpha}(\tau) \equiv \hat{\boldsymbol{R}}_{\boldsymbol{yx}}^{-\alpha H}(-\tau)e^{j2\pi\alpha\tau}$.

D-9 The cyclic cross-correlation coefficient between $u(n)$ and $v(n)$ for cycle frequency α and lag τ is defined as

$$\rho_{uv}^{\alpha}(\tau) \triangleq R_{uv}^{\alpha}(\tau) \Big/ \sqrt{R_{uu}(0)R_{vv}(0)}\ ,$$

where it is noted that $\left|\rho_{uv}^{\alpha}(\tau)\right| \leq 1$ for all α and τ (e.g., [30]). The magnitude of $\rho_{ss}^{\alpha}(\tau)$ is sometimes referred to as the feature strength (at cycle frequency α and lag τ) of the signal $s(n)$. The feature strength is a normalized measure that can be interpreted as indicating the degree to which a signal is correlated with a frequency-shifted and delayed copy of itself.

D-10 The singular value decomposition of the cyclic correlation matrix $\boldsymbol{R}_{\boldsymbol{xv}}^{\alpha}(\tau)$ is denoted by

$$\boldsymbol{R}_{\boldsymbol{xy}}^{\alpha}(\tau) = [\boldsymbol{S}\ \ \boldsymbol{G}]\begin{bmatrix} \Sigma_S & 0 \\ 0 & \Sigma_G \end{bmatrix}[\boldsymbol{V}_S\ \ \boldsymbol{V}_G]^{H} = \boldsymbol{S}\Sigma_S\boldsymbol{V}_S^{H},$$

where $[\boldsymbol{S}\ \ \boldsymbol{G}]$ and $[\boldsymbol{V}_S\ \ \boldsymbol{V}_G]$ are unitary, the diagonal elements of the diagonal matrix Σ_S are nonnegative, and $\Sigma_G = 0$. The SVD of the cyclic correlogram matrix $\hat{\boldsymbol{R}}_{\boldsymbol{xy}}^{\alpha}(\tau)$ is denoted similarly except that all quantities have hats (ˆ) on them.

3.2 Mean and Covariance

Here, three results on the mean and covariance of cyclic correlation matrices are summarized. More detailed discussion can be found in the appendix of this overview, where the assumptions under which these results are true are given.

First, $\hat{\boldsymbol{R}}_{\boldsymbol{xy}}^{\alpha}(\tau)$ is found to be asymptotically unbiased.

Result 1 *The mean of* $\hat{\boldsymbol{R}}_{\boldsymbol{xy}}^{\alpha}(\tau)$ *for* $\tau > 0$ *can be expressed as*

$$\begin{aligned} E\left\{\hat{\boldsymbol{R}}_{\boldsymbol{xy}}^{\alpha}(\tau)\right\} &= \sum_{\beta} \boldsymbol{R}_{\boldsymbol{xy}}^{\beta}(\tau) Z_N(\alpha-\beta,\tau) \\ &= \left(1-\frac{\tau}{N}\right)\boldsymbol{R}_{\boldsymbol{xy}}^{\alpha}(\tau) + \sum_{\beta\neq\alpha} \boldsymbol{R}_{\boldsymbol{xy}}^{\beta}(\tau)\, Z_N(\alpha-\beta,\tau) \\ &= \boldsymbol{R}_{\boldsymbol{xy}}^{\alpha}(\tau) + O\left(\frac{1}{N}\right). \end{aligned}$$

Second, the covariance is found.

Result 2 *Given two arbitrary vectors* **g** *and* **w**, *then*

$$cov\left\{\hat{\boldsymbol{R}}_{\boldsymbol{xy}}^{\alpha H}(\tau)\boldsymbol{g}, \hat{\boldsymbol{R}}_{\boldsymbol{xy}}^{\alpha T}(\tau)\boldsymbol{w}^*\right\} = J_1 + J_2 + O\left(\frac{1}{N^2}\right)$$

where

$$J_1 = \frac{1}{N}\sum_{\beta}\sum_{k=-\infty}^{\infty} \boldsymbol{w}^T \boldsymbol{R}_{\boldsymbol{xx}^*}^{\beta H}(k)\boldsymbol{g}\boldsymbol{R}_{\boldsymbol{yy}^*}^{\beta-2\alpha}(k)\, e^{j2\pi\alpha k} e^{-j2\pi\beta\tau}$$

$$J_2 = \frac{1}{N}\sum_{\beta}\sum_{k=-\infty}^{\infty} \boldsymbol{R}_{\boldsymbol{yx}}^{\beta-2\alpha}(k-\tau)\boldsymbol{w}\boldsymbol{g}^T \boldsymbol{R}_{\boldsymbol{xy}}^{\beta *}(k+\tau) e^{j2\pi\alpha k} e^{j2\pi(\beta-2\alpha)\tau}.$$

Result 3 *Given two arbitrary vectors* **g** *and* **w**, *then*

$$cov\left\{\hat{\boldsymbol{R}}_{\boldsymbol{xy}}^{\alpha H}(\tau)\boldsymbol{g}, \hat{\boldsymbol{R}}_{\boldsymbol{xy}}^{\alpha H}(\tau)\boldsymbol{w}\right\} = K_1 + K_2 + O\left(\frac{1}{N^2}\right)$$

where

$$K_1 = \frac{1}{N}\sum_{\beta}\sum_{k=-\infty}^{\infty} \boldsymbol{w}^H \boldsymbol{R}_{\boldsymbol{xx}}^{\beta H}(k)\boldsymbol{g}\boldsymbol{R}_{\boldsymbol{yy}}^{\beta}(k) e^{j2\pi\alpha k} e^{-j2\pi\beta\tau}$$

$$K_2 = \frac{1}{N}\sum_{\beta}\sum_{k=-\infty}^{\infty} \boldsymbol{R}_{\boldsymbol{yx}^*}^{\beta}(k-\tau)\boldsymbol{w}^*\boldsymbol{g}^T \boldsymbol{R}_{\boldsymbol{xy}^*}^{\beta *}(k+\tau) e^{j2\pi\alpha k} e^{j2\pi\beta\tau}.$$

Under the assumptions given in the Appendix, it is easily shown that $\hat{\boldsymbol{R}}_{\boldsymbol{xy}}^{\alpha}(\tau)$ converges in mean square to $\boldsymbol{R}_{\boldsymbol{xy}}^{\alpha}(\tau)$ since its bias and covariance are both $O\left(\frac{1}{N}\right)$.

3.3 Implications for Sensor Array Processing

Under the assumption that the narrowband model holds well and L_α signals of interest (SOI) $s_1(n), \ldots, s_{L_\alpha}(n)$ having cycle frequency α arrive at the array,

$$\boldsymbol{x}(n) = \sum_{l=1}^{L_\alpha} \boldsymbol{a}(\theta_l)\, s_l(n) + \boldsymbol{i}(n),$$

then the results on the mean and covariance of $\hat{\boldsymbol{R}}_{\boldsymbol{xx}}^{\alpha}(\tau)$ imply that

$$\hat{\boldsymbol{R}}_{\boldsymbol{xx}}^{\alpha}(\tau) \longrightarrow \boldsymbol{A}(\Theta)\boldsymbol{R}_{\boldsymbol{ss}}^{\alpha}(\tau)\boldsymbol{A}^H(\Theta)$$

$$\hat{\boldsymbol{R}}_{\boldsymbol{xx}^*}^{\alpha}(\tau) \longrightarrow \boldsymbol{A}(\Theta)\boldsymbol{R}_{\boldsymbol{ss}^*}^{\alpha}(\tau)\boldsymbol{A}^T(\Theta)$$

with bias and covariance $O\left(\frac{1}{N}\right)$. This in turn implies that a measurement of the spatial characteristics (the vector space spanned by the SOIs' array response vectors) of the L_α SOIs can be made, even if the SOI waveforms and directions of arrival are unknown, and even if $\boldsymbol{i}(n)$ has completely arbitrary and unknown spatial characteristics. However, note that additional restrictions on the ambiguity of the manifold and on the rank of $\boldsymbol{R}_{\boldsymbol{ss}}^{\alpha}(\tau)$ are needed to determine Θ and/or $\boldsymbol{A}(\Theta)$ from $\boldsymbol{R}_{\boldsymbol{xx}}^{\alpha}(\tau)$.

In subsequent sections it is shown how this signal selectivity can be exploited to solve waveform and parameter estimation problems in sensor array processing.

Furthermore, Results 1–3 are needed, for example, in the analytical performance evaluation of direction-finding methods that exploit cyclostationarity, as summarized in Section 6.4.

$\boxed{\alpha}$ Another implication of these results is that error in the knowledge of the cycle frequency manifests itself as cycle leakage in the sense that the regenerated spectral line at the true cycle frequency α_0 leaks into the estimate at α with weight $Z_N(\alpha_0 - \alpha)$. Thus, the allowable error in the estimated cycle frequency should be less than $1/2N$. Methods for estimating the cycle frequencies of a signal are available, although they are either computationally intensive (requiring computation of the cyclic spectrum over a potentially large portion of the (f, α) plane as in [45]) (cf. [65] in this volume for efficiently implemented cyclic spectrum analyzers), or they are somewhat susceptible to error and require a potentially large number of data samples [13, 75]. However, it has been observed in some cases [68] that the CRLB on the variance of cycle frequency estimates decreases as $1/N^3$ (in contrast to $1/N$ for estimates of direction of arrival, carrier phases, and so forth), so there it may be possible to develop extremely reliable cycle frequency estimators. Even in the case of scalar data this remains an open problem.

4 SPATIAL FILTERING

In this section the problem of using a sensor array to spatially filter the received signals without knowing a training signal or direction of arrival is addressed.

4.1 Structures for Spatial Filtering

Spatial filtering is used for purposes similar to those of temporal filtering: to enhance desired signal components, to attenuate undesired signal components, and to minimize noise.

The simplest spatial filter considered here linearly combines the signals from the sensors to yield an output signal $\hat{s}(n)$:

$$\hat{s}(n) = \mathbf{w}^H \boldsymbol{x}(n).$$

If the narrowband model (6) holds well then

$$\hat{s}(n) = \sum_{l=1}^{L_\alpha} \left[\mathbf{w}^H \boldsymbol{a}(\theta_l)\right] s_l(n) + \mathbf{w}^H \boldsymbol{i}(n),$$

which clearly shows that the gain applied to a signal arriving from angle θ is $\mathbf{w}^H \boldsymbol{a}(\theta)$. Thus, $\mathbf{w}^H \boldsymbol{a}(\theta)$ is analogous to the transfer function (Fourier-transformed impulse response) of a linear time-invariant (LTI) finite-impulse-response (FIR) temporal filter and is referred to as the spatial transfer function or antenna pattern of the spatial filter. When multiple signals arrive from different directions, a carefully chosen spatial filter can extract one of these signals while rejecting the others, as depicted by the antenna pattern in Fig. 3, where $\boldsymbol{w}$ has been chosen to extract the signal arriving from 30 degrees and to reject (or null) the signals arriving from -20 degrees and 0 degrees.

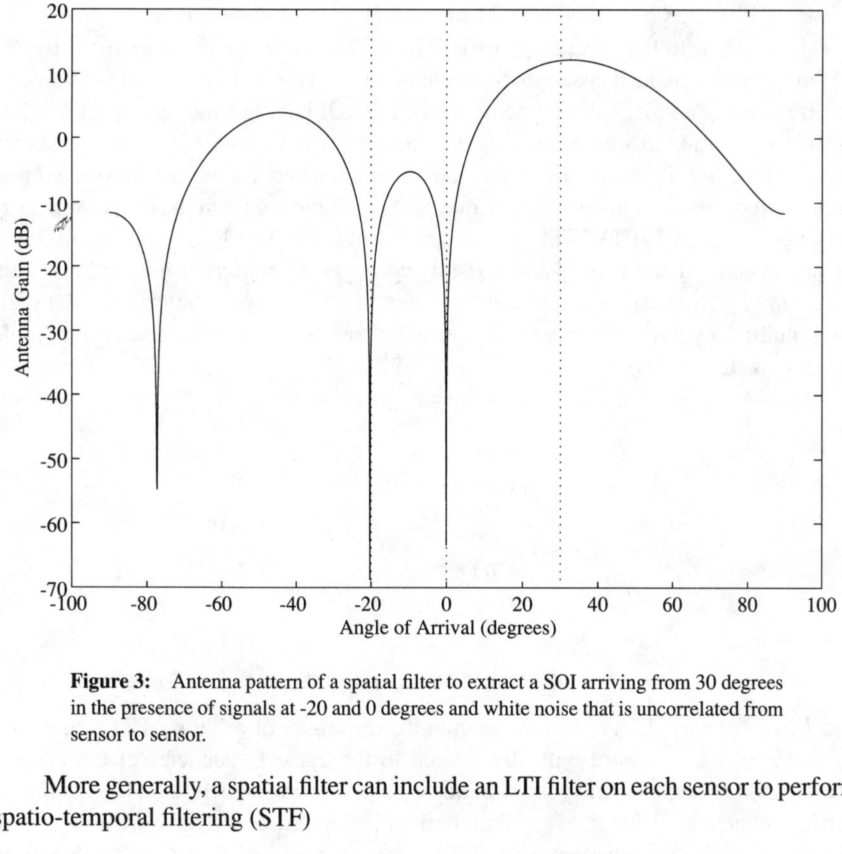

Figure 3: Antenna pattern of a spatial filter to extract a SOI arriving from 30 degrees in the presence of signals at -20 and 0 degrees and white noise that is uncorrelated from sensor to sensor.

More generally, a spatial filter can include an LTI filter on each sensor to perform spatio-temporal filtering (STF)

$$\hat{s}(n) = \sum_{m=1}^{M} w_m^*(n) \star x_m(n) = \boldsymbol{w}^H(n) \star \boldsymbol{x}(n),$$

for which the antenna pattern is now a function of frequency (i.e., $\boldsymbol{w}^H(-f)\boldsymbol{a}(\theta, f)$), and the frequency dependence of $\boldsymbol{a}(\theta)$ is included to emphasize that the STF is appropriate even if the narrowband model does not hold.

Yet more generally, a spatial filter can include a linear-conjugate-linear polyperiodic time-variant (LCL-PTV) filter on each sensor to perform LCL-PTV STF, which is the generalization to multiple inputs of the scalar LCL-PTV temporal filter. An extensive discussion of LCL-PTV filters appears in [23] where they are referred to as polyperiodic linear filters. As discussed there and in [31], the LCL-PTV filter structure can implement the Cyclic Wiener filter, which is the generalization of the Wiener filter from stationary to cyclostationary signals. The fractionally spaced equalizer structure that is ubiquitous in digital communication systems is also a restricted implementation of an LCL-PTV filter. Since LCL-PTV filter structures have been shown to be capable of separating two cyclostationary signals even if they are completely overlapping both temporally and spectrally, it is reasonable to expect there to be substantial benefits of LCL-PTV STFs over LTI STFs in some cases.

In addition to the preceding, LCL-PTV STFs can be motivated simply by observing that they make it possible to simultaneously exploit multiple cyclostationarity features. For example, a digital communication SOI having baud rate f_{baud} exhibits useful cyclostationarity at multiple cycle frequencies, such as $\pm f_{baud}$ (and harmonics thereof if the SOI is not bandwidth-efficient), and the cyclic autocorrelations at these cycle frequencies contain useful information at multiple values of the lag parameter τ.

Thus, the LCL-PTV STF can be interpreted as linearly combining multiple frequency-shifted and filtered and possibly conjugated versions of the received data, or as simply providing a signal, which when correlated with the original data, allows multiple cyclostationarity properties to be manifested simultaneously in a single measurement. That is,

$$\hat{s}(n) = \boldsymbol{w}^H \boldsymbol{y}(n) \tag{7}$$

where

$$\boldsymbol{y}(n) = \begin{bmatrix} (\boldsymbol{x}(n) \star h_1(n))\, e^{j2\pi\alpha_1 n} \\ \vdots \\ (\boldsymbol{x}(n) \star h_J(n))\, e^{j2\pi\alpha_J n} \\ (\boldsymbol{x}(n)^* \star h_{J+1}(n))\, e^{j2\pi\alpha_{J+1} n} \\ \vdots \\ (\boldsymbol{x}(n)^* \star h_K(n))\, e^{j2\pi\alpha_K n} \end{bmatrix} \tag{8}$$

and $h_k(n)$ for $k = 1, \ldots, K$ are the impulse responses of arbitrary LTI filters, and α_k for $k = 1, \ldots, K$ are typically related to the cycle frequencies of the desired cyclostationary signals (and possibly undesired interfering signals) $\boldsymbol{s}(n)$ (e.g., doubled carrier frequencies, baud rates and their harmonics, and sums and differences of these). Figure 4 depicts an implementation of (7), without the branches that use conjugation.

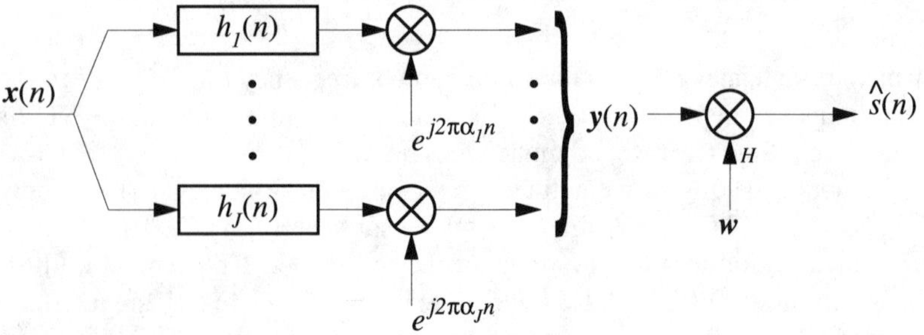

Figure 4: Block diagram of an LCL-PTV STF. The conjugate branches are omitted for simplicity.

$\boxed{\alpha}$ Several open problems relate to LCL-PTV STFs. In particular, practical methods of adapting them, their convergence time and SINR at convergence, methods for *blindly* adapting them (i.e., without the use of training signals), and their use in such applications as equalization and/or demodulation of severely corrupted digital communication signals (e.g., see Section 5) and interception (blind despreading) of spread spectrum communication signals are of significant interest.

4.2 Performance Criteria

The performance measures most commonly used to evaluate waveform estimators such as spatial filters are summarized here.

Perhaps the most obvious of these measures is the mean-squared error (MSE) of the estimated waveform relative to the desired waveform,

$$MSE(\hat{s}, s) = \left\langle \left\| \hat{s}(n) - s(n) \right\|^2 \right\rangle_N . \tag{9}$$

A closely related measure is the signal-to-interference-and-noise ratio (SINR) which can be expressed as

$$\begin{aligned} SINR(\hat{s}, s) &\triangleq \frac{\text{power of desired signal components in } \hat{s}}{\text{power of everything else in } \hat{s}} \\ &= \frac{\left|\hat{R}_{\hat{s}s}\right|^2 / \hat{R}_{ss}}{\hat{R}_{\hat{s}\hat{s}} - \left|\hat{R}_{\hat{s}s}\right|^2 / \hat{R}_{ss}} = \frac{\left|\hat{\rho}_{\hat{s}s}\right|^2}{1 - \left|\hat{\rho}_{\hat{s}s}\right|^2} . \end{aligned} \tag{10}$$

The SINR increases without bound as $\hat{s}$ becomes arbitrarily highly correlated with s (i.e., as $\left|\hat{\rho}_{\hat{s}s}\right| \to 1$). All three of these measures ($MSE(\hat{s}, s)$, $SINR(\hat{s}, s)$, and $\left|\hat{\rho}_{\hat{s}s}\right|^2$) are applicable to single time-series, but are often averaged over multiple realizations (e.g., sample paths of a stochastic process) or multiple data segments of length N comprising a much longer time series.

Minimizing $MSE(g\hat{s}, s)$ with respect to the complex scalar g and estimated waveform $\hat{s}$ is exactly equivalent to maximizing $SINR(\hat{s}, s)$ and then computing $g = \hat{R}_{s\hat{s}} / R_{\hat{s}\hat{s}}$.

In this overview SINR is distinct from SNR, which is used here to specify the power levels of received signals relative to the power of Gaussian noise.

Finally, in the special, though increasingly important, case in which the desired signal is a digital communication signal, the ultimate goal is typically to obtain a good estimate of the underlying bit or symbol stream. Thus, the relevant performance measure is the bit error rate (BER), which is strongly dependent both on the type of interference, noise, and channel distortion that corrupts the signal at the input to the demodulator and on the demodulator itself. Simple expressions for BER as a function of SNR (relative to white Gaussian noise) in the absence of channel distortion are well known and can be found in any standard text on digital communications (e.g., [60]).

4.3 Conventional Methods

Conventional approaches to adapting a spatial filter have been well understood for more than two decades, and a tutorial that includes them and more recent advances can be found in [95]. Comprehensive treatments can also be found in [20] and [54]. For the purposes of this overview, three of the most popular methods are summarized here.

The most direct method simply minimizes $MSE(\hat{s}, s)$ to obtain the minimum MSE (MMSE) solution

$$w_{MMSE} = \hat{\boldsymbol{R}}_{xx}^{-1}\hat{\boldsymbol{R}}_{xs}, \tag{11}$$

for which the corresponding SINR (referred to in this overview as the maximum SINR SINR_{max} attainable by a memoryless time-invariant spatial filter) is given by (11) with

$$\left|\hat{\rho}_{\hat{s}s}\right|^2 = \hat{\boldsymbol{R}}_{xs}^H\hat{\boldsymbol{R}}_{xx}^{-1}\hat{\boldsymbol{R}}_{xs}/\hat{R}_{ss}.$$

As can be inferred from (11), a spatial filter attaining this maximum SINR can also be found by maximizing $\left|\hat{\rho}_{\hat{s}s}\right|^2$. For applications in which it is undesirable, expensive, or impossible for a known training signal to be sent by the transmitter and a local copy of it used by the receiver for adaptation, this method is unsuitable.

An alternative to the MMSE method makes use of prior knowledge of the direction of arrival (DOA) of the signal of interest (SOI) to form a beam on the SOI and to null any interferers. This method is referred to in the literature as minimum variance distortionless response (MVDR) beamforming and is a special case of a more general framework called linearly constrained minimum variance beamforming. Yet more general variants exist which accommodate quadratic, derivative, and eigenvector constraints (cf. [95]). The spatial filter weights are chosen so as to minimize the average output power while maintaining unity array gain in the direction θ_0 of the SOI, since doing so necessarily minimizes the contributions of interferers that arrive from other directions:

$$\max_{\boldsymbol{w} \text{ s.t. } \boldsymbol{w}^H\boldsymbol{a}(\theta_0)=1} \hat{\boldsymbol{R}}_{\hat{s}\hat{s}} \Longleftrightarrow \max_{\boldsymbol{w} \text{ s.t. } \boldsymbol{w}^H\boldsymbol{a}(\theta_0)=1} \boldsymbol{w}^H\hat{\boldsymbol{R}}_{xx}\boldsymbol{w}.$$

It can be shown (e.g., via the method of Lagrange multipliers) that the solution is given by

$$w_{MVDR} = \left[\boldsymbol{a}^H(\theta_0)\hat{\boldsymbol{R}}_{xx}^{-1}\boldsymbol{a}(\theta_0)\right]^{-1}\hat{\boldsymbol{R}}_{xx}^{-1}\boldsymbol{a}(\theta_0). \tag{12}$$

If the SOI arrives only from angle θ_0, then $\hat{\boldsymbol{R}}_{xs}$ converges to $\boldsymbol{a}(\theta_0)R_{ss}$, which with (11) and (12) implies that the output SINR of the MVDR spatial filter converges to SINR_{max}. For applications in which it is impossible to know θ_0 and difficult to estimate it (e.g., using one of the direction-finding methods described in Section 6), this method is unsuitable.

As yet another alternative, when neither a known training signal nor a known direction of arrival is available, a direction-finding algorithm can be applied. For example, given known array calibration data, the method of [55] estimates the directions of arrival of all signals and then uses these to compute the weights of the spatial filter for each signal. However, in some applications, the computational load of estimating all of the directions of arrival and post processing the spatially filtered signals to select those of interest may be prohibitive. Also, accurate array calibration data can be difficult or impossible to obtain in some applications.

4.4 SCORE Algorithms

$\boxed{\alpha}$ The general problem of blind adaptive filtering is by nature lacking in specificity, since it is desired to extract a good quality estimate of a signal by exploiting whatever knowledge (if any) is available in the event that conventional methods are unsuitable. This lack of specificity requires investigation of possibly numerous approaches, each tailored to a different set of knowable facts about the SOI, the interference, the noise, the channel, etc.

This section describes a class of algorithms for one such set of facts, namely knowledge that the desired signal exhibits cyclostationarity, where it is assumed that the specific cyclostationarity properties are either known at the receiver or can be estimated. The spectral coherence restoral (SCORE) algorithms, which exploit cyclostationarity properties, require neither a known training signal nor knowledge of the spatial characteristics of the SOIs, interference or noise, and don't require array calibration data (i.e., knowledge of the array manifold). The SCORE algorithms are explained in this section from several different perspectives with the intent of demonstrating that there are multiple approaches to the blind adaptive spatial filtering problem, and that some of the approaches yield identical algorithms. Finally, the SCORE framework is generalized to exploit other signal properties.

4.4.1 Least-Squares SCORE

The challenge of blind adaptive spatial filtering is to minimize $MSE(\hat{s}, s)$ without knowing s, without knowing $\boldsymbol{a}(\theta_0)$, and without resorting to methods based on direction finding which require knowledge of the array calibration data. At first this challenge may seem insurmountable. However, if s arrives from only one direction (i.e., no multipath or smart jamming is present), then $\hat{\boldsymbol{R}}_{xs}$ converges to $\boldsymbol{a}(\theta_0)R_{ss}$; indeed, any vector that converges to a vector that is proportional to $\boldsymbol{a}(\theta_0)$ can reasonably serve as a substitute for $\hat{\boldsymbol{R}}_{xs}$ in (11). In particular, if $u(n)$ is any signal that is correlated with $s(n)$ but uncorrelated with the remaining signals comprising $\boldsymbol{x}(n)$, then $\hat{\boldsymbol{R}}_{xu} = \boldsymbol{a}(\theta_0)\hat{R}_{su} + \hat{\boldsymbol{R}}_{iu}$ converges to $\boldsymbol{a}(\theta_0)R_{su}$, which is clearly proportional to $\boldsymbol{a}(\theta_0)$ as desired. This implies that $\boldsymbol{w} = \hat{\boldsymbol{R}}_{xx}^{-1}\hat{\boldsymbol{R}}_{xu}$ might be a suitable adaptive spatial filtering algorithm. If possible, the convergence time should be reduced by choosing the signal $u(n)$ so as to maximize the ratio of the desired contributions $\boldsymbol{a}(\theta_0)\hat{R}_{su}$ to the residual contributions $\hat{\boldsymbol{R}}_{iu}$.

If it is known that $s(n)$ is cyclostationary with cycle frequency α, but $\boldsymbol{i}(n)$ is not, then a reference signal $u(n)$ can be derived directly from the received data,

$$u(n) = \boldsymbol{c}^H\boldsymbol{x}(n-\tau)\,e^{j2\pi\alpha n}, \tag{13}$$

where $\boldsymbol{c}$ is any vector of spatial filter weights such that $\boldsymbol{c}^H\boldsymbol{a}(\theta_0) \neq 0$ (e.g., $\boldsymbol{c} =$

$[1,\ 0, \ldots,\ 0]^T$), and τ is chosen so as to maximize (if possible) $R_{ss}^{\alpha}(\tau)$. The key observation is that $u(n)$ is correlated with $s(n)$ but uncorrelated with $\boldsymbol{i}(n)$. This observation leads directly to the *Least-Squares SCORE* (LS-SCORE) optimization problem

$$\min_{\boldsymbol{w}} MSE(\hat{s}, u) \Longleftrightarrow \max_{\boldsymbol{w}} \left|\hat{\rho}_{\hat{s}u}\right|^2 \tag{14}$$

which has solution

$$\boldsymbol{w} = \hat{\boldsymbol{R}}_{\boldsymbol{xx}}^{-1}\hat{\boldsymbol{R}}_{\boldsymbol{x}u} = \hat{\boldsymbol{R}}_{\boldsymbol{xx}}^{-1}\hat{\boldsymbol{R}}_{\boldsymbol{xx}}^{\alpha}(\tau)\boldsymbol{c} \tag{15}$$

referred to as the LS-SCORE spatial filter. Alternatively, if $s(n)$ exhibits conjugate cyclostationarity, then $u(n)$ is given by

$$u(n) = \boldsymbol{c}^H\boldsymbol{x}^*(n-\tau)\, e^{j2\pi\alpha n}, \tag{16}$$

which yields

$$\boldsymbol{w} = \hat{\boldsymbol{R}}_{\boldsymbol{xx}}^{-1}\hat{\boldsymbol{R}}_{\boldsymbol{x}u} = \hat{\boldsymbol{R}}_{\boldsymbol{xx}}^{-1}\hat{\boldsymbol{R}}_{\boldsymbol{xx}^*}^{\alpha}(\tau)\boldsymbol{c}.$$

4.4.2 Cross-SCORE

The preceding derivation of the LS-SCORE algorithm leaves some questions unanswered, one of which is "How is $\boldsymbol{c}$ chosen so as to minimize convergence time?" Without resorting to prior knowledge of the spatial characteristics of $s(n)$ or $\boldsymbol{i}(n)$, a direct method is to maximize $\left|\hat{\rho}_{\hat{s}u}\right|^2$ with respect to both $\boldsymbol{w}$ and $\boldsymbol{c}$,

$$\max_{\boldsymbol{w},\boldsymbol{c}} \left|\hat{\rho}_{\hat{s}u}\right|^2 \Longleftrightarrow \max_{\boldsymbol{w},\boldsymbol{c}} \frac{\left|\boldsymbol{w}^H\hat{\boldsymbol{R}}_{\boldsymbol{xx}}^{\alpha}(\tau)\boldsymbol{c}\right|^2}{\left[\boldsymbol{w}^H\hat{\boldsymbol{R}}_{\boldsymbol{xx}}\boldsymbol{w}\right]\left[\boldsymbol{c}^H\hat{\boldsymbol{R}}_{\boldsymbol{xx}}\boldsymbol{c}\right]} \tag{17}$$

for which the solutions are the dominant eigenvectors $\boldsymbol{w}_1$ and $\boldsymbol{c}_1$, respectively:

$$\begin{aligned} \hat{\boldsymbol{R}}_{\boldsymbol{xx}}^{\alpha}(\tau)\hat{\boldsymbol{R}}_{\boldsymbol{xx}}^{-1}\hat{\boldsymbol{R}}_{\boldsymbol{xx}}^{\alpha H}(\tau)\boldsymbol{w}_m &= \lambda_m\hat{\boldsymbol{R}}_{\boldsymbol{xx}}\boldsymbol{w}_m \\ \hat{\boldsymbol{R}}_{\boldsymbol{xx}}^{\alpha H}(\tau)\hat{\boldsymbol{R}}_{\boldsymbol{xx}}^{-1}\hat{\boldsymbol{R}}_{\boldsymbol{xx}}^{\alpha}(\tau)\boldsymbol{c}_m &= \lambda_m\hat{\boldsymbol{R}}_{\boldsymbol{xx}}\boldsymbol{c}_m, \end{aligned} \tag{18}$$

where $m = 1, \ldots, M$ and $\lambda_1 \geq \cdots \geq \lambda_M$.

Since this algorithm maximizes the cyclic cross-correlation coefficient (or cross spectral coherence) between $\hat{s}(n)$ and $\boldsymbol{c}^H\boldsymbol{x}(n-\tau)$, it is called the *Cross-SCORE* algorithm. Loosely speaking, it adapts $\boldsymbol{w}$ and $\boldsymbol{c}$ to maximize the degree of cyclostationarity exhibited by $\hat{s}$ at cycle frequency α and lag τ.

Another question unanswered in the discussion of LS-SCORE is, "How can multiple SOI waveforms be estimated (i.e., when more than one signal has the chosen cycle frequency α)?" In the following, assume that L_α SOIs have cycle frequency α. In [7] it is shown that the L_α most dominant eigenvectors $\boldsymbol{w}_1, \ldots, \boldsymbol{w}_{L_\alpha}$ in (19), which correspond to the L_α highest stationary points of $\left|\hat{\rho}_{\hat{s}u}\right|^2$ (i.e., locations at which the gradients with respect to $\boldsymbol{w}$ and $\boldsymbol{c}$ are both equal to zero), are spatial filter weights that can be used to obtain estimates of the L_α signals having cycle frequency α and lag

τ. It is also shown that each of these signal estimates has nearly the same SINR as the corresponding MMSE signal estimate, provided that a sufficient number of data samples are available for adaptation, that input SNR is not too low, that fewer signals than sensors arrive at the array, the array response vectors are linearly independent, no multipath is present, and that each signal has a different feature strength. The condition on the feature strengths can be easily derived in the absence of noise and effects due to finite-averaging by simplifying (19) to yield

$$\boldsymbol{R}_{ss}^{\alpha}(\tau)\boldsymbol{R}_{ss}^{-1}\boldsymbol{R}_{ss}^{\alpha H}(\tau)\boldsymbol{g} = \lambda \boldsymbol{R}_{ss}\boldsymbol{g} \tag{19}$$

where $\boldsymbol{g} = \boldsymbol{A}(\Theta)^H\boldsymbol{w}$ and it is assumed that the L_α columns of $\boldsymbol{A}(\Theta)$ are linearly independent, and noticing that perfect signal separation implies that the eigenvectors $[\boldsymbol{g}_1 \ \cdots \ \boldsymbol{g}_{L_\alpha}]$ are a permutation of the identity matrix $\boldsymbol{I}$, which occurs only when $\boldsymbol{R}_{ss}$ and $\boldsymbol{R}_{ss}^{\alpha}(\tau)$ are diagonal and $\left|\rho_{s_n s_n}^{\alpha}(\tau)\right| \neq \left|\rho_{s_m s_m}^{\alpha}(\tau)\right|$ for all $m \neq n$. Thus, the SOIs must be uncorrelated and have distinct feature strengths for perfect signal separation in Cross-SCORE. In practice, noise is present (so (19) does not apply), and thus signal separation is less than perfect. Nonetheless, signal separation has been observed in simulations to be adequate, provided that the input SNR (but not necessarily the input SINR) is positive.

Alternatively, as in LS-SCORE, conjugate cyclostationarity (e.g., doubled carrier features) may be exploited by solving

$$\begin{aligned} \hat{\boldsymbol{R}}_{xx^*}^{\alpha}(\tau)\hat{\boldsymbol{R}}_{xx}^{-*}\hat{\boldsymbol{R}}_{xx^*}^{\alpha H}(\tau)\boldsymbol{w}_m &= \lambda_m \hat{\boldsymbol{R}}_{xx}\boldsymbol{w}_m \\ \hat{\boldsymbol{R}}_{xx^*}^{\alpha H}(\tau)\hat{\boldsymbol{R}}_{xx}^{-1}\hat{\boldsymbol{R}}_{xx^*}^{\alpha}(\tau)\boldsymbol{c}_m &= \lambda_m \hat{\boldsymbol{R}}_{xx}^{*}\boldsymbol{c}_m . \end{aligned} \tag{20}$$

In practical terms, Cross-SCORE avoids the need for a known training signal, albeit at the expense of increased convergence time relative to the MMSE method. The benefits of this to cellular communication system design are discussed in Section 4.5.1. For example, using only the knowledge of the baud rate of desired bauded digital communications signals (e.g., PSK, QAM, etc.) or only the doubled carrier of communications signals that exhibit cyclic conjugate correlation at the doubled carrier frequency (e.g., BPSK, DSB-AM, NBFM), the Cross-SCORE and conjugate Cross-SCORE algorithms, respectively, can nearly achieve the maximum SINR attainable by an LTI beamformer [7].

A brief illustration of the performance of LS-SCORE and Cross-SCORE is given here. In the computer simulations, 100 independent trials are performed and the average SINR at the output of the SCORE processor is computed as a function of the number N of time samples used for adaptation. In each trial, independent BPSK signals having baud rates 0.25 and 0.2, carrier frequencies 0 and 0.1, and angles of arrival 0 and 20 degrees, respectively, arrive at a 4-element ULA in the presence of additive white Gaussian noise that is uncorrelated from sensor to sensor. Both signals use Nyquist-shaped pulses with 100% excess bandwidth and have SNR equal to 10 dB. LS-SCORE and Cross-SCORE are both applied to the data, first with $\alpha = 0.25$ and then using the conjugate form with $\alpha = 0$; that is, first the baud-rate feature and then the doubled-carrier feature of the first signal is exploited. Similar results are obtained using the cycle frequencies of the second BPSK signal. As shown in Fig. 5, Cross-SCORE adapts more quickly than LS-SCORE because Cross-SCORE adapts both $\boldsymbol{w}$ and $\boldsymbol{c}$. Convergence is much quicker for doubled-carrier features than for baud-rate features because the feature strength at the doubled carrier is much stronger than at the baud rate; a stronger feature enables more reliable discrimination between the desired signal and the interference, thus reducing convergence time. This is analogous to the fact that the MMSE method adapted with a slightly noisy training signal converges more quickly than the same method adapted with a very noisy training signal. Indeed, the results in Fig. 5 show that Cross-SCORE, using the doubled-carrier feature, converges at nearly the same rate as the MMSE method, which uses perfect knowledge of the signal waveform.

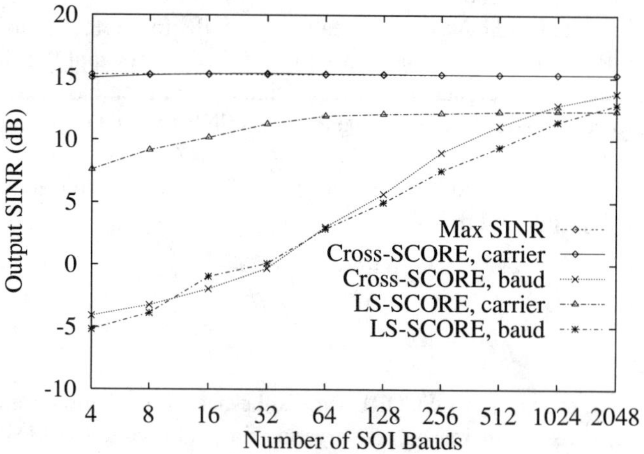

Figure 5: Average output SINR of LS-SCORE and Cross-SCORE for different feature choices.

Other performance results are given in Section 4.4.8, and in [5, 6, 7, 72], where it is also shown that, in certain environments in which a baud-rate feature is being used, Cross-SCORE can converge much more quickly than LS-SCORE, in contrast to the results shown in Fig. 5.

4.4.3 A Maximum-Likelihood Interpretation

In [81] a constrained conditional maximum-likelihood (CCML) STF is found for unknown cyclostationary signals in stationary zero-mean complex white Gaussian noise having arbitrary unknown spatial covariance matrix. Since the publication of [81], a more complete interpretation, described here, has been developed.

The derivation of the algorithm consists of two steps: 1) the CCML STF is developed to obtain a set of optimal (in the CCML sense) data-derived training signals for the SOIs; and 2) the resulting estimates of the training signals replace the ideal training signal in (11) to yield the spatial filter weights used to estimate the SOI waveforms. The two-step procedure is adopted because it will be seen that the data-derived training signals cannot be high-quality estimates of the SOIs, but they *can* be effective training signals.

Step 1: CCML Data-Derived Training Signal Before becoming embroiled in the details of the algorithm, the justifications for the qualifiers *constrained* and *conditional* are presented.

The STF can be constrained to be LCL-PTV to directly generalize the single cycle frequency transformations (13) and (16) used in LS-SCORE. The general LCL-PTV structure provides a way to exploit multiple cyclostationarity properties simultaneously. Another motivation for using LCL-PTV structures is that it is known that this structure, when properly adapted, yields the minimum MSE for scalar cyclostationary signals in noise [31] (and it's conjectured to yield the MMSE for vector-valued cyclostationary signals). Also, the LCL-PTV filtering structure is both tractable to work with analytically and feasible to implement.

However, other signal properties could also be exploited by simply choosing $\boldsymbol{y}(n)$ to be some other transformation of $\boldsymbol{x}(n)$. For example, either of the Kronecker products $\boldsymbol{x}(n) \otimes \boldsymbol{x}(n-\tau)$, $\boldsymbol{x}(n) \otimes \boldsymbol{x}(n-\tau_1) \otimes \boldsymbol{x}(n-\tau_2)$, variations of these involving conjugation of some terms, and further variations in which frequency shifts are included, could be useful transformations of $\boldsymbol{x}(n)$ that yield data-derived training signals. These examples could be appropriate for exploitation of higher-order stationarity or higher-order cyclostationarity. These statistical properties are discussed in depth in [87].

The ML problem is conditional (on the unknown waveforms of the cyclostationary signals) because noise can often be accurately modeled as being stationary and Gaussian, whereas communication signals of interest are almost never Gaussian nor stationary, and their probability distribution functions are typically unknown or virtually intractable to work with. In particular, in some signal interception and signal classification applications, almost nothing might be known about the signals of interest. Also, the conditional ML problem easily admits the user-programmable constraints just described.

The vector of sampled complex envelopes at the output of an M-element sensor array is denoted by $\boldsymbol{x}(n)$, which can be modeled under the narrowband assumption by

$$\boldsymbol{x}(n) = \boldsymbol{A}\boldsymbol{s}(n) + \boldsymbol{i}(n)$$

where $\boldsymbol{s}(n)$ denotes the vector of L unknown signals, $\boldsymbol{A}$ denotes the $M \times L$ matrix of unknown array response vectors of the signals, and $\boldsymbol{i}(n)$ denotes the noise. The estimates of the data-derived training signals are constrained to be the linear combinations

$$\hat{\boldsymbol{d}}(n) = \boldsymbol{W}_y^H \boldsymbol{y}(n) \tag{21}$$

where $\boldsymbol{y}(n)$ is almost any specific realization of the general LCL-PTV form in (8) or another suitable transformation as discussed previously. It is necessary to constrain the processor structure to something other than the usual linear processor $\hat{\boldsymbol{d}}(n) = \boldsymbol{W}^H\boldsymbol{x}(n)$ since the resulting solution for $\boldsymbol{W}$ in this case could be any arbitrary matrix having column-rank equal to L, as will be seen in Section 4.4.6. Furthermore, for the same reason, $\boldsymbol{x}(n)$ cannot appear in unmodified form in $\boldsymbol{y}(n)$. Thus, $\hat{\boldsymbol{d}}(n)$ cannot in general be a high-quality estimate of $s(n)$, but it will be seen to be a useful training signal.

The constrained conditional likelihood function for the estimates $\hat{\boldsymbol{A}}$, $\hat{\boldsymbol{D}}$, and $\hat{\boldsymbol{R}}_{ii}$ of $\boldsymbol{A}$, $\boldsymbol{S} \triangleq \{s(n) : 1 \leq n \leq N\}$, and $\boldsymbol{R}_{ii} \triangleq \langle \boldsymbol{i}(n)\boldsymbol{i}(n)^H \rangle_\infty$, respectively, is given by

$$L(\hat{\boldsymbol{A}}, \hat{\boldsymbol{D}}, \hat{\boldsymbol{R}}_{ii}) = \left|\pi\hat{\boldsymbol{R}}_{ii}\right|^{-N} \exp\left\{-N\left\langle \boldsymbol{e}(n)^H\hat{\boldsymbol{R}}_{ii}^{-1}\boldsymbol{e}(n)\right\rangle_N\right\}$$

where $\boldsymbol{e}(n) = \boldsymbol{x}(n) - \hat{\boldsymbol{A}}\hat{\boldsymbol{d}}(n)$.

By application of results from matrix calculus [41] and complex gradients of nonanalytic functions [14], it can be shown that the CCML estimates are given by

$$\begin{aligned}\hat{\boldsymbol{R}}_{ii}^{(ML)} &= \hat{\boldsymbol{R}}_{xx} - \hat{\boldsymbol{A}}^{(ML)}\hat{\boldsymbol{R}}_{x\hat{d}}^H - \hat{\boldsymbol{R}}_{x\hat{d}}\left[\hat{\boldsymbol{A}}^{(ML)}\right]^H + \hat{\boldsymbol{A}}^{(ML)}\hat{\boldsymbol{R}}_{\hat{d}\hat{d}}\left[\hat{\boldsymbol{A}}^{(ML)}\right]^H, \\ \hat{\boldsymbol{A}}^{(ML)} &= \hat{\boldsymbol{R}}_{x\hat{d}}\hat{\boldsymbol{R}}_{\hat{d}\hat{d}}^{-1},\end{aligned}$$

and $\hat{\boldsymbol{d}}(n)$ is given by (21) where $\boldsymbol{W}$ has L columns given by any full-column-rank linear combination of the L most dominant eigenvectors $\boldsymbol{E}$ of the matrix $\boldsymbol{T}_y$ defined by

$$\boldsymbol{T}_y = \hat{\boldsymbol{R}}_{yy}^{-1}\hat{\boldsymbol{R}}_{xy}^H\hat{\boldsymbol{R}}_{xx}^{-1}\hat{\boldsymbol{R}}_{xy}. \tag{22}$$

That is, $\boldsymbol{W}_y = \boldsymbol{EQ}$ for any full-rank $L \times L$ matrix $\boldsymbol{Q}$.

If, in addition to the constraint on the processor structure, it is required that the estimated signals be uncorrelated, $\boldsymbol{W}_y^H\hat{\boldsymbol{R}}_{yy}\boldsymbol{W}_y = \boldsymbol{I}$, then the log-likelihood is maximized *only* by the L most dominant eigenvectors of $\boldsymbol{T}_y$, instead of any linear combination thereof; that is, $\boldsymbol{W}_y = \boldsymbol{E}$. A mathematical result related to the solution (22) was proposed in [112] and proven in [86]. Also, the result of this section includes that in [4] as the special case for which one signal is present ($L = 1$) and $\boldsymbol{y}(n) = \boldsymbol{x}^*(n)$.

An alternate proof of (22) can be obtained by noting that maximizing $L(\hat{\boldsymbol{A}}^{(ML)}, \hat{\boldsymbol{D}}, \hat{\boldsymbol{R}}_{ii}^{(ML)})$ is equivalent to maximizing

$$f(\boldsymbol{W}_y) = \prod_{m=1}^{M} \lambda_m(\boldsymbol{W}_y^H\hat{\boldsymbol{R}}_{xy}^H\hat{\boldsymbol{R}}_{xx}^{-1}\hat{\boldsymbol{R}}_{xy}\boldsymbol{W}_y, \boldsymbol{W}_y^H\hat{\boldsymbol{R}}_{yy}\boldsymbol{W}_y),$$

where $\lambda_m(\cdot, *)$ denotes the mth generalized eigenvalue of the matrix pair $(\cdot, *)$, and then applying the Poincaré Separation Theorem for generalized eigenvalues of a pair of Hermitian matrices (e.g., [62]). In any case, the additional constraint that the estimated signals be uncorrelated yields a unique solution $\boldsymbol{W}_y$ that can also be obtained from a different perspective as discussed in Section 4.4.4.

Step 2: Using the Data-Derived Training Signal In this step, $\hat{\boldsymbol{d}}(n)$ is used as a training signal to minimize $MSE(\hat{s}, \hat{\boldsymbol{d}})$.

Direct substitution of $\hat{\boldsymbol{d}} = (\boldsymbol{EQ})^H \boldsymbol{y}(n)$ into

$$\boldsymbol{W}_x = \hat{\boldsymbol{R}}_{\boldsymbol{xx}}^{-1} \hat{\boldsymbol{R}}_{\boldsymbol{x}\hat{\boldsymbol{d}}}$$

and algebraic manipulation reveal that $\boldsymbol{W}_x$ can be any full-rank linear combination of the L most dominant eigenvectors of $\boldsymbol{T}_x$,

$$\boldsymbol{T}_x = \hat{\boldsymbol{R}}_{\boldsymbol{xx}}^{-1} \hat{\boldsymbol{R}}_{\boldsymbol{xy}} \hat{\boldsymbol{R}}_{\boldsymbol{yy}}^{-1} \hat{\boldsymbol{R}}_{\boldsymbol{xy}}^H. \tag{23}$$

As before, if the elements of $\hat{\boldsymbol{s}}(n) = \boldsymbol{W}_x^H \boldsymbol{x}(n)$ must be uncorrelated, then the additional constraint $\boldsymbol{W}_x^H \hat{\boldsymbol{R}}_{\boldsymbol{xx}} \boldsymbol{W}_x = \boldsymbol{I}$ implies that $\boldsymbol{W}_x$ is exactly the L most dominant eigenvectors of $\boldsymbol{T}_x$.

4.4.4 Canonical Correlation Analysis

In this section the processor derived in Section 4.4.3 is derived, from a different perspective, as the solution to a canonical correlation analysis (CCA) problem.

The CCML beamformer maximizes the conditional likelihood of the received data subject to the constraint that the training signal estimates of the unknown signals are obtained by means of an LCL-PTV processor (or possibly other processor, as discussed previously). An alternative interpretation is that the CCML beamformer obtains optimal (in the CCML sense) estimates of the unknown signals appearing in the received data set as functions of another data set (which happens to be derived from the received data set in a constrained optimal way). That is, the CCML beamformer obtains estimates of the signals that are common to both the received data $\boldsymbol{x}(n)$ and the derived data $\boldsymbol{y}(n)$. In multivariate statistics, this task is referred to as *canonical correlation analysis* (e.g., [15, 48, 86]).

In the canonical correlation analysis of two data sets $\boldsymbol{x}(n)$ and $\boldsymbol{y}(n)$ that are believed to share some number L of signals jointly denoted by $\boldsymbol{s}(n)$, it is desired to minimize the mean-squared error between the estimates of $\boldsymbol{s}(n)$ linearly obtained from $\boldsymbol{x}(n)$ by a processor $\boldsymbol{W}_x^H \boldsymbol{x}(n)$ and the estimates of $\boldsymbol{s}(n)$ linearly obtained from $\boldsymbol{y}(n)$ by a processor $\boldsymbol{W}_y^H \boldsymbol{y}(n)$, subject to the constraints that $\boldsymbol{W}_x^H \hat{\boldsymbol{R}}_{\boldsymbol{xx}} \boldsymbol{W}_x = \boldsymbol{I}$ and $\boldsymbol{W}_y^H \hat{\boldsymbol{R}}_{\boldsymbol{yy}} \boldsymbol{W}_y = \boldsymbol{I}$. Mathematically, this can be accomplished by minimizing

$$\left\langle \left\| \boldsymbol{W}_x^H \boldsymbol{x}(n) - \boldsymbol{W}_y^H \boldsymbol{y}(n) \right\|^2 \right\rangle_N \tag{24}$$

subject to the constraints that $\boldsymbol{W}_x^H \hat{\boldsymbol{R}}_{\boldsymbol{xx}} \boldsymbol{W}_x = \boldsymbol{I}$ and $\boldsymbol{W}_y^H \hat{\boldsymbol{R}}_{\boldsymbol{yy}} \boldsymbol{W}_y = \boldsymbol{I}$. The resulting weight matrices $\boldsymbol{W}_x$ and $\boldsymbol{W}_y$ are given by the L most dominant eigenvectors of

$$\boldsymbol{T}_x = \hat{\boldsymbol{R}}_{\boldsymbol{xx}}^{-1} \hat{\boldsymbol{R}}_{\boldsymbol{xy}} \hat{\boldsymbol{R}}_{\boldsymbol{yy}}^{-1} \hat{\boldsymbol{R}}_{\boldsymbol{xy}}^H \tag{25}$$

and

$$\boldsymbol{T}_y = \hat{\boldsymbol{R}}_{\boldsymbol{yy}}^{-1} \hat{\boldsymbol{R}}_{\boldsymbol{xy}}^H \hat{\boldsymbol{R}}_{\boldsymbol{xx}}^{-1} \hat{\boldsymbol{R}}_{\boldsymbol{xy}}, \tag{26}$$

respectively.

4.4.5 Contrast Between CCML and CCA

Clearly, $\boldsymbol{T}_y$ in (26) and $\boldsymbol{T}_y$ in (22) are identical, which shows that the CCML and CCA frameworks yield identical algorithms when the signal estimates are constrained to be uncorrelated. Thus, a primary utility of having these two approaches is simply a conceptual one, providing the engineer with some flexibility in how the application of interest is modeled or conceptualized.

The CCML conceptual framework is obviously applicable when the spatial covariance of the interference and noise $\boldsymbol{i}(n)$ is unknown, but it requires that $\boldsymbol{i}(n)$ be stationary, white, and Gaussian. In contrast, the CCA conceptual framework is not obviously optimum when $\boldsymbol{R}_{ii}(0)$ is unknown (although from its equivalence with CCML we know it is), but its simple least-squares formulation does not require that $\boldsymbol{i}(n)$ be stationary, white, or Gaussian.

The other contrast of interest here is that CCML can be formulated without the constraint that the signal estimates be uncorrelated, whereas CCA cannot so easily admit this flexibility. In many applications, uncorrelated signal estimates are required. However, in the presence of multipath it is conjectured (based on results for Cross-SCORE in [7]) that multiple signal estimates, one for each propagation path of a SOI, are obtained by CCML. In this case, the desired action of the processor is first to reject other SOIs, interference, and noise from the multiple signal estimates of a single SOI, and then to linearly combine these multiple estimates to mitigate the multipath. Thus, a benefit of the CCML framework, without the constraint that signal estimates be uncorrelated, is that linearly combining the multipath estimates is the optimal type of processing (in the CCML sense). Note, however, that the CCML framework does not suggest *how* to choose the weights in the linear combiner. Another adaptive processor (e.g., one of the blind equalization algorithms reviewed in [21], or either of the algorithms in Section 5) could be configured to perform this task.

4.4.6 Special Cases of CCML/CCA

The Cross-SCORE and conjugate Cross-SCORE [7] blind adaptive beamformers are specific realizations of the CCML/CCA beamformer. Two special choices of $\boldsymbol{y}(n)$ merit consideration. The first choice is $\boldsymbol{y}(n) = \boldsymbol{x}(n - \tau)e^{j2\pi\alpha n}$ for which each of $\boldsymbol{W}_x^H\boldsymbol{x}(n)$ and $\boldsymbol{W}_y^H\boldsymbol{y}(n)$ is a CCML and CCA beamformer. Also, $\boldsymbol{W}_x^H\boldsymbol{x}(n)$ in this case is exactly the Cross-SCORE blind adaptive STF (19). The block diagram of the signal processor for this case is shown in Fig. 6, although the general CCML/CCA processor can be obtained by replacing the dash-boxed section with the diagram for the appropriate LCL-PTV transformation of $\boldsymbol{x}(t)$ (i.e., the relevant part of Fig. 4).

Similarly, the second choice is $\boldsymbol{y}(n) = \boldsymbol{x}(n-\tau)^* e^{j2\pi\alpha n}$ for which each of $\boldsymbol{W}_x^H\boldsymbol{x}(n)$ and $\boldsymbol{W}_y^H\boldsymbol{y}(n)$ is a CCML and CCA beamformer. Also, $\boldsymbol{W}_x^H\boldsymbol{x}(n)$ in this case is exactly the *conjugate* Cross-SCORE blind adaptive beamformer (21).

The observation that Cross-SCORE and conjugate Cross-SCORE are CCA beamformers was first made in [74] where it was used only to motivate the Cyclic Correlation Significance Test for estimating the number of signals having a specified cycle frequency (see also Section 7.2.1).

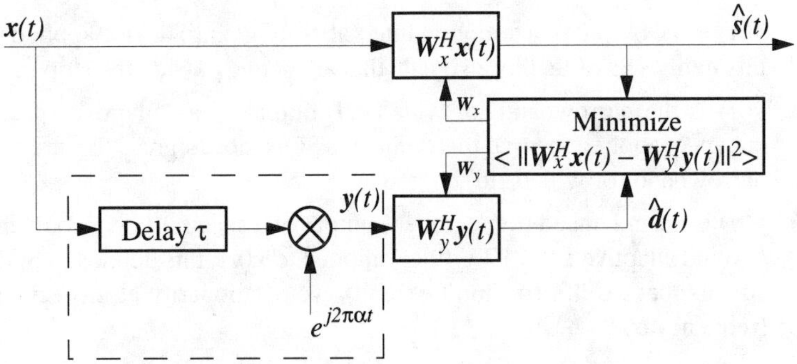

Figure 6: Block diagram of Cross-SCORE processor.

4.4.7 Programmable CCA

Here the restriction on the choice of $\boldsymbol{y}(n)$ mentioned earlier is discussed, and it is noted that the CCML/CCA framework can exploit a wider variety of signal properties than just those related to cyclostationarity.

From (24) it can be seen that $\boldsymbol{y}(n)$ should not contain $\boldsymbol{x}(n)$ as a literal element, since any solution of the form

$$\boldsymbol{W}_y = \begin{bmatrix} \boldsymbol{W}_x \\ \boldsymbol{W} \end{bmatrix}$$

would minimize (24); this observation implies that this CCML/CCA framework cannot directly yield a blind Cyclic Wiener filter.

As alluded to earlier, the CCML/CCA framework need not use an LCL-PTV transformation to obtain $\boldsymbol{y}(n)$. Indeed, the transformation is entirely up to the user, provided that the restriction just discussed is met. To emphasize this flexibility, the *programmable canonical correlation analyzer* (PCCA) is proposed, wherein the transformation used to obtain $\boldsymbol{y}(n)$ is completely programmable by the user. Thus, the PCCA can use many types of signal properties to distinguish between desired signals and interference. A nonexhaustive list of transformations is proposed here:

1. $\boldsymbol{y}(n)$ is a frequency-shifted (by α) and delayed (by τ) version of $\boldsymbol{x}(n)$ or $\boldsymbol{x}^*(n)$, which yield the Cross-SCORE and conjugate Cross-SCORE algorithms, respectively; this defines as SOIs those signals that exhibit cyclostationarity or conjugate cyclostationarity with cycle frequency α, and can be generalized to multiple frequency shifts, multiple delays, and prefiltering.
2. $\boldsymbol{y}(n)$ is the output of a band-stop (or band-pass) LTI filter applied to $\boldsymbol{x}(n)$; this defines as SOIs those signals that have spectral support outside (or inside) the stop-band (or pass-band), and can be generalized to more complicated regions of spectral support.
3. $\boldsymbol{y}(n)$ is a delayed version of $\boldsymbol{x}(n)$; this defines as SOIs those signals for which the coherence time is greater than or equal to τ.

4. $y(n)$ is the output of a temporal interval-stop (gating) device applied to $x(n)$; this defines as SOIs those signals that are active outside the stop intervals.
5. $y(n)$ is the narrowband (or wideband) output of an adaptive spectral-line enhancer applied to $x(n)$; this defines as SOIs those signals that are relatively narrowband (or wideband).
6. $y(n)$ is the enhanced (or degraded) output of a spectral-correlation enhancer (a blind adaptive LCL-PTV filter) applied to $x(n)$; this defines as SOIs those signals that exhibit (or don't exhibit) cyclostationarity at a specified cycle frequency α.
7. $y(n)$ is the constant modulus (or nonconstant modulus) output of an LTI filter (or LTI canceller) adapted by the constant modulus (CM) algorithm (CMA); this defines as SOIs those signals that have (or do not have) constant modulus.
8. $y(n)$ is the output of a demodulation/remodulation device that is applied to $x(n)$ and is structured to select FM, PM, FSK, or PSK signals.
9. $y(n)$ is the output of a nonlinear transformation such as $x(n) \odot x(n) \odot x(n)$, $x(n) \odot x(n) \odot x^*(n)$, $x(n) \otimes x(n) \otimes x(n)$, $x(n) \otimes x(n) \otimes x^*(n)$, or time-variant nonmemoryless generalizations thereof, where $\odot$ denotes the elementwise product and $\otimes$ denotes the Kronecker product; this defines as SOIs those signals that have the higher-order stationarity or higher-order cyclostationarity properties selected for by the chosen transformation.

$\boxed{\alpha}$ Although empirical results on the performance of Cross-SCORE for finite numbers of time samples are available in [5, 6, 7, 35, 72, 81], no analytical results are available, such as the mean and variance of the output SINR as a function of the number N of time samples. It is likely that existing results in the multivariate statistics literature on the behavior of canonical correlation analyzers could be applicable to the performance analysis of Cross-SCORE and the CCML/CCA generalizations thereof. Also, as mentioned, it is of interest to determine whether the coherent combination of $\boldsymbol{W}_x^H \boldsymbol{x}(n)$ and $\boldsymbol{W}_y^H \boldsymbol{y}(n)$ is useful as an LCL-PTV STF.

4.4.8 Simulation Results

Here the performance of the CCA beamformer is briefly illustrated via computer simulations as in [81].

In the simulated environment, a 4-element ULA having half-wavelength sensor-spacing receives 2 BPSK signals in the presence of stationary complex white Gaussian noise. The two signals are spectrally overlapping and have equal signal-to-noise ratios (SNRs) but arrive from different directions (0 and 20 degrees) and have different baud rates (0.25 and 0.33 times the sampling rate). The average output SINR was obtained by averaging the output SINRs from one hundred independent trials. Two different input SNRs (0 dB and 10 dB) are considered. In each case, the average output SINR is evaluated for different numbers of cycle frequencies α and filters $h_k(n) = \delta_{n-\tau_k}$ in (8). In particular, 6 combinations of one or two values of α (0.25 and -0.25)

and 3 values of τ (0, 1, and -1) are considered; since these values of α correspond to cycle frequencies of the BPSK signal with baud rate 0.25 (because the cyclic autocorrelations at these cycle frequencies and lags τ are nonzero), this signal is extracted by the spatial filter. If α were chosen to be 0.33 and/or -0.33, then the other BPSK signal would be chosen. If only one value of α and one value of τ are used, then the CCA beamformer is equivalent to the Cross-SCORE beamformer. As seen in Figs. 7 and 8, the use of multiple values of α and τ substantially improves convergence, with between one-half and one-eighth as many data samples required to achieve the same performance as Cross-SCORE.

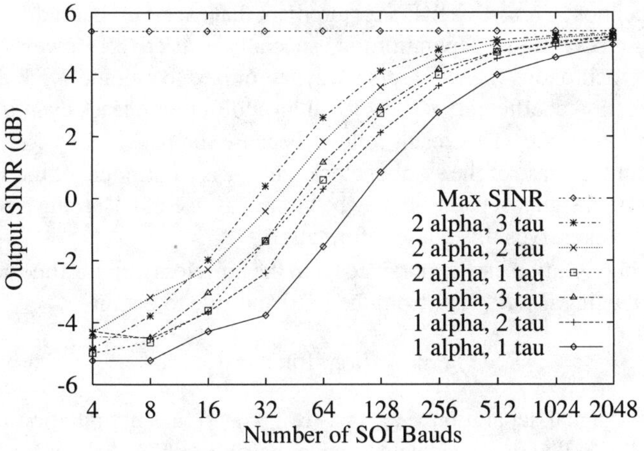

Figure 7: Average output SINR for different numbers of cycle frequencies and lags. The input SNR is 0 dB.

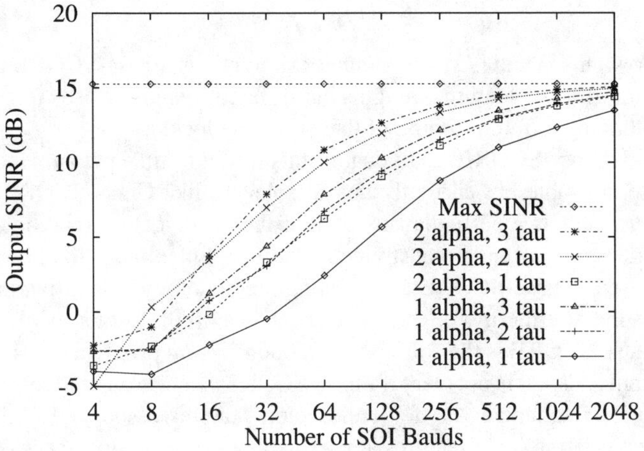

Figure 8: Average output SINR for different numbers of cycle frequencies and lags. The input SNR is 10 dB.

4.4.9 Phase-SCORE

One very important occurrence of multiple spectrally overlapping SOIs having the same cycle frequency α is in modern communication networks based on code-division multiple-access (CDMA). In CDMA, multiple signals can share the same spectral band because each has the unique code used by the transmitter to spread the message signal across a wide spectral band which is also used by the receiver to despread it. However, this code does not despread the signals of other users but instead spreads them even further. Depending upon the choice of spreading codes, the feature strengths of the CDMA signals of the multiple users can be identical. Thus, in this application, Cross-SCORE's requirement that each SOI has a different feature strength (to enable signal separation) is especially restrictive. However, unless the network is synchronous (i.e., all users are synchronized to a single symbol clock), the cyclic correlation coefficients $\rho^{\alpha}_{s_m s_m}(\tau)$ will have different phases even if the feature strengths are the same. This result follows because the phase of $\rho^{\alpha}_{s_m s_m}(\tau)$ is linearly dependent on the phase of the symbol clock when α is a harmonic of the symbol rate. Thus, the task of signal separation may be viewed as one of designing a SCORE-like algorithm that preserves this phase information.

An ad hoc solution can be obtained from the noiseless infinite-time analysis (19) by simply deleting $\boldsymbol{R}_{ss}^{-1}\boldsymbol{R}_{ss}^{\alpha H}(\tau)$ from the left-hand side to obtain

$$\boldsymbol{R}^{\alpha}_{ss}(\tau)\boldsymbol{g}_l = \lambda_l \boldsymbol{R}_{ss}\boldsymbol{g}_l \text{ for } l = 1, \ldots, L_{\alpha}. \tag{27}$$

Here, perfect signal separation (i.e., $[\boldsymbol{g}_1, \ \ldots \ , \ \boldsymbol{g}_{L_\alpha}]$ is a permutation of $\boldsymbol{I}$) occurs only when the SOIs are uncorrelated and $\rho^{\alpha}_{s_m s_m}(\tau) \neq \rho^{\alpha}_{s_n s_n}(\tau)$ for all $m \neq n$. That is, the phase information of the feature at α is preserved and is used to distinguish each signal from the others. Working backwards from (27) yields the *Phase-SCORE* algorithm

$$\hat{\boldsymbol{R}}^{\alpha}_{xx}(\tau)\boldsymbol{w} = \lambda \hat{\boldsymbol{R}}_{xx}\boldsymbol{w}. \tag{28}$$

As shown in [7] and [72] via computer simulations, Phase-SCORE can separate SOIs having the same feature strengths but different feature phases (e.g., caused by the signals having different timing of their symbol clocks).

$\boxed{\alpha}$ As with Cross-SCORE, analytical results for finite-time performance of Phase-SCORE are desirable but currently unavailable. Unlike Cross-SCORE, which has been reinterpreted as a particular instance of the canonical correlation analyzer, or as the solution to a particular constrained conditional maximum-likelihood problem, no link between Phase-SCORE and existing methodology in multivariate statistics has been found. Furthermore, it is unclear (at least to the author as of this writing) whether Phase-SCORE is the solution to an optimization problem or it is a purely ad hoc algorithm. Also, it is not clear if Phase-SCORE can be extended to accommodate multiple cycle frequencies or if it can exploit features associated with the cyclic conjugate correlation (e.g., features at the doubled carrier frequency).

4.4.10 Wideband SCORE

SCORE can also be applied to wideband environments in which the received data $\boldsymbol{x}(t)$ can be modeled in terms of finite-time Fourier transforms (FTFT):

$$\begin{aligned}\tilde{\boldsymbol{x}}(t,f) &\approx \sum_{l=1}^{L_\alpha} \tilde{\boldsymbol{a}}(\theta_l,f)\tilde{s}_l(t,f)+\tilde{\boldsymbol{i}}(t,f) \\ &= \tilde{\boldsymbol{A}}(\Theta,f)\tilde{\boldsymbol{s}}(t,f)+\tilde{\boldsymbol{i}}(t,f),\end{aligned} \tag{29}$$

where $\tilde{\boldsymbol{x}}(t,f) \triangleq \left(1/\sqrt{\Delta}\right)\int_{t-\Delta/2}^{t+\Delta/2} \boldsymbol{x}(t)\,e^{-j2\pi ft}\,dt$ is the FTFT of the received data, $\tilde{\boldsymbol{a}}(\theta,f)$ is the transfer function of the array for a signal arriving from angle θ, $\tilde{\boldsymbol{s}}(t,f)$ is the FTFT of the L_α cyclostationary signals impinging on the array from angles $\theta_1,\ldots,\theta_{L_\alpha}$, respectively, and having cycle frequency α, and $\tilde{\boldsymbol{i}}(t,f)$ is the FTFT of all other signals and noise that do not have cycle frequency α. The approximation in (30) holds well if the FTFT integration time Δ is greater than the duration of the impulse response of the array. Similarly, the FTFT of the lth extracted signal is given by $\tilde{y}_l(t,f)=\tilde{\boldsymbol{w}}_l(t,f)^H\tilde{\boldsymbol{x}}(t,f)$ for $l=1,\ldots,L$.

Application of the SCORE concept to this problem can be pursued by reexpressing it as a CCA problem. Notice that the FTFT $\tilde{\boldsymbol{y}}(t,f)$ of the auxiliary signal defined by $\boldsymbol{y}(t) \triangleq \boldsymbol{x}(t)\,e^{j2\pi\alpha t}$ is given by

$$\begin{aligned}\tilde{\boldsymbol{r}}(t,f) &= \tilde{\boldsymbol{x}}(t,f-\alpha) \\ &= \tilde{\boldsymbol{A}}(\Theta,f-\alpha)\,\tilde{\boldsymbol{s}}(t,f-\alpha)+\tilde{\boldsymbol{i}}(t,f-\alpha).\end{aligned} \tag{30}$$

Since $\boldsymbol{s}(t)$ is cyclostationary with cycle frequency α, $\tilde{\boldsymbol{s}}(t,f-\alpha)$ is correlated (shares a common component) with $\tilde{\boldsymbol{s}}(t,f)$. Consequently, a solution based on canonical correlation analysis can be found for each value of f:

$$\hat{\boldsymbol{S}}_{xr}(f)\,\hat{\boldsymbol{S}}_{rr}(f)^{-1}\,\hat{\boldsymbol{S}}_{rx}(f)\tilde{\boldsymbol{w}}_l(f) = \lambda_l^{(w)}(f)\hat{\boldsymbol{S}}_{xx}(f)\tilde{\boldsymbol{w}}_l(f) \tag{31}$$

$$\hat{\boldsymbol{S}}_{rx}(f)\,\hat{\boldsymbol{S}}_{xx}(f)^{-1}\,\hat{\boldsymbol{S}}_{xr}(f)\tilde{\boldsymbol{c}}_l(f) = \lambda_l^{(c)}(f)\hat{\boldsymbol{S}}_{rr}(f)\tilde{\boldsymbol{c}}_l(f), \tag{32}$$

where $\hat{\boldsymbol{S}}_{xr}(f)$ is the estimated cross-correlation matrix of $\tilde{\boldsymbol{x}}(t,f)$ and $\tilde{\boldsymbol{r}}(t,f)$ (and can also be interpreted as the cross-periodogram matrix of $\boldsymbol{x}(t)$ and $\boldsymbol{r}(t)$).

Other details of the method, including the regions over which $\tilde{\boldsymbol{w}}_l(f)$ and $\tilde{\boldsymbol{c}}_l(f)$ are defined, are given in [79]. Also, as discussed in [79], Wideband SCORE differs from the frequency-dependent SCORE method presented in [3], which typically requires that a complicated set of coupled matrix equations be solved, which hinders practical application.

4.5 Application Examples

In this section two specific applications of blind-adaptive spatial filtering to digital communication systems are summarized. In the first application, the Cross-SCORE algorithm forms the basis for a completely new architecture for a land mobile digital cellular radio system that accommodates a substantially greater number of users than existing or proposed systems. In the second application, the Wideband SCORE algorithm is applied to a direct-sequence spread spectrum communication system to reduce the amount of processing gain required from spreading alonc, thcrcby allowing an increase in channel capacity and/or a decrease in bit error rate.

α Both of the examples to be presented here are based on the results of computer simulations. A standing open problem in this regard (and, indeed, in the case of many signal-processing algorithms in sensor array processing) is to test the algorithms on real data from an actual sensor array.

4.5.1 Digital Cellular Radio

Demand for mobile communication continues to increase as it becomes easier to use, is more widely available, and offers a greater variety of services. The need for new mobile communication systems having increased spectral efficiency relative to current systems is compounded by increasing demand for radio spectrum allocations from other communication services. Various temporal processing schemes have been proposed for increasing spectral efficiency, including time division multiple access (TDMA) and frequency division multiple access (FDMA), although code division multiple access (CDMA) might offer the greatest potential increase in capacity (e.g., see [37, 51, 57]) in addition to inherently mitigating the effects of multipath. However, except for modifications of these schemes that use fixed multibeam or multisector antennas to further increase capacity, which is roughly equivalent to subdividing each cell into smaller cells, none of these schemes fully exploits the multiplicity of spatial channels that arises because each mobile user occupies a unique spatial location.

All schemes use a large-scale form of space division multiple access (SDMA) by dividing a large geographic area into cells. The mobile users within each cell are served by a base station, shown in Fig. 9 as being at the center of each cell.

Some proposed schemes use SDMA via spatial filtering within each cell to separate spectrally overlapping signals from different users, and have potential increases in spectral efficiency over conventional analog FM-FDMA schemes of a factor of 30 or more (cf. [35, 90]). These schemes adapt the antenna array either by estimating the directions of arrival of the spectrally overlapping signals and then using these estimates to compute appropriate weights for the spatial filter [8], or by minimizing the time-averaged squared error between a known training signal and the output of the spatial filter [42, 90, 105, 111]. In either case, the properly adapted array at the base station can spatially separate spectrally overlapping users in the same cell, as shown in Fig. 10 (where multipath is neglected for clarity).

The schemes based on direction estimation have numerous disadvantages, including computationally intensive algorithms, poor performance in the presence of multipath signals arriving from different directions, the need to measure, store, and update array calibration data, and considerable sensitivity to errors in the array calibration data. The schemes that require a training signal have different disadvantages,

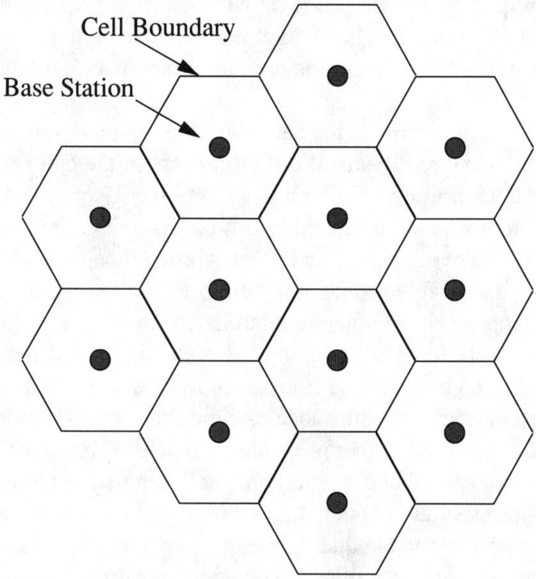

Figure 9: Typical cell structure in a cellular radio system.

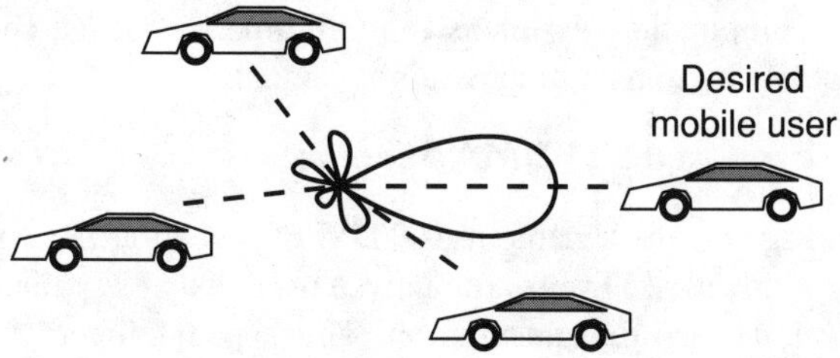

Figure 10: Antenna pattern of array at base station extracting one of several spectrally overlapping signals from mobile users.

including the need to use capacity to periodically transmit the training signal, the need to synchronize the received and locally generated copies of the training signal, and the need to adaptively increase or decrease the duration of the training signal to accommodate varying levels of interference.

The scheme proposed in [35] and described here uses spatial, temporal, and frequency-division multiplexing to allow multiple users to communicate simultaneously, and so is referred to here as a space-time-frequency division multiple access (STFDMA) scheme. This STFDMA scheme has some things in common with the scheme proposed in [42], but the former does not require any reference (training) signal because Cross-SCORE is used instead to adapt the array. Unlike schemes that rely solely on frequency, time, or code division multiplexing and thus use only one spatial channel, the proposed scheme exploits space as well as partial time and frequency division multiplexing and thus uses multiple spatial, temporal, and spectral channels.

An adaptive antenna array at the base station separates the temporally and spectrally overlapping received signals of different users in the cell and transmits directively to each user, exploiting multipath when present. Users whose signals arriving at the base station are spatially separable can be assigned to spectral bands that are completely spectrally overlapping, and users whose signals arriving at the base station are spatially inseparable can be assigned to disjoint spectral bands. In practice, channel assignments might be made without determining spatial separability; detection of substantial interference in another user's signal could trigger an automatic reassignment. Also, signals coming from the individual users can be assigned to time intervals that are interleaved with those assigned to signals coming from the base station. Under the assumption that users are sufficiently well distributed throughout the cell, all available spatial and spectral channels can be used effectively. Since the number of multiple spatial channels that can be separated from each other by the antenna array is approximately equal to the number of antenna elements in the array (which can be quite large), overall capacity can be much greater than schemes using a single spatial channel. Also, unlike adaptive array schemes that require direction estimation processors or known training signals, the proposed scheme uses the conjugate Cross-SCORE algorithm (cf. Section 4.4.2), and thus does not require array calibration data or computationally intensive multidimensional searches nor does it waste channel capacity by transmitting a training signal.

4.5.2 The Proposed STFDMA Scheme

As shown in Fig. 11, the scheme uses TDM of reception (Rx, or uplink) and transmission (Tx, or downlink) frames for a given user so that a spatial filter adapted during reception can be used for transmission. Thus, spatial directivity is preserved for both. Up to 10 users share the same carrier frequency and are separable by the TDM scheme shown in Fig. 11. Thus, the data rate of each user's vocoded speech is 8 kb/s, which must then be doubled due to TDM of Rx and Tx, and multiplied further by a factor of 10 due to TDM of 10 users, to yield an instantaneous data rate of 160 kb/s. The signals are BPSK using 100% excess-bandwidth Nyquist-shaped pulses, for a signal bandwidth of 320 kHz. This data rate is high enough that a sufficient number of independent time samples can be collected for adaptation of the spatial filter over a period of roughly 500 μs, which is short enough that the propagation environment is approximately stationary, given a typical fast fading rate of 100 fades/s for land mobile radio.

As shown in Fig. 12, carrier assignments for different groups of 10 TDM'd users can be such that their signals are almost completely spectrally overlapping, provided that the users assigned to the first TDM time slot in different groups are spatially separable, and similarly for the users assigned to the second through the tenth TDM

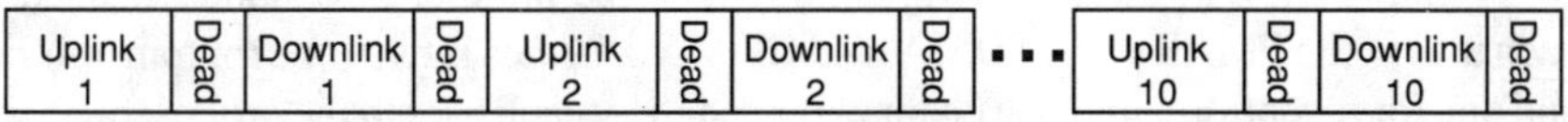

Figure 11: One complete TDM cycle for the uplink (reception) and downlink (transmission) phases at the base station for 10 mobile users sharing the same carrier frequency.

time slots. Since only one user having a particular carrier frequency is active at any given time, conjugate Cross-SCORE is used to adapt a spatial filter to separate that user from all other users. The number of spatially separable users is limited by the number of antennas in the array, which in turn is limited by convergence-time considerations.

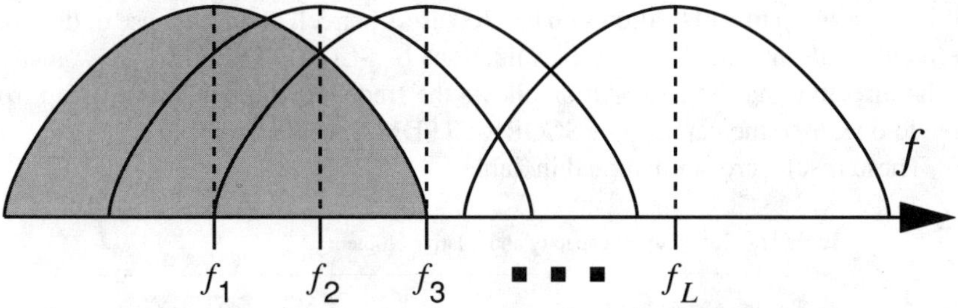

Figure 12: Frequency allocation in the proposed scheme. The spectrum of TDM group number one is shaded to enhance clarity.

In general, for a total system-bandwidth B_t, single-user channel-bandwidth B_c, frequency-reuse factor r, and minimum carrier separation f_{sep}, the maximum number L of users that can be accommodated in one cell by the frequency allocation scheme in Fig. 12 is $L = 10(B_t/r - B_c)/f_{sep} + 10$. The number M of antenna elements required to separate these signals is bounded from below by the number K of users whose signals are spectrally overlapping with any given user's signal, where $M > K = 2(B_c/f_{sep} - 1)$. Using $B_c = 320$ kHz and $f_{sep} = 10$ kHz, at least 63 antenna elements (which can be omnidirectional) are needed to separate the signals of all users, assuming that the energy from each user arrives at the base station from a single direction, and assuming that the users are uniformly distributed throughout the cell. In practice, more antenna elements might be required to achieve adequate performance at full capacity in the presence of spatially separable multipath, although fewer antenna elements can suffice if a lower capacity is opted for and appropriate channel allocation is done. Also, adaptive equalization should follow the spatial filtering to mitigate the time-smearing effects of multipath.

4.5.3 Capacity Example

For the purpose of comparing the potential increase in capacity due to the proposed SCORE-STFDMA scheme relative to the analog FM-FDMA, TDMA, and CDMA schemes as considered in [51], consider a total system-bandwidth of

$B_t = 12.5$ MHz. In the following comparison, the number of channels needed by each user in the FM-FDMA, TDMA, and CDMA schemes is two (one for transmission and one for reception), and one channel is needed by each user in the proposed scheme because transmission and reception are multiplexed in time. With FM-FDMA, using a channel bandwidth of 30 kHz, 2 channels per user, and a frequency reuse factor of 7 yields 30 users per cell. With TDMA, using a channel bandwidth of 30 kHz with three time slots for TDMA, 2 channels per user, and a frequency reuse factor of 4 yields 150 users per cell. With CDMA, using 2 channels per user, a frequency reuse factor of 1, sectorization of 3, and voice activity factor of $3/8$ yields 1200 channels per cell [51] or 600 users per cell. With the proposed SCORE-STFDMA scheme, using a channel bandwidth of 320 kHz, 1 channel per user, a frequency reuse factor of 3, and a carrier separation of 10 kHz yields up to 3850 users per cell. Decreasing the carrier separation to 5 kHz allows up to 7700 users per cell at the expense of doubling the number of antennas. If future evaluations of SCORE-STFDMA show that the spatial directivity at the base station allows the frequency reuse factor to drop from three to one, then the capacity of SCORE-STFDMA would triple.

These results are summarized in Table 1.

Table 1: Summary of capacity and relative efficiency.

Scheme	FM-FDMA	TDMA	CDMA	SCORE-STFDMA $f_{sep} = 10$ kHz
Users/cell	30	150	600	3850
Rel. Eff.	1	5	20	128

4.5.4 Performance

In this section results of computer simulations of the proposed SCORE-STFDMA scheme illustrate its potential.

The antenna array consists of 64 omnidirectional elements spaced one-half wavelength apart on a circle that is coplanar with the mobile units (i.e., the elevation angle is assumed to be negligible). This would require an array diameter of 3 meters for a system center frequency of 1 GHz, which is reasonable for a base station. The noise is zero-mean complex temporally-white Gaussian noise that is independent and identically distributed for each sensor. The signal-to-noise ratio (SNR) of each signal is defined to be the power of this signal received along the direct path divided by the power of the noise on a single sensor. The signal-to-interference-and-noise ratio (SINR) of the output is defined to be the power of the desired signal components in the output divided by the power of everything else in the output. The signals from the mobile units are independent BPSK signals having Nyquist-shaped pulses with 100% excess bandwidth and baud rate of 160 kb/s. The carrier separation f_{sep} is 10 kHz, and the data collection interval is 500 μs. The average output SINR is computed for the signal having zero carrier offset, which is referred to hereafter as the desired signal. Although the direction of arrival of the desired signal is 0 degrees in these simulations, any direction of arrival is allowable since SCORE does not require knowledge of the directions of arrival of any of the signals.

Fifty independent trials are performed for each combination of one and two multipath reflections having nominal angular separation of one and five degrees from the direct path. The direct-path signals and multipath reflections of 48 spectrally overlapping users are present, with each of the direct-path signals having SNR of 10 dB. The directions of arrival of the direct-path signals from the mobile units are nominally equally spaced on the interval $[-180, 180]$ degrees with a random perturbation in each direction of up to 2 degrees to more accurately model a typical environment. The multipath reflections each have SNR 0 dB (10 dB less than the direct-path signal) and have phases that are independent and uniformly distributed on the interval $[0, 2\pi]$ radians. Each reflection is nominally separated by 1 or 5 degrees from its direct-path signal with a random perturbation of up to 1 degree to more accurately model a typical environment. Even in the presence of this multipath, the SCORE algorithm attains a high average output SINR as shown in Table 2. The results in Table 2 demonstrate that SCORE coherently combines the direct and reflected paths of the desired signal to increase the output SINR. A typical antenna pattern of the spatial filter in this environment that extracts the desired signal and rejects the signals of other users is shown in Fig. 13. More extensive results showing the bit error rate of this scheme can be found in [82].

Table 2: Average output SINR of SCORE-SDMA scheme for different numbers of multipath reflections and angular separations. Forty-eight spectrally overlapping users are present.

# of multipath reflections	0	1		2	
Angular sep. of multipath	—	1	5	1	5
Average output SINR (dB)	19.3	19.6	20.0	21.0	20.8

$\boxed{\alpha}$ Much work remains to be done on the use of SCORE-adapted antenna arrays for increasing capacity in wireless communications. In particular, simulations are needed that use more realistic models of the propagation environment (i.e., to properly model the multipath "seen" by the base station as it receives the signals from the mobile units, and to account for cochannel interference from nearby cells), and that evaluate the bit error rate (BER) after spatial filtering and equalization is performed. It also remains to be investigated whether SCORE could be applied to a preexisting standard such as the North American standard IS-54 [22] (cf. [40]) for digital cellular TDMA to either increase capacity or improve signal quality. Further work evaluating and improving the methods of Section 5 for use in existing and proposed systems is also needed. And, the more general problem of purposely designing a communication system to exploit LCL-PTV STF remains open.

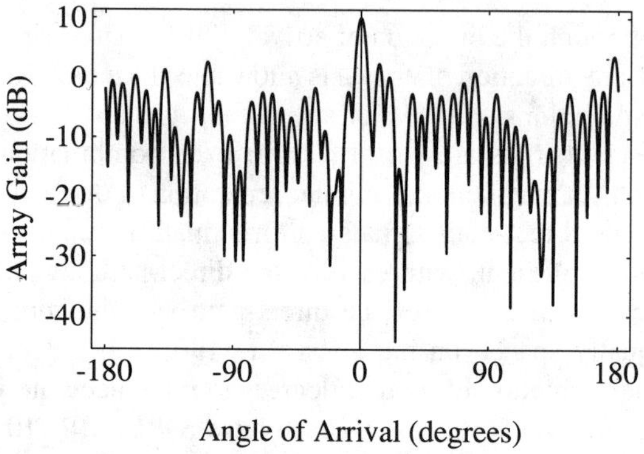

Figure 13: Typical antenna pattern of spatial filter found by SCORE. The desired signal arrives from 0 degrees and the signals of the remaining 47 users are roughly equally spaced on $[-180, 180]$ degrees.

4.5.5 Performance Improvement for a Spread Spectrum System

In this example, the Wideband SCORE method of Section 4.4.10 is used to extract a low-SNR BPSK direct-sequence spread spectrum signal of interest hidden beneath five high-SNR narrowband BPSK signals.

The array is linear and consists of four identical equi-spaced omnidirectional sensors; the sensor spacing is equal to one-half of the wavelength corresponding to the highest frequency in the receiver band. The lowest RF frequency is 0.75 times the highest, yielding a relative bandwidth (defined here as the ratio of the width to the center frequency of the receiver band) of about 28%. The noise consists of stationary complex white Gaussian noise that is uncorrelated from sensor to sensor. The signals not of interest (SNOIs) are identical BPSK signals having bit rate $1/10$ (normalized to the sampling rate) and independent and identically distributed (i.i.d.) random bit sequences and Nyquist-shaped pulses with 100% excess bandwidth. Five such SNOIs having carrier offsets (relative to the center frequency of the receiver band and normalized to the sampling rate) $-2/5$, $-1/5$, 0, 1/5, and 2/5 arrive from -20 degrees, 10 degrees, -10 degrees, 40 degrees, and 0 degrees, respectively. Each SNOI has inband SNR (defined here as the ratio of the signal power to the power of the noise within the band occupied by the signal) equal to 15 dB. The signal of interest (SOI) is a BPSK signal having chip rate 1/2 with chip sequence modeled as an i.i.d. random bit sequence, and rectangular chip pulses; it has zero carrier frequency offset and arrives from 25 degrees. The SNR of the SOI is -5 dB, for a total SINR of -20 dB.

One hundred independent trials are conducted in which the Wideband SCORE algorithm is applied using $\alpha = 1/2$ to extract the SOI. The wideband data is decomposed into data from eight spectrally disjoint subbands using an FFT channelizer, such that each subband closely follows the narrowband model. These subbands are spatially filtered by Wideband SCORE to obtain a channelized estimate of the SOI. The average SINR (averaged over the 100 trials) of the extracted signal in each of the eight subbands is plotted in Fig. 14 for different numbers N of data samples used to compute the weight vector.

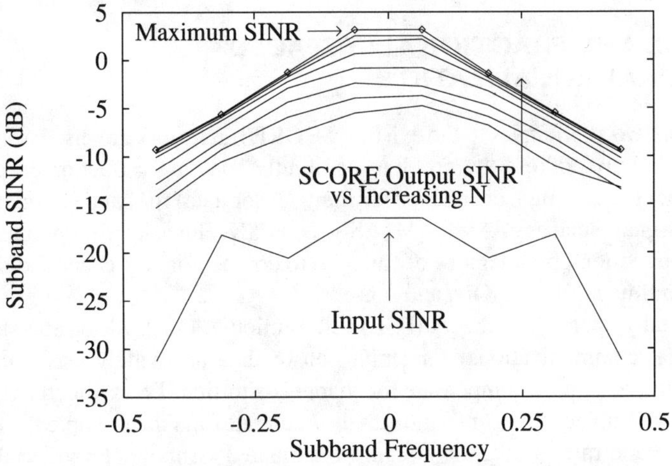

Figure 14: Average SINR (averaged over 100 trials) in each of the 8 subbands of SCORE processor output, plotted for different numbers N of data samples: 128, 256, 512, ..., 8192. The input SINR at the first sensor and the maximum-attainable output SINR are shown at the extreme bottom and top of the graph, respectively.

These results indicate that the use of a sensor array for this cochannel interference problem could be exploited in the design of a spread spectrum communication system to accomplish either of two objectives without requiring a training signal or other potentially troublesome prior knowledge: 1) reduce the bit error rate in the despread SOI while keeping the spreading factor constant, or 2) allow a much smaller spreading factor to be used at the transmitter while keeping the bit error rate constant. For example, in a typical application a spreading factor (processing gain) of about 1000 might be needed without the use of the array to overcome the −20 dB SINR of the spread signal. In contrast, the increase in output SINR of between 10 dB and 20 dB due to the SCORE processor implies that a much lower spreading factor between 100 and 10, respectively, would be needed.

The results also indicate that in signal interception applications, the SINR of the SOI at the output of the SCORE processor could be sufficiently high that a blind despreading technique might be applied successfully.

4.6 Summary

Several blind adaptive spatial filtering algorithms are available, although their performance is not thoroughly understood and none appears to be statistically optimum and devoid of undesirable behavior. Nonetheless, their performance and capabilities suggest that blind adaptive spatial filtering and blind adaptive LCL-PTV STF may significantly improve the performance (i.e., signal quality and capacity) of communication systems relative to that obtained from conventional methods.

5 SPATIAL AND FRACTIONALLY SPACED TEMPORAL EQUALIZATION

In this section two alternatives to the existing SCORE framework are discussed. Both alternatives directly address the problem of blindly adapting a fractionally spaced equalizer (for either a single channel or for multisensor data) for bauded digital communication signals such as PAM, QAM, ASK, and PSK. However, the emphasis here is on using spatio-temporal degrees of freedom to compensate for channel distortion rather than multiple SOIs, SNOIs, and noise.

Fractionally spaced equalizers (FSE) (cf. Section 7.4 of [38], or any standard text on digital communications) for single-sensor data are widely used in digital communication systems to compensate for channel distortion. FSEs filter the received data (typically sampled at two or four times the baud rate) and then sample the filtered output at the baud rate to obtain a stream of estimated symbols. Provided that the LTI filter in the FSE is chosen properly, it can be shown that this structure is the optimum linear receiver for PAM signals corrupted by linear channel distortion and additive stationary noise. The FSE structure can also be understood in terms of LCL-PTV filtering (also called LCL frequency-shift filtering or LCL-FRESH filtering), as discussed in [23, 32].

Tutorial overviews of equalization in digital communication systems can be found in [59, 61], and additional discussion of blind equalization can be found in [21] and references therein. Most of the blind channel identification and equalization methods studied to date have exploited the fact that well-equalized baud-sampled data should have samples drawn from a finite alphabet, or that the higher-order statistics of the baud-sampled data can be used to infer the effective channel seen by the symbol stream as it propagates through the modulator, transmitter, propagation channel, and receiver. In contrast, the methods considered in [21, 94] and the new method proposed in Section 5.2 exploit only the second-order cyclostationarity of the fractionally sampled data, a property which is destroyed by the baud-sampling typically performed prior to application of most blind equalization methods. Unlike the second-order statistics of stationary data, the second-order statistics of cyclostationarity data have been shown to be sufficient to perform identification and equalization of nonminimum-phase channels [21, 94].

5.1 The Blind FSE of Tong, Xu, and Kailath

In this section the blind channel identification and equalization method of Tong, Xu, and Kailath, referred to here as the TXK method for the sake of brevity, is reinterpreted here as the solution to a least-squares problem. The method and an algebraic derivation is presented in [93] and is generalized in a straightforward way in [94] to accommodate multiple sensors. It accommodates arbitrary time-limited pulse shapes and channel distortion but requires that the noise be spatially and temporally white (i.e., it has no interference rejection capability).

The TXK method performs blind adaptive temporal equalization of a noisy distorted QAM signal,

$$x(n) = \sum_{k=-\infty}^{\infty} h(n - kT)s(k) + i(n) \tag{33}$$

received at a single sensor, where $\{s(k)\}$ is the stream of independent symbols, T is the symbol period, and $h(n)$ is the unknown distorted pulse. In [93, 94], the TXK method is derived by exploiting various signal subspace properties of the ideal correlation matrices of the received data to yield an algebraically motivated solution. In [94], the straightforward extension to sensor array or multichannel data is proposed by replacing $x(\cdot)$ in (33) by the received data $\boldsymbol{x}(\cdot)$ at the output of the sensor array, and replacing $h(\cdot)$ by the multisensor impulse response $\boldsymbol{h}(\cdot)$. This general case is used in this section.

The method operates on the vector-stationarized data

$$z(k) = \begin{bmatrix} \boldsymbol{x}(kT) \\ \vdots \\ \boldsymbol{x}(kT + KT - 1) \end{bmatrix} = \boldsymbol{Hs}(k) \tag{34}$$

where

$$\boldsymbol{H} = \begin{bmatrix} \boldsymbol{h}(-(k_0 + 0)T) & \cdots & \boldsymbol{h}(-(k_0 + d - 1)T) \\ \vdots & & \vdots \\ \boldsymbol{h}(KT - 1 - (k_0 + 0)T) & \cdots & \boldsymbol{h}(KT - 1 - (k_0 + d - 1)T) \end{bmatrix}$$

and

$$\boldsymbol{s}(k) = \begin{bmatrix} s(k_0 + k) \\ \vdots \\ s(k_0 + k + d - 1) \end{bmatrix},$$

with K equal to any integer, k_0 equal to an arbitrary integer (e.g., zero for simplicity), and d equal to a best guess of the number of symbols that contribute significantly to any given sample of the signal (i.e., the estimated number of samples in the distorted pulse $\boldsymbol{h}(n)$, divided by T). In addition to reexpressing the data as the stationary output of a multidimensional LTI system, this representation is useful because the autocorrelations $\boldsymbol{R}_{zz}(0)$ and $\boldsymbol{R}_{zz}(1)$ exhibit a special structure, namely $\boldsymbol{R}_{zz}(0) = \boldsymbol{HH}^H + \sigma^2\boldsymbol{I}$ and $\boldsymbol{R}_{zz}(1) = \boldsymbol{HJH}^H + \sigma^2\boldsymbol{J}^T$, where the superscript T here temporarily denotes exponentiation, not transposition. This structure is exploited by the TXK method as described in [94].

The objective is to identify the channel $\boldsymbol{h}(n)$ through which the symbols are sent, equalize the channel, and estimate the symbol values. The equalization structure is constrained to be linear, with the estimated symbols $\hat{s}(k)$ given by

$$\hat{s}(k) = W^H z(k).$$

Thus, not only must d be long enough that the estimated channel can be a reasonable approximation of the true channel, but K must be large enough that the equalization filter W can adequately compensate for the distortion.

Here, the TXK method for blind adaptive spatio-temporal equalization is rederived as the solution to a least-squares problem for a finite data set:

$$\min_{W,\hat{H}} \left\langle \left\| z(k) - \hat{H}\hat{s}(k) \right\|^2 \right\rangle_{N/T}, \tag{35}$$

where the time average $\langle\cdot\rangle$ is taken over the symbol index k, the estimated symbols are $\hat{s}(k) = W^H z(k)$, and the autocorrelations of $\hat{s}$ are restricted to obey $\hat{R}_{\hat{s}\hat{s}}(0) = I$ and $\hat{R}_{\hat{s}\hat{s}}(1) = J$, with

$$J = \begin{bmatrix} 0 & 1 & 0 & \cdots & 0 \\ & 0 & 1 & \ddots & \vdots \\ \vdots & & \ddots & \ddots & 0 \\ & & & 0 & 1 \\ 0 & & \cdots & & 0 \end{bmatrix}.$$

Reexpressing the cost function (35) in terms of estimated correlations and minimizing with respect to W yields the solution

$$W = \hat{H}\left(\hat{H}^H\hat{H}\right)^{-1}.$$

Substituting this solution into (35) and minimizing with respect to $\hat{H}$ can be shown to lead to the maximization problem

$$\max_{\hat{H}} \operatorname{tr}\left\{P_{\hat{H}}\hat{R}_{zz}(0)\right\}$$

where $P_{\hat{H}} = \hat{H}\left(\hat{H}^H\hat{H}\right)^{-1}\hat{H}^H$, which in turn is equivalent to

$$\max_{W} \operatorname{tr}\left\{P_W\hat{R}_{zz}(0)\right\}.$$

Expressing the problem in terms of the SVD of $W = U\Sigma V^H$ yields the constrained optimization problem

$$\max_{W} \operatorname{tr}\left\{UU^H\hat{R}_{zz}\right\}$$

subject to

$$\begin{aligned} V\Sigma U^H\hat{R}_{zz}(0)U\Sigma V^H &= I \\ V\Sigma U^H\hat{R}_{zz}(1)U\Sigma V^H &= J. \end{aligned}$$

Straightforward manipulations reveal that maximization subject to the first constraint requires that $\boldsymbol{U}$ equals the d most dominant eigenvectors of $\hat{\boldsymbol{R}}_{zz}$ and Σ equals the reciprocal of the square root of the corresponding eigenvalues. The second constraint then defines $\boldsymbol{V}$.

This derivation leads to the same algorithm as obtained in [93, 94] by a different, algebraically motivated argument for the single-channel case. However, the second constraint can be satisfied only if the matrix $\Sigma \boldsymbol{U}^H \hat{\boldsymbol{R}}_{zz}(1)\boldsymbol{U}\Sigma$ has Jordan form $\boldsymbol{J}$, which occurs with probability zero due to finite averaging time. A possible solution to this dilemma is given without proof (but motivated by standard practices in the sensor array processing literature) in [93, 94].

5.2 Another Blind FSE

Partially motivated by the least-squares derivation of the TXK method, another algorithm that blindly adapts an FSE is proposed here.

As before, let

$$\boldsymbol{x}(kT+n) = \sum_{l=-\infty}^{\infty} \boldsymbol{h}(n-lT)s(l+k) + \boldsymbol{i}(kT+n)$$

be a bauded communication signal distorted by an unknown multisensor channel having impulse response $\boldsymbol{h}(n)$, where T is the baud period, $s(k)$ is the kth symbol (unknown), and $\boldsymbol{i}(n)$ is unknown noise. Provided that $\boldsymbol{h}(n)$ decays sufficiently fast as $|n| \to \infty$, $\boldsymbol{x}(kT+n)$ can be approximated by

$$\boldsymbol{x}(kT+n) \approx \sum_{l=-L}^{L} \boldsymbol{h}(n-lT)s(l+k) + \boldsymbol{i}(kT+n). \tag{36}$$

For most physical channels of interest, this simply means that L is sufficiently large, or equivalently, that only $2L+1$ symbols contribute significantly to any particular baud period of the signal. Defining

$$\boldsymbol{y}(k) = \begin{bmatrix} \boldsymbol{x}(kT) \\ \boldsymbol{x}(kT+1) \\ \vdots \\ \boldsymbol{x}(kT+T-1) \end{bmatrix}, \quad \check{\boldsymbol{h}}(k) = \begin{bmatrix} \boldsymbol{h}(-kT) \\ \boldsymbol{h}(-kT+1) \\ \vdots \\ \boldsymbol{h}(-kT+T-1) \end{bmatrix},$$

$$\check{\boldsymbol{i}}(k) = \begin{bmatrix} \boldsymbol{i}(kT) \\ \boldsymbol{i}(kT+1) \\ \vdots \\ \boldsymbol{i}(kT+T-1) \end{bmatrix},$$

and

$$\boldsymbol{s}(k) = [s(-L+k),\ s(-L+1+k), \cdots,\ s(L+k)]^T,$$

(36) can be expressed as

$$y(k) = Hs(k) + \check{i}(k), \tag{37}$$

where $H \triangleq \left[\check{h}(-L) \;\; \check{h}(-L+1) \;\cdots\; \check{h}(L)\right]$. It is noted that H as defined here is unstructured because $y(k)$ contains only a single baud period of the signal, whereas H as defined in the TXK method is a block Toeplitz matrix.

The signal is filtered by an FIR filter of length $(2K+1)T$ samples to yield an estimate of the symbol stream,

$$\hat{s}(k) = \begin{bmatrix} w_{-K} \\ w_{-K+1} \\ \vdots \\ w_K \end{bmatrix}^H \begin{bmatrix} y(k-K) \\ y(k-K+1) \\ \vdots \\ y(k+K) \end{bmatrix} = w^H \begin{bmatrix} y(k-K) \\ y(k-K+1) \\ \vdots \\ y(k+K) \end{bmatrix}, \tag{38}$$

and thus the vector $\hat{s}(k)$ of symbol estimates is given by

$$\hat{s}(k) = W^H \begin{bmatrix} y(k-K-L) \\ y(k-K-L+1) \\ \vdots \\ y(k+K+L) \end{bmatrix} = W^H z(k), \tag{39}$$

where $z(k)$ is defined in the obvious way and W is a block Toeplitz matrix of dimension $(2(K+L)+1) \times (2L+1)$ with block size $T \times 1$ (or $MT \times 1$ if $x(n)$ comes from an M-sensor array) and first column equal to $[w_{-K}^H \; w_{-K+1}^H \cdots w_K \; 0 \cdots 0]^H$ and first row equal to $[w_{-K} \;\; 0 \cdots 0]$.

In a vein similar to (35) but having different constraints, a least-squares minimization problem can be posed,

$$\min_{W,\hat{H}} \left\langle \left\| y(k) - \hat{H}\hat{s}(k) \right\|^2 \right\rangle_{N/T}, \tag{40}$$

subject only to the constraint that W is block Toeplitz. Note that there is no requirement that the symbols be uncorrelated with each other or that the noise be spatio-temporally white (although these assumptions might be needed to insure consistency; this is an open problem). Equivalently,

$$\min_{W,\hat{H}} \operatorname{tr} \left\{ \hat{R}_{yy} - \hat{H}\hat{R}_{y\hat{s}}^H - \hat{R}_{y\hat{s}} \hat{H}^H + \hat{H}\hat{R}_{\hat{s}\hat{s}} \hat{H}^H \right\}. \tag{41}$$

Alternatively, if $i(n)$ is zero-mean stationary spatio-temporally white Gaussian noise, then it is easily shown that (40)–(41) maximize the conditional likelihood of the received data, with H being an unknown parameter matrix to be estimated and the conditioning performed over the unknown symbol values, whose estimates are then constrained to be given by (38)–(39).

Minimization of (41) with respect to $\hat{H}$ is a standard problem whose solution is well known. However, in preparation for the subsequent optimization with respect to W which does not seem to be a standard problem, the basics of complex matrix calculus are summarized here. It is shown in [14] that an appropriate definition of the complex gradient of a real-valued function with respect to a complex matrix and its conjugate can be found simply by treating the matrix and its conjugate as independent variables. This step is necessary because the usual definition of differentiation with respect to a complex variable is not valid unless the function is analytic. Consequently, it is also shown in [14] that the stationary point of a real-valued function of a complex matrix can be found by setting the complex gradient with respect to the conjugate of the matrix equal to zero (equivalently, the gradient with respect to the matrix could be equated with zero). Then, it is easily shown that

$$\nabla_{\hat{\boldsymbol{H}}^*} \operatorname{tr}\left\{\hat{\boldsymbol{H}}^H \boldsymbol{A}\right\} = \boldsymbol{A} \quad \text{and} \quad \nabla_{\hat{\boldsymbol{H}}^*} \operatorname{tr}\left\{\hat{\boldsymbol{H}}^H \boldsymbol{A} \hat{\boldsymbol{H}}\right\} = \boldsymbol{A}\hat{\boldsymbol{H}}.$$

Consequently, the gradient of the cost function in (41) is

$$\nabla_{\hat{\boldsymbol{H}}^*} (\text{cost func}) = -\hat{\boldsymbol{R}}_{y\hat{s}} + \hat{\boldsymbol{H}}\hat{\boldsymbol{R}}_{\hat{s}\hat{s}}.$$

Equating to zero and solving yields the standard solution,

$$\hat{\boldsymbol{H}} = \hat{\boldsymbol{R}}_{y\hat{s}} \hat{\boldsymbol{R}}_{\hat{s}\hat{s}}^{-1}. \tag{42}$$

Substituting (42) into (41) yields

$$\min_{\boldsymbol{W}} \operatorname{tr}\left\{\hat{\boldsymbol{R}}_{yy} - \hat{\boldsymbol{R}}_{y\hat{s}} \hat{\boldsymbol{R}}_{\hat{s}\hat{s}}^{-1} \hat{\boldsymbol{R}}_{y\hat{s}}^H\right\}, \tag{43}$$

which can be reexpressed as

$$\max_{\boldsymbol{W}} \operatorname{tr}\left\{\boldsymbol{W}^H \hat{\boldsymbol{R}}_{yz}^H \hat{\boldsymbol{R}}_{yz} \boldsymbol{W} \left(\boldsymbol{W}^H \hat{\boldsymbol{R}}_{zz} \boldsymbol{W}\right)^{-1}\right\}. \tag{44}$$

It is crucial to note that W is a block Toeplitz matrix. However, (44) can be reexpressed in terms of the unstructured vector w as follows. Note that the (m, n)th element of a quadratic form of W can be expressed as

$$\left[\boldsymbol{W}^H \boldsymbol{R} \boldsymbol{W}\right]_{m,n} = \sum_{l=-K}^{K} \sum_{k=-K}^{K} \boldsymbol{w}_l^H \boldsymbol{R}_{l+m,k+n} \boldsymbol{w}_k \tag{45}$$

where $\boldsymbol{R}_{l+m,k+n}$ is the block element (matrix) at block-row $l + m$ and block-column $k + n$ in $\boldsymbol{R}$, and $l + m$ and $k + n$ each range from $-(K + L)$ to $(K + L)$. Denoting

$$\boldsymbol{R}^{m,n} = \begin{bmatrix} \boldsymbol{R}_{m-K,n-K} & \cdots & \boldsymbol{R}_{m-K,n+K} \\ \vdots & & \vdots \\ \boldsymbol{R}_{m+K,n-K} & \cdots & \boldsymbol{R}_{m+K,n+K} \end{bmatrix},$$

(45) can be expressed as

$$\left[\boldsymbol{W}^H \boldsymbol{R} \boldsymbol{W}\right]_{m,n} = \boldsymbol{w}^H \boldsymbol{R}^{m,n} \boldsymbol{w}. \tag{46}$$

Thus, (44) is equivalent to

$$\max_{\boldsymbol{w}} \operatorname{tr}\left\{\boldsymbol{A}(\boldsymbol{w})\,[\boldsymbol{B}(\boldsymbol{w})]^{-1}\right\} \tag{47}$$

where

$$[\boldsymbol{A}(\boldsymbol{w})]_{m,n} = \boldsymbol{w}^H \left[\hat{\boldsymbol{R}}_{yz}^H \hat{\boldsymbol{R}}_{yz}\right]^{m,n} \boldsymbol{w} \quad \text{and} \quad [\boldsymbol{B}(\boldsymbol{w})]_{m,n} = \boldsymbol{w}^H \left[\hat{\boldsymbol{R}}_{zz}\right]^{m,n} \boldsymbol{w}$$

with m and n each ranging from $-L$ to L. As of this writing, no further simplification of (47) has been found, except in the special case where $L = 0$ (i.e., no symbol's contribution outside its own symbol period is deemed to be significant). In this special case, $\boldsymbol{A}(\boldsymbol{w})$ and $\boldsymbol{B}(\boldsymbol{w})$ are scalars, and the standard solution for $\boldsymbol{w}$ is simply the most dominant eigenvector $\boldsymbol{w}_{max}$ of the system

$$\hat{\boldsymbol{R}}_{yz}^H \hat{\boldsymbol{R}}_{yz} \boldsymbol{w}_{max} = \lambda_{max} \hat{\boldsymbol{R}}_{zz} \boldsymbol{w}_{max}. \tag{48}$$

This special case provides some hope that (47) can be solved without resort to a gradient-based search method. However, this remains an open problem.

5.2.1 Algorithm Summary

In general, the algorithm can be summarized in the following steps:

1. Estimate $\hat{\boldsymbol{R}}_{yz}$ and $\hat{\boldsymbol{R}}_{zz}$.
2. If $L = 0$, then solve (48) and exit.
3. If $L \neq 0$, initialize the weight vector $\boldsymbol{w}$ with a Kronecker delta function, or prior knowledge of a suitable equalizer for the multichannel impulse response (e.g., in single-channel communication systems for which the transmitter filter has transfer function equal to the square root of the raised cosine function, the initial guess for $\boldsymbol{w}$ could be the impulse response of this filter).
4. Iteratively update the guess $\boldsymbol{w}^{(k)}$ at the kth iteration using the simple gradient method
$$\boldsymbol{w}^{(k)} = \boldsymbol{w}^{(k-1)} + \mu \nabla_{\boldsymbol{w}^*} \tag{49}$$
where $\nabla_{\boldsymbol{w}^*}$ is the gradient with respect to $\boldsymbol{w}^*$ of the objective function in (47) evaluated at $\boldsymbol{w}^{(k-1)}$, and can be shown using results in [14, 41] to be equal to the following,
$$\nabla_{\boldsymbol{w}^*} = \sum_{m,n} \left\{ C_{n,m}\ \tilde{\boldsymbol{A}}_{m,n}\boldsymbol{w} - \left(\boldsymbol{w}^H \tilde{\boldsymbol{A}}_{m,n}\boldsymbol{w}\right) \sum_{p,q} C_{n,p}\ \tilde{\boldsymbol{B}}_{p,q}\boldsymbol{w}\ C_{q,m} \right\} \tag{50}$$
where each summation index ranges from $-L$ to L and
$$\tilde{\boldsymbol{A}}_{m,n} = \left[\hat{\boldsymbol{R}}_{yz}^H \hat{\boldsymbol{R}}_{yz}\right]^{m,n}, \quad \tilde{\boldsymbol{B}}_{m,n} = \left[\hat{\boldsymbol{R}}_{zz}\right]^{m,n}, \quad \text{and} \quad \boldsymbol{C} = [\boldsymbol{B}(\boldsymbol{w})]^{-1}. \tag{51}$$

$\boxed{\alpha}$ Cauchy's method (cf. [9]) could be an alternative to the simple gradient method (49), and consists simply of (49) with μ replaced by a scalar variable over which maximization occurs at each iteration. Another alternative is a limited-step Newton's method, although this requires that the Hessian matrix be computed and inverted at each iteration, which significantly increases the computational load at each iteration. These issues are left as open problems, as are those concerning the choice of the values for the parameters L and K.

5.3 A Simulation Example

Here the TXK method and the method of Section 5.2 are simulated in the example environment considered in [94]. That is, the transmitter filter and the propagation channel together have impulse response $(c(n) - 0.8c(n-0.5) + 0.4c(n-3))u_{6T}(n)$, where $c(n)$ is a Nyquist-shaped pulse having 11% excess bandwidth and $u_{6T}(n)$ is a rectangular window of duration 6 symbol intervals. The symbol stream is binary, and the sampling rate is 4 samples per symbol. The TXK method is implemented according to the description in [94], with the parameter settings $d = 10$ and $K = 5$. The new method is implemented as described at the end of Section 5.2, with $K = L = 0$ chosen to enable the simple optimization based on the eigenvalue problem (48). One hundred independent trials of each method at each SNR were performed. In each trial, 100 symbols were collected and used to adapt the equalizer, and the performance of this equalizer was then measured by processing 1000 symbols. As shown in Fig. 15, the BER of the new method is significantly less than the BER of the TXK method over a wide range of SNRs in this environment.

5.4 Summary

A comprehensive evaluation of these second-order cyclostationarity-exploiting methods and the higher-order/finite-alphabet methods referred to at the beginning of this section has yet to be performed, so no firm conclusions can be drawn as to their relative merits. Issues such as relative convergence time, robustness to deviation from underlying assumptions, and computational complexity must all be considered. However, neither approach by itself can be optimum since the signals in question exhibit both properties, collectively described by their nth-order cyclostationarity properties for $n = 2, 3, \ldots$.

$\boxed{\alpha}$ Many open problems exist in blind spatio-temporal equalization, even if the scope is restricted to a single, distorted, digital communication signal in noise. For example, the consistency properties of the method of Section 5.2 remain to be determined. Finite-time analyses and complete simulation studies of the two methods considered in this section remain to be done. Also, more computationally efficient methods of solving (47) need to be found. Both the TXK method and the new method of Section 5.2 could be easily extended to constrain the estimated symbols to be real, which is appropriate for signals such as BPSK and ASK having real constellations. On the speculative side, extension of these algorithms to the new capability of simul-

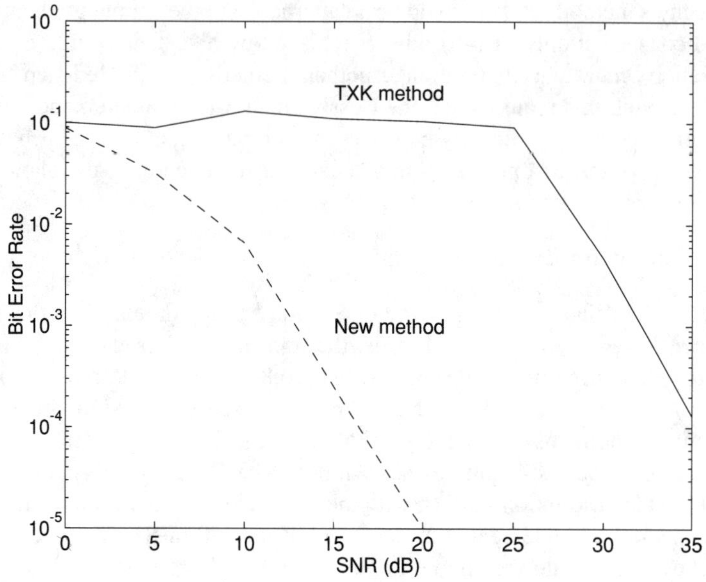

Figure 15: Bit error rate after blind equalization by the TXK method and the new method.

taneously separating and equalizing multiple spectrally overlapping signals could be used to increase communication capacity in wireless cellular networks. Except for the temporal equalization step, such a scheme based on Cross-SCORE has already been proposed in [35, 82] (cf. Section 4.5.1).

6 DIRECTION FINDING

In this section the problem of estimating the directions of arrival of signals received at an array of sensors is addressed. Several existing algorithms that exploit cyclostationarity to simplify this task and/or to outperform conventional methods are summarized.

6.1 Objectives of Direction Finding

The objective of direction finding (DF) is the estimation of the directions of arrival (DOAs) of signals of interest at the sensor array. With reference to Fig. 1 and surrounding discussion in Section 2, the array manifold maps these DOAs to vectors of gains and phases that determine the spatial structure of the received data $\boldsymbol{x}(n)$. Restated, the objective is to determine the DOAs for which the corresponding array response vectors best describe the received data. Desirable performance attributes of superior algorithms include the resolution of DOAs of closely spaced signals and the accommodation of correlation between signals, which can arise from multipath propagation or smart jamming. Thus, in addition to the usual parameter-estimator performance measures such as bias and MSE, the probability of resolution is also used to characterize performance. However, because this probability can be difficult to calculate, one convention is to declare that estimators $\hat{\theta}_1$ and $\hat{\theta}_2$ of DOAs θ_1 and θ_2 resolve the DOAs if

$$\sqrt{MSE(\hat{\theta}_1, \theta_1)} \leq \frac{1}{2}|\theta_1 - \theta_2|$$

$$\sqrt{MSE(\hat{\theta}_2, \theta_2)} \leq \frac{1}{2}|\theta_1 - \theta_2|.$$

6.2 Conventional Algorithms

The Multiple Signal Classification (MUSIC) algorithm [83, 84, 85] is perhaps the single most popular DF algorithm in existence, at least from a research standpoint. In the literature it is shown to exploit the signal and/or noise subspace structure (eigenstructures) of the autocorrelation matrix $\boldsymbol{R}_{xx}$, under the following assumptions:

1. The autocorrelation matrix $\boldsymbol{R}_{ii}$ of the interference and noise is proportional to the identity matrix: $\boldsymbol{R}_{ii} = \sigma^2 \boldsymbol{I}$.
2. $L < M$ signals $\boldsymbol{s}(n) = [s_1(n), \cdots, s_L(n)]^T$ arrive at the array with unknown DOAs $\Theta = [\theta_1, \cdots, \theta_L]^T$.
3. The columns of the array response matrix $\boldsymbol{A}(\Theta) = [\boldsymbol{a}(\theta_1), \cdots, \boldsymbol{a}(\theta_L)]$ are linearly independent.

Under these assumptions, it can be shown that the $M - L$ smallest eigenvalues found from

$$\boldsymbol{R}_{xx}\boldsymbol{e}_m = \lambda_m \boldsymbol{e}_m$$

are all equal to σ^2, and that the $M - L$ corresponding eigenvectors $[\boldsymbol{e}_{L+1}, \cdots, \boldsymbol{e}_M]$ are orthogonal to the columns of $\boldsymbol{A}(\Theta)$. The MUSIC algorithm is implemented by simply replacing the ideal correlation matrices with their estimates:

1. Estimate $\hat{\boldsymbol{R}}_{xx} = \langle \boldsymbol{x}(n)\boldsymbol{x}^H(n) \rangle_N$.
2. Solve $\hat{\boldsymbol{R}}_{xx}\boldsymbol{e}_m = \lambda_m \boldsymbol{e}_m$, and let $\hat{\sigma}^2 = \frac{1}{M-L}\sum_{m=L+1}^{M} \lambda_m$.
3. Find the L minima of $\|[\boldsymbol{e}_{L+1}, \cdots, \boldsymbol{e}_M]^H \boldsymbol{a}(\theta)\|^2$ or the L maxima of $\|[\boldsymbol{e}_1, \cdots, \boldsymbol{e}_L]^H \boldsymbol{a}(\theta)\|^2$, whichever is easier.

An alternative formulation in [80] (cf. Section 4.1 of [67] for related material) shows that MUSIC finds the set $\boldsymbol{W} = [\boldsymbol{w}_1, \cdots, \boldsymbol{w}_L]$ of orthogonal spatial filters that maximize the total average power of the spatially filtered data, and then finds the L DOAs corresponding to the peaks in the resulting antenna gain pattern. Equivalently, MUSIC finds the set $\boldsymbol{W} = \left[\boldsymbol{w}_{L+1}, \cdots, \boldsymbol{w}_M\right]$ of orthogonal spatial filters that minimize the total average power of the spatially filtered data and then finds the L DOAs corresponding to the nulls in the antenna pattern. Thus, it is clear that there is no inherent disadvantage to using beamforming or null-steering as the basis of a high-resolution DF method, provided that both are used in the generalized sense that applies to multiple signals.

Other popular DF algorithms include ESPRIT [56, 66, 67] and conditional maximum likelihood (cf. [88, 113]), which together with MUSIC, are special cases of weighted subspace fitting (WSF) methods [97, 98]. All of these methods have been shown to have low (or no) bias, low MSE, and high probability of resolution under the three assumptions listed above for MUSIC.

However, in some applications there may be more signals present than sensors in the array, or the number and angular separation of signals may be such that the methods cannot resolve the signals in a reasonable number of time samples (cf. [24]). Also, if only a subset of the signals are SOIs, then resources are wasted on SNOIs, and postprocessing may be required to classify each estimated DOA as belonging to a SOI or a SNOI. Unknown spatial characteristics of the noise also degrade performance and/or cause these methods to fail, although this can be partially mitigated [52, 63, 91, 102, 106, 108].

6.3 Cyclic DF Algorithms

The four primary algorithms for DF of cyclostationary signals (as of this writing) are described here. Three of them have interpretations based on subspace fitting and its usual complementary perspective of least-squares optimization (cf. [80]), and a different three of them can be understood, loosely, as cyclostationarity-exploiting versions of conventional algorithms.

In the following, the notational convention employed is that $\boldsymbol{s}(n)$ contains only the L_α signals that have cycle frequency α, and all of the remaining signals (of which there are $L - L_\alpha$) and the noise are lumped into $\boldsymbol{i}(n)$.

6.3.1 Cyclic MUSIC

As shown in [29], MUSIC can be simplified if it is known that exactly one signal arriving at the array has cycle frequency α, and $\hat{\boldsymbol{R}}_{\boldsymbol{xx}}^{\alpha}(\tau)$ is analyzed instead of $\hat{\boldsymbol{R}}_{\boldsymbol{xx}}$. This observation can be generalized to accommodate multiple signals having cycle frequency α, yielding the Cyclic MUSIC algorithm [73] (cf. [68, 71, 75, 76, 80]). In the subspace-fitting interpretation of Cyclic MUSIC, it is noted that $\boldsymbol{R}_{\boldsymbol{xx}}^{\alpha}(\tau) = \boldsymbol{A}(\Theta)\boldsymbol{R}_{\boldsymbol{ss}}^{\alpha}(\tau)\boldsymbol{A}^H(\Theta)$ has the same column space as $\boldsymbol{A}(\Theta)$ and its left null space is orthogonal to $\boldsymbol{A}(\Theta)$. For a finite number of time samples, the algorithm can be implemented as the following:

1. Estimate $\hat{\boldsymbol{R}}_{\boldsymbol{xx}}^{\alpha}(\tau) = \left\langle \boldsymbol{x}(n)\boldsymbol{x}^H(n-\tau)\, e^{-j2\pi\alpha n}\right\rangle_N$.
2. Compute SVD

$$\hat{\boldsymbol{R}}_{\boldsymbol{xx}}^{\alpha}(\tau) = \begin{bmatrix}\hat{\boldsymbol{S}} & \hat{\boldsymbol{G}}\end{bmatrix}\begin{bmatrix}\hat{\Sigma}_S & 0 \\ 0 & \hat{\Sigma}_G\end{bmatrix}\begin{bmatrix}\hat{\boldsymbol{V}}_S & \hat{\boldsymbol{V}}_G\end{bmatrix}^H,$$

where $\begin{bmatrix}\hat{\boldsymbol{S}} & \hat{\boldsymbol{G}}\end{bmatrix}$ and $\begin{bmatrix}\hat{\boldsymbol{V}}_S & \hat{\boldsymbol{V}}_G\end{bmatrix}$ are unitary, and the diagonal elements of the diagonal matrices $\hat{\Sigma}_S$ and $\hat{\Sigma}_G$ are nonnegative and appear in decreasing order, with $\hat{\Sigma}_G \to 0$ as $N \to \infty$.

3. Find the L_α minima of $\|\hat{G}^H a(\theta)\|^2$ or the L_α maxima of $\|\hat{S}^H a(\theta)\|^2$, whichever is easier.

Cyclic MUSIC (and other cyclic DF methods to be discussed in the next subsections) can be less complicated than MUSIC and other conventional methods because: 1) R_{ii} need not be known or estimated; 2) the number L_α of DOAs for Cyclic MUSIC to estimate is never larger than the number L of DOAs for MUSIC to estimate (and $L_\alpha \ll L$ in some cases); 3) the need for postprocessing and classification of DOAs is inherently reduced by signal selectivity; and 4) if closely spaced signals have different cycle frequencies, then the purely spatial resolution problem confronted by a conventional algorithm is replaced by the easier (potentially much easier) problem of resolving jointly in space and cycle frequency. Brief illustrations of these capabilities are provided in Section 6.4.

6.3.2 Generalizations of Cyclic MUSIC

$\boxed{\alpha}$ Several generalizations of Cyclic MUSIC have been proposed and are summarized very briefly here. This entire subsection is designated as being descriptive of open problems because some of the generalizations are not yet well understood and others may be suggested.

Just as Wideband SCORE extended SCORE to wideband data, Cyclic MUSIC can be generalized in much the same way. As suggested in [29] and pursued in [53], either the cyclic autocorrelation matrix or the cyclic spectrum matrix contains the desired signal-selective measurements of the SOIs' spatial characteristics. Applying this suggestion to the Cyclic MUSIC algorithm of Section 6.3.1 requires some care because the appropriate array response vector to use with the left singular vectors of $\hat{S}_{xx}^\alpha(f)$ is $a(\theta, f + \alpha/2)$. This association follows because

$$\hat{S}_{xx}^\alpha(f) \to A(\Theta, f + \alpha/2)\, S_{ss}^\alpha(f) A^H(\Theta, f - \alpha/2).$$

A more serious complication arises as in wideband generalizations of conventional methods (e.g., [104]) because the algorithm yields a set of DOA estimates for each value of f. Potentially complicated postprocessing and classification may be needed to yield a single DOA estimate for each signal whose spectral support extends over multiple values of f. Nonetheless, a simulation-based study [53] shows that Wideband Cyclic MUSIC exhibits promising performance.

In addition to this possible wideband method, a promising method [110] is summarized in Section 6.3.4. Alternate cyclic DF methods for wideband data may be obtained by generalizing existing conventional methods for wideband data, including those in [17, 49, 99, 100].

In [73] the Phase-SCORE algorithm for signal extraction was applied to the DF problem by recognizing that the $M - L_\alpha$ least dominant eigenvectors of the Phase-SCORE eigenvalue equation converge to spatial filters that null the SOIs and, therefore, could be used instead of the least dominant left singular vectors of $\hat{\boldsymbol{R}}_{\boldsymbol{xx}}^{\alpha}(\tau)$ in step 3 of the Cyclic MUSIC algorithm. Another extension [3] uses all of the eigenvectors of (28) and searches for minima of the function

$$\frac{\left\| \left[\boldsymbol{w}_{L_\alpha+1} \ \cdots \ \boldsymbol{w}_M \right]^H \boldsymbol{a}(\theta) \right\|^2}{\left\| \left[\boldsymbol{w}_1 \ \cdots \ \boldsymbol{w}_{L_\alpha} \right]^H \boldsymbol{a}(\theta) \right\|^2},$$

although simulation results in [68] show that these two Phase-SCORE-based Cyclic MUSIC methods do not perform as well as Cyclic MUSIC in many cases.

In [110], a modification to Cyclic MUSIC is proposed, whereby $\hat{\boldsymbol{R}}_{\boldsymbol{xx}}^{\alpha}(\tau)$ is replaced by a matrix containing contributions from multiple lags:

$$\left[\hat{\boldsymbol{R}}_{\boldsymbol{xx}}^{\alpha}(\tau_1) \ \cdots \ \hat{\boldsymbol{R}}_{\boldsymbol{xx}}^{\alpha}(\tau_K) \right]$$

for some suitable number K of different lags $\tau_1, \ldots, \tau_K$ for which $\hat{\boldsymbol{R}}_{\boldsymbol{xx}}^{\alpha}(\tau_k)$ are all significant. In general, in light of the developments in Section 4 and [81], $\hat{\boldsymbol{R}}_{\boldsymbol{xx}}^{\alpha}(\tau)$ in Cyclic MUSIC could be replaced by $\hat{\boldsymbol{R}}_{\boldsymbol{xy}}$ where $\boldsymbol{y}(n)$ is any LCL-PTV transformation given by (8). This gives added flexibility and may substantially improve performance (e.g., reduce bias and MSE) of the Cyclic MUSIC DF method, just as convergence time of Cross-SCORE is reduced by the same technique (cf. Section 4.4.8 and [81]). However, no empirical or analytical results are currently available for this generalization of Cyclic MUSIC.

Finally, in [13, 75], three extensions of Cyclic MUSIC for estimating the cycle-frequency parameter α are proposed. The rudimentary method in [75] simply searches for significant peaks in the FFT of the lag-product waveform $x_1(n)x_1^*(n - \tau)$ to determine the frequencies at which quadratically regenerated sine waves are present. These cycle-frequency estimates are then used to compute the corresponding cyclic autocorrelation matrices, to which Cyclic MUSIC is then applied one at a time. Two more sophisticated methods are investigated in [13].

6.3.3 Method of Izzo, Paura, and Poggi

The method described in [47] can be interpreted as a cyclic version of the method discussed in [12, 58]. That is, since $\boldsymbol{R}_{\boldsymbol{xx}^*}^{\alpha}(\tau)$ has rank L_α in the absence of perfectly correlated signals, its $(L_\alpha + 1) \times (L_\alpha + 1)$ principal submatrix has a one-dimensional null space that is orthogonal to the array response vectors of the cyclostationary signals. This observation also applies to $\boldsymbol{R}_{\boldsymbol{xx}}^{\alpha}(\tau)$ and forms the basis of a method for estimating the number of cyclostationary signals arriving at the array, as summarized in Section 7.2.3.

However, because the correlations involving the remaining $M - L_\alpha - 1$ sensors are ignored, this DF method is not expected to perform as well as Cyclic MUSIC.

6.3.4 Method of Xu and Kailath

In [110], a method that analyzes the signal-subspace of a collection of cyclic autocorrelations is presented as a solution to the wideband DF problem. This method is based on a data model in which uncorrelated signals $s_1(n), \cdots, s_{L_\alpha}(n)$ arrive at the array, yielding

$$x_m(n) = \sum_{l=1}^{L_\alpha} g_{ml}\, s_l(n - d_m(\theta_l)),$$

where $d_m(\theta) = (q_m/c)\sin\theta$, and q_m is the coordinate of the mth sensor as in Section 2. A key observation is made, which shows that the vector of cyclic autocorrelations follows an induced narrowband model,

$$\boldsymbol{R}^\alpha(\tau) = \begin{bmatrix} R^\alpha_{x_1x_1}(\tau) \\ \vdots \\ R^\alpha_{x_Mx_M}(\tau) \end{bmatrix} = \begin{bmatrix} \boldsymbol{a}^\alpha(\theta_1) & \cdots & \boldsymbol{a}^\alpha(\theta_{L_\alpha}) \end{bmatrix} \begin{bmatrix} R^\alpha_{s_1s_1}(\tau) \\ \vdots \\ R^\alpha_{s_{L_\alpha}s_{L_\alpha}}(\tau) \end{bmatrix},$$

where $a^\alpha_m(\theta_l) = |g_{ml}|^2 \exp(-j2\pi\alpha d_m(\theta_l))$. That is, even though $\boldsymbol{x}(n)$ need not follow the narrowband model, the vector $\boldsymbol{R}^\alpha(\tau)$ does follow a narrowband model having induced manifold $\{\boldsymbol{a}^\alpha(\theta)\}$. Standard signal-subspace methods can then be applied to estimate the DOAs from the matrix

$$\begin{bmatrix} \boldsymbol{R}^\alpha(\tau_1) & \cdots & \boldsymbol{R}^\alpha(\tau_K) \end{bmatrix},$$

where the K values of τ should be chosen such that $\boldsymbol{R}^\alpha_{ss}(\tau_1), \ldots, \boldsymbol{R}^\alpha_{ss}(\tau_K)$ are significant.

As noted in [110], the signals must be uncorrelated (i.e., no multipath or smart jamming), and the true sensor responses $a_m(\theta, f)$ must be factorizable as $a_m(\theta, f) = g_m(f)\, a_m(\theta)$ in order for the method to operate properly. Also, it is important to note that the induced manifold $\{\boldsymbol{a}^\alpha(\theta)\}$ collapses to a single vector which is independent of θ and of the true DOAs in the limit as $(\alpha\, q_m)/c \to 0$. That is, if the array is designed for high-performance signal extraction in which the sensors are separated by half the wavelength of the highest frequency in the band, and α is equal to a fixed value (e.g., a baud rate), then the method's performance degrades toward complete failure as the carrier frequency increases without bound. An example of this degradation appears in [53]. Alternatively, for a fixed set of sensor separations, α must be sufficiently large that the induced manifold varies appreciably over the allowable range of θ.

As discussed in [110], this difficulty can be avoided by exploiting conjugate cyclostationarity by using $R^\alpha_{x_mx_m^*}(\tau)$, although this is naturally limited to those signals that exhibit such features (e.g., BPSK or SQPSK, but not QPSK or QAM).

Another requirement of this method, noted recently in [53], is that the sequence of cyclic autocorrelations $R^\alpha_{s_ms_m}(\tau_1), \cdots, R^\alpha_{s_ms_m}(\tau_K)$, from the mth signal must not be a linear combination of the corresponding sequences from the other signals having cycle frequency α. Otherwise, the "signals" (i.e., these sequences of cyclic autocorrelations) in the induced narrowband model are perfectly correlated, which requires that multidimensional subspace-fitting techniques be used. This requirement may help to explain the rather odd choice of pulse shapes used for the two SOIs in simulations in [110].

Simulations in [110] show that the method can outperform Cyclic MUSIC (i.e., exhibit less MSE) over a large range of relative bandwidths (the ratio of bandwidth to carrier frequency ranged from 1% to 40%), provided that the aforementioned requirements on sensor spacing for the method of Xu and Kailath are met. Since the model assumptions for the method are met in these simulations, whereas those for Cyclic MUSIC are not met exactly, this result is reasonable. However, it is unknown whether the performance advantage would be maintained if Cyclic MUSIC also exploited K multiple lag values as summarized in 6.3.2. It is also interesting to note that the performance of Cyclic MUSIC in these simulations is insensitive to the relative bandwidth, which supports the claim of Section 2 that the narrowband model may still be acceptable even at large relative bandwidths (e.g., 40%).

Finally, it is noted that an alternate formulation of this method is obtained by posing the least-squares minimization problem

$$\min_{\boldsymbol{C},\hat{\Theta}} \left\| \hat{\boldsymbol{R}}^{\alpha}\boldsymbol{C} - \boldsymbol{A}^{\alpha}(\hat{\Theta}) \right\|_F^2 ,$$

where $\| \cdot \|_F$ denotes the Frobenius norm (cf. [39]). This problem leads to the SVD-based solution of [110]. One benefit of this formulation is that it shows that the method can be interpreted as an M-sensor generalization of the 2-sensor Cyclic Phase Difference (CPD) time-difference-of-arrival (TDOA) method of [34].

6.3.5 Cyclic Least Squares

Although Cyclic MUSIC and the method of [47] can accommodate partial correlation among signals (e.g., arising from multipath propagation or smart jamming), they fail in the presence of perfect correlation (although in practice high correlation is sufficient to cause failure) because the rank of $\hat{\boldsymbol{R}}^{\alpha}_{\boldsymbol{xx}}(\tau)$ or $\hat{\boldsymbol{R}}^{\alpha}_{\boldsymbol{xx}^*}(\tau)$ is no longer equal to L_α. Also, the method of [110] cannot accommodate any correlation.

One alternative that addresses this problem is the Cyclic Least Squares (CLS) method,

$$\min_{\hat{\Theta},\boldsymbol{W}} \left\langle \left\| \boldsymbol{x}(n) - \boldsymbol{A}(\hat{\Theta})\hat{\boldsymbol{s}}(n), \right\|^2 \right\rangle_N \tag{52}$$

where $\hat{\boldsymbol{s}}(n) = \boldsymbol{W}^H\boldsymbol{x}(n-\tau)\exp(j2\pi\alpha n)$. Clearly, the CLS method can be interpreted as a cyclic version of the conventional least squares method. In particular, if $\boldsymbol{i}(n)$ is white, stationary, and Gaussian with $\boldsymbol{R}_{\boldsymbol{ii}} \propto \boldsymbol{I}$ and $s(n)$ is unknown, then (52) with $\alpha = 0$ is the conditional maximum-likelihood DF method (cf. [88, 89, 113] and references therein). The solution to CLS can be shown [68] to be given by

$$\begin{aligned} \boldsymbol{W} &= \hat{\boldsymbol{R}}^{-1}_{\boldsymbol{xx}}\hat{\boldsymbol{R}}^{\alpha H}_{\boldsymbol{xx}}(\tau)\boldsymbol{A}(\hat{\Theta})\left(\boldsymbol{A}^H(\hat{\Theta})\boldsymbol{A}(\hat{\Theta})\right)^{-1} \\ &\min_{\hat{\Theta}} \operatorname{tr}\left\{\boldsymbol{P}^{\perp}_{\boldsymbol{A}}(\hat{\Theta})\hat{\boldsymbol{R}}^{\alpha}_{\boldsymbol{xx}}(\tau)\hat{\boldsymbol{R}}^{-1}_{\boldsymbol{xx}}\hat{\boldsymbol{R}}^{\alpha H}_{\boldsymbol{xx}}(\tau)\right\}, \end{aligned} \tag{53}$$

where $P_A^{\perp}(\hat{\Theta})$ is the orthogonal complement of the projection matrix for $A(\hat{\Theta})$. Analytical and simulation-based results in [68, 76] show that CLS retains all of the benefits of signal selectivity exhibited by the other cyclic DF methods, in addition to operating properly in the presence of perfectly correlated multipath.

$\boxed{\alpha}$ As with Cyclic MUSIC, CLS can be generalized to exploit multiple cyclostationarity properties by replacing $\hat{s}(n)$ with $\hat{s}(n) = W^H y(n)$, where $y(n)$ is any LCL-PTV transformation (8), yielding

$$W = \hat{R}_{yy}^{-1}\hat{R}_{xy}^{H}A(\hat{\Theta})\left(A^H(\hat{\Theta})A(\hat{\Theta})\right)^{-1}$$
$$\min_{\hat{\Theta}} \operatorname{tr}\left\{P_A^{\perp}(\hat{\Theta})\hat{R}_{xy}\hat{R}_{yy}^{-1}\hat{R}_{xy}^{H}\right\}. \tag{54}$$

However, no simulation results or analytical performance evaluation are available yet for this generalization.

As reported in [68], a minor computational simplification of CLS can be obtained by simply deleting $\hat{R}_{xx}^{-1}$ from the argument of the minimization in (53) without any significant performance degradation. A similar result may be expected to hold true for the generalization, in which $\hat{R}_{yy}^{-1}$ would be deleted.

In all cases, the L_α-dimensional search can be implemented using the method of [113] solely, or as an initialization for the modified variable projection method described in [96, 98].

6.4 Cyclic DF Performance

As with the SCORE algorithms, numerous simulation results are available [1, 13, 18, 68, 71, 73, 75–78, 109, 110]. However, unlike SCORE, analytical results are available for Cyclic MUSIC [71], based on the results summarized in Section A and the method of proof used in [88].

In addition to the assumptions A-1–A-4 listed in Section A.1, additional assumptions discussed in [71] are used to obtain simple expressions for the asymptotic mean and covariance of Cyclic MUSIC DOA estimates:

- A-6 A cycle frequency α of the signals of interest, a suitable value of the lag parameter τ, and the number L_α of these signals are known.
- A-7 The narrowband model $x(n) = As(n) + i(n)$ is assumed to hold exactly.
- A-8 The array manifold is unambiguous; that is, any $M \times L$ matrix $A(\Theta)$ of array response vectors has rank L if $L \leq M$.
- A-9 The cyclic autocorrelation matrix $R_{ss}^{\alpha}(\tau)$ has rank L_α.
- A-10 A-10 No component of $i(n)$ exhibits conjugate cyclostationarity: $R_{ii^*}^{\beta}(k) = 0$ for all β, k.
- A-11 No two cycle frequencies of $i(n)$ differ by 2α.
- A-12 The process $i(n)$ is substantially temporally white in the sense that

$$\sum_{\beta\neq\alpha}\sum_{n=-\infty}^{\infty} g_l^H R_{ii}^{\beta H}(n) g_k R_{xx}^{\beta}(n)\, e^{j2\pi\alpha n} \approx g_l^H R_{ii}(0) g_k R_{xx}(0),$$

where g_k are the least-dominant left singular vectors of $R_{xx}^{\alpha}(\tau)$.

Of the several results in [71], one is especially useful for it describes the behavior of Cyclic MUSIC in the presence of SNOIs. In particular, assume that the SOIs $s(n)$ are independent and that $\boldsymbol{i}(n)$ consists of spatially white noise plus SNOIs $\boldsymbol{q}(n)$ having DOAs $\phi_1, \ldots, \phi_{L-L_\alpha}$, and denote $\boldsymbol{B} = \left[\boldsymbol{a}(\phi_1) \cdots \boldsymbol{a}(\phi_{L-L_\alpha})\right]$, $\boldsymbol{A} = \boldsymbol{A}(\Theta)$, $\boldsymbol{D} = \left[\boldsymbol{d}(\theta_1) \cdots \boldsymbol{d}(\theta_{L_\alpha})\right]$ where $\boldsymbol{d}(\theta) = \partial \boldsymbol{a}(\theta)/\partial\theta$, and $h(\theta) = \boldsymbol{d}^H(\theta)\boldsymbol{G}\boldsymbol{G}^H\boldsymbol{d}(\theta)$. Remaining symbols follow the notation of Section 3.1. In [71] it is shown that the bias and MSE of Cyclic MUSIC DOA estimates are given by

$$E\left\{\hat{\theta}_i - \theta_i\right\} = -\frac{\mathrm{Re}\left\{\sum_{\beta\neq\alpha}\left[\boldsymbol{R}_{ss}^{\alpha H}(\tau)^{-1}\boldsymbol{A}^{+}\boldsymbol{B}\boldsymbol{R}_{qq}^{\beta}(\tau)\boldsymbol{B}^H\boldsymbol{P}_A^{\perp}\boldsymbol{D}\right]_{ii} Z_N(\alpha-\beta,\tau)\right\}}{h(\theta_i)} + O\left(\frac{1}{N^2}\right)$$

and

$$E\left\{(\hat{\theta}_i - \theta_i)^2\right\} = \frac{1}{2Nh^2(\theta_i)}\left[\boldsymbol{D}^H\boldsymbol{P}_A^{\perp}\left[\boldsymbol{I}+\boldsymbol{B}\frac{\boldsymbol{R}_{qq}(0)}{\sigma^2}\boldsymbol{B}^H\right]\boldsymbol{P}_A^{\perp}\boldsymbol{D}\right]_{ii} \times\left[\frac{1+\left(\boldsymbol{A}^H\boldsymbol{A}\right)_{ii}^{-1}\big/\mathrm{SNR}_i}{\left|\rho_{s_is_i}^{\alpha}(\tau)\right|^2\mathrm{SNR}_i} + \sigma^2\left[\boldsymbol{R}_{ss}^{\alpha H}(\tau)^{-1}\boldsymbol{A}^{+}\boldsymbol{B}\boldsymbol{R}_{qq}(0)\boldsymbol{B}^H\boldsymbol{A}^{+H}\boldsymbol{R}_{ss}^{\alpha}(\tau)^{-1}\right]_{ii}\right].$$

It should be noted that in the absence of SNOIs these expressions reduce to

$$E\left\{\hat{\theta}_i - \theta_i\right\} = O\left(\frac{1}{N^2}\right)$$

and

$$E\left\{(\hat{\theta}_i - \theta_i)^2\right\} \approx \frac{1}{2N\left|\rho_{s_is_i}^{\alpha}(\tau)\right|^2\mathrm{SNR}_i}\left[1+\frac{\left(\boldsymbol{A}^H\boldsymbol{A}\right)_{ii}^{-1}}{\mathrm{SNR}_i}\right]\frac{1}{h(\theta_i)} + O\left(\frac{1}{N^2}\right).$$

These expressions in turn reduce to the expressions for conventional MUSIC in [88] when $\alpha = 0$, and thus, when SNOIs are absent and the noise is spatially white, the ratio of MSEs is given by

$$\frac{E\left\{(\hat{\theta}_i - \theta_i)^2\right\}_{Cyclic\ MUSIC}}{E\left\{(\hat{\theta}_i - \theta_i)^2\right\}_{MUSIC}} = \left|\rho_{s_is_i}^{\alpha}(\tau)\right|^{-2} \geq 1,$$

which implies that MUSIC is preferred over Cyclic MUSIC in this type of environment. Since the cyclic correlation coefficient, evaluated at the baud rate, for most signals such as PAM, M-ary QAM, and various PSK signals is $1/3$ or less, the preference of MUSIC over Cyclic MUSIC in these simple environments (that is, all signals have cycle frequency α and the noise is spatially white) can be strong.

On the other hand, when SNOIs are present, Cyclic MUSIC may be strongly preferred over all conventional algorithms. This is predicted by the analytical results because the MSE of Cyclic MUSIC remains bounded as the DOAs of the SNOIs approach those of the SOIs (i.e., as the smallest nonzero singular value of the matrix $[\boldsymbol{A}\ \boldsymbol{B}]$ approaches zero), whereas the MSE of MUSIC approaches infinity. This behavior is demonstrated in this simulation, in which the dependence of MSE on angular separation between a SOI and a SNOI is investigated. The SOI, which is a Gaussian PAM signal (i.e., the symbol stream is Gaussian) having 100% excess bandwidth, arrives from 0 degrees, and the SNOI, which is also Gaussian PAM with 100% excess bandwidth but has a different baud rate than the SOI, arrives from a different angle. The SNR is 10 dB, and the number of data samples is 8192. The Cramér-Rao lower bound (CRLB) for the cyclostationary case was computed using the method of [77], although the proposed generalization of Whittle's Theorem to cyclostationary signals [70] could also be used.

As shown in Fig. 16, the RMSE of Cyclic MUSIC is nearly insensitive with respect to the angular separation, as predicted, and it should be noted that these results are nearly identical to those obtained for non-Gaussian signals (i.e., BPSK). The RMSE of Cyclic MUSIC using α equal to the baud rate of the SNOI is comparable and is not shown. The results show that MUSIC performs much worse than Cyclic MUSIC when the signals are close together (e.g., 3 degrees apart or less), and better when the signals are farther apart. No data on the RMSE of MUSIC is shown for

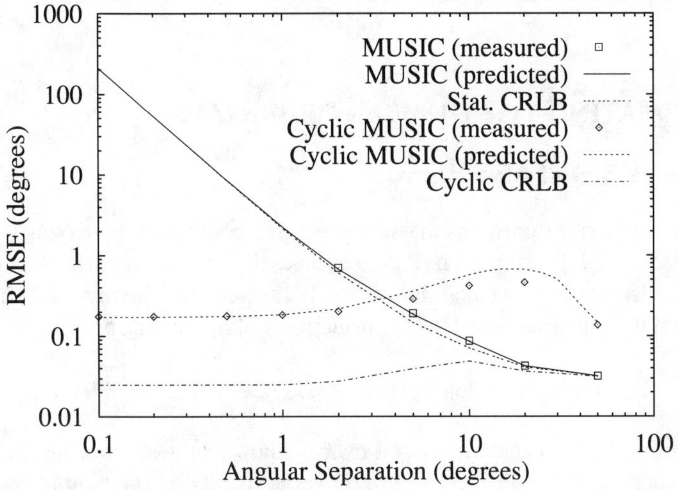

Figure 16: RMSE as a function of angular separation for one SOI with 10 dB SNR in the presence of 1 equal-powered SNOI and complex white Gaussian noise.

angular separation less than 2 degrees since MUSIC obtained two DOA estimates in fewer than 10% of these trials. Extrapolating from these results, it is interesting to note that conventional MUSIC would require more than 10^6 times as many data samples as Cyclic MUSIC to achieve the same RMSE when the angular separation is 0.1 degrees. It should also be noted that it was shown empirically in [68, 77] that conjugate Cyclic MUSIC and conjugate Cyclic Least Squares nearly achieved the CRLB for double-sideband suppressed-carrier amplitude modulation (DSB-SC AM) signals.

Recent results in [78] show that Cyclic MUSIC can operate properly in the presence of much more error in the knowledge of the array manifold than MUSIC can when two signals having different cycle frequencies are present. This capability may prove to be significant in practical applications, where sensor positions and characteristics may be perturbed by vibration, thermal expansion, drift in the values of electronic components, and so forth.

6.5 Summary

As with the spatial filtering methods of Section 4, several cyclic DF methods are available. Analytical and simulation-based results demonstrate their superiority over conventional methods, especially those in which the number of signals exceeds the number of sensors but the number of signals having cycle frequency α is less than the number of sensors, or those in which noise is unknown, or those in which signals having different cycle frequencies have similar DOAs.

$\boxed{\alpha}$ Significant open problems remain, including the development of statistically optimum cyclic DF methods (e.g., maximum-likelihood or WSF), implementation and testing of algorithms on real data, and development of better methods for wideband data.

7 ESTIMATING THE NUMBER OF SIGNALS

7.1 Conventional Algorithms

Of the conventional algorithms that detect the number of signals arriving at the array, the method of [103] is summarized here as an easily implemented, high-performance detector. Also, its performance shows clearly the power of the minimum description length (MDL) principle of [64]. This principle yields a criterion

$$MDL(k) = -\log f(\boldsymbol{x}(0), \cdots, \boldsymbol{x}(N-1) \mid \hat{z}) + \frac{1}{2} k \log N,$$

that selects the model (parameterized by k unknown values $\hat{z}$) having the minimum length, where $f(\boldsymbol{x}(0), \cdots, \boldsymbol{x}(N-1) \mid \hat{z})$ is the likelihood function of the observed data given a guess $\hat{z}$ of the parameter values. For independent stationary temporally white Gaussian signals arriving at a sensor array in the presence of white noise, it can be shown that the estimated number $\hat{L}$ of signals is the value of k that minimizes

$$MDL(k) = -(M-k)N \log \left[\frac{\prod_{i=k+1}^{M} \lambda_i^{1/(M-k)}}{\frac{1}{M-k}\sum_{i=k+1}^{M} \lambda_i} \right] + \frac{1}{2}k(2M-k)\log N$$

over $k = 0, \ldots, M-1$, where $\lambda_1 \geq \cdots \lambda_M$ are the eigenvalues of $\hat{\boldsymbol{R}}_{xx}$. More recent methods that perform joint detection and DF are described in [98, 101, 107] and references therein. To address the degradation of these methods in the presence of unknown noise, more robust methods are described in [19, 102, 108] and references therein.

7.2 Algorithms for Cyclostationary Signals

$\boxed{\alpha}$ In this section the problem of estimating the number of signals having particular cyclostationarity properties is discussed. In any practical implementation or simulation of the spatial filtering methods and direction-finding methods discussed in this overview, this number must be either known or estimated. The algorithms developed to date attempt to estimate this number either by estimating the rank of the cyclic autocorrelation matrix of the received data or by estimating the number of signal components that are common to the received data and a frequency-shifted and delayed version thereof. Given the lack of theoretical justification for some parts of the algorithms and the less-than-outstanding performance, all of the algorithms for cyclostationary signals discussed here seem to be suboptimal solutions, and much remains to be done toward finding superior methods.

7.2.1 Canonical Correlation Significance Test (CCST)

As noted in Section 4.4.4, the task of estimating the waveforms of signals having cycle frequency α can be posed as a canonical correlation analysis problem. This observation suggests that the methods from multivariate statistics for estimating the number of common signals between two data sets may be useful in the present problem. However, the standard methods [10, 11, 50] use sequential hypothesis testing, which requires the subjective setting of thresholds and related parameters.

Following the success of information-theoretic criteria such as MDL, summarized in Section 7.1, work on applying these criteria to the canonical correlation significance problem has appeared recently (e.g., see [25, 92], although no empirical evaluations of algorithms are provided there). In [74] the MDL principle is (loosely) applied to this problem to yield the canonical correlation significance test (CCST), in which the estimated value of $\hat{L}_\alpha$ is declared to be the value of k that minimizes the expression

$$-N \log \prod_{i=k+1}^{M} (1-\lambda_i) + \frac{1}{2}k(2M-k)\log N$$

where $\lambda_1 \geq \cdots \geq \lambda_M$ are the eigenvalues of the Cross-SCORE equation (19).

Although several simulation results of CCST appear in [74], the figures there are incorrect. A very brief simulation-based study of MDL, CCST, and the methods to be summarized next appears in Section 7.2.5.

$\boxed{\alpha}$ The development of CCST makes the usual assumptions of canonical correlation analysis, namely that two jointly stationary, jointly Gaussian, temporally white time-series are being analyzed, when in fact the time-series in question are certainly not stationary and are typically neither Gaussian nor white. Thus, CCST is necessarily suboptimal. A more rigorous application of the MDL principle, in which the cyclostationarity is properly accounted for in the likelihood function, should yield a superior method.

7.2.2 Canonical Correlation Test (CCT)

In [19], canonical correlation analysis is applied to detect the number L of unknown signals arriving at two spatially separated arrays in the presence of unknown noise and is based on the same test statistic as CCST. In [19] the data are assumed to be stationary, Gaussian, and temporally white, in which case Bartlett's test statistic [11]

$$C(k) = -2\,[N - M - 1/2]\log \prod_{i=k}^{M} (1 - \lambda_i)$$

is approximately distributed as χ^2 with $2(M - k + 1)^2$ degrees of freedom if there are $k - 1$ signals present. Thus, $C(k)$, for $k = M - 1, M - 2, \ldots, 0$ is tested against a sequence of threshold values $t(k)$. As described in [19] the threshold values can be chosen to yield a constant probability of false detection, and these values can be obtained from a table of percentage points of the χ^2 distribution (cf. [2]). The estimate $\hat{L}$ is declared to be equal to the first value of k such that $C(k) > t(k)$.

It is proposed here that CCT be applied to the problem of estimating the number of signals having cycle frequency α by using the eigenvalues $\lambda_1, \cdots, \lambda_M$ obtained from the Cross-SCORE equation (19). The performance of this ad hoc method is evaluated in Section 7.2.5.

7.2.3 A Determinant-Based Method (DBM)

In the method of [47], the sequence of determinants of principal submatrices of $\hat{\boldsymbol{R}}^{\alpha}_{\boldsymbol{xx}^*}(\tau)$ is analyzed to estimate the probable rank of the matrix and thus the number of cyclostationary signals. This method is referred to here as the determinant-based method (DBM). Let D_k denote the determinant of the $k \times k$ principal submatrix of $\frac{1}{b}\hat{\boldsymbol{R}}^{\alpha}_{\boldsymbol{xx}^*}(\tau)$, where

$$b = \frac{1}{M^2} \sum_{i,j=1}^{M} \left| R^{\alpha}_{x_i x_j^*}(\tau) \right|^2$$

is a normalization constant. Alternately, $\hat{\boldsymbol{R}}^{\alpha}_{\boldsymbol{xx}}(\tau)$ could be analyzed to estimate the number of signals exhibiting cyclostationarity rather than conjugate cyclostationarity. Based on the observation that $D_1, \cdots, D_{L_\alpha} \neq 0$ and $D_{L_\alpha+1}, \cdots, D_M \to 0$ as the averaging time N approaches infinity, the method attempts to find the "edge" between the significant and insignificant determinants. This detection is accomplished in DBM by declaring $\hat{L}_\alpha$ equal to the value of k that maximizes

$$\frac{D_k^2}{D_{k-1}D_{k+1}}, \quad \text{where } D_0 \triangleq 1.$$

Simulation results are available in [47] (where a very large number of data samples was required due to the very narrow bandwidth of the SOI) and in Section 7.2.5 (where the method is shown to perform comparably to the other cyclic DF methods).

7.2.4 Variable Coefficient-Based Method (VCBM)

In the method of [36], the singular values $\hat{\sigma}_1 \geq \cdots \geq \hat{\sigma}_M \geq 0$ of $\hat{\boldsymbol{R}}_{\boldsymbol{xx}}^{\alpha}(\tau)$ are analyzed to estimate L_α. This method is referred to here as the variable coefficient-based method (VCBM). Theoretically, VCBM should be superior to DBM because it is known that the kth-largest singular value of an arbitrary matrix equals the minimum 2-norm distance of this matrix to the set of all matrices having rank $k-1$ [39], and thus a superior rank estimator could be constructed by analyzing the singular values. The open question remains "How small should the k-th largest singular value be for it to be declared as insignificant and the rank of the matrix as $k-1$?" In VCBM this question is answered by declaring $\hat{L}_\alpha$ equal to the value of k that maximizes v_k/m_k^2, where

$$m_k = \frac{1}{M-k+1}\sum_{i=k}^{M}\hat{\sigma}_i$$

$$v_k = \frac{1}{M-k+1}\sum_{i=k}^{M}(\hat{\sigma}_i - m_k)^2.$$

Simulation results in [36] and Section 7.2.5 show that VCBM can be superior to both DBM and CCST in some environments, although it is very sensitive to differences in the received power of the SOIs.

7.2.5 Performance

The simulated environment duplicates one of those used in [74]. In addition to correcting erroneously plotted results, the results presented here also include evaluations of the other methods discussed in this section. Three BPSK signals having Nyquist-shaped pulses with 100% excess bandwidth are received at a 4-element ULA. Two of these signals have baud rate 0.25 and arrive from 10 and -15 degrees, and the third signal has baud rate 0.2 and arrives from 0 degrees. The CCST, CCT (using thresholds determined by a probability of false alarm equal to 0.1), DBM, VCBM, and MDL methods are evaluated. The cyclic methods are applied using the baud rates of the desired signals. For the case in which all three signals have equal SNR of 10 dB, the numbers of correct estimates are tabulated in Table 3. For a different case in which the two signals having baud rate 0.25 have SNRs of 10 and 5 dB, respectively, and the SNR of the signal having baud rate 0.2 is 10 dB, the numbers of correct estimates are tabulated in Table 4.

As can be seen from these results, only MDL performs consistently well. Among the cyclic methods, VCBM is superior when SNRs are equal but degrades severely

Table 3: Number of correct results in 100 trials for 3 signals having equal SNR of 10 dB. Two signals share the same baud rate.

	1 signal: $\alpha = 0.2$				2 signals: $\alpha = 0.25$				
N	CCST	CCT	VCBM	DBM	CCST	CCT	VCBM	DBM	MDL
64	78	4	81	55	12	0	59	37	91
128	78	7	92	59	20	0	93	62	100
256	75	10	95	71	26	1	100	66	100
512	96	43	100	76	65	53	100	80	100
1024	100	95	100	88	100	98	100	92	100
2048	95	95	100	94	100	98	100	95	100
4096	99	95	100	95	100	97	100	99	100
8192	97	96	100	98	100	91	100	100	100

Table 4: Number of correct results in 100 trials for 3 signals having unequal SNR. Two signals share the same baud rate and have SNRs of 10 and 5 dB, respectively. The SNR of the third signal is 10 dB.

	1 signal: $\alpha = 0.2$				2 signals: $\alpha = 0.25$				
N	CCST	CCT	VCBM	DBM	CCST	CCT	VCBM	DBM	MDL
64	80	4	92	62	6	0	3	18	69
128	80	8	95	72	8	0	6	25	100
256	78	13	98	74	8	2	13	26	100
512	94	45	100	90	38	35	22	43	100
1024	100	93	100	97	96	93	32	46	100
2048	99	94	100	98	100	98	67	80	100
4096	100	94	100	100	100	97	85	95	100
8192	100	96	100	100	100	92	100	98	100

when the SNRs differ. CCST outperforms CCT in all cases considered. Also, although DBM occasionally outperforms CCST for some smaller values of N, CCST generally outperforms DBM, especially for longer averaging times N.

However, MDL fails in the presence of more signals than sensors or when spatially correlated noise is present (e.g., see [74] for simulations). Also, if a cyclic DF method is to be applied, then the number of signals having a given cycle frequency α is needed, and this cannot be supplied by the conventional MDL for stationary signals. In contrast, the cyclic methods are relatively insensitive to the spatial correlation of the noise, and can accommodate more signals than sensors ($L > M$) provided that the number of signals sharing a cycle frequency of interest is less than the number of sensors ($L_\alpha < M$).

$\boxed{\alpha}$ Nonetheless, the high performance of MDL when its requirements are met, and recent extensions to perfectly correlated sources and unknown noise suggest that application of the general MDL principle to the case of cyclostationary signals may be successful. Similarly, correct generalization of other high-performance algorithms for stationary signals may also yield superior methods for cyclostationary signals.

7.3 Summary

In contrast to recent work on methods for stationary signals, it is clear that all methods currently available for cyclostationary signals are ad hoc and exhibit undesirable performance attributes such as long convergence times and sensitivity to small differences in received signal power. Rigorous formulation of the likelihood function for cyclostationary signals arriving at an array, followed by careful application of principles applied to stationary signals, such as MDL, should yield superior detectors having better performance than existing methods.

8 CONCLUSIONS

Work on algorithms for spatial filtering and DF of cyclostationary signals has shown that exploitation of cyclostationarity can offer substantially improved performance over that of conventional algorithms in some signal environments and can offer reasonably good performance in other environments where conventional algorithms fail completely. However, some of these signal-selective algorithms are suboptimal and exhibit undesirable performance attributes. Thus, numerous open problems, some of which are identified in this overview, exist and range from application-oriented to more fundamental theoretical problems. These technical challenges and the growing appreciation among industrial, federal, and military organizations for the benefits of cyclostationarity exploitation in sensor array processing bode well for researchers and engineers in this area.

ACKNOWLEDGMENTS

The author gratefully acknowledges the encouragement and support received from Professor William A. Gardner and the helpful comments on this and previous work made by Professor A. Lee Swindlehurst, Dr. Chad M. Spooner, and the members of the α-group at the University of California, Davis.

This overview is based on the plenary lecture given by the author at the 1992 Workshop on Cyclostationary Signals in Yountville, CA. Much of the work surveyed in this chapter was supported by the following programs: the Office of Naval Research under contract N00014-92-J-1218 (PIs: W.A. Gardner and S.V. Schell); E-Systems, Inc., Greenville Div.; the Army Research Office under contract DAAL03-89-C-0035 sponsored by the U.S. Army Communications Electronics Command Center for Signals Warfare (PI: W. A. Gardner); the National Science Foundation under grant MIP-88-12902 (PI: W. A. Gardner); and ESL, Inc. and the California State MICRO Program (PI: W. A. Gardner).

APPENDIX MOMENTS OF CYCLIC CORRELATION MATRICES

Recent results from [69] on the mean and covariance of cyclic correlation matrices are summarized. Although the results were derived in the stochastic-process framework where $E\{\}$ denotes the ensemble average, they are also true in the nonstochastic time-series framework where $E\{\}$ denotes the polyperiodic component extractor [32]. Relevant results can also be found in [44].

A.1 Assumptions

First, the assumptions about the processes $\boldsymbol{x}(n)$ and $\boldsymbol{y}(n)$ to be used at various points in the derivations are listed. A brief discussion of these assumptions follows.

A-1 The time-variant means of $\boldsymbol{x}(n)$ and $\boldsymbol{y}(n)$ are identically equal to zero, $E\{\boldsymbol{x}(n)\} \equiv E\{\boldsymbol{y}(n)\} \equiv 0$.

A-2 The time-variant correlations of $\boldsymbol{x}(n)$ and $\boldsymbol{y}(n)$ have Fourier series representations consisting of a finite number of finite-magnitude terms; thus, for all τ, $\left|\boldsymbol{R}^{\beta}_{\boldsymbol{xx}}(\tau)\right|$, $\left|\boldsymbol{R}^{\beta}_{\boldsymbol{xx}^*}(\tau)\right|$, $\left|\boldsymbol{R}^{\beta}_{\boldsymbol{yy}}(\tau)\right|$, $\left|\boldsymbol{R}^{\beta}_{\boldsymbol{yy}^*}(\tau)\right|$, $\left|\boldsymbol{R}^{\beta}_{\boldsymbol{xy}}(\tau)\right|$, and $\left|\boldsymbol{R}^{\beta}_{\boldsymbol{xy}^*}(\tau)\right|$ are all finite for all β and each is nonzero for only a finite number of cycle frequencies β.

A-3 The processes $\boldsymbol{x}(n)$ and $\boldsymbol{y}(n)$ satisfy the mixing condition

$$\sum_{m=0}^{\infty}\sum_{k=m}^{\infty}\left|R^{\beta}_{uv}(k+\tau)R^{\gamma}_{pq}(k-\tau)\right| < \infty$$

for all possible choices of elements $u(n)$, $v(n)$, $p(n)$, and $q(n)$ of $\boldsymbol{x}(n)$, $\boldsymbol{x}^*(n)$, $\boldsymbol{y}(n)$, and $\boldsymbol{y}^*(n)$, and all cycle frequencies β and γ, and all lag-values τ.

A-4 The processes $\boldsymbol{x}(n)$ and $\boldsymbol{y}(n)$ are complex and jointly Gaussian.

Some comments on the appropriateness of these assumptions are in order. A-1 is true for many communication signals such as DSB-SC AM, and most digital signals such as BPSK, QPSK, M-ary QAM, FSK, and spread spectrum signals. A-2 is true for most persistent communication signals (e.g., see [28, 30] for explanations of this from the stochastic-process and nonstochastic time-average frameworks, respectively; see also [27, 33]). It is straightforward to verify that A-2 implies that the processes $\boldsymbol{x}(n)$ and $\boldsymbol{y}(n)$ are uniformly almost periodically correlated (UAPC) in the sense of [46]. Also, as originally explained in [26] using the alternative terminology *almost cyclostationary*, most communication signals are UAPC because their time-variant autocorrelations are expressible as Fourier series having a finite number of terms (each of which corresponds to a cycle frequency such as a baud rate or a doubled-carrier frequency). A-3 can be easily shown to be true if, for example, $|R_{uv}(t+\tau,t)| < ce^{-\tau}$ for all choices of elements $u(n)$ and $v(n)$ of $\boldsymbol{x}(n)$, $\boldsymbol{x}^*(n)$, $\boldsymbol{y}(n)$, and $\boldsymbol{y}^*(n)$ and for all t and some choice of positive constant c; loosely speaking, it requires that the statistical dependence between any two time samples of the processes decays fast enough as the temporal separation between them increases. If $\boldsymbol{x}(n)$ can be shown to be a phi-mixing process, which is simply a process for which the statistical dependence between two time samples is uniformly controlled by a function solely of the temporal separation (cf. [46]), and A-2 is true, then it is proposed that the estimator $\hat{\boldsymbol{R}}^{\alpha}_{\boldsymbol{xy}}(\tau)$ is strongly consistent and asymptotically normal (the comparable result for the continuous-time case was proven in [46] under the assumptions that the process was phi-mixing and UAPC, the latter of the two being implied by A-2). A-4 is true for $\boldsymbol{x}(n)$ and $\boldsymbol{y}(n)$ each consisting of, for example, sums of Gaussian noises, DSB-SC AM signals having Gaussian basebands, and/or PAM and QAM signals having Gaussian symbol sequences.

A.2 Moments of Cyclic Correlation Estimates

The mean and second-order moments of the cyclic correlogram $\hat{\boldsymbol{R}}_{\boldsymbol{xy}}^{\alpha}(\tau)$ of the cyclostationary processes $\boldsymbol{x}(n)$ and $\boldsymbol{y}(n)$ as derived in [69] are summarized here under the assumptions A-1 through A-4 described in Section A-1. Additionally, in light of the last part of D-8, only expressions for $\tau > 0$ are derived directly. These results generalize (and make a typographical correction to) those in [43] (cf. [44]).

A.3 Mean

Lemma 1 *Given A-2, the mean of* $\hat{\boldsymbol{R}}_{\boldsymbol{xy}}^{\alpha}(\tau)$ *for* $\tau > 0$ *can be expressed as*

$$E\left\{\hat{\boldsymbol{R}}_{\boldsymbol{xy}}^{\alpha}(\tau)\right\} = \sum_{\beta} \boldsymbol{R}_{\boldsymbol{xy}}^{\beta}(\tau) Z_N(\alpha - \beta, \tau).$$

Corollary 1 *Given A-2, the cyclic correlogram* $\hat{\boldsymbol{R}}_{\boldsymbol{xy}}^{\alpha}(\tau)$ *is an asymptotically unbiased estimator of* $\boldsymbol{R}_{\boldsymbol{xy}}^{\alpha}(\tau)$*:*

$$\begin{aligned} E\left\{\hat{\boldsymbol{R}}_{\boldsymbol{xy}}^{\alpha}(\tau)\right\} &= \left(1 - \frac{\tau}{N}\right) \boldsymbol{R}_{\boldsymbol{xy}}^{\alpha}(\tau) + \sum_{\beta \neq \alpha} \boldsymbol{R}_{\boldsymbol{xy}}^{\beta}(\tau)\, Z_N(\alpha - \beta, \tau) \\ &= \boldsymbol{R}_{\boldsymbol{xy}}^{\alpha}(\tau) + O\left(\frac{1}{N}\right). \end{aligned}$$

A.4 Covariance

The covariances of the elements of $\hat{\boldsymbol{R}}_{\boldsymbol{xy}}^{\alpha}(\tau)$ are derived here by first considering the covariance $\operatorname{cov}\left\{\hat{R}_{uv}^{\eta}(\tau), \hat{R}_{pq}^{\lambda}(\tau)\right\}$ where each of $u(n)$, $v(n)$, $p(n)$, and $q(n)$ is any element of $\boldsymbol{x}(n)$, $\boldsymbol{x}^*(n)$, $\boldsymbol{y}(n)$, or $\boldsymbol{y}^*(n)$.

Lemma 2 *Given A-1, A-2, A-3, and A-4, then*

$$cov\left\{\hat{R}_{uv}^{\eta}(\tau), \hat{R}_{pq}^{\lambda}(t)\right\} = I_1 + I_2 + O\left(\frac{1}{N^2}\right)$$

where

$$I_1 = \frac{1}{N} \sum_{\beta} \left[\sum_{k=-\infty}^{\infty} R_{up}^{\beta}(k + \tau - t) R_{vq}^{(\beta-\eta+\lambda)*}(k) e^{-j2\pi\eta k} \right] e^{j2\pi\beta t}$$

$$I_2 = \frac{1}{N} \sum_{\beta} \left[\sum_{k=-\infty}^{\infty} R_{uq^*}^{\beta}(k + \tau) R_{vp^*}^{(\beta-\eta+\lambda)*}(k - t) e^{-j2\pi\eta k} \right] e^{-j2\pi(\beta-\eta+\lambda)t}.$$

As noted in [69], Lemma 2 is a generalization of the result obtained in [43], where the latter is derived for the cyclic autocorrelogram of a real-valued continuous-time scalar process for which all cycle frequencies are integer multiples of a fundamental cycle frequency, whereas the former is derived for the cyclic cross-correlogram of complex-valued discrete-time processes having cycle frequencies that need not be integer multiples of a fundamental. However, with reference to results and notations in [43], namely Lemma 1 and equations (42a) and (45) of Proposition 3, a minor correction is obtained by careful application of Lemma 1 to equation (45) which yields

$$F(u, \tau_1, \tau_2) = \sum_{p=-\infty}^{\infty} B_p(u + \tau_1 - \tau_2) B_{j-k-p}(u) e^{-i2\pi(ju - p\tau_2)/T}$$

rather than $F(u, \tau_1, \tau_2)$ as defined in (42a). With reference to Lemma 2 of this overview, in the special case in which all cycle frequencies are harmonics of a fundamental cycle frequency, $u(n) = v(n) = p(n) = q(n)$ are real, and noting that $R_{qq}^{\beta *}(k) = R_{qq}^{-\beta}(k)$ for all β, k when $q(n)$ is real, it is easily verified that Lemmas 1 and 2 of this overview yield a covariance that is identical in form to that obtained from Proposition 3 (after the minor correction) in [43].

The next two corollaries, which follow directly from Lemma 2 after algebraic substitution and rearrangement, characterize the covariances $\mathrm{cov}\left\{\hat{\boldsymbol{R}}_{\boldsymbol{xy}}^{\alpha H}(\tau)\boldsymbol{g}, \hat{\boldsymbol{R}}_{\boldsymbol{xy}}^{\alpha T}(\tau)\boldsymbol{w}^*\right\}$ and $\mathrm{cov}\left\{\hat{\boldsymbol{R}}_{\boldsymbol{xy}}^{\alpha H}(\tau)\boldsymbol{g}, \hat{\boldsymbol{R}}_{\boldsymbol{xy}}^{\alpha H}(\tau)\boldsymbol{w}\right\}$. These covariances are needed, for example, in the expressions for the second moments of the singular vectors of $\hat{\boldsymbol{R}}_{\boldsymbol{xy}}^{\alpha}(\tau)$ which arise in the performance analysis [71] of the Cyclic MUSIC direction-finding method (see also Section 6.3.1).

Corollary 2 *Given A-1, A-2, A-3, A-4, and two arbitrary vectors* **g** *and* **w**, *then*

$$\mathrm{cov}\left\{\hat{\boldsymbol{R}}_{\boldsymbol{xy}}^{\alpha H}(\tau)\boldsymbol{g}, \hat{\boldsymbol{R}}_{\boldsymbol{xy}}^{\alpha T}(\tau)\boldsymbol{w}^*\right\} = J_1 + J_2 + O\left(\frac{1}{N^2}\right)$$

where

$$J_1 = \frac{1}{N} \sum_{\beta} \sum_{k=-\infty}^{\infty} \boldsymbol{w}^T \boldsymbol{R}_{\boldsymbol{xx}^*}^{\beta H}(k) \boldsymbol{g} \boldsymbol{R}_{\boldsymbol{yy}^*}^{\beta - 2\alpha}(k) e^{j2\pi\alpha k} e^{-j2\pi\beta\tau}$$

$$J_2 = \frac{1}{N} \sum_{\beta} \sum_{k=-\infty}^{\infty} \boldsymbol{R}_{\boldsymbol{yx}}^{\beta - 2\alpha}(k - \tau) \boldsymbol{w} \boldsymbol{g}^T \boldsymbol{R}_{\boldsymbol{xy}}^{\beta *}(k + \tau) e^{j2\pi\alpha k} e^{j2\pi(\beta - 2\alpha)\tau}$$

Corollary 3 *Given A-1, A-2, A-3, A-4, and two arbitrary vectors* **g** *and* **w**, *then*

$$\mathrm{cov}\left\{\hat{\boldsymbol{R}}_{\boldsymbol{xy}}^{\alpha H}(\tau)\boldsymbol{g}, \hat{\boldsymbol{R}}_{\boldsymbol{xy}}^{\alpha H}(\tau)\boldsymbol{w}\right\} = K_1 + K_2 + O\left(\frac{1}{N^2}\right)$$

where

$$K_1 = \frac{1}{N}\sum_{\beta}\sum_{k=-\infty}^{\infty} \boldsymbol{w}^H \boldsymbol{R}_{\boldsymbol{xx}}^{\beta H}(k)\boldsymbol{g}\boldsymbol{R}_{\boldsymbol{yy}}^{\beta}(k)e^{j2\pi\alpha k}e^{-j2\pi\beta\tau}$$

$$K_2 = \frac{1}{N}\sum_{\beta}\sum_{k=-\infty}^{\infty} \boldsymbol{R}_{\boldsymbol{yx^*}}^{\beta}(k-\tau)\boldsymbol{w}^*\boldsymbol{g}^T\boldsymbol{R}_{\boldsymbol{xy^*}}^{\beta *}(k+\tau)e^{j2\pi\alpha k}e^{j2\pi\beta\tau}.$$

Clearly, $\hat{\boldsymbol{R}}_{\boldsymbol{xy}}^{\alpha}(\tau)$ converges in mean square to $\boldsymbol{R}_{\boldsymbol{xy}}^{\alpha}(\tau)$ since its bias and covariance are both $O\left(\frac{1}{N}\right)$.

It should be noted that Corollaries 2 and 3 simplify in the following two special cases: 1) If $\boldsymbol{y}(n) = \boldsymbol{x}(n)$ and $\boldsymbol{x}(n)$ does not exhibit conjugate cyclostationarity ($\boldsymbol{R}_{\boldsymbol{xx^*}}^{\beta}(k) \equiv 0$ for all β), then $J_1 = K_2 = 0$; and 2) If $\boldsymbol{y}(n) = \boldsymbol{x}^*(n)$ and $\boldsymbol{x}(n)$ does not exhibit nonconjugate cyclostationarity ($\boldsymbol{R}_{\boldsymbol{xx}}^{\beta}(k) \equiv 0$ for all $\beta \neq 0$), then only the $\beta = 0$ terms remain in J_2 and K_1.

REFERENCES

[1] T. J. Abatzoglou and B. F. Rice, "Application of CTLS for estimating the direction of arrival of cyclostationary signals," in *Proc. Twenty-Fifth Asilomar Conf. on Signals, Systems and Computers*, pp. 247–251, Pacific Grove, CA, November 1991.

[2] M. Abramowitz and I. A. Stegun, eds., *Handbook of Mathematical Functions*, New York: Dover, 1965.

[3] B. G. Agee, *The Property-Restoral Approach to Blind Adaptive Signal Extraction*, Ph.D. thesis, Dept. of Electrical Engineering and Computer Science, University of California, Davis, CA, 1989.

[4] B. G. Agee, "Maximum-likelihood approaches to blind adaptive signal extraction using narrowband antenna arrays," in *Proc. Twenty-Fifth Asilomar Conf. on Signals, Systems, and Computers*. IEEE Computer Society Press, 1991.

[5] B. G. Agee, S. V. Schell, and W. A. Gardner, "Self-coherence restoral: A new approach to blind adaptation of antenna arrays," in *Proc. of the Twenty-First Asilomar Conf. on Signals, Systems, and Computers*, pp. 589–593, Pacific Grove, CA, November 1987.

[6] B. G. Agee, S. V. Schell, and W. A. Gardner, "The SCORE approach to blind adaptive signal extraction: An application of the theory of spectral correlation," in *Proc. of the Fourth ASSP Workshop on Spectrum Estimation and Modelling*, pp. 277–282, Minneapolis, MN, August 1988.

[7] B. G. Agee, S. V. Schell, and W. A. Gardner, "Spectral self-coherence restoral: A new approach to blind adaptive signal extraction," *Proc. IEEE*, vol. 78, pp. 753–767, April 1990.

[8] S. Anderson, M. Millnert, M. Viberg, and B. Wahlberg, "An adaptive array for mobile communication systems," *IEEE Trans. Vehic. Tech.*, vol. 40, pp. 230–236, February 1991.

[9] M. Avriel, *Nonlinear Programming: Analysis and Methods*, Englewood Cliffs, NJ: Prentice-Hall, 1976.

[10] M. S. Bartlett, "Further aspects of the theory of multiple regression," *Proc. Cambridge Philos. Soc.*, vol. 34, pp. 33–40, 1938.

[11] M. S. Bartlett, "Multivariate analysis," *J. Roy. Stat. Soc.*, vol. 9, pp. 176–197, 1947.

[12] A. J. Berni, "Target identification by natural resonance estimation," *IEEE Trans. Aerospace Electron. Syst.*, vol. AES-11, pp. 147–154, March 1975.

[13] T. E. Biedka and B. G. Agee, "Subinterval cyclic MUSIC—robust DF with error in cyclic frequency knowledge," in *Proc. Twenty-Fifth Asilomar Conf. on Signals, Systems and Computers*, pp. 262–266, Pacific Grove, CA, November 1991.

[14] D. H. Brandwood, "A complex gradient operator and its application in adaptive array theory," *IEE Proc.*, vol. 130(Pts. F and H, No. 1), pp. 11–16, February 1983.

[15] D. R. Brillinger, *Time Series: Data Analysis and Theory*, San Francisco, CA: Holden-Day, expanded edition, 1981.

[16] K. M. Buckley, "Spatial/spectral filtering with linearly constrained minimum variance beamformers," *IEEE Trans. Acoust., Speech, Signal Processing*, vol. ASSP-35, pp. 249–266, March 1987.

[17] K. M. Buckley and L. J. Griffiths, "Broadband signal-subspace spatial-spectrum (BASS-ALE) estimation for sensor array processing," *IEEE Trans. Acoust., Speech, Signal Processing*, vol. ASSP-36, pp. 953–964, July 1988.

[18] R. A. Calabretta, "On cyclic MUSIC algorithms for signal-selective direction estimation," Master's thesis, Dept. of Electrical Engineering and Computer Science, University of California, Davis, CA, 1989.

[19] W. G. Chen, J. P. Reilly, and K. M. Wong, "Application of canonical correlation analysis in detection in presence of spatially correlated noise," in *SPIE Vol. 1566 Advanced Signal Processing Algorithms, Architecures, and Implementations II*, pp. 464–473, 1991.

[20] R. T. Compton, Jr., *Adaptive Antennas*, Englewood Cliffs, NJ: Prentice-Hall, 1988.

[21] Z. Ding, "Blind channel identification and equalization using spectral correlation measurements, Part I: Frequency domain analysis," in W. A. Gardner, ed., *Cyclostationarity in Communications and Signal Processing*, New York: IEEE Press, 1994.

[22] Electronic Industries Association, "Dual-mode subscriber equipment–network equipment compatibility specification," *Interim Standard 54*, December 1989.

[23] L. E. Franks, "Polyperiodic linear filtering," in W. A. Gardner, ed., *Cyclostationarity in Communications and Signal Processing*, New York: IEEE Press, 1994.

[24] B. Friedlander and A. J. Weiss, "On the number of signals whose directions can be estimated by an array," *IEEE Trans. Signal Processing*, vol. 39, pp. 1686–1689, July 1991.

[25] Y. Fujikoshi, "Selection of variables in discriminant analysis and canonical correlation analysis," in P. R. Krishnaiah, ed., *Multivariate Analysis—VI*, pp. 219–236. Amsterdam: Elsevier Science Pub. B.V., 1985.

[26] W. A. Gardner, "Stationarizable random processes," *IEEE Trans. Inform. Theory*, vol. IT-24, pp. 8–22, January 1978.

[27] W. A. Gardner, "Spectral correlation of modulated signals: Part I—Analog modulation," *IEEE Trans. Comm.*, vol. COM-35, pp. 584–594, June 1987.

[28] W. A. Gardner, *Statistical Spectral Analysis: A Nonprobabilistic Theory*, Englewood Cliffs, NJ: Prentice-Hall, 1987.

[29] W. A. Gardner, "Simplification of MUSIC and ESPRIT by exploitation of cyclostationarity," *Proc. IEEE*, vol. 76, pp. 845–847, July 1988.

[30] W. A. Gardner, *Introduction to Random Processes with Applications to Signals and Systems,* 2nd ed., New York: McGraw-Hill, 1989.

[31] W. A. Gardner, "Cyclic Wiener filtering: Theory and method," *IEEE Trans. Comm.*, vol. 41, pp. 151–163, January 1993.

[32] W. A. Gardner, "An introduction to cyclostationary signals," in W. A. Gardner, ed., *Cyclostationarity in Communications and Signal Processing*, New York: IEEE Press, 1994.

[33] W. A. Gardner, W. A. Brown, III, and C.-K. Chen, "Spectral correlation of modulated signals: Part II—Digital modulation," *IEEE Trans. Comm.*, vol. COM-35, pp. 595–601, June 1987.

[34] W. A. Gardner and C.-K. Chen, "Signal-selective time-difference-of-arrival estimation for passive location of manmade signal sources in highly corruptive environments. Part I: Theory and method," *IEEE Trans. Signal Processing*, vol. 40, pp. 1168–1184, May 1992.

[35] W. A. Gardner, S. V. Schell, and P. A. Murphy, "Multiplication of cellular radio capacity by blind adaptive spatial filtering," in *Proc. of IEEE Int'l. Conf. on Selected Topics in Wireless Comm.*, pp. 102–106, Vancouver, B.C., Canada, June 1992.

[36] G. Gelli, L. Izzo, L. Paura, and G. Poggi, "A cyclic SVD-based algorithm for multiple source localization," in *13th GRETSI Symposium*, Nice, Sept. 16–20 1991.

[37] K. S. Gilhousen, I. M. Jacobs, R. Padovani, A. J. Viterbi, L. A. Weaver, Jr., and C. E. Wheatley III, "On the capacity of cellular CDMA system," *IEEE Trans. Vehic. Tech.*, vol. 40, pp. 303–312, May 1991.

[38] R. D. Gitlin, J. F. Hayes, and S. B. Weinstein, *Data Communication Principles*, New York: Plenum, 1992.

[39] G. H. Golub and C. F. van Loan, *Matrix Computations*, Baltimore, Maryland: Johns Hopkins University Press, 2nd ed., 1989.

[40] D. J. Goodman, "Trends in cellular and cordless communications," *IEEE Communications Mag.*, vol. 29, pp. 31–40, June 1991.

[41] A. Graham, *Kronecker Products and Matrix Calculus: with Applications*, England: Ellis Horwood, Ltd., 1981.

[42] P. S. Henry and B. S. Glance, "A new approach to high-capacity digital mobile radio," *Bell Syst. Tech. J.*, vol. 60, pp. 1891–1904, October 1981.

[43] H. L. Hurd, "Nonparametric time series analysis for periodically correlated processes," *IEEE Trans. Inform. Theory*, vol. 35, pp. 350–359, March 1989.

[44] H. L. Hurd and D. Dehay, "Performance analysis of estimators for the cyclic autocorrelation and cyclic spectrum," in W. A. Gardner, ed., *Cyclostationarity in Communications and Signal Processing*, New York: IEEE Press, 1994.

[45] H. L. Hurd and N. L. Gerr, "Graphical methods for determining the presence of periodic correlation," *J. Time Series Anal.*, vol. 12, pp. 337–350, 1991.

[46] H. L. Hurd and J. Leskow, "Strongly consistent and asymptotically normal estimation of the covariance for almost periodically correlated processes," Technical Report 160, Dept. of Statistics and Applied Probability, University of California, Santa Barbara, February 1991.

[47] L. Izzo, L. Paura, and G. Poggi, "An interference-tolerant algorithm for localization of cyclostationary-signal sources," *IEEE Trans. Signal Processing*, vol. 40, pp. 1682–1686, July 1992.

[48] R. A. Johnson and D. W. Wichern, *Applied Multivariate Statistical Analysis*, 2nd ed., Englewood Cliffs, NJ: Prentice Hall, 1988.

[49] J. Krolik and D. Swingler, "Multiple broad-band source location using steered covariance matrices," *IEEE Trans. Acoust., Speech, Signal Processing*, vol. 37, pp. 1481–1494, October 1989.

[50] D. N. Lawley, "Tests of significance in canonical analysis," *Biometrika*, vol. 46, pp. 59–66, 1959.

[51] W. C.Y. Lee, "Overview of cellular CDMA," *IEEE Trans. Vehic. Tech.*, vol. 40, pp. 291–302, May 1991.

[52] F. Li and R. J. Vaccaro, "Performance degradation of DOA estimators due to unknown noise fields," *IEEE Trans. Signal Processing*, vol. 40, pp. 686–690, March 1992.

[53] K. Mauck, "Wideband cyclic MUSIC," in *Proc. IEEE Int. Conf. Acoust., Speech, Signal Processing*, vol. IV, pp. 288–291, Minneapolis, MN, April 1993.

[54] R. A. Monzingo and T. W. Miller, *Introduction to Adaptive Arrays*, New York: John Wiley and Sons, 1980.

[55] B. Ottersten, R. Roy, and T. Kailath, "Signal waveform estimation in sensor array processing," in *Proc. Twenty-Third Asilomar Conf. on Signals, Systems and Computers*, pp. 787–791, Pacific Grove, CA, 1989.

[56] A. Paulraj, R. Roy, and T. Kailath, "Estimation of signal parameters via rotational invariance techniques—ESPRIT," in *Proc. Nineteenth Asilomar Conf. on Circuits, Systems and Computers*, pp. 83–89, Pacific Grove, CA, November 1985.

[57] R. L. Pickholtz, L. B. Milstein, and D. L. Schilling, "Spread spectrum for mobile communications," *IEEE Trans. Vehic. Tech.*, vol. 40, pp. 313–322, May 1991.

[58] V. F. Pisarenko, "The retrieval of harmonics from a covariance function," *Geophys. J. Royal Astronomical Soc.*, vol. 33, pp. 347–366, 1973.

[59] J. G. Proakis, "Advances in equalization for intersymbol interference," in A. J. Viterbi, ed., *Advances in Communication Systems*, New York: Academic, 1975, vol. 4.

[60] J. G. Proakis, *Digital Communications*, 2nd ed., New York: McGraw-Hill, 1989.

[61] S. U. H. Qureshi, "Adaptive equalization," *Proc. IEEE*, vol. 53, pp. 1349–1387, September 1985.

[62] C. R. Rao, "Separation theorems for singular values of matrices and their applications in multivariate analysis," *J. Multiv. Anal.*, vol. 9, pp. 362–377, 1979.

[63] J. P. Reilly and K. M. Wong, "Estimation of the directions of arrival of signals in unknown correlated noise, Part II: Asymptotic behavior and performance of the MAP approach," *IEEE Trans. Signal Processing*, vol. 40, pp. 2018–2028, August 1992.

[64] J. Rissanen, "Modeling by shortest data description," *Automatica*, vol. 14, pp. 465–471, 1978.

[65] R. S. Roberts, W. A. Brown, and H. H. Loomis, Jr., "A review of digital spectral correlation analysis: Theory and implementation," in W. A. Gardner, ed., *Cyclostationarity in Communications and Signal Processing*, New York: IEEE Press, 1994.

[66] R. Roy and T. Kailath, "ESPRIT—Estimation of signal parameters via rotational invariance techniques," *IEEE Trans. Acoust., Speech, Signal Processing*, vol. ASSP-37, pp. 984–995, July 1989.

[67] R. H. Roy, *ESPRIT: Estimation of Signal Parameters Via Rotational Invariance Techniques*, Ph.D. thesis, Stanford University, Stanford, CA, August 1987.

[68] S. V. Schell, *Exploitation of Spectral Correlation for Signal-Selective Direction Finding*, Ph.D. thesis, Dept. of Electrical Engineering and Computer Science, University of California, Davis, CA, 1990.

[69] S. V. Schell, "Asymptotic moments of estimated cyclic correlation matrices," Technical report TR001-93, Dept. of Electrical Engineering, Pennsylvania State University, 1993.

[70] S. V. Schell, "Generalization of Whittle's theorem to cyclostationary signals," in *Proc. of the IEEE Sixth SP Workshop on Statistical Signal and Array Processing*, pp. 110–113, Victoria, B.C., Canada, October 1992.

[71] S. V. Schell, "Performance analysis of the cyclic MUSIC method of direction estimation for cyclostationary signals," Technical report TR002-93, Dept. of Electrical Engineering, Pennsylvania State University, 1993.

[72] S. V. Schell and B. G. Agee, "Application of the SCORE algorithm and SCORE extensions to sorting in the rank-l self-coherence environment," in *Proc. of the Twenty-Second Annual Asilomar Conf. on Signals, Systems, and Computers*, pp. 274–278, Pacific Grove, CA, November 1988.

[73] S. V. Schell, R. A. Calabretta, W. A. Gardner, and B. G. Agee, "Cyclic MUSIC algorithms for signal-selective DOA estimation," in *Proc. IEEE Int. Conf. Acoust., Speech, Signal Processing*, pp. 2278–2281, Glasgow, Scotland, May 1989.

[74] S. V. Schell and W. A. Gardner, "Detection of the number of cyclostationary signals in unknown interference and noise," in *Proc. of the Twenty-Fourth Annual Asilomar Conf. on Signals, Systems, and Computers*, pp. 473–477, Pacific Grove, CA, November 1990.

[75] S. V. Schell and W. A. Gardner, "Progress on signal-selective direction finding," in *Proc. of the Fifth ASSP Workshop on Spectrum Estimation and Modeling*, pp. 144–148, Rochester, NY, October 1990.

[76] S. V. Schell and W. A. Gardner, "Signal-selective high-resolution direction finding in multipath," in *Proc. IEEE Int. Conf. Acoust., Speech, Signal Processing*, pp. 2667–2670, Albuquerque, NM, April 1990.

[77] S. V. Schell and W. A. Gardner, "Cramer-Rao lower bound for directions of arrival of Gaussian cyclostationary signals," *IEEE Trans. Inform. Theory*, vol. 38, pp. 1418–1422, July 1992.

[78] S. V. Schell and W. A. Gardner, "Robustness of direction-finding methods for cyclostationary signals in the presence of array calibration error," in *Proc. of the IEEE Sixth SP Workshop on Statistical Signal and Array Processing*, pp. 346–349, Victoria, B.C., Canada, October 1992.

[79] S. V. Schell and W. A. Gardner, "Blind adaptive spatio-temporal filtering for wideband cyclostationary signals," *IEEE Trans. Signal Processing*, vol. 41, pp. 1961–1964, May 1993.

[80] S. V. Schell and W. A. Gardner, "High-resolution direction finding," in N. K. Bose and C. R. Rao, eds., *Handbook of Statistics, Vol. 10*, Amsterdam: North-Holland, 1993.

[81] S. V. Schell and W. A. Gardner, "Maximum likelihood and common factor analysis-based blind adaptive spatial filtering for cyclostationary signals," in *Proc. IEEE Int. Conf. Acoust., Speech, Signal Processing*, vol. IV, pp. 292–295, Minneapolis, MN, April 1993.

[82] S. V. Schell, W. A. Gardner, and P. A. Murphy, "Blind adaptive antenna arrays in cellular communications for increased capacity," in *Proc. of 3rd Virginia Tech. Symp. on Wireless Personal Comm.*, June 1993.

[83] R. O. Schmidt, "Multiple emitter location and signal parameter estimation," in *Proc. RADC Spectrum Estimation Workshop*, Rome Air Develop. Center, NY, October 1979.

[84] R. O. Schmidt, *A Signal Subspace Approach to Multiple Source Location and Spectral Estimation*, Ph.D. thesis, Stanford Univ., Stanford, CA, 1981.

[85] R. O. Schmidt, "Multiple emitter location and signal parameter estimation," *IEEE Trans. Antennas Propagat.*, vol. AP-34, pp. 276–280, March 1986.

[86] M. Siotani, T. Hayakawa, and Y. Fujikoshi, *Modern Multivariate Statistical Analysis: A Graduate Course and Handbook*, Columbus, OH: American Sciences Press, 1985.

[87] C. M. Spooner, "Higher-order statistics for nonlinear processing of cyclostationary signals," in W. A. Gardner, ed., *Cyclostationarity in Communications and Signal Processing*, New York: IEEE Press, 1994.

[88] P. Stoica and A. Nehorai, "MUSIC, maximum likelihood, and Cramer–Rao bound," *IEEE Trans. Acoust., Speech, Signal Processing*, vol. 37, pp. 720–741, May 1989.

[89] P. Stoica and A. Nehorai, "Performance study of conditional and unconditional direction-of-arrival estimates," *IEEE Trans. Signal Processing*, vol. 38, pp. 1783–1795, October 1990.

[90] S. C. Swales, M. A. Beach, D. J. Edwards, and J. P. McGeehan, "The performance enhancement of multibeam adaptive base-station antennas for cellular land mobile radio systems," *IEEE Trans. Vehic. Tech.*, vol. 39, pp. 56–67, February 1990.

[91] A. L. Swindlehurst and T. Kailath, "A performance analysis of subspace-based methods in the presence of model errors—Part 2: Multidimensional algorithms," *IEEE Trans. Signal Processing*, vol. 41, September 1993.

[92] M. Taniguchi and P. R. Krishnaiah, "Asymptotic distributions of functions of the eigenvalues of the sample covariance matrix and canonical correlation matrix in multivariate time series," *J. Multivariate Anal.*, vol. 22, pp. 156–176, 1987.

[93] L. Tong, G. Xu, and T. Kailath, "A new approach to blind identification and equalization of multipath channels," in *Proc. Twenty-Fifth Asilomar Conf. on Signals, Systems and Computers*, pp. 856–860, Pacific Grove, CA, November 1991.

[94] L. Tong, G. Xu, and T. Kailath, "Blind channel identification and equalization using spectral correlation measurements, Part II: A time-domain approach," in W. A. Gardner, ed., *Cyclostationarity in Communications and Signal Processing*, New York: IEEE Press, 1994.

[95] B. D. van Veen and K. M. Buckley, "Beamforming: A versatile approach to spatial filtering," *IEEE ASSP Magazine*, vol. 5, pp. 4–24, April 1988.

[96] M. Viberg, *Subspace Fitting Concepts in Sensor Array Processing*, Ph.D. thesis, Linköping University, 1989.

[97] M. Viberg and B. Ottersten, "Sensor array processing based on subspace fitting," *IEEE Trans. Signal Processing*, vol. 39, pp. 1110–1121, May 1991.

[98] M. Viberg, B. Ottersten, and T. Kailath, "Detection and estimation in sensor arrays using weighted subspace fitting," *IEEE Trans. Signal Processing*, vol. 39, pp. 2436–2449, November 1991.

[99] H. Wang and M. Kaveh, "Coherent signal-subspace processing for the detection and estimation of angles of arrival of multiple wide-band sources," *IEEE Trans. Acoust., Speech, Signal Processing*, vol. ASSP-33, pp. 823–831, August 1985.

[100] H. Wang and M. Kaveh, "On the performance of signal-subspace processing—Part II: Coherent wide-band systems," *IEEE Trans. Acoust., Speech, Signal Processing*, vol. ASSP-35, pp. 1583–1591, November 1987.

[101] M. Wax, "Detection and localization of multiple sources via the stochastic signals model," *IEEE Trans. Signal Processing*, vol. 39, pp. 2450–2456, November 1991.

[102] M. Wax, "Detection and localization of multiple sources in noise with unknown covariance," *IEEE Trans. Signal Processing*, vol. 40, pp. 245–249, January 1992.

[103] M. Wax and T. Kailath, "Detection of signals by information theoretic criterion," *IEEE Trans. Acoust., Speech, Signal Processing*, vol. ASSP-33, pp. 387–392, April 1985.

[104] M. Wax, T.-J. Shan, and T. Kailath, "Spatio-temporal spectral analysis by eigenstructure methods," *IEEE Trans. Acoust., Speech, Signal Processing*, vol. ASSP-32, pp. 817–827, August 1984.

[105] J. H. Winters, "Optimum combining in digital mobile radio with cochannel interference," *IEEE Trans. Vehic. Tech.*, vol. VT-33, pp. 144–155, August 1984.

[106] K. M. Wong, J. P. Reilly, Q. Wu, and S. Qiao, "Estimation of the directions of arrival of signals in unknown correlated noise, Part I: The MAP approach and its implementation," *IEEE Trans. Signal Processing*, vol. 40, pp. 2007–2017, August 1992.

[107] Q. Wu and D. R. Fuhrmann, "A parametric method for determining the number of signals in narrow-band direction finding," *IEEE Trans. Signal Processing*, vol. 39, pp. 1848–1857, August 1991.

[108] Q. Wu, K. M. Wong, and J. P. Reilly, "Optimal array signal processing in unknown noise environments via parametric approaches," in *Proc. of the IEEE Sixth SP Workshop on Statistical Signal and Array Processing*, pp. 504–507, Victoria, B.C., Canada, October 1992.

[109] G. Xu and T. Kailath, "Array signal processing via exploitation of spectral correlation—A combination of temporal and spatial processing," in *Proc.*

Twenty-Third Asilomar Conf. on Signals, Systems and Computers, pp. 945–949, Pacific Grove, CA, November 1989.

[110] G. Xu and T. Kailath, "Direction of arrival estimation via exploitation of cyclostationarity—A combination of temporal and spatial processing," *IEEE Trans. Signal Processing*, vol. 40, pp. 1775–1786, July 1992.

[111] Y.-S. Yeh and D. O. Reudnik, "Efficient spectrum utilization for mobile radio systems using space diversity," *IEEE Trans. Comm.*, vol. COM-30, pp. 447–455, March 1982.

[112] V. J. Yohai and M. S. G. Ben, "Canonical variables as optimal predictors," *Ann. Statist.*, vol. 8, pp. 865–869, 1980.

[113] I. Ziskind and M. Wax, "Maximum likelihood localization of multiple sources by alternating projection," *IEEE Trans. Acoust., Speech, Signal Processing*, vol. ASSP-36, pp. 1553–1560, October 1988.

Chapter 4

Polyperiodic Linear Filtering

L. E. Franks
University of Massachusetts at Amherst
Amherst, MA 01003

1 INTRODUCTION

This chapter evolved from a tutorial lecture presented at the 1992 Workshop on Cyclostationary Signals. The lecture addressed the basic theory of periodically- and polyperiodically-time-varying linear systems and, as such, was complementary to the presentations on cyclostationary signals, which constituted the main thrust of the workshop. Time-varying systems are extensively employed as signal extraction and parameter estimation filters for such signals as well as models for the generation of cyclostationary processes. We shall consider various forms of input/output and state-variable descriptions of the systems and the nature of the implementation configurations suggested by these descriptions. Inasmuch as periodicity is a certain form of constancy, or persistence, in time, we expect periodically-varying systems to exhibit more similarities to time-invariant systems than time-varying systems in general. We show that all the system descriptions lead to implementation configurations having an embedded time-invariant component. The periodically-varying component consists solely of zero-memory operations; i.e., multiplications by periodic waveforms on an instantaneous basis. We refer to the devices for performing these multiplications as modulators. From a historical perspective, the earlier work in electrical periodically-varying systems dealt primarily with circuits such as the parametric amplifier, which contained reactive elements, such as a capacitor, which were "pumped" by a local oscillator. Thus the prevailing view of the configuration of such systems was that of a few variable-parameter elements located at various points within a predominantly time-invariant system. The practical utility of such devices inspired a great deal of research into the basic properties of periodically-time-varying linear systems, leading to canonical realizations which today form the basis for design of signal processors having the desired signal transformation characteristics.

We first discuss system configurations based on Fourier series expansions of system response functions. Then we consider systems that are characterized by a finite-order differential equation, or equivalently by a finite-dimensional state vector. This approach gives the designer more control over the overall complexity or implementation cost of the filter. It also reveals classes of systems that yield equivalent responses so that a certain degree of design flexibility is afforded. Finally, we present a discussion of optimum linear filtering for cyclostationary processes with simple examples that illustrate how periodic filtering can provide signal extraction results that could not be obtained using purely time-invariant signal processing.

2 PERIODIC AND POLYPERIODIC LINEAR SYSTEMS

A continuous-time, linear, time-varying system H is characterized, in general, by the integral transformation,

$$y(t) = H \circ x(t) = \int_{-\infty}^{\infty} h(t, s)x(s)ds \quad \text{for } -\infty < t < \infty, \tag{1}$$

relating the input signal (continuous-time waveform) $x(t)$ to the output signal $y(t)$. Since we are interested here in the signal-transforming properties of the system, we consider only the zero-state response; that is, the part that depends linearly on the input signal as in (1). The kernel function in the integral transform is called the impulse response of the system.

Much of the theory of filter design, or of system analysis in general, is limited to time-invariant systems. Such systems are characterized by how response is affected by time shifts of the input. Let D_T denote a T-second time-shift (delay) operator; $x(t - T) = D_T \circ x(t)$. A system H is said to be time-invariant if and only if it commutes with the delay operator for each value of the T parameter: $D_T \circ (H \circ x(t)) = H \circ (D_T \circ x(t))$. It is well known, of course, that for a time-invariant linear system, the impulse response function in (1) depends only on the single variable $t - s$, and the integral transform becomes a convolution integral. For a periodically-time-varying (PTV) system, on the other hand, D_T commutes with H only for specific values of the T parameter. If commutativity holds for a particular value of T, then it also holds for any integer multiple of that value. Hence The Periodicity of the system is designated as the smallest positive value of T for which the commutative relation holds. It is easy to show that the commutativity condition for the PTV system implies the following joint periodicity of the impulse response function:

$$h(t + T, s + T) = h(t, s) \quad \text{for all } t, s. \tag{2}$$

For some purposes, it is convenient to define an alternative impulse response function by means of a simple change of time variable. Let

$$p(t,\tau) = h(t, t-\tau); \tag{3}$$

then $p(t,\tau)$ is periodic (T) in t and is assumed to have a Fourier series expansion so that we can write

$$h(t,s) = \sum_k p_k(t-s)e^{j2\pi kt/T}, \tag{4a}$$

where

$$p_k(\tau) = \frac{1}{T}\int_0^T h(t, t-\tau)e^{-j2\pi kt/T}dt \tag{4b}$$

and we have

$$y(t) = \sum_k e^{j2\pi kt/T}\int_{-\infty}^{\infty} p_k(t-s)x(s)ds. \tag{4c}$$

Equation (4c) gives our first canonical configuration for the PTV system by direct inspection. Each term in the sum represents a time-invariant system, with impulse response $p_k(t-s)$, followed by an exponential-type modulator. Hence the overall system is a parallel bank of subsystems, each driven by the common input signal, and the output signal is formed by summing the outputs of the subsystems. A shorthand graphical representation for this parallel configuration, which will be generally useful for all configurations, is shown in Fig. 1.

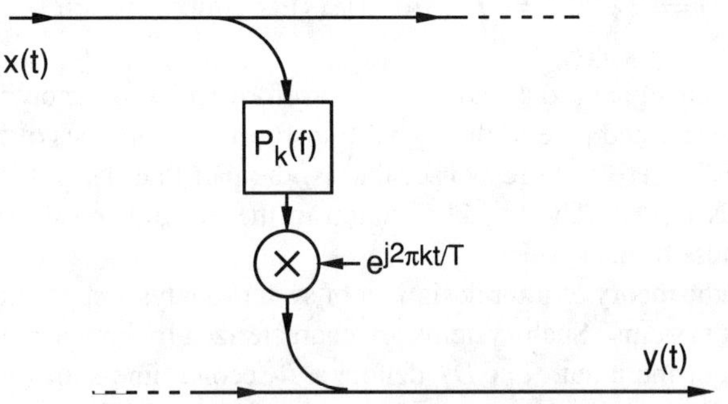

Figure 1: Graphical representation to be used to portray parallel-path configurations. For this particular example, the k-th path consists of a time-invariant filter, with transfer function $P_k(f)$, followed by an exponential modulator having a frequency of k/T.

The periodicity expressed in (2) would also permit another kind of Fourier expansion, in s rather than t:

$$h(t,s) = \sum_k q_k(t-s)e^{j2\pi ks/T}, \tag{5}$$

which leads directly to an equivalent parallel-path configuration involving exponential input modulators followed by time-invariant filters, with transfer functions $Q_k(f) = P_k(f-k/T)$. A convenient general equivalence between systems that use modulators of the exponential type is shown in Fig. 2. This equivalence can be used in a variety of ways to derive alternative configurations. For example, if a particular analysis technique results in a configuration using both input and output exponential modulators, then it can easily be converted to a configuration using only input modulators or only output modulators.

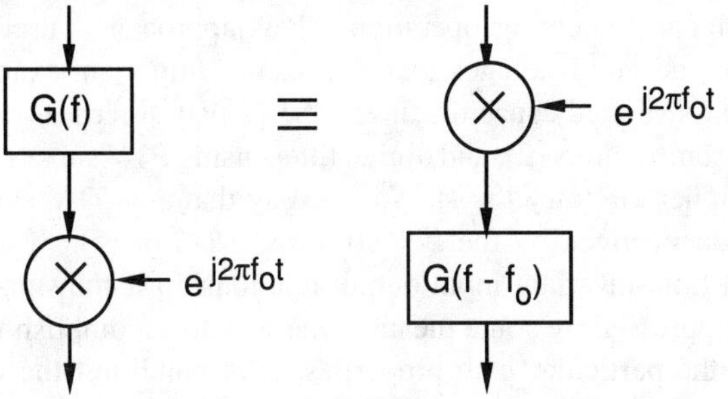

Figure 2: Input/output equivalent subsystems that employ exponential modulators.

An important generalization of the PTV system is the polyperiodic linear system wherein the restriction that the exponential functions all have harmonically related frequencies is removed. If the set of frequencies $\{m/T; m \text{ integer}\}$ for the PTV system is replaced by an arbitrary discrete set $\{f_m\}$, then $p(t, \tau)$ in (3) can be regarded as an almost-periodic function in t, which leads to the expansion

$$h(t, s) = \sum_m p_m(t - s)e^{j2\pi f_m t}, \tag{6a}$$

where

$$p_m(\tau) = \lim_{T_o \to \infty} \frac{1}{2T_o} \int_{-T_o}^{T_o} h(t, t - \tau)e^{-2\pi f_m t} dt. \tag{6b}$$

An example of a common situation where the polyperiodic filter response is required arises in the extraction of a bandpass carrier signal at frequency f_o carrying a baseband synchronous digital data signal with a symbol rate of $1/T$. Such a signal is most naturally regarded as polycyclostationary since there is usually no practical reason to regard f_o and $1/T$ as harmonically related; they could possibly be incommensurate numbers.

The class of polyperiodic systems corresponds closely with the concept of "stationary linear time-varying" systems introduced by Claasen and Mecklenbrauker [1]. They define this class of systems as that which maps almost-periodic inputs into almost-periodic outputs. They show that such systems are partially characterized by a set of frequency-mapping functions $\{f_m(\lambda)\}$, which reflects the fact that each input of the form $\exp[j2\pi\lambda t]$ is mapped into a linear combination of the set $\{\exp[j2\pi f_m(\lambda)t]\}$. They then proceed to show that the constraint of causality, $h(t, s) = 0$ for $t < s$, implies that each of these frequency-mapping functions is linear; that is, $f_m(\lambda) = \lambda + \lambda_{0m}$ for each m. Then we have

$$h(t, s) = \sum_m p_m(t - s)e^{j2\pi\lambda_{0m} t} \tag{7}$$

so that, with the constraint of causality, the impulse responses have the same form as that in (6) for the polyperiodic filter.

Another useful application for the Fourier-series type representations examined here is in the analysis of PTV systems that are designed to implement a time-invariant (convolution) signal-processing operation. This approach is used because of the unique implementation advantages that can result. Important examples are analog integrated-circuit switched capacitor filters, the N-path filter for realization of multiple passband (comb) filters [2], and digital filters using PTV subsystems to eliminate expensive multiplier circuitry [3, 4]. We can say that any PTV system has a time-invariant component given by the $k = 0$ term in (4) or (5). The overall system would exhibit a time-invariant input/output relationship if transmission through all other paths is suppressed by some means. One way to accomplish this, in a manner independent of the particular path properties, is to bandlimit the input and output signals to the frequency range $|f| \leq 1/2T$. Thus, using the appropriate antialiasing filters at the input and output of a PTV system makes it equivalent to a time-invariant system.

3 FINITE-ORDER PTV SYSTEMS

The previous configurations are not physically realizable unless the Fourier series expansions (4) or (5) can be replaced by finite sums, or unless suitable bandwidth constraints can be placed on the input and output signals. For example, if $p(t, \tau)$ in (3) exhibits discontinuities in t, then an infinite number of time-invariant filters (and modulators) would be required for an exact realization. Even considering a finite series, the number of terms in the series is not necessarily a useful measure of implementation costs. Note, for example, that it is much easier to implement multiplication of a signal by a square wave than to implement multiplication by a sine wave. For a discrete-time system, the expansions corresponding to (4) or (5) will be finite, that is, L terms (L parallel paths) for a system having a period L. However, the parameter L can also be regarded as a characterization of a limited signal bandwidth since it must be chosen large enough to avoid aliasing in the sampling process.

For many purposes, the order of the differential equation (or difference equation for the discrete-time case) describing the system dynamics is a more accurate indication of implementation complexity. It is certainly more closely related to the number of components in the time-invariant subsystem of the overall PTV filter. For a digital system, the system order is a direct indication of memory requirements for the processor.

The general M-th order, single-input, single-output, linear, continuous-time system is described by an M-dimensional, first-order state equation and an output equation,

$$\dot{\mathbf{w}}(t) = \mathbf{A}(t)\mathbf{w}(t) + \mathbf{B}(t)x(t) \quad (8a)$$

$$y(t) = \mathbf{C}(t)\mathbf{w}(t) + \mathbf{D}(t)x(t), \quad (8b)$$

where $\mathbf{w}(t)$ is the M-dimensional state vector, $\dot{\mathbf{w}}(t)$ is its time derivative, $\mathbf{A}(t)$, $\mathbf{B}(t)$, $\mathbf{C}(t)$, $\mathbf{D}(t)$ are $M \times M$, $M \times 1$, $1 \times M$, and 1×1 time-varying matrices, respectively. For the PTV system, they will vary periodically with time. In the following discussion, we assume $\mathbf{D}(t) = 0$ because it simply represents a parallel input/output path consisting only of a single modulator.

The zero-state response of the system (assume $\mathbf{w}(-\infty) = 0$) is given by

$$y(t) = \int_{-\infty}^{t} \mathbf{C}(t)\Phi(t, s)\mathbf{B}(s)x(s)ds, \tag{9}$$

where $\Phi(t, s)$ is the $M \times M$ state transition matrix for the system [5, 6]. Its columns are the M linearly independent solutions of the homogeneous part of (8a); that is,

$$\frac{d}{dt}\Phi(t, t_0) = \mathbf{A}(t)\Phi(t, t_0) \tag{10}$$

having the transitional properties

$$\begin{aligned} \mathbf{\Phi}(t, t_0) &= \mathbf{\Phi}(t, t_1)\mathbf{\Phi}(t_1, t_0) && \text{(11a)} \\ \mathbf{\Phi}(t, t) &= \mathbf{I} \quad \text{(the } M \times M \text{ identity matrix)} && \text{(11b)} \\ \mathbf{\Phi}^{-1}(t, t_0) &= \mathbf{\Phi}(t_0, t). && \text{(11c)} \end{aligned}$$

For the PTV system the transition matrix has a very special form. We can use (10) and the periodicity to show that it can be expressed as the product of two matrices: one nonsingular, periodic matrix and another matrix having the typical exponential form of the transition matrix for a time-invariant system,

$$\mathbf{\Phi}(t, 0) = \mathbf{J}(t)e^{\mathbf{R}t}, \tag{12}$$

where $\mathbf{J}(t)$ is periodic (T) with $\mathbf{J}(0) = \mathbf{J}(T) = \mathbf{I}$ and $\mathbf{R}$ is a constant matrix [6]. Now if $\mathbf{J}(t)$ is used as a state transformation matrix, the system equations (8) can be rewritten to reveal that the overall PTV system can be implemented as a configuration consisting of an M-th order, M-input, M-output time-invariant subsystem with banks of M-element input and output modulators (but not necessarily of the exponential form) [7]. Define a new state variable by

$$\mathbf{z}(t) = \mathbf{J}^{-1}(t)\mathbf{w}(t); \tag{13}$$

then (8) becomes

$$\begin{aligned} \dot{\mathbf{z}}(t) &= \mathbf{R}\mathbf{z}(t) + \mathbf{J}^{-1}(t)\mathbf{B}(t)x(t) && \text{(14a)} \\ y(t) &= \mathbf{C}(t)\mathbf{J}(t)\mathbf{z}(t) && \text{(14b)} \end{aligned}$$

where $\mathbf{J}^{-1}(t)\mathbf{B}(t)$ is a column matrix whose elements are the input modulators and $\mathbf{C}(t)\mathbf{J}(t)$ is a row matrix giving the output modulators. An additional state transformation permits the time-invariant subsystem (characterized by $\mathbf{R}$) to be implemented as M parallel (uncoupled) paths, thus minimizing the complexity of the system. Let $\mathbf{K}$ be the $M \times M$ matrix that diagonalizes $\mathbf{R}$,

$$\mathbf{K}^{-1}\mathbf{R}\mathbf{K} = \mathbf{\Lambda} = \text{diag}\{\lambda_m\}_1^M. \tag{15}$$

Then with

$$\mathbf{v}(t) = \mathbf{K}^{-1}\mathbf{z}(t) \tag{16}$$

the system equations become

$$\dot{\mathbf{v}}(t) = \mathbf{\Lambda}\mathbf{v}(t) + \mathbf{B}'(t)x(t) \tag{17a}$$

$$y(t) = \mathbf{C}'(t)\mathbf{v}(t), \tag{17b}$$

where $\mathbf{B}'(t) = \mathbf{K}^{-1}\mathbf{J}^{-1}(t)\mathbf{B}(t)$ and $\mathbf{C}'(t) = \mathbf{C}(t)\mathbf{J}(t)\mathbf{K}$ give the modified input and output modulators, respectively. Using (9) for the state-transformed system, we see that the impulse response of any M-th order PTV system has the separable form,

$$h(t,s) = \begin{cases} \sum_{i=1}^{M} f_i(t)g_i(s) & \text{for } t \geq s \\ 0 & \text{otherwise} \end{cases} \tag{18a}$$

where

$$\begin{aligned} f_i(t) &= c_i'(t)e^{\lambda_i t} \\ g_i(s) &= b_i'(s)e^{-\lambda_i s} \end{aligned} \tag{18b}$$

and $c_i'(t)$ and $b_i'(t)$ denote the ith elements of the transformed matrices $\mathbf{C}'(t)$ and $\mathbf{B}'(t)$, respectively, in (17). For each of the real eigenvalues (poles) λ_i of the system, the corresponding path is a first-order (single integrator with feedback) system with periodic input modulator $b_i'(t)$ and periodic output modulator $c_i'(t)$. For each complex eigenvalue there is a corresponding conjugate eigenvalue and these can be paired in (18a) and implemented as a second-order subsystem. Typically, when this diagonalization procedure is applied in the analysis of a time-invariant system to put the system in its "normal" or "state-variable" form, the complex poles are implemented with two cross-coupled first-order systems. The PTV system affords an attractive alternative that avoids the cross-coupling. This is often a significant practical advantage when the eigenvalue has a large imaginary part in comparison to the real part (high-Q poles). To show this alternative, consider the typical complex eigenvalue pair $\alpha + j\beta$ and $\alpha - j\beta$, along with the associated input modulators $b(t)$ and $b^*(t)$ and output modulators $c(t)$ and $c^*(t)$. From (18a), the impulse response of this subsystem is

$$\begin{aligned} h(t,s) &= 2\mathcal{R}e\left[c(t)e^{(\alpha+j\beta)(t-s)}b(s)\right] \\ &= 2\mathcal{R}e\left[c(t)e^{j\beta t}e^{\alpha(t-s)}e^{-j\beta s}b(s)\right] \\ &= 2\mathcal{R}e\left[c'(t)e^{\alpha(t-s)}b'(s)\right] \end{aligned} \tag{19a}$$

where

$$c'(t) = c(t)e^{j\beta t} \quad \text{and} \quad b'(t) = b(t)e^{-j\beta t}. \tag{19b}$$

Hence by appropriately modifying the modulating functions, the entire M-th order PTV system can be implemented by M parallel paths as shown in Fig. 3. Each path contains a first-order system with a pole given by the real part of the corresponding eigenvalue in (15). Since the frequencies $\beta/2\pi$ and $1/T$ are not necessarily harmonically related, we have here an example of how a polyperiodic system can be used to advantage to realize a periodic system. The situation is analogous to using a periodic system to realize a time-invariant system, as discussed previously.

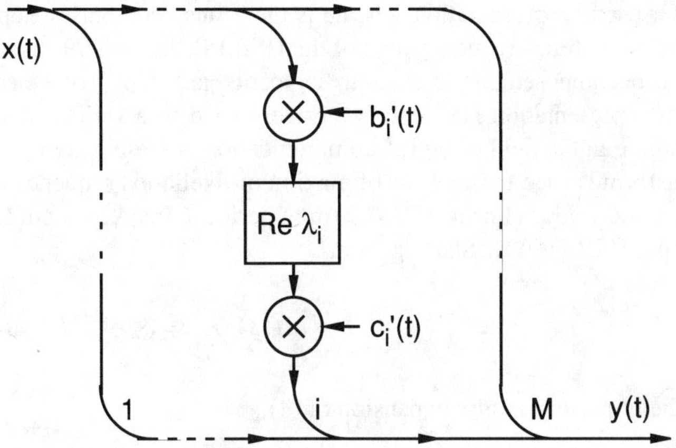

Figure 3: Canonical configuration for M-th order PTV system.

4 PERIODIC FILTERING OF CYCLOSTATIONARY SIGNALS

When dealing with cyclostationary signals, various signal representation schemes help to characterize linear signal processing operations. These representations tend to be of a mixed discrete/continuous nature and they can be useful to reveal the time-invariant aspects of periodic filtering of cyclostationary signals. One such scheme is the Harmonic Series Representation (HSR) whereby an arbitrary continuous-time signal is represented by a sequence of bandlimited signals [8]. We can write

$$x(t) = \sum_n x_n(t)e^{j2\pi nt/T}, \tag{20a}$$

where

$$x_n(t) = \int_{-\infty}^{\infty} v(t-\tau)e^{-j2\pi n\tau/T}x(\tau)d\tau \tag{20b}$$

and $v(t) = (1/T)\,\mathrm{sinc}(t/T)$ is the impulse response of an ideal lowpass filter with passband in $|f| \le 1/2T$. The nature of the decomposition (20a) is quite transparent in the frequency domain. The Fourier transforms of the $x_n(t)$ signals are simply contiguous segments, of width $1/T$, of the Fourier transform of $x(t)$. Since $V(f) = 1$

in $|f| \leq 1/2T$ and $V(f) = 0$ elsewhere, we can write

$$X(f) = \sum_n X_n(f - n/T), \tag{21a}$$

where

$$X_n(f) = V(f)X(f + n/T) \tag{21b}$$

and taking inverse Fourier transforms of (21) gives (20) directly.

Another useful representation scheme is the Translation Series Representation (TSR) which is the time-frequency dual of the HSR [8]. In the TSR, the continuous-time signal is decomposed into contiguous segments (or "chips") of T-second width. This type of representation has been extensively used in a variety of fields. One notable example in the field of digital communication is presented in [9], where the chip D-transform is used for analysis of maximum-likelihood sequence estimation.

We now show how a linear PTV system of period T maps an input HSR $\{x_n(t)\}$ into the output HSR $\{y_n(t)\}$. Starting from

$$y(t) = \sum_n y_n(t)e^{j2\pi nt/T} = \int_{-\infty}^{\infty} h(t, s) \sum_m x_m(s)e^{j2\pi ms/T} ds \tag{22}$$

and using the impulse response expansion of (4),

$$h(t, s) = \sum_k p_k(t - s)e^{j2\pi kt/T} \tag{23}$$

and the HSR for each of the terms in (23), we obtain

$$p_k(t - s) = \sum_m p_{km}(t - s)e^{j2\pi m(t-s)/T}, \tag{24}$$

which gives a double Fourier expansion of $h(t, s)$:

$$h(t, s) = \sum_n \sum_m h_{nm}(t - s)e^{j2\pi nt/T} e^{-j2\pi ms/T}, \tag{25}$$

where

$$h_{nm}(t) \triangleq p_{(n-m)m}(t).$$

Now since both the $x_m(t)$ and the $h_{nm}(t)$ functions are bandlimited, the result of substituting (25) into (22) can be greatly simplified. Equating terms having the same exponential factors, we get the convolution

$$y_n(t) = \sum_m \int_{-\infty}^{\infty} h_{nm}(t - s)x_m(s)ds \quad \text{for all } n \tag{26}$$

or, taking Fourier transforms and using matrix notation,

$$\mathbf{Y}(f) = \mathbf{H}(f)\mathbf{X}(f), \tag{27}$$

where $\mathbf{X}(f) = \mathrm{col}\{X_n(f)\}$, $\mathbf{Y}(f) = \mathrm{col}\{Y_n(f)\}$, and the elements of the $\mathbf{H}(f)$ matrix,

$$H_{nm}(f) = P_{n-m}(f + m/T) \quad \text{for all } |f| \le 1/2T, \tag{28}$$

are transfer functions of lowpass filters (bandlimited to the frequency range $|f| \le 1/2T$). This representation leads to an implementation configuration consisting of a bandlimited (lowpass) time-invariant subsystem with banks of input and output exponential modulators. A physical interpretation of the signal processing is that the input modulators perform frequency shifting of various portions of the input spectrum to a set of "baseband" signals in the band $|f| \le 1/2T$. Then lowpass filtered versions of these signals are linearly combined and the output modulators perform frequency shifting back into various spectral regions to form the output signal. This configuration is easily converted to the parallel-path (but not bandlimited) configuration of Fig. 1 using only output modulators. From (24) and (25), we have

$$P_k(f) = \sum_m H_{(m+k)m}(f - m/T) \tag{29}$$

for the transfer functions of the filters in the parallel-path/output-modulator configuration. The equivalence shown in Fig. 2 can be used to convert to a parallel-path/input-modulator configuration.

For a cyclostationary random process, the HSR exhibits a very useful kind of stationarity. If $x(t)$ is cyclostationary with period T, then its autocorrelation has a periodicity expressed by

$$E[x(t)x^*(s)] = k_{xx}(t, s) = k_{xx}(t + T, s + T) \quad \text{for all } s, t. \tag{30}$$

The HSR for $x(t)$ has elements that are jointly stationary (in the wide sense) [8]. Hence, a scalar cyclostationary process can be treated as a stationary vector process. Let the cross-correlations of the elements of the HSR be denoted by

$$r_{nm}(t - s) = E[x_n(t)x_m^*(s)]. \tag{31}$$

Then the autocorrelation for x(t) is expressed as a double Fourier series,

$$k_{xx}(t, s) = \sum_n \sum_m r_{nm}(t - s)e^{j2\pi nt/T}e^{-j2\pi ms/T}. \tag{32}$$

Assuming that $x(t)$ is a zero-mean process, we note from (32) that it is wide-sense stationary (WSS) if and only if the HSR elements are uncorrelated; that is, $\{r_{nm}(\cdot)\}$ is a diagonal matrix. In fact, the diagonal entries of this matrix form the HSR for the stationary autocorrelation function. The off-diagonal entries represent the correlation between spectral components in nonoverlapping frequency regions that is characteristic of cyclostationary processes. If the signal process is subjected to severe bandlimiting, or if phase randomizing parameters are introduced, then the spectral correlation is reduced or eliminated and signal recovery schemes that take advantage of the spectral correlation will suffer in performance.

To illustrate how the HSR is used to solve signal recovery problems, we consider the class of problems known as minimum mean-squared-error continuous waveform estimation. We seek the optimum linear filter to produce an output waveform $y(t)$, which is nearest (in a mean-squared-error sense) to a desired waveform $w(t)$. The input signal $z(t)$ is frequently a received signal that results from modulation of a periodic "carrier" signal by a message process $w(t)$ and subsequent transmission over a noisy and dispersive communication channel. The receiver is diagrammed in Fig. 4.

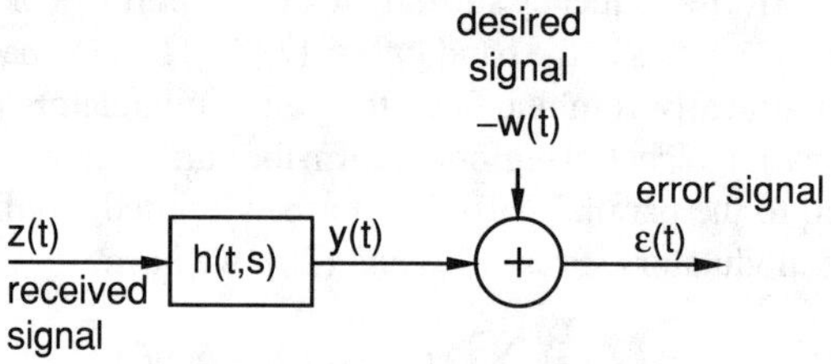

Figure 4: Continuous waveform estimation for real signals.

To minimize $E[|\epsilon(t)|^2]$, for each instant t, we select the filter that satisfies the orthogonality condition, which says that the correlation between the error and all of the observed data should vanish [10].

$$E[\{y(t) - w(t)\}z^*(s)] = 0 \quad \text{for all } t, s. \tag{33}$$

The resulting condition on the impulse response of the optimum filter can be obtained directly from (33):

$$\int_{-\infty}^{\infty} h(t, s)k_{zz}(\sigma, s)d\sigma = k_{wz}(t, s) \quad \text{for all } t, s. \tag{34}$$

This is the orthogonality condition for the case of real signals. For complex signal-processing problems, typical of bandpass carrier-type signals, the orthogonality condition needs to be extended. This is discussed later in Section 5.

Now use representations of the form (25) for the impulse response $h(t, s)$, and the form (32) for the autocorrelation $k_{zz}(t, s)$ in (34). Using the fact that both the $h_{nm}(t - \sigma)$ and the $r_{nm}(\sigma - s)$ functions are bandlimited to the interval $|f| \leq 1/2T$, the LHS of (34) can be simplified and written in terms of the Fourier transforms of these functions as

$$\sum_n \sum_m \sum_k \int_{-\infty}^{\infty} H_{nm}(f)R_{mk}(f)e^{j2\pi f(t-s)}df e^{j2\pi nt/T} e^{-j2\pi ks/T}. \tag{35}$$

Using the same format for the cross-correlation of $w(t)$ and $z(t)$, the RHS of (34) can be written as

$$k_{wz}(t, s) = \sum_n \sum_k w_{nk}(t - s)e^{j2\pi nt/T} e^{-j2\pi ks/T}. \tag{36}$$

Since the condition is to hold for all t and s, we can equate factors of corresponding exponential terms. Then taking Fourier transforms of the functions of $(t-s)$, we get

$$\sum_m H_{nm}(f)R_{mk}(f) = W_{nk}(f) \quad \text{for all } n, k \tag{37}$$

or, in matrix notation, the solution for the optimum PTV filter is

$$\mathbf{H}(f) = \mathbf{W}(f)\mathbf{R}^{-1}(f) \tag{38}$$

which closely resembles the form of the solution for the Wiener filter for stationary signals. In fact, it is simply a multiinput/multioutput Wiener filter for the HSR representor vector [8].

We illustrate the application of this result by considering a familiar problem in digital data communication. The problem is to recover a synchronous pulse-amplitude-modulation (PAM) signal which is received with additive, zero-mean, stationary noise $u(t)$, characterized by the power spectral density function $K_{uu}(f)$. A baseband carrier pulse $g(t)$ is used to transmit the stationary data sequence $\{c_n\}$. Hence the optimum filtering problem can be stated in terms of the quantities shown in Fig. 4 as

$$z(t) = x(t) + u(t) \tag{39a}$$

$$w(t) = x(t) = \sum_n c_n g(t - nT). \tag{39b}$$

The relevant correlation functions are

$$k_{zz}(t,s) = k_{xx}(t,s) + k_{uu}(t-s) \tag{40a}$$

$$\begin{aligned} k_{wz}(t,s) &= k_{xz}(t,s) = k_{xx}(t,s) \\ &= \sum_n \sum_m w_{nm}(t-s) e^{j2\pi nt/T} e^{-j2\pi ms/T} \end{aligned} \tag{40b}$$

where, using (20) for the HSR $\{x_n(t)\}$ for $x(t)$,

$$w_{nm}(t-s) = E[x_n(t)x_m^*(s)]$$

and some straightforward calculations give the Fourier transform of the HSR cross-correlation matrix elements for PAM as

$$W_{nm}(f) = \frac{1}{T} M(f) G(f + n/T) G^*(f + m/T) V(f), \tag{41}$$

where $V(f)$ is the rectangular bandlimiting factor used in (20) and (21). The $M(f)$ factor in (41) is the power spectral density of the discrete-time data sequence; that is,

$$M(f) \triangleq \sum_n E[c_{k+n} c_k^*] e^{-j2\pi nTf}. \tag{42}$$

Note that $M(f)$ is periodic $(1/T)$ in f. The key step in solving for the optimum filter is inversion of the autocorrelation matrix $\mathbf{R}(f)$ for the received signal $z(t)$ in (38). Although the $\mathbf{R}(f)$ matrix is, in general, of infinite dimension, its inversion is simple because the matrix has the form of the sum of a diagonal matrix and a rank-one matrix. To see this, we define the following matrices in terms of the HSRs for $k_{uu}(t)$ and $g(t)$, respectively,

$$\begin{aligned}\mathbf{K}(f) &= \operatorname{diag}\{K_{uu}(f+n/T)\}\\ \mathbf{G}(f) &= \operatorname{col}\{G(f+n/T)\}.\end{aligned} \tag{43}$$

Then from (40) and (41), we have

$$\mathbf{R}(f) = \mathbf{K}(f) + \frac{1}{T} M(f)\mathbf{G}(f)\mathbf{G}^{*T}(f) \quad \text{for all } |f| \le 1/2T. \tag{44}$$

Now, using the Matrix Inversion Lemma [11], we get

$$\begin{aligned}\mathbf{R}^{-1}(f) &= \mathbf{K}^{-1}(f)\\ &\quad -\mathbf{K}^{-1}(f)\mathbf{G}(f)\left[\frac{T}{M(f)}\mathbf{I} + \mathbf{G}^{*T}(f)\mathbf{K}^{-1}(f)\mathbf{G}(f)\right]^{-1}\mathbf{G}^{*T}(f)\mathbf{K}^{-1}(f).\end{aligned} \tag{45}$$

In (45), $\mathbf{G}^{*T}(f)\mathbf{K}^{-1}(f)\mathbf{G}(f)$ is a scalar,

$$\mathbf{G}^{*T}(f)\mathbf{K}^{-1}(f)\mathbf{G}(f) = TL(f), \tag{46}$$

where $L(f)$ is defined as the periodic $(1/T)$ scalar function

$$L(f) = \frac{1}{T}\sum_n \frac{|G(f+n/T)|^2}{K_{uu}(f+n/T)}. \tag{47}$$

Now premultiplying (45) by the matrix $\mathbf{W}(f)$, as in (38), and performing the matrix multiplications, we express the solution to the problem as

$$H_{nm}(f) = \frac{1}{T} D(f)\left[\frac{G(f+n/T)G^*(f+m/T)}{K_{uu}(f+m/T)}\right] \quad \text{for all } |f| \le 1/2T, \tag{48}$$

where we define the periodic $(1/T)$ function $D(f)$ by

$$D(f) = \frac{M(f)}{1+M(f)L(f)}. \tag{49}$$

To convert this solution to the parallel-path structure of Fig. 1, we use (29) to get

$$\begin{aligned}P_k(f) &= \sum_m H_{(m+k)m}(f-m/T)\\ &= \frac{1}{T} D(f)\frac{G^*(f)}{K_{uu}(f)}G(f+k/T).\end{aligned} \tag{50}$$

For this example, a great deal more simplification can be accomplished. First, extract the common factor $D(f)G^*(f)/K_{uu}(f)$ in (50) as an input filter preceding all the parallel paths. Next, use the equivalence shown in Fig. 2 on each path, giving an exponential modulator followed by a time-invariant filter $G(f)$. This filter is common to all the paths, hence it can be extracted as an output filter following the path summor. Now the separate paths contain only modulators, which can be summed to give a single-path modulator equal to

$$\frac{1}{T}\sum_k e^{j2\pi kt/T} = \sum_n \delta(t - nT); \tag{51}$$

that is, the sum is an impulse modulator (IM) with impulses occurring every T seconds. The simplified result is shown in Fig. 5, where MF denotes the classical matched filter for the pulse $g(t)$ in additive noise $u(t)$ [10]. The block marked TDL represents a tapped-delay-line filter having transfer function $D(f)$. Since $D(f)$, in general, will not be a polynomial in $\exp[-j2\pi Tf]$, the TDL realization might require an infinite number of taps; however, a finite structure using delay lines in a feedback configuration could be used to realize a more general $D(f)$ transfer function.

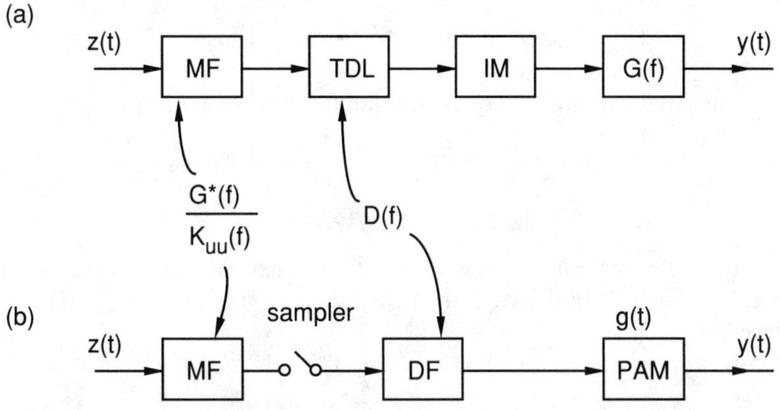

Figure 5: Optimum PTV filter for recovery of baseband PAM waveform. (a) Matched-filter/tapped-delay-line/impulse-modulator implementation. (b) Equivalent Matched-filter/digital-filter/PAM implementation.

The TDL/IM combination in Fig. 5a can be replaced by a sampler and digital (discrete-time) filter (DF) combination as shown in Fig. 5b. The discrete-time signal at the output of the DF is used to form a new PAM signal using the carrier pulse $g(t)$ at the output of the continuous waveform estimator. The optimum PTV filter for the PAM signal can be interpreted in terms of known results for the optimum linear receiver/demodulator for PAM [10]. This result states that the sampled output of a time-invariant receiver filter, $H(f) = G^*(f)D(f)/K_{uu}(f)$, provides the minimum mean-squared-error estimate of the data sequence $\{c_n\}$. Given that the pulse shape $g(t)$ is known, it is clear that reconstruction of the PAM signal from these estimates gives the best output waveform. Other equivalent structures can be obtained when the carrier pulses are bandlimited (usually to a frequency that is some fraction in excess of the Nyquist bandwidth, $1/2T$). Then the received signal can be sampled at its Nyquist rate, or higher, and the MF/TDL combination can be implemented as a single digital filter operating at the elevated rate; a result which corresponds to the use of "fractionally-spaced equalizers" for the data channel. This is discussed in more detail in [12].

As a second example to illustrate continuous waveform estimation on a cyclostationary signal process, we consider a spread-spectrum communication system where a stationary message signal process $x(t)$ is multiplied by an arbitrary periodic spreading signal $c(t)$. The resulting signal is transmitted over an additive white-noise channel. The receiver filter is required to transform the received signal $z(t)$ into a minimum mean-squared-error estimate of the original waveform $x(t)$. Thus, the problem can be stated as

$$z(t) = c(t)x(t) + n(t) \tag{52a}$$

$$w(t) = x(t) \tag{52b}$$

and

$$c(t) = c(t+T) = \sum_p C_p e^{j2\pi pt/T}, \tag{53}$$

which results in the following expressions for the correlation functions:

$$k_{zz}(t,s) = c(t)c^*(s)k_{xx}(t-s) + N_o\delta(t-s) \tag{54a}$$

$$k_{wz}(t,s) = c^*(s)k_{xx}(t-s). \tag{54b}$$

Consider first the very simple case where $x(t)$ is bandlimited to frequencies in the interval $|f| \le 1/2T$. In this case, the HSR for $c(t)x(t)$ is simply $\{C_n x(t)\}$, so that we have

$$R_{nm}(f) = K_{xx}(f)C_nC_m^* + N_o\delta_{nm} \tag{55a}$$

$$W_{nk}(f) = \delta_{n0}C_k^* K_{xx}(f). \tag{55b}$$

The solution for the **H** matrix in (38) has only a single nonzero row, and

$$H_{0m}(f) = \frac{C_m^* K_{xx}(f)}{N_o + P_o K_{xx}(f)}, \tag{56}$$

where

$$P_o = \sum_i |C_i|^2$$

is the power in the spreading function. Then using this solution for **H** in (25) and performing the summation, we see that the entire PTV filter is simply

$$h(t,s) = c^*(s)h_o(t-s), \tag{57}$$

where

$$H_o(f) = \frac{K_{xx}(f)}{N_o + P_o K_{xx}(f)}.$$

That is, multiplication of the received signal $z(t)$ by the conjugate of the spreading function $c(t)$, followed by time-invariant filtering with a filter that corresponds to the optimum (Wiener) filter for $\sqrt{P_o}x(t)$ in additive white noise. As signal-to-noise ratio increases ($N_0 \to 0$), this filter becomes independent of the particular shape of the signal spectral density function and approaches an ideal lowpass filter with a cutoff frequency that does not exceed $1/2T$. The simplicity of the solution in this case can be attributed to the fact that we assumed $x(t)$ to be bandlimited to the extent that the spectrum of $c(t)x(t)$ consists of translated nonoverlapping versions of the spectrum of $x(t)$, as shown in Fig. 6. Considering, for the moment, that $x(t)$ has a Fourier transform and that it is represented by the triangular shape in the figure, we can interpret the spread signal as a set of double-sideband (DSB) signals centered on each of the harmonics of $c(t)$. Since the individual DSB signals do not overlap in frequency, we could use a normal DSB demodulation technique, regarding one of the harmonics as a carrier, and $x(t)$ could be recovered without interference from aliased versions of itself. However, this approach would not realize the processing gain inherent in a spread-spectrum system. Assuming $H_o(f)$ to be an ideal lowpass filter of bandwidth $1/2T$, the output signal-to-noise ratio (SNR) of the optimum filter of (57) is $P_x P_o(T/N_o)$, where P_x is the power in the $x(t)$ signal. On the other hand, if DSB recovery of $x(t)$ from the carrier at frequency n/T was used, the SNR would be $P_x|C_n|^2(T/N_o)$. In other words, the optimum receiver provides a processing gain equal to the ratio of the total power of the spreading signal to the power in its n-th harmonic.

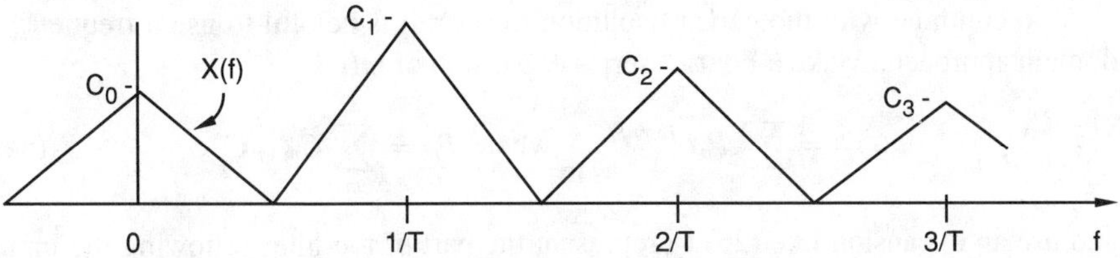

Figure 6: Spectrum of spread signal.

For a more general treatment of this example, we consider a wider bandwidth signal $x(t)$ (or a longer period T for $c(t)$) such that there is some degree of spectral overlap (aliasing) of the DSB components of the spread signal. Let $\{s_n(\cdot)\}$ be the HSR for the autocorrelation of the message process $x(t)$, that is,

$$k_{xx}(\tau) = \sum_n s_n(\tau)e^{j2\pi n\tau/T} \tag{58}$$

so that

$$S_n(f) = K_{xx}(f + n/T) \quad \text{for all } |f| \leq 1/2T. \tag{59}$$

Then the more general expressions for (55a) and (55b) become

$$R_{nm}(f) = \sum_{\ell} C_{n-\ell} C^*_{m-\ell} S_\ell(f) + N_o \delta_{nm} \tag{60a}$$

and

$$W_{nk}(f) = C^*_{k-n} S_n(f). \tag{60b}$$

The form of the orthogonality condition can be simplified by taking a hint from the previous result and assuming a general solution of the form

$$h(t, s) = c^*(s) g(t, s). \tag{61}$$

Then (34) can be reexpressed for this problem as

$$\int_{-\infty}^{\infty} |c(\sigma)|^2 g(t, \sigma) k_{xx}(\sigma - s) d\sigma + N_o g(t, s) = k_{xx}(t - s). \tag{62}$$

At this point, the asymptotic solution for low SNR ($N_o \to \infty$) can be easily seen for the general case of arbitrary spectral overlap of the signal components. In this case, the second term on the LHS of (62) dominates and we can solve directly for $g(t, s)$. The asymptotic result for low SNR is

$$h(t, s) = \frac{1}{N_o} c^*(s) k_{xx}(t - s). \tag{63}$$

So the optimum PTV filter is a $c^*(t)$ multiplier followed by a time-invariant filter with transfer function $K_{xx}(f)/N$, which is simply the low-SNR Wiener filter for the original signal in white noise.

To continue with the general solution to (62), it is helpful to use a frequency-domain approach. Make a Fourier series expansion of $|c(t)|^2$,

$$|c(t)|^2 = \sum_n P_n e^{j2\pi nt/T} \quad \text{where } P_n \triangleq \sum_m C_{m+n} C^*_m, \tag{64}$$

and use an expansion like (25) to represent the part of the filter following the input multiplier. Then, using (58) for $k_{xx}(\tau)$, the expression equivalent to (37) for this problem is

$$\sum_m P_{m-k} G_{nm}(f) S_k(f) + N_o G_{nk}(f) = S_k(f) \delta_{nk} \tag{65}$$

or, in matrix notation,

$$\mathbf{G}(f)\mathbf{P}\mathbf{S}(f) + N_o \mathbf{G}(f) = \mathbf{S}(f), \tag{66}$$

where

$$\mathbf{S}(f) = \text{diag}\{S_n(f)\} \quad \text{and} \quad \mathbf{P} = \{P_{n-m}\}.$$

We illustrate the solution of (66) by considering the case where there is spectral overlap only between adjacent DSB components of the spread signal, that is,

$K_{xx}(f) = 0$ for $|f| \geq 1/T$. In this case, $S_n(f) = 0$ for $|n| > 1$ and the diagonal matrix on the RHS of (66) has only three nonzero entries. Hence the solution is easily manageable without a computer. In fact, solving (66) separately in the regions $f < 0$ and $f > 0$ requires inversion of only 2×2 matrices. As in the simpler case examined previously, the solution for the optimum filter becomes independent of the shape of the message spectrum as $N_0 \to 0$. Noting that $P_{-\ell} = P_\ell^*$ since $|c(t)|^2$ is real, the solution can be expressed as

$$\mathbf{G}(f) = \frac{1}{P_0^2 - |P_1|^2} \begin{bmatrix} P_o U(f) & -P_1^* U(f) & 0 \\ -P_1 U(f) & P_o V(f) & -P_1^* U(-f) \\ 0 & -P_1 U(-f) & P_o U(-f) \end{bmatrix}, \tag{67}$$

where we have defined the rectangular filtering functions

$$U(f) = \begin{cases} 1 & \text{in } 0 \leq f < 1/2T \\ 0 & \text{otherwise} \end{cases} \tag{68}$$

and

$$V(f) = U(f) + U(-f).$$

This solution represents a structure having multiple paths between the three inputs and three outputs ($G(f)$ is not diagonal). It can be converted to a parallel-path structure using a summation as in (29). Finally, the equivalence shown in Fig. 2 can be used to make all the path filters symmetric about $f = 0$ and the implementation shown in Fig. 7 is the optimum receiver filter at high SNR. Note that the time-invariant (center) path has an ideal lowpass filter with twice the bandwidth of the filters in the other two paths.

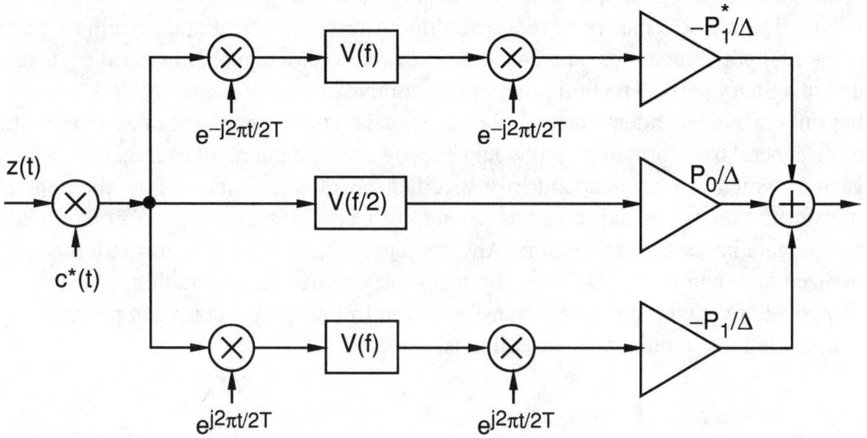

Figure 7: Optimum demodulation filter for spread-spectrum signal at high SNR. $\Delta = P_0^2 - |P_1|^2$.

Calculation of the time-averaged mean-squared error for the optimum filter gives

$$\frac{1}{T}\int_0^T E[|\epsilon(t)|^2]dt = N_o \sum_n \int_{-1/2T}^{1/2T} G_{nn}(f)df \tag{69}$$

indicating that the aliasing error is completely eliminated as $N_0 \to 0$ and the output SNR approaches

$$P_x[(P_0^2 - |P_1|^2)/P_o](T/2N_o).$$

5 POLYPERIODIC FILTERING OF POLYCYCLOSTATIONARY SIGNALS

As a final example of a waveform extraction or signal recovery problem, we consider optimum linear demodulation of a bandpass carrier signal when one or more interfering carrier signals are present, each having significant spectral overlap with the signal of interest, as well as additive noise. This example introduces two important aspects of signal recovery that have not been addressed so far: 1) Some situations are best modeled in terms of polycyclostationary processes, leading to polyperiodic filters, for which some of the foregoing analysis techniques are inappropriate. In this example, the carrier frequencies of the various bandpass signals need not be harmonically related to each other, nor to the possibly cyclostationary message processes (e.g. digital data) which modulate the various carriers; 2) complex envelope representation has many advantages for dealing with signals of a bandpass nature; however, unlike the case of time-invariant bandpass filtering, a general time-varying signal-processing operation cannot be characterized by a single complex transformation on the complex envelope.

Addressing the second issue first, we observe that a transformation like 1) with a complex impulse response $h(t, s)$ operating on a complex signal $x(t)$ can also be regarded as a 2×2 matrix of real operators (given by the real and imaginary parts of $h(t, s)$) which operates on a 2-dimensional real vector signal composed of its real and imaginary parts. The limitation of this complex operator is that this 2×2 matrix has only two independent entries, whereas signal recovery problems often require the more general transformation that would be provided by a matrix of four real operators, each of which can be independently specified. Brown [13] shows how the general transformation can be expressed as the sum of a complex linear transformation and a conjugate linear transformation. Any conjugate linear (CL) transformation can be realized as a conjugator followed by a complex linear transformation. It can also be realized by a complex linear transformation followed by a conjugator, simply by conjugating the impulse response, that is,

$$\int g(t,s)x^*(s)ds = \left[\int g^*(t,s)x(s)ds\right]^*. \tag{70}$$

Hence, for the analysis of optimum filtering in terms of processing a complex envelope signal $z(t)$, we use a transformation characterized by

$$y(t) = \int_{-\infty}^{\infty} [h(t,s)z(s) + g(t,s)z^*(s)]ds, \tag{71}$$

which is implemented as the two-path structure shown in Fig. 8. Such a transformation is referred to as linear/conjugate linear (LCL) [12–15].

The conventional definition for the complex envelope $z(t)$ of a real bandpass signal $x(t)$, relative to a "center" frequency f_o, is [10]

$$x(t) = \mathcal{R}e[z(t)e^{j2\pi f_o t}] \tag{72a}$$

$$z(t) = [x(t) + j\hat{x}(t)]e^{-j2\pi f_o t} \tag{72b}$$

where $\hat{x}(t)$ denotes the Hilbert transform of $x(t)$. Thus we see that the transformation (72a) from $z(t)$ to $x(t)$ is LCL, while the transformation (72b) from $x(t)$ to $z(t)$ is simply complex linear and can be characterized by the single impulse response function

$$c(t,s) = u_o(t-s)e^{-j2\pi f_o t}, \tag{73}$$

where the transfer function of the time-invariant part is the step-function

$$U_o(f) = 1 + j(-j \operatorname{sgn} f) = 1 + \operatorname{sgn} f$$

and the output of $U_o(f)$ is often called the "analytic signal" or "pre-envelope" for $x(t)$ [10]. Typically, a bandpass time-varying filtering problem will be solved in terms of the operations on $z(t)$, then the result can be combined with the structure for forming the complex envelope, shown to the left in Fig. 8, to form the overall processor operating on the real bandpass signal.

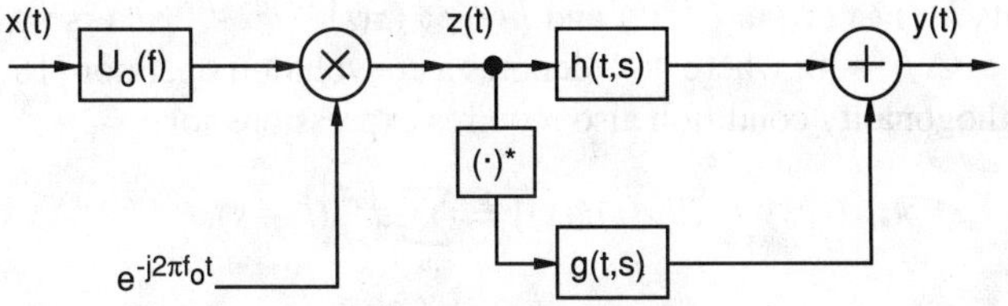

Figure 8: Complex envelope transformation followed by LCL processing.

For optimum filtering problems based on complex envelopes, the impulse response $h(t,s)$ in Fig. 4 must be replaced by the LCL transformation in Fig. 8 and the orthogonality condition needs to be restated to reflect the requirement that the error $\epsilon(t)$ should be orthogonal to both $z(s)$ and $z^*(s)$ for all t and s in order to achieve minimum mean-squared-error. Hence the orthogonality condition requires that the following pair of equations be simultaneously satisfied:

$$\int_{-\infty}^{\infty} h(t,\sigma)k_{zz}(\sigma,s)d\sigma + \int_{-\infty}^{\infty} g(t,\sigma)k_{zz^*}^*(\sigma,s)d\sigma = k_{wz}(t,s) \tag{74a}$$

$$\int_{-\infty}^{\infty} h(t,\sigma)k_{zz^*}(\sigma,s)d\sigma + \int_{-\infty}^{\infty} g(t,\sigma)k_{zz}^*(\sigma,s)d\sigma = k_{wz^*}(t,s), \tag{74b}$$

where we have used the facts that

$$k_{z^*z}(t,s) = E[z^*(t)z^*(s)] = k^*_{zz^*}(t,s)$$

and

$$k_{z^*z^*}(t,s) = E[z^*(t)z(s)] = k^*_{zz}(t,s).$$

We now examine the nature of the orthogonality condition when $z(t)$ is a polycyclostationary process, that is, its autocorrelation can be expressed in terms of an almost-periodic function in s

$$k_{zz}(t,s) = E[z(t)z^*(s)] = \sum_m c^{\alpha_m}(t-s)e^{j2\pi\alpha_m s}, \tag{75a}$$

where

$$c^{\alpha_m}(\tau) = \lim_{T\to\infty} \frac{1}{2T}\int_{-T}^{T} k_{zz}(t+\tau,t)e^{-j2\pi\alpha_m t}\,dt \tag{75b}$$

for some set of cycle frequencies $\{\alpha_m\}$. The coefficient functions $c^{\alpha_m}(\cdot)$ are called cyclic autocorrelation functions, although Gardner, who originated the terminology, uses a slightly different definition involving time variables $t+\tau/2$ and $t-\tau/2$ instead of t and s in (75a) [16, 17]. The Fourier transforms of the cyclic autocorrelations, which will be used in this approach to the satisfaction of the orthogonality condition, are known as cyclic spectral density functions. These functions characterize the spectral correlation exhibited by polycyclostationary signals, which has a straightforward physical interpretation. Let $h(t)$ be the impulse response of a lowpass analysis filter having a bandwidth parameter Δf. Then it is easy to show that the cross-correlation of the signals $h(t) \otimes (z(t)e^{-j2\pi ft})$ and $h(t) \otimes (z(t)e^{-j2\pi(f+\alpha_m)t})$ is proportional to $\Delta f C^{\alpha_m}(f)$ as $\Delta f \to 0$, where "$\otimes$" denotes a convolution operator [16, 17].

The orthogonality condition also requires expressions for

$$k_{zz^*}(t,s) = E[z(t)z(s)] = \sum_m d^{\alpha_m}(t-s)e^{j2\pi\alpha_m s} \tag{76}$$

and similar expressions for $k_{wz}(\cdot,\cdot)$ and $k_{wz^*}(\cdot,\cdot)$, which will be characterized by the cyclic cross-correlations $\{e^{\alpha_m}(\cdot)\}$ and $\{f^{\alpha_m}(\cdot)\}$, respectively. We assume that the set $\{\alpha_m\}$, called the cycle spectrum, is closed under all additions and subtractions. Then the optimum filter can be represented by frequency-shift operations followed by time-invariant filtering using this same set of cycle frequencies, that is,

$$h(t,s) = \sum_k q_{\alpha_k}(t-s)e^{j2\pi\alpha_k s} \tag{77a}$$

$$g(t,s) = \sum_k \bar{q}_{\alpha_k}(t-s)e^{j2\pi\alpha_k s}. \tag{77b}$$

Substituting all the above series expansions into the orthogonality condition (74a,b), equating terms with like exponential factors, and taking Fourier transforms of the result gives the new expressions to be solved for the optimum filter [12, 14, 15, 17]:

$$\sum_k Q_{\alpha_k}(f)C^{\beta_\ell-\alpha_k}(f-\alpha_k)+\sum_k \bar{Q}_{\alpha_k}(f)D^{-\beta_\ell+\alpha_k}(-f+\alpha_k)^* = E^{\beta_\ell}(f) \quad (78a)$$

$$\sum_k Q_{\alpha_k}(f)D^{\beta_\ell-\alpha_k}(f-\alpha_k)+\sum_k \bar{Q}_{\alpha_k}(f)C^{-\beta_\ell+\alpha_k}(-f+\alpha_k)^* = F^{\beta_\ell}(f) \quad (78b)$$

for each $\beta_\ell \in \{\alpha_m\}$.

We illustrate the solution of (78a,b) by an example simple enough to be worked out without the aid of a computer. Consider the situation, shown in Fig. 9, where two DSB signals, with carriers separated by frequency f_o, have the same bandwidth and total overlap of one sideband. Let $z(t)$ be the complex envelope relative to the carrier frequency of the signal of interest, which is modulated by a real WSS process $x(t)$, bandlimited to $|f| \leq f_0$. The interfering carrier signal is modulated with an independent real WSS process $v(t)$, having the same bandwidth. There is an additional interference caused by additive white noise with spectral density N_0. Therefore we have

$$z(t) = x(t)e^{j\theta} + v(t)e^{j\phi}e^{j2\pi f_o t} + n(t) \quad (79a)$$
$$w(t) = x(t), \quad (79b)$$

where θ and ϕ are the known phases of the signal of interest and the interfering signal, respectively. For this case the relevant correlation functions are

$$k_{zz}(t,s) = k_{xx}(t-s) + k_{vv}(t-s)e^{j2\pi f_0(t-s)} + 4N_o\delta(t-s) \quad (80a)$$
$$k_{zz^*}(t,s) = k_{xx}(t-s)e^{j2\theta} + k_{vv}(t-s)e^{j2\phi}e^{j2\pi f_o(t+s)} \quad (80b)$$
$$k_{wz}(t,s) = k_{xx}(t-s)e^{-j\theta} \quad (80c)$$
$$k_{wz*}(t,s) = k_{xx}(t-s)e^{j\theta}. \quad (80d)$$

Note that no noise term appears in (80b) since $n(t)$ is the complex envelope of a WSS real process, and so $E[n(t)n(s)]$ vanishes [10]. The cyclic spectral densities are obtained directly from (80a)–(80d). Taking Fourier transforms, we have

$$\begin{aligned} C^o(f) &= K_{xx}(f) + K_{vv}(f-f_0) + 4N_o \\ D^o(f) &= K_{xx}(f)e^{j2\theta} \\ D^{2f_o}(f) &= K_{vv}(f-f_o)e^{j2\phi} \\ E^o(f) &= K_{xx}(f)e^{-j\theta} \\ F^o(f) &= K_{xx}(f)e^{j\theta} \end{aligned} \quad (81)$$

and all other cyclic spectral density terms vanish. This means that only three terms appear in each equation on the LHS of (78a,b). Using $\beta_\ell = 0, -2f_o$, and $2f_o$, we find that the equations can be satisfied by

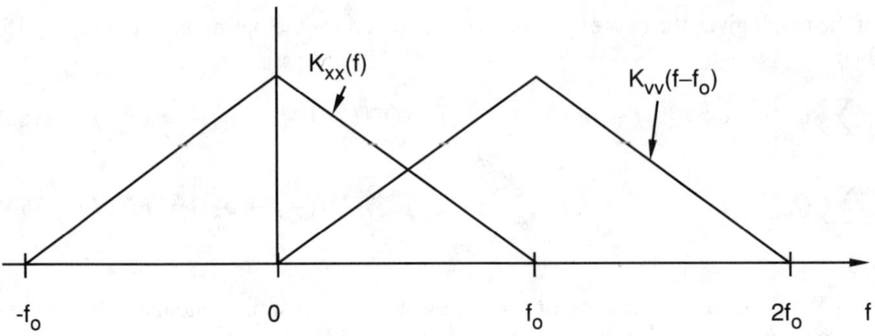

Figure 9: Example involving two interfering DSB signals.

$$
\begin{aligned}
Q_o(f) &= [\bar{Q}_o(-f)]^* \\
&= \frac{1}{\Delta(f)}[4N_o K_{xx}(f)e^{-j\theta}] \\
&\quad \times (2K_{vv}(f+f_0)+4N_o)U(-f)+(4N_0+K_{vv}(f-f_o))U(f)] \\
\bar{Q}_{2f_o}(f) &= [Q_{-2f_o}(-f)]^* \\
&= \frac{1}{\Delta(f)}[-4N_o K_{xx}(f)K_{vv}(f-f_o)e^{-j(\theta-2\phi)}]U(f),
\end{aligned}
\tag{82}
$$

where

$$
\begin{aligned}
\Delta(f) &= \Delta_+(f)U(f)+\Delta_+(-f)U(-f) \\
\Delta_+(f) &= 12N_o K_{xx}(f)K_{vv}(f-f_o)+32N_o^2(K_{xx}(f)+K_{vv}(f-f_0))+64N_o^3
\end{aligned}
\tag{83}
$$

and $U(f)$ is the nonsymmetric rectangular filtering function: $U(f)=1$ in $0 \le f < f_o$ and $U(f)=0$ otherwise.

Of course, the solution is greatly simplified as $N_o \to 0$ and also the mean-squared-error approaches zero as $N_o \to 0$, indicating that perfect cancellation of the interference from the other carrier signal can be attained. If the complex envelope transformation shown in Fig. 8 is combined with the asymptotically optimum LCL filter for high SNR, we can display the entire receiver filter in terms of real processing operations. This is done in Fig. 10, where $G(f) = U(f)+U(-f)$ is an ideal lowpass filter with a cutoff frequency of f_o and $H(f) = -j[U(f)-U(-f)]$ is the corresponding bandlimited Hilbert transformer.

The fact that cochannel interference can be completely eliminated in this simple example is easily understood by noting that one of the sidebands in Fig. 9 is free from interference from the other carrier signal. Therefore, for example, a highly selective bandpass filter could be used to extract the lower sideband of the signal of interest and coherent demodulation used on the resulting single-sideband signal to recover

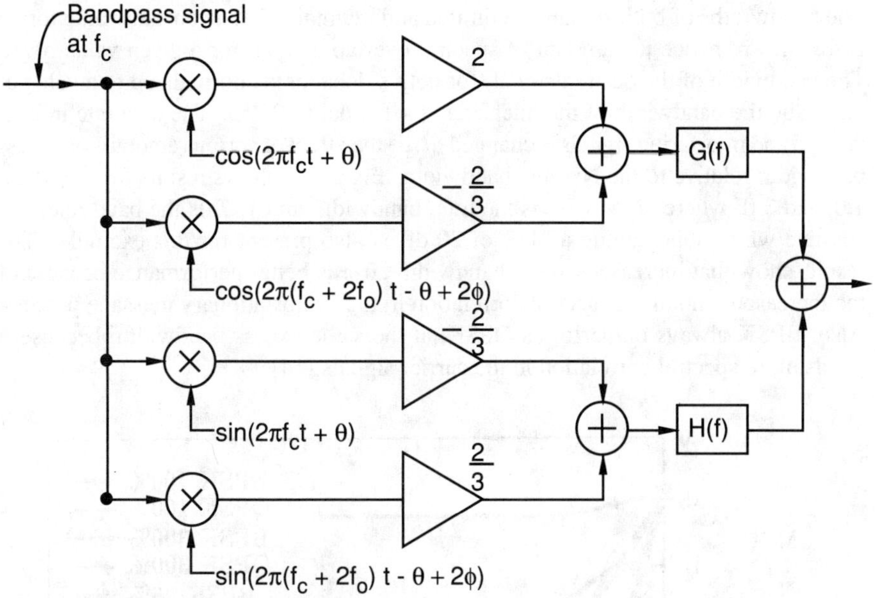

Figure 10: Optimum linear demodulator for DSB signal with upper-sideband co-channel interference and high SNR.

the message process $x(t)$. Alternatively, a bandpass filter could extract the upper sideband of the interfering signal and coherent demodulation used to recover a noisy version of $v(t)$ for later cancellation from the demodulated signal of interest. The solution given by (82) and (83) uses both these techniques in the proper combination to minimize the total effect of additive noise and cochannel interference.

The complexity of the solution to (78a,b) grows rapidly as the problem is made more interesting by increasing the amount of spectral overlap, by increasing the number of interfering carrier signals, and by letting the message process have its own cycle frequencies that might also be different from the cycle frequencies modulated onto the other carriers. Finding the optimum filter becomes extremely tedious or even impossible without some form of bandwidth constraints. This has led to the investigation of constrained optimum receivers, where the number of frequency-shift operations permitted by the receiver is set to some prescribed value. Even with a limited number of frequency shifts, there still is the problem of determining which set of frequency shifts will produce the best results. The problem has been examined in detail for a number of situations of practical interest. Results are reported in [12, 14, and 15]. We reproduce here some of the results presented in [14], where the computed minimum mean-squared-error is shown graphically as a function of the allowed number of frequency shifts, as shown in Fig. 11. In this example there is a single interferer with a carrier frequency offset by $f_o = 0.2257$ from the carrier of the desired signal. Both carriers are modulated with synchronous digital data, either in a binary phase-shift-keyed (BPSK) or quarternary phase-shift-keyed (QPSK) format.

The bandwidths of both signals are limited and maintained at a constant value for all cases shown, hence the amount of spectral overlap is fixed for a given value of f_o. The bandwidth of the desired signal (for both sidebands) is normalized to a value of 1.0 while the bandwidth of the interferer is set equal to 0.753. The data rate in both desired and interfering signals is changed to see the effect of various amounts of excess bandwidth relative to the Nyquist bandwidth. Figure 11 shows results for $e = 0.25$, 1.0, and 3.0; where $(1 + e)/T$ is the total bandwidth and $1/T$ is the baud rate. An additive white noise giving a SNR of 20 dB is also present for this example. The results show that increased excess bandwidth affords better performance because of the increased amount of spectral correlation in the cyclostationary message process. Also, BPSK always outperforms QPSK, at the same excess bandwidth, because it yields more spectral correlation in the carrier signals [14].

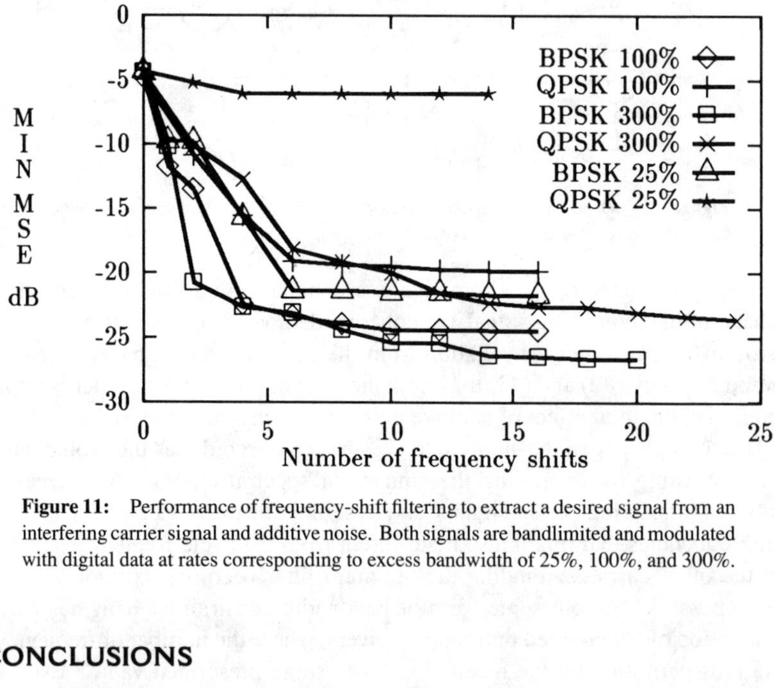

Figure 11: Performance of frequency-shift filtering to extract a desired signal from an interfering carrier signal and additive noise. Both signals are bandlimited and modulated with digital data at rates corresponding to excess bandwidth of 25%, 100%, and 300%.

6 CONCLUSIONS

We have examined a variety of input/output and state-variable characterizations for periodically- and polyperiodically-time-varying linear filters. The state-variable descriptions have the advantage that, unlike the Fourier series methods, the filter complexity does not depend directly on the harmonic complexity of the time-varying parameters. On the other hand, the Fourier series methods are more adaptable to the solution of optimum filtering problems. In this paper we have investigated such problems by applying the orthogonality condition, which expresses the optimum filter response implicitly in terms of the autocorrelation of its input and the cross-correlation between input and desired output. We consider some simple examples to illustrate how the orthogonality condition is solved for problems involving extraction of a cyclostationary signal process from a received signal having stationary and/or cyclostationary interference. We also present an extended form of the orthogonality condition for processing of complex-valued signals.

REFERENCES

[1] T. Claasen and W. Mecklenbrauker, "On Stationary Linear Time-Varying Systems," *IEEE Trans.*, vol. CAS-29, pp. 169–184, March 1982.

[2] L. E. Franks and I. W. Sandberg, "An Alternative Approach to the Realization of Network Transfer Functions: The N-Path Filter," *Bell Sys. Tech. Jour.*, vol. 39, pp. 1321–1350, September 1960.

[3] F. L. Kitson and L. J. Griffiths, "Design and Analysis of Recursive Periodically Time-Varying Digital Filters with Highly Quantized Coefficients," *IEEE Trans.*, vol. ASSP-36, pp. 674–685, May 1988.

[4] S. Ghanekar, S. Tantaratana, and L. E. Franks, "Implementation of Recursive Filters Using Highly Quantized Periodically Time-Varying Coefficients," *Proc. 1991 ICASSP*, pp. 1625-1628, May 1991.

[5] R. W. Brockett, *Finite Dimensional Linear Systems*, New York: Wiley, 1970.

[6] W. J. Rugh, *Linear System Theory*, Englewood Cliffs, NJ: Prentice-Hall, 1993.

[7] H. E. Meadows, L. M. Silverman, and L. E. Franks, "A Canonical Network for Periodically-Variable Linear Systems," *Proc. Fourth Annual Allerton Conference on Circuit and Systems Theory*, pp. 649–658, October 1966.

[8] W. A. Gardner and L. E. Franks, "Characterization of Cyclostationary Random Signal Processes," *IEEE Trans.*, vol. IT-21, pp. 4–14, January 1975.

[9] G. D. Forney, Jr., "Maximum-Likelihood Sequence Estimation of Digital Sequences in the Presence of Intersymbol Interference," *IEEE Trans.*, vol. IT-18, pp. 363–378, May 1972.

[10] L. E. Franks, *Signal Theory*, Rev. ed., Dowden and Culver, Inc., 1981.

[11] A. P. Sage and J. L. Melsa, *Estimation Theory with Applications to Communications and Control*, Appendix A, New York: McGraw-Hill, 1971.

[12] W.A. Gardner and W.A. Brown, "Frequency-Shift Filtering Theory for Adaptive Co-Channel Interference Removal," *Proc. 23rd Annual Asilomar Conference on Signals, Systems & Computers*, November 1989.

[13] W. M. Brown, *Analysis of Linear Time-invariant Systems*, New York: McGraw-Hill, Inc., 1963.

[14] W. A. Gardner, "Cyclic Wiener Filtering: Theory and Method," *IEEE Trans. Communicat.*, vol. 41, pp. 151–163, January 1993.

[15] W. A. Brown, "On the Theory of Cyclostationary Signals," Ph.D. Dissertation, 1987, Department of Electrical Engineering and Computer Science, University of California, Davis.

[16] W. A. Gardner, *Introduction to Random Processes*, 2nd Ed., New York: McGraw-Hill, 1986.

[17] W. A. Gardner, *Statistical Spectral Analysis: A Nonprobabilistic Theory*, Englewood Cliffs, NJ: Prentice-Hall, 1988.

Chapter 5

Representation, Prediction, and Identification of Cyclostationary Processes—A State-Space Approach

Sergio Bittanti Paolo Bolzern Luigi Piroddi
Dipartimento di Elettronica e Informazione
Politecnico di Milano, 20133 - Milano

Giuseppe De Nicolao
Dipartimento di Informatica e Sistemistica
Università di Pavia, 27100 - Pavia

1 INTRODUCTION

Recent years have witnessed a spate of research devoted to cyclostationary (CS) processes and their application in several fields, including telecommunications, signal processing, biomedicine, natural sciences, and economics, see e.g., [1–4] and references quoted therein. Starting with early investigations in the late 1950s, most contributions have focused on frequency domain and input-output descriptions. Only a few contributions were made to the state-space theory of such processes [5–9]. In this framework, a CS process is modeled as the output of a linear dynamical system with periodic coefficients excited by white noise inputs (*state-space Markovian* or simply *Markovian* model). This kind of description is well suited to clarify many important issues related to the representation, prediction, and identification of CS processes. Consider for instance the optimal linear prediction problem. As is well known, in the stationary case this issue is intimately related to spectral factorization theory and can be tackled either in an input-output context (by Wiener-Kolmogorov techniques) or in a state-space one (by Kalman-Bucy theory). Now, among other things, spectral factorization calls for the notions of poles, zeros, stability, minimum phase representation. At present, a full characterization of these notions for CS processes relies on state-space concepts, the necessary technical background being provided by the recent developments in the theory of deterministic linear periodic systems [8]. Furthermore, state-space models for CS signals are particularly useful in dealing with control problems in the presence of cyclostationary disturbances. In fact many existing control design techniques take advantage of a state-space description of (possibly nonstationary) disturbances affecting the controlled plant, see e.g., [10], where a problem of active control of vibrations in helicopters is tackled by means of LQG (Linear Quadratic Gaussian) methods.

This chapter offers an extensive view of state-space theory of Gaussian CS processes in discrete time (some partial results in continuous time are given in [5] and [6]). The connections among the three alternative descriptions depicted in Fig. 1 are investigated: (i) jointly periodic autocovariance functions, (ii) state-space stochastic models (Markovian representations), (iii) Autoregressive Moving Average models with Periodic coefficients (PARMA). In particular, we focus on the links (i)$\leftrightarrow$(ii) and (ii)$\leftrightarrow$(iii). First, it is shown that, under the stability assumption, a linear periodic system fed by white noise asymptotically generates a CS process, whose statistics are independent of the initial conditions ((i)$\leftarrow$(ii)). The inverse problem of autocovariance generation, i.e., the problem of finding a state-space model yielding a prescribed autocovariance ((i)$\rightarrow$(ii)) is far more involved. In particular, we show that the derivation of a Markovian model from covariance data calls for the solution of a deterministic realization problem and the computation of the periodic solution of a suitable Riccati equation. To the authors' knowledge, this periodic stochastic realization problem has not been adequately dealt with in the previous literature. As for the relationship between PARMA models and state-space representations, it is easy to derive a state-space model equivalent to a given PARMA ((ii)$\leftarrow$(iii)). We exploit this fact to introduce the notions of poles and zeros of a PARMA from the analogous state-space concepts. The passage from state-space to input-output ((ii)$\rightarrow$(iii)) is more intriguing. It turns out that a given linear periodic system admits an equivalent PARMA representation of the same order if and only if the system matrices can be put in the so-called Observer Canonical Form. This is possible only if a suitable observability condition is fulfilled. When this is not the case, the transformation from state-space to input-output can still be performed, but the order of the PARMA is greater than that of the original state-space model.

Another topic treated here is the classification of the main state-space representations. A first family is given by *innovationslike* representations. In some sense, these are the most parsimonious Markovian representations, since they are the only ones whose input is a white noise vector of the same dimension as the CS process. The *canonical innovations* representation enjoys the additional property of being minimum-phase, i.e., all its zeros belong to the open unit circle. Finally, *prediction error* representations are special innovationslike representations and are obtained from the equations of the Kalman filter. A main result is that, given a generic Markovian model and the associated Riccati equation for the prediction problem, a bijective

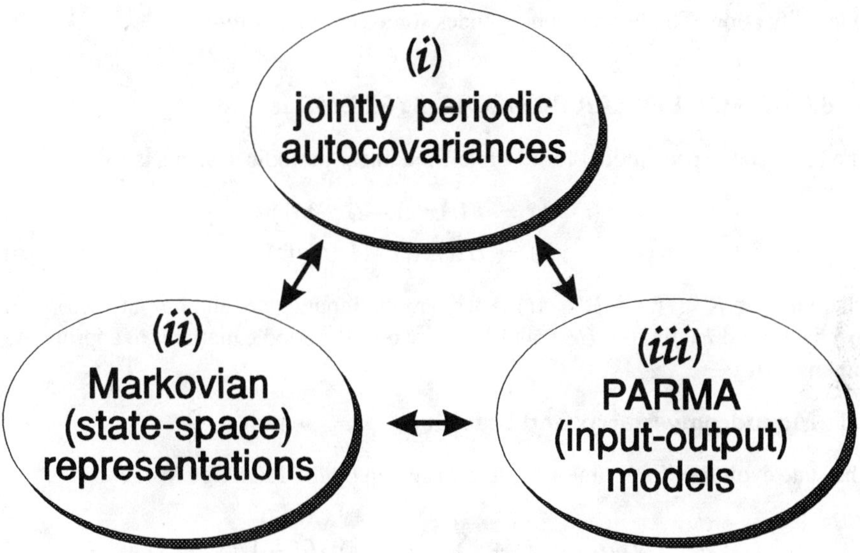

Figure 1: Alternative representations of CS processes.

correspondence exists between the innovationslike representations and the symmetric periodic solutions of the Riccati equation. Moreover, under mild assumptions, the unique canonical innovations representation corresponds to the maximal periodic solution of the same Riccati equation.

In the last part of the chapter, canonical factorization, optimal prediction, and identification of PARMA models are tackled by state-space tools. In particular, we show that, if a given CS process admits a nonminimum-phase PARMA representation, a canonical minimum-phase PARMA representation can always be computed under the assumption that no zero of the original PARMA lies on the unit circle. As for the prediction problem for PARMA models, in the minimum-phase case the optimal predictor is given an explicit formula, which extends the well-known formula for stationary ARMA models. This solution applies also to the nonminimum-phase case, provided that the original model is replaced by its canonical representation. Finally, it is shown that canonical factorization methods are essential in order to guarantee stability of the filtering operations involved in iterative identification of PARMA models.

The chapter is organized as follows: Section 2 is devoted to the introduction of the necessary background regarding deterministic periodic systems. In Section 3, the properties of state-space representations are investigated. In particular, state-space and covariance representations are related to each other. Moreover, a rather complete classification of the various types of state-space representations is provided. In Section 4, PARMA models are investigated from a state-space point of view. We address the equivalence between input-output and state-space representations, the linear prediction problem, the canonical factorization of PARMA models, and its role in identification. The last section provides some concluding remarks.

2 BASICS OF LINEAR PERIODIC SYSTEMS

The basic state-space model for discrete-time linear periodic systems is

$$x(t+1) = F(t)x(t) + G(t)u(t) \tag{1a}$$

$$y(t) = H(t)x(t) + D(t)u(t), \tag{1b}$$

where $u(t) \in \mathbf{R}^m$, $x(t) \in \mathbf{R}^n$, $y(t) \in \mathbf{R}^p$, are the input, state, and output vectors respectively, and $F(\cdot)$, $G(\cdot)$, $H(\cdot)$, and $D(\cdot)$, are real T-periodic matrices of appropriate dimensions.

2.1 Monodromy Matrix and Stability

The state evolution from time τ to time t is given by the Lagrange formula

$$x(t) = \Psi_F(t, \tau)x(\tau) + \sum_{j=\tau+1}^{t} \Psi_F(t, j)G(j-1)u(j-1)$$

where $\Psi_F(t, \tau) = F(t-1)F(t-2)\ldots F(\tau)$ is the transition matrix of the system over the interval $[\tau, t]$. It is easily seen that periodicity entails:

$$\Psi_F(t+T, \tau+T) = \Psi_F(t, \tau). \tag{2}$$

The transition matrix over one period, viz.

$$\Phi_F(t) = \Psi_F(t+T, t),$$

is named *monodromy matrix* at time t and is T-periodic. Apparently, this matrix determines the system behavior from one period (starting at t) to the subsequent one (starting at $t+T$). In particular, the T-sampled free motion is given by $x(t+kT) = \Phi_F(t)^k x(t)$. This indicates that the system or, equivalently, the matrix sequence $F(\cdot)$ is asymptotically stable if and only if the eigenvalues of $\Phi_F(t)$ belong to the open unit disk. Such eigenvalues, referred to as *characteristic multipliers*, are independent of t, see e.g., [8]. Thus, the characteristic multipliers can be regarded as the periodic equivalent of the *poles* of a time-invariant system.

2.2 Reachability and Observability

In the state-space theory of linear systems, the notions of *reachability* and *observability* play a major role. As is well known [11], reachability deals with the possibility of driving $x(t)$ to any desired point in the state-space by using a proper input sequence, while observability is connected with the possibility of uniquely determining $x(t)$ from the observation of the subsequent outputs. When any state at any time can be reached [observed] in an interval of length k, we speak of *k-step reachability* [*k-step observability*]. Moreover, a system is said to be *reachable* [*observable*] if it is *k-step reachable* [*k-step observable*] for some k. If not all points in the state-space can be reached [observed], one can define the *reachability* [*unobservability*] *subspace* as the set of states that are reachable [unobservable]. For periodic systems, a thorough body of results regarding these notions is available, see e.g., [8] and references quoted therein. We also refer to [8] for more precise definitions of the reachability and unobservability subspaces. Among the various characterizations, the following are worth mentioning.

Reachability Criterion. *System (1) is k-step reachable iff rank* $[\boldsymbol{R}_k(t)] = n$ $\forall t$, *where*

$$\boldsymbol{R}_k(t) = [G(t-1) \;\; \Psi_F(t, t-1)G(t-2) \ldots \Psi_F(t, t-k+1)G(t-k)].$$

Moreover, system (1) is reachable iff it is nT-step reachable.

Observability Criterion. *System (1) is k-step observable iff rank* $[\boldsymbol{O}_k(t)] = n$ $\forall t$, *where*

$$\boldsymbol{O}_k(t) = [H(t)' \;\; \Psi_F(t+1, t)'H(t+1)' \ldots \Psi_F(t+k-1, t)'H(t+k-1)'].$$

Moreover, system (1) is observable iff it is nT-step observable.

For time-invariant systems, $\boldsymbol{R}_n(t)$ and $\boldsymbol{O}_n(t)$ become the celebrated Kalman reachability and observability matrices [11]:

$$\boldsymbol{R} = [G \; FG \ldots F^{n-1}G]$$
$$\boldsymbol{O} = [H' \; F'H' \ldots (F')^{n-1}H'].$$

In the periodic case, even if $\boldsymbol{R}_{nT}(t)$ [or $\boldsymbol{O}_{nT}(t)$] has maximum rank for some t, it may fail to exhibit the same property at a different time point. This corresponds to the fact that the dimensions of reachability and unobservability subspaces of system (1) are, in general, time-varying.

It is also said that the pair $(F(\cdot), G(\cdot))$ is reachable [$(F(\cdot), H(\cdot))$ is observable] to mean that system (1) is reachable [observable].

An alternative characterization of reachability and observability of periodic systems refers to the characteristic multipliers as follows.

Reachability Modal Characterization. *A characteristic multiplier* λ *of* $F(\cdot)$ *is said to be* $(F(\cdot), G(\cdot))$*-unreachable at time* τ, *if there exists* $\eta \neq 0$, *such that*

$$\Phi_F(\tau)'\eta = \lambda\eta \quad \text{and} \quad G(j-1)'\Psi_F(\tau, j)'\eta = 0 \qquad \forall j \in [\tau - T + 1, \tau].$$

System (1) is reachable if no characteristic multiplier is unreachable.

Observability Modal Characterization. *A characteristic multiplier* λ *of* $F(\cdot)$ *is said to be* $(F(\cdot), H(\cdot))$*-unobservable at time* τ, *if there exists* $\zeta \neq 0$, *such that*

$$\Phi_F(\tau)\zeta = \lambda\zeta \quad \text{and} \quad H(j)\Psi_F(j, \tau)\zeta = 0 \qquad \forall j \in [\tau, \tau + T - 1].$$

System (1) is observable if no characteristic multiplier is unobservable.

These "modal" notions are the periodic counterpart of the so-called PBH (Popov-Belevitch-Hautus) characterization of reachability and observability in the time invariant case, see e.g., [12]. For the following developments, it should be noted that, if a characteristic multiplier $\lambda \neq 0$ is unreachable at time t, it is also unreachable at any time point. Somewhat unexpectedly, the same is not true for null multipliers, as discussed in [13].

2.3 Zeros

As is well known, the zeros of time-invariant systems can be characterized in terms of the so-called blocking property: Associated with any zero there is an input function such that the output is identically zero for a suitable choice of the initial state. A consistent definition of zeros of discrete-time periodic systems has been first given in [14]. In this respect, it is advisable to introduce the following Time-Invariant Representation (TIR) of the periodic system (1), which dates back to 1959 [15]. Letting

$$\begin{aligned}
&F_\tau = \phi_F(\tau),\\
&G_\tau = [\Psi_F(\tau+T,\tau+1)G(\tau)\ \ \Psi_F(\tau+T,\tau+2)G(\tau+1)\ldots G(\tau+T-1)],\\
&H_\tau = [H(\tau)'\ \ \Psi_F(\tau+1,\tau)'H(\tau+1)'\ldots\Psi_F(\tau+T-1,\tau)'H(\tau+T-1)']',\\
&D_\tau = \{(D_\tau)_{ij}\}, \qquad i,j=1,2,\ldots,T,
\end{aligned} \tag{3}$$

$$(D_\tau)_{ij} = \begin{cases} 0, & i<j\\ D(\tau+i-1), & i=j\\ H(\tau+i-1)\ \Psi_F(\tau+i-1,\tau+j)G(\tau+j-1), & i>j \end{cases}$$

$$u_\tau(k) = [u(\tau+kT)'\ u(\tau+kT+1)'\ldots u(\tau+(k+1)T-1)']',$$

where $F_\tau \in \boldsymbol{R}^{n\times n}$, $G_\tau \in \boldsymbol{R}^{n\times mT}$, $H_\tau \in \boldsymbol{R}^{pT\times n}$, and $D_\tau \in \boldsymbol{R}^{pT\times mT}$, consider the time-invariant system

$$x_\tau(k+1) = F_\tau x_\tau(k) + G_\tau u_\tau(k) \tag{4a}$$

$$y_\tau(k) = H_\tau x_\tau(k) + D_\tau u_\tau(k). \tag{4b}$$

It is easy to see that, if $u_\tau(\cdot)$ is constructed according to (3), and $x_\tau(0)$ is taken equal to $x(\tau)$, then

$$\begin{aligned}
&x_\tau(k) = x(\tau+kT)\\
&y_\tau(k) = [y(\tau+kT)'\ y(\tau+kT+1)'\ldots y(\tau+(k+1)T-1)']'.
\end{aligned}$$

Therefore, the time-invariant system (4) can be seen as a state-sampled representation of system (1), fed by an augmented input vector and producing an augmented output vector. Such vectors $u_\tau(k)$ and $y_\tau(k)$ are obtained by stacking the values of the input $u(\cdot)$ and the output $y(\cdot)$ over each period. This kind of representation is also named *lifted representation*. For an extension of lifting techniques to periodic continuous-time systems the interested reader is referred to [16].

The definition of zeros of a periodic system relies on the TIR defined above. For what matters herein, we restrict consideration to the case $m = p$.

Definition 1 (Zeros of a Periodic System) [14]: Assume $m = p$. Then z is a zero at time τ of system (1), if it is a zero of the corresponding TIR (4), i.e., if $\det N(z) = 0$, where

$$N(z) = \begin{bmatrix} zI - F_\tau & -G_\tau \\ H_\tau & D_\tau \end{bmatrix}$$

is *Rosenbrock's system matrix* [17] for system (4).

As discussed in [14, 18], the nonnull zeros together with their multiplicities are in fact independent of τ. In [14] it has been proven that one can associate to any zero of the system a suitable initial state and an input function that result in the null output (transmission-blocking property). Precisely, given a zero $\bar{z} \neq 0$, there exists an initial state $x(0)$ such that, for any T-th root m of $\bar{z}$, there exists a T-periodic function $a(t)$ such that the input $u(t) = a(t)m^t$ yields $y(t) = 0 \; \forall t$.

When all zeros and poles belong to the open unit disk, system (1) is said to be *minimum phase*, in analogy with the corresponding notion for time invariant systems.

An important property of zeros is their invariance with respect to output injection, as precisely stated in the following lemma.

Lemma 1 [19] *Consider the following periodic system generated from system (1) by means of a periodic output injection*

$$\tilde{x}(t+1) = F(t)\tilde{x}(t) + G(t)u(t) + E(t)\tilde{y}(t)$$
$$\tilde{y}(t) = H(t)\tilde{x}(t) + D(t)u(t),$$

where $E(\cdot)$ is an arbitrary periodic matrix of suitable dimensions. The zeros at time τ of this system coincide with the zeros at τ of system (1).

2.4 Change of Basis

A change of basis in the state-space amounts to transforming the state vector $x(t)$ into a new vector $\tilde{x}(t) = S(t)x(t)$, where $S(t)$ is T-periodic and nonsingular for each t. In the new basis, the system is characterized by the 4-tuple $(\tilde{F}(\cdot), \tilde{G}(\cdot), \tilde{H}(\cdot), \tilde{D}(\cdot))$:

$$\tilde{F}(t) = S(t+1)F(t)S(t)^{-1},$$
$$\tilde{G}(t) = S(t+1)G(t),$$
$$\tilde{H}(t) = H(t)S(t)^{-1},$$
$$\tilde{D}(t) = D(t).$$

The original and the new 4-tuple are said to be *algebraically equivalent*. A given system admits infinitely many algebraically equivalent state-space representations sharing the same input-output behavior. Furthermore, stability, reachability, observability, poles, and zeros are not affected by a change of state-coordinates.

When comparing different models, we also make use of the weaker notion of *equivalence* to indicate that two representations yield the same input-output behavior. Of course, two algebraically equivalent models are also equivalent, but the converse is not true in general. In particular, equivalent models may well have different orders. For more details on reachability, observability, change of basis, equivalence, etc., the interested reader is referred to [20, 21].

3 STATE-SPACE REPRESENTATION OF CYCLOSTATIONARY PROCESSES

3.1 Stochastic Periodic Systems

We now focus on the state-space representation of stochastic periodic systems, i.e., systems of the type ation

$$x(t+1) = F(t)x(t) + G(t)v(t) \tag{5a}$$

$$y(t) = H(t)x(t) + w(t), \tag{5b}$$

where $v(\cdot)$ and $w(\cdot)$ are m- and p-dimensional white noises with Gaussian distribution and T-periodic moments. Precisely, denoting by

$$\begin{bmatrix} \bar{v}(t) \\ \bar{w}(t) \end{bmatrix} \stackrel{\text{def}}{=} \mathrm{E} \begin{bmatrix} v(t) \\ w(t) \end{bmatrix},$$

the expected value and by

$$\begin{bmatrix} Q(t) & S(t) \\ S(t)' & R(t) \end{bmatrix} \stackrel{\text{def}}{=} \mathrm{Var} \begin{bmatrix} v(t) \\ w(t) \end{bmatrix},$$

the covariance matrix, we assume that $\bar{v}(t) = \bar{v}(t+T)$, $\bar{w}(t) = \bar{w}(t+T)$, $Q(t) = Q(t+T)$, $S(t) = S(t+T)$, and $R(t) = R(t+T)$, $\forall t$. The initial state $x(\tau)$ is assumed to be independent of $v(\cdot)$ and $w(\cdot)$, and normally distributed with $\mathrm{E}\,[x(\tau)] = \bar{x}_\tau$ and $\mathrm{Var}\,[x(\tau)] = \Sigma_\tau$.

Throughout this chapter, we make the following assumptions:

- $F(\cdot)$ is asymptotically stable.
- $(F(\cdot), H(\cdot))$ is observable.

We show that the stability assumption implies the convergence of the state and output processes to CS processes. The observability assumption is useful to avoid unnecessary redundancy in the state-space representation.

3.2 Evolution of the Moments of a Stochastic Periodic System

3.2.1 Periodic Mean

The expected values $\bar{x}(t)$ and $\bar{y}(t)$ of the state and output processes trivially satisfy the following deterministic equations:

$$\bar{x}(t+1) = F(t)\bar{x}(t) + G(t)\bar{v}(t) \tag{6a}$$
$$\bar{y}(t) = H(t)\bar{x}(t) + \bar{w}(t). \tag{6b}$$

Since $F(\cdot)$ is asymptotically stable, it is easy to see that there exists a unique $\bar{x}_\tau$ such that

$$\bar{x}(\tau+T) = \Phi_F(\tau)\bar{x}_\tau + \sum_{j=\tau+1}^{t} \Psi_F(t,j)G(j-1)\bar{v}(j-1)$$

is equal to $\bar{x}_\tau$. With such an initial condition, the state and the output of system (6) are T-periodic. Due to the stability assumption, such a cyclic equilibrium turns out to be globally attractive. Consequently, given any initial condition $\bar{x}_\tau$, the cyclic equilibrium is asymptotically recovered for $\tau \to -\infty$.

3.2.2 Periodic Variance

We denote the state and output autocovariance functions by

$$\sigma(t,\tau) = \mathrm{E}\,[(x(t)-\bar{x}(t))(x(\tau)-\bar{x}(\tau))']$$
$$\gamma(t,\tau) = \mathrm{E}\,[(y(t)-\bar{y}(t))(y(\tau)-\bar{y}(\tau))'].$$

Moreover, for the sake of conciseness, the covariance matrices $\sigma(t,t)$ and $\gamma(t,t)$ are respectively denoted by $\Sigma(t)$ and $\Gamma(t)$.

It is easy to see that the covariance matrix $\Sigma(\cdot)$ of the state of system (5) obeys the periodic Lyapunov equation (PLE)

$$\Sigma(t+1) = F(t)\Sigma(t)F(t)' + G(t)Q(t)G(t)', \tag{7}$$

with initial condition $\Sigma(\tau) = \Sigma_\tau$. Now, it is straightforward to verify that the solution of the PLE is given by

$$\Sigma(t) = \Psi_F(t,\tau)\Sigma(\tau)\Psi_F(t,\tau)' + W(\tau,t),$$

where

$$W(\tau,t) = \sum_{j=\tau+1}^{t} \Psi_F(t,j)G(j-1)Q(j-1)G(j-1)'\Psi_F(t,j)'$$

is the *reachability Grammian matrix*. Imposing the periodicity constraint $\Sigma(\tau+T) = \Sigma(\tau)$ yields

$$\Sigma(\tau) = \Phi_F(\tau)\Sigma(\tau)\Phi_F(\tau)' + W(\tau, \tau + T). \tag{8}$$

Equation (8) is a discrete-time Algebraic Lyapunov Equation. Since $F(\cdot)$ is stable, all the eigenvalues of $\Phi_F(\tau)$ lie in the open unit disk, so that (8) admits a unique solution $\overline{\Sigma}_\tau$ (which is in fact positive semidefinite). Such a solution $\overline{\Sigma}_\tau$ acts as a periodic generator for the PLE, i.e., the solution $\overline{\Sigma}(\cdot)$ of the PLE with initial condition $\overline{\Sigma}(\tau) = \overline{\Sigma}_\tau$ is T-periodic. Moreover, it is easy to see that

$$\Sigma(t) - \overline{\Sigma}(t) = \Psi_F(t, \tau)\left[\Sigma(\tau) - \overline{\Sigma}_\tau\right]\Psi_F(t, \tau)',$$

where $\Sigma(\cdot)$ is any solution of the PLE. Obviously, the stability of system (5) implies the convergence of the difference $\Sigma(t) - \overline{\Sigma}(t)$ to zero regardless of the initial condition $\Sigma(\tau)$.

Turning now to the state autocovariance function, it is easy to see that

$$\sigma(t, \tau) = \Psi_F(t, \tau)\Sigma(\tau), \quad t > \tau.$$

Since $\sigma(t + T, \tau + T) = \Psi_F(t + T, \tau + T)\Sigma(\tau + T)$ and $\sigma(t, \tau) = \Psi_F(t, \tau)\Sigma(\tau)$, equality (2) together with $\Sigma(\tau + T) = \Sigma(\tau)$ imply $\sigma(t + T, \tau + T) = \sigma(t, \tau)$.

As for the system output, an easy computation shows that

$$\gamma(t, \tau) = H(t)\Psi_F(t, \tau + 1)\left[F(\tau)\Sigma(\tau)H(\tau)' + G(\tau)S(\tau)\right], \quad t > \tau \tag{9a}$$

$$\Gamma(t) = \gamma(t, t) = H(t)\Sigma(t)H(t)' + R(t). \tag{9b}$$

Therefore, $\Sigma(\tau + T) = \Sigma(\tau)$ implies $\Gamma(\tau + T) = \Gamma(\tau)$, so that $\gamma(t + T, \tau + T) = \gamma(t, \tau)$, as well.

To summarize, under the stability assumption of system (5), whatever the initial condition, the stochastic process $y(\cdot)$ asymptotically tends to a process with the following characteristics:

i. The expected value $\bar{y}(\cdot)$ is T-periodic.

ii. The autocovariance function $\gamma(t, \tau)$ is *jointly periodic* in its arguments, i.e., it possesses the property $\gamma(t + T, \tau + T) = \gamma(t, \tau)\ \forall t, \tau$.

In other words, the output (and also the state) of system (5) asymptotically converges to a CS process. Furthermore, the asymptotic CS process is independent of the initial condition.

3.3 Markovian Representation of a CS Autocovariance Function

The periodic stochastic realization problem consists of finding a state-space model whose output asymptotically exhibits a given expected value $\bar{y}(t)$ and a prescribed CS autocovariance function $\gamma(t, \tau)$ (see arrow (i)→(ii) in Fig. 1). Whenever this happens, the stochastic system will be said to be a *Markovian representation* or a *Markovian realization* of the specified CS process. Since the expected value requirement can be trivially fulfilled by a suitable choice of $\bar{w}(t)$, without any loss of generality we consider only the zero-mean case, $\bar{y}(t) = 0 \ \forall t$.

Obviously, not all CS processes admit a finite dimensional Markovian representation. For this reason, we focus on *rational* CS processes, according to the following definition. A CS process with autocovariance $\gamma(t, \tau) = \gamma(t+T, \tau+T)$ is said to be rational of order n if there exist T-periodic matrices $F(\cdot)$, $\tilde{G}(\cdot)$, $H(\cdot)$, of orders $n \times n, n \times m, p \times n$, such that

- $F(\cdot)$ is asymptotically stable
- $(F(\cdot), \tilde{G}(\cdot))$ is reachable and $(F(\cdot), H(\cdot))$ is observable
- $\gamma(t, \tau) = H(t)\Psi_F(t, \tau+1)\tilde{G}(\tau), \ t > \tau.$

Now, the stochastic realization problem can be split into two steps. The first one consists of finding a triple $(F(\cdot), \tilde{G}(\cdot), H(\cdot))$ from a given $\gamma(t, \tau)$; the second one consists of passing from this triple to a complete Markovian model. As for the first step, notice that $\gamma(t, \tau) = H(t)\Psi_F(t, \tau+1)\tilde{G}(\tau)$ can be interpreted as the impulse response of a linear deterministic system characterized by the triple $(F(\cdot), \tilde{G}(\cdot), H(\cdot))$. In the systems and control literature, the problem of associating a state-space triple to a given impulse response is known as the (deterministic) *realization problem.* A complete treatment of such a problem in the periodic case has been recently provided in [22], where a realization algorithm is given, provided that a certain Hankel matrix associated with $\gamma(t, \tau)$ satisfies some rank requirements.

Turning now to the second step, given the triple $(F(\cdot), \tilde{G}(\cdot), H(\cdot))$, a complete Markovian representation can be found by looking for a periodic 4-tuple $(\Sigma(\cdot), \tilde{Q}(\cdot), \tilde{S}(\cdot), R(\cdot))$, with $\Sigma(t) \geq 0 \ \forall t$, such that

$$\Sigma(t+1) - F(t)\Sigma(t)F(t)' = \tilde{Q}(t) \tag{10a}$$

$$\tilde{G}(t) - F(t)\Sigma(t)H(t)' = \tilde{S}(t) \tag{10b}$$

$$\Gamma(t) - H(t)\Sigma(t)H(t)' = R(t) \tag{10c}$$

$$\begin{bmatrix} \tilde{Q}(t) & \tilde{S}(t) \\ \tilde{S}(t)' & R(t) \end{bmatrix} \geq 0 \ \forall t. \tag{10d}$$

Indeed, from (7), (9a), and (9b), it is easy to see that if (10a–d) are fulfilled, system (5) with $G(t) = I$, $Q(t) = \tilde{Q}(t)$, $S(t) = \tilde{S}(t)$ is a Markovian representation of the CS process whose autocovariance function is $\gamma(t, \tau)$.

A direct inspection of (10a–c) reveals that, once a periodic matrix $\Sigma(\cdot)$ is selected, the remaining matrices $\tilde{Q}(\cdot)$, $\tilde{S}(\cdot)$, $R(\cdot)$ of the 4-tuple are uniquely determined. However, there is no guarantee that condition (10d) is met. We denote by $\mathcal{S}$ the set of Symmetric Periodic Positive Semidefinite (SPPS) matrices $\Sigma(\cdot)$ such that (10a–d) hold. A significant characterization of such a set is given by the following lemma.

Lemma 2 (Periodic Positive Real Lemma) *Let $\gamma(t, \tau)$ be any function such that*

- $\gamma(t, \tau) = \gamma(\tau, t)'$
- $\gamma(t, \tau) = \gamma(t + T, \tau + T)$
- *there exists a triple* $(F(\cdot), \tilde{G}(\cdot), H(\cdot))$ *satisfying*

$$\gamma(t, \tau) = H(t)\Psi_F(t, \tau + 1)\tilde{G}(\tau), \ t > \tau.$$

Then $\gamma(t, \tau)$ *is the autocovariance function of a CS process if and only if the associated set* $\mathcal{S}$ *is not empty.*

The "if" part of the lemma is easily demonstrated in view of the previous discussion. As for the "only if" part, it can be obtained by specializing to the periodic case the Positive Real Lemma for nonstationary processes (see e.g., [23]). The lemma owes its name to the fact that, for a function to be an autocovariance, an associated operator must satisfy a nonnegativity condition (in the stationary case, for instance, the spectrum must be nonnegative real).

The computation of an element $\Sigma(\cdot)$ of $\mathcal{S}$ can be performed by using the following algorithm based on the periodic Riccati equation

$$\begin{aligned}\Sigma(t+1) = {} & F(t)\Sigma(t)F(t)' \\ & +[\tilde{G}(t) - F(t)\Sigma(t)H(t)'][\Gamma(t) - H(t)\Sigma(t)H(t)']^{\dagger} \\ & \times[\tilde{G}(t) - F(t)\Sigma(t)H(t)']',\end{aligned} \tag{11}$$

where the symbol † denotes the Moore-Penrose pseudoinverse.

Algorithm *Consider the solution of (11) with initial condition* $\Sigma(\tau) = 0$. *As* $\tau \to -\infty$, *such a solution tends to a periodic equilibrium denoted by* $\Sigma_*(t)$. *It can be proven that* $\Sigma_*(\cdot)$ *is the minimal element of set* $\mathcal{S}$ *(here minimality makes reference to the usual ordering of positive semidefinite matrices; in other words,* $\Sigma(t) - \Sigma_*(t) \geq 0$, *for every* $\Sigma(\cdot) \in \mathcal{S}$).

This procedure can be seen to be a special case of the general algorithm provided in [23] for the time-varying case. As a matter of fact, again following the ideas of [23], it can be shown that the set $\mathcal{S}$ is convex and that all the SPPS solutions of (11) are extremal points for such a set.

To make things simpler, it is hereafter assumed that $R(t) > 0 \ \forall t$ (regularity condition), so that (11) can be written as follows:

$$\begin{aligned}\Sigma(t+1) = {} & F(t)\Sigma(t)F(t)' \\ & +[\tilde{G}(t) - F(t)\Sigma(t)H(t)'][\Gamma(t) - H(t)\Sigma(t)H(t)']^{-1} \\ & \times[\tilde{G}(t) - F(t)\Sigma(t)H(t)']'.\end{aligned} \tag{12}$$

3.4 Kalman Predictor

The classical linear prediction problem for a stochastic process $y(\cdot)$ consists of finding a linear functional of past data yielding optimal estimates of the present and future values of the process. When optimality is in the mean-square sense and the process is given a Markovian representation, the natural framework for the solution of the prediction problem is provided by Kalman filtering theory [24, 25]. Precisely, denoting by $x(t+1|t)$, the optimal one-step-ahead prediction of the state $x(t)$ of system (5) based on $\{y(j), j \leq t\}$, the optimal predictor is

$$x(t+1|t) = F(t)x(t|t-1) + K(t)\eta(t) \quad (13a)$$

$$y(t|t-1) = H(t)x(t|t-1) \quad (13b)$$

$$\eta(t) = y(t) - y(t|t-1), \quad (13c)$$

with

$$x(\tau|\tau-1) = \bar{x}_\tau.$$

The Kalman gain matrix $K(t)$ appearing in (13a) is given by

$$K(t) = [G(t)S(t) + F(t)P(t)H(t)'][H(t)P(t)H(t)' + R(t)]^{-1} \quad (14)$$

where $P(t)$ is the solution of the Riccati equation with periodic coefficients

$$\begin{aligned} P(t+1) = {} & F(t)P(t)F(t)' + G(t)Q(t)G(t)' \\ & -[G(t)S(t) + F(t)P(t)H(t)'][H(t)P(t)H(t)' + R(t)]^{-1} \\ & \times[G(t)S(t) + F(t)P(t)H(t)']', \end{aligned} \quad (15)$$

with initial condition $P(\tau) = \Sigma_\tau =$ Var$[x(\tau)]$. Matrix $P(t)$ is the covariance matrix of the state prediction error $x(t) - x(t|t-1)$. Therefore, it is not surprising that, if $\Sigma_\tau \geq 0$, then $P(t) \geq 0\ \forall t$.

Using (13b) and (13c) in (13a), it turns out that

$$x(t+1|t) = [F(t) - K(t)H(t)]x(t|t-1) + K(t)y(t).$$

Hence, the stability of the predictor is determined by the matrix $\hat{A}(t) = F(t) - K(t)H(t)$. In particular, when $K(\cdot)$ is T-periodic, the predictor is stable iff all the characteristic multipliers of $\hat{A}(\cdot)$ lie inside the unit circle.

Equation (15) can also be given the equivalent expression

$$\begin{aligned} P(t+1) = {} & A(t)P(t)A(t)' + B(t)B(t)' \\ & -A(t)P(t)H(t)'[H(t)P(t)H(t)' + R(t)]^{-1}H(t)P(t)A(t)', \end{aligned} \quad (16)$$

where

$$A(t) = F(t) - G(t)S(t)R(t)^{-1}H(t) \quad (17a)$$

$$B(t)B(t)' = G(t)[Q(t) - S(t)R(t)^{-1}S(t)']G(t)'. \quad (17b)$$

This alternative formulation can be associated with the prediction problem relative to the stochastic system

$$x(t+1) = A(t)x(t) + B(t)\tilde{v}(t) \quad (18a)$$

$$y(t) = H(t)x(t) + \tilde{w}(t), \quad (18b)$$

where $\tilde{v}(t)$ and $\tilde{w}(t) = w(t)$ are *independent* white noises with zero mean and covariance matrices given by I and $R(t)$, respectively.

The asymptotic behavior of the solutions of a periodic Riccati equation can include a variety of cases, such as asymptotic convergence to a periodic equilibrium of period T, asymptotic convergence to a periodic equilibrium of period kT or convergence to an almost-periodic solution (see [26] for the continuous-time case). Our attention is focused on the first case. Note that, if, as in the present case, the initial condition $P(\tau) = \Sigma_\tau$ is symmetric positive semidefinite, then, whenever the solution converges to a periodic equilibrium, it actually converges to a SPPS solution. Then, the gain tends to a periodic matrix $K(\cdot)$ and the prediction error $\eta(\cdot)$ defined by (13c) converges to a white noise with periodic variance $H(\cdot)P(\cdot)H(\cdot)' + R(\cdot)$. Correspondingly, if in (13) $\eta(t)$ is regarded as a stochastic input and $y(t)$ as the output, we obtain

$$x(t+1|t) = F(t)x(t|t-1) + K(t)\eta(t) \tag{19a}$$

$$y(t) = H(t)x(t|t-1) + \eta(t), \tag{19b}$$

$$\eta(\cdot) \sim \text{WN}(0, H(\cdot)P(\cdot)H(\cdot)' + R(\cdot)), \tag{19c}$$

which can be seen to be an alternative Markovian model of the CS process $y(\cdot)$. (WN $(\cdot,\cdot)$ denotes white noise with mean and covariance specified by the first and second arguments, respectively.) This representation, which is referred to as the *Prediction Error* (PE) representation, enjoys the distinctive feature of being fed by a p-dimensional noise signal $\eta(\cdot)$, whereas the general Markovian model (5) is fed by a $(m+p)$-dimensional white noise $[v(\cdot)'w(\cdot)']'$. Summing up, any Markovian representation of a CS process can be transformed into a PE-representation, via the solution of the Riccati equation (15).

Before addressing the problem of characterizing the set of PE-representations, it is useful to overview some basic facts concerning the Riccati equation. For more details on the properties of this equation, the interested reader is referred to [27, 28, 29].

3.5 Properties of the Riccati Equation

With reference to the predictor dynamic matrix $\hat{A}(t) = F(t) - K(t)H(t)$, where $K(t)$ is defined by (14), a symmetric periodic solution $\bar{P}(\cdot)$ of (16) is said to be

- *maximal* if $P(t) \leq \bar{P}(t)\ \forall t$, for all symmetric periodic solutions $P(\cdot)$
- *strong* if the characteristic multipliers of $\hat{A}(\cdot)$ belong to the closed unit disk
- *stabilizing* if the characteristic multipliers of $\hat{A}(\cdot)$ belong to the open unit disk.

In the following theorem, the existence of the maximal, strong, and stabilizing solution is analyzed. The existence conditions are stated in terms of structural properties of system (18).

Theorem 1 ([28, 29, 30])

i. *If $(A(\cdot), H(\cdot))$ is observable, then (16) admits a maximal symmetric periodic solution, which is in fact positive semidefinite and strong.*

ii. *If $A(\cdot)$ is stable and $(A(\cdot), H(\cdot))$ is observable, then (16) admits a unique SPPS solution, which is in fact stabilizing.*

iii. *Assume that $(A(\cdot), H(\cdot))$ is observable. Then, (16) admits a stabilizing solution iff no $(A(\cdot), B(\cdot))$-unreachable characteristic multiplier lies on the unit circle. Moreover, such a stabilizing solution is unique and coincides with the maximal solution.*

It is advisable to reinterpret these results in terms of the original stochastic representation (5). Observe that the stability of $F(\cdot)$ in no way implies the stability of $A(\cdot)$. However, it is not difficult to see that the assumed observability of the pair $(F(\cdot), H(\cdot))$ entails that of $(A(\cdot), H(\cdot))$ too. Therefore, in our context, one can conclude from Theorem 1 that the Riccati equation (16) admits a maximal solution. However, there may exist other SPPS solutions and the maximal solution itself does not necessarily yield a stable predictor. To achieve uniqueness of the SPPS solution and to guarantee predictor stability, additional conditions on the pair $(A(\cdot), B(\cdot))$ are to be met, as stated in points (ii) and (iii) of Theorem 1.

As for the numerical computation of the periodic solutions of the Riccati equation, several approaches have been proposed in the literature, see e.g., [28, 29, 31]. These methods include direct integration, quasi-linearization and time-invariant reformulations.

3.6 Periodic Innovations Representation of CS Processes

Given a Markovian representation (5) of a CS process, one can obtain the alternative PE-representation (19) associated with an SPPS solution of (16). Below, we go even farther by showing that there is a one-to-one correspondence between all symmetric periodic solutions (not necessarily positive semidefinite) of the periodic Riccati equation (16) and a certain class of Markovian representations. Precisely, consider the stochastic system

$$x(t+1) = F(t)x(t) + L(t)\varepsilon(t) \tag{20a}$$
$$y(t) = H(t)x(t) + \varepsilon(t), \tag{20b}$$

where $\varepsilon(t)$ is a zero-mean white noise with periodic covariance matrix $\tilde{R}(t) = \tilde{R}(t+T)$, and $L(t)$ is a T-periodic matrix. If, for some $\tilde{R}(\cdot)$ and $L(\cdot)$, system (20) is a representation of a given CS process, it will be said to be an *innovationslike (IL) representation* of the process. It is immediately seen that the PE-representations (19) are just a subclass within the family of IL-representations. If the zeros of (20) belong to the open unit disk, the representation is termed the *(canonical) innovations representation.*

We have just observed that, for any given SPPS solution $P(\cdot)$ of the Riccati equation (16), the associated PE-representation (19) is an IL-representation. Below, we show that a model of the type (19) is still an IL-representation even if the Kalman gain $K(\cdot)$ is computed from (14) in correspondence to any (not necessarily positive semidefinite) periodic solution P($\cdot$) of the Riccati equation (16). In this respect, the problem of whether $H(t)P(t)H(t)' + R(t)$ still represents a covariance matrix in accordance with expression (19c) arises. In [32] it is shown that, for any symmetric periodic solution $P(\cdot)$ of (16), $H(t)P(t)H(t)' + R(t)$ is indeed positive definite at each time point t.

The next results proven in [19] establish a one-to-one correspondence between all symmetric periodic solutions of (16) and the class of IL-representations.

Theorem 2 [19] *Suppose that $F(\cdot)$ is stable and $(F(\cdot), H(\cdot))$ is observable.*

i. *Consider any symmetric periodic solution $P(\cdot)$ of (16), with the associated gain $K(\cdot)$ defined by (14). Then, letting $L(t) = K(t)$ and $\tilde{R}(t) = H(t)P(t)H(t)' + R(t)$, (20) provides an IL-representation of the CS process $y(\cdot)$ defined by (5). Moreover, different solutions lead to different IL-representations.*

ii. *Assume that (20) is an IL-representation of the CS process generated by (5). Then, there exists a symmetric periodic solution $P(\cdot)$ of (16) such that $\tilde{R}(t) = R(t) + H(t)P(t)H(t)'$ and $L(t) = K(t)$, where $K(t)$ is the associated Kalman gain.*

This theorem tells us that, for system (20) to be an IL-representation, matrix $L(t)$ must coincide with the Kalman gain $K(t)$ associated with a periodic solution of the Riccati equation.

3.7 Canonical Innovations Representation

By adding and subtracting $K(t)y(t)$ to the right-hand side of (20a) and letting $L(t) = K(t)$, the IL-representation (20) can be rewritten as

$$\begin{aligned} x(t+1) &= \hat{A}(t)x(t) + K(t)y(t) \\ y(t) &= H(t)x(t) + \varepsilon(t), \end{aligned}$$

where $\hat{A}(t) = F(t) - K(t)H(t)$ is the closed-loop matrix. Regarding $y(\cdot)$ as input and $\varepsilon(\cdot)$ as output, the following *whitening filter* is obtained:

$$\begin{aligned} x(t+1) &= \hat{A}(t)x(t) + K(t)y(t) \\ \varepsilon(t) &= -H(t)x(t) + y(t). \end{aligned}$$

We now show that the zeros of the IL-representation coincide with the poles of the whitening filter, i.e., with the poles of $\hat{A}(\cdot)$. Indeed, application of the output injection $-K(t)y(t)$ to (20) (with $L(t) = K(t)$) leads to the system

$$x(t+1) = \hat{A}(t)x(t) \tag{21a}$$

$$y(t) = H(t)x(t) + \varepsilon(t). \tag{21b}$$

According to Def. 1, the zeros of system (21) are the complex numbers z such that

$$\det \begin{bmatrix} zI - \Phi_{\hat{A}}(t) & 0 \\ \Omega & \Xi \end{bmatrix} = 0,$$

where Ξ turns out to be a nonsingular matrix and Ω is a matrix we do not consider explicitly. Hence, the zeros of (21) coincide with the characteristic multipliers of $\hat{A}(\cdot)$, which are the poles of the periodic whitening filter. On the other hand, by Lemma 1, the zeros of (21) coincide with those of (20).

In view of the above discussion, it is apparent that an IL-representation is canonical (minimum-phase) if and only if $\hat{A}(\cdot)$ is stable. This is equivalent to saying that the canonical innovations representation exists if and only if the periodic Riccati equation (16) admits a stabilizing solution. In conclusion, by Theorem 1 (iii) a canonical innovations representation exists if and only if the unit circle is free of $(A(\cdot), B(\cdot))$-unreachable characteristic multipliers. Precisely, such a representation is generated by the unique stabilizing periodic solution of the Riccati equation.

3.8 Mid-chapter Conclusions

As already pointed out in the Introduction, CS processes admit at least three alternative representations: (i) jointly periodic autocovariance functions, (ii) state-space stochastic models, (iii) Autoregressive Moving Average models with Periodic coefficients (PARMA). So far, we have analyzed the connections between (i) and (ii). In particular, it has been shown that a stable state-space model fed by white noise generates a process whose covariance is jointly periodic, and explicit formulas for such a covariance have been given ((i)$\leftarrow$(ii)). The inverse problem of finding a state-space model yielding a prescribed autocovariance ((i)$\rightarrow$(ii)), is more involved. First of all, the autocovariance has to satisfy certain "rationality" requirements. When these are fulfilled, the realization algorithm consists of two nontrivial tasks: the solution of a deterministic periodic realization problem and the computation of the periodic solution of the realization Riccati equation. This procedure yields one Markovian representation of the autocovariance. However, not all Markovian representations are of equal value: innovationslike representations, for instance, are more parsimonious than others. Rather interestingly, once an arbitrary Markovian representation is available, all and only innovationslike representations are obtained from the periodic solutions of the prediction Riccati equation. In particular, the stabilizing solution, if any, provides the (unique) minimum-phase innovations model.

Two different Riccati equations, (12) and (16), are involved in the derivation of a state-space model from the autocovariance function and in the search for the innovationslike representations, respectively. It is perhaps unsurprising that a strict connection exists between the periodic solutions of the two Riccati equations. Indeed, it is a matter of mere computation to show that, in view of (10a)–(10c), $\widetilde{\Sigma}(\cdot)$ is a periodic solution of (12) if and only if $P(\cdot) = \Sigma(\cdot) - \widetilde{\Sigma}(\cdot)$ satisfies (16), where $\Sigma(\cdot)$

is the periodic solution of the Lyapunov equation (7). Thus, given $\Sigma(\cdot)$, the minimal solution of (16) leads to the maximal solution of (12) and vice versa.

4 PARMA REPRESENTATION OF CYCLOSTATIONARY PROCESSES

A typical input-output representation of a CS process is provided by Autoregressive Moving Average models with Periodic coefficients (PARMA):

$$y(t) = \sum_{i=1}^{q(t)} A_i(t-i)y(t-i) + \sum_{i=1}^{r(t)} C_i(t-i)e(t-i) + e(t)$$

where $y(t), e(t) \in \mathbf{R}^p$, $e(\cdot)$ is a white Gaussian noise with $\mathrm{E}[e(t)] = \bar{e}(t)$, $\mathrm{Var}[e(t)] = R(t)$, and $q(\cdot)$, $r(\cdot)$, $A_i(\cdot)$, $C_i(\cdot)$, $\bar{e}(\cdot)$, and $R(\cdot)$ are periodic with period T. Letting

$$s = \max(\max_t q(t), \max_t r(t))$$

be the *order* of the PARMA model, and putting $A_i(t-i) = 0, i > q(t), C_i(t-i) = 0, i > r(t)$, the above model can be equivalently written as

$$y(t) = \sum_{i=1}^{s} A_i(t-i)y(t-i) + \sum_{i=1}^{s} C_i(t-i)e(t-i) + e(t). \tag{22}$$

The theory of PARMA models has its roots in the early contributions [33, 34] on PAR (Periodic Autoregressive) models, and has found many applications in connection with time-series prediction and identification. Some examples of real data analysis can be found in [35, 36, 37].

The purpose of this section is to discuss the connections between periodic state-space models and PARMA representations ((ii)↔(iii) in Fig. 1), completing the picture concerning the possible characterizations of CS processes.

4.1 From PARMA to State-Space Model

The first problem we address consists of finding a state-space representation of a given PARMA model (22). It is easily verified that the system

$$x(t+1) = F(t)x(t) + G(t)e(t), \quad x(t) \in \mathbf{R}^{ps} \tag{23a}$$
$$y(t) = H(t)x(t) + e(t), \tag{23b}$$

where

$$F(t) = \begin{bmatrix} 0 & 0 & \dots & 0 & A_s(t) \\ I & 0 & \dots & 0 & A_{s-1}(t) \\ 0 & I & \dots & 0 & A_{s-2}(t) \\ \vdots & & & & \vdots \\ 0 & 0 & \dots & I & A_1(t) \end{bmatrix}, \quad G(t) = \begin{bmatrix} A_s(t) + C_s(t) \\ A_{s-1}(t) + C_{s-1}(t) \\ A_{s-2}(t) + C_{s-2}(t) \\ \vdots \\ A_1(t) + C_1(t) \end{bmatrix}, \tag{23c}$$

$$H(t) = [\ 0 \ \ 0 \ \ \dots \ \ 0 \ \ I\ ], \tag{23d}$$

is equivalent to model (22) in that the input-output relationship is the same. Interestingly enough, independently of the PARMA coefficients, the pair $(F(\cdot), H(\cdot))$ is observable (just apply the observability criterion). On the contrary, there is no guarantee that $(F(\cdot), G(\cdot))$ is reachable. Hence, such a state-space representation is not necessarily minimal, i.e., there might exist equivalent Markovian representations with a number of state variables lower than ps.

4.2 Poles, Zeros, and Minimum Phase

A simple way of defining the poles and zeros of the PARMA model is to identify them with those of the associated state-space model (23). The poles of the PARMA are then the characteristic multipliers of $F(\cdot)$. If all the poles lie inside the unit disk, then the process generated by the PARMA model asymptotically tends to a CS process. As for the zeros, observe that by Lemma 1 the zeros of (23) coincide with those of the system:

$$\begin{aligned} x(t+1) &= A(t)x(t) \\ y(t) &= H(t)x(t) + e(t), \end{aligned}$$

with

$$A(t) = F(t) - G(t)H(t) = \begin{bmatrix} 0 & 0 & \dots & 0 & -C_s(t) \\ I & 0 & \dots & 0 & -C_{s-1}(t) \\ 0 & I & \dots & 0 & -C_{s-2}(t) \\ \vdots & & & & \vdots \\ 0 & 0 & \dots & I & -C_1(t) \end{bmatrix}.$$

By a rationale similar to that employed in Section 3.7, it is seen that the zeros of the above system are just the characteristic multipliers of $A(\cdot)$, which will be hereafter referred to as zeros of the PARMA model. In particular, we say that the PARMA model is *minimum phase* if both poles and zeros belong to the open unit disk.

4.3 From State-Space to PARMA Model

As for the transformation from a given Markovian representation of a CS process to an equivalent PARMA model, we focus, for simplicity, on the zero-mean univariate case ($\bar{e}(t) = 0\ \forall t,\ p = 1$). Consider the system

$$\begin{aligned} x(t+1) &= F(t)x(t) + G(t)e(t), \quad && x(t) \in \mathbf{R}^n \\ y(t) &= H(t)x(t) + e(t), && y(t) \in \mathbf{R}^1 \end{aligned}$$

with $F(\cdot)$ stable and $(F(\cdot), H(\cdot))$ observable. Denote by k the minimum integer such that $(F(\cdot), H(\cdot))$ is k-step observable ($n \le k \le nT$, in view of the observability criterion). Now, two cases are possible. If $k = n$, it can be shown (see [38, 39]) that there exists a change of basis in the state-space leading to a representation of the form

$$\begin{aligned} \tilde{x}(t+1) &= \tilde{F}(t)\tilde{x}(t) + \tilde{G}(t)e(t), \qquad && \tilde{x}(t) \in \mathbf{R}^n \\ y(t) &= \tilde{H}(t)\tilde{x}(t) + e(t), \end{aligned}$$

with

$$\tilde{F}(t) = \begin{bmatrix} 0 & 0 & \dots & 0 & f_n(t) \\ 1 & 0 & \dots & 0 & f_{n-1}(t) \\ 0 & 1 & \dots & 0 & f_{n-2}(t) \\ \vdots & & & & \vdots \\ 0 & 0 & \dots & 1 & f_1(t) \end{bmatrix}, \quad \tilde{G}(t) = \begin{bmatrix} g_n(t) \\ g_{n-1}(t) \\ g_{n-2}(t) \\ \vdots \\ g_1(t) \end{bmatrix},$$

$$\tilde{H}(t) = [\ 0 \ \ 0 \ \ \dots \ \ 0 \ \ 1\ ].$$

This is known as the *Observer Canonical Form* (OCF). Then, an equivalent univariate PARMA model of order $s = n$ is easily obtained by letting $A_i(t) = f_i(t)$ and $C_i(t) = g_i(t) - f_i(t)$ (compare with (23c)).

If $k > n$, it can be shown that the system does not admit any algebraically equivalent OCF representation. However, it is still possible to find a system of order k in OCF which preserves the input-output behavior [39]. Then, one can obtain a PARMA model of order $s = k$, as above.

In conclusion, an nth-order Markovian representation of a univariate CS process can always be transformed into a PARMA model, but the order of the PARMA model can be greater than n.

4.4 Prediction for PARMA Models

The state-space theory for CS processes provides valuable insights in connection with prediction and identification of PARMA models. For instance, the solution of the prediction problem is better understood if reference is made to the notion of *canonical* PARMA representations, to which the first part of the subsection is devoted. For the sake of simplicity only univariate ($p = 1$) PARMA models are considered (symbols $a_i(\cdot)$ and $c_i(\cdot)$ are used in place of $A_i(\cdot)$ and $C_i(\cdot)$ to denote the scalar coefficients of the PARMA model).

4.4.1 Canonical Representation of PARMA Models

The state-space representation (23) of the original PARMA model (22) coincides with system (5), provided that $F(t)$, $G(t)$, $H(t)$ are defined according to (23c–d). As for the variances $Q(t)$, $R(t)$, and the cross-covariance $S(t)$ of the disturbances $v(\cdot)$ and $w(\cdot)$ appearing in (5), one has to take $Q(t) = S(t) = R(t)$, where $R(t)$ is the variance of $e(t)$. Next, we try to derive the expression for the prediction Riccati equation (16) associated with (23). Matrix $B(t)$ defined in (17b) turns out to be the zero matrix $\forall t$. Therefore, the prediction Riccati equation relative to the state-space representation (23) of the PARMA model is given by

$$\begin{aligned} P(t+1) = {} & A(t)P(t)A(t)' \\ & -A(t)P(t)H(t)'[H(t)P(t)H(t)' + R(t)]^{-1}H(t)P(t)A(t)', \end{aligned} \tag{24}$$

with $A(t) = F(t) - G(t)H(t)$. Notice that, since $B(t) = 0$, the pair $(A(\cdot), B(\cdot))$ is, so to say, "completely unreachable." Conversely, as already pointed out, observability of $(F(\cdot), H(\cdot))$ entails that of $(A(\cdot), H(\cdot))$.

If all the poles of the PARMA model lie inside the unit circle, its output converges to a CS process. However, in analogy with the stationary case, a CS process can be described by several PARMA representations. Under suitable assumptions, one of these representations is minimum-phase.

Theorem 3 (Periodic Spectral Factorization Theorem) *Consider a PARMA model having all its poles inside the unit circle and the associated CS process $y(\cdot)$. Given any symmetric periodic solution $P(\cdot)$ of (24), let $\tilde{c}_i(t) = c_i(t) + \bar{k}_{s-i+1}(t)$, where $\bar{k}_i(t)$ is the i-th entry of the Kalman gain $\bar{K}(t)$:*

$$\bar{K}(t) = A(t)P(t)H(t)'[H(t)P(t)H(t)' + R(t)]^{-1}. \tag{25}$$

Then, the new PARMA model

$$y(t) = \sum_{i=1}^{s} a_i(t-i)y(t-i) + \sum_{i=1}^{s} \tilde{c}_i(t-i)\tilde{e}(t-i) + \tilde{e}(t), \tag{26}$$

where $\tilde{e}(\cdot) \sim WN(0, R(\cdot) + H(\cdot)P(\cdot)H(\cdot)')$, provides an alternative representation of the CS process $y(\cdot)$.

Furthermore, there exists a minimum-phase representation iff the additional assumption is made that no zero of the PARMA model (22) lies on the unit circle. Then, the minimum-phase representation is unique and is obtained in correspondence of the unique stabilizing solution of (24).

To prove the theorem, consider the state-space representation (23) of the PARMA model. In view of Theorem 2, letting $\tilde{R}(t) = R(t) + H(t)P(t)H(t)'$ and $L(t) = K(t)$, system (20) provides an IL-representation of the CS process $y(\cdot)$. Note, that, since $S(t) = R(t)$, from (17a) and (14) it follows that $K(t) = G(t) + \bar{K}(t)$. Recall that (22) is the input-output representation of the triple $(F(\cdot), G(\cdot), H(\cdot))$ given in (23); by comparing the triples $(F(\cdot), G(\cdot), H(\cdot))$ and $(F(\cdot), G(\cdot) + \bar{K}(\cdot), H(\cdot))$, it is easy to see that (26) is in fact the input-output representation of the IL-system (20).

Finally, the zeros of the PARMA model (22) coincide with the characteristic multipliers of $\hat{A}(\cdot) = F(\cdot) - K(\cdot)H(\cdot) = A(\cdot) - \bar{K}(\cdot)H(\cdot)$. Therefore, the canonical minimum-phase representation is obtained in correspondence with the stabilizing solution of the Riccati equation (24). Since $(A(\cdot), H(\cdot))$ is completely observable and $(A(\cdot), B(\cdot))$ completely unreachable, Theorem 1(iii) implies that such a stabilizing solution exists iff $A(\cdot)$ has no unit-modulus characteristic multiplier, i.e., iff the original PARMA model has no zero on the unit circle.

From the above theorem it is clear that in order to compute the minimum-phase PARMA representation one has to find the periodic stabilizing solution of a Riccati equation. However, a different (possibly more efficient) computational scheme, based on the multivariate Rissanen factorization algorithm [40], can be pursued as illustrated below.

First, notice that, in view of Theorem 3, in the derivation of the canonical representation only the MA part of the PARMA model is involved. Hence, without loss of generality, we restrict our attention to the following periodic MA model:

$$y(t) = \sum_{i=1}^{s} c_i(t-i)e(t-i) + e(t). \tag{27}$$

Now, order the parameters of (27) in a matrix as follows:

$$C = \begin{bmatrix} 1 & 0 & \dots & 0 \\ c_1(0) & 1 & & 0 \\ \vdots & \vdots & & \vdots \\ c_{s-1}(0) & c_{s-2}(1) & & 1 \\ c_s(0) & c_{s-1}(1) & & c_1(T-1) \\ 0 & c_s(1) & & c_2(T-1) \\ \vdots & \vdots & & \vdots \\ 0 & 0 & & c_s(T-1) \\ 0 & 0 & \dots & 0 \end{bmatrix}, \qquad C \in \mathbf{R}^{(r+1)T\times T},$$

where r is the minimum integer greater than or equal to s/T. Furthermore, partition C into matrices of dimension $T \times T : C = [C_0' C_1' \dots C_r']'$, where $C_i \in \mathbf{R}^{T\times T}$. Note that C_0, C_{r-1}, and C_r have zero entries in some predetermined locations (for instance C_0 is lower triangular). Now, define the matrix polynomial $C(q) = C_0 + C_1 q^{-1} + \dots + C_r q^{-r}$, where q^{-1} is the unit-delay operator. Then the univariate periodic relationship (27) can be reformulated as a multivariate stationary MA model:

$$y_0(k) = C(q)e_0(k), \tag{28}$$

where

$$y_0(k) = [y(kT)\; y(kT+1) \dots y((k+1)T-1)]',$$
$$e_0(k) = [e(kT)\; e(kT+1) \dots e((k+1)T-1)]'.$$

The covariance matrix of $e_0(k)$ is given by $R_e \stackrel{\text{def}}{=} \text{Var}[e_0(k)] = \text{diag}(R(0), R(1), \dots, R(T-1))$, where $R(\cdot)$ is the periodic (scalar) variance of $e(t)$. Moreover, it can be shown that the zeros of the (invariant) MA model (28) (i.e., the complex values q such that $\det C(q) = 0$) coincide with the zeros of the (periodic) MA (27).

Now, assume that (27), and therefore (28), is not minimum phase. Then, the Rissanen projection algorithm allows the computation of a Hurwitz polynomial matrix $\bar{C}(q) = I + \bar{C}_1 q^{-1} + \dots + \bar{C}_r q^{-r}$ and a covariance matrix $\bar{R}_e$ such that

$$C(q)R_e C(q^{-1})' = \bar{C}(q)\bar{R}_e\bar{C}(q^{-1})'.$$

As far as the computation of the canonical factor $\bar{C}(q)$ is concerned, the actual value of R_e is immaterial so that an arbitrary diagonal positive matrix can be chosen, e.g., $R_e = I$. Matrix $\bar{R}_e$ provided by the Rissanen algorithm is not necessarily diagonal. Hence, the multivariate MA model defined by $\bar{C}(q)$ and $\bar{R}_e$ cannot be directly interpreted as a periodic MA model. To cope with this problem, consider the factorization $\bar{R}_e = \hat{C}_0\hat{R}_e\hat{C}_0'$, where $\hat{C}_0$ is lower triangular with unit-entries on the main diagonal and $\hat{R}_e$ is diagonal. This is just a standard UDU factorization. Then, $\hat{C}(q) = \bar{C}(q)\hat{C}_0$ is seen to satisfy all the requirements in order to be reinterpreted in terms of a periodic MA model and provides the desired canonical factor. The parameters $\hat{c}_i(t)$ of the canonical periodic representation are finally obtained from the entries of the matrix $\hat{C} = [\hat{C}_0'\hat{C}_1' \dots \hat{C}_r]'$.

4.4.2 Optimal One-Step-Ahead Prediction for PARMA Models

The solution of the optimal prediction problem for PARMA models is obtained by applying Kalman prediction theory to the state-space representation of the PARMA model. Then the obtained state-space predictor is more conveniently reformulated in input-output form. Along these lines it is not difficult to derive the following results, the proofs of which can be found in [19].

Theorem 4 *Consider the PARMA model given in (22), and assume that it has no unit-modulus zero.*

i. *If the PARMA model is minimum phase, the optimal periodic predictor is*

$$\begin{aligned} y(t|t-1) = &- \sum_{i=1}^{s} c_i(t-i)y(t-i|t-i-1) \\ &+ \sum_{i=1}^{s} [c_i(t-i) + a_i(t-i)]y(t-i). \end{aligned} \tag{29}$$

Moreover, such a predictor is stable.

ii. *If the PARMA model is nonminimum phase, the stable optimal periodic predictor is*

$$\begin{aligned} y(t|t-1) = &- \sum_{i=1}^{s} \tilde{c}_i(t-i)y(t-i|t-i-1) \\ &+ \sum_{i=1}^{s} [\tilde{c}_i(t-i) + a_i(t-i)]y(t-i), \end{aligned} \tag{30}$$

where $\tilde{c}_i(t) = c_i(t) + \bar{k}_{s-i+1}(t)$, $\bar{k}_i(t)$ being the i-th entry of the Kalman gain $\bar{K}(t)$ associated with the stabilizing solution of (24).

In view of the Periodic Spectral Factorization Theorem, it is apparent that solving the prediction problem for a stable nonminimum-phase PARMA model is equivalent to finding a minimum-phase PARMA realization of the associated CS process $y(\cdot)$ and then applying (29) to such a canonical PARMA model.

4.5 PARMA Identification

Identification of PARMA models from time-series data has been studied from different viewpoints. PAR models do not pose particular difficulties since they can be identified by solving a set of linear equations that generalize the well-known Yule-Walker equations, see [34]. A maximum-likelihood approach for the estimation of PARMA models was proposed in [37]. In this subsection, we present a prediction-error algorithm for the identification of the PARMA model (22) with $p = 1$, starting from a set of data $Y = \{y(1), y(2), \ldots, y(N)\}$. It is assumed that the data have been pretreated in order to obtain a zero-mean realization of a CS process. For instance, a nonzero periodic mean could have been removed by means of standard deseasoning techniques.

Denoting with $a_i(\cdot)$ and $c_i(\cdot)$ the scalar coefficients of the PARMA model, the identification problem consists of computing an estimate of the parameter vector:

$$\theta = [\alpha_1\, \alpha_2\, \ldots\, \alpha_s\, \gamma_1\, \gamma_2\, \ldots\, \gamma_s]',$$

where

$$\alpha_i = [a_i(0)\; a_i(1) \ldots a_i(T-1)], \qquad i = 1, 2, \ldots, s,$$

$$\gamma_i = [c_i(0)\; c_i(1) \ldots c_i(T-1)], \qquad i = 1, 2, \ldots, s.$$

For any value of θ corresponding to a minimum-phase PARMA model, consider the optimal prediction rule given in (29). Then, it is easy to verify that the prediction error $\varepsilon(t) = y(t) - y(t|t-1)$ satisfies the following recursion:

$$\varepsilon(t) = -\sum_{i=1}^{s} c_i(t-i)\varepsilon(t-i) + y(t) - \sum_{i=1}^{s} a_i(t-i)y(t-i). \tag{31}$$

According to the general viewpoint of prediction-error identification [41], the estimation of θ is carried out by minimizing with respect to θ the quadratic performance index

$$J(\theta) = \frac{1}{N}\sum_{t=1}^{N} \varepsilon(t)^2.$$

By analogy with the case of stationary ARMA models, the minimization can be performed by means of the following Gauss-Newton iteration:

$$\hat{\theta}^{(h+1)} = \hat{\theta}^{(h)} - \left[\sum_{t=1}^{N} \Psi(t)'\Psi(t)\right]^{-1}_{\theta=\hat{\theta}^{(h)}} \left[\sum_{t=1}^{N} \varepsilon(t)\Psi(t)'\right]_{\theta=\hat{\theta}^{(h)}},$$

where $\hat{\theta}^{(h)}$ represents the parameter estimate at the h-th iteration and

$$\Psi(t) = \frac{\partial \varepsilon(t)}{\partial \theta} = [\beta_1(t)\; \beta_2(t) \ldots \beta_s(t)\; \zeta_1(t)\; \zeta_2(t) \ldots \zeta_s(t)]$$

where

$$\beta_i(t) = [\beta_{i1}(t)\; \beta_{i2}(t) \ldots \beta_{iT}(t)] = \left[\frac{\partial \varepsilon(t)}{\partial a_i(0)} \frac{\partial \varepsilon(t)}{\partial a_i(1)} \cdots \frac{\partial \varepsilon(t)}{\partial a_i(T-1)}\right]$$

$$\zeta_i(t) = [\zeta_{i1}(t)\; \zeta_{i2}(t) \ldots \zeta_{iT}(t)] = \left[\frac{\partial \varepsilon(t)}{\partial c_i(0)} \frac{\partial \varepsilon(t)}{\partial c_i(1)} \cdots \frac{\partial \varepsilon(t)}{\partial c_i(T-1)}\right].$$

As for the elements of vector $\Psi(t)$, in view of (31), they can be recursively computed as

$$\beta_{jk}(t) = -\sum_{i=1}^{s} c_i(t-i)\beta_{jk}(t-i) - \delta_T(t, j+k-1)y(t-j) \qquad (32a)$$

$$\zeta_{jk}(t) = -\sum_{i=1}^{s} c_i(t-i)\zeta_{jk}(t-i) - \delta_T(t, j+k-1)\varepsilon(t-j), \qquad (32b)$$

where $\delta_T(i, j)$ is a T-periodic Kronecker delta, i.e., $\delta_T(i, j) = 1$ for $i = j + mT$, m integer, and $\delta_T(i, j) = 0$ otherwise.

In conclusion, at each iteration, recursions (31) and (32) are needed to compute $\varepsilon(t)$ and $\Psi(t)$, $t = 1, \ldots, N$. At each step, the coefficients $a_i(\cdot)$ and $c_i(\cdot)$ involved in these computations are just the entries of the current parameter estimate $\hat{\theta}^{(h)}$.

Observe that such filtering operations are stable if and only if the PARMA model parameterized by $\hat{\theta}^{(h)}$ is minimum phase. Therefore, at each step of the identification algorithm, the minimum-phase property of the currently estimated model must be checked. In the negative case, the current estimate of the PARMA model has to be suitably replaced by the corresponding canonical representation according to the projection techniques discussed in the previous subsection.

An obvious drawback of the outlined procedure is that the number of unknown parameters increases linearly with the period T. For instance, if $T = 12$ and $s = 4$, then 96 independent parameters have to be estimated. In such cases, it might be more convenient to resort to a (low order) Fourier series expansion of the coefficients $a_i(t)$ and $c_i(t)$. In such a case, the minimization algorithm can be easily extended as shown in [42]. Unfortunately, there is no obvious way of adapting the Rissanen projection technique to such a Fourier parameterization. A simple heuristic procedure to overcome this problem is provided in [42].

5 CONCLUSIONS

A comprehensive survey of state-space theory of cyclostationary processes is presented in this chapter. The assumption of Gaussian processes was made so that first- and second-order moments suffice to completely characterize a CS process. However, most results straightforwardly extend to the case of non-Gaussian (wide-sense) cyclostationary processes. A point to be noted is that in the non-Gaussian case the Kalman predictor is optimal *only within the class of linear prediction rules*.

Attention has been restricted to discrete-time CS processes having integer period. In particular this corresponds to the case of continuous-time CS processes sampled with an integer number of samples per period. The extension of state-space theory to more general models encompassing noninteger periods or multiple incommensurate periods (polycyclostationary processes) is an open problem.

ACKNOWLEDGMENTS

This work was supported by MURST project *Model Identification, System Control, Signal Processing* and CNR *Centro di Teoria dei Sistemi* and special project *Algorithms and Architectures for Identification and Robust and Adaptive Control.*

REFERENCES

[1] W. A. Gardner, "Characterization of cyclostationary random signal processes," *IEEE Trans. on Information Theory*, vol. IT-21, no. 1, pp. 4–14, 1975.

[2] *IEEE Signal Processing Magazine*, Special Issue on Cyclostationary Signals, W. A. Gardner, ed., April 1991.

[3] *Proc. of Workshop on Cyclostationary Signals*, W. A. Gardner, ed., Yountville CA, August 16–18, 1992.

[4] W. A. Gardner, *Statistical Spectral Analysis: A Nonprobabilistic Theory*, Englewood Cliffs, NJ: Prentice-Hall, 1987.

[5] J. Rootenberg and S. A. Ghozati, "Stability properties of periodic filters," *Int. J. Systems Sci.*, vol. 8, pp. 953–959, 1977.

[6] J. Rootenberg and S. A. Ghozati, "Generation of a class of non-stationary random processes," *Int. J. Systems Sci.*, vol. 9, pp. 935-947, 1978.

[7] S. Bittanti, "The periodic prediction problem for cyclostationary processes—an introduction," *Proc. NATO Adv. Res. Workshop*, Groningen (NL) 1986, pp. 239–249, 1987.

[8] S. Bittanti, "Deterministic and stochastic linear periodic systems," in *Time Series and Linear Systems*, S. Bittanti ed., Berlin: Springer-Verlag, LNCIS Series, vol. 86, pp. 141–182, 1986.

[9] S. Bittanti and G. De Nicolao, "Markovian representations of cyclostationary processes," in *Topics in Stochastic Systems: Modelling, Estimation and Adaptive Control*, L. Gerencsér and P. E. Caines, eds., Berlin: Springer-Verlag, LNCIS Series, vol. 161, pp. 31–46, 1991.

[10] S. Bittanti, F. Lorito, and S. Strada, "An LQG disturbance modeling approach to active control of vibrations in helicopters," *Proc. 19th European Rotorcraft Forum*, Cernobbio, Italy, vol. 2, 914 1–7,1993.

[11] R. E. Kalman, P. L. Falb, and M. A. Arbib, *Topics in Mathematical System Theory*, New York: McGraw-Hill, 1969.

[12] T. Kailath, *Linear Systems*, Englewood Cliffs, NJ: Prentice-Hall, 1980.

[13] S. Bittanti and P. Bolzern, "Discrete-time linear periodic systems: Gramian and modal criteria for reachability and controllability," *Int. J. Control*, vol. 41, pp. 909–928, 1985.

[14] P. Bolzern, P. Colaneri, and R. Scattolini, "Zeros of discrete-time linear periodic systems," *IEEE Trans. Autom. Control*, vol. AC-31, pp. 1057–1058, 1986.

[15] E. J. Jury and F. J. Mullin, "The analysis of sampled-data control systems with a periodically time-varying sampling rate," *IRE Trans. Autom. Control*, vol. 5, pp. 15-21, 1959.

[16] B. A. Bamieh and J. B. Pearson, "A general framework for linear periodic systems with applications to H^{∞} sampled-data control," *IEEE Trans. Autom. Control*, vol. 37, pp. 418–435, 1992.

[17] H. H. Rosenbrock, *State-Space and Multivariable Theory*, New York: Wiley, 1970.

[18] O. M. Grasselli and S. Longhi, "Zeros and poles of linear periodic multivariable discrete-time systems," *Circuits Systems Signal Processing*, vol. 7, pp. 361–380, 1988.

[19] S. Bittanti and G. De Nicolao, "Spectral factorization of linear periodic systems with application to the optimal prediction of periodic ARMA models," *Automatica*, vol. 29, pp. 517–522, 1993.

[20] F. M. Callier and C. A. Desoer, *Multivariable Feedback Systems*, Berlin: Springer Verlag, 1982.

[21] S. Pinzoni, "Stabilization and control of linear time-varying systems," Ph. D. Dissertation, Arizona State University, December 1989.

[22] P. Colaneri and S. Longhi, "The realization problem for linear periodic systems," *Proc. IFAC World Congress*, Sydney, 1993, vol. 5, pp. 69-76.

[23] P. Faurre, M. Clerget, and F. Germain, *Operateurs Rationnels Positifs*, Paris: Dunod, 1979.

[24] R. E. Kalman, "New methods in Wiener filtering," *Proc. of First Symp. on Eng. Applications of Random Function Theory and Probability* (J. Bogdanoff and F. Kozin, eds.), New York: Wiley, pp. 270-388, 1963.

[25] B. D. O. Anderson and J. B. Moore, *Optimal Filtering*, Englewood Cliffs, NJ: Prentice-Hall, 1979.

[26] M. A. Shayman, "On the phase portrait of the matrix Riccati equation arising from the periodic control problem," *SIAM J. Control and Optimization*, vol. 23, no. 5, pp. 717–751, 1985.

[27] S. Bittanti, P. Colaneri, and G. De Nicolao, "The periodic Riccati equation," *The Riccati Equation* (S. Bittanti, A. J. Laub, J. C. Willems, eds.), Berlin: Springer-Verlag, Communication and Control Series, 1991.

[28] S. Bittanti, P. Colaneri, and G. De Nicolao, "The difference periodic Riccati equation for the periodic prediction problem," *IEEE Trans. Autom. Control*, vol. AC-33, pp. 706–712, 1988.

[29] S. Bittanti, P. Colaneri, and G. De Nicolao, "An algebraic Riccati equation for the discrete-time periodic prediction problem," *Syst. & Control Letters*, vol. 14, pp. 71–78, 1990.

[30] C. E. de Souza, "Existence conditions and properties for the maximal periodic solution of periodic Riccati difference equations," *Int. J. Control*, vol. 50, pp. 731–742, 1989.

[31] B. Lennartson, "Periodic solutions of Riccati equations applied to multirate sampling," *Int. J. of Control*, vol. 48, pp. 1025–1042, 1988.

[32] S. Bittanti and G. De Nicolao, "A note on the periodic Riccati equation and innovations representations of cyclostationary processes," *Proc. 31st Conference on Decision and Control*, Tucson, Arizona, Dec. 1992, pp. 1243–1244.

[33] W. M. Brelsford, "Probability predictions and time series with periodic structure," Ph.D. Dissertation, Johns Hopkins Univ., Baltimore, Maryland, 1967.

[34] M. Pagano, "On periodic and multiple autoregressions," *Ann. Statist.*, vol. 6, pp. 1310–1317, 1978.

[35] E. Parzen and M. Pagano, "An approach to modeling seasonally stationary time-series," *J. Econometrics*, vol. 9, pp. 137–153, 1979.

[36] T. P. Barnett, "Interaction of the monsoon and Pacific trade wind system at interannual time scales. I. The equatorial zone," *Mon. Weather Rev.*, vol. 3, pp. 756–773, 1983.

[37] A. V. Vecchia, "Maximum likelihood estimation for periodic autoregressive moving average models," *Technometrics*, vol. 27, pp. 375–384, 1985.

[38] D. Malah and B. A. Shenoi, "Reduction and transformation of linear discrete-time-varying systems," *Int. J. of Control*, vol. 16, pp. 1127–1136, 1972.

[39] S. Bittanti, P. Bolzern, and G. Guardabassi, "Some citical issues on the state-representation of time-varying ARMA models," *Preprints IFAC Symp. on Ident. and Sys. Par. Est.*, York (UK): pp. 1479–1483, 1985.

[40] J. Rissanen, "Algorithms for triangular decomposition of block Hankel and Toeplitz matrices with application to factoring positive matrix polynomials," *Math. Comput.*, vol. 27, pp. 147–154, 1973.

[41] P. E. Caines, *Linear Stochastic Systems*, New York: Wiley, 1988.

[42] S. Bittanti, P. Bolzern, G. De Nicolao, L. Piroddi, and D. Purassanta, "A minimum prediction error algorithm for estimation of periodic ARMA models,"*Proc. European Control Conference*, Grenoble, France, July 1991, pp. 1200–1203.

Chapter 6

Representation and Estimation for Periodically and Almost Periodically Correlated Random Processes

Dominique Dehay
IRMAR, Campus de Beaulieu
35042 Rennes, France

Harry L. Hurd*
Harry L. Hurd Assoc.
and Center for Stochastic Processes
University of North Carolina
Chapel Hill, NC

1 INTRODUCTION

This chapter contains a review of the spectral theory of *periodically correlated* (PC) and *almost periodically correlated* (APC) stochastic processes, and of results concerning the consistent estimation of the Fourier coefficients of the correlation function and their corresponding densities. We also review process representations and, in particular, the role of groups of unitary operators in the spectral and representation theory for these processes. Finally, some topics for future research are given.

The theory of PC and APC processes begun by Gladyshev [37, 38] has developed along probabilistic lines and the emphasis has been on the underlying structure, but significant consideration has been given to applications, such as spectral estimation.

*Sponsored by ONR contract N00014-92-C-0057.

A few years prior to Gladyshev's initial work, Bennett [4] introduced *cyclostationary* (synonymous with *periodically correlated*) processes in the context of communications theory. The theory of *cyclostationary* (CS) and *almost cyclostationary* (ACS) processes that evolved from this beginning has emphasized applications but significant consideration has been given to the underlying structure.

The CS theory includes probabilistic models as well as nonprobabilistic notions of CS functions or time-series [29] that may be understood as an extension and generalization of Wiener's generalized harmonic analysis [84]. Some of the results obtained by these two paths of development are similar, particularly those concerning estimation of the coefficient functions (cyclic correlations) and their (spectral correlation) densities, which are discussed later in Section 5 (see [31] for a clear description of these similarities). The similarities are due to the fact that in both cases, the coefficient functions are related to spectral densities through Fourier transforms and because similar estimation formulas are used. On the other hand, the issues of difference appear mainly to be structural. One deals with random processes and defines correlation through expectation while the other deals with nonrandom functions belonging to a certain class and defines correlation through time averages. The understanding of both views is of value and the existence of this volume will certainly facilitate this understanding.

In this overview, we primarily use the terms PC and APC.

A random process $X(t)$ taking values in the L_2 random variables of a probability space $(\Omega, \mathbf{F}, P)$ and indexed on $\mathbf{I} = \mathbf{Z}$ or $\mathbf{R}$ is called periodically correlated or PC if there exists a smallest $T > 0$ such that

$$\mu(t) = E\{X(t)\} = \mu(t+T) \tag{1}$$

$$R(t_1, t_2) = E\{[X(t_1) - \mu(t_1)]\overline{[X(t_2) - \mu(t_2)]}\} = R(t_1 + T, t_2 + T) \tag{2}$$

for every t, t_1, and t_2 in $\mathbf{I}$ ([37, 38]). For discrete time ($\mathbf{I} = \mathbf{Z}$) it is required that $T > 1$, otherwise the process is stationary. For continuous time ($\mathbf{I} = \mathbf{R}$) Gladyshev also requires continuity of the correlation function.

Similarly, a process $X: \mathbf{I} \to L_2(\Omega)$ is called almost periodically correlated if

$$\mu(t) = E\{X(t)\}$$

and

$$R(t_1 + t, t_2 + t) = E\{[X(t_1 + t) - \mu(t_1 + t)]\overline{[X(t_2 + t) - \mu(t_2 + t)]}\} \tag{3}$$

are almost periodic functions with respect to the variable t (in the sense of Bohr) for every t_1 and t_2 in $\mathbf{I}$ [37]. In this case, Gladyshev treats only processes indexed on $\mathbf{R}$ and requires the covariance to be uniformly continuous.

The theory of PC and APC processes reviewed in this chapter is substantially based on the theory of weakly stationary processes. The spectral theory for stationary processes began with the spectral theory for the covariance, based on the work of Khintchine [65] and Bochner [6]. This was followed by the spectral representation of the process, based on the work of Cramér [12] and Kolmogorov [66] (who uses a theorem of Stone [83] on the spectral representation of groups of unitary operators). Early contributors to the spectral theory for multidimensional stationary processes were Cramér [12], Kolmogorov [66], Rozanov [79], Wiener and Masani [85]. A particularly interesting fact about PC processes is that they can be viewed as stationary processes that take values in a larger space. For example, second-order PC sequences of period T are exactly the same (there is a bijection) as T-dimensional stationary vector sequences. It is therefore of particular interest that Gladyshev worked on multivariate stationary processes prior to his work on PC processes [36]. In the case of continuous-time PC processes, the larger space may need to be infinite dimensional, thus requiring the theory for stationary Hilbert-space valued random processes (see Gangolli [24] and Kallianpur and Mandrekar [63]).

Finally, a group of researchers in the former USSR, principally Y. P. Dragan and his co-workers, have made substantial progress in many of the same topics. The books by Dragan [21], Dragan and Yavorskii [22] and Dragan, Rozhkov, and Yavorskii [23] (all in Russian) contain a large volume of work that was not reviewed for this survey. The book by Yaglom [86] contains a review of PC processes and gives additional references to Dragan's work.

2 SPECTRAL THEORY FOR THE COVARIANCE

2.1 PC Processes

We begin the spectral theory for the covariance (we assume henceforth that $\mu(t) \equiv 0$) with the observation that the diagonal periodicity or almost periodicity of (2) and (3) can be transformed into periodicity in one variable through a simple transformation. Following Gladyshev [37], we thus define the rotated covariance

$$B(t, \tau) = R(t + \tau, t), \tag{4}$$

and so the property (2) becomes $B(t, \tau) = B(t + T, \tau)$ for every t and $\tau \in \mathbf{I}$. This suggests that for every τ, $B(t, \tau)$ can be expressed as a Fourier series

$$B(t, \tau) \sim \sum_k B_k(\tau) \exp(i2\pi kt/T) \tag{5}$$

where the sense of convergence depends on the smoothness of $B(t, \tau)$ in the variable t, and

$$B_k(\tau) = \frac{1}{T} \int_0^T \exp(-i2\pi kt/T) B(t, \tau) dt. \tag{6}$$

In the discrete-time case, the representation (5) is a pointwise equality and the integral (6) is a sum. In the continuous-time case, since $B(t, \tau)$ is continuous with respect to t for each fixed τ, $B(t, \tau)$ is represented by its Fourier series in the sense that the Cesàro means of the partial sums converge at each t, τ. Under stronger conditions, such as B is Lipschitz in t at (t, τ), the partial sums themselves converge to $B(t, \tau)$.

Part of the general spectral theory for the covariance is thus provided through the Fourier series representation (5) and to address the nature of the coefficient functions $B_k(\tau)$, Gladyshev answered the following question, first for sequences [37], and then for continuous time [38]. Given a continuous function $B(t, \tau)$, what conditions on the coefficient functions $\{B_k(\tau)\}$ are implied by the fact that B arises from a PC covariance R (which is nonnegative definite); and conversely, what conditions on the coefficients imply that B will correspond to a covariance R. This is Gladyshev's answer for continuous time: A sequence of coefficient functions $\{B_k(\tau), k \in \mathbf{Z}, \tau \in \mathbf{R}\}$ arises from some PC process if and only if for every N and every sequence of complex numbers $\{c_1, c_2, \ldots, c_N\}$, real numbers $\{\tau_1, \tau_2, \ldots, \tau_N\}$, and integers $\{k_1, k_2, \ldots, k_N\}$ it follows that

$$\sum_{p=1}^{N}\sum_{q=1}^{N} c_p \bar{c}_q \beta_{k_p k_q}(\tau_p - \tau_q) \geq 0 \tag{7}$$

where $\beta_{jk}(\tau) = B_{k-j}(\tau)\exp(i2\pi j\tau/T)$. We note that Gladyshev's result is in the spirit of Khintchine [65] who obtains the spectral representation of the covariance of a stationary process by using the fact that covariances must be nonnegative definite (followed by an application of Bochner's theorem).

In [47] it is shown that Gladyshev's result also holds for a class of PC processes that need not be continuous in quadratic mean.

The condition (7) is necessary and sufficient (see [12] or [80], p. 20) for each $\beta_{jk}(\tau)$ to be a cross-correlation function belonging to some weakly stationary vector process[1] and hence each $\beta_{jk}(\tau)$ is a Fourier transform of a complex measure and thus the functions $B_k(\tau)$ are also. This remains true in a general sense; if $R(t, t) = B(t, 0) = B(t + T, 0)$ is Lebesgue integrable on $[0, T]$, then the following are equivalent [47]:

(a) $B_0(\tau)$ is continuous at $\tau = 0$;

(b) $B_k(\tau) = \int_{-\infty}^{\infty} \exp(i\lambda\tau) m_k(d\lambda); \int_{-\infty}^{\infty} |m_k(d\lambda)| \leq B_0(0) = \int_{-\infty}^{\infty} m_0(d\lambda)$, (8)

where m_0 is a nonnegative measure on the Borel sets of $\mathbf{R}$.

Recently [20] it has been shown that m_k is absolutely continuous with respect to m_0. The Lebesgue decomposition of the measure m_k implies that there will always be a unique (spectral) density g_k (although it may be null) associated with every m_k. But when and only when m_k is absolutely continuous with respect to Lebesgue measure, may we write

$$B_k(\tau) = \int_{-\infty}^{\infty} \exp(i\lambda\tau) g_k(\lambda) d\lambda. \tag{9}$$

Although (8) can be inverted to obtain m_k (see [69], p. 186), the direct inversion $g_k(\lambda) = \frac{1}{2\pi}\int_{-\infty}^{\infty} B_k(\tau)\exp(-i\lambda\tau)d\tau$ requires that $B_k \in L_1(\mathbf{R})$. For an example of a

---

[1] We will subsequently show how to obtain a stationary vector process having this correlation structure.

PC process that is not continuous in quadratic mean, consider the process formed by amplitude modulation $X(t) = f(t)Y(t)$ where Y is wide sense stationary and $f(t) = f(t+T)$ is $L_2[0, T]$. It is clear that $X(t)$ is PC but not necessarily continuous in quadratic mean, but from (6) we obtain $B_0(\tau) = T^{-1}R_Y(\tau)\int_0^T f(t+\tau)\overline{f(t)}dt$ which is continuous everywhere and so the coefficient functions $B_k(\tau)$ are each Fourier transforms. Further results on the representation theory of such processes and their relation to $L_2[0, T]$-valued stationary sequences are given in [56]. In the preceding example, if we set $f(t) = f(t+T) = 1$ for $t \in [0, T/2)$ and $f(t) = f(t+T) = -1$ for $t \in [T/2, T)$, then from (6) it follows that $B_k(0) = m_k(\mathbf{R}) = 0$ for $k \neq 0$, but yet it is not true that $B_k(\tau) \equiv 0$ or equivalently $m_k \equiv 0$. Evidently for $k \neq 0$, the measure m_k can "oscillate" to produce $m_k(\mathbf{R}) = 0$.

There is an intimate connection between PC processes and processes that can be made stationary by an independent uniform time shift. If the random variable Θ is independent of a q.m. (quadratic mean) continuous second-order process X, then setting $Y(t) = X(t+\Theta)$ leads to

$$\begin{aligned} R_Y(s,t) &= E_\Theta E_X\{[X(s+\Theta) - \mu(s+\Theta)]\overline{[X(t+\Theta) - \mu(t+\Theta)]}\} \\ &= \int R_X(s+\theta, t+\theta)d\mu(\theta) \end{aligned} \tag{10}$$

where μ is the distribution function for Θ. If we add to the general hypothesis of independence that Θ is uniformly distributed over $[0, T]$, then Y is weakly stationary if and only if X is PC with period T [46].

Conditions of independence that imply (10), along with some results for the strict-sense case are given in [48]. Gardner [25, 26] showed that a vector stationary process $\{X_j(t), j \in \mathbf{Z}\}$ whose cross correlations are given by $\beta_{jk}(\tau)$ appearing in (7) can be obtained by

$$X_j(t) = X(t+\Theta)\exp[i2\pi j(t+\Theta)/T]. \tag{11}$$

2.2 APC Processes

Results similar to those above hold for APC processes. Since for every τ the function $B(t,\tau)$ is almost periodic with respect to the variable t, it follows from $E\{|X(t)|^2\} = B(t, 0)$ that APC processes have bounded second moments. Further, for each τ the function B possesses a Fourier series

$$B(t,\tau) \sim \sum_{\alpha\in\alpha_\tau} B_\alpha(\tau)\exp(i\alpha t) \tag{12}$$

whose Fourier coefficients are determined by

$$B_\alpha(\tau) = \lim_{A\to\infty}\frac{1}{2A}\int_{-A}^{A}\exp(-i\alpha t)B(t,\tau)dt \tag{13}$$

and the set $\alpha_\tau = \{\alpha : B_\alpha(\tau) \neq 0\}$ is countable. The sense of representation expressed by (12) is that the frequencies α_τ and the coefficients $B_\alpha(\tau)$ are uniquely determined by $B(t, \tau)$, and that there exists a sequence of trigonometric polynomials whose frequencies are in $\boldsymbol{\alpha}_\tau$ that converges uniformly in t to $B(t, \tau)$. If $\boldsymbol{\alpha}_\tau$ is a finite set, then the representation (12) is equality. In the language of ACS processes and functions, the frequencies $\boldsymbol{\alpha}_\tau$ are called the *cycle* frequencies [27, 28, 29].

The assumed uniform continuity of B (which follows from that of R) implies that $B_\alpha(\tau)$ is continuous with respect to τ and this can be used to prove [51] a conjecture of Gardner [28] that

$$\boldsymbol{\alpha} = \bigcup_\tau \boldsymbol{\alpha}_\tau = \{\alpha : B_\alpha(\tau) \neq 0 \text{ for some } \tau\} \tag{14}$$

is also countable and hence $\boldsymbol{\alpha}_\tau$ appearing in (12) can be replaced with $\boldsymbol{\alpha}$.

Gladyshev [38] also gave a condition for APC processes corresponding to (7), and it also remains true that all the functions $\{B_\alpha(\tau), \alpha \in \boldsymbol{\alpha}\}$ are Fourier transforms of complex measures [51]

$$B_\alpha(\tau) = \int_{-\infty}^{\infty} \exp(i\lambda\tau) m_\alpha(d\lambda); \int_{-\infty}^{\infty} |m_\alpha(d\lambda)| \leq B_0(0) = \int_{-\infty}^{\infty} m_0(d\lambda). \tag{15}$$

As in the PC case, m_α is absolutely continuous with respect to m_0 [20] and

$$B_\alpha(\tau) = \int_{-\infty}^{\infty} \exp(i\lambda\tau) g_\alpha(\lambda) d\lambda. \tag{16}$$

when and only when m_α is absolutely continuous with respect to Lebesgue measure. The remarks made for the PC case also apply to the inversion of (15) and (16).

We note that the frequencies $\boldsymbol{\alpha}$ and the coefficients together have constraints imposed by the requirement that $B(t, \tau)$ must be almost periodic in t for every τ and those imposed by the requirement (Gladyshev's conditions) that R is a covariance. But these conditions still do not prevent the occurrence of other strange conditions, such as the set $\boldsymbol{\alpha}$ containing limit points. This issue is mentioned again subsequently.

Extensions of the ideas of APC processes include the following: First, rather than considering $B(t, \tau)$ as a collection of almost periodic functions, it may be viewed as a function from **R** to the Banach space of bounded continuous functions. Gardner introduces this idea in [27] where he calls them almost cyclostationary (ACS). In order to distinguish them from the (Gladyshev) APC processes, they are also called *uniformly* APC or UAPC [16, 20]; it is also pointed out that UAPC $\subset$ APC and that the two notions coincide when X is harmonizable.

The idea of random time shifts can also be applied to APC processes. Gardner [27] shows that if X is quasi-CS ($\boldsymbol{\alpha}$ has a finite number of elements), the process Y is stationary if $\Theta = \sum \eta_j$ where the $\{\eta_j\}$ are mutually independent, independent of X, and uniformly distributed over the interval $[-\pi/\alpha_j, \pi/\alpha_j]$. The result is also true for almost CS [27] and almost PC processes [51] having countable $\boldsymbol{\alpha}$, but in both cases $\sum 1/\alpha_j^2 < \infty$ ensures $\mathrm{Var}[\Theta] < \infty$. Gardner [27] also points out that the vector stationary process $\{X_j(t)\}$ for PC processes generalizes to almost CS processes by replacing the cycle frequency k/T with α_k and taking $\Theta = \sum \eta_j$ as above.

Finally, the ideas of APC processes can be extended [58] to processes that are not continuous in quadratic mean by using the notion of almost periodic functions in the sense of Stepanov (see [68]), and in another sense of generalization, to APC processes indexed on locally compact abelian groups [59]. In another direction, Gardner [27] also introduces the asymptotically ACS processes, which intuitively are the same as ACS except that they differ by a "transient" part that does not affect the cyclic correlations, which are determined by long-term averages. Isokawa [61] investigated an identification problem for APC and asymptotically APC processes formed by the output of a linear time invariant filter driven by a point process.

There also exist results connected with ergodic theory and the laws of large numbers. Boyles and Gardner [7] introduce the asymptotically mean CS processes, which are nonstationary processes whose covariances contain additive periodic components (an additive trigonometric series), and hence contain the asymptotically ACS processes. They introduce the notion of cycloergodicity, and give both weak and strong types of results for CS sequences. Honda [42] obtains pointwise results for the continuous-time strictly CS processes. Results concerning both weak and strong laws of large numbers are given for continuous-time PC and APC processes in [9]; the strong-law results rely on the random time shift.

3 SPECTRAL THEORY FOR THE HARMONIZABLE CASE

The notion of a harmonizable process can be motivated from the fact that every wide-sense stationary process has an integral spectral representation of the form

$$X(t) = \int_{-\infty}^{\infty} \exp(i\lambda t)\, dZ(\lambda) \tag{17}$$

where the frequency-indexed random process $Z(\lambda)$ has orthogonal or uncorrelated increments and the equality is in $L_2(\Omega)$. This theory is treated in many texts on stationary second-order random processes (or sequences); for example, see [14, 69, 81].

Corresponding to (17) is the Fourier integral representation for the correlation function

$$E\{X(s)\overline{X(t)}\} = R(s-t) = \int_{-\infty}^{\infty} \exp[i\lambda(s-t)]\, dF_Z(\lambda) \tag{18}$$

where the real-valued nonnegative spectral distribution function $F_Z(\lambda)$ is defined by $F_Z(b) - F_Z(a) = E\{|Z(b) - Z(a)|^2\}$ for $a \le b$.

The notion of *harmonizable* (in the sense of Loève [69]) processes preserves the spectral representation (17) but the increments of $Z(\lambda)$ need not be orthogonal. For such processes the correlation (covariance) function is given by a two-dimensional Fourier integral

$$E\{X(s)\overline{X(t)}\} = R(s,t) = \int_{-\infty}^{\infty}\int_{-\infty}^{\infty} \exp(i\lambda_1 s - i\lambda_2 t)\, M_Z(d\lambda_1, d\lambda_2) \tag{19}$$

where the *spectral correlation* measure defined by

$$M_Z([a,b),[c,d)) = E\{[Z(b) - Z(a)]\overline{[Z(d) - Z(c)]}\} \tag{20}$$

is a finite complex measure on $\mathbf{B(R)} \times \mathbf{B(R)}$ for which (i) $M_Z(A, B) = \overline{M_Z(B, A)}$ for every $A, B \in \mathbf{B(R)}$, and (ii) $\sum_{i,j=1}^{n} a_i \bar{a}_j M(A_i, A_j) \geq 0$ for every $n, a_1, a_2, \ldots, a_n \in \mathbf{C}$ and $A_1, A_2, \ldots, A_n \in \mathbf{B(R)}$. We note that the notion of *weakly* harmonizable processes, described in [10, 74] still requires (i) and (ii) but only that M_Z is a *bimeasure*; that is, for every fixed $A \in \mathbf{B(R)}$, the set function $M_Z(A, \cdot)$ is a measure with respect to the second set variable and conversely. Even the weakly harmonizable processes have bounded second moments and are uniformly continuous in quadratic mean. Other interesting concepts of harmonizability may be found in [44, 45].

The exact manner in which the increments of $Z(\lambda)$ are *not* orthogonal gives some information about the nature of the nonstationarity of $X(t)$. In the case when $X(t)$ is *stationary*, the spectral correlation measure is concentrated on the diagonal and is determined completely by $M_Z([a,b),[a,b)) = F_Z(b) - F_Z(a)$. A principal nonstationary example is given by the case of PC and APC processes.

If $X(t)$ is *harmonizable and periodically correlated*, then the increments of the spectral process $Z(\lambda)$ are correlated in a characteristic manner. Specifically, a pair of increments of $Z(\lambda)$, written here as $dZ(\lambda_1)$ and $dZ(\lambda_2)$ have nonzero correlation only when $\lambda_2 = \lambda_1 - 2\pi k/T$. This fact may also be described in terms of the support set of the spectral correlation measure given by (20); a harmonizable $X(t)$ is periodically correlated if and only if the support of M_Z is contained in the union of the diagonal lines $D_k = \{(\lambda_1, \lambda_2) : \lambda_2 = \lambda_1 - 2\pi k/T\}$ [46]. In this case the coefficient functions can be determined from

$$B_k(\tau) = \int\int_{D_k} \exp(i\lambda_1\tau) M_Z(d\lambda_1, d\lambda_2) \tag{21}$$

or, equivalently, the measure m_k can be identified with the restriction of M_Z to the line D_k; this restricted measure is denoted by $M_Z|_{D_k}$. It is possible for harmonizable processes to have support lines that are not of unity slope (see [3]).

If $X(t)$ is *harmonizable and almost periodically correlated*, then the pair of increments $dZ(\lambda_1)$ and $dZ(\lambda_2)$ have nonzero correlation if and only if $\lambda_2 = \lambda_1 - \alpha$ for $\alpha \in \boldsymbol{\alpha}$. In other words, a harmonizable $X(t)$ is APC if and only if M_Z is concentrated on the union of the diagonal lines $D_\alpha = \{(\lambda_1, \lambda_2) : \lambda_2 = \lambda_1 - \alpha\}$ [49]. The coefficient functions can be determined from

$$B_\alpha(\tau) = \int\int_{D_\alpha} \exp(i\lambda_1\tau) M_Z(d\lambda_1, d\lambda_2), \tag{22}$$

and the measure m_α can be identified with $M_Z|_{D_\alpha}$. This identification is also true [20] in a more general sense for weakly harmonizable APC processes having a σ−finite bimeasure [15].

Gladyshev [37] showed that all PC *sequences* are harmonizable by arguing that $B_k(\tau)$ being a Fourier transform implies, through the use of (5), that $B(t, \tau)$ can be represented as a two-dimensional Fourier transform and, therefore, the covariance $R(s, t)$ also can be. The support of $M_Z(d\lambda_1, d\lambda_2)$ is then characteristically constrained to be in the intersection of the $2T - 1$ diagonal lines $\{(\lambda_1, \lambda_2) : \lambda_2 = \lambda_1 - 2\pi k/T, k = -(T-1), \ldots, T-1\}$ with the square $[0, 2\pi) \times [0, 2\pi)$. This support set is illustrated in Fig. 1.

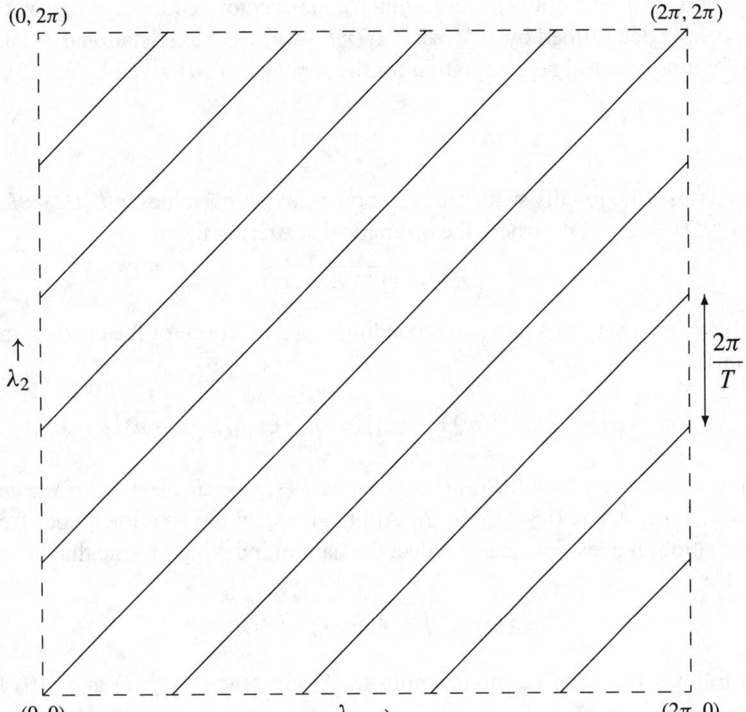

Figure 1: Spectral mass locations for periodically correlated sequences.

Gladyshev also gave an example [38] of a continuous-time PC process that is q.m. continuous and yet not harmonizable. The example is essentially based on the existence of continuous periodic functions whose Fourier coefficients are not (absolutely) summable. Subsequent examples have been given in [20]. Since q.m. continuous PC processes are APC, these same examples show that continuous-time APC processes are not all harmonizable.

Because of the identification of the measures m_k with $M_Z|_{D_k}$ and m_α with $M_Z|_{D_\alpha}$, it follows that PC and APC processes are harmonizable if and only if $\sum_k |m_k|(\mathbf{R}) < \infty$ or $\sum_\alpha |m_\alpha|(\mathbf{R}) < \infty$ [20].

4 PROCESS REPRESENTATIONS

Several representations, usually based in some way on spectral theory, have been developed for PC sequences and processes. We first discuss the harmonizable PC case and then the unitary operator of a PC process and its connection to the representation theory. Finally, we examine the representation issue for APC processes.

4.1 Harmonizable PC Processes

Gladyshev made the important observation that a second-order sequence $X(t)$ is PC with period T if and only if the T-dimensional vector sequence $Y(n)$ whose pth component is determined by $Y^{(p)}(n) = X(nT + p)$ is weakly stationary. There is a corresponding spectral representation for the sequence $Y(n)$ given by

$$Y(n) = \int_0^{2\pi} \exp(i\lambda n)\, d\zeta(\lambda), \tag{23}$$

where ζ is an orthogonally scattered vector measure taking values in $L_2(\Omega) \times L_2(\Omega) \times \ldots \times L_2(\Omega) = L_2(\Omega)^T$, where the orthogonal scattering means

$$E\{\zeta_p(A)\overline{\zeta_q(B)}\} = 0 \tag{24}$$

if $A \bigcap B = \emptyset$. This leads to a corresponding representation for the cross-correlation functions

$$R_{pq}(m-n) = E\{Y^{(p)}(m)\overline{Y^{(q)}(n)}\} = \int_0^{2\pi} \exp[i\lambda(m-n)]\, dF_{pq}(\lambda), \tag{25}$$

and the correspondence with $R(s,t)$ is given by $R(s,t) = R_{pq}(m-n)$ if $s = mT + p$ and $t = nT + q$ where $0 \le p, q < T$. Although useful information comes from this representation, it does not clearly reflect the natural indexing of time that is obtained in

$$X(t) = \int_0^{2\pi} \exp(i\lambda t)\, dZ(\lambda), \tag{26}$$

which follows from the harmonizability of X and rewriting (17) and (19) for sequences.

Gladyshev proceeded further with the representation (26) to show that PC sequences can also be expressed in terms of another T-dimensional stationary vector sequence. To see this, the interval $[0, 2\pi)$ is partitioned into T consecutive subintervals of length $2\pi/T$, so that (26) becomes

$$\begin{aligned} X(t) &= \sum_{k=0}^{T-1} \int_{k2\pi/T}^{(k+1)2\pi/T} \exp(i\lambda t)\, dZ(\lambda) \\ &= \sum_{k=0}^{T-1} \int_0^{2\pi/T} \exp(i\lambda t + i2\pi kt/T) dZ(\lambda + 2\pi k/T) \\ &= \sum_{k=0}^{T-1} \exp(i2\pi kt/T) \int_0^{2\pi/T} \exp(i\lambda t) dZ(\lambda + 2\pi k/T) \\ &= \sum_{k=0}^{T-1} \exp(i2\pi kt/T) a_k(t), \end{aligned} \tag{27}$$

where

$$a_k(t) = \int_0^{2\pi/T} \exp(i\lambda t) dZ(\lambda + 2\pi k/T) = \int_0^{2\pi} \exp(i\gamma t) d\tilde{Z}_k(\gamma) \quad (28)$$

and the measures $\tilde{Z}_k$ are obtained from Z by

$$\tilde{Z}_k([a, b)) = Z([a/T, b/T + 2\pi k/T)).$$

The measures $\tilde{Z}_k$ can now be seen, with the help of Fig. 1, to be mutually orthogonally scattered in the same sense as (24), so finally we see from (28) that $X(t)$ is represented by a sum of products of harmonically spaced complex exponentials with the components of a vector stationary sequence[2]. In the case of sequences, the sense of equality in (28) is that of $L_2(\Omega)$ and the notion of limit is not an issue since the sum is finite. See [37] and [72] for the relationship between the cross-spectral distribution function $F_{pq}(\lambda)$ appearing in (25) and those of the family $\{a_k(t)\}$.

Harmonic series representations

$$X(t) \sim \sum_{k=-\infty}^{\infty} \exp(i2\pi kt/T) a_k(t), \quad (29)$$

have also been obtained for continuous-time PC processes [26, 42, 49, 75] but now there is the additional issue of the sense of convergence of the sum (29). In the case of harmonizable PC processes, the proof follows from the partitioning of the real line into intervals of length $2\pi/T$, as for sequences, along with the fact that the support set for M_Z is contained in $D = \bigcup_k \{(\lambda_1, \lambda_2) : \lambda_2 = \lambda_1 - 2\pi k/T\}$. *The proof of the harmonic series representation for harmonizable PC processes clearly illustrates the close connection between the periodic partition applied to* (17) *and the periodic spacing of the support lines D_k*. The processes $a_k(t)$ whose existence and properties are established by the representation theorems can also be expressed, for harmonizable X, directly in terms of X by [26]

$$a_k(t) = \int_{-\infty}^{\infty} w(t - \sigma) X(\sigma) \exp(-i2\pi k\sigma/T) d\sigma \quad (30)$$

where $w(t) = \sin(\pi t/T)/\pi t$.

It is not clear that a representation (29) in terms of stationary processes holds for APC processes even if they are harmonizable. However, such a representation has been shown [70] for bounded correlation autoregressive sequences (which are a subset of the harmonizable APC sequences with α a finite set).

4.2 The Unitary Operator of PC Processes

An important notion in the theory of wide-sense stationary processes is the naturally associated group of unitary operators. It is, in essence, the notion underlying the

[2] In the terminology of cyclostationary processes, this is a harmonic series representation [26].

spectral representation (17) for stationary processes. We first mention a few pertinent facts about unitary operators and groups of unitary operators, then we discuss their importance to stationary and PC processes (where they naturally occur) and how they can be used to generate an important subclass of the APC processes.

A unitary operator on a Hilbert space H is a continuous linear operator U for which $\langle Ux, Uy\rangle = \langle x, y\rangle$ for every $x, y \in H$; that is, unitary operators preserve inner products. Every unitary operator can be written as an integral (read as a sum) with respect to orthogonal projections; that is

$$U = \int_0^{2\pi} \exp(i\lambda) dE(\lambda), \tag{31}$$

where $E(\lambda)$, $\lambda \in [0, 2\pi)$ is an increasing family of orthogonal projections such that $E(2\pi^-) = I$, the identity operator [82]. A family of unitary operators $\{U(\tau), \tau \in \mathbf{I}\}$, where $\mathbf{I}$ is an index set (here $\mathbf{Z}$ or $\mathbf{R}$), is a group (under the operation of composition) if $U(s) \circ U(t)$ is given by $U(s+t)$ for every $s, t \in \mathbf{I}$ and $U(0)$ is the identity; this is often shortened to $U(s+t) = U(s)U(t)$.

If $\mathbf{I} = \mathbf{Z}$, then any $U(\tau)$ has the representation

$$U(\tau) = U^\tau = \int_0^{2\pi} \exp(i\lambda\tau) dE(\lambda); \tag{32}$$

if $\mathbf{I} = \mathbf{R}$ and the family is strongly continuous, $||[U(h) - U(0)]x|| \to 0$ as $h \to 0$, (see [82], p. 380) then

$$U(\tau) = U^\tau = \int_{-\infty}^{\infty} \exp(i\lambda\tau) dE(\lambda). \tag{33}$$

Beginning only with the view that a second-order wide-sense stationary process $X(t)$ is a curve in a Hilbert space for which $\langle X(s), X(t)\rangle = R(s-t)$, one can demonstrate (see [80], p. 14) the existence of a family of unitary operators $\{U(\tau), \tau \in \mathbf{I}\}$ that form a group under composition and satisfy

$$X(t+\tau) = U(\tau)[X(t)] \tag{34}$$

for every t, τ. Since the group $\{U(\tau), \tau \in \mathbf{I}\}$ through (34) "explains" any time shift of the process, it is often called the shift group. Setting $\tau = 0$ gives $X(t) = U(t)[X(0)]$ for every t, or we see that the entire process (curve) is determined by one random variable (we chose $X(0)$) and the family of shift operators $\{U(\tau), \tau \in \mathbf{I}\}$. Taking $\mathbf{I} = \mathbf{R}$ and using (33) gives the representation (17) where the random measure Z is determined by $Z([a, b)) = E([a, b))X(0)$, the application of the projection $E([a, b))$ to the element $X(0)$.

We now turn to the unitary operator of a PC process (or sequence) [56, 71]. If X is PC with period T we denote $H(X) = \overline{sp}\{X(t), t \in \mathbf{I}\}$ as the Hilbert space generated by X. For vectors of the form $z = \sum_{j=1}^n a_j X(t_j)$ in $sp\{X(t), t \in \mathbf{I}\}$, define the operator U_X by $U_X z = \sum_{j=1}^n a_j X(t_j + T)$. It is easy to show that U_X is linear and, due to (2), is unitary on $sp\{X(t), t \in \mathbf{I}\}$. It follows from the continuity of U_X that it extends to the closure $H(X)$.

As a consequence, we may write

$$X(t+T) = U_X[X(t)] \tag{35}$$

for every $t \in \mathbf{I}$, so U_X is a shift operator for X *but only for shifts of length* T. If X is a PC sequence ($\mathbf{I} = \mathbf{Z}$) and U_X is a shift operator for $T = 1$, then X is stationary. Also in the case of sequences there is a close relationship between U_X and the unitary operator associated with the vector $Y(n)$ appearing in relation (23). Weakly stationary vector sequences have the property that there is a single unitary operator U_Y that acts as the shift operator for each component of Y; that is, $Y^{(p)}(n+1) = U_Y[Y^{(p)}(n)]$ for every p (see [80], p. 14). It follows that $U_Y = U_X$.

In the continuous-time case when X is q.m. continuous, then we can construct a similar Y; indeed, define the infinite dimensional stationary sequence Y by $Y^{(p)}(n) = X(nT + r_p)$ where r_p is the pth rational in the interval $[0, T)$ (with any ordering). Then $U_Y = U_X$ and any $X(t)$ is a limit along some subset of values of $Y^{(p)}(n)$. We have already seen, for the harmonizable case, one representation of continuous-time PC processes in terms of infinite dimensional vector stationary processes, and we shall subsequently see some others.

The existence of U_X leads to another characterization of PC sequences and processes: *X is PC with period T if and only if there exists a group of unitary operators $\{U(\tau), \tau \in \mathbf{I}\}$ and a periodic function (process) $P(t)$ taking values in $L_2(\Omega)$ for which*

$$X(t) = U(t)[P(t)] \tag{36}$$

for every t [56]. If $\mathbf{I} = \mathbf{R}$ and X is q.m. continuous, then $U(\tau)$ is strongly continuous and $P(t)$ is a continuous periodic function.[3] The proof is straightforward. If X is given by (36) then denoting inner product by $\langle \cdot, \cdot \rangle$,

$$\begin{aligned} \langle X(s), X(t) \rangle &= \langle U(s)[P(s)], U(t)[P(t)] \rangle \\ &= \langle U(T)U(s)[P(s+T)], U(T)U(t)[P(t+T)] \rangle \\ &= \langle X(s+T), X(t+T) \rangle \end{aligned} \tag{37}$$

so X is PC. Conversely, if X is PC, the spectral representation of U_X given by (31) permits us to construct a group of unitary operators by

$$U(\tau) = \int_0^{2\pi} \exp(i\lambda\tau/T) dE(\lambda), \tag{38}$$

where $U(T) = U_X$. Therefore $P(t) = U(-t)[X(t)]$ is periodic with period T and so P and the group U satisfy (36).

The expression (36) provides the clear contrast to the stationary case, in which $P(t) = X(0)$, a fixed random variable. We also note that the characterization is not

[3] Even when X is not q.m. continuous, there is a sense in which (36) holds; see [56] for the details.

unique since $\tilde{U}(t) = U(t)\exp(i2\pi t/T)$ and $\tilde{P}(t) = P(t)\exp(-i2\pi t/T)$ will also solve (36). See [56] for additional discussion.

4.3 Representations of X Based on the Unitary Operator

Representations for X follow naturally from (36) by employing representations for U and P. We begin with a representation of the *harmonic series* [26] type. Under various conditions (see [64], p. 51), the partial sums

$$S_N(t) = \sum_{k=-N}^{N} P_k \exp(i2\pi kt/T)$$

converge to $P(t)$ where $P_k = T^{-1}\int_0^T P(t)\exp(i2\pi kt/T)dt$; it follows that

$$X_N(t) = U(t)\left[\sum_{k=-N}^{N} P_k \exp(i2\pi kt/T)\right] = \sum_{k=-N}^{N} a_k(t)\exp(i2\pi kt/T), \quad (39)$$

converges in $L_2(\Omega)$ to $X(t)$. Since $a_k(t) = U(t)[P_k]$ (note that $a_k(0) = U(0)[P_k] = P_k$) given in this manner are jointly wide-sense stationary, harmonic series representations are obtained without the assumption of harmonizability. If $X_N(t)$ were formed from $U(t)$ operating on the Cesàro means of $S_N(t)$, then $X_N(t) \to X(t)$ at every t. The issue of expressing $a_k(t)$ directly in terms of a nonharmonizable X, as in (30), is not completely settled. Some results are given in [60]. The representation (39), with the common group $U(t)$ acting on the terms $P_k \exp(i2\pi kt/T)$, provides additional meaning to the spectral redundancy of cyclostationary processes [32].

Another similar representation is obtained for all q.m. continuous PC processes and the $L_2(\Omega)$ convergence is uniform in t. First write $P(t) = \sum_{k=1}^{\infty}\langle P(t), \xi_k\rangle\xi_k$ where $\{\xi_k\}$ is a complete orthonormal set (CONS) in $H(X)$; such a CONS exists because of the q.m. continuity, and the convergence is unconditional (in a sense, absolute). Since $U(t)$ is unitary for each fixed t it follows from the unconditional convergence that

$$X(t) = U(t)\left[\sum_{k=-\infty}^{\infty} f_k(t)\xi_k\right] = \sum_{k=-\infty}^{\infty} f_k(t)U(t)[\xi_k] = \sum_{k=-\infty}^{\infty} f_k(t)a_k(t) \quad (40)$$

where $f_k(t) = \langle P(t), \xi_k\rangle$ is a scalar periodic function for every k and the sequence of processes $a_k(t) = U(t)[\xi_k]$ are jointly stationary in the wide sense. As in the preceding paragraph, we observe that $a_k(0) = U(0)[\xi_k] = \xi_k$.

Finally, using the spectral representation (31) for $U(t)$ which is applied at each t to the vector $P(t)$ gives

$$X(t) = U(t)[P(t)] = \int_{-\infty}^{\infty} \exp(i\lambda t)dE(\lambda)[P(t)] \quad (41)$$

$$= \int_{-\infty}^{\infty} \exp(i\lambda t)dZ(\lambda, t) \quad (42)$$

where the *time-dependent spectral measure* Z is defined through the application of the projection-valued measure appearing in (33) to the vector $P(t)$. To be precise, for the interval $[a, b)$, we have

$$Z([a, b), t) = E([a, b))[P(t)].$$

In this case the family of measures are mutually orthogonally scattered in the sense

$$\langle Z([a, b), t_1), Z([c, d), t_2)\rangle = 0$$

whenever $[a, b) \cap [c, d) = \emptyset$. This is in contrast to the harmonizable representation in which the measure is independent of t but is not orthogonally scattered.

The unitary operator also has a role in the representations of the *translation series* type, which are based on the Karhunen-Loève expansion. This representation was introduced by Jordan [62] and was clarified and applied to a filtering problem by Gardner and Franks [26]. The following statement appearing in [56], p. 267, summarizes the basic representation result: *A second-order process $X(t)$ is continuous and PC if and only if*

$$X(t) = \sum_{j=1}^{\infty} a_{jn}\phi_j(t) \qquad nT \le t < (n+1)T \tag{43}$$

where (*a*) *the sense of convergence of* (43) *is in q.m. and uniform with respect to t;* (*b*) *the sequence $\{\phi_j(t), t \in \mathbf{R}, j \in \mathbf{N}_+\}$ are continuous periodic functions $\phi_j(t) = \phi_j(t+T)$ that satisfy for all $j, k \in \mathbf{N}_+$ and $0 \le t < T$,*

$$\int_0^T R(s, t)\overline{\phi_j(s)}ds = \lambda_j\phi_j(t) \tag{44}$$

and

$$\int_0^T \phi_j(t)\overline{\phi_k(t)}dt = \delta_{jk}; \tag{45}$$

and (*c*) *for fixed n, the random variables $\{a_{jn}, j \in \mathbf{N}_+\}$ are orthogonal, $E\{a_{jn}\overline{a_{kn}}\} = \lambda_j\delta_{j,k}$, and the countable collections $A_n = \{a_{jn}, j \in \mathbf{N}_+\}$ are weakly stationary in the sense that*

$$E\{a_{jm}\overline{a_{kn}}\} = r_{jk}(m-n) \tag{46}$$

for every $j, k \in \mathbf{N}_+, m, n \in \mathbf{Z}$.

The connection between the unitary operator U_X and the sequence $\{A_k, k \in \mathbf{Z}\}$ can be expressed by $a_{j(n+1)} = U_X a_{jn}$ for every j, n. This follows from writing

$$a_{jn} = \int_{nT}^{(n+1)T} X(t)\phi_j(t)dt \tag{47}$$

which is a Riemann integral because $X(t)$ is q.m. continuous. Then U_X can be brought inside the integral (consider approximating Riemann sums) and hence $a_{j(n+1)} =$

$U_X a_{jn}$ follows. Thus U_X is the same unitary operator that is naturally associated with the vector sequence $\{A_k\}$.

4.4 Almost Periodically Unitary Processes

Since unitary operators arise naturally in stationary and PC processes, do they also arise in APC processes? This question motivates the *almost periodically unitary* (APU) processes, which can be defined as the second-order processes X for which there exists a strongly continuous group of unitary operators $\{U(\tau), \tau \in \mathbf{I}\}$ for which the set of "almost periods"

$$S(\epsilon, X, U) = \left\{\tau : \sup_{t\in\mathbf{R}} ||X(t+\tau) - U(\tau)X(t)||_{L_2} < \epsilon\right\} \tag{48}$$

is a *relatively dense* set for every $\epsilon > 0$; a set E of real numbers is relatively dense if for some $L > 0$ every interval of length L has a nonnull intersection with E. This definition of APU processes is motivated by the Bohr definition of almost periodic functions. There is an easier equivalent definition. A q.m. continuous process $X(t)$ is APU if and only if it can be given by (36) where $P(t)$ is an almost periodic function taking values in $L_2(\Omega)$ [57]. We observe that PC processes can always be called periodically unitary.

The standard examples of APC processes that begin with periodic modulations of stationary processes are all APU as well. We do not know of a nice characterization of processes that are APC and not APU.

The representations based on (36) each have a corresponding version for the APU case, although a few qualifications are needed. In the case of the *harmonic series representation*, the sense of convergence of

$$X_N(t) = U(t)\left[\sum_{k=-N}^{N} P_k \exp(i\lambda_k t)\right] = \sum_{k=-N}^{N} a_k(t)\exp(i\lambda_k t), \tag{49}$$

must be taken in a sense corresponding to almost periodic functions (see [57]), and the processes $a_k(t)$ are no longer necessarily bandlimited although they are still jointly stationary. The representation (40) is roughly the same except the functions $f_j(t)$ are almost periodic in the usual Bohr sense, and $\sum_{j>N} |f_j(t)|^2$ must converge to 0 uniformly in t. The time-dependent spectral representation (42) follows in the same manner except the $L_2(\Omega)$-valued *time-dependent spectral measure* $Z([a,b),t) = E([a,b))[P(t)]$ is almost periodic in the sense of Bohr. We do not expect a *translation series representation* for these processes based on the Karhunen-Loève expansion because the rigid constraint of periodicity is needed to get (35) from which everything follows. Nor do we expect to be able to generally express $a_k(t)$ directly in terms of $X(t)$ via (30) because for the APU case, the $a_k(t)$ need not be bandlimited, another fact that follows from (30).

5 ESTIMATION OF THE COEFFICIENT FUNCTIONS AND THEIR SPECTRAL DENSITIES

We first review the stationary case, then the PC case, for which results are relatively straightforward. When X is APC, some additional complication can arise because generally there is nothing to prevent the set of cycle frequencies $\boldsymbol{\alpha}$ from possessing cluster points. This issue was first discussed by Gardner [27] in his discussion of ACS processes. In our work it arose in attempting to extend the approach of density estimation for PC processes to the APC case. To get some appreciation of the potential difficulty, suppose α_0 is a limit point from $\boldsymbol{\alpha}$ so there is a subsequence $\{\alpha_n\}$ with $\alpha_n \to \alpha_0$. If we wish to estimate $B_{\alpha_0}(\tau)$ or $g_{\alpha_0}(\lambda)$ from averaging operations whose resolution in the dimension α is imperfect (> 0) for any finite averaging interval A, then for any A there will be a countable number of α_n that are unresolvable from α_0. This problem is also discussed by Gardner [29] under the topic of cycle leakage. Several types of restrictions can be made that diminish this difficulty. First, it is possible for the amplitudes corresponding to the subsequence $\{\alpha_n\}$ to diminish sufficiently fast that they do not matter. Both harmonizability (absolute summability of the spectral correlation measures) and certain mixing conditions cause this to occur. And of course, the set $\boldsymbol{\alpha}$ may not have cluster points and still be interesting; for example, the condition $\sum 1/\alpha_j^2 < \infty$ (discussed earlier in connection with the random shift for APC processes) prevents the occurrence of cluster points in the set $\boldsymbol{\alpha}$.

5.1 The Stationary Case

In the case of a stationary complex-valued process with covariance $R(\tau)$, the usual estimator for R

$$\hat{R}(A, \tau) = \frac{1}{A} \int_0^{A-|\tau|} X(t+\tau)\overline{X(t)}dt \tag{50}$$

is consistent under conditions such as $\int_{-\infty}^{\infty} |R(\tau)|d\tau < \infty$ that control the correlation between distant events. This may be considered a form of mixing, which will appear again subsequently. If X is Gaussian, then $\hat{R}(A, \tau)$ is consistent if and only if F_Z appearing in (18) has no discrete component [41].

Pointwise consistency (for fixed τ or λ) is usually taken to be in the sense of quadratic mean or in probability. But since we are actually trying to estimate functions, $R(\tau)$ or $g(\lambda)$ for the stationary case, function-space consistency is also of interest. The remainder of this preliminary discussion treats pointwise consistency.

A related problem is the estimation of the spectral density function g where $R(\tau) = \int_{-\infty}^{\infty} \exp(i\lambda\tau)g(\lambda)d\lambda$. The typical approach to the nonparametric[4] estima-

[4] No constraints, other than $g \geq 0$ and $\int_{-\infty}^{\infty} g(\lambda)d\lambda < \infty$, are imposed upon g.

tion of g (pointwise) is through the use of the finite-time Fourier transform

$$\tilde{X}_A(\lambda) = \int_0^A X(t) \exp(-i\lambda t) dt \tag{51}$$

from which the periodogram

$$I_A(\lambda) = \frac{1}{2\pi A} |\tilde{X}_A(\lambda)|^2 \tag{52}$$

is formed. The periodogram is an asymptotically unbiased estimator for g although it is not consistent. Consistent estimators for g are obtained by tapering the correlation estimator by a multiplicative kernel $k(\cdot) \in L_1(-\infty, \infty)$ which is bounded and even with $k(0) = 1$. This estimator can equivalently be expressed as a smoothing of the periodogram by K, the Fourier transform of k; thus

$$\begin{aligned} \hat{g}(A, \lambda) &= \frac{1}{2\pi} \int_{-A}^{A} k(B_A v) \hat{R}(A, v) \exp(-i\lambda v) dv \\ &= \frac{1}{B_A} \int_{-\infty}^{\infty} K\left(\frac{\sigma - \lambda}{B_A}\right) I_A(\sigma) d\sigma \end{aligned} \tag{53}$$

where $B_A \to 0$ and $AB_A \to \infty$ as $A \to \infty$. This procedure gives kernels that are dependent upon A in a manner that produces the consistency of $\hat{g}$. For example, if a stationary process X is Gaussian and $R \in L_1(-\infty, \infty)$, then $\hat{g}$ is a consistent estimator at every λ. For additional details see [81] and [76]. Gardner [29] provides a unified treatment of these and other methods, such as time averaging (with respect to the parameter s) of the sliding periodogram obtained by replacing the interval of integration $[0, A]$ in (51) with $[s, s + A]$.

Parzen [77] obtains consistency results for $\hat{R}(A, \tau)$ and $\hat{g}(A, \lambda)$ for the *asymptotically stationary* processes for which the limiting quantities are the asymptotic covariance and its corresponding spectral density; these are $B_0(\tau)$ and $g_0(\lambda)$ for PC or APC processes.

We now review the corresponding results that have been obtained for both PC and APC processes. In summary, modified versions of (50) and (53) produce consistent estimators for B_k and g_k, but there are additional considerations.

5.2 Estimation of $B_k(\tau)$ for PC Case

In the case of PC processes, estimators for $B_k(\tau)$ are given by

$$\hat{B}_k(A, \tau) = \frac{1}{A} \int_{I(A,\tau)} X(t + \tau) \overline{X(t)} \exp(-i2\pi kt/T) dt \tag{54}$$

where the interval $I(A, \tau) = [0, A - \tau]$ for $\tau \geq 0$ and $[-\tau, A]$ for $\tau < 0$. The need for two such intervals disappears by defining $B(t, \tau) = R(t + \tau/2, t - \tau/2)$, then $I(A, \tau) = [|\tau|/2, A - |\tau|/2]$ [28, 46]. But this presents an obvious problem in the discrete-time case (consider odd values of τ).

In examination of q.m. consistency it is often useful to explicitly separate the bias and variance of the estimator:

$$\begin{aligned} E\{|\hat{B}_k(A,\tau) - B_k(\tau)|^2\} \\ &= |E\{\hat{B}_k(A,\tau)\} - B_k(\tau)|^2 + E\left\{|\hat{B}_k(A,\tau) - E\{\hat{B}_k(A,\tau)\}|\right\} \\ &= |\text{bias}[\hat{B}_k(A,\tau)]|^2 + \text{var}[\hat{B}_k(A,\tau)]. \end{aligned}$$

It follows that [50]

$$E\{\hat{B}_k(A,\tau)\} = B_k(\tau)\left[1 - \frac{|\tau|}{A}\right] + \frac{\epsilon(A,\tau)}{A} \tag{55}$$

where $\epsilon(A,\tau)$ is bounded with respect to both A and τ and hence the bias is $O(1/A)$. See [35] for a similar result in the discrete-time case.

Sufficient conditions for $\lim_{A\to\infty} \text{var}[\hat{B}_k(A,\tau)] = 0$ (and hence the desired consistency) are now easily stated in terms of the process $Z_\tau(t) = X(t+\tau)\overline{X(t)} - B(t,\tau)$ and its correlation, $R_{Z,\tau}(t_1,t_2) = E\{Z(t_1,\tau)\overline{Z(t_2,\tau)}\}$. Whenever X is real and Isserlis's condition,

$$\begin{aligned} E\{X(t_1)X(t_2)X(t_3)X(t_4)\} &= E\{X(t_1)X(t_2)\}E\{X(t_3)X(t_4)\} \\ &+ E\{X(t_1)X(t_3)\}E\{X(t_2)X(t_4)\} + E\{X(t_1)X(t_4)\}E\{X(t_2)X(t_3)\} \end{aligned} \tag{56}$$

is satisfied, as when $X(t)$ is real and Gaussian, $R_{Z,\tau}(t_1,t_2)$ can be expressed in terms of pairwise correlations. This is used in the immediately following and subsequent results. If X is PC with bounded absolute fourth moments, then the following are sufficient for the desired consistency:

- Ia. $\int_{-\infty}^{\infty}\int_{-\infty}^{\infty} |R_{Z,\tau}(t_1,t_2)|dt_1dt_2 < \infty$;
- Ib. $\lim_{u\to\infty} R_{Z,\tau}(t+u,t) = 0$ uniformly in t;
- Ic. if $Z_\tau(t)$ is PC and $\int_{-\infty}^{\infty}\int_0^T |R_{Z,\tau}(t_1,t_2)|dt_1dt_2 < \infty$;
- Id. if X is real, satisfies Isserlis's condition and $\int_{-\infty}^{\infty}\int_0^T B^2(t,\tau)dtd\tau < \infty$;
- Ie. if X is real, satisfies Isserlis's condition and $\lim_{\tau\to\infty} B(t,\tau) = 0$ uniformly in t.

For real-valued discrete-time Gaussian PC sequences, Genossar, Lev-Ari, and Kailath [35] show that consistency for every k and τ occurs if and only if one of the following hold:

- IIa. the measure m_0 has no discrete component;
- IIb. $\lim_{A\to\infty} A^{-1}\sum_{\tau=0}^{A-1} |B_0(\tau)|^2 = 0$.

These results should be compared to the remarks concerning F_Z for the stationary case discussed in Section 5.1.

The study of consistency of $\hat{B}_k(A,\tau)$ seems to have begun with Gudzenko [40] who essentially gave condition (Ib). Proofs of the others may be found in [46, 50]. A necessary and sufficient condition (see [28]) for q.m. consistency for APC and, more generally, asymptotically mean cyclostationary processes, can also be expressed in terms of the process $Z_\tau(t)$ by using

$$\mathrm{Var}[\hat{B}_k(A,\tau)] = \frac{1}{A^2}\int_{I(A,\tau)}\int_{I(A,\tau)} K_{Z,\tau}(s,t)\exp[-i2\pi k(s-t)]dsdt \quad (57)$$

where

$$K_{Z,\tau}(s,t) = R_{Z,\tau}(s,t) - B(s,\tau)\overline{B(t,\tau)}. \quad (58)$$

Conditions for the a.s. convergence of $\hat{B}_k(A,\tau)$ can be obtained from ([43], Theorem 5.1) for Gaussian PC processes.

Additional consistency results for PC processes can be inferred from recent results for APC processes (discussed in Section 5.4).

5.3 Estimation of $g_k(\lambda)$ for the PC Case

The estimator for g_k considered here is formed by tapering $\hat{B}_k(A,\tau)$ or equivalently by smoothing the biperiodogram

$$J_A(\lambda_1,\lambda_2) = \frac{1}{2\pi A}\tilde{X}_A(\lambda_1)\overline{\tilde{X}_A(\lambda_2)} \quad (59)$$

along the line D_k. As in the stationary case (note $I_A(\lambda) = J_A(\lambda,\lambda)$) the "raw" (unsmoothed) shifted periodogram $I_A(k,\lambda) = J_A(\lambda, \lambda-2\pi k/T)$ is an asymptotically unbiased but inconsistent estimator for $g(\lambda)$, but appropriate smoothing can produce consistency. Specifically, using the same notation for the kernels appearing in the stationary case, $\hat{g}_k(A,\lambda)$ can be expressed in two ways,

$$\begin{aligned}\hat{g}_k(A,\lambda) &= \frac{1}{2\pi}\int_{-A}^{A} k(B_A v)\hat{B}_k(A,v)\exp(-i\lambda v)dv \\ &= \frac{1}{B_A}\int_{-\infty}^{\infty} K\left(\frac{\sigma-\lambda}{B_A}\right) J_A(\sigma,\sigma-2\pi k/T)d\sigma.\end{aligned} \quad (60)$$

Qualitatively, the "bandwidth" of the estimator becomes small as $A \to \infty$ at a rate proportional to B_A; but the sample size (degrees-of-freedom) of the biperiodogram is increasing at a faster rate, yielding a compromise between bias and variance that produces consistency.

The exact analysis of both the bias and variance of $\hat{g}_k(A,\lambda)$ requires too much space for this survey, but the following considerations for the bias illuminate the problem. First, assuming that $B_k \in L_1(\mathbf{R})$ so that $g_k(\lambda) = \frac{1}{2\pi}\int_{-\infty}^{\infty} B_k(\tau)\exp(-i\lambda\tau)d\tau$, and that X is jointly measurable with bounded second moments (so we can use Fubini's theorem) we compute

$$E\{\hat{g}_k(A,\lambda)\} = \frac{1}{2\pi}\int_{-A}^{A} k(B_A v)E\{\hat{B}_k(A,v)\}\exp(-i\lambda v)dv \quad (61)$$

and so

$$
\begin{aligned}
|E\{\hat{g}_k(A,\lambda)\} - g_k(\lambda)| \leq \frac{1}{2\pi}\int_{-A}^{A} |k(B_A v)E\{\hat{B}_k(A,v)\} - B_k(v)|dv \\
+\frac{1}{2\pi}\int_{|v|>A} |B_k(v)|dv.
\end{aligned}
\tag{62}
$$

So the manner of convergence of the bias to zero depends on the difference between $B_k(v)$ and $k(B_A v)E\{\hat{B}_k(A,v)\}$ and on the tails of $B_k(v)$. Parzen [76] has developed a method by which rate of convergence can be given in terms of the properties of $k(\cdot)$ and $B_k(\cdot)$; it is applied to the PC case in [46].

Turning now to the issue of the variance, if X is real, Gaussian, and $\int_{-\infty}^{\infty}\left[\int_0^T B^2(t,\tau)dt\right]^{1/2} d\tau < \infty$, which can be interpreted as a sort of mixing condition (see [81]), then there exists a positive constant M for which

$$
AB_A|\text{cov}\left[\hat{g}_j(A,\lambda_1), \hat{g}_k(A,\lambda_2)\right]| \leq M \tag{63}
$$

or, in other words, $|\text{cov}\left[\hat{g}_j(A,\lambda_1), \hat{g}_k(A,\lambda_2)\right]|$ is $O((AB_A)^{-1})$ and hence q.m. consistency follows [46, 50].

In a work mainly concerned with ergodicity, Honda [43] gives a condition for the a.s. convergence of the smoothed shifted periodogram to the smoothed density g_k; but the consistency question is not completely answered because smoothing is taken as a fixed function with respect to A.

An alternative approach to the problem for real Gaussian PC processes has been given by Alekseev [1] who first forms the estimator

$$
\hat{B}^{(N)}(t,\tau) = \frac{1}{2N+1}\sum_{n=-N}^{N} X(t+nT+\tau)X(t+nT) \tag{64}
$$

and then

$$
\hat{B}_k^{(N)}(\tau) = \frac{1}{T}\int_0^T \hat{B}^{(N)}(t,\tau)\exp(-i2\pi kt/T)dt. \tag{65}
$$

Then the kth periodogram

$$
I_k^{(N)}(\lambda) = \frac{1}{2\pi}\int_{-\infty}^{\infty} \hat{B}_k^{(N)}(\tau)\exp(-i\lambda\tau)d\tau \tag{66}
$$

is smoothed by a kernel to produce an estimator $\hat{g}_k^{(N)}(\lambda)$ for the density $g_k(\lambda)$. We shall not pursue the apparent correspondence between the formulations here, but will summarize Alekseev's main result. Suppose the densities g_k satisfy $\sup_\lambda |g_k(\lambda)| \leq M_k$ with $\sum_k M_k < \infty$, and that $\text{Re}[g_k(\lambda)]$ and $\text{Im}[g_k(\lambda)]$ have bounded derivatives of all relevant orders, then

$$
E|\hat{g}_k^{(N)}(\lambda) - g_k(\lambda)|^2 = O(N^{-2r/(2r+1)}) \tag{67}
$$

where r is a parameter of the smoothing window, and the window scaling factor $h(N)$ must be chosen proportional to $N^{-1/(2r+1)}$ as $N \to \infty$.

Gardner [29] introduces a method that time-averages a sliding biperiodogram and unifies this method along with (60) and other methods in terms of the single representation

$$w_{\lambda}^{\alpha}(t) = \int_{-\infty}^{\infty}\int_{-\infty}^{\infty} k_{\lambda}^{\alpha}(u, v) X(t-u)\overline{X(t-v)}\,du\,dv \exp(-i2\pi\alpha t) \tag{68}$$

where the kernel $k_{\lambda}^{\alpha}(u, v)$ can be chosen to produce estimators of the time-dependent analogs of the quantities we have been discussing above. For example, by appropriate choice of $k_{\lambda}^{\alpha}(u, v)$ one can obtain $w_{\lambda}^{\alpha}(t) = S_{X_{\Delta t}}^{\alpha}(t, \lambda)_{\Delta\lambda}$, which is the time-average version of the estimator for the cyclic spectral density (at cycle frequency α) at frequency λ at "local" time t. The time resolution of the average is Δt and the frequency resolution is $\Delta\lambda$. General yet explicit formulas for the bias and variance of all estimators in this family are obtained by using Isserlis's condition; although these results are obtained for time-average definitions of bias and variance (rather than expected-value definitions), the results agree with those obtainable using expected values [29, 31]. The connection between our notation and that of [29] can be facilitated by identifying the quantity A (time duration of the estimator) in (63) with Δt and B_A (the frequency resolution of the estimator) with $\Delta\lambda$. Gardner also discusses the problem of "cycle leakage" that occurs with finite time estimators and is thus often not treated in the discussions of consistency.

5.4 Estimation of $B_\alpha(\tau)$ for the APC Case

The basic form of the estimator considered here is a natural extension of that for the PC case:

$$\hat{B}_\alpha(A, \tau) = \frac{1}{A}\int_{I(A,\tau)} X(t+\tau)\overline{X(t)} \exp(-i\alpha t)\,dt, \tag{69}$$

where $I(A, \tau)$ is the same as that given in (54).

As in the PC case we first consider bias$[\hat{B}_\alpha(A, \tau)]$. Since APC processes have bounded second moments and are continuous in q.m.,

$$E\{\hat{B}_\alpha(A, \tau)\} = \frac{1}{A}\int_{I(A,\tau)} B(t, \tau) \exp(-i\alpha t)\,dt. \tag{70}$$

To be specific, if we take $\tau \geq 0$ so that $I(A, \tau) = [0, A - \tau]$, then

$$\begin{aligned} E\{\hat{B}_\alpha(A, \tau)\} &= \frac{1}{A}\int_0^{A-\tau} B(t, \tau) \exp(-i\alpha t)\,dt \\ &= \left[\frac{A-\tau}{A}\right]\frac{1}{A-\tau}\int_0^{A-\tau} B(t, \tau)\exp(-i\alpha t)\,dt. \end{aligned} \tag{71}$$

For any fixed τ, the term in brackets converges to 1 as $A \to \infty$; and since $B(t,\tau)$ is almost periodic in t for each τ, the second term converges to the Fourier coefficient $B_\alpha(\tau)$; hence $\hat{B}_\alpha(A,\tau)$ is asymptotically unbiased.

Although for the PC case we were able to show generally that the bias was $O(A^{-1})$, we cannot do so here since the basic results on almost periodic functions do not provide rates of convergence of such integrals to their limits. But rates can be determined for specific subclasses of almost periodic functions. For example, Cambanis et al. [9] show that if the frequencies $\{\alpha_j\}$ of an almost periodic function satisfy $\sum_j 1/\alpha_j^2 < \infty$, then the convergence is $O(A^{-1})$; note this condition prevents the occurrence of cluster points in the sequence $\{\alpha_j\}$. If this condition applies to $\boldsymbol{\alpha}$ (excluding $\alpha = 0$) then the bias will be $O(A^{-1})$ for every τ.

Again, sufficient conditions for $\lim_{A\to\infty} \text{var}[\hat{B}_\alpha(A,\tau)] = 0$ can be given in terms of the process $Z_\tau(t) = X(t+\tau)\overline{X(t)} - B(t,\tau)$; any of these conditions, along with the preceding remarks on the asymptotic bias, yields consistency. If X is AMCS[5] with bounded absolute fourth moments, then conditions (Ia) and (Ib) of Section 5.2 are sufficient. But the rest require modification to

IIIc. if $Z_\tau(t)$ is APC and $\int_{-\infty}^{\infty} \lim_{A\to\infty} A^{-1} \int_0^A |R_{Z,\tau}(t_1,t_2)| dt_1 dt_2 < \infty$;
IIId. if X is APC, real, satisfies Isserlis's condition and $\int_{-\infty}^{\infty} \lim_{A\to\infty} A^{-1} \int_0^A B^2(t,\tau) dt d\tau < \infty$;
IIIe. if X is APC, real, satisfies Isserlis's condition and $\lim_{u\to\infty} B(u,t) = 0$ uniformly in t.

By restricting consideration to Gaussian processes having bounded fourth moments, Genossar, Lev-Ari, and Kailath [35] refine the earlier work of Boyles and Gardner [7] on discrete-time AMCS processes with the following collection of equivalent statements (paraphrased to our notation):

IVa. $\hat{B}_\alpha(A,\tau)$ is consistent for all α and τ;
IVb. $\hat{B}_0(A,0)$ is consistent;
IVc. $(1/A^2)\sum_{t=1}^{A}\sum_{s=1}^{A} R^2(t,s) \to 0$ as $A \to \infty$;
IVd. $(1/A)\sum_{s=1}^{A} R^2(A,s) \to 0$ as $A \to \infty$.
IVe. If the process is harmonizable, (all PC sequences are) $\hat{B}_\alpha(A,\tau)$ is consistent for all α and τ if and only if the spectral correlation measure M_Z has no discrete components and this occurs if and only if m_0 has no discrete component.

The summability conditions on the correlations can be seen to be a form of mixing, as they control the correlation between separated events. It is shown in [7] that either weak mixing or uniformly strong mixing guarantees the *cycloergodicity*

[5]The class of *asymptotically mean cyclostationary* processes introduced in [7] contain the PC and APC processes.

property which then ensures the consistency of the estimator (69) for AMCS discrete-time processes. The following results use the idea of ϕ−mixing. A continuous-time process X is called ϕ−mixing ([5], p. 166) if for every t,

$$\sup_{\substack{A\in \mathbf{F}_X(-\infty,t)\\ B\in \mathbf{F}_X(t+s,\infty)}} |P(A|B) - P(A)| \leq \phi(s) \tag{72}$$

where $\mathbf{F}_X(a, b)$ denotes the sub σ−algebra of events generated by $\{X(t), a \leq t \leq b\}$. It follows (see [5], p. 170) that the correlation is controlled by ϕ in the sense

$$\left[E\{|X(t)\overline{X(t+s)}|\}\right]^2 \leq C\phi(s)E\{|X(t)|^2\}E\{|X(t+s)|^2\} \tag{73}$$

where $C = 2$. If the APC process X is ϕ−mixing with $\phi^{1/2}(\cdot) \in L_1(\mathbf{R})$ and $B_0(\tau)$ is continuous at $\tau = 0$, then there exist constants C_1 and C_2 for which

$$\mathrm{Var}[\hat{B}_\alpha(A,\tau)] \leq (C_1|\tau| + C_2)/A, \tag{74}$$

for every α and τ. This is clearly sufficient for q.m. consistency [53]; see the discussion of (57). In [54] it is shown that adding only the hypothesis that X has uniformly bounded fourth moments gives almost sure consistency of $\hat{B}_\alpha(A,\tau)$ for every α and τ. Further, if X is UAPC and

$$\phi^{1/2}(t) = \mathrm{O}(t^{-(3+\delta)})$$

for $\delta > 0$, and if α is not a cluster point of $\boldsymbol{\alpha}$, then strong consistency in $L_1(\mathbf{R})$ is obtained; that is,

$$\int_{-\infty}^{\infty} |\hat{B}_\alpha(A,\tau) - E\{\hat{B}_\alpha(A,\tau)\}|d\tau \to 0$$

almost surely as $A \to \infty$. And finally, if the function $V(t, \tau_1, \tau_2, \tau_3)$ is almost periodic in t uniformly (see [11], p. 55) with respect to $\tau_i, i = 1, 2, 3$, then the estimator is asymptotically normal.

The following results [20] show that if $\mathrm{Var}[\hat{B}_\alpha(A,\tau)]$ is $\mathrm{O}(A^{-\gamma})$ for $\gamma > 0$, and assuming some variants of the fourth moment property, then almost sure (and q.m.) consistency is obtained. Generally it is assumed that X is measurable, has finite fourth moments for all t, and one of the following is satisfied for some C and $\gamma > 0$:

Va. $E\{|X(t+\tau)X(t)|^2\} \leq C$ for fixed τ and all t, and

$$\int_0^A \int_0^A \mathrm{Cov}[Z_\tau(u), Z_\tau(v)] \exp[-i\alpha(u-v)]dudv = \mathrm{O}(A^{2-\gamma});$$

Vb. $|X(t+\tau)X(t)| \leq C$ for fixed τ and all t a.s. (P), and

$$\int_a^\infty A^{-3} \int_0^A \int_0^A |\mathrm{Cov}[Z_\tau(u), Z_\tau(v)]|dudvdA < \infty$$

for some $a > 0$;

Vc. X satisfies Isserlis's condition (56) and

$$\int_0^A \int_0^A |R(u,v)|^2 du dv = O(A^{2-\gamma});$$

Vd. X satisfies Isserlis's condition, $|X(t)| \le C$ for all t a.s. (P) and

$$\int_a^\infty A^{-3} \int_0^A \int_0^A |R(u,v)|^2 du dv dA < \infty$$

for some $a > 0$.

A variation of the a.s. result that uses mixing appears in [18]. Condition (Vd), while similar to condition (IVc) for discrete-time AMCS processes discussed above, shows that a.s. consistency is obtained if for $\gamma > 0$,

$$\frac{1}{A^2} \int_0^A \int_0^A |R(u,v)|^2 du dv = O(\frac{1}{A^\gamma}).$$

5.5 Estimation of $g_\alpha(\lambda)$ for the APC Case

The estimator for the density $g_\alpha(\lambda)$ is formed in the same manner as for the PC case, by tapering $\hat{B}_\alpha(A,\tau)$ or equivalently by smoothing the biperiodogram (59) along the line D_α. That is,

$$\begin{aligned} \hat{g}_\alpha(A,\lambda) &= \frac{1}{2\pi} \int_{-A}^{A} k(B_A v) \hat{B}_\alpha(A,v) \exp(-i\lambda v) dv \\ &= \frac{1}{B_A} \int_{-\infty}^{\infty} K\left(\frac{\sigma-\lambda}{B_A}\right) I_A(\sigma, \sigma-\alpha) d\sigma. \end{aligned} \tag{75}$$

As previously stated, the approach to proving consistency followed for Gaussian PC processes does not extend in a simple manner to the APC case. The difficulty can be related to $\boldsymbol{\alpha}$ possibly containing cluster points. However, if the APC process is assumed to be ϕ−mixing with $\phi^{1/2}(\cdot) \in L_1(\mathbf{R})$, $B_0(\tau)$ is continuous at $\tau = 0$, and $\int_{-\infty}^{\infty} u^{1/2}|k(u)|du < \infty$, then for every $\alpha_1, \alpha_2, \lambda_1, \lambda_2$

$$\lim_{A\to\infty} \text{Cov}[\hat{g}_{\alpha_1}(A,\lambda_1), \hat{g}_{\alpha_2}(A,\lambda_2)] = 0$$

if $B_A \to 0$ and $AB_A^3 \to \infty$ as $A \to \infty$ [53]. It would appear that such mixing conditions are strong enough to prevent any problem from clustering cycle frequencies.

Analogous results on the estimation of spectral densities, based on mixing conditions have been found for bivariate APC processes [17].

The asymptotic normality of $\hat{g}_\alpha(A,\lambda)$ has been shown [67] under the following assumptions:

VIa. X is ϕ–mixing with $\phi(t) = O(t^{-(2+\delta)})$ for $\delta > 0$;

VIb. $V(t, \tau_1, \tau_2, \tau_3)$ is almost periodic in t uniformly with respect to $\tau_i, i = 1, 2, 3$;

VIc. $B(t, \tau)$ is uniformly almost periodic in t with respect to $\tau \in \mathbf{R}$;

VId. $\int_{-\infty}^{\infty} |u| K^2(u) du < \infty$, where K is the Fourier transform of the covariance tapering kernel k in (53).

The problem of clustering of the cycle frequencies can also be addressed through explicit conditions [20]. First, define

$$K_1(\tau) = \sum_{\alpha} |B_\alpha(\tau)|, \tag{76}$$

$$K_2(\tau) = \sum_{\alpha} |B_\alpha(\tau)|^2 \tag{77}$$

and

$$R_\alpha(\tau) = \sum_{\beta \neq \alpha} |B_\beta(\tau)||\beta - \alpha|^{-1}. \tag{78}$$

Note $K_1(\tau)$ is bounded if X is harmonizable. Suppose X is a zero-mean real measurable APC process that satisfies Isserlis's condition (56). If for any specific α, $B_\alpha(\cdot) \in L_1(\mathbf{R})$ and

$$\int_{A^\delta < |\tau|} |B_\alpha(\tau)| d\tau = o(A^{-\epsilon}),$$

and either of the conditions (VIIa) and (VIIb) is satisfied, then

$$\lim_{A \to \infty} A^\epsilon [\hat{g}_\alpha(A, \lambda) - g_\alpha(\lambda)] = 0 \quad \text{a.s. and in q.m.}$$

The conditions are:

VIIa. for some ϵ, γ, and δ, $4\epsilon \leq 1 - \gamma - 4\delta$ and there is a $C > 0$ such that $|R_\alpha(\tau)| < C$ for almost all τ; in addition, $\int_{-A}^{A} K_1(\tau) d\tau = o(A^\gamma)$ for some $\gamma < 1$ and $B_A = A^\delta$ for $4\delta \leq 1 - \gamma$;

VIIb. for some ϵ, γ, and δ, $4\epsilon \leq 1 - 2\gamma - 3\delta$ and there is a $C > 0$ such that $|K_1(\tau)| < C$ for almost all τ; in addition, $\int_{-A}^{A} [K_2(\tau)]^{1/2} d\tau = o(A^\gamma)$ for some $\gamma < 1/2$ and $B_A = A^\delta$ for $3\delta \leq 1 - 2\gamma$.

A variation of (VIIb) that uses mixing appears in [18].

The general family of estimators (see Gardner [29]) given by (68) can also be applied to the APC case, where the discussion of cycle leakage along with the possibility for α to have cluster points become particularly relevant.

6 CONCLUSIONS AND NEW DIRECTIONS OF RESEARCH

This survey concentrates on the structural properties of second-order PC and APC processes, and on the estimation of the coefficient functions and their spectral densities. Many related issues, which are treated in other contributions to this volume, are omitted.

Interesting problems related to these processes remain open. To name a few, we begin with the problem of PC and APC random fields. Alekseev [2] has begun the problem of spectral density estimation of two-dimensional Gaussian PC fields. Ramanathan and Zeitouni [78] introduce a connection between wavelets and cyclostationary processes; Cambanis and Houdré [8] take this idea a little farther. The determination of the presence of the PC or APC property in a time-series is another problem of interest; a test based on Goodman's [39] spectral coherence is given in [52]. Results for the second-order case can also be found in the work of Dandawate and Giannakis [13] on testing for the presence of kth-order cyclostationarity. The weak-signal maximum-likelihood approach for parametric detection, and a maximum signal-to-noise ratio approach for nonparametric detection are pursued by Gardner and Spooner in [30, 33, 34]. The estimation of the period of a PC process (see [73]), and of the frequencies α of an APC process, and the determination of their relative importance is also of current interest. The works of Dragan and Yavorskiy [21, 22, 23, 87] also treat these problems. For a discussion of computationally efficient algorithms for estimation of $B_\alpha(\tau)$ and $g_\alpha(\lambda)$, see Article 6 in this volume. Additional problems for APC processes include their representation (the existence of a stationary representation as in the PC case) and prediction. Finally, we believe the connection between the probabilistic (Kolmogorov) and nonprobabilistic (Wiener) theories is a problem of great interest and challenge. See Chapter 1 of this volume for further discussion.

REFERENCES

[1] V. G. Alekseev, "Estimating the Spectral Densities of a Gaussian Periodically Correlated Stochastic Process," *Problems of Information Transmission*, vol. 24, no. 2, pp. 109–115, 1988.

[2] V. G. Alekseev, "On Spectral Density Estimates of a Gaussian Periodically Correlated Random Field," *Probability and Mathematical Statistics*, vol. 11, no. 2, pp. 157–167, 1991.

[3] J. Allen and S. Hobbs, "Detecting Target Motion by Frequency-Plane Smoothing," *Proc. of the Twenty-Sixth Asilomar Conference on Systems and Computers*, Pacific Grove, CA, pp. 1042–1047, 1992.

[4] W. R. Bennett, "Statistics of Regenerative Digital Transmission," *Bell System Tech. J.*, vol. 37, pp. 1501–1542, 1958.

[5] P. Billingsley, *Convergence of Probability Measures*, New York: John Wiley, 1968.

[6] S. Bochner, *Verlesengen uber Fouriersche Integrale*, Leipzig, 1932.

[7] R. A. Boyles and W. A. Gardner, "Cycloergodic Properties of Discrete-Parameter Nonstationary Stochastic Processes," *IEEE Trans. on Information Theory*, vol. IT-29, no. 1, pp. 105–114, 1983.

[8] S. Cambanis and C. H. Houdré, "On the Continuous Wavelet Transform of Second Order Random Processes," Technical Report No. 390, Center for Stochastic Processes, Dept. of Statistics, UNC at Chapel Hill, 1993.

[9] S. Cambanis, C. H. Houdré, H. L. Hurd, and J. Leskow, "Laws of Large Numbers for Periodically and Almost Periodically Correlated Processes," Technical Report No. 334, Center for Stochastic Processes, Dept. of Statistics, UNC at Chapel Hill, 1992.

[10] D. K. Chang and M. M. Rao, "Bimeasures and nonstationary processes," in *Real and Stochastic Analysis*, M. M. Rao, ed., New York: John Wiley, 1987.

[11] C. Corduneanu, *Almost Periodic Functions*, New York: Chelsea, 1989.

[12] H. Cramér, "On the Theory of Stationary Random Processes," *Math. Ann.*, vol. 41, pp. 215–230, 1940.

[13] A. V. Dandawate and G. B. Giannakis, "Statistical Tests for the Presence of Cyclostationarity," to appear in *Proc. Intl. Workshop on Higher-Order Statistics*, June, 1993.

[14] J. L. Doob, *Stochastic Processes*, New York: John Wiley, 1953.

[15] D. Dehay, "On a Class of Asymptotically Stationary Harmonizable Processes," *J. Multivariate Anal.*, vol. 22, pp. 251–257, 1987.

[16] D. Dehay, "Non Linear Analysis for Almost Periodically Correlated Strongly Harmonizable Processes," presented at *2nd World Congress of the Bernoulli Society at Uppsala, Sweden*, August 13–18, 1990.

[17] D. Dehay, "Processus bivariés presque périodiquement corrélés: analyse spectrale et estimation des densités spectrales croisées," presented at *XXXIII Journées de Statistiques à Strasbourg*, Mai 1991.

[18] D. Dehay, "Estimation de paramètres fonctionnels spectraux de certains processus non-nécessairement stationnaires," *Comptes Rendus de l'Académie des Sciences de Paris*, vol. 314(4), pp. 313–316, 1992.

[19] D. Dehay and R. Moché, "Trace Measures of a Positive Definite Bimeasure," *J. Multivariate Anal.*, vol. 40, pp. 115–131, 1992.

[20] D. Dehay, "Spectral Analysis of the Covariance of the Almost Periodically Correlated Processes," to appear in *Stochastic Processes and their Application*.

[21] Y. P. Dragan, *Structure and Representation of Stochastic Signal Models* (in Russian), Kiev: Naukova Dumka, 1980.

[22] Y. P. Dragan and I. N. Yavorskiy, *Rhythmics of Sea Waves and Underwater Acoustic Signals* (in Russian), Kiev: Naukova Dumka, 1982.

[23] Y. P. Dragan, V. A. Rozhkov, and I. N. Yavorskiy, *Methods of Probabilistic Analysis of Rhythms of Oceanological Processes* (in Russian), Leningrad: Gidrometeoizdat, 1987.

[24] R. Gangolli, "Wide Sense Stationary Sequences of Distributions on Hilbert Space and the Factorization of Operator-Valued Functions," *J. Math. Mech.*, vol. 12, pp. 893–910, 1963.

[25] W. A. Gardner, "Representation and Estimation of Cyclostationary Processes," Ph.D. Dissertation, Department of Electrical and Computer Engineering, University of Mass., 1972, reprinted as *Signal and Image Processing Lab Technical Report No. SIPL-82-1*, Department of Electrical and Computer Engineering, University of California at Davis, 1982.

[26] W. A. Gardner and L. E. Franks, "Characterization of Cyclostationary Random Signal Processes," *IEEE Trans. on Information Theory*, vol. IT-21, no. 1, pp. 4–14, 1975.

[27] W. A. Gardner, "Stationarizable Random Processes," *IEEE Trans. on Information Theory*, vol. IT-24, no. 1, pp. 8–22, 1978.

[28] W. A. Gardner, *Introduction to Random Processes with Application to Signals and Systems*, New York: Macmillan, 1985.

[29] W. A. Gardner, *Statistical Spectral Analysis: A Nonprobabilistic Theory*, Englewood Cliffs, NJ: Prentice-Hall, 1987.

[30] W. A. Gardner, "Signal Interception: A Unifying Theoretical Framework for Feature Detection," *IEEE Trans. on Communications*, vol. COM-36, pp. 897–906, 1988.

[31] W. A. Gardner, "Two Alternative Philosophies for Estimation of the Parameters of Time-Series," *IEEE Trans. on Information Theory*, vol. 37, no. 1, pp. 216–218, 1991.

[32] W. A. Gardner, "Exploitation of Spectral Redundancy in Cyclostationary Signals," *IEEE ASSP Magazine*, vol. 8, no. 2, pp. 14–36, 1991.

[33] W. A. Gardner and C. M. Spooner, "Signal Interception: Performance Advantages of Cyclic Feature Detectors," *IEEE Trans. on Communications*, vol. 40, pp. 149–159, 1992.

[34] W. A. Gardner and C. M. Spooner, "Detection and Source Location of Weak Cyclostationary Signals: Simplification of the Maximum Likelihood Receiver," *IEEE Trans. on Communications*, vol. 41, no. 6, pp. 905–916, 1993.

[35] M. J. Genossar, H. Lev-Ari, and T. Kailath, "Consistent Estimation of the Cyclic Autocorrelation," to appear in *IEEE Trans. Sig. Proc.*, March 1994.

[36] E. G. Gladyshev, "On Multi-Dimensional Stationary Random Processes," *Theory Prob. and Appl.*, vol. 3, pp. 425–428, 1958.

[37] E. G. Gladyshev, "Periodically Correlated Random Sequences," *Sov. Math.*, vol. 2, pp. 385–388, 1961.

[38] E. G. Gladyshev, "Periodically and Almost Periodically Correlated Random Processes with Continuous Time Parameter," *Theory Prob. and Appl.*, vol. 8, pp. 173–177, 1963.

[39] N. R. Goodman, "Statistical Tests for Nonstationarity within the Framework of Harmonizable Processes," Rocketdyne Research Report No. 65-28, AD619270, August 2, 1965.

[40] L. I. Gudzenko, "On Periodically Nonstationary Processes," *Radiotekhnika i elektronika*, vol. 4, no. 6, pp. 1062–1064, 1959.

[41] E. J. Hannan, *Multiple Time Series*, New York: John Wiley, 1970.

[42] I. Honda, "On the Spectral Representation and Related Properties of Periodically Correlated Stochastic Processes," *Trans. IECE Japan*, vol. E65, no. 12, pp. 723–729, 1982.

[43] I. Honda, "On the Ergodicity of Gaussian Periodically Correlated Stochastic Processes," *Trans. IEICE Japan*, vol. E73, no. 10, pp. 1729–1737, 1990.

[44] C. H. Houdré, "Harmonizability, V-Boundedness, (2,p)-Boundedness of Stochastic Processes," *Prob. Theory and Related Fields*, vol. 84, pp. 39–54, 1987.

[45] C. H. Houdré, "Linear Fourier and Stochastic Analysis," *Prob. Theory and Related Fields*, vol. 87, pp. 167–188, 1990.

[46] H. L. Hurd, "An Investigation of Periodically Correlated Stochastic Processes," Ph.D. dissertation, Duke Univ. Dept. of Electrical Engineering, November, 1969.

[47] H. L. Hurd, "Periodically Correlated Processes with Discontinuous Correlation Functions," *Theory Prob. and Appl.*, vol. 19, no. 4, pp. 834–838, 1974.

[48] H. L. Hurd, "Stationarizing Properties of Random Shifts," *SIAM J. Appl. Math.*, vol. 26, pp. 203–211, 1974.

[49] H. L. Hurd, "Representation of Strongly Harmonizable Periodically Correlated Processes and Their Covariances," *J. Multivariate Anal.*, vol. 29, no. 1, pp. 53–67, 1989.

[50] H. L. Hurd, "Nonparametric Time Series Analysis for Periodically Correlated Processes," *IEEE Trans. on Information Theory*, vol. 35, pp. 350–359, 1989.

[51] H. L. Hurd, "Correlation Theory of Almost Periodically Correlated Processes," *J. Multivariate Anal.*, vol. 37, no. 1, pp. 24–45, 1991.

[52] H. L. Hurd and N. L. Gerr, "Graphical Methods for Determining the Presence of Periodic Correlation in Time Series," *J. Time Series Anal.*, vol. 12, no. 4, pp. 337–350, 1991.

[53] H. L. Hurd and J. Leskow, "Estimation of the Fourier Coefficient Functions and Their Spectral Densities for $\phi-$Mixing Almost Periodically Correlated Processes," *Statistics and Probability Letters*, vol. 14, no. 4, pp. 299–306, 1992.

[54] H. L. Hurd and J. Leskow, "Strongly Consistent and Asymptotically Normal Estimation of the Covariance for Almost Periodically Correlated Processes," *Statistics and Decisions*, vol. 10, pp. 201–225, 1992.

[55] H. L. Hurd and V. Mandrekar, "Spectral Theory of Periodically and Quasi-Periodically Stationary $S\alpha S$ Sequences," Technical Report No. 349, Center for Stochastic Processes, Dept. of Statistics, UNC at Chapel Hill, Sept. 1991.

[56] H. L. Hurd and G. Kallianpur, "Periodically Correlated and Periodically Unitary Processes and Their Relationship to $L_2[0, T]$-valued Stationary Sequences," in *Nonstationary Stochastic Processes and Their Application*, J. C. Hardin and A. G. Miamee, eds., Singapore: World Scientific Publishing Co., 1992.

[57] H. L. Hurd, "Almost Periodically Unitary Stochastic Processes," *Stoch. Processes and Their Appl.*, vol. 43, pp. 99–113, 1992.

[58] H. L. Hurd and A. Russek, "Almost Periodically Correlated and Almost Periodically Unitary Processes in the Sense of Stepanov," Technical Report No. 368, Center for Stochastic Processes, Dept. of Statistics, UNC at Chapel Hill, July 1992.

[59] H. L. Hurd and A. Russek, "Almost Periodically Correlated Processes in LCA Groups," Technical Report No. 369, Center for Stochastic Processes, Dept. of Statistics, UNC at Chapel Hill, 1992.

[60] H. L. Hurd, "Representation of Discontinuous Processes by (C,1) Means of a Stochastic Fourier Series," Technical Report, Center for Stochastic Processes, Dept. of Statistics, UNC at Chapel Hill, Oct. 1993.

[61] Y. Isokawa, "An Identification Problem in Almost and Asymptotically Almost Periodically Correlated Processes," *J. Appl. Probability*, vol. 19, no. 1, pp. 53–67, 1982.

[62] K. L. Jordan, "Discrete Representations of Random Signals," Technical Report No. 378, MIT Research Laboratory of Electronics, 1961.

[63] G. Kallianpur and V. Mandrekar, "Spectral Theory of Stationary H-Valued Processes," *J. Multivariate Anal.*, vol. 1, no. 1, pp. 1–16, 1971.

[64] Y. Katznelson, *An Introduction to Harmonic Analysis*, New York: Dover, 1976.

[65] A. Khintchine, "Korrelations theorie de stationaren stochastischen prozesse," *Math. Ann.*, vol. 109, pp. 604–615, 1934.

[66] A. Kolmogorov, "Stationary Sequences in Hilbert Space," (in Russian), *Bull. Math. Univ. Moscow*, vol. 2, 1941. Translated in report CN/74/2, J. F. Barrett, trans., Department of Engineering, Cambridge Univ., pp. 1–24, 1974.

[67] J. Leskow, "Asymptotically Normality of the Spectral Density Estimators for Almost Periodically Correlated Processes," *Tech. Report No. 170*, Dept. of Statistics and Applied Probability, UC at Santa Barbara, 1991.

[68] B. M. Levitan and V. V. Zhikov, *Almost Periodic Functions and Differential Equations*, London: Cambridge Univ. Press, 1982.

[69] M. Loève, *Probability Theory*, New York: Van Nostrand, 1965.

[70] A. Makagon and A. G. Miamee, "Correlation Autoregressive Sequences: Their Representations and Harmonizability," preprint.

[71] A. Makagon, A. G. Miamee, and H. Salehi, "Continuous Time Periodically Correlated Processes: Spectrum and Prediction," to appear in *Stoch. Processes and Their Appl.*

[72] A. G. Miamee and H. Salehi, "On the Prediction of Periodically Correlated Stochastic Processes," in *Multivariate Anal. V*, P. R. Krishnaiah, ed., pp. 167–179, Amsterdam: North-Holland, 1980.

[73] D. E. K. Martin, "Estimation of the Minimal Period of Periodically Correlated Sequences," Ph.D. dissertation, Dept. Mathematics, Univ. Maryland at College Park, 1990.

[74] H. Niemi, "Stochastic Processes as Fourier Transforms of Stochastic Measures," *Ann. Acad. Sci. Fenn. AI Math.*, vol. 591, pp. 1–47, 1975.

[75] H. Ogura, "Spectral Representation of Periodic Nonstationary Random Processes," *IEEE Trans. on Information Theory*, vol. IT-17, pp. 143–149, 1971.

[76] E. Parzen, "On Consistent Estimates of the Spectrum of a Stationary Time Series," *Ann. Math. Statist.*, vol. 28, pp. 24–43, 1957.

[77] E. Parzen, "Spectral Analysis of Asymptotically Stationary Time Series," *Bull. Int. Statist. Inst.*, vol. 39, no. 2, pp. 87–103, 1962.

[78] J. Ramanathan and O. Zeitouni, "On the Wavelet Transform of Fractional Brownian Motion," *IEEE Trans. on Information Theory*, vol. 37, pp. 1156–1158, 1991.

[79] Y. A. Rozanov, "Spectral Theory of Multi-Dimensional Stationary Random Processes with Discrete Time," *Uspekhi Mat. Nauk*, XIII, vol. 2, pp. 93–142, 1958.

[80] Y. A. Rozanov, *Stationary Random Processes*, San Francisco: Holden Day, 1967.

[81] M. Rosenblatt, *Stationary Sequences and Random Fields*, Boston: Birkhäuser, 1985.

[82] F. Riesz and B. Sz.-Nagy, *Functional Analysis*, New York: Fredrick Ungar, 1965.

[83] M. H. Stone, "On One Parameter Unitary Groups in Hilbert Space," *Annals of Math.*, vol. 33, pp. 643–648, 1932.

[84] N. Wiener, "Generalized Harmonic Analysis," *Acta Math.*, vol. 55, pp. 117–258, 1930.

[85] N. Wiener and P. Masani, "The Prediction Theory of Multivariate Stochastic Processes," *Acta. Math.*, vol. 98, pp. 111–150, 1950.

[86] A. M. Yaglom, *Correlation Theory of Stationary and Related Random Functions*, New York: Springer-Verlag, 1987.

[87] I. N. Yavorskii, "The Statistical Analysis of Periodically Correlated Random Processes," (in Russian), *Radiotekhnika i elektronika*, vol. 30, no. 6, pp. 1096–1104, 1985.

PART II

Article 1

Joint Transmitter/Receiver Optimization for Multiuser Communications

S. Roy J. Yang P. S. Kumar
Department of Electrical Engineering
University of Pennsylvania
Philadelphia, PA 19104

1 INTRODUCTION

The classical communication system design problem has centered around single-user communication in additive noise. Within the last decade, increasing attention has been centered on multiuser communication having qualitatively different features than the single-user problem. While some of these developments have been prompted by specific multiaccess communication technologies (such as time code-division multiaccess in mobile radio), improved system design in the presence of cochannel interference (and not merely additive noise) is increasingly the canonical issue that must be addressed in many future system design and optimization tasks.

For *linear* communication systems, a common approach to system design is based on the joint minimization (with respect to the transmit-receive filter pair) of the mean-squared error (MSE) between the transmitted symbol and its estimate at the output of the receiver, subject to a power constraint at the transmitter output. Traditionally, receiver optimization (for a fixed transmit filter) has been overwhelmingly emphasized compared to joint transmit-receive optimization even for the single-user problem (as is evident by the extensive literature on the subject, summarized in standard references such as [1]), even though the optimum transmit filter has been known since 1967 [2]. This lack of emphasis on joint transmitter-receiver optimization for single-user communications (represented by a single-input, single-output (SISO) system) perturbed solely by additive noise may be partially explained by the availability of implementationally simpler alternatives such as forward error correction (FEC) to

This work was partially supported by NSF under NCR 89-14431.

improve system reliability. However, for high performance multiuser communications in interference-limited environments, all avenues of system optimization must be explored, particularly since in some situations, FEC coding may not be effective. Increasingly, evidence in recent work [3, 4, 5] appears to support the hypothesis that *transmitter* optimization (in addition to receiver optimization) contributes significantly to cross-talk suppression in multiuser systems.

The purpose of this paper is two-fold: (i) to present a coherent view of system design approaches that include different but related multiinput, multioutput (MIMO) models representing multiuser scenarios, and (ii) to summarize results that constitute the present state-of-knowledge and underscore the potential for cochannel interference suppression by these methods.

A generic block diagram of a MIMO system is shown in Fig. 1 that transmits a *vector* of symbols $\mathbf{a}_l$ every T seconds, which is subsequently detected as $\hat{\mathbf{a}}_l$ at the receiver output. The dimension of the input stream (corresponding to the number of users) is N, while the dimension of the transmitter output is P, implying that the transmit matrix is $P \times N$. The channel matrix is assumed to be $P \times P$ in this work; a more general case is treated in [6].

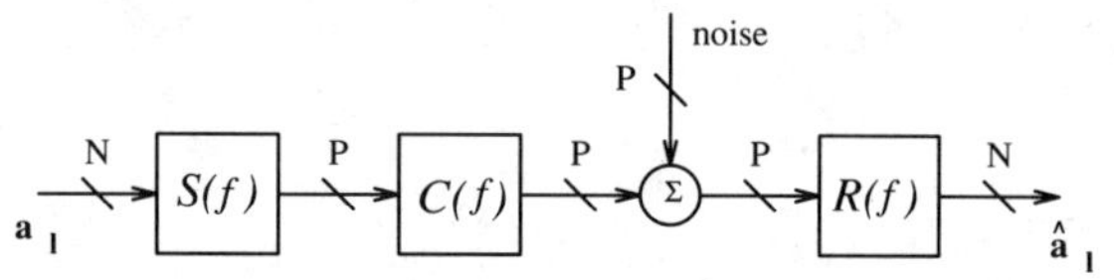

Figure 1: MIMO system representation.

1.1 MIMO System Optimization: Prelude

The problems of MIMO system optimization may superficially appear to be a straightforward vector generalization of the scalar (SISO) case. Such a view is misleading due to the singular fact that cross-talk from other users (that consists of data-modulated signals such as that from the desired user) is not wide-sense stationary (WSS) but wide-sense cyclostationary (WSCS) [7, 8] assuming synchronous transmission whenever the system responses have excess bandwidth, i.e., the spectral support exceeds the first Nyquist zone $(-\frac{1}{2T}, \frac{1}{2T})$ corresponding to a symbol interval T, as is usually true. A key feature of such (interfering) WSCS processes is that they exhibit spectral correlation as characterized by the (nonzero) *cycle frequencies* of the cyclic correlation function (or spectral correlation density) [7–9]. An optimized system seeks to exploit this spectral redundancy inherent in WSCS signals to achieve cancellation of the unwanted signals. Thus, the performance of the conventional single-user receiver (i.e., a matched-filter followed by a tapped-delay line equalizer and detector) that is optimum for the single source in additive Gaussian noise (AGN), but which treats such cross-talk interference as WSS noise, can be significantly suboptimal compared to an optimized MIMO system.

A consistent understanding and appreciation of the implications of the WSCS nature of data-modulated signals began with the seminal contributions of [10]. Of the many results in [10], a key one (for our purposes) was the solution to the time-domain linear MMSE estimation problem for jointly WSCS processes (akin to the standard Wiener filtering problem for jointly WSS processes). The essential results were summarized in [11] and applied to the single-user receiver design problem. In particular, it was shown that the optimal receiver belongs to the class of linear periodically time-varying systems. Some structural implications of this were elucidated in [8, 12], particularly with a view to adaptive implementations. In more recent work [13, 14, 15], frequency-domain interpretations of the above as embodied in the concept of FREquency-SHift (FRESH) filtering have been developed. Examples presented there underscore the potential for separation of cochannel WSCS interfering signals based on the exploitation of spectral redundancy. In some cases, this can involve exploiting differences in the cycle frequencies of the desired and interfering signals in the design of the receiver. For digitally modulated signals, such differences typically occur due to (i) different symbol (or keying) rates and/or (ii) different carrier frequencies for separate users.

However, in many scenarios, the necessity for standardization may imply that several cochannel users communicate using a common symbol rate and (nominal) carrier frequency. In these situations, exploitation of the differences in the *shapes* of the spectral correlation densities of the WSCS processes, achieved by appropriate pulse shaping at the transmitter, becomes critical to cochannel interference rejection. This naturally leads to investigations of the joint optimization of the receiver and the (transmit) shaping filter, which has been relatively unexplored since [16]. Some evidence of the importance of the spectral support of the system toward interference suppression was presented in [3, 15], and enhanced by the more recent analysis and results concerning transmit pulse shape optimization in [4, 5, 6, 17]. In this article, an overview of these new results is provided. Taken together, the results reported here and those in [4, 5, 14, 15, 18] offer a comprehensive picture of the state-of-the-art in co- and adjacent-channel interference rejection via the exploitation of WSCS.

As with the SISO case, the joint optimization of MIMO systems can be viewed as a two-step procedure:

1. The receive (matrix) filter is first optimized for a given transmit filter;
2. Substituting for the optimal receiver filter, the MSE is then minimized with respect to the transmit (matrix) filter, subject to an average transmit power constraint to ensure a well-posed problem.

In consonance with SISO optimization, step (1) is analytically simpler and has consequently gained considerable attention, beginning with [19] and developed more comprehensively by [4, 18, 20].

It is useful to highlight here some aspects of the MIMO system optimization problem. In a generic multipoint to multipoint communication scenario, all the input data streams constitute "desired" signals and need to be detected simultaneously. Such a goal could be embodied in the simultaneous minimization of the total (i.e., over all N users) MSEs,

$$\min_{\mathbf{R}} \xi = E\{\|\mathbf{a}_l - \hat{\mathbf{a}}_l\|\}^2. \tag{1}$$

However, the design of any one receiver (nominally the ith) may be based on minimization of the MSE only for its desired input to the exclusion of the others,

$$\min_{R_i} \xi_i = E\left\{a_l^{(i)} - \hat{a}_l^{(i)}\right\}^2. \tag{2}$$

It deserves to be emphasized (as was indeed noted by [4]) that the optimum *receiver* for these two different problems is identical, due to the fact that the first problem of global minimization over all the users' data decouples into *independent* subproblems at each of the receiver locations and can be viewed as interference rejection problems. In contrast, no such decoupling is feasible for transmitter optimization, since the MSE ξ_i for the ith user is a function of *all* the transmit filters. Consequently, transmitter optimization (once the receiver has already been optimized for a given transmitter) is a problem of greater analytical complexity, with fewer results to date.

Finally, physically different multiuser scenarios impose *structural restrictions* on MIMO system models used for optimization. In this work, we describe two such models that lead to somewhat different analytical results. These distinctions are based on assumptions about the nature of cooperation between the different users. In applications such as the high-speed Digital Subscriber Lines (DSL), coordinated two-pair transmission has been proposed [21] implying a cross-coupled transmit matrix. Such a system with a fully cross-coupled receiver is shown in Fig. 2 for the two-user case. In contrast, in a mobile communications scenario, different users do not coordinate with each other and this is represented by an (uncoupled) diagonal transmit matrix. Assuming an *uncoupled receiver* leads to a fully uncoordinated system, which is shown in Fig. 5(a) for the two-user case.

The rest of the paper concentrates on the optimization problem for these two situations and is organized as follows. Sec. 2 deals with results for a fully coordinated MIMO system beginning with the *Nyquist bandlimited* case in Sec. 2.1. These results can then be applied to the situation of interest—systems with excess bandwidth, where the cross-talk is indeed cyclostationary—as is shown in Sec. 2.2. The optimization problem for uncoordinated systems is investigated in Sec. 3. Following the major developments in Sec. 3.2 assuming cyclostationary interference, Sec. 3.3 revisits the problem but with a stationary cross-talk assumption, to provide a baseline for appreciation of the additional suppression of synchronous cross-talk that is achievable. The paper concludes with representative computational examples in Sec. 4 to compare system performance as a function of system *structure* (i.e., uncoordinated vs. coordinated) and critical system *parameters* such as bandwidth (measured in number of Nyquist zones available at the symbol rate) vis-à-vis number of users.

2 MIMO SYSTEM OPTIMIZATION: COORDINATED USERS

To provide a perspective to our developments for the MIMO optimization problem, it is necessary to revisit the pioneering results of Berger and Tufts [2] for the SISO case (i.e., additive noise only). Their key observation regarding the optimum transmit filter can be summarized as follows:

Consider all frequencies in the set $\mathcal{I} \equiv \{f : f_0 + \frac{k}{T},\ k \in \mathcal{Z}\}$ where T is the symbol period, and $f_0 \in \mathcal{I}_0 \equiv [-\frac{1}{2T}, \frac{1}{2T}]$. The optimal transmitter has support at $f \in \mathcal{I}$ only if (i) a minimum-allowable spectral signal-to-noise ratio (SNR) at the channel output is preserved; and (ii) within this subset of $\mathcal{I}$, it allocates power only at the unique k for which this SNR is maximum (in case of nonuniqueness, any k that maximizes the SNR can be chosen).

As a direct corollary, this implies that, in general, the optimal transmitter is *not* bandlimited to the first Nyquist zone and its support is nonzero only on a set (termed a Nyquist set in [2]) of measure (at most) $\frac{1}{T}$ [22]. Note that the support set, in general, is not connected, and exceeds the basic Nyquist interval $\mathcal{I}_0$.

The first attempt to extend the above to multiuser or MIMO systems appeared in [20]. However, the results obtained were restricted to the case where the system is bandlimited to the first Nyquist zone $(-\frac{1}{2T}, \frac{1}{2T})$. Its significance is thereby limited in view of [2]. Nevertheless, a notable contribution of [20] was that the optimization problem for a set of mutually cross-coupled channels was shown to be reducible to that of an equivalent *decoupled* system consisting of parallel subchannels. As will be apparent, the developments presented here revolve around two key decompositions (of the transmit and receive filters, respectively) that allow a similar transformation of the original coupled problem into an equivalent decoupled one.

A basic coordinated MIMO two-user system is shown in Fig. 2 for $P = N = 2$ (merely for convenience); the analytical results are valid however for all P, N [6]. The derivation presented focuses on the $P \geq N$ case, since this will be subsequently used (in Sec. 2.2) to derive optimum systems with excess bandwidth.

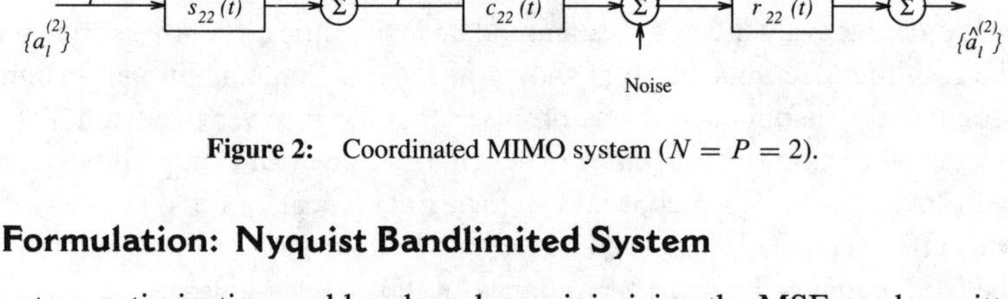

Figure 2: Coordinated MIMO system ($N = P = 2$).

2.1 Formulation: Nyquist Bandlimited System

The system optimization problem based on minimizing the MSE can be written as

$$\min \xi = tr\left\{\mathbf{E}\,\mathbf{e}_l\mathbf{e}_l^\dagger\right\} = \mathbf{E}\left\{\|\hat{\mathbf{a}}_l - \mathbf{a}_l\|^2\right\}, \tag{3}$$

where $\mathbf{e}_l = \hat{\mathbf{a}}_l - \mathbf{a}_l$ is the error sequence corresponding to the transmitted and the received (vector) symbols. The above can be rewritten in the frequency domain as

$$\min_{\mathbf{S}(f),\,\mathbf{R}(f)} \xi = tr\, T\int_{-\frac{1}{2T}}^{\frac{1}{2T}} \left\{[\mathbf{I} - \frac{1}{T}\mathbf{RCS}]\boldsymbol{\Sigma}_{\mathbf{a}}[\mathbf{I} - \frac{1}{T}\mathbf{RCS}]^\dagger + \frac{1}{T}\mathbf{R}\boldsymbol{\Sigma}_{\mathbf{n}}\mathbf{R}^\dagger\right\} df \tag{4}$$

subject to the transmit power constraint

$$\frac{1}{T}\int_{-\frac{1}{2T}}^{\frac{1}{2T}} tr\{\mathbf{S}(f)\mathbf{\Sigma_a}(f)\mathbf{S}^{\dagger}(f)\}df = NP_0, \tag{5}$$

where $\mathbf{\Sigma_a}(f)$ and $\mathbf{\Sigma_n}(f)$ are the power spectral density matrices of the (vector) input data and noise processes, respectively; $\mathbf{R}(f), \mathbf{C}(f), \mathbf{S}(f)$ represent the receiver, channel, and transmitter's frequency responses, and $\dagger$ denotes conjugate transposition. Note that in this section the support of all (frequency domain) quantities is assumed restricted to $\mathcal{I}_0$.

In the interests of analytical tractability, it is useful to transform the above problem into an equivalent one defined in terms of the new variables $\mathbf{F}(f)$ and $\mathbf{E}(f)$,

$$\mathbf{S}(f) = \mathbf{V}(f)\mathbf{F}(f)\mathbf{U}^{\dagger}(f), \tag{6}$$

$$\mathbf{R}(f) = \mathbf{U}(f)\mathbf{E}(f)\mathbf{V}^{\dagger}(f)\mathbf{C}^{\dagger}(f)\mathbf{\Sigma_n}^{-1}(f) \tag{7}$$

where $\mathbf{U}(f)$ and $\mathbf{V}(f)$ are unitary matrices that diagonalize the input data covariance matrix $\mathbf{\Sigma_a}(f)$ and the Hermitian matrix $\mathbf{C}^{\dagger}(f)\mathbf{\Sigma_n}^{-1}(f)\mathbf{C}(f)$, respectively, i.e.,

$$\mathbf{U}^{\dagger}(f)\mathbf{\Sigma_a}(f)\mathbf{U}(f) = \mathbf{K_a}(f) \tag{8}$$

$$\frac{1}{T}\mathbf{V}^{\dagger}(f)\mathbf{C}^{\dagger}(f)\mathbf{\Sigma_n}^{-1}(f)\mathbf{C}(f)\mathbf{V}(f) = \mathbf{\Lambda_n}^{-1}(f). \tag{9}$$

It is apparent that $\mathbf{K_a}$ and $\mathbf{\Lambda_n}^{-1}$ are diagonal matrices that contain the eigenvalues of $\mathbf{\Sigma_a}(f)$ and $\frac{1}{T}\mathbf{C}^{\dagger}(f)\mathbf{\Sigma_n}^{-1}(f)\mathbf{C}(f)$, respectively, and that such a decomposition always exists. In addition, we assume without loss of generality that the elements of $\mathbf{K_a}$ and $\mathbf{\Lambda_n}$ are ordered as follows

$$\kappa_1(f) \geq \kappa_2(f) \geq \cdots \geq \kappa_N(f) > 0 \tag{10}$$

$$0 < \lambda_1(f) \leq \lambda_2(f) \geq \cdots \leq \lambda_P(f). \tag{11}$$

The motivation for this ordering will be made apparent subsequently. The case of singular (zero) eigenvalues can be accommodated with little difficulty as shown in [6].

The resulting decomposition is shown in Fig. 3. Some additional insight into this decomposition is obtained if one observes that the receiver front end $\mathbf{V}^{\dagger}\mathbf{C}^{\dagger}\mathbf{\Sigma_n}^{-1}$ is simply the whitening (matrix) filter matched to the channel $\mathbf{C}$ in additive Gaussian noise with covariance $\mathbf{\Sigma_n}$, which is known to be a canonical lossless (i.e., information preserving) front end [18].

The MSE can now be rewritten in terms of the new variables as

$$\begin{aligned}\xi &= trT\int_{-\frac{1}{2T}}^{\frac{1}{2T}} \left\{[\mathbf{I} - \mathbf{E}\mathbf{\Lambda_n}^{-1}\mathbf{F}]\mathbf{K_a}[\mathbf{I} - \mathbf{E}\mathbf{\Lambda_n}^{-1}\mathbf{F}]^{\dagger} + \mathbf{E}\mathbf{\Lambda_n}^{-1}\mathbf{E}^{\dagger}\right\} df \\ &= trT\int_{-\frac{1}{2T}}^{\frac{1}{2T}} \left\{[\mathbf{I} - \mathbf{D}(f)\mathbf{F}(f)]\mathbf{K_a}(f)[\mathbf{I} - \mathbf{D}(f)\mathbf{F}(f)]^{\dagger} + \mathbf{D}(f)\mathbf{\Lambda_n}(f)\mathbf{D}^{\dagger}(f)\right\} df\end{aligned} \tag{12}$$

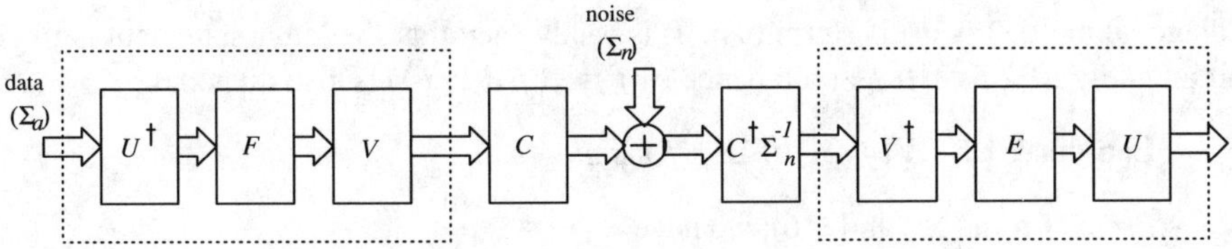

Figure 3: Transmitter and receiver decomposition.

where

$$\mathbf{D}(f) = \mathbf{E}(f)\boldsymbol{\Lambda}_{\mathbf{n}}^{-1}(f). \tag{13}$$

The transformed transmit power constraint is now given by

$$\frac{1}{T}\int_{-\frac{1}{2T}}^{\frac{1}{2T}} tr\{\mathbf{F}(f)\mathbf{K_a}(f)\mathbf{F}^\dagger(f)\}df \;=\; NP_0. \tag{14}$$

Comparing (12) and (14) with (4) and (5), it is clear that the original problem has been converted to that of finding the optimal transmit-receive pair for mutually uncorrelated input data and noise sequences with uncorrelated components in the presence of an ideal channel.

The method of Lagrange multipliers can now be invoked to solve the constrained optimization problem given by (12) and (14). Using standard techniques from variational calculus and setting the first variation with respect to **D** and **F** to zero, yields the following necessary conditions:

$$\mathbf{D}(f) = \mathbf{K_a}(f)\mathbf{F}^\dagger(f)[\mathbf{F}(f)\mathbf{K_a}(f)\mathbf{F}^\dagger(f) + \boldsymbol{\Lambda}_{\mathbf{n}}(f)]^{-1} \tag{15}$$

$$\mathbf{D}^\dagger(f)\mathbf{D}(f)\mathbf{F}(f) + \gamma\mathbf{F}(f) = \mathbf{D}^\dagger(f)\,, \tag{16}$$

where the constant γ is chosen to satisfy the power constraint (14).

Lemma 2.1.1 $\mathbf{F}(f)\mathbf{K_a}(f)\mathbf{F}^\dagger(f)$ *is diagonal.*

Proof: From (16), we have upon postmultiplying by $\mathbf{K_a}(f)\mathbf{F}^\dagger(f)$

$$\mathbf{D}^\dagger(f)\mathbf{D}(f)\mathbf{F}(f)\mathbf{K_a}(f)\mathbf{F}^\dagger(f) + \gamma\mathbf{F}(f)\mathbf{K_a}(f)\mathbf{F}^\dagger(f) = \mathbf{D}^\dagger(f)\mathbf{K_a}(f)\mathbf{F}^\dagger(f)\,, \tag{17}$$

while (15) straightforwardly gives

$$\mathbf{D}(f)\mathbf{F}(f)\mathbf{K_a}(f)\mathbf{F}^\dagger(f) = \mathbf{K_a}(f)\mathbf{F}^\dagger(f) - \mathbf{D}(f)\boldsymbol{\Lambda}_{\mathbf{n}}(f)\,. \tag{18}$$

Premultiplying (18) by $\mathbf{D}^\dagger(f)$ and subtracting from (17) leads to

$$\gamma\mathbf{F}(f)\mathbf{K_a}(f)\mathbf{F}^\dagger(f) = \mathbf{D}^\dagger(f)\mathbf{D}(f)\boldsymbol{\Lambda}_{\mathbf{n}}(f). \tag{19}$$

Then, since the left-hand side of (19) is clearly Hermitian, it implies that $\mathbf{D}^\dagger(f)\mathbf{D}(f)\boldsymbol{\Lambda}_{\mathbf{n}}(f)$ (which is the product of a Hermitian matrix and a nondegenerate diagonal matrix) is itself Hermitian. It is easily seen that the latter statement is true if and only if $\mathbf{D}^\dagger(f)\mathbf{D}(f)$ (and hence $\mathbf{F}(f)\mathbf{K_a}(f)\mathbf{F}^\dagger(f)$) is also diagonal.

Lemma 2.1.2 $\mathbf{F}(f)\mathbf{K}_\mathbf{a}^2(f)\mathbf{F}^\dagger(f)$ *is diagonal.*

Proof: From (15) and (16), we have

$$[\mathbf{F}(f)\mathbf{K_a}(f)\mathbf{F}^\dagger(f) + \mathbf{\Lambda_n}(f)][\mathbf{D}^\dagger(f)\mathbf{D}(f) + \gamma\mathbf{I}]\mathbf{F}(f)\mathbf{K_a}(f)\mathbf{F}^\dagger(f) \\ = \mathbf{F}(f)\mathbf{K}_\mathbf{a}^2(f)\mathbf{F}^\dagger(f). \tag{20}$$

The diagonality of $\mathbf{F}(f)\mathbf{K}_\mathbf{a}^2(f)\mathbf{F}^\dagger(f)$ follows from the fact that the left-hand side of (20) is the product of three diagonal matrices.

Proposition 2.1.1 *The optimal* $\mathbf{F}(f)$ *(*$P \times N$*) is diagonal; i.e., only* $\mathbf{F}_{ii}(f)$ *are nonzero.*

This central result is established in three steps. In Parts 1 and 2, it is shown that the optimum $\mathbf{F}(f)$ has at most *one* nonzero element in each *row* and *column* respectively. Part 3 then demonstrates that the optimal $\mathbf{F}(f)$ within this class is diagonal. In the sequel, the frequency variable f is omitted, wherever possible, without sacrificing clarity.

Part 1: To show there is at most *one* nonzero element in each *row*, note that the MSE of (12) can be rewritten using (15) as

$$\xi = tr\ T \int_{-\frac{1}{2T}}^{\frac{1}{2T}} \mathbf{K_a}(f)\left\{\mathbf{I} - \mathbf{F}^\dagger(f)[\mathbf{F}(f)\mathbf{K_a}(f)\mathbf{F}^\dagger(f) + \mathbf{\Lambda_n}(f)]^{-1}\mathbf{F}(f)\mathbf{K_a}(f)\right\} df. \tag{21}$$

Thus minimization of ξ is equivalent to

$$\max\ tr \int_{-\frac{1}{2T}}^{\frac{1}{2T}} \mathbf{F}(f)\mathbf{K}_\mathbf{a}^2(f)\mathbf{F}^\dagger(f)\,[\mathbf{F}(f)\mathbf{K_a}(f)\mathbf{F}^\dagger(f) + \mathbf{\Lambda_n}(f)]^{-1}df. \tag{22}$$

Using Lemma 2.1.1 and Lemma 2.1.2, the integrand of (22) simplifies to

$$\sum_i \frac{[\mathbf{F}(f)\mathbf{K}_\mathbf{a}^2(f)\mathbf{F}^\dagger(f)]_{ii}}{[\mathbf{F}(f)\mathbf{K_a}(f)\mathbf{F}^\dagger(f) + \mathbf{\Lambda_n}(f)]_{ii}}. \tag{23}$$

Note that the ith diagonal element of both $\mathbf{F}(f)\mathbf{K_a}(f)\mathbf{F}^\dagger(f)$ and $\mathbf{F}(f)\mathbf{K}_\mathbf{a}^2(f)$ $\mathbf{F}^\dagger(f)$ depend *only* on the ith row of $\mathbf{F}(f)$,

$$[\mathbf{F}\mathbf{K_a}\mathbf{F}^\dagger]_{ii} = \sum_j |F_{ij}|^2\kappa_j \text{ and } [\mathbf{F}\mathbf{K}_\mathbf{a}^2\mathbf{F}^\dagger]_{ii} = \sum_j |F_{ij}|^2\kappa_j^2\,. \tag{24}$$

Assume now that the ith row of $\mathbf{F}(f)$ has more than one nonzero element. Consider the matrix $\hat{\mathbf{F}}(f)$ obtained from $\mathbf{F}(f)$ in the following manner:

For the index set $U_j^{(i)} \triangleq \{j : F_{ij} \neq 0\}$ for each i, find $l = \arg\max_{j \in U_j^{(i)}} \kappa_j$ and choose $|\hat{F}_{il}|^2$ such that

$$\sum_j |F_{ij}|^2 \kappa_j = |\hat{F}_{il}|^2 \kappa_l \text{ and } \hat{F}_{ij} = 0,\ j \neq l. \tag{25}$$

Note that this operation preserves the diagonality of $\mathbf{F}(f)\mathbf{K_a}(f)\mathbf{F}^\dagger(f)$ and $\mathbf{F}(f)\mathbf{K}_\mathbf{a}^2(f)\mathbf{F}^\dagger(f)$, implying that (23) is still valid for minimization of the mean-squared error, with $\hat{\mathbf{F}}(f)$ substituted for $\mathbf{F}(f)$. Further, note that $\hat{\mathbf{F}}(f)$ also satisfies the power constraint (14) due to (25), and that $\hat{\mathbf{F}}(f)$ contains only one nonzero element in each row. Now

$$\sum_j |F_{ij}|^2 \kappa_j^2 \leq \kappa_l \sum_j |F_{ij}|^2 \kappa_j = |\hat{F}_{il}|^2 \kappa_l^2, \tag{26}$$

where the last equality follows using (25). It is evident by using (26) in (22) that (23) is optimized for $\mathbf{F}(f) = \hat{\mathbf{F}}(f)$. The above establishes that the optimal $\mathbf{F}(f)$ can have at most one nonzero element in each row for nondegenerate $\mathbf{K_a}(f)$.

Remark 2.1.1: If $\mathbf{K_a}(f)$ is degenerate, e.g., $\mathbf{K_a}(f) = \sigma_a^2 I$ (as in [20]), it follows from the above that the optimal $\mathbf{F}(f)$ is unique only up to a *unitary* transformation. However, since a unitary transformation leaves the MSE unchanged while satisfying the power constraint, the minimum MSE is always achievable by an $\mathbf{F}(f)$ with at most one nonzero element in each row.

Part 2: We now proceed to show that there can be at most one nonzero element in each *column* of $\mathbf{F}(f)$ as well. Note that $\mathbf{F}(f)\mathbf{K_a}(f)\mathbf{F}^\dagger(f)$ can be written as the outer-product decomposition

$$\mathbf{F}(f)\mathbf{K_a}(f)\mathbf{F}^\dagger(f) = \sum_k \kappa_k(f)\mathbf{f}_k(f)\mathbf{f}_k^\dagger(f), \tag{27}$$

where $\mathbf{f}_k(f)$ is the kth column vector of $\mathbf{F}(f)$. Now each off-diagonal element of (27) can have contribution from at most *one* of the terms in the summation, because of the just-proven fact that there is at most *one* nonzero element in each row of $\mathbf{F}(f)$. Assuming that $\mathbf{K_a}(f)$ is nonsingular with probability 1, it follows that in order for the right-hand side of (27) to be diagonal, there can be at most one nonzero element in each $\mathbf{f}_i$ or, equivalently, there is at most one nonzero element in each column of $\mathbf{F}(f)$. Thus the important property that the optimal $\mathbf{F}(f)$ contains at most *one* nonzero element in each *row* and *column* of $\mathbf{F}(f)$ is established.

Part 3: Based on the results in Parts 1 and 2, the optimal $\mathbf{F}(f)$ can be represented as

$$\mathbf{F}(f) = \mathcal{P}(f)\tilde{\mathbf{F}}(f), \tag{28}$$

where $\tilde{\mathbf{F}}(f)$ is a $P \times N$ matrix with nonzero elements on the main diagonal, and $\mathcal{P}(f)$ is a $P \times P$ permutation matrix. We are now left with the problem of showing that $\mathcal{P}(f) = I$ corresponds to the desired result, and this will be achieved by a constructive demonstration of the fact that any other $\mathcal{P}(f)$ is not optimal.

From (15), it is clear that $\mathbf{D}(f)$ can have nonzero elements only at those locations where $\mathbf{F}^{\dagger}(f)$ has a nonzero element since $\mathbf{D}(f)$ is equal to $\mathbf{F}^{\dagger}(f)$ pre- and postmultiplied by diagonal matrices. Hence we can conclude that $\mathbf{D}(f)$ also has at most one nonzero element per row and column, allowing the representation

$$\mathbf{D}(f) = \tilde{\mathbf{D}}(f)\mathcal{P}^{\dagger}(f) \tag{29}$$

where, $\tilde{\mathbf{D}}(f)$ is an $N \times P$ matrix with nonzero elements along the main "diagonal."

Using the property of permutation matrices that

$$\mathcal{P}^{\dagger}(f)\mathcal{P}(f) = \mathcal{P}(f)\mathcal{P}^{\dagger}(f) = \mathbf{I}, \tag{30}$$

the MSE in (12) can be reexpressed as

$$\xi = trT\int_{-\frac{1}{2T}}^{\frac{1}{2T}} \left\{[\mathbf{I} - \tilde{\mathbf{D}}(f)\tilde{\mathbf{F}}(f)]\mathbf{K_a}(f)[\mathbf{I} - \tilde{\mathbf{D}}(f)\tilde{\mathbf{F}}(f)]^{\dagger} + \tilde{\mathbf{D}}(f)\tilde{\mathbf{\Lambda}}_{\mathbf{n}}(f)\tilde{\mathbf{D}}^{\dagger}(f)\right\} df \tag{31}$$

where

$$\tilde{\mathbf{\Lambda}}_{\mathbf{n}}(f) = \mathcal{P}^{\dagger}(f)\mathbf{\Lambda}_{\mathbf{n}}(f)\mathcal{P}(f) \tag{32}$$

is a diagonal matrix of the additive noise variances with its elements permuted vis-à-vis $\mathbf{\Lambda}_{\mathbf{n}}(f)$. The transformed power constraint (14) now becomes

$$\frac{1}{T}\int_{-\frac{1}{2T}}^{\frac{1}{2T}} tr\left\{\tilde{\mathbf{F}}(f)\mathbf{K_a}(f)\tilde{\mathbf{F}}^{\dagger}(f)\right\} df = \frac{1}{T}\int_{-\frac{1}{2T}}^{\frac{1}{2T}} \sum_i |\tilde{F}(f)_{ii}|^2\kappa_i(f)df = NP_0 \tag{33}$$

and a necessary condition commensurate with (15) now holds for the permuted receiver matrix $\tilde{\mathbf{D}}(f)$, i.e.,

$$\tilde{\mathbf{D}}(f) = \mathbf{K_a}(f)\tilde{\mathbf{F}}^{\dagger}(f)[\tilde{\mathbf{F}}(f)\mathbf{K_a}(f)\tilde{\mathbf{F}}^{\dagger}(f) + \tilde{\mathbf{\Lambda}}_{\mathbf{n}}(f)]^{-1}. \tag{34}$$

Using (34), the MSE of (31) can be expressed as

$$\xi = T\int_{-\frac{1}{2T}}^{\frac{1}{2T}} \xi(f)df = T\int_{-\frac{1}{2T}}^{\frac{1}{2T}} \sum_{i=1}^{N} \frac{\kappa_i(f)}{|\tilde{F}_{ii}(f)|^2\frac{\kappa_i(f)}{\tilde{\lambda}_i(f)} + 1} df \tag{35}$$

where $\tilde{F}_{ii}$ and $\tilde{\lambda}_i$ denote the ith elements on the diagonals of $\tilde{\mathbf{F}}$ and $\tilde{\mathbf{\Lambda}}_{\mathbf{n}}$, respectively. Since (35) is monotonic nondecreasing in $\{\tilde{\lambda}_i\}_{i=1}^{N}$'s, it is clear that the permutation $\mathcal{P}$ is one that selects the smallest N diagonal elements of $\mathbf{\Lambda}_{\mathbf{n}}$ $\{\lambda_1, \ldots, \lambda_N\}$. Thus, the minimal MSE does not depend on the remaining $P - N$ largest λ_i's, and we need only examine the *ordering* of the first N diagonal elements of $\tilde{\mathbf{\Lambda}}_{\mathbf{n}}(f)$.

Note from (11) that the ordering of the elements in $\mathbf{\Lambda_n}(f)$ satisfies, for any $j < k$,

$$\lambda_j(f) \leq \lambda_k(f), \ \forall f. \tag{36}$$

Thus any $\tilde{\mathbf{\Lambda}}_{\mathbf{n}}(f)$ that is not identical to $\mathbf{\Lambda}(f)$ in the ordering of the first N diagonal elements, contains at least one pair of indices $j < k \leq N$ such that for some f_0,

$$\tilde{\lambda}_j(f_0) \geq \tilde{\lambda}_k(f_0). \tag{37}$$

Consider $\overline{\mathbf{\Lambda}}_{\mathbf{n}}(f)$ that is identical to $\tilde{\mathbf{\Lambda}}_{\mathbf{n}}(f)$ except for swapping the jth and kth diagonal elements at f_0, i.e.,

$$\overline{\lambda}_j(f_0) = \tilde{\lambda}_k(f_0) \text{ and } \overline{\lambda}_k(f_0) = \tilde{\lambda}_j(f_0). \tag{38}$$

The difference between the integrand of (35) for these two permutations is easily shown to be

$$\xi(f) - \overline{\xi}(f) = \begin{cases} \left(\dfrac{\kappa_j \tilde{\lambda}_j}{|\tilde{F}_{jj}|^2 \kappa_j + \tilde{\lambda}_j} + \dfrac{\kappa_k \tilde{\lambda}_k}{|\tilde{F}_{kk}|^2 \kappa_k + \tilde{\lambda}_k}\right) \\ \qquad - \left(\dfrac{\kappa_j \overline{\lambda}_j}{|\overline{F}_{jj}|^2 \kappa_j + \overline{\lambda}_j} + \dfrac{\kappa_k \overline{\lambda}_k}{|\overline{F}_{kk}|^2 \kappa_k + \overline{\lambda}_k}\right), & f = f_0 \\ 0 & \text{otherwise.} \end{cases} \tag{39}$$

It is shown in [6] that the right-hand side above is nonnegative while satisfying

$$|\overline{F}_{jj}|^2 + |\overline{F}_{kk}|^2 = |\tilde{F}_{jj}|^2 + |\tilde{F}_{kk}|^2. \tag{40}$$

This implies that $\overline{\mathbf{\Lambda}}_{\mathbf{n}}(f)$ is at least as good as $\tilde{\mathbf{\Lambda}}_{\mathbf{n}}(f)$ while preserving the same transmit power due to (40). This process can be repeated until no such j, k pairs exist, thereby proving that the ordering of the elements of $\mathbf{\Lambda_n}(f)$ is optimal and, by implication, the diagonality of $\mathbf{F}(f)$. With the diagonality of the optimal $\mathbf{F}(f)$ established, the final expression for the minimal MSE becomes

$$\xi = T \int_{-\frac{1}{2T}}^{\frac{1}{2T}} \sum_{i=1}^{N} \frac{\kappa_i(f)}{|F_{ii}(f)|^2 \frac{\kappa_i(f)}{\lambda_i(f)} + 1} \, df. \tag{41}$$

The final step in the proof is the specification of the elements $F_{ii}(f)$'s. This is achieved by observing that (41) is convex in $\{|F_{ii}(f)|^2\}$, implying that the optimal $|F_{ii}(f)|^2$ that minimize (35) subject to the convex set (33) is obtained as the solution to a well-known allocation problem [23] popularly illustrated by a "water-pouring" argument [24]. The solution is

$$|F_{ii}(f)|^2 = \max\left(0, \sqrt{\frac{\lambda_i(f)}{\mu \kappa_i(f)}} - \frac{\lambda_i(f)}{\kappa_i(f)}\right), \qquad i = 1, 2, \ldots, N, \tag{42}$$

where μ is chosen so as to satisfy the power constraint (14).

From (31) and the diagonality of $F(f)$, it is evident that the original system has been converted into the one shown in Fig. 4 consisting of N parallel independent subsystems, each communicating over an ideal (unity gain, zero-phase) channel.

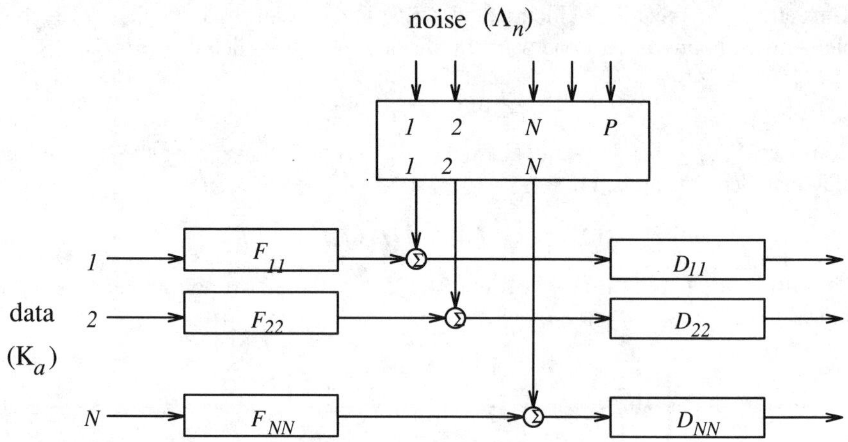

Figure 4: Optimal transmitter and receiver filter structure.

Once the optimal $\mathbf{F}(f)$ is determined by the preceding procedure, one can obtain $\mathbf{D}(f)$ from (15). Finally, the optimal coupled transmitter and receiver filter $\mathbf{S}(f)$ and $\mathbf{R}(f)$ for the original coupled MIMO system can be determined from (6), (7), and (13).

Remark 2.1.2: It is interesting to note that κ_i/λ_i is an effective SNR for the ith subchannel of the decomposed MIMO system. The optimal ordering given by (10) and (11) simply attempts to utilize the N channels with the best SNR, in accordance with intuition.

Remark 2.1.3: Note that (42) specifies only the magnitude of the optimal F_{ii}, implying that the transmit phase can be freely chosen without affecting the optimality as long as the phase of the receive filter is adjusted accordingly, i.e., as long as (34) is satisfied. This parallels the result in [2] for the SISO optimization problem.

2.2 Application to Excess Bandwidth Systems

We reconsider the original MIMO system shown in Fig. 1, but now allow the system to have excess bandwidth. This case is of considerable interest due to the capability for superior cross-talk rejection by systems with excess bandwidth, as demonstrated in [3, 4, 5, 15]. It will be shown that the optimization problem can be converted to that of an equivalent MIMO system bandlimited to the basic Nyquist interval $(-\frac{1}{2T}, \frac{1}{2T})$ with matrix transmit and receive filters of augmented dimensions. Hence, the results of the previous section can be directly applied to obtain a solution to this important problem.

We assume that the transmit filter $\mathbf{S}(f)$, receive filter $\mathbf{R}(f)$, channel response $\mathbf{C}(f)$, and the noise power spectral density (psd) $\mathbf{\Sigma_n}(f)$ are all bandlimited to M Nyquist zones. Let $\mathbf{\Omega}_i(f)$ denote the bandlimited filter restricted to the first Nyquist zone whose response is the same as that of $\mathbf{\Omega}(f)$ in ith Nyquist zone, i.e.,

$$\mathbf{\Omega}_i(f) = \begin{cases} \mathbf{\Omega}(f - \frac{i-1}{T}) & -\frac{1}{2T} \leq f \leq \frac{1}{2T} \\ 0 & \text{elsewhere} \end{cases}. \tag{43}$$

Utilizing this representation, the folded spectrum $\mathbf{\Phi}(f)$ of the cascade of the transmit-channel-receive filters can be expressed as

$$\begin{aligned} \mathbf{\Phi}(f) &= \tfrac{1}{T} \textstyle\sum_{i=1}^{M} \mathbf{R}(f - \frac{i-1}{T})\mathbf{C}(f - \frac{i-1}{T})\mathbf{S}(f - \frac{i-1}{T}) \\ &= \frac{1}{T} \sum_{i=1}^{M} \mathbf{R}_i(f)\mathbf{C}_i(f)\mathbf{S}_i(f) \\ &= \frac{1}{T} \vec{\mathbf{R}}(f)\vec{\mathbf{C}}(f)\vec{\mathbf{S}}(f) \end{aligned} \tag{44}$$

and, similarly, the noise psd at the sampler output in the receiver is given by

$$\begin{aligned} \mathbf{\Psi}(f) &= \frac{1}{T} \sum_{i=1}^{M} \mathbf{R}_i(f)\mathbf{\Sigma}_{\mathbf{n},i}(f)\mathbf{R}_i^{\dagger}(f) \\ &= \frac{1}{T} \vec{\mathbf{R}}(f)\vec{\mathbf{\Sigma}}_{\mathbf{n}}(f)\vec{\mathbf{R}}^{\dagger}(f), \end{aligned} \tag{45}$$

where the quantities with subscript i are Nyquist bandlimited filters defined as in (43), and we have introduced the concatenated matrices

$$\vec{\mathbf{R}}(f) = \left(\mathbf{R}_1(f) \quad \mathbf{R}_2(f) \quad \ldots \quad \mathbf{R}_M(f) \right), \tag{46}$$

$$\vec{\mathbf{S}}(f) = \left(\mathbf{S}_1^T(f) \quad \mathbf{S}_2^T(f) \quad \cdots \quad \mathbf{S}_M^T(f) \right)^T, \tag{47}$$

$$\vec{\mathbf{\Sigma}}_{\mathbf{n}}(f) = \mathrm{diag}\left(\mathbf{\Sigma}_{\mathbf{n},1}(f) \quad \mathbf{\Sigma}_{\mathbf{n},2}(f) \quad \ldots \quad \mathbf{\Sigma}_{\mathbf{n},\mathrm{M}}(f) \right), \tag{48}$$

$$\vec{\mathbf{C}}(f) = \mathrm{diag}\left(\mathbf{C}_1(f) \quad \mathbf{C}_2(f) \quad \ldots \quad \mathbf{C}_M(f). \right) \tag{49}$$

Assuming the usual uncorrelatedness of the input data and noise, the transmitter-receiver joint optimization problem can be reformulated as

$$\begin{aligned} \min \xi &= tr\, T \int_{-\frac{1}{2T}}^{\frac{1}{2T}} \left\{ [\mathbf{I} - \mathbf{\Phi}(f)]\, \mathbf{\Sigma_a}(f)\, [\mathbf{I} - \mathbf{\Phi}(f)]^{\dagger} + \mathbf{\Psi}(f) \right\} df \\ &= tr\, T \int_{-\frac{1}{2T}}^{\frac{1}{2T}} \left\{ [\mathbf{I} - \frac{1}{T}\vec{\mathbf{R}}\vec{\mathbf{C}}\vec{\mathbf{S}}]\mathbf{\Sigma_a}[\mathbf{I} - \frac{1}{T}\vec{\mathbf{R}}\vec{\mathbf{C}}\vec{\mathbf{S}}]^{\dagger} + \frac{1}{T}\vec{\mathbf{R}}\vec{\mathbf{\Sigma}}_{\mathbf{n}}\vec{\mathbf{R}}^{\dagger} \right\} df \end{aligned}$$

subject to the power constraint at the transmitter output

$$tr\left\{\sum_{i=1}^{M}\frac{1}{T}\int_{-\frac{1}{2T}}^{\frac{1}{2T}}\mathbf{S}_i(f)\mathbf{\Sigma_a}(f)\mathbf{S}_i^{\dagger}(f)\right\}df$$
$$= tr\left\{\frac{1}{T}\int_{-\frac{1}{2T}}^{\frac{1}{2T}}\vec{\mathbf{S}}(f)\mathbf{\Sigma_a}(f)\vec{\mathbf{S}}^{\dagger}(f)\right\}df = NP_0. \tag{50}$$

The problem is now converted to one of transmitter/receiver optimization of a strictly Nyquist bandlimited system in terms of the matrix variables $\vec{\mathbf{S}}(f)$ $(MN \times N)$ and $\vec{\mathbf{R}}(f)$ $(N \times MN)$. The results of Section 2.1 can now be applied to obtain a solution, by identifying $P = MN$. The optimal transmitter structure in the equivalent decoupled system thus allows the following intuitive interpretation: for each of the N decoupled channels, choose only the optimum (in the sense of the SNR defined earlier in Sec. 2.1) frequencies in the set $\{f : f_0 + \frac{k}{T}, k = 0, 1, \ldots, M-1\}$ to allocate transmit power, in a manner analogous to [2] for the scalar case.

3 MIMO SYSTEM OPTIMIZATION: UNCOORDINATED USERS

In Section 2, the design of the transmit matrix $\mathbf{S}(f)$ and receive matrix $\mathbf{R}(f)$ was based on the premise that the transmitter has access to data from all users. Similarly, it was assumed that the receiver acts on all the channels, outputs to simultaneously detect all the users' data. In many practical situations, such as in mobile radio, the users operate independently implying that transmitter coordination is infeasible. Analogously, in some situations, the receiver may have access to the output of only one channel (more generally, a subset of the channel outputs) to decode the corresponding data. It follows that structural constraints such as the lack of full coordination at either the transmitter or receiver leads to different MSE performance. For a simple qualitative illustration of one instance of this, consider the situation of two users using two *uncoupled* channels to transmit data, where the passband of one channel is superior (in its gain and/or phase response) to the other. In this case, transmit coordination would assure that the better channel is shared by both the users for a given system bandwidth. This superior allocation of system resources (and hence improved system performance) would not be possible when the users are uncoordinated.

A *fully uncoordinated* system in which both the transmit and receive matrices are uncoupled is shown in Fig. 5(a) (for a 2×2 system for simplicity) in contrast to the fully coordinated situation. Here, we exclusively concentrate on a special subproblem of the above that is characterized by the following constraints:

- All users implement identical transmit pulse shaping filters.
- Each user sees the same direct and cross-talk channel responses, i.e., the matrix of channel responses is *circulant*.

This is a natural generalization of the two-user version of the same problem investigated by [5]. It now follows readily from the previous structural assumptions

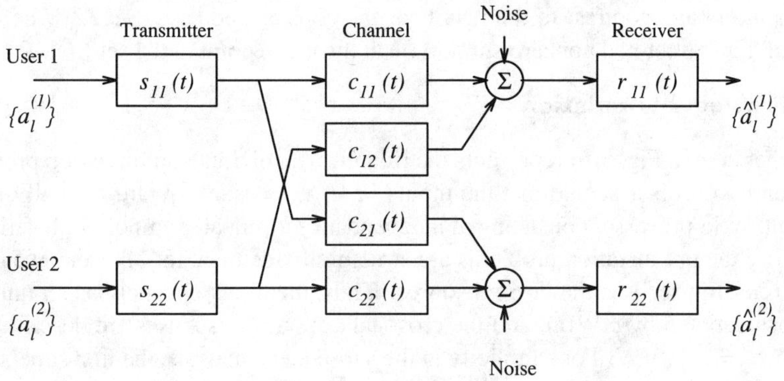

(a) 2-user Uncoordinated MIMO System

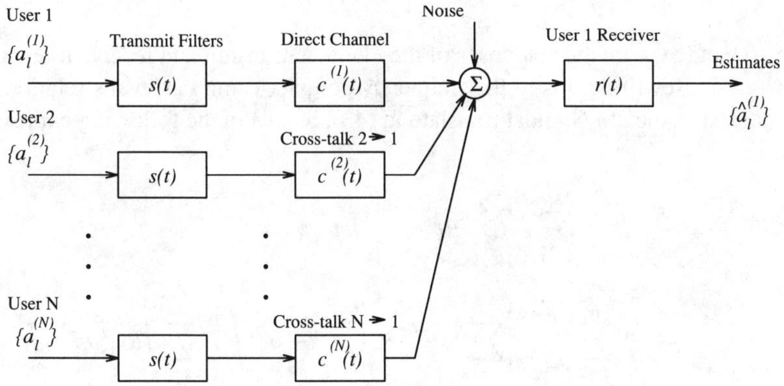

(b) Individual MMSE Problem (Symmetric Channel, N-users)

Figure 5: Uncoordinated MIMO systems.

that the global MMSE problem (i.e., over all users) leads to *identical decoupled* MMSE subproblems for each user, i.e., we need only to consider the single-user in the interference problem shown in Fig. 5(b). Clearly, *identical*, *uncoupled* receive filters are employed by each user.

While the above formulation is clearly not the most general, it may be justified due to the resulting simplicity in system implementation. Also, this model may be tenable for a cluster of subscriber loop cables arrayed with circular symmetry. We summarize the following key result, concerning the optimal transmitter for this problem, which is striking due to its fundamental distinction from that obtained for the optimal transmitter for the case of coordinated users presented in Section 2:

- The optimal transmitter has nonzero support on at most $L \leq 2N - 1$ points in the set $\{f : f_0 + \frac{k}{T}\}$.

This is in distinct contrast to the result for the coordinated case (Sec. 2) where the optimal transmitter had nonzero support on at most one point in the set $\{f : f_0 + \frac{k}{T}\}$.

3.1 Problem Formulation

The schematic in Fig. 5(b) represents the (identical) individual optimization problem for each user. It is assumed that the inputs $\{a_l^{(i)}\}$, $i = 1, \dots, N$ are mutually independent, WSS processes consisting of independent zero-mean, symbols with variance σ_a^2. Since the optimization problems are structurally identical for all users, we nominally refer to user 1 as the desired source. Thus, the direct channel has an impulse response denoted by $c^{(11)}(t)$, and the cross-talk impulse responses are described by $c^{(1i)}(t)$, $i = 2, \dots, N$. For simplicity in the subsequent analysis, the first superscript will be omitted and the channel responses identified simply by $c^{(i)}(t)$, $i = 1, \dots, N$.

The optimization entails minimization of the MSE for the 1st user

$$\min_{s(t),r(t)} \xi_1 = E\left\{|\hat{a}_l^{(1)} - a_l^{(1)}|^2\right\}, \tag{51}$$

where $s(t)$ and $r(t)$ are the responses of the identical transmit and receive filters used by each user. Rewriting this in the frequency domain, using Parseval's relation and the definition of the kth Nyquist translate in (43), results in the following expression for MSE:

$$\begin{aligned}\xi_1 &= \frac{\sigma_a^2}{T}\int_{-\frac{1}{2T}}^{\frac{1}{2T}}\left|\sum_{k=1}^{M} S_k C_k^{(1)} R_k - T\right|^2 df \\ &+ \frac{\sigma_a^2}{T}\sum_{i=2}^{N}\int_{-\frac{1}{2T}}^{\frac{1}{2T}}\left|\sum_{k=1}^{M} S_k C_k^{(i)} R_k\right|^2 df + \sigma_n^2\int_{-\frac{1}{2T}}^{\frac{1}{2T}}\sum_{k=1}^{M}|R_k|^2 df\end{aligned} \tag{52}$$

subject to the usual power constraint

$$\frac{\sigma_a^2}{T}\int_{-\frac{1}{2T}}^{\frac{1}{2T}}\sum_{k=1}^{M}|S_k|^2 df = P_0. \tag{53}$$

Note that, because of the structural assumptions concerning the problem, P_0 is effectively an *individual* power constraint at each of the transmit filter outputs in distinction to NP_0 in (50), which represents the *total* power at the output of the transmit matrix.

Introducing a Lagrange multiplier λ, the preceding problem can be transformed to the following unconstrained optimization problem

$$\begin{aligned}\xi_1 &= \frac{\sigma_a^2}{T}\int_{-\frac{1}{2T}}^{\frac{1}{2T}}\left|\sum_{k=1}^{M} S_k C_k^{(1)} R_k - T\right|^2 df + \frac{\sigma_a^2}{T}\sum_{i=2}^{N}\int_{-\frac{1}{2T}}^{\frac{1}{2T}}\left|\sum_{k=1}^{M} S_k C_k^{(i)} R_k\right|^2 df \\ &+ \sigma_n^2\int_{-\frac{1}{2T}}^{\frac{1}{2T}}\sum_{k=1}^{M}|R_k|^2 df + \lambda\left[\frac{\sigma_a^2}{T}\int_{-\frac{1}{2T}}^{\frac{1}{2T}}\sum_{k=1}^{M}|S_k|^2 df - P_0\right].\end{aligned} \tag{54}$$

Remark 3.1.1: Note from (54) that the MSE depends on R_k and S_k only through their product and, hence, via the *sum* of their phases. Consequently, the phase can be arbitrarily distributed between the transmit and receive filter pair S_k and R_k, while maintaining a constant sum.

3.2 Analysis

We now consider the joint optimization of the transmit and receiver filters. We note that (54) can also be compactly written in two complementary forms:

$$\xi_1 = \frac{\sigma_a^2}{T} \int_{-\frac{1}{2T}}^{\frac{1}{2T}} \|\mathbf{A}_1\mathbf{r} - T\mathbf{b}\|^2 df + \sigma_n^2 \int_{-\frac{1}{2T}}^{\frac{1}{2T}} \|\mathbf{r}\|^2 df + \lambda \left[\frac{\sigma_a^2}{T} \int_{-\frac{1}{2T}}^{\frac{1}{2T}} \|\mathbf{s}\|^2 df - P_0 \right] \tag{55}$$

$$= \frac{\sigma_a^2}{T} \int_{-\frac{1}{2T}}^{\frac{1}{2T}} \|\mathbf{A}_2\mathbf{s} - T\mathbf{b}\|^2 df + \sigma_n^2 \int_{-\frac{1}{2T}}^{\frac{1}{2T}} \|\mathbf{r}\|^2 df + \lambda \left[\frac{\sigma_a^2}{T} \int_{-\frac{1}{2T}}^{\frac{1}{2T}} \|\mathbf{s}\|^2 df - P_0 \right]. \tag{56}$$

where $\| \, . \, \|$ is the L_2 norm of the corresponding vector; the vectors and matrices written in boldface are understood to be functions of f and are defined as follows

$$\mathbf{A} \equiv \begin{bmatrix} C_1^{(1)} & C_2^{(1)} & \cdots & C_M^{(1)} \\ C_1^{(2)} & C_2^{(2)} & \cdots & C_M^{(2)} \\ \cdot & \cdot & & \cdot \\ \cdot & \cdot & & \cdot \\ \cdot & \cdot & & \cdot \\ C_1^{(N)} & C_2^{(N)} & \cdots & C_M^{(N)} \end{bmatrix} \in C^{N \times M} \tag{57}$$

$$\mathbf{A}_1 \equiv \mathbf{A}\,\mathbf{S}_d \in C^{N \times M} \quad ; \quad \mathbf{A}_2 \equiv \mathbf{A}\,\mathbf{R}_d \in C^{N \times M} \tag{58}$$

$$\mathbf{S}_d \equiv \text{diag}\,[S_1, \ldots, S_M] \in C^{M \times M} \quad ; \quad \mathbf{R}_d \equiv \text{diag}\,[R_1, \ldots, R_M] \in C^{M \times M} \tag{59}$$

$$\mathbf{s} \equiv [S_1, \ldots, S_M]^T \in C^{M \times 1} \quad ; \quad \mathbf{r} \equiv [R_1, \ldots, R_M]^T \in C^{M \times 1} \tag{60}$$

$$\mathbf{b} \equiv [1, 0, \ldots, 0]^T \in R^{N \times 1}. \tag{61}$$

For a fixed transmitter $s(t)$, the optimal receiver can be found by optimizing (55) at each frequency. In a dual manner, (56) can be used to obtain the optimal transmitter for a fixed receiver $r(t)$. Using standard variational techniques, these yield the two necessary conditions for the jointly optimal transmit-receive pair:

$$\mathbf{r}_o = T \left[\mathbf{A}_1^\dagger \mathbf{A}_1 + \eta^{-1} \mathbf{I} \right]^{-1} \mathbf{A}_1^\dagger \mathbf{b} \tag{62}$$

$$\mathbf{s_o} = T \left[\mathbf{A}_2^\dagger \mathbf{A}_2 + \beta^{-1} \mathbf{I} \right]^{-1} \mathbf{A}_2^\dagger \mathbf{b} \tag{63}$$

where we have introduced (for notational simplicity) $\eta = \sigma_a^2/\sigma_n^2 T$, $\beta = 1/\lambda$. Note that (62) and (63) actually represent two *coupled* conditions due to the implicit dependence of $\mathbf{A}_1$ and $\mathbf{A}_2$ on $\mathbf{s}$ and $\mathbf{r}$, respectively.

Proposition 3.2.1 $|S(f)|^2 = \dfrac{\sigma_n^2 T}{\lambda \sigma_a^2} |R(f)|^2 \ \forall f.$

Proof: This follows in a direct manner upon observing the following relation in (54):

$$\xi_1(|S_k|, |R_k|) = \xi_1\left(|R_k|\sqrt{\frac{T\sigma_n^2}{\lambda\sigma_a^2}}, |S_k|\sqrt{\frac{\lambda\sigma_a^2}{T\sigma_n^2}}\right). \tag{64}$$

Equating the respective arguments of $\xi_1(.,.)$ on both sides of (64) directly leads to the desired result.

Remark 3.2.1: A direct corollary of the above is that $S(f) = 0$ if and only if $R(f) = 0$.

The optimum N-user receiver structure is well known [18] to consist of a bank of N filters matched to the respective overall channels [19], followed by T-spaced tapped delay lines, the outputs of which are summed together to produce an estimate of the desired symbol; i.e.,

$$\begin{aligned} R(f) = {} & S^*(f)C^{(1)*}(f)Y_1(f) \\ & + S^*(f)C^{(2)*}(f)Y_2(f) + \cdots + S^*(f)C^{(N)*}(f)Y_N(f) \end{aligned} \tag{65}$$

where $*$ denotes the complex conjugate and $Y_1(f), \ldots, Y_N(f)$ are periodic filters with period $1/T$. It follows that

$$\begin{aligned} |R(f)|^2 = {} & |S(f)|^2 \big|C^{(1)*}(f)Y_1(f) \\ & + C^{(2)*}(f)Y_2(f) + \cdots + C^{(N)*}(f)Y_N(f)\big|^2. \end{aligned} \tag{66}$$

Using Proposition 3.2.1, this becomes

$$\begin{aligned} \frac{\lambda\sigma_a^2}{\sigma_n^2 T} = {} & |C^{(1)*}(f)Y_1(f) + C^{(2)*}(f)Y_2(f) \\ & + \cdots + C^{(N)*}(f)Y_N(f)|^2, \ \text{if } S(f) \neq 0. \end{aligned} \tag{67}$$

It is easily shown [17] that $Y_1(f)$ is real, but $\{Y_2(f), \ldots, Y_N(f)\}$ are, in general, complex. Evaluating (67) at $\left\{f_0 + \frac{k}{T}\right\}$, and denoting $Y_i(f_0) \equiv Y_{i0}$, we obtain

$$\frac{\lambda\sigma_a^2}{\sigma_n^2 T} = |C_k^{(1)*}Y_{10} + C_k^{(2)*}Y_{20} + \cdots + C_k^{(N)*}Y_{N0}|^2, \ \forall\, k \text{ for which } S_k \neq 0. \tag{68}$$

Proposition 3.2.2 *Let the number of indices k for which S_k is nonzero be L. Then (68) represents a system of L equations in $2N - 1$ variables (one unknown* *corresponding to the real Y_{10} and two each for the complex $Y_{20}, \ldots, Y_{n0}$). This implies that for a solution to exist generically, the number of nonzero elements S_k satisfies $L \leq 2N - 1$.*

Remark 3.2.2: Note that for $L > 2N - 1$, a solution could still exist for some special cases, but over the ensemble of all possible channels, this event is of measure zero. For $N = 2$ which corresponds to the case of one interferer, this reduces to the result obtained in [5] that $L \leq 3$. Further, if the direct and cross-talk responses are assumed to be real (i.e., $Y_{10}, \ldots, Y_{n0}$ are all real), (68) reduces to a system of L equations in N variables. By similar arguments as above, we obtain $L \leq N$, which, for $N = 2$, corroborates the claim in [5] that $L \leq 2$.

Proposition 3.2.3

$$\lambda = \frac{\sigma_n^2}{P_0} \int_{-\frac{1}{2T}}^{\frac{1}{2T}} \|\mathbf{r}_o\|^2 df > 0.$$

Proof: This follows by integrating both sides of the expression in Proposition 3.2.1 and using the power constraint.

Proposition 3.2.4 $\sum_{k=1}^{M} S_k C_k^{(1)} R_k$ *is real and nonnegative.*

Proof: From (63), we get

$$\left[\mathbf{A}_2^\dagger \mathbf{A}_2 + \beta^{-1}\mathbf{I}\right] \mathbf{x}_o = T\mathbf{A}_2^\dagger \mathbf{b} \tag{69}$$

$$\Rightarrow \mathbf{R}_d^\dagger \mathbf{A}^\dagger \mathbf{A}\, \mathbf{R}_d \mathbf{x}_o + \beta^{-1}\mathbf{x}_o = T\mathbf{R}_d^\dagger \mathbf{A}^\dagger \mathbf{b}. \tag{70}$$

On multiplying both sides of (70) by $\mathbf{s}_o^\dagger$, we obtain

$$\Rightarrow \mathbf{s}_o^\dagger \mathbf{R}_d^\dagger \mathbf{A}^\dagger \mathbf{A}\, \mathbf{R}_d \mathbf{s}_o + \beta^{-1}\mathbf{s}_o^\dagger \mathbf{s}_o = T\mathbf{s}_o^\dagger \mathbf{R}_d^\dagger \mathbf{A}^\dagger \mathbf{b}. \tag{71}$$

Since the left-hand side of (71) is nonnegative and real, it follows that the right-hand side $\mathbf{s}_o^\dagger \mathbf{R}_d^\dagger \mathbf{A}^\dagger \mathbf{b} = \sum_{k=1}^{M} S_k^\dagger C_k^{(1)\dagger} R_k^\dagger$ is also real and nonnegative.

Using Proposition 3.2.4, (71) can be rewritten as

$$\|\mathbf{A}\, \mathbf{R}_d \mathbf{s}_o\|^2 + \beta^{-1}\|\mathbf{s}_o\|^2 = T \sum_{k=1}^{M} S_k C_k^{(1)} R_k \tag{72}$$

$$\Rightarrow \left(\sum_{k=1}^{M} S_k C_k^{(1)} R_k\right)^2 + \sum_{i=2}^{N} \left|\sum_{k=1}^{M} S_k C_k^{(i)} R_k\right|^2 + \beta^{-1} \sum_{k=1}^{M} |S_k|^2 = T \sum_{k=1}^{M} S_k C_k^{(1)} R_k. \tag{73}$$

On substituting (73) in (54) and simplifying, we arrive at the expression for the MSE as

$$\xi_1 = \sigma_a^2 \left[1 - \int_{-\frac{1}{2T}}^{\frac{1}{2T}} \sum_{k=1}^{M} S_k C_k^{(1)} R_k \, df\right]. \tag{74}$$

At each frequency f, let the set of indices j, for which S_j is nonzero, be denoted $\mathcal{J}(f) = \{k_1, k_2, \ldots, k_L\}$, i.e.,

$$R_i - \begin{cases} \sqrt{\frac{\eta}{\beta}}\ S_i \exp(j\phi_i), \text{ for some } \phi_i, \text{ for } i \in \mathcal{J} \\ 0, \ i \in \mathcal{J}. \end{cases} \tag{75}$$

We form the matrix **B** by deleting those columns of **A** whose indices are not in $\mathcal{J}$. Let $\mathbf{S}_{dp}$ and $\mathbf{R}_{dp}$ be defined as

$$\mathbf{S}_{dp} \equiv \operatorname{diag}\left[S_{k_1}, \ldots, S_{k_L}\right] \in C^{L\times L} \ ; \ \mathbf{R}_{dp} \equiv \operatorname{diag}\left[R_{k_1}, \ldots, R_{k_L}\right] \in C^{L\times L}. \tag{76}$$

$$\mathbf{s}_p \equiv \left[S_{k_1}, \ldots, S_{k_L}\right]^T \in C^{L\times 1} \ ; \ \mathbf{r}_p \equiv \left[R_{k_1}, \ldots, R_{k_L}\right]^T \in C^{L\times 1} \tag{77}$$

$$\mathbf{B}_1 \equiv \mathbf{B}\,\mathbf{S}_{dp} \in C^{N\times L} \ ; \ \mathbf{B}_2 \equiv \mathbf{B}\,\mathbf{R}_{dp} \in C^{N\times L}. \tag{78}$$

Using these definitions, (70) can be rewritten as

$$\Rightarrow \mathbf{R}_{dp}^{\dagger}\mathbf{B}^{\dagger}\mathbf{B}\,\mathbf{R}_{dp}\mathbf{s}_{po} + \beta^{-1}\mathbf{s}_{po} = T\mathbf{R}_{dp}^{\dagger}\mathbf{B}^{\dagger}\mathbf{b}. \tag{79}$$

Note that $\mathbf{R}_{dp}$ has an inverse because it is formed by taking the nonzero elements of $\mathbf{R}_d$. Multiplying (79) on the left by $\mathbf{R}_{dp}^{\dagger^{-1}}$, we obtain,

$$\mathbf{B}^{\dagger}\mathbf{B}\,\mathbf{R}_{dp}\mathbf{s}_{po} + \beta^{-1}\mathbf{R}_{dp}^{\dagger^{-1}}\mathbf{s}_{po} = T\mathbf{B}^{\dagger}\mathbf{b}. \tag{80}$$

We consider the special case $L = 1$, which corresponds to the transmit filter having nonzero support at a solitary point in the Nyquist set $\{f : f_0 + \frac{k}{T}\}$. The MSE expression (74) simplifies to

$$\xi_1 = \sigma_a^2\left[1 - \int_{-\frac{1}{2T}}^{\frac{1}{2T}} S_l C_l^{(1)} R_l \, df\right] \tag{81}$$

$$= \sigma_a^2\left[1 - \sqrt{\frac{\eta}{\beta}}\int_{-\frac{1}{2T}}^{\frac{1}{2T}} |S_l|^2 |C_l^{(1)}| \, df\right] \tag{82}$$

by using (75) and Proposition 3.2.4. To obtain the optimum l (which is a function of frequency), we first solve (80) for $|S_l|^2$ to obtain

$$|S_l|^2 = \frac{\beta}{\eta}\,\frac{\sqrt{\frac{\eta}{\beta}}\,T|C_l^{(1)}| - \beta^{-1}}{|C_l^{(1)}|^2 + \sum_{i=2}^{N}|C_l^{(i)}|^2}. \tag{83}$$

Substituting this in (82), we deduce that the MSE is minimized if l is chosen according to

$$l = \arg\max_k \frac{\beta}{\eta}|C_k^{(1)}|\frac{\sqrt{\frac{\eta}{\beta}}T|C_k^{(1)}| - \beta^{-1}}{|C_k^{(1)}|^2 + \sum_{i=2}^{N}|C_k^{(i)}|^2}. \tag{84}$$

Note that if (84) yields $l = 1 \, \forall f$, it implies that the optimum transmit filter is strictly bandlimited to the first Nyquist zone $\left(-\frac{1}{2T}, \frac{1}{2T}\right)$. The more general case $L \geq 2$ does not appear to be analytically tractable, and numerical techniques must be resorted to. Toward this end, we write

$$S_i = |S_i| \exp(j\theta_i) \;\Rightarrow\; R_i = \sqrt{\frac{\eta}{\beta}} |S_i| \exp\{j(\theta_i + \phi_i)\}. \tag{85}$$

Substituting (85) in (80), gives us

$$\sqrt{\frac{\eta}{\beta}} \mathbf{B}^{\dagger}\mathbf{B} \begin{bmatrix} \vdots \\ |S_i|^2 \exp\{j(\phi_i + 2\theta_i)\} \\ \vdots \end{bmatrix} + \frac{1}{\sqrt{\beta\eta}} \begin{bmatrix} \vdots \\ \exp\{j(\phi_i + 2\theta_i)\} \\ \vdots \end{bmatrix} = T\mathbf{B}^{\dagger}\mathbf{b}. \tag{86}$$

Let $z_i = |S_i|^2 \exp\{j(\phi_i + 2\theta_i)\}$ and define $\mathbf{Z} = [z_1, \ldots, z_L]^T$, $|\mathbf{Z}|_d = \text{diag}\,[|z_1|, \ldots, |z_L|]$ as the diagonal matrix containing $|z_i|$ as the elements. Then, (86) can be rewritten as

$$\sqrt{\frac{\eta}{\beta}} \mathbf{B}^{\dagger}\mathbf{B} \begin{bmatrix} \vdots \\ z_i \\ \vdots \end{bmatrix} + \frac{1}{\sqrt{\beta\eta}} \begin{bmatrix} \vdots \\ \frac{z_i}{|z_i|} \\ \vdots \end{bmatrix} = T\mathbf{B}^{\dagger}\mathbf{b} \tag{87}$$

$$\Rightarrow \sqrt{\frac{\eta}{\beta}} \mathbf{B}^{\dagger}\mathbf{B}\mathbf{Z} + \frac{1}{\sqrt{\beta\eta}} |\mathbf{Z}|_d^{-1}\mathbf{Z} = T\mathbf{B}^{\dagger}\mathbf{b}. \tag{88}$$

Clearly, (88) is a set of coupled nonlinear equations in $|z_i|$, $i = 1, \ldots, M$. An iterative approach that is guaranteed to produce a local optimum proceeds by introducing the squared error ϵ between the left- and right-hand sides of (88) and updating the current estimate $\mathbf{Z}_n$ via a gradient descent algorithm.

Proposition 3.2.5 *A necessary and sufficient condition for uniqueness of the solution to* (88) *for the case* $M \leq N$ (*number of Nyquist zones less than or equal to total number of users*) *is that the matrix* **B** *have full column rank for each* f.

Proof: We first note that the assumption that the channel vectors, formed by the elements of all the frequency translates, are linearly independent implies that the matrix **B** is of full column rank. This in turn implies that $\mathbf{B}^{\dagger}\mathbf{B}$ is nonsingular and hence is positive definite. The proof follows by contradiction. Let us assume that (88) has two distinct solutions $\mathbf{Z}_1 \neq \mathbf{Z}_2$. It then follows using (88) that

$$\sqrt{\frac{\eta}{\beta}} \mathbf{B}^{\dagger}\mathbf{B}\mathbf{Z}_1 + \frac{1}{\sqrt{\beta\eta}} \mathbf{Z}_{d_1}^{-1}\mathbf{Z}_1 = \sqrt{\frac{\eta}{\beta}} \mathbf{B}^{\dagger}\mathbf{B}\mathbf{Z}_2 + \frac{1}{\sqrt{\beta\eta}} \mathbf{Z}_{d_2}^{-1}\mathbf{Z}_2 \tag{89}$$

$$\Rightarrow \mathbf{B}^{\dagger}\mathbf{B}(\mathbf{Z}_1 - \mathbf{Z}_2) = \frac{1}{\eta}\left[\mathbf{Z}_{d_2}^{-1}\mathbf{Z}_2 - \mathbf{Z}_{d_1}^{-1}\mathbf{Z}_1\right]. \tag{90}$$

Since $\mathbf{B}^\dagger\mathbf{B}$ is positive definite, we have

$$(\mathbf{Z}_1 - \mathbf{Z}_2)^\dagger \mathbf{B}^\dagger \mathbf{B}(\mathbf{Z}_1 - \mathbf{Z}_2) > 0\,,\ \mathbf{Z}_1 \neq \mathbf{Z}_2. \tag{91}$$

Left-multiplying (90) by $(\mathbf{Z}_1 - \mathbf{Z}_2)^\dagger$ and using (91), we obtain

$$(\mathbf{Z}_1 - \mathbf{Z}_2)^\dagger(\mathbf{Z}_{d_2}^{-1}\mathbf{Z}_2 - \mathbf{Z}_{d_1}^{-1}\mathbf{Z}_1) > 0. \tag{92}$$

This can be expanded to obtain

$$\mathbf{Z}_1^\dagger \mathbf{Z}_{d_2}^{-1}\mathbf{Z}_2 + \mathbf{Z}_2^\dagger \mathbf{Z}_{d_1}^{-1}\mathbf{Z}_1 > \mathbf{Z}_1^\dagger \mathbf{Z}_{d_1}^{-1}\mathbf{Z}_1 + \mathbf{Z}_2^\dagger \mathbf{Z}_{d_2}^{-1}\mathbf{Z}_2. \tag{93}$$

Writing this component-wise gives us

$$\frac{Z_{1i}^* Z_{2i}}{|Z_{2i}|} + \frac{Z_{2i}^* Z_{1i}}{|Z_{1i}|} > |Z_{1i}| + |Z_{2i}|. \tag{94}$$

From (94), we have

$$\begin{aligned} \text{LHS} &\leq \left| \frac{Z_{1i}^* Z_{2i}}{|Z_{2i}|} + \frac{Z_{2i}^* Z_{1i}}{|Z_{1i}|} \right| \\ &\leq \frac{|Z_{1i}^* Z_{2i}|}{|Z_{2i}|} + \frac{|Z_{2i}^* Z_{1i}|}{|Z_{1i}|} \\ &= |Z_{1i}| + |Z_{2i}|. \end{aligned} \tag{95}$$

Combining (94) and (95), we obtain

$$|Z_{1i}| + |Z_{2i}| \geq \frac{Z_{1i}^* Z_{2i}}{|Z_{2i}|} + \frac{Z_{1i}^* Z_{2i}}{|Z_{1i}|} > |Z_{1i}| + |Z_{2i}|\,, \tag{96}$$

implying a contradiction. Hence, $\mathbf{Z}_1 = \mathbf{Z}_2$.

Remark 3.2.3: A *sufficient* condition for Proposition 3.2.5 to hold is that the matrix $\mathbf{A}$ be of full column rank, which obviously implies that $\mathbf{B}$ is also of full column rank.

Remark 3.2.4: Proposition 3.2.5 together with Remark 3.2.2 implies that the optimal solution for the transmitter is unique for channels with real frequency responses.

3.3 Interference Modeled as Stationary Noise

In [3], MMSE calculations based on a "stationarized" interference model were included to provide a baseline for comparison of the results for the cyclostationary case. As is well known, in the case of a large number of interferers with random initial phases, the net interference can be well approximated as WSS [8]. Consequently, the primary purpose of this section is to provide an *analytical* foundation to the optimization problem when the interference is *assumed* to be WSS, and compare the results obtained thereby to the case when the interference is (more accurately) modelled as wide-sense cyclostationary (WSCS).

Assuming WSS interference, the MSE expression (51) now becomes

$$\xi_1 = \frac{\sigma_a^2}{T}\int_{-\frac{1}{2T}}^{\frac{1}{2T}} |\sum_{k=1}^{M} S_k C_k^{(1)} R_k - T|^2 df$$
$$+\frac{\sigma_a^2}{T}\int_{-\frac{1}{2T}}^{\frac{1}{2T}} \sum_{k=1}^{M} |S_k R_k|^2 \sum_{i=2}^{N} |C_k^{(i)}|^2 df + \sigma_n^2 \int_{-\frac{1}{2T}}^{\frac{1}{2T}} \sum_{k=1}^{M} |R_k|^2 df. \tag{97}$$

We introduce

$$|I(f)|^2 = \sum_{i=2}^{N} |C^{(i)}(f)|^2 \tag{98}$$

in (97) to obtain

$$\xi_1 = \frac{\sigma_a^2}{T}\int_{-\frac{1}{2T}}^{\frac{1}{2T}} |\sum_{k=1}^{M} S_k C_k^{(1)} R_k - T|^2 df$$
$$+\frac{\sigma_a^2}{T}\int_{-\frac{1}{2T}}^{\frac{1}{2T}} \sum_{k=1}^{M} |S_k R_k I_k|^2 df + \sigma_n^2 \int_{-\frac{1}{2T}}^{\frac{1}{2T}} \sum_{k=1}^{M} |R_k|^2 df. \tag{99}$$

As in Sec. 3.2, we define the matrices

$$\hat{\mathbf{A}} \equiv \begin{bmatrix} C_1^{(1)} & C_2^{(1)} & C_3^{(1)} & \cdots & C_M^{(1)} \\ I_2 & 0 & 0 & \cdots & 0 \\ 0 & I_3 & 0 & \cdots & 0 \\ \cdot & \cdot & & & \cdot \\ \cdot & \cdot & & & \cdot \\ \cdot & \cdot & & & \cdot \\ 0 & 0 & 0 & \cdots & I_M \end{bmatrix} \tag{100}$$

$$\hat{\mathbf{A}}_1^{M\times M} \equiv \mathbf{A}\,\mathbf{S}_d \tag{101}$$

$$\hat{\mathbf{A}}_2^{M\times M} \equiv \mathbf{A}\,\mathbf{R}_d. \tag{102}$$

The MSE (99) can now be written as

$$\xi_1 = \frac{\sigma_a^2}{T}\int_{-\frac{1}{2T}}^{\frac{1}{2T}} \|\hat{\mathbf{A}}_1\mathbf{r} - T\mathbf{b}\|^2 df$$
$$+\sigma_n^2\int_{-\frac{1}{2T}}^{\frac{1}{2T}} \|\mathbf{r}\|^2 df + \lambda\left[\frac{\sigma_a^2}{T}\int_{-\frac{1}{2T}}^{\frac{1}{2T}} \|\mathbf{s}\|^2 df - P_0\right] \tag{103}$$

$$= \frac{\sigma_a^2}{T}\int_{-\frac{1}{2T}}^{\frac{1}{2T}} \|\hat{\mathbf{A}}_2\mathbf{s} - T\mathbf{b}\|^2 df$$
$$+\sigma_n^2\int_{-\frac{1}{2T}}^{\frac{1}{2T}} \|\mathbf{r}\|^2 df + \lambda\left[\frac{\sigma_a^2}{T}\int_{-\frac{1}{2T}}^{\frac{1}{2T}} \|\mathbf{s}\|^2 df - P_0\right]. \tag{104}$$

Setting the first variation of (104) to zero yields the necessary conditions for the optimal transmitter and receiver filters:

$$\mathbf{r}_o = T\left[\hat{\mathbf{A}}_1^\dagger\hat{\mathbf{A}}_1 + \eta^{-1}\mathbf{I}\right]^{-1}\hat{\mathbf{A}}_1^\dagger\mathbf{b} \tag{105}$$

$$\mathbf{s}_o = T\left[\hat{\mathbf{A}}_2^\dagger\hat{\mathbf{A}}_2 + \beta^{-1}\mathbf{I}\right]^{-1}\hat{\mathbf{A}}_2^\dagger\mathbf{b}. \tag{106}$$

Following arguments identical to those in Sec. 3.2 leads to a similar relation between $S(f)$ and $R(f)$ (see Proposition 3.2.1):

$$\Rightarrow |S(f)|^2 = \frac{\sigma_n^2 T}{\lambda\sigma_a^2}|R(f)|^2. \tag{107}$$

Using (107) in (106) leads (repeating the derivation in Sec. 3) to the final form of the equation for the optimal transmit filter

$$\sqrt{\frac{\eta}{\beta}}\hat{\mathbf{B}}^\dagger\hat{\mathbf{B}}\mathbf{Z} + \frac{1}{\sqrt{\beta\eta}}|\mathbf{Z}|_d^{-1}\mathbf{Z} = T\hat{\mathbf{B}}^\dagger\mathbf{b} \tag{108}$$

where $\hat{\mathbf{B}}$ is obtained from $\hat{\mathbf{A}}$ by deleting those columns of $\hat{\mathbf{A}}$ that correspond to the elements in the Nyquist set $\{f : f_0 + \frac{k}{T}\}$ where the transmitter has no support. Now, the spectra of the total WSS cross-talk and noise is given by

$$N(f) = \sigma_n^2 + \frac{\sigma_a^2}{T}|S(f)|^2\sum_{i=2}^{N}|C^{(i)}(f)|^2. \tag{109}$$

The optimal MMSE receiver in the presence of additive WSS interference is well known to be given by

$$R(f) = \frac{X^*(f)C^{(1)*}(f)}{N(f)}Y(f) \tag{110}$$

$$\Rightarrow |R(f)|^2 = \frac{|S(f)|^2|C^{(1)}(f)|^2}{|N(f)|^2}|Y(f)|^2 \tag{111}$$

$$\Rightarrow |Y(f)|^2 = \frac{\lambda\sigma_a^2}{\sigma_n^2 T}\frac{|N(f)|^2}{|C^{(1)}(f)|^2}, \tag{112}$$

where $Y(f)$ is a periodic filter (corresponding to the tapped delay line), and (107) was used to obtain (112) from (111). Using the periodicity

$$Y(f) = Y(f + \frac{k}{T}),\ k = 1, \ldots, M-1 \tag{113}$$

in (112) yields

$$\frac{N(f)}{C^{(1)}(f)} = \frac{N(f+\frac{k}{T})}{C^{(1)}(f+\frac{k}{T})},\ k = 1, \ldots, M-1. \tag{114}$$

Substituting (109) into (114) results in the following set of equations

$$\begin{bmatrix} a_1 & -1 & 0 & 0 & \cdots \\ a_2 & 0 & -1 & 0 & \cdots \\ \vdots & \vdots & \vdots & & \\ a_{L-1} & 0 & 0 & \cdots & -1 \end{bmatrix} \begin{bmatrix} |S_{i_1}|^2 \\ |S_{i_2}|^2 \\ \vdots \\ |S_{i_L}|^2 \end{bmatrix} = \begin{bmatrix} b_1 \\ b_2 \\ \vdots \\ b_{M-1} \end{bmatrix} \tag{115}$$

where $i_1, \ldots, i_L$ correspond to the indices of the Nyquist translates for which $S(f)$ is nonzero. Note that in obtaining (115), only (106) and the structure of the receiver (110) was used. Hence, it follows that (115) and (108) represent two independent sets of constraints that the transmit filter must satisfy. This implies a total of $2L - 1$ equations ($L - 1$ from (115) and L from (108), respectively) in L variables. For a solution to exist generically, it follows that $2L - 1 \leq L \Leftrightarrow L \leq 1$. Neglecting the case $L = 0$ (which yields a trivial solution), only $L = 1$ remains as a bona-fide candidate. For this situation, we can obtain an explicit closed form expression for the transmit filter:

$$|S_l|^2 = \frac{\beta}{\eta} \frac{\sqrt{\frac{\eta}{\beta}} T|C_l^{(1)}| - \beta^{-1}}{|C_l^{(1)}|^2 + \sum_{i=2}^{N} |C_l^{(i)}|^2} \tag{116}$$

where l is chosen such that

$$l = \arg\max_k \frac{\beta}{\eta} |C_k^{(1)}| \frac{\sqrt{\frac{\eta}{\beta}} T|C_k^{(1)}| - \beta^{-1}}{|C_k^{(1)}|^2 + \sum_{i=2}^{N} |C_k^{(i)}|^2}. \tag{117}$$

Note that these are identical to (83) and (84), which were obtained for the $L = 1$ case but when the interference was modelled as WSCS. Clearly, the optimum transmit filter under the assumption of WSS interference is identical to the optimum transmit filter in the presence of WSCS interference only when the latter is restricted to one generalized Nyquist zone, i.e., when the support for the transmit filter is (at most) of measure $\frac{1}{T}$, and is such that it is nonzero on at most one of $\{f : f_0 + \frac{k}{T}\}$ [2]. This is in accordance with the fact that digitally modulated signals are WSS when the pulse-shaping filter responses have support restricted to one generalized Nyquist zone. Finally, this result demonstrates analytically that greatly improved cross-talk rejection is achievable when the spectral correlation properties of WSCS signals are exploited, as described in [14], [15]. The computations in Sec. 4 quantify the improvement in the MSE as a function of L (number of points in the Nyquist set $\{f : f_0 + \frac{k}{T}\}$ where the transmit filter has nonzero support) and M (total number of Nyquist zones in the given system bandwidth). Note that for a fixed system bandwidth, varying M is equivalent to changing the symbol duration T.

4 NUMERICAL EXAMPLES AND DISCUSSIONS

Some representative numerical results are presented to highlight the MSE performance of jointly optimized MIMO systems for both coordinated and uncoordinated cases. The selection of the examples was motivated primarily by the need to investigate comparative system performance for coordinated and uncoordinated transmission as a qualitative function of the *shape* and *gain* of the channels. All examples assume i.i.d. input data with additive white noise. The noise power is chosen to be 45 dB below the transmit power ($P_0/\sigma_n^2 = 45$ dB), so that system performance is essentially cross-talk limited. Only two-user examples are included; a more exhaustive catalog of numerical results for $N > 2$ users is given in [6, 17].

Example 1: A 2×2 symmetric MIMO channel that spans two Nyquist zones is assumed. Fig. 6(a) shows the direct and cross-channel frequency response, where $\mathbf{C}_{ij}$ represents the channel responses from the jth input to the ith output. The direct paths ($\mathbf{C}_{11} = \mathbf{C}_{22}$) exhibit low-pass characteristics and the cross channels ($\mathbf{C}_{12} = \mathbf{C}_{21}$) a high-pass response, representing typical HDSL loop characteristics [4, 5]. The optimized coordinated transmitter matrix response is shown in Fig. 6(b) where, similar to above, $\mathbf{S}_{ij}$ is the transmit filter response from the jth input data source to the ith channel input. The uncoordinated transmitter response is shown in Fig. 7(a) for $L = 1$ and in Fig. 7(b) for $L = 2$.

The average MSE per user for the coordinated case was computed as was the MSE for identical channel responses when the receiver only is optimized to serve as a base line for comparison. The latter assumed commonly employed uncoupled transmit filters that have raised-cosine responses with 100% excess bandwidth (note that this is clearly nonoptimal for this case). These results and those for the uncoordinated case (for $L = 1, 2$, respectively) are shown in Table 1. A gain of 7.1 dB in MSE performance is noted for the jointly optimized coordinated system vis-à-vis one that employs only the optimal receiver. Similarly, the gain for the coordinated system over the uncoordinated system is about 5.5 dB. Note that $L = 1$, which corresponds implicitly to system design based on WSS interference, is inferior by nearly 18.5 dB compared to $L = 2$ when the system is not coordinated. The best uncoordinated system ($L = 2$) achieves marginally better MSE performance (by 1.6 dB) than the coordinated system with only an optimized receiver and uncoupled raised-cosine transmit filters.

Table 1: MSE for Example 1

	Coordinated	Uncoordinated($L = 1$)	Uncoordinated($L = 2$)	Receiver Only
MSE(dB)	−40.65	−16.62	−35.14	−33.54

Example 2: Figure 8(a) shows another set of channel responses that differ from those used in Example 1, in that both direct and cross paths now have similar low-pass

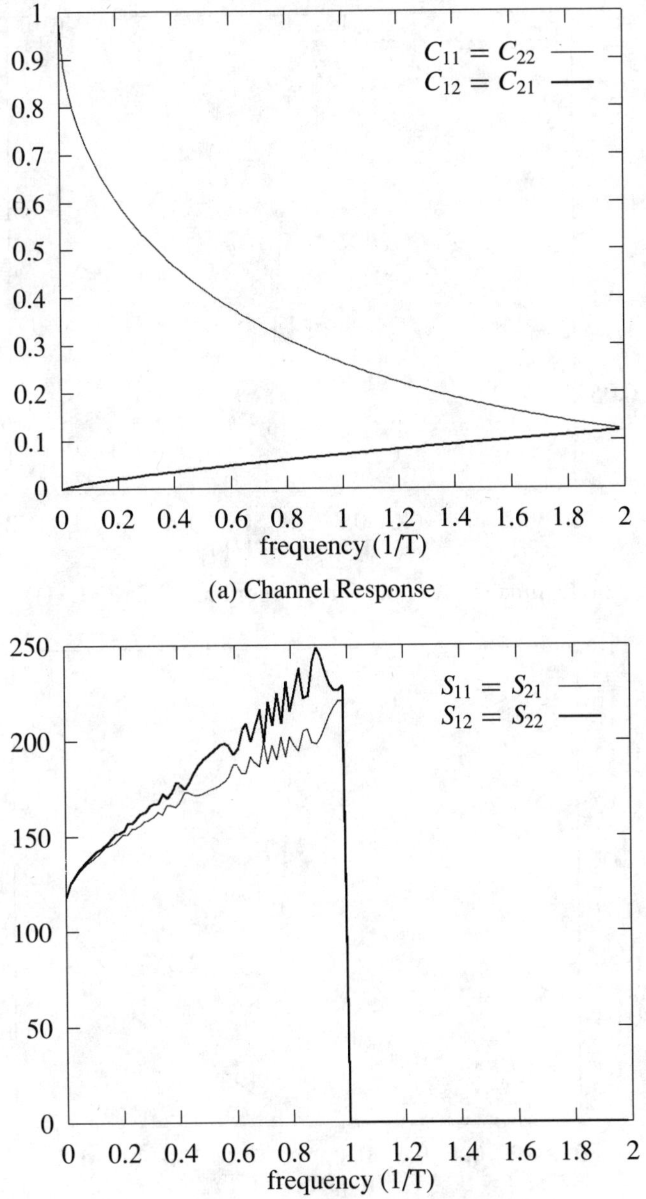

(a) Channel Response

(b) Optimal Transmitter for Co-Ordinated Case

Figure 6: Example 1.

characteristics. The optimal transmit filter is shown in Fig. 8(b) for the coordinated case and in Figs. 9(a) and (b) for uncoordinated cases $L = 1$ and $L = 2$, respectively. The MSE for the several designs discussed above are shown below in Table 2.

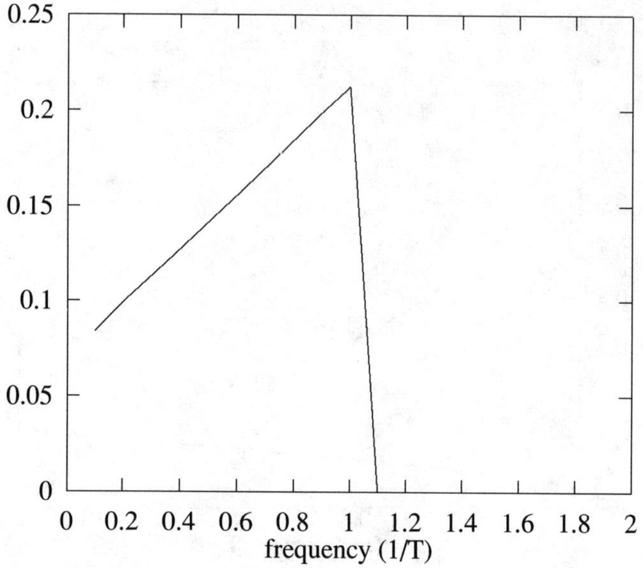

(a) Optimal Transmitter for Un-coordinated Case (L=1)

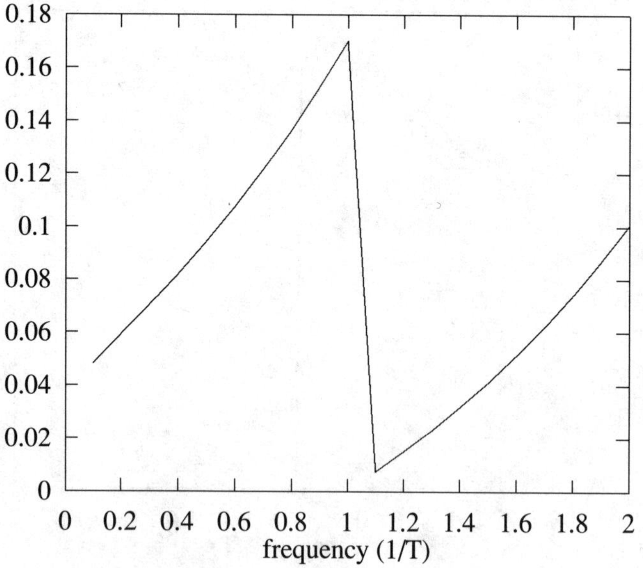

(b) Optimal Transmitter for Un-coordinated Case (L=2)

Figure 7: Example 1 (continued).

The gain of the fully coordinated design vis-à-vis the optimized-receiver-only system is now 5.9 dB (which is 1.2 dB less than in Example 1) whereas the advantage over an optimized noncoordinated system is 6.6 dB. This qualitatively highlights a

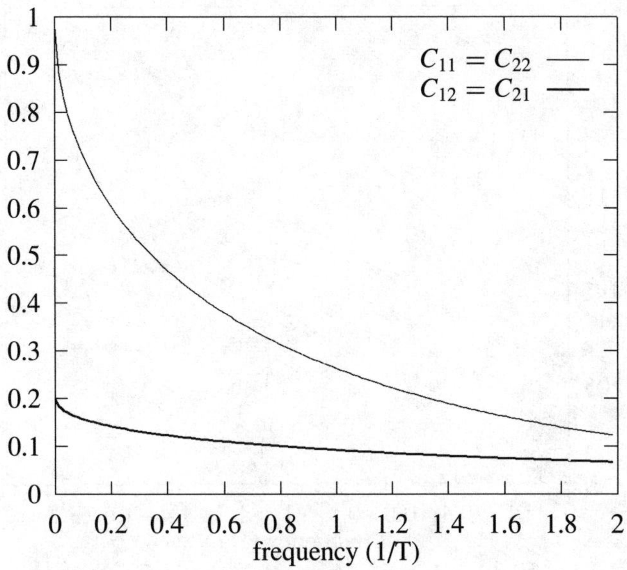

(a) Channel Response

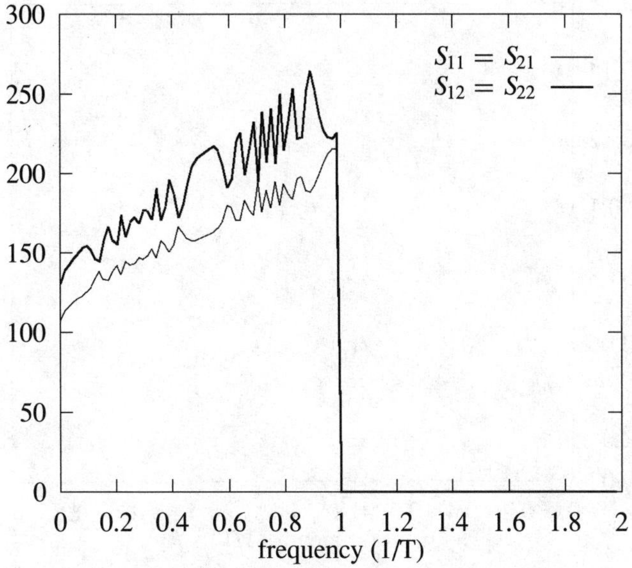

(b) Optimal Transmitter for Co-Ordinated Case

Figure 8: Example 2.

fundamental aspect of system performance—that a system with a fixed (or unoptimized) transmit filter will suffer poor performance depending on the relative shape of the direct and cross-talk channels. In contrast, a coordinated system with

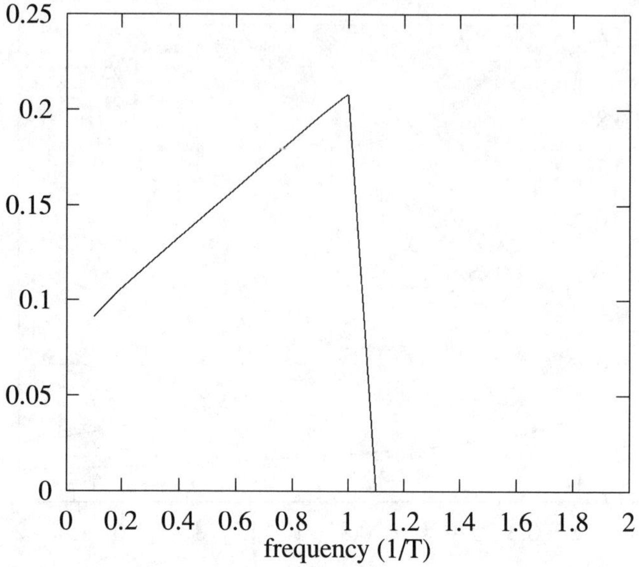

(a) Optimal Transmitter for Un-coordinated Case (L=1)

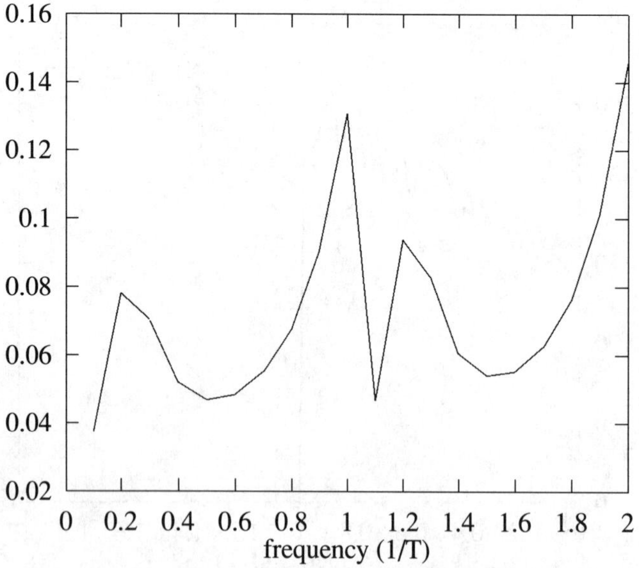

(a) Optimal Transmitter for Un-coordinated Case (L=2)

Figure 9: Example 2 (continued).

an optimized transmitter adjusts to the channel characteristics. Thus, *transmitter optimization is crucial in systems that are interference limited.* In both Examples 1

Table 2: MSE Results of Example 2

	Coordinated	Uncoordinated($L = 1$)	Uncoordinated($L = 2$)	Receiver Only
MSE(dB)	−40.49	−9.34	−33.87	−34.91

and 2, the optimal transmitter exhibits the following symmetry

$$\mathbf{S}_{11}(f) = \mathbf{S}_{21}(f),\ \mathbf{S}_{12}(f) = \mathbf{S}_{22}(f) \tag{118}$$

that follows directly from the symmetry of the channel response matrix. Finally, we observe that in both examples, the WSS assumption results in severely degraded performance (by 24 dB and 31 dB compared to the jointly optimized and coordinated systems, respectively).

5 CONCLUSIONS

The joint transmit/receive optimization problem for multiuser communication is investigated for fully coordinated and uncoordinated scenarios. The optimum transmitters are derived and system performance is obtained by simple numerical procedures. The performance improvements achievable vis-à-vis system design based on a stationary interference model when the system is uncoordinated is also quantified analytically. The numerical examples further substantiate the consensus that each additional interferer may be effectively suppressed by the addition of a Nyquist zone [3, 15].

ACKNOWLEDGMENTS

The authors gratefully acknowledge Glenn Golden, J. Mazo, and J. Salz for stimulating discussion and for providing a preprint of [5], an anonymous reviewer, and W. A. Gardner for several helpful suggestions.

REFERENCES

[1] J. G. Proakis, *Digital Communications*. New York: McGraw-Hill, 1989.

[2] T. Berger and D. W. Tufts, "Optimum pulse amplitude modulation—Pt. 1: Transmitter-receiver design and bounds from information theory," *IEEE Trans. Info. Theory*, vol. IT-13, pp. 196–208, Apr. 1967.

[3] B. R. Petersen and D. D. Falconer, "Minimum mean square equalization in cyclostationary and stationary interference-analysis and subscriber line calculations," *IEEE J. Sel. Areas Comm.*, vol. 9, pp. 931–940, Aug. 1991.

[4] M. L. Honig, P. Crespo, and K. Steiglitz, "Optimization of pre- and post-filters in the presence of near- and far-end crosstalk," *IEEE J. Sel. Areas Comm.*, vol. 10, pp. 614–629, Apr. 1992.

[5] G. D. Golden, J. E. Mazo, and J. Salz, "Transmitter design for data transmission in the presence of data-like interferer," *IEEE Trans. Comm.* (*to appear*), 1993.

[6] J. Yang and S. Roy, "On joint transmitter and receiver optimization for multiple-input multiple-output (MIMO) transmission system," *IEEE Trans. Comm.*, 1993.

[7] W. A. Gardner, *Statistical Spectral Analysis–A Nonprobabilistic Theory*. Englewood Cliffs, NJ: Prentice-Hall, 1988.

[8] W. A. Gardner, *Introduction to Random Processes with Applications to Signals and Systems*, 2nd ed. New York: McGraw-Hill, 1989.

[9] W. A. Gardner, "Exploitation of spectral redundancy in cyclostationary signals," *IEEE Sig. Proc. Magazine*, vol. 9, pp. 14–36, 1992.

[10] W. A. Gardner, "Representation and estimation for cyclostationary processes," Ph. D. Dissertation, Dept. of Electrical and Computer Engineering, University of Massachusetts, Amherst, 1972 (updated and reprinted as *U.C. Davis Tech. Rept. SIPL-82-1*, 1982).

[11] W. A. Gardner, "The structure of least-mean-square linear estimators for synchronous m-ary signals," *IEEE Trans. Info. Theory*, vol. IT-19, pp. 240–243, 1973.

[12] J. H. Reed and T. C. Hsia, "The performance of time-dependant adaptive filters for interference rejection," *IEEE Trans. Acoust. Speech Sig. Proc.*, vol. 38, pp. 1373–1385, Aug. 1990.

[13] W. A. Gardner and W. A. Brown, "Frequency-shift filtering theory for adaptive co-channel interference removal," *Proc. 23rd Asilomar Conf. Sig., Syst. Comp.*, pp. 562–567, 1989.

[14] W. A. Gardner and S. Venkataraman, "Performance of optimum and adaptive frequency-shift filters for co-channel interference and fading," *Proc. 24th Asilomar Conf. Sig., Syst. Comp.*, pp. 242-247, 1990.

[15] W. A. Gardner, "Cyclic Wiener filtering: Theory and method," *IEEE Trans. Comm.*," vol. 41, pp. 151–163, Jan. 1993.

[16] F. Graef, "Joint optimization of transmitter and receiver for cyclostationary random signal processes," *Proc. NATO Adv. Study Institute on Nonlinear Stochastic Problems*, pp. 581–592, Sept. 1982.

[17] P. S. Kumar and S. Roy, "Optimization for crosstalk suppression with non-coordinating transmitters," *IEEE Trans. Comm.* (*to appear*), 1993.

[18] A. Duel-Hallen, "Equalizers for multiple input/output channels and PAM system with cyclostationary input sequences," *IEEE J. Sel. Areas Comm.*, vol. 10, pp. 630–639, Apr. 1992.

[19] A. R. Kaye and D. A. George, "Transmission of multiplexed PAM signals over multiple channel and diversity systems," *IEEE Trans. Comm. Tech.*, vol. COM-18, pp. 520–526, Oct. 1970.

[20] J. Salz, "Digital transmission over cross-coupled linear channels," *AT&T Technical Journal*, vol. 64, pp. 1147–1159, July-Aug. 1985.

[21] J. W. Lechleider, "Coordinated transmission for two-pair digital subscriber lines," *IEEE J. Sel. Areas Comm.*, vol. 9, pp. 920–930, Aug. 1991.

[22] T. Ericson, "Optimum PAM filters are always band-limited," *IEEE Trans. Info. Theory*, vol. IT-19, pp. 570–572, July 1973.

[23] D. G. Luenberger, *Optimization by Vector Space Methods*. New York: Wiley, 1969.

[24] R. G. Gallagher, *Information Theory and Reliable Communication*. New York: Wiley, 1968.

Article 2

Beneficial Effects of Spectral Correlation on Synchronization

Nelson M. Blachman
GTE Government Systems Corporation
Mountain View, California 94039-7188

Abstract

The clock waveform of a PAM signal may be recovered by feeding a nonlinear function (e.g., the square) of that signal to a narrowband filter. The deterministic relationship between pairs of spectral components of the PAM signal whose frequencies differ by or sum to a multiple of the clock frequency has opposite effects on the in-phase and the quadrature components of the output of such a narrowband filter centered on a harmonic $m/2T$ of half the clock frequency. The mean squared value of the component in quadrature with the clock harmonic determines the mean squared phase error (jitter) in the clock waveform regenerated in this way when m is even (usually $m = 2$). This mean squared value is studied as a function of the narrowband filter's shape, bandwidth, and center frequency—first for the PAM signal itself and then for its square, whose behavior is remarkably similar—thus clarifying and extending previous work on this topic.

1 INTRODUCTION

Discovering and characterizing periodicities is a very widespread and fundamental problem in time-series analysis. In the field of communications it is particularly important in regard to sampling, demodulation, demultiplexing, and decoding. Because of its crucial role in the proper operation of coherent communication systems, the synchronization problem has been the subject of considerable research and development for several decades, especially in recent years with the increasing demands on system performance that result from rapidly growing usage.

The synchronization of the phase and frequency of an analogue oscillator with a signal's carrier and the synchronization of those of a digital clock with the symbols conveyed by that signal are two applications of this very general problem. A popular approach has been to break the synchronization problem down into two tasks: (i) using a nonlinearity to transform the time-series to be synchronized into a timing waveform whose spectrum contains a line (representing a periodicity of interest) along with a continuous spectrum (representing additive noise), and (ii) locking the frequency and phase of an oscillator or clock to the sinusoid represented by that spectral line with the help of a phase-locked loop (PLL) or a similar device such as a narrow bandpass filter or zero-crossing detector. Both the design and the performance of such synchronizers can be characterized in terms of spectral correlation and higher-order spectral-moment functions [1]. In particular, a timing sinusoid of frequency α can be generated from a random time-series by using an nth-degree nonlinearity if and only if the time-series exhibits cyclostationarity of order n with cycle frequency α, and this occurs if and only if the nth-order spectral-moment function for spectral components whose frequencies sum algebraically to α is not identically zero [2].

The spectrum of this timing wave at the output of an nth-degree nonlinearity is given explicitly in terms of the $2n$th-order spectral-moment function of the input, which is determined by the $2n$th-order and lower-order spectral cumulant functions. For the class of digital quadrature-amplitude-modulated (QAM) signals and pulse-amplitude-modulated (PAM) signals, the spectral cumulants can be expressed simply and explicitly in terms of the Fourier transform of the pulse shape and the cumulant of the distribution of the random amplitude factor representing the transmitted symbol [2]. Moreover, the mean squared value of the phase jitter of the PLL that locks onto the timing sinusoid is characterized by the continuous spectrum accompanying this sinusoid and the associated autocorrelation function [1]. For synchronizing with a QAM or PAM signal in the presence of strong additive Gaussian noise, it has been shown that the quadratic transformation generating the timing sinusoid with the greatest signal-to-noise ratio and the maximum-likelihood timing-phase estimator is determined by the spectral correlation function of the digital signal and the power spectrum of the additive noise.

These general results demonstrate the central role played by spectral correlation and higher-order spectral-moment and cumulant functions in the design and analysis of synchronizers. However, to obtain more refined results showing more specifically how to design synchronizers and evaluate their performance, we must focus on particular classes of signals and nonlinearities. This article treats the class of amplitude-shift-keyed signals and their baseband counterparts, the pulse-amplitude-modulated signals, along with quadratic nonlinearities consisting of a squaring device preceded and followed by fixed linear filters, and it carries out a detailed study of the statistical characteristics of the resulting waveforms with a view to their optimization.

Reference [3] calculated the power spectrum of the square of the binary phase-shift-keyed or PAM signal

$$u(t) = \sum_{j=-\infty}^{\infty} c_j g(t - jT) \tag{1}$$

with statistically independent, identically distributed real coefficients $\{c_j\}$ having mean 0, mean squared value 1, and fourth moment M:

$$\mathbf{E}\{c_j\} = 0, \quad \mathbf{E}\{c_j^2\} = 1, \quad M \triangleq \mathbf{E}\{c_j^4\}. \tag{2}$$

That spectrum consists of a clock-frequency line and a self-noise spectrum that results from the random nature of the signal. The part of the latter representing self-noise components in quadrature with the clock-frequency line can cause jitter in the phase of the regenerated clock. However, (1) is cyclostationary and cycloergodic [4–6], and, as mentioned in [3], [7], and elsewhere, there are correlations [8] between Fourier components of any single realization of (1) separated in frequency by any harmonic of $1/T$ that remain even when the signal ensemble is stationarized as in [3] by randomizing the signal's epoch, although this randomization destroys the signal's cyclostationarity [5]. These spectral correlations profoundly affect the magnitude of the quadrature component of the self-noise.

When the transmitted waveform is $u(t)$ times a sinusoidal carrier, its upper and lower sidebands are well known to be deterministically related. Because $\mathbf{E}\{c_j\} = 0$, it has no carrier-frequency spectral component, but its square contains a term that is $u(t)^2$ times the second harmonic of the carrier. Since $u(t)^2 \geq 0$, this term is the sum of a pure second-harmonic sinusoid plus in-phase noise produced by the fluctuations of $u(t)^2$ around its mean value; there is no self-noise in quadrature with the second harmonic of the carrier to introduce any error in the times of the zero-crossings or in the phase of the carrier regenerated from this harmonic. The problem of clock-regeneration error, however, is far more complicated; to study it we suppose henceforth that $u(t)$ is available without multiplication by a carrier.

W. R. Bennett [6] determined the in-phase and quadrature spectra of the self-noise in special cases, and Manley [9] showed that, in the case of on–off keying, the self-noise arising from the signal may resemble a double-sideband suppressed-carrier (DSBSC) waveform whose carrier is in phase with the clock. Franks and Bubrouski [10] (see also [1], [11]) carried out a general analysis of the rms clock-regeneration phase error for the case of a bandlimited $g(t)$ and found conjugate-symmetry conditions on the Fourier transform of $g(t)$ and the narrowband clock-regenerating filter $H(f)$ (Figures 1 and 5) that ensure the vanishing of the quadrature self-noise. F. M. Gardner [12], Franks [13, p. 310], and others previously cited have stressed that the rms jitter of the regenerated clock as recovered by nonlinear processing (e.g., squaring) of the signal $u(t)$ can thus be far smaller than might be mistakenly inferred from the spectrum of the total self-noise [3], and a number of books [14–16] point out the benefits that can thus be obtained from the proper design of $g(t)$ or from its effective modification by a "prefilter" preceding the squaring of $u(t)$. Fang [17] has obtained similar results for square-law clock regenerators, and Mengali et al. [18–20] have found similar effects when nonquadratic nonlinearities are used.

With some simplifying assumptions, Franks [21] and F. M. Gardner [12] have found for the component of the self-noise of $u(t)^2$ in quadrature with the clock a spectral density that vanishes at the clock frequency. It was inferred that the mean squared phase jitter in the regenerated clock should therefore be proportional to the cube of the bandwidth B of the filter $H(f)$ used in the clock regeneration. In fact, if the signal pulse and the narrowband filter have the requisite conjugate symmetry, the phase jitter should remain zero as B varies up to half the clock frequency. It is shown later, however, that the term proportional to frequency in the departure from conjugate symmetry of the spectrum of the signal pulse causes the mean squared jitter to be indeed proportional to B^3 when $BT \ll 1$ if $H(f)$ possesses conjugate symmetry around the clock frequency. It is also found later that the mean squared jitter also generally includes a term proportional to the square of the offset of the center frequency of the narrowband filter $H(f)$ from the clock frequency $1/T$.

This article studies the complex spectrum of the self-noise output of the filter $H(f)$ in order to make clear the relationships between pairs of its spectral components whose frequencies differ by or sum to a multiple of the clock frequency and to determine the mean squared value of the quadrature self-noise component, which is responsible for clock-regeneration phase errors. As a result, some extensions of the work of Franks and Bubrouski [10] are possible. Straightforward analysis of (1) in Section 2 and of its square in Section 3 suffices to provide both insight into the nature of the self-noise and a quantitative description of the mean squared clock-phase jitter showing its dependence upon $g(t)$ and $H(f)$. Section 2 reveals the amplitude and phase relationships between different Fourier components of the signal which, when the signal is squared, result in spectral relationships very similar to those of the signal itself, producing, among other things, the clock-frequency spectral line discussed in Section 3. Finally, the latter section's results are applied to several signal-pulse waveforms $g(t)$ and filter responses $H(f)$.

2 SPECTRAL ANALYSIS OF THE PAM SIGNAL ITSELF

To understand the nature of the PAM signal $u(t)$, it is useful to notice that it can be generated by feeding the PAM train of impulses

$$v(t) = \sum_{j=-\infty}^{\infty} c_j \delta(t - jT) \tag{3}$$

to a filter with impulse response

$$g(t) = \int_{-\infty}^{\infty} G(f) e^{2\pi i f t} df, \tag{4}$$

where $G(f)$ is the Fourier transform of $g(t)$. Thus, the PAM signal can be described as the convolution

$$u(t) = v(t) \star g(t). \tag{5}$$

In the frequency domain, this equation takes the form

$$U(f) = V(f)G(f), \tag{6}$$

where

$$V(f) = \sum_{j=-\infty}^{\infty} c_j e^{2j\pi i f T}. \tag{7}$$

Here it is evident that

$$V(f) = V\left(f + \frac{1}{T}\right) \tag{8}$$

is a periodic function of f with period $1/T$ and that

$$V(-f) = V^*(f) \tag{9}$$

is its complex conjugate. Hence, $V(f) = V^*(\frac{m}{T} - f)$ for every integer m and, therefore, $V(\cdot)$ has the conjugate symmetry

$$V\left(\frac{m}{2T} + f\right) = V^*\left(\frac{m}{2T} - f\right) \tag{10}$$

around every multiple $m/2T$ of half the clock frequency.

These amplitude and phase relationships between the Fourier component of $v(t)$ of frequency f and those of frequencies $n/T \pm f$ for all integers n would persist even if the summations over j were, for the sake of mathematical rigor, truncated to a finite number of terms, which would at a later stage be allowed to grow infinite. The amplitude and phase relationships among the corresponding Fourier components of $u(t)$ are exactly those for $v(t)$ but modified in a deterministic way by the amplitude and phase responses $|G(f)|$ and $\arg G(f)$ of the pulse-shaping filter.

When f is a multiple of $1/2T$, $V(f)$ is real. Otherwise, $V(f)$ is complex, and, by the central limit theorem, it has a circular normal distribution with equal-variance zero-mean uncorrelated components in any two orthogonal directions. Thus, for any f that is not a multiple of $1/2T$, the contributions of $V(f)$ to the in-phase and orthogonal components of the self-noise have equal mean squared values. Because of the correlations between the values of $V(f)$ at different frequencies, however, the net results of combining different frequencies passed by the narrowband filter $H(f)$ are different for the in-phase and quadrature components.

These correlations are described by the expectations $\mathbf{E}\{V(f_1)V^*(f_2)\}$ and $\mathbf{E}\{V(f_1)V(f_2)\}$, which will be needed in the next subsection, the latter being called a *conjugate correlation* [22, 23]. Making use of the fact that $\mathbf{E}\{c_j c_k\}$ is 0 for $j \neq k$ and is 1 for $j = k$ and of the fact that a train of unit impulses (Dirac delta functions) with unit spacing (known as the shah function [24, pp. 77–79] or comb function [25, p. 28]) is its own Fourier transform, we can express these expectations as

$$\begin{aligned}\mathbf{E}\{V(f_1)V^*(f_2)\} &= \sum_{j=-\infty}^{\infty} e^{-2j\pi i(f_1-f_2)T} \\ &= \frac{1}{T}\sum_{n=-\infty}^{\infty} \delta\Big(f_1 - f_2 - \frac{n}{T}\Big)\end{aligned} \tag{11}$$

and

$$\mathbf{E}\{V(f_1)V(f_2)\} = \frac{1}{T}\sum_{n=-\infty}^{\infty} \delta\Big(f_1 + f_2 - \frac{n}{T}\Big). \tag{12}$$

These two equations merely reflect the relationships revealed more fully by (8) and (9).

If a random constant τ distributed uniformly between 0 and T is added to t in order to randomize the epoch of the signal and thus to make its ensemble statistically stationary, the phases of the Fourier components of frequencies $\{f + n/T\}$ will no longer be the same, but they will depend linearly on n, thus continuing to show fixed phase relations although in a slightly disguised form. However, in order to maintain a fixed clock phase as a basis for our investigation of the in-phase and quadrature self-noise in $u(t)$ and in $u(t)^2$, such a time shift is not introduced in this article.

2.1 Variances and Covariance of In-Phase and Quadrature Components

To understand the more complicated case in which a narrowband filter with impulse response $h(t)$ and frequency response

$$H(f) = \int_{-\infty}^{\infty} h(t)e^{-2\pi ift}\,dt \tag{13}$$

passing only frequencies in the neighborhood of a harmonic $m/2T$ of half the clock frequency is fed the *square* of the signal $u(t)$, we shall first consider the result of passing $u(t)$ itself through such a filter, whose output $w(t)$ in this case has Fourier transform

$$W(f) = V(f)G(f)H(f). \tag{14}$$

To obtain the complex form of the filter's output (the "analytic signal"), which directly exhibits the amplitude and phase of this output, we suppose that $H(f) = 0$ for $f < 0$. Multiplying the resulting output

$$w(t) = \int_{-\infty}^{\infty} V(f)G(f)H(f)e^{2\pi ift}\,df$$

of the filter by $e^{-m\pi it/T}$, we get the complex amplitude of the output expressed with reference to the frequency $m/2T$, whose in-phase and quadrature components $x(t)$ and $y(t)$ (Fig. 1) are the real and imaginary parts of

$$w(t)e^{-m\pi it/T} = x(t) + iy(t) = e^{-m\pi it/T}\int_{-\infty}^{\infty} V(f)G(f)H(f)e^{2\pi ift}\,df. \tag{15}$$

Since the expected value of the square of the absolute value of (15) can be seen to be independent of t and since the expected value of the square of (15) can similarly be seen to be independent of t whenever the filter $H(f)$ has a bandwidth $B < 1/2T$ and thus passes only frequencies in the neighborhood of a single harmonic $m/2T$ of half the clock frequency, we can keep our notation simple by dealing with the second moments of (15) at the convenient time $t = 0$, i.e., the moments of

$$w(0) = x + iy = \int_{-\infty}^{\infty} V(f)G(f)H(f)\,df, \tag{16}$$

where $x = x(0)$ and $y = y(0)$. Because $\mathbf{E}\{c_j\} = 0$, the mean value of (16) is zero. If the impulse response $h(t)$ of the filter is very long because of a small bandwidth, the central limit theorem is applicable, and hence the joint probability distribution of x and y is very nearly the bivariate normal distribution [26, ch. 1].

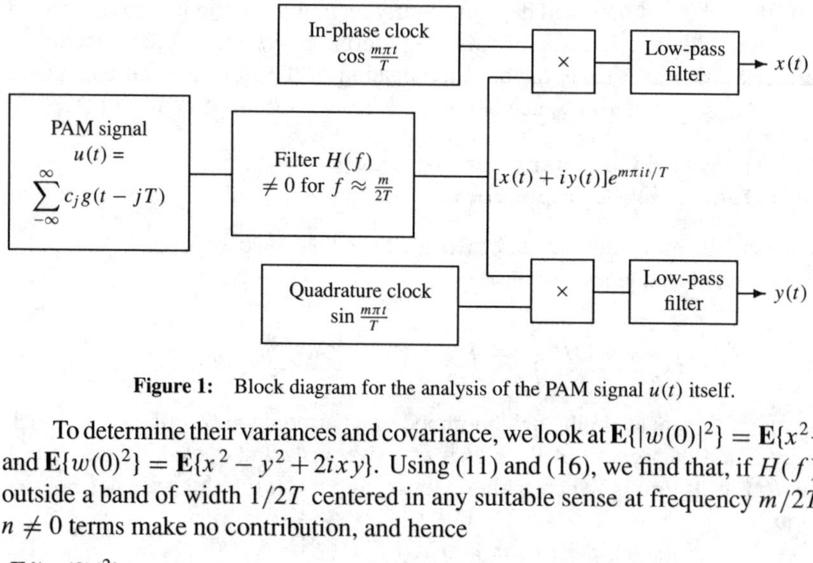

Figure 1: Block diagram for the analysis of the PAM signal $u(t)$ itself.

To determine their variances and covariance, we look at $\mathbf{E}\{|w(0)|^2\} = \mathbf{E}\{x^2+y^2\}$ and $\mathbf{E}\{w(0)^2\} = \mathbf{E}\{x^2 - y^2 + 2ixy\}$. Using (11) and (16), we find that, if $H(f) = 0$ outside a band of width $1/2T$ centered in any suitable sense at frequency $m/2T$, the $n \neq 0$ terms make no contribution, and hence

$$\begin{aligned}
\mathbf{E}\{|w(0)|^2\}& \\
&= \mathbf{E}\{x^2 + y^2\} \\
&= \frac{1}{T}\int_{-\infty}^{\infty}\int_{-\infty}^{\infty}\sum_{n=-\infty}^{\infty} G(f_1)H(f_1)G^*(f_2)H^*(f_2)\delta\Big(f_1 - f_2 - \frac{n}{T}\Big)\,df_2\,df_1 \\
&= \frac{1}{T}\int_{-\infty}^{\infty} G(f)H(f)G^*(f)H^*(f)\,df \\
&\approx \frac{1}{T}\left|G\Big(\frac{m}{2T}\Big)\right|^2 \int_{-\infty}^{\infty} |H(f)|^2\,df.
\end{aligned} \tag{17}$$

The approximation in the last line of (17) is valid if the variation in $G(f)$ over the passband of $H(f)$ can be neglected. Its first factor represents the PAM total power spectrum [3] $|G(f)|^2/T$; the integral factor expresses the effect of the analyzing filter.

Similarly, if the filter's passband again lies in the neighborhood of the harmonic $m/2T$ of half the clock frequency, we have from (12) and (16)

$$\begin{aligned}
&\mathbf{E}\{w(0)^2\}\\
&= \mathbf{E}\{x^2 - y^2 + 2ixy\}\\
&= \frac{1}{T}\int_{-\infty}^{\infty}\int_{-\infty}^{\infty}\sum_{n=-\infty}^{\infty} G(f_1)H(f_1)G(f_2)H(f_2)\delta\Big(f_1 + f_2 - \frac{n}{T}\Big)\,df_2\,df_1\\
&= \frac{1}{T}\int_{-\infty}^{\infty}\sum_{n=-\infty}^{\infty} G\Big(\frac{n}{T} - f\Big)H\Big(\frac{n}{T} - f\Big)G(f)H(f)\,df \qquad (18)\\
&= \frac{1}{T}\int_{-\infty}^{\infty} G\Big(\frac{m}{2T} - f\Big)H\Big(\frac{m}{2T} - f\Big)G\Big(\frac{m}{2T} + f\Big)H\Big(\frac{m}{2T} + f\Big)\,df\\
&\approx \frac{1}{T}G\Big(\frac{m}{2T}\Big)^2\int_{-\infty}^{\infty} H\Big(\frac{m}{2T} - f\Big)H\Big(\frac{m}{2T} + f\Big)\,df.
\end{aligned}$$

This is $G(m/2T)^2/T$ times the convolution of $H(f)$ with itself evaluated at $f = m/T$. Notice that this convolution is zero if, for example, $m/2T$ is outside the passband of the filter $H(f)$. When (52) is zero, its first line implies that $x(t)$ and $y(t)$ are uncorrelated and that they have equal mean squared values.

Adding and subtracting (17) with f replaced by $\frac{m}{2T} \pm f$ and the real part of (52), we find that the mean squared value of the in-phase part of the filter's output can be expressed as

$$\begin{aligned}
\mathbf{E}\{x^2\} = \frac{1}{4T}\int_{-\infty}^{\infty}\Big|&G\Big(\frac{m}{2T} + f\Big)H\Big(\frac{m}{2T} + f\Big)\\
&+G^*\Big(\frac{m}{2T} - f\Big)H^*\Big(\frac{m}{2T} - f\Big)\Big|^2\,df
\end{aligned} \qquad (19)$$

and that of the quadrature part of the filter's output is

$$\begin{aligned}
\mathbf{E}\{y^2\} = \frac{1}{4T}\int_{-\infty}^{\infty}\Big|&G\Big(\frac{m}{2T} + f\Big)H\Big(\frac{m}{2T} + f\Big)\\
&-G^*\Big(\frac{m}{2T} - f\Big)H^*\Big(\frac{m}{2T} - f\Big)\Big|^2\,df
\end{aligned} \qquad (20)$$

if the filter $H(f)$ passes frequencies in the neighborhood of no more than one harmonic $m/2T$ of half the clock frequency. Thus, (20) vanishes if and only if $G(\frac{m}{2T} + f)\,H(\frac{m}{2T} + f) \equiv G^*(\frac{m}{2T} - f)\,H^*(\frac{m}{2T} - f)$, e.g., only if $G(\cdot)$ and $H(\cdot)$ possess conjugate symmetry around the clock harmonic $m/2T$ or, equivalently, if $g(t)$ and $h(t)$ are, respectively, $\cos m\pi t/T$ and $e^{m\pi it/T}$ times a real function of t.

2.2 The Vanishing of the Quadrature Self-Noise

From (20) we see that $\mathbf{E}\{y^2\} = 0$, i.e., there is no quadrature self-noise, if

$$G\Big(\frac{m}{2T} + f\Big)\, H\Big(\frac{m}{2T} + f\Big) = G^*\Big(\frac{m}{2T} - f\Big)\, H^*\Big(\frac{m}{2T} - f\Big) \qquad (21)$$

for almost all values of f. This condition can be achieved by tailoring either $G(f)$ or $H(f)$ to make their product satisfy (21). When this is the case, the impulse response of the two filters $G(f)$ and $H(f)$ in tandem is $e^{m\pi it/T}$ times a real function of time, and the real part of the latter product is therefore a double-sideband amplitude-modulated suppressed-carrier waveform with carrier frequency $m/2T$ that passes through zero at exactly the same times as the clock-harmonic waveform $\cos m\pi t/T$, with no error due to self-noise. For the case of the squared signal $u(t)^2$, we shall find a set of similar conditions in Section 3 that involve $H(f)$ and each of a set of functions $\{G_k(f)\}$.

To satisfy (21) it suffices to ensure that

$$H\Big(\frac{m}{2T} + f\Big) = H^*\Big(\frac{m}{2T} - f\Big) \quad \text{and} \quad G\Big(\frac{m}{2T} + f\Big) = G^*\Big(\frac{m}{2T} - f\Big), \qquad (22)$$

i.e., that $H(\cdot)$ has conjugate symmetry around the clock harmonic $m/2T$ and that the part of $G(\cdot)$ lying within the passband of $H(\cdot)$ has the same symmetry. In this case, the impulse response $h(t)$ of the $H(f)$ filter is $e^{m\pi it/T}$ times a real function of time, and the part of $g(t)$ whose spectrum lies within the passband of $H(f)$ is $\cos m\pi t/T$ times a real function of time—again a DSBSC waveform. The latter condition is not hard to meet when that passband is very narrow and includes the frequency $m/2T$.

Since, when $G(f)$ is real, i.e., when $g(t)$ is symmetric, any two Fourier components of $u(t)$ whose frequencies sum to a multiple of $1/T$ have equal amplitudes and opposite phases on account of the $V(f)$ factor in (14), they are like the sidebands of a suppressed-carrier AM signal whose carrier frequency is the average of their two frequencies and whose carrier phase is the same as that of the clock. If the filter $H(f)$ passes the two of them with equal attenuation and opposite phase shifts, there is no quadrature clock-harmonic component in their sum.

When the condition (21) is satisfied, (52) is real, and so x and y, which have mean 0, are uncorrelated. In this case all of the power represented by (17) lies in the in-phase component. From (20) we see that quadrature components of the filter's output result from any differences between $G(\frac{m}{2T} + f)H(\frac{m}{2T} + f)$ and $G^*(\frac{m}{2T} - f)H^*(\frac{m}{2T} - f)$, which destroy the totally in-phase addition of $V(\frac{m}{2T} + f)$ and $V(\frac{m}{2T} - f)$ that occurs when (21) is satisfied. If only one of the two frequencies $m/2T \pm f$ falls within the narrowband filter's passband, it contributes equally to the in-phase and the quadrature components of the filter output.

2.3 Examples: Ideal Bandpass and Gaussian $H(f)$

Ideal Bandpass Filter: When $H(f)$ is a normalized ideal bandpass filter with bandwidth $B \leq 1/2T$ centered F above the clock harmonic $m/2T$ (because of mistuning, F may fail to be zero), i.e., when $H(\frac{m}{2T} + F) = 1$ and

$$H(f) = \text{rect}\,\frac{f - \frac{m}{2T} - F}{B}, \tag{23}$$

where [25, p. 29]

$$\text{rect}\, z \triangleq \begin{cases} 1 & \text{for } |z| < \frac{1}{2} \\ 0 & \text{otherwise} \end{cases} \tag{24}$$

(rect z is sometimes denoted by $\Pi(z)$ [24, p. 52]), the integral appearing in (17) is

$$\int_{-\infty}^{\infty} |H(f)|^2\, df = B. \tag{25}$$

The convolution of $H(f)$ with itself appearing in (52) is conveniently expressed in terms of the triangle function [24, p. 53]

$$\begin{aligned} \Lambda(z) &\triangleq \text{rect}\, z \star \text{rect}\, z \\ &= \begin{cases} 1 - |z| & \text{for } |z| \le 1 \\ 0 & \text{otherwise} \end{cases} \end{aligned} \tag{26}$$

(Fig. 2). This convolution,

$$B\Lambda\left(\frac{f - \frac{m}{T} - 2F}{B}\right),$$

has the shape of an isosceles triangle of height B and total width $2B$ centered at $m/2T + F$. Thus, if $G(m/2T)$ is real, (20) becomes

$$\mathbf{E}\{y^2\} \approx \frac{B}{2T} G\left(\frac{m}{2T}\right)^2 \left[1 - \Lambda\left(\frac{2F}{B}\right)\right], \tag{27}$$

and the quadrature component of the filter output is reduced in mean squared value as shown in Fig. 3b when the detuning $|F|$ is less than $B/2$, i.e., whenever a multiple $m/2T$ of half the clock frequency lies within the narrowband filter's passband.

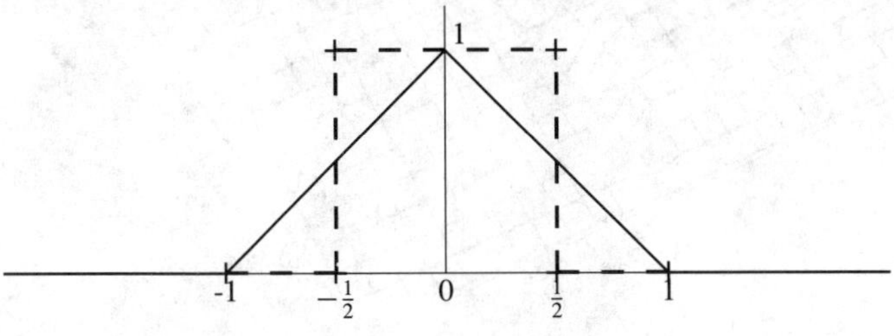

Figure 2: The rect (dashed) and Λ (solid) functions.

$G(m/2T)$ may not be real if the signal pulse $g(t)$ is not even, as $G(f)$ is then not always real, and (52) may therefore not be real. If it is not, x and y are correlated. It is then appropriate to shift the phase of the reference clock as described later in Section 3.3 in order to permit the quadrature component of the self-noise to vanish.

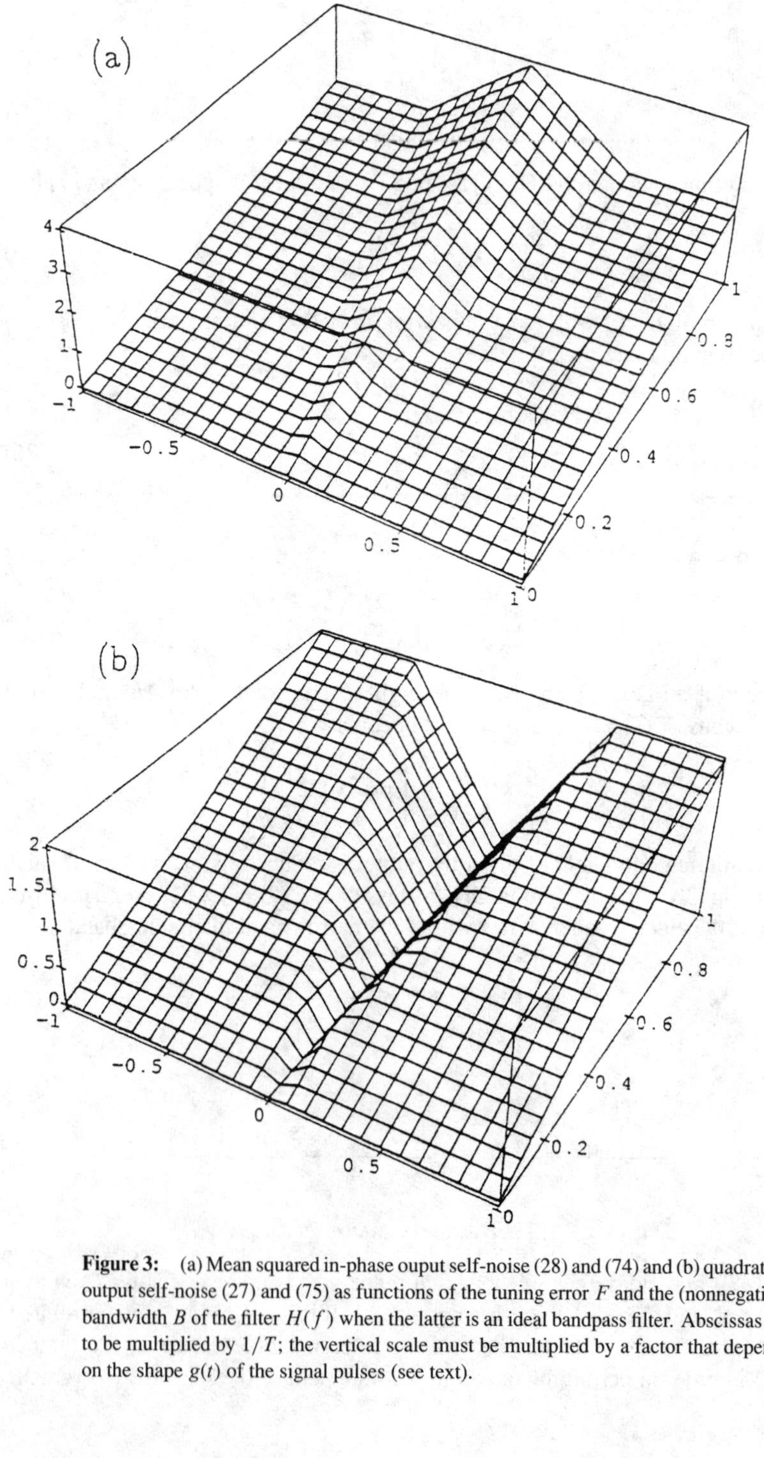

Figure 3: (a) Mean squared in-phase ouput self-noise (28) and (74) and (b) quadrature output self-noise (27) and (75) as functions of the tuning error F and the (nonnegative) bandwidth B of the filter $H(f)$ when the latter is an ideal bandpass filter. Abscissas are to be multiplied by $1/T$; the vertical scale must be multiplied by a factor that depends on the shape $g(t)$ of the signal pulses (see text).

The power of the in-phase component (19) is compensatingly increased, becoming

$$\mathbf{E}\{x^2\} \approx \frac{B}{2T} G\left(\frac{m}{2T}\right)^2 \left[1 + \Lambda\left(\frac{2F}{B}\right)\right] \tag{28}$$

(Fig. 3a), as the sum $E\{x^2\} + \mathbf{E}\{y^2\}$ is hardly affected by any small detuning F even if $F > B$. When the ideal bandpass filter is not centered at $m/2T$, it will pass a band of width $|F|$ on one side of $m/2T$ that is not paired with the corresponding band on the opposite side to make a DSBSC waveform because the filter does not pass the latter band. The self-noise power in the former band therefore contributes equally to in-phase and quadrature components, thus accounting for the fact that the quadrature self-noise power is proportional to $|F|$ for $|F| \leq B/2$. When $|F| \geq B/2$, *none* of the self-noise passed by the filter is paired with self-noise equally far on the opposite of $m/2T$, and *all* of the self-noise is divided equally between in-phase and quadrature components, just as if there were no spectral correlations.

Gaussian Filter: When

$$H(f) = \exp\left[-\frac{(f - \frac{m}{2T} - F)^2}{2B^2}\right], \tag{29}$$

(20) becomes

$$\mathbf{E}\{y^2\} \approx \frac{\sqrt{\pi} B}{2T} G\left(\frac{m}{2T}\right)^2 \left(1 - e^{-F^2/B^2}\right), \tag{30}$$

which is proportional to F^2/BT when $|F| \ll B$, is proportional to B/T when $F \gg B$, and is zero regardless of the value of B when $F = 0$. See Fig. 4b; for application to the present section, ordinates in Figs. 3 and 4 are to be multiplied by $BG\left(\frac{m}{2T}\right)^2/4T$. Likewise (19) becomes

$$\mathbf{E}\{x^2\} \approx \frac{\sqrt{\pi} B}{2T} G\left(\frac{m}{2T}\right)^2 \left(1 + e^{-F^2/B^2}\right) \tag{31}$$

(Fig. 4a), which, when added to (30), gives a total self-noise power proportional to B that is independent of any small mistuning F.

3 SPECTRAL ANALYSIS OF $u(t)^2$

Because of the correlation (11) and conjugate correlation (12) between spectral components whose frequencies respectively differ by (if one of the components is conjugated) or sum to (if neither component is conjugated) any multiple of the clock frequency, the squaring of $u(t)$ (Fig. 5) offers a natural way to bring about the heterodyning of pairs of correlated components to yield spectral lines at harmonics of the clock frequency. If the squaring precedes narrowband filtering, all of the received frequencies are able to participate in the process.

When $v(t)$ is squared, pairs of components like these add together to produce spectral lines in the $k = 0$ portion of (32) at multiples of the clock frequency. As

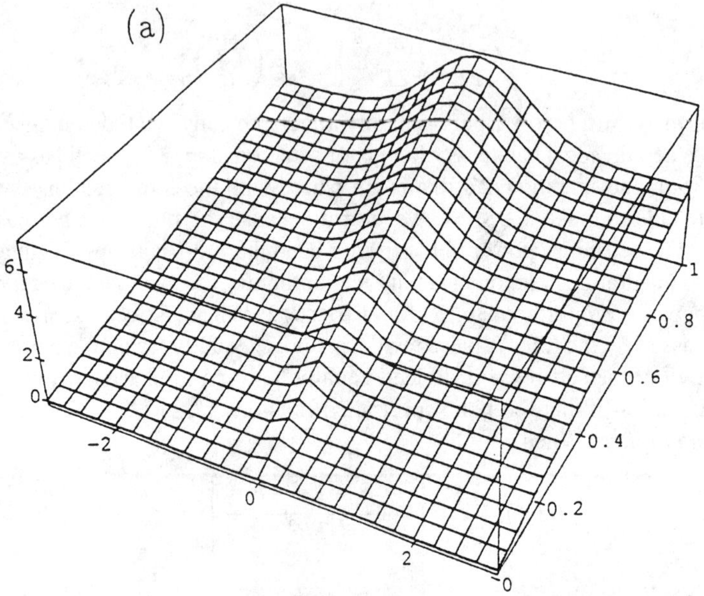

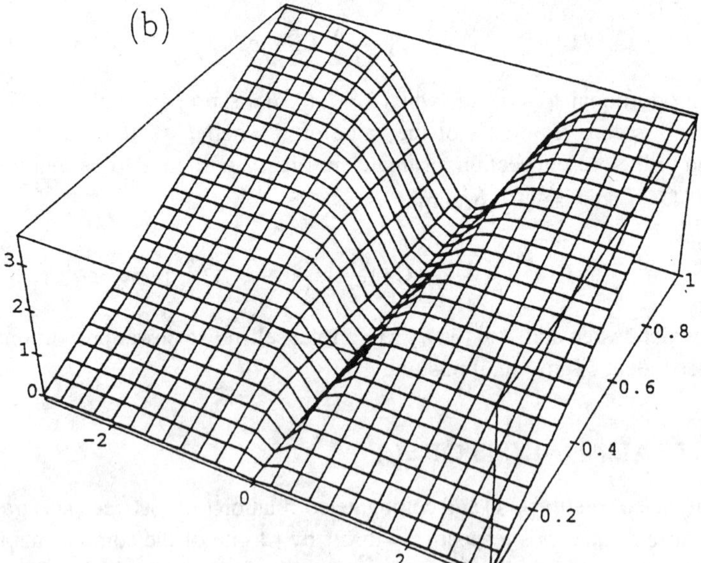

Figure 4: (a) Mean squared in-phase output self-noise (31) and (76) and (b) quadrature output self-noise (30) and (77) as functions of the tuning error F and the (nonnegative) bandwidth B of the filter $H(f)$ when this filter has a Gaussian frequency response. Abscissas are to be multiplied by $1/T$; the vertical scale must be multiplied by a factor that depends on the shape $g(t)$ of the signal pulses (see text).

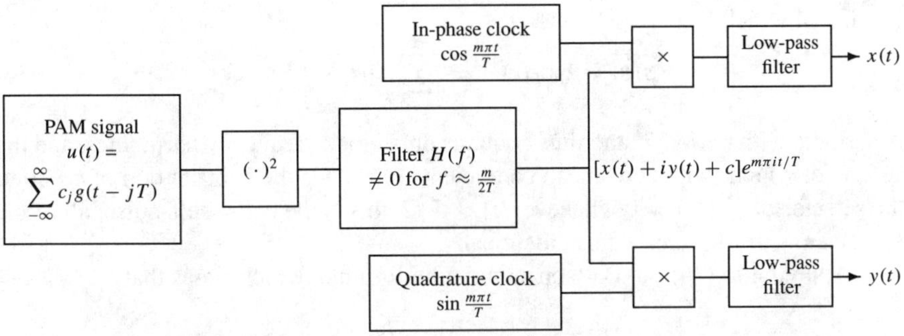

Figure 5: Block diagram for the analysis of the square $u(t)^2$ of the PAM signal. In addition to the self-noise components $x(t)$ and $y(t)$, there will be constant output components $\Re\{c\}$ and $\Im\{c\}$ from the low-pass filter that result from the clock-harmonic output $ce^{m\pi it/T}$ of the filter $H(f)$.

(52)–(55) show, each pair whose frequencies do not sum to a multiple of the clock frequency combines additively with the pair whose frequencies are their reflections in a multiple of half the clock frequency, to produce self-noise spectral components of the squared signal that, to the extent the filters $G(f)$ and $H(f)$ allow these correlated combining components to have equal amplitudes and opposite phases, exhibit the same sort of spectral correlations (44) and (45) as the signal itself.

The square of the signal (1) can be expressed in the form

$$\begin{aligned} u(t)^2 &= \sum_{k=0}^{\infty} \sum_{j=-\infty}^{\infty} \epsilon_k c_j c_{j+k} g(t - jT) g(t - (j+k)T) \\ &= \sum_{k=0}^{\infty} \epsilon_k v_k(t) \star g_k(t), \end{aligned} \tag{32}$$

where the Neumann factor ϵ_k is 1 for $k = 0$ and is 2 otherwise,

$$v_k(t) = \sum_{j=-\infty}^{\infty} c_j c_{j+k} \delta\left(t - \frac{2j+k}{2} T\right), \tag{33}$$

and

$$g_k(t) = g\left(t + \frac{kT}{2}\right) g\left(t - \frac{kT}{2}\right). \tag{34}$$

Because the $\{c_j\}$ are statistically independent, zero-mean random variables, the terms of (32) with different values of k are completely uncorrelated, regardless of any time shift, as noted by Fang [17].

The $k = 0$ part of (32) can be separated into two terms,

$$\sum_{j=-\infty}^{\infty} c_j^2 g(t - jT)^2 = s(t) + \sum_{j=-\infty}^{\infty} (c_j^2 - 1) g(t - jT)^2, \tag{35}$$

the first of which,

$$s(t) \triangleq \mathbf{E}\{u(t)^2\} = \sum_{j=-\infty}^{\infty} g(t - jT)^2, \tag{36}$$

is periodic with period T and thus contains only clock-frequency harmonics, and the second of which (with zero-mean coefficients $c_j^2 - 1$) is the contribution of $k = 0$ to the self-noise. If $g(t) = 0$ whenever $|t| > T/2$, this is the entire self-noise; all of the $k > 0$ terms of (32) then vanish identically.

Applying to (36) the Poisson summation formula, which states that

$$\sum_{k=-\infty}^{\infty} q(kT) = \frac{1}{T} \sum_{j=-\infty}^{\infty} Q\left(\frac{j}{T}\right) \tag{37}$$

for any Fourier-transform pair $\{q(t), Q(f)\}$ (because the shah function is its own Fourier transform), we can express $s(t)$ as the Fourier series

$$s(t) = \frac{1}{T} \sum_{k=-\infty}^{\infty} e^{2k\pi it/T} \int_{-\infty}^{\infty} G\left(\frac{k}{2T} - f\right) G\left(\frac{k}{2T} + f\right) df. \tag{38}$$

This series contains only a finite number of nonzero terms when $G(f)$ is bandlimited. If the convolution integral in (38) is not real for $k = m/2$, i.e., for the clock harmonic $m/2T$ being regenerated, and, instead has a phase angle ψ, the self-noise component in the direction $\psi + \frac{1}{2}\pi$ rather than the component $y(t)$ in the direction $\frac{1}{2}\pi$ will produce errors in the phase of the regenerated clock. When $g(t)$ is an even function of t and $G(f)$ is therefore real, however, ψ will be zero, and the mean squared value of $y(t)$ will determine the mean squared phase error of the regenerated clock. The case of $\psi \neq 0$ is discussed later in Section 3.3.

If, instead of being real, the coefficients $\{c_j\}$ in (3) were equally likely to be each of the nth roots of unity with $n > 2$, (3) would not be real, and (9) would not be valid, thus invalidating (10). Moreover, because $\mathbf{E}\{c_j\} = 0$, the right-hand side of (12) would be replaced by zero and, hence, that of (52) also. Thus, $\mathbf{E}\{x^2\} = \mathbf{E}\{y^2\}$ and $\mathbf{E}\{xy\} = 0$; i.e., the in-phase and quadrature components of $u(t)$ would then always have identical spectra and would be uncorrelated. In addition, (36) would vanish, and there would be no periodic component from which to regenerate the clock. However, such a periodic component is then obtainable, for example, from $|u(t)|^2$ because there is still a correlation between spectral components of $u(t)$ whose frequencies differ by any multiple of the clock frequency.

From (32)–(34) it follows that the Fourier transform of $u(t)^2$ is

$$\sum_{k=0}^{\infty} \epsilon_k V_k(f) G_k(f), \tag{39}$$

where

$$V_k(f) = \sum_{j=-\infty}^{\infty} c_j c_{j+k} e^{-(2j+k)\pi ifT}, \tag{40}$$

and

$$G_k(f) = \int_{-\infty}^{\infty} g\left(t + \frac{kT}{2}\right) g\left(t - \frac{kT}{2}\right) e^{-2\pi i f t}\, dt$$
$$= \int_{-\infty}^{\infty} G\left(\frac{f}{2} + \phi\right) G\left(\frac{f}{2} - \phi\right) e^{\pm 2k\pi i \phi T}\, d\phi \tag{41}$$

are the Fourier transforms of $v_k(t)$ and $g_k(t)$. The second line of (41) shows that $G_k(f)$ comes from pairs of Fourier components of $u(t)$ whose frequencies sum to f. By comparing (38) with (41), notice that $G_0(m/2T)$ is the complex amplitude of the Fourier component of $s(t)$ of frequency $m/2T$ for even m, from which the clock can be recovered if $G_0(m/2T) \neq 0$.

From (40) it follows that

$$V_k^*(-f) = V_k(f) = V_k\left(f + \frac{1}{T}\right) \tag{42}$$

and, hence,

$$V_k\left(\frac{m}{2T} + f\right) = V_k^*\left(\frac{m}{2T} - f\right) \tag{43}$$

for every integer m, as is the case for $V(f)$ as given by (7) [cf. (8)–(10)]; i.e., every $V_k(f)$, too, has conjugate symmetry about each multiple of half the clock frequency. Whenever k is odd and f is a multiple of $1/T$ or k is even and f is a multiple of $1/2T$, (40) is real; otherwise, it has a circular normal distribution like that of $V(f)$, and so it too results in contributions to $x(t)$ and to $y(t)$ that are, for each f, uncorrelated, zero-mean random variables.

Just as we needed the correlation properties of $V(f)$ in Section 2, we shall shortly need those of the $\{V_k(f)\}$. Again recognizing the Fourier series for the shah function [24, pp. 77–79], we find that

$$\mathbf{E}\{V_{k_1}(f_1)V_{k_2}^*(f_2)\} = \frac{\delta_{k_1 k_2}}{T} \sum_{n=-\infty}^{\infty} \delta\left(f_1 - f_2 - \frac{n}{T}\right) \tag{44}$$

and

$$\mathbf{E}\{V_{k_1}(f_1)V_{k_2}(f_2)\} = \frac{\delta_{k_1 k_2}}{T} \sum_{n=-\infty}^{\infty} \delta\left(f_1 + f_2 - \frac{n}{T}\right). \tag{45}$$

In contrast with the Dirac delta function $\delta(\cdot)$, the Kronecker delta $\delta_{k_1 k_2}$ is 1 for $k_1 = k_2$ and is 0 otherwise. It arises here because the $\{c_j\}$ are zero-mean independent random variables, of which just the first power of at least one is a factor in each term for which $k_1 \neq k_2$. Thus, the $\{V_k(f)\}$ have the same autocorrelation properties [17] as $V(f)$ and are completely uncorrelated with one another. Moreover, the sum of the terms of (32) for each $k > 0$ behaves just like $u(t)$ with $g(t)$ replaced by $g_k(t)$, and so our analysis of $u(t)$ can be applied directly to $u(t)^2$ except for the $k = 0$ terms.

Because all three express the correlation between signal values displaced in time by kT or by radar range difference and shifted in frequency by heterodyning or by the Doppler effect, (41) and Woodward's [25, sec. 7.2] complex radar ambiguity function $\chi(\tau, \phi)$ and W. A. Gardner's cyclic finite autocorrelation function [5, p. 356, eq. (12.149)] are identical in form whenever $g(t)$ is real for all t and, consequently, $G(\frac{1}{2}f - \phi) = G^*(\phi - \frac{1}{2}f)$ for all values of f. Otherwise, (41) must be regarded as a conjugate correlation [22] and [23], which can be regarded as a form of cross-ambiguity function.

From the first line of (41) we see that, if $g(t)$ is an even function, $G_k(f)$ is real, and so this factor in (39) will leave the proper phase relations between (39) at frequency f and (39) at frequency $l/2T \pm f$ for every integer l. The amplitude relationships will generally be altered, however, preventing full cancellation of the quadrature self-noise unless the bandwidth B of the filter $H(f)$ is small enough that $|G(f)|$ varies negligibly over its passband, since the Fourier transform of the narrowband filter's output is

$$W(f) = H(f) \sum_{k=0}^{\infty} \epsilon_k V_k(f) G_k(f). \tag{46}$$

Subtraction from (32) of the periodic component (36) of $u(t)^2$, which gives rise to the clock-harmonic spectral lines, replaces the $k = 0$ term of (32) by the last term of (35). Multiplying the time-domain version of (39) by $e^{-m\pi i t/T}$ as in Section 2.2 in order to get its phasor representation in terms of the reference frequency $m/2T$ and again setting $t = 0$ for convenience, we now find that, at $t = 0$, the phasor representing the self-noise near this frequency that is passed by the filter $H(f)$ is

$$x + iy = w(0) = \int_{-\infty}^{\infty} H(f) \left[V_{00}(f) G_0(f) + 2 \sum_{k=1}^{\infty} V_k(f) G_k(f) \right] df, \tag{47}$$

where

$$V_{00}(f) = \sum_{j=-\infty}^{\infty} (c_j^2 - 1) e^{-2\pi i f T}. \tag{48}$$

Notice that (47) resembles (16). The autocorrelations of (48), which are completely uncorrelated with $V_k(f)$ for $k > 0$, is

$$\mathbf{E}\{V_{00}(f_1) V_{00}^*(f_2)\} = \frac{M-1}{T} \sum_{n=-\infty}^{\infty} \delta\left(f_1 - f_2 - \frac{n}{T}\right) \tag{49}$$

and

$$\mathbf{E}\{V_{00}(f_1) V_{00}(f_2)\} = \frac{M-1}{T} \sum_{n=-\infty}^{\infty} \delta\left(f_1 + f_2 - \frac{n}{T}\right), \tag{50}$$

where M is the fourth moment (2) of the coefficients $\{c_j\}$.

3.1 The Statistics of the In-Phase and Quadrature Components

From (44) and (47) we find that, if the bandwidth B of the $H(f)$ filter does not exceed $1/2T$,

$$\begin{aligned}
&\mathbf{E}\{x^2+y^2\}\\
&\quad=\mathbf{E}\{|w(0)|^2\}\\
&\quad=\frac{M-1}{T}\int_{-\infty}^{\infty}|G_0(f)|^2|H(f)|^2\,df+\frac{4}{T}\sum_{k=1}^{\infty}\int_{-\infty}^{\infty}|G_k(f)|^2|H(f)|^2\,df \qquad (51)\\
&\quad\approx\left[\frac{M-1}{T}\left|G_0\left(\frac{m}{2T}\right)\right|^2+\frac{4}{T}\sum_{k=1}^{\infty}\left|G_k\left(\frac{m}{2T}\right)\right|^2\right]\int_{-\infty}^{\infty}|H(f)|^2\,df.
\end{aligned}$$

Likewise, the square of the self-noise phasor has expectation

$$\begin{aligned}
&\mathbf{E}\{x^2-y^2+2ixy\}\\
&\quad=\mathbf{E}\{w(0)^2\}\\
&\quad=\frac{M-1}{T}\sum_{n=-\infty}^{\infty}\int_{-\infty}^{\infty}G_0\left(\frac{n}{2T}+f\right)G_0\left(\frac{n}{2T}-f\right)H\left(\frac{n}{2T}+f\right)H\left(\frac{n}{2T}-f\right)df\\
&\quad\quad+\frac{4}{T}\sum_{n=-\infty}^{\infty}\sum_{k=1}^{\infty}\int_{-\infty}^{\infty}G_k\left(\frac{n}{2T}+f\right)G_k\left(\frac{n}{2T}-f\right) \qquad (52)\\
&\quad\quad\times H\left(\frac{n}{2T}+f\right)H\left(\frac{n}{2T}-f\right)df\\
&\quad\approx\left[\frac{M-1}{T}G_0\left(\frac{m}{2T}\right)^2+\frac{4}{T}\sum_{k=1}^{\infty}G_k\left(\frac{m}{2T}\right)^2\right]\\
&\quad\quad\times\int_{-\infty}^{\infty}H\left(\frac{m}{2T}+f\right)H\left(\frac{m}{2T}-f\right)df
\end{aligned}$$

if the passband of $H(f)$ lies in the neighborhood of the harmonic $m/2T$ of half the clock frequency; (52) is zero if $m/2T$ does not lie within the passband of this filter. Equations (51) and (52) are convenient when the signal pulse $g(t)$ has a finite duration, as the summation over k then includes only a finite number of nonzero terms.

Notice that the terms of (52) for each n involve two pairs of Fourier components of $g(t)$ whose frequencies sum respectively to $(n/2T)\pm f$, i.e., with n/T as the sum of the four frequencies, while each term of (51) comes from a tetrad of Fourier components of $g(t)$, pairs of whose frequencies have arbitrary equal sums f. This is related to the fact that the fourth-order spectral-moment function studied by Spooner [2] is nonzero only when the sum of the four frequencies equals a clock harmonic. Notice also that, if $M=1$, as in the case where each c_j in (1) takes only the values ±1, the terms of (51) and of (52) involving G_0 disappear, and only the terms involving G_k with $k\geq 1$ remain.

Combining (51) and (52) just as we did (17) and (52), we find that

$$\begin{aligned}\mathbf{E}\{x^2\} = \frac{M-1}{4T} \int_{-\infty}^{\infty} \Big| G_0\Big(\frac{m}{2T}+f\Big) H\Big(\frac{m}{2T}+f\Big) \\ + G_0^*\Big(\frac{m}{2T}-f\Big) H^*\Big(\frac{m}{2T}-f\Big)\Big|^2 \, df \\ + \frac{1}{T}\sum_{k=1}^{\infty} \int_{-\infty}^{\infty} \Big| G_k\Big(\frac{m}{2T}+f\Big) H\Big(\frac{m}{2T}+f\Big) \\ + G_k^*\Big(\frac{m}{2T}-f\Big) H^*\Big(\frac{m}{2T}-f\Big)\Big|^2 \, df\end{aligned} \tag{53}$$

and that

$$\begin{aligned}\mathbf{E}\{y^2\} = \frac{M-1}{4T} \int_{-\infty}^{\infty} \Big| G_0\Big(\frac{m}{2T}+f\Big) H\Big(\frac{m}{2T}+f\Big) \\ - G_0^*\Big(\frac{m}{2T}-f\Big) H^*\Big(\frac{m}{2T}-f\Big)\Big|^2 \, df \\ + \frac{1}{T}\sum_{k=1}^{\infty} \int_{-\infty}^{\infty} \Big| G_k\Big(\frac{m}{2T}+f\Big) H\Big(\frac{m}{2T}+f\Big) \\ - G_k^*\Big(\frac{m}{2T}-f\Big) H^*\Big(\frac{m}{2T}-f\Big)\Big|^2 \, df.\end{aligned} \tag{54}$$

When $g(t)$ is bandlimited and thus has long tails, (51)–(54) involve summations over many values of k. To recast (54) in a form that then involves only a finite number of nonzero terms, we add to and subtract from (54) half of what would be the $k = 0$ term if the summation over k began with $k = 0$. Then applying the Poisson summation formula, (37), to the resulting summation over k, we can express (54) in the form

$$\begin{aligned}&\mathbf{E}\{y^2\} \\ &= \frac{M-3}{4T} \int_{-\infty}^{\infty} \Big| G_0\Big(\frac{m}{2T}+f\Big) H\Big(\frac{m}{2T}+f\Big) - G_0^*\Big(\frac{m}{2T}-f\Big) H^*\Big(\frac{m}{2T}-f\Big)\Big|^2 \, df \\ &+ \frac{1}{2T^2} \int_{-\infty}^{\infty}\int_{-\infty}^{\infty} \Big[G\Big(\frac{m}{4T}+f_1\Big) G\Big(\frac{m}{4T}+f_2\Big) H\Big(\frac{m}{2T}+f_1+f_2\Big) \\ &- G^*\Big(\frac{m}{4T}-f_1\Big) G^*\Big(\frac{m}{4T}-f_2\Big) H^*\Big(\frac{m}{2T}-f_1-f_2\Big)\Big] \\ &\times \sum_{j=-\infty}^{\infty} \Big[G^*\Big(\frac{m}{4T}+f_1+\frac{j}{T}\Big) G^*\Big(\frac{m}{4T}+f_2-\frac{j}{T}\Big) H^*\Big(\frac{m}{2T}+f_1+f_2\Big) \\ &- G\Big(\frac{m}{4T}-f_1-\frac{j}{T}\Big) G\Big(\frac{m}{4T}-f_2+\frac{j}{T}\Big) H\Big(\frac{m}{2T}-f_1-f_2\Big)\Big] \, df_1\, df_2.\end{aligned} \tag{55}$$

(The corresponding form for (53) is the same except for + signs between products where − signs appear here.) As observed in [2], when the signal pulse $g(t)$ contains no frequencies higher than half that of the clock, i.e., when $G(f) = 0$ for $|f| > 1/2T$, the $j \neq 0$ terms of (55) vanish because one factor or another is zero for every f_1 and f_2. For $j = 0$ the integrand is the square of an absolute value.

3.2 The Vanishing of the Quadrature Self-Noise

The first term in (55), with the factor $M - 3$, vanishes whenever the distribution of the $\{c_j\}$ is mesokurtic, e.g., if it is Gaussian, as M is then 3; this results from the vanishing of the fourth-order spectral cumulant [2]. If $M \neq 3$, this term will generally not vanish, but we shall shortly see that, when the rest of (55) vanishes, the first term does also. These other terms vanish when their common factor is zero, i.e., when

$$G\left(\frac{m}{4T} + f_1\right) G\left(\frac{m}{4T} + f_2\right) H\left(\frac{m}{2T} + f_1 + f_2\right) = G^*\left(\frac{m}{4T} - f_1\right) G^*\left(\frac{m}{4T} - f_2\right) H^*\left(\frac{m}{2T} - f_1 - f_2\right) \tag{56}$$

for all values of f_1 and f_2. From (54) we obtain a different but equivalent condition for the vanishing of the mean squared quadrature self-noise:

$$G_k\left(\frac{m}{2T} + f\right) H\left(\frac{m}{2T} + f\right) = G_k^*\left(\frac{m}{2T} - f\right) H^*\left(\frac{m}{2T} - f\right) \tag{57}$$

for all values of k and all values of f. Multiplying (57) by $e^{2k\pi i \psi T}$, summing over k from $-\infty$ to ∞, and integrating with respect to the ϕ in the definition (41) of G_k, we find, after substituting f_1 for $f/2 - \psi$ and f_2 for $f/2 + \psi$, that (57) implies

$$\sum_{j=-\infty}^{\infty} G\left(\frac{m}{4T} + f_1 + \frac{j}{T}\right) G\left(\frac{m}{4T} + f_2 - \frac{j}{T}\right) H\left(\frac{m}{2T} + f_1 + f_2\right) = \sum_{j=-\infty}^{\infty} G^*\left(\frac{m}{4T} - f_1 - \frac{j}{T}\right) G^*\left(\frac{m}{4T} - f_2 + \frac{j}{T}\right) H^*\left(\frac{m}{2T} - f_1 - f_2\right) \tag{58}$$

for all values of f_1 and f_2, which is similar to (56) apart from the shifts by j/T and the summation over j and represents the vanishing of the *second* factor in the last integrand of (55).

The condition (57) is clearly satisfied when

$$H\left(\frac{m}{2T} + f\right) = H^*\left(\frac{m}{2T} - f\right) \tag{59}$$

for all f and

$$G_k\left(\frac{m}{2T} + f\right) = G_k^*\left(\frac{m}{2T} - f\right) \tag{60}$$

for all values of k and for all values of f for which $H(f) \neq 0$. The conjugate-symmetry condition (59) on $H(f)$ means that the impulse response $h(t)$ of the narrow-band filter is $e^{m\pi i t/T}$ times a real function of time. The conjugate symmetry condition

(60) on $G_k(f)$ means that the part of $g_k(t)$ whose spectrum lies within the passband of $H(f)$ is $\cos m\pi t/T$ times a real function of time. For $g_k(t) \triangleq g(t - kT/2)g(t + kT/2)$ to include the factor $\cos m\pi t/T$ after exclusion of frequencies not passed by $H(f)$, it is sufficient that $g(t)$ be $\cos m\pi t/2T$ times a real function of time whose spectral density vanishes at all frequencies above $m/4T$ and that $H(f)$ should pass no frequencies below $m/4T$ so that the dc component of $\cos^2 m\pi t/T$ will have no effect. Thus, we find that the same conjugate-symmetry condition

$$G\left(\frac{m}{4T} + f\right) = G^*\left(\frac{m}{4T} - f\right) \tag{61}$$

that has been found by other authors ([10], [17], and others) suffices to eliminate quadrature self-noise.

The conditions (59) and (61) together clearly suffice to satisfy (56), and they ensure the vanishing of the first term on the right-hand side of (55) as well as the second. They are not, however, the most general conditions under which $\mathbf{E}\{y^2\} = 0$, as any $G(f)$ and $H(f)$ satisfying (59) and (61) can be multiplied, respectively, by a^{-b-cf} and a^{b+cf}, where a, b, and c are any real or complex constants, and will still satisfy (56) and (57). The resulting asymmetric $G(f)$ and $H(f)$ represent the most general solution to (56) and, probably, to (58) also, as can be seen by letting $\gamma(\phi) \triangleq G(\frac{m}{4T} + \phi)/G^*(\frac{m}{4T} + \phi)$ and $\eta(\phi) \triangleq H(\frac{m}{2T} + \phi)/H^*(\frac{m}{2T} + \phi)$. Then (56) becomes $\gamma(\phi)\gamma(\psi)\eta(\phi + \psi) = 1$ for all values of ϕ and ψ. If the two sides of this equation are divided by the values they have when $\psi = 0$ and if the two sides of the resulting equation are divided by the values they have when $\phi = 0$, the equation $\eta(0)\eta(\phi + \psi) = \eta(\phi)\eta(\psi)$ is obtained, whose only continuous solution is readily found (by putting ψ equal to successive multiples of ϕ) to be an $\eta(\phi)$ that is the exponential of a linear function of ϕ, and it follows from setting $\psi = 0$ that $\gamma(\phi)$ must be the reciprocal of that exponential. The foregoing factors a^b and a^{-b} representing this exponential will take care of any phase mismatch at the clock frequency. The factor a^{-cf}, however, may destroy the conjugate symmetry of $G(f)$ around $f = 0$, which is necessary if $g(t)$ is to be real.

When $H(f)$ possesses conjugate symmetry around the clock harmonic $m/2T$ but $G_k(f)$ does not, suitable filtering before squaring ("prefiltering") may give it the requisite symmetry within the band of width B passed by $H(f)$. Otherwise, the dominant effect of the asymmetry will generally come from the linear Taylor-series term of $G_k(f)$, say $C \cdot (f - m/2T)$, which will, for $BT \ll 1$, make the $k > 0$ term of (54) proportional to $(4/T)\int_{-\infty}^{\infty} |C \cdot (f - m/2T)H(m/2T + f)|^2 df$, which varies as the cube of the bandwidth B of $H(\cdot)$ if its tails fall more than 9 dB per octave. $\mathbf{E}\{y^2\}$ will be proportional to B^3 even if $G_k(m/2T) = 0$, as, for example, in the case of (64) substituted into (20). This cubic dependence of the mean squared jitter on bandwidth is like that of the output noise in high-SNR high-modulation-index FM reception.

3.3 Other Reference Phases and the Case of Correlated x and y

If the signal pulse $g(t)$ is not symmetric, the phase angle of the clock harmonic m/T in (38) is likely to be, say $\beta \neq 4$, and, instead of $\mathbf{E}\{y(t)^2\}$, we need to know the mean squared value of the self-noise component $y'(t)$ in the direction $\beta + \frac{1}{2}\pi$, which is responsible for the errors in the phase of the clock-frequency component of the output of the narrowband filter. The mean squared value of $y'(t)$ can be determined by reference to Fig. 6, which shows the circle [26, prob. 1-9*a*] having as a diameter the line from $(\mathbf{E}\{y^2\}, -\mathbf{E}\{xy\})$ to $(\mathbf{E}\{x^2\}, \mathbf{E}\{xy\})$, whose inclination is 2θ, where

$$2\theta = \arctan \frac{2\mathbf{E}\{xy\}}{\mathbf{E}\{x^2\} - \mathbf{E}\{y^2\}} \tag{62}$$

is the phase angle of (52). As complex numbers, this diameter and the center of the circle are, respectively, (52) and half of (51). When (52) is not real, it is because $\mathbf{E}\{xy\}$ is not zero, and x and y are correlated; $\mathbf{E}\{y^2\}$ can then be made smaller by this sort of shift in the phase reference.

If the reference phase is changed from zero to β, the diameter seen in Fig. 6 turns clockwise through the angle 2β, with the abscissas of its ends becoming $\mathbf{E}\{x'(t)^2\}$ and $\mathbf{E}\{y'(t)^2\}$, and their ordinates becoming $\pm\mathbf{E}\{x'(t)y'(t)\}$. When $\beta = \theta$, $\mathbf{E}\{x'(t)^2\}$ and $\mathbf{E}\{y'(t)^2\}$ reach their extreme values, $\overline{x^2}_{\max}$ and $\overline{y^2}_{\min}$, which are the abscissas of the rightmost and leftmost points of the circle, and $x'(t)$ and $y'(t)$ are then uncorrelated. The effect of this rotation can be found by replacing H and H^* in (52)–(54) by $He^{-i\beta}$ and $H^*e^{i\beta}$, respectively.

3.4 Examples

Raised Cosine: When the signal pulse has the raised-cosine form

$$g(t) = \cos^2 \frac{\pi t}{2T} \operatorname{rect} \frac{t}{2T}, \tag{63}$$

with Fourier transform

$$\begin{aligned} G(f) &= \frac{T}{2} \sum_{j=-1}^{1} \binom{2}{j+1} \operatorname{sinc}(2fT - j) \\ &= \frac{\sin 2\pi fT}{2\pi f(1 - 4f^2T^2)}, \end{aligned} \tag{64}$$

the expectation of $u(t)^2$, (36), is a constant plus a clock-frequency sinusoid,

$$s(t) = \frac{3}{4} + \frac{1}{4} \cos \frac{2\pi t}{T}; \tag{65}$$

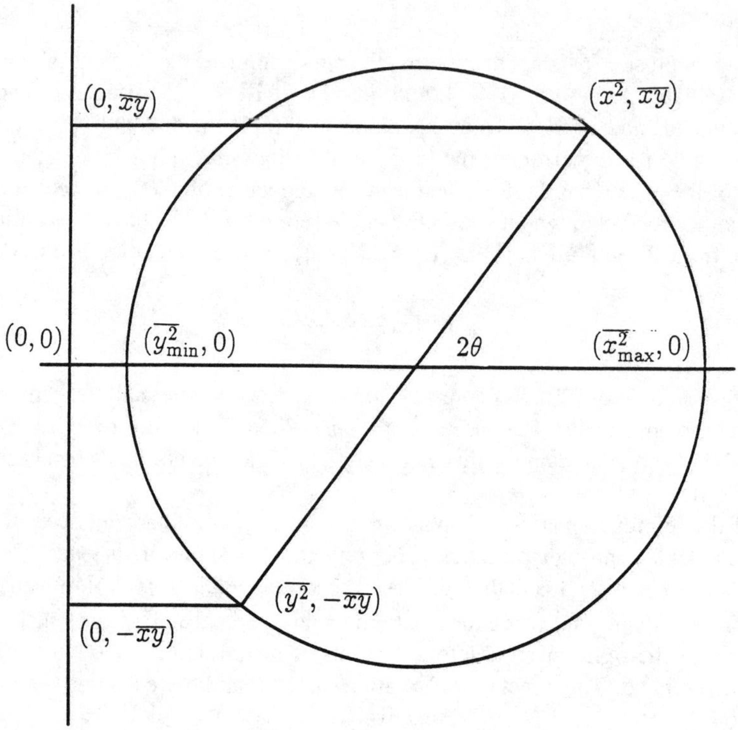

Figure 6: Relationship between the variances $\overline{x^2} = \mathbf{E}\{x^2\}$ and $\overline{y^2} = \mathbf{E}\{y^2\}$ and covariance $\overline{xy} = \mathbf{E}\{xy\}$ of the amplitudes of the in-phase and quadrature components and the extrema of the variances $\overline{x^2}_{\max} = \mathbf{E}\{x^2\}_{\max}$ and $\overline{y^2}_{\min} = \mathbf{E}\{y^2\}_{\min}$ attainable by shifting the reference phase of the filter output.

no other clock harmonic is present. Because (63) is zero for $|t| \geq T$, $G_k(f)$ vanishes for $k > 1$, and for the smaller values of k we have (Fig. 7)

$$\begin{aligned} G_0(f) &= \frac{T}{8} \sum_{j=-2}^{2} \binom{4}{j+2} \operatorname{sinc}(2fT - j) \\ &= \frac{3 \sin 2\pi fT}{8\pi f(1 - f^2T^2)(1 - 4f^2T^2)} \end{aligned} \tag{66}$$

and

$$\begin{aligned} G_1(f) &= \frac{T}{16} \sum_{j=-1}^{1} \binom{2}{j+1} \operatorname{sinc}(fT - j) \\ &= \frac{\sin \pi fT}{8\pi f(1 - f^2T^2)} \end{aligned} \tag{67}$$

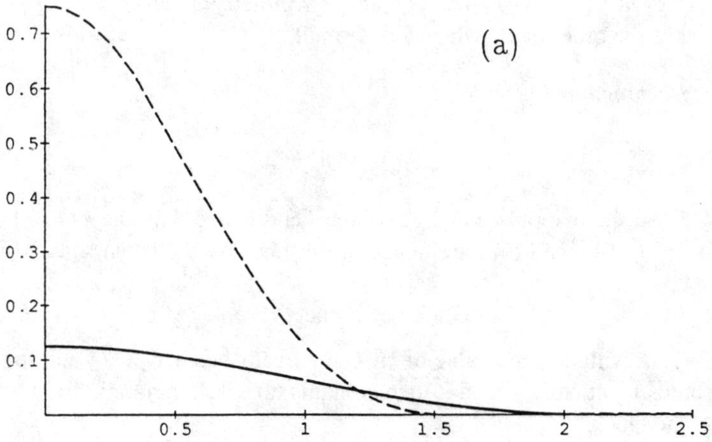

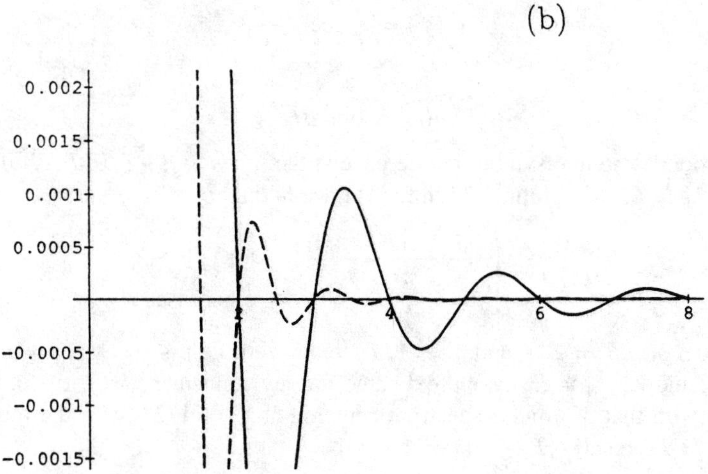

Figure 7: $G_0(f)$ (dashed) and $G_1(f)$ (solid) when the signal pulse $g(t)$ is the raised cosine (63): (a) main lobe and (b) sidelobes.

where

$$\operatorname{sinc} fT \triangleq \frac{\sin \pi f T}{\pi f T} \tag{68}$$

is the Fourier transform of $T \operatorname{rect}(t/T)$. At the clock frequency their values are $G_0(1/T) = T/8$ and $G_1(1/T) = T/16$. Hence, the quadrature self-noise output of the $H(f)$ filter will tend to vanish only if the response of this filter has conjugate symmetry about the clock frequency and its passband is very narrow. Because $G_0(f)$ and $G_1(f)$ are not flat near $f = 1/T$, their variations with f will prevent perfect cancellation even when $H(f)$ has conjugate symmetry about $f = 1/T$, causing $\mathbf{E}\{y^2\}$ to be of the order of B^3 when BT is small.

Short Rectangular Pulse: When

$$g(t) = \text{rect}\,\frac{t}{\tau} \tag{69}$$

with $\tau \leq T$, successive pulses do not overlap, and hence only the $k = 0$ terms of (41), (46), (47), (51), and (52) are nonzero. In this case the mean squared jitter is determined by

$$G_0(f) = \tau \text{ sinc } f\tau. \tag{70}$$

For $f = m/2T$ with an even value of m, $G_0(f)$ vanishes if $\tau = T$, but, for this τ, $u(t)^2$ becomes identically 1, and it then contains no clock harmonic from which to regenerate the clock.

Ideal Bandlimited $G(f)$: When

$$G(f) = \frac{1}{B}\,\text{rect}\,\frac{f}{B}, \tag{71}$$

the pulse shape is

$$g(t) = \text{sinc } Bt, \tag{72}$$

which produces no intersymbol interference at the times $\ldots, -T, 0, T, \ldots$ if B is a multiple of $1/T$. Substituting (72) into (41), we find that

$$G_k(f) = \begin{cases} \dfrac{\sin k\pi(B - |f|)T}{k\pi B^2 T} & \text{for } |f| \leq B \\ 0 & \text{otherwise.} \end{cases} \tag{73}$$

If, for example, $B = 2/T$ and $f = 1/T$, this is zero unless $k = 0$, in which case it is $T/4$, and $\mathbf{E}\{y^2\}$ is easily approximated for any convenient $H(f)$. This case is identical with that of a pulse spectrum centered at $f = 1/2T$ with the form [10] $G(f) = (T/2)\,\text{rect}\,(|f|T - \frac{1}{2})$.

Ideal Bandpass $H(f)$: When $H(f)$ is an ideal bandpass filter (23) with bandwidth $B \leq 1/2T$ centered F above the clock harmonic $m/2T$, where m is any positive, even integer, we find, upon substitution into the last line of (53), that, if the $\{G_k(m/2T)\}$ are all real, as in the case when (60) is satisfied,

$$\mathbf{E}\{x^2\} \approx \frac{2B}{T}\left[\frac{M-1}{4}G_0\left(\frac{m}{2T}\right)^2 + \sum_{k=1}^{\infty} G_k\left(\frac{m}{2T}\right)^2\right]\left[1 + \Lambda\left(\frac{2F}{B}\right)\right] \tag{74}$$

(Fig. 3a) and, upon substituting into the last line of (54), that

$$\mathbf{E}\{y^2\} \approx \frac{2B}{T}\left[\frac{M-1}{4}G_0\left(\frac{m}{2T}\right)^2 + \sum_{k=1}^{\infty} G_k\left(\frac{m}{2T}\right)^2\right]\left[1 - \Lambda\left(\frac{2F}{B}\right)\right] \tag{75}$$

(Fig. 3b). For a fixed F, as B increases from zero, (75) increases linearly until $B = 2F$, and it remains constant as B increases further because up to that point the band of width B passed by the filter is not balanced by anything on the opposite side of the clock harmonic $m/2T$, but thereafter there is only a band of fixed width $2F$ that is not balanced by its mirror image in the clock-harmonic frequency.

If B is large enough that the $\{G_k(f)\}$ must be treated as linear functions of f in the neighborhood of $f = m/2T$ instead of as constants, $\mathbf{E}\{y^2\}$ will include higher-order (up to cubic) terms in B and F in addition to the linear term in B and the constant and linear terms in F. $\mathbf{E}\{y^2\}$ is then no longer zero for $F = 0$, but, as mentioned in the Introduction, suitable prefiltering of the signal before squaring and narrowband filtering can alleviate this problem.

Gaussian Filter H(f): When $H(f)$ has the Gaussian form (29), the last line of (53) becomes

$$\mathbf{E}\{x^2\} \approx \frac{2\sqrt{\pi}B}{T}\left[\frac{M-1}{4}G_0\left(\frac{m}{2T}\right)^2 + \sum_{k=1}^{\infty} G_k\left(\frac{m}{2T}\right)^2\right]\left(1 + e^{-F^2/B^2}\right) \quad (76)$$

(Fig. 4a), and the last line of (54) becomes

$$\mathbf{E}\{y^2\} \approx \frac{2\sqrt{\pi}B}{T}\left[\frac{M-1}{4}G_0\left(\frac{m}{2T}\right)^2 + \sum_{k=1}^{\infty} G_k\left(\frac{m}{2T}\right)^2\right]\left(1 - e^{-F^2/B^2}\right) \quad (77)$$

(Fig. 4b), which is proportional to F^2/BT when $|F| \ll B$ and which is proportional to B/T when $F \gg B$. If $\{G_k(f)\}$ must be treated as linear functions of f in the neighborhood of $f = m/2T$, additional terms proportional to BF, B^3T, BF^2T, and $B^3Te^{-F^2/B^2}$ are required in (77). These result, for example, in $\mathbf{E}\{y^2\}$'s being proportional to B^3T when $F = 0$.

4 CONCLUSIONS

The quadrature self-noise $y(t)$ introduces a phase error of rms value $2y_{\rm rms}/m|G_0(2m/T)|$ in the regenerated clock, where $G_0(2m/T)$ is the magnitude of the Fourier coefficient of (36) found in (38) that corresponds to the clock harmonic $m/2T$. Because of the amplitude and phase relationships between the pairs of Fourier components of (39) having frequencies whose sums or differences are multiples of the clock frequency $1/T$, the mean squared clock-regeneration jitter is far smaller than might be predicted on the basis of the spectrum of the total self-noise [3]. It is given by an integral (20) in the case of the signal $u(t)$ itself (Fig. 1) whose integrand involves *two* frequencies, since the quadrature self-noise on one of the foregoing sort of pair of frequencies can cancel the quadrature self-noise on the other while the in-phase self-noise components on these two frequencies add constructively. In the case of $u(t)^2$ (Fig. 5), (54) has an integrand involving *four* frequencies that combine in this way.

The phase and amplitude correlations between the Fourier components of frequencies $m/2T \pm f$, where m is any integer, cause them to combine like the sidebands of a suppressed-carrier AM signal of frequency $m/2T$ whose carrier is in phase with the clock if the two frequencies fall within the passband of the filter $H(f)$ and suffer equal attenuation and opposite phase shifts, thus producing no jitter. If, as a result of asymmetry of $H(f)$, only one of the two falls within the passband, it contributes equally and uncorrelatedly to the in-phase and the quadrature filter outputs. The foregoing analysis should help to clarify the source and the nature of the relationships between spectral components of the square of a PAM signal and the effects that these relationships have upon clock recovery. These foregoing results show explicitly how the design-and-performance analysis of square-law synchronizers is characterized by the spectral correlation function and the fourth-order spectral-moment function of a PAM signal.

ACKNOWLEDGMENTS

I am grateful to Dr. Floyd M. Gardner of the Gardner Research Company, Palo Alto, California, for pointing out the important difference in the behavior of the in-phase and quadrature components of the self-noise of a PAM signal, for suggestions regarding the improvement of a draft of this article, and for calling my attention to some of the relevant literature. I want to express my appreciation to Professor Lewis E. Franks of the University of Massachusetts for very helpful comments on the draft and for supplying numerous references and copies of some of his unpublished work. In addition, I am greatly indebted to Professor W. A. Gardner of the University of California at Davis for his careful reading of that draft, for his comments on it, and for further relevant references.

REFERENCES

[1] W. A. Gardner, "The role of spectral correlation in design and performance analysis of synchronizers," *IEEE Trans. Commun.*, vol. COM-34, no. 11, pp. 1089–1095, Nov. 1986.

[2] C. M. Spooner, "Higher-order statistics for nonlinear processing of cyclostationary signals," chap. 2 in *Cyclostationarity in Communications and Signal Processing*, W. A. Gardner, ed., New York: IEEE Press, 1994.

[3] N. M. Blachman and S. H. Mousavinezhad, "The spectrum of the square of a synchronous random pulse train," *IEEE Trans. Commun.*, vol. 38, no. 1, pp. 13–17, Jan. 1990.

[4] L. E. Franks, *Signal Theory*. Englewood Cliffs, NJ: Prentice-Hall, 1969.

[5] W. A. Gardner, *Introduction to Random Processes with Applications to Signals and Systems*, 2nd ed. New York: McGraw-Hill, 1990, ch. 12.

[6] W. R. Bennett, "Statistics of regenerative digital transmission," *Bell Sys. Tech. J.*, vol. 37, pp. 1501–1542, Nov. 1958.

[7] L. E. Franks, "Carrier and bit synchronization in data communication—a tutorial review," *IEEE Trans. Commun.*, vol. COM-28, no. 8, pp. 1107–1121, Aug. 1980.

[8] W. A. Gardner, "Spectral correlation of modulated signals: Part I—Analogue modulation," *IEEE Trans. Commun.*, vol. COM-35, pp. 584–594, June 1987.

[9] J. M. Manley, "The generation and accumulation of timing noise in PCM systems—an experimental and theoretical study," *Bell Sys. Tech. J.*, vol. 48, no. 3, pp. 541–613, March 1969.

[10] L. E. Franks and J. P. Bubrouski, "Statistical properties of timing jitter in a PAM timing recovery scheme," *IEEE Trans. Commun.*, vol. COM-22, no. 7, pp. 913–920, July 1974.

[11] M. Moeneclaey, "Linear phase-locked loop theory for cyclostationary input disturbances," *IEEE Trans. Commun.*, vol. COM-30, no. 10, pp. 2253–2259, Oct. 1982.

[12] F. M. Gardner, "Self-noise in synchronizers," *IEEE Trans. Commun.*, vol. COM-28, no. 8, pp. 1159–1163, Aug. 1980.

[13] L. E. Franks, "Synchronization subsystems: Analysis and design," ch. 7, pp. 294–335, in *Digital Communications*, K. Feher, ed., Englewood Cliffs, NJ: Prentice-Hall, 1983.

[14] E. A. Lee and D. G. Messerschmitt, *Digital Communication*. Boston, Mass.: Kluwer, 1988.

[15] R. E. Blahut, *Digital Transmission of Information*. Reading, Mass.: Addison-Wesley, 1990.

[16] J. G. Proakis, *Digital Communications*, 2nd ed. New York: McGraw-Hill, 1991.

[17] T. T. Fang, "I and Q decomposition of self-noise in square-law clock regenerators," *IEEE Trans. Commun.*, vol. 36, no. 9, pp. 1044–1052, Sept. 1988.

[18] U. Mengali, "A new approach to the pulse shaping problem in timing recovery," paper no. A3 (14 pp.), *International Workshop on Digital Communications*, Tirrenia, Italy, 30 Aug.–2 Sept. 1983.

[19] A. M. Abdulsatar and G. Maral, "Self-noise spectral measurements in clock synchronizers," *IEEE Trans. Commun.*, vol. COM-31, no. 10, pp. 1204–1207, Oct. 1983.

[20] N. A. d'Andrea and U. Mengali, "A simulation study of clock recovery in QPSK and 9 QPRS systems," *IEEE Trans. Commun.*, vol. COM-33, no. 10, pp. 1139–1142, Oct. 1985.

[21] L. E. Franks, Dept. of Electrical and Computer Engineering, University of Mass., Amherst, MA, course notes, 1990.

[22] W. A. Gardner, "An introduction to cyclostationary signals," chap. 1 in *Cyclostationarity in Communications and Signal Processing*, W. A. Gardner, ed., New York: IEEE Press, 1994.

[23] W. A. Gardner, *Statistical Spectral Analysis: A Nonprobabilistic Theory*. Englewood Cliffs, NJ: Prentice-Hall, 1987, ch. 10, sec. C.

[24] R. N. Bracewell, *The Fourier Transform and Its Applications*. New York: McGraw-Hill, 1965; 2nd ed., 1986.

[25] P. M. Woodward, *Probability and Information Theory, with Applications to Radar*. London: Pergamon, 1953.

[26] N. M. Blachman, *Noise and Its Effect on Communication*. New York: McGraw-Hill, 1966; 2nd ed., Malabar, Fla.: Krieger, 1982.

Article 3

Cyclostationarity-Exploiting Methods for Multipath-Channel Identification

Luciano Izzo   Antonio Napolitano   Luigi Paura
Università di Napoli "Federico II"
Dipartimento di Ingegneria Elettronica
via Claudio, 21 I-80125 Napoli, Italy

Abstract

Recently proposed cyclostationarity-exploiting methods for multipath-channel identification are reviewed. They are inherently tolerant to both noise and (possibly correlated) interference and can, in principle, provide arbitrarily accurate multipath-parameter estimates when a sufficiently long collection time is available. The principles of operation of these techniques are explained and their algorithmic implementations are discussed. Computer simulation results show the capability of the new methods to correctly estimate the multipath parameters in noise and interference environments where the conventional techniques (based on power spectrum measurements) fail. Finally, a comparison among the methods, in terms of estimation accuracy, arrival-time resolution, and operating range, is made.

1 INTRODUCTION

The problem of multipath-channel identification (MCI) is of great interest in many applications including radar, sonar, seismics, ocean acoustics, communications, and others. For example, the use of digital communication systems characterized by higher and higher bit rates requires detailed channel modeling to counteract the degrading effects of intersymbol interference and selective fading arising from the presence of delayed replicas of the transmitted signal.

The standard cross-correlation method [1, 2] identifies the channel, that is, estimates the arrival times (ATs) and the scaling amplitudes (SAs) of the multipath signal, by locating and measuring the peaks of the temporal cross-correlation function of noisy measurements of the channel input and output, provided that the individual pulses in the impulse response of the channel are separated by time intervals greater than the duration of the autocorrelation function of the uncorrupted exciting signal (ES).

A variety of techniques, referred to as the generalized cross-correlation (GCC) methods [1, 2], reduce the degrading effects of noise and interference by filtering the input and output measurements prior to estimating their cross-correlation function. Some of these methods, proposed in the time-difference-of-arrival (TDOA) estimation context, with the PHAT technique as an example, cannot be utilized for MCI since the cross-correlation function between the filtered measurements does not necessarily exhibit peaks in correspondence of the ATs. Moreover, some GCC techniques, with the Roth method as an example, allow one to estimate the ATs but are not able to perform SA estimates. In any case, all GCC algorithms that are appropriate for MCI exhibit poor arrival-time resolution.

Greater arrival-time resolution can be achieved by an algorithm based on the autoregressive spectral analysis technique that is referred to as the Prony algorithm [3]. The method, however, turns out to be very sensitive to measurement errors due to the presence of noise and interference.

All the above-mentioned estimation methods have been found to perform very poorly when the measurements of the channel output and, especially, input are highly corrupted by noises and/or interferences, particularly if the input and output undesired signals are correlated with each other. Such a correlation typically occurs when the channel input and output cannot be reached or fully controlled by the experimenter.

Recently, a new approach to the identification of linear time-invariant systems [4, 5] has been utilized to obtain multipath-parameter estimates [6–10]. It is applicable as long as the ES exhibits cyclostationarity with a cycle frequency for which the input undesired signals do not exhibit cyclostationarity and, moreover, the input and output undesired signals do not exhibit joint cyclostationarity. In short, the new methods exploit the spectral correlation properties, as well as the temporal correlation properties, in order to gain immunity to noise and interference. In particular, they can assure satisfactory performances when the input and output undesired signals exhibit a high degree of temporal correlation.

This article provides a tutorial review of recently proposed cyclostationarity-based (say cyclic) methods for multipath-channel identification. After a brief introduction to cyclostationarity, Section 2 describes the general cyclostationary model for the MCI problem. Then, the cyclic methods so-far proposed [6–10] are derived and their principles of operation are explained in Section 3, whereas problems arising from their algorithmic implementations are presented and discussed in Section 4. Section 5 presents numerical results of Monte Carlo simulation experiments aimed at comparing the cyclic methods with each other and with the corresponding conventional (i.e., noncyclic) estimation techniques. Finally, conclusions are drawn in Section 6 and some analytical details on the derivation of two of the described methods are given in the Appendix.

2 CYCLOSTATIONARY MODEL FOR MCI

Before introducing the cyclostationarity-exploiting methods for MCI, let us present here the necessary background on cyclostationarity and the cyclostationary model for MCI.

A real signal $s(t)$ exhibits cyclostationarity [4] with cycle frequency $\alpha \neq 0$ if the cyclic autocorrelation function

$$R_s^\alpha(\tau) \triangleq \langle s\left(t+\frac{\tau}{2}\right) s\left(t-\frac{\tau}{2}\right) e^{-j2\pi\alpha t}\rangle \tag{1}$$

is not identically zero; that is, for some lag values τ, the lag product $s(t+\frac{\tau}{2})s(t-\frac{\tau}{2})$ contains a finite additive sinewave component with frequency $\alpha \neq 0$. In (1), $\langle\cdot\rangle$ denotes infinite-time average. For $\alpha = 0$, (1) is recognized as the conventional autocorrelation function $R_s(\tau)$.

Analogously, two real signals $s_1(t)$ and $s_2(t)$ exhibit joint cyclostationarity with cycle frequency $\alpha \neq 0$ if the cyclic cross-correlation function

$$R_{s_1 s_2}^\alpha(\tau) \triangleq \langle s_1\left(t+\frac{\tau}{2}\right) s_2\left(t-\frac{\tau}{2}\right) e^{-j2\pi\alpha t}\rangle \tag{2}$$

is not identically zero. For $\alpha = 0$, (2) is the conventional cross-correlation function $R_{s_1 s_2}(\tau)$.

The multipath-channel identification problem can be described by the following model:

$$x(t) = s(t) + n(t) \tag{3a}$$

$$y(t) = \sum_{k=1}^{N} A_k s(t - D_k) + m(t) \tag{3b}$$

where $x(t)$ and $y(t)$ are noisy versions of the input and output signals (respectively) of the multipath channel. Both signals $n(t)$ and $m(t)$ take into account the presence of noise-plus-interference on the real exciting signal $s(t)$ and on its multipath version. We desire to estimate the scaling amplitudes $A_1, A_2, \ldots, A_N$ and the arrival times $D_1, D_2, \ldots, D_N$.

Assuming that $s(t)$, $n(t)$, and $m(t)$ have zero-mean (time-average) values and that $s(t)$ is statistically independent (over time) of $n(t)$ and $m(t)$ and exhibits cyclostationarity with cycle frequency α, then, from (3), accounting for (1) and (2), it follows that

$$R_x^\alpha(\tau) = R_s^\alpha(\tau) + R_n^\alpha(\tau) \tag{4a}$$

$$R_{yx}^\alpha(\tau) = \sum_{k=1}^{N} A_k R_s^\alpha(\tau - D_k) e^{-j\pi\alpha D_k} + R_{mn}^\alpha(\tau). \tag{4b}$$

If the undesired signal $n(t)$ does not exhibit cyclostationarity with the considered cycle frequency α ($R_n^\alpha(\tau) \equiv 0$) and, moreover, $m(t)$ and $n(t)$ do not exhibit joint cyclostationarity with this cycle frequency ($R_{mn}^\alpha(\tau) \equiv 0$), (4a) and (4b) become

$$R_x^\alpha(\tau) = R_s^\alpha(\tau) \tag{5a}$$

$$R_{yx}^\alpha(\tau) = \sum_{k=1}^{N} A_k R_s^\alpha(\tau - D_k) e^{-j\pi\alpha D_k}. \tag{5b}$$

By Fourier transforming (5), one finds that the cyclic spectrum $S_x^\alpha(f)$ and the cyclic cross-spectrum $S_{yx}^\alpha(f)$ (also referred to as spectral correlation density functions) are given by

$$S_x^\alpha(f) = S_s^\alpha(f) \tag{6a}$$

$$S_{yx}^\alpha(f) = S_s^\alpha(f) \sum_{k=1}^{N} A_k e^{-j2\pi(f+\alpha/2)D_k}. \tag{6b}$$

Equations (5) and (6) show the potential immunity of AT and SA estimation methods based on the spectral correlation properties ($\alpha \neq 0$) against noise and interference, regardless of the extent of the spectral overlap of $s(t)$, $n(t)$, and $m(t)$ and of the degree of temporal correlation between $n(t)$ and $m(t)$. In contrast to this, the GCC methods, which are based on the temporal correlation only (i.e., $\alpha = 0$ in (4)), are affected by the presence of noise and interference. Even if $n(t)$ and $m(t)$ are uncorrelated, the presence of $n(t)$ affects (see (4a)) the measurement of the autocorrelation function of the ES.

3 CYCLOSTATIONARITY-EXPLOITING METHODS FOR MCI

In this section the recently proposed [6–10] cyclostationarity-exploiting methods for multipath-channel identification are presented and their principles of operation are explained.

3.1 CYCCOR Method

On the basis of (5), and using estimates of $R_{yx}^\alpha(\tau)$ and $R_x^\alpha(\tau)$, the ATs and SAs can be estimated by assuming that the replicas in (5b) do not overlap, i.e., that the true values D_k satisfy the resolution relations

$$|D_h - D_k| \geq d_\alpha \qquad h \neq k; \qquad h, k = 1, 2, \ldots, N \tag{7}$$

where d_α is the width of $R_s^\alpha(\tau)$. In fact, on this assumption, from (5b) it follows that

$$\left|R_{yx}^\alpha(\tau)\right| = \sum_{k=1}^{N} A_k \left|R_s^\alpha(\tau - D_k)\right| \tag{8}$$

and, therefore, the ATs D_k can be estimated by evaluating the shifts of the replicas of $\left|R_s^\alpha(\tau)\right|$ and the SAs A_k can be determined by considering the values of $|R_{yx}^\alpha(\tau)| / \left|R_x^\alpha(\tau - D_k)\right|$ corresponding to values of τ such that $|\tau - D_k| \leq d_\alpha/2$, provided that $\left|R_x^\alpha(\tau - D_k)\right| \neq 0$. The evaluation of replica shifts benefits from the fact that $R_s^\alpha(\tau)$ is an even function of τ. This method, referred to as the CYCCOR (CYclic Cross-CORrelation) method [7, 9], reduces for $\alpha = 0$ to the conventional cross-correlation technique. Moreover, for $\alpha \neq 0$, it includes the CCC (Cyclic Cross-Correlation) method of TDOA estimation as the special case for which only one delay exists ($N = 1$) and no estimate of the SA is sought [4, 11].

3.2 SPECCOA Method

An alternative method can be obtained using an ad hoc least-squares optimization procedure. More specifically, the SAs and ATs are estimated by determining the values of the amplitudes $\gamma \triangleq (\gamma_1, \gamma_2, \ldots, \gamma_N)$ and the arrival times $\beta \triangleq (\beta_1, \beta_2, \ldots, \beta_N)$ that minimize the norm of the difference of the two sides of (6b) where, accounting for (6a), $S_s^\alpha(f)$ has been replaced by $S_x^\alpha(f)$:

$$\min_{\gamma,\beta} \left\{ \int_{-\infty}^{+\infty} \left| S_{yx}^\alpha(f) - S_x^\alpha(f) \sum_{k=1}^{N} \gamma_k e^{-j2\pi(f+\alpha/2)\beta_k} \right|^2 df \right\}. \tag{9}$$

Although this minimization problem is in general very difficult to solve, it can be simplified by assuming that the true values D_k satisfy the resolution relations

$$|D_h - D_k| \geq 2d_\alpha \qquad h \neq k; \qquad h, k = 1, 2, \ldots, N. \tag{10}$$

In fact, on this assumption, as is shown in the Appendix, the problem can be solved considering that the function

$$g_\alpha(\tau) \triangleq \frac{\mathrm{Re}\left\{ e^{j\pi\alpha\tau} \int_{-\infty}^{+\infty} S_{yx}^\alpha(f) S_x^\alpha(f)^* e^{j2\pi f\tau} df \right\}}{\int_{-\infty}^{+\infty} \left|S_x^\alpha(f)\right|^2 df} \tag{11}$$

is equivalent to the superposition of N nonoverlapping replicas

$$\frac{\sum_{k=1}^{N} A_k \mathrm{Re}\left\{ e^{j\pi\alpha(\tau - D_k)} \int_{-\infty}^{+\infty} \left|S_s^\alpha(f)\right|^2 e^{j2\pi f(\tau - D_k)} df \right\}}{\int_{-\infty}^{+\infty} \left|S_s^\alpha(f)\right|^2 df} \tag{12}$$

each of which presents a peak of amplitude A_k at $\tau = D_k$. Therefore, the ATs and SAs can be estimated by locating and measuring the maximum of each replica in (11) where the cyclic spectra are replaced by their estimates.

This method maximizes the correlation over f of two cyclic spectra by phase alignments through adjustment of the linear phase-versus-frequency term $2\pi(f + \alpha/2)\tau$. It has been referred to as the SPECCOA (SPECtral COherence Alignment) method [6, 7, 9] in accordance with [11, 12] in the TDOA estimation context. For $\alpha = 0$, this technique reduces to the GCC method with the power spectrum $S_x(f)$ as weighting function. Moreover, for $\alpha \neq 0$, it includes the SPECCOA method of [11, 12] as the special case for which $N = 1$ and the estimate of the SA is ignored (in which case the denominator in (11) and (12) can be replaced with unity).

3.3 BL-SPECCORR Method

A further estimation method can be obtained by the least-squares minimization procedure

$$\min_{\gamma,\beta} \left\{ \int_{||f|-f_\alpha| \le B_\alpha/2} \left| \frac{S_{yx}^\alpha(f)}{S_x^\alpha(f)} - \sum_{k=1}^{N} \gamma_k e^{-j2\pi(f+\alpha/2)\beta_k} \right|^2 df \right\} \tag{13}$$

as suggested by (6). In (13), f_α and B_α are the center and width of the support band for $S_s^\alpha(f)$.

The minimization problem can be reasonably simplified by assuming that (7) holds. In fact, on this assumption, as is shown in the Appendix, the problem can be solved recognizing that the function

$$v_\alpha(\tau) \triangleq \frac{1}{\displaystyle\int_{||f|-f_\alpha| \le B_\alpha/2} df} \operatorname{Re}\left\{ \int_{||f|-f_\alpha| \le B_\alpha/2} \frac{S_{yx}^\alpha(f)}{S_x^\alpha(f)} e^{j2\pi(f+\alpha/2)\tau} df \right\} \tag{14}$$

is equivalent to the superposition of N nonoverlapping replicas

$$\sum_{k=1}^{N} A_k \frac{\sin[\pi B_\alpha(\tau - D_k)]}{\pi B_\alpha(\tau - D_k)} \cos[\pi\alpha(\tau - D_k)] \cos[2\pi f_\alpha(\tau - D_k)] \tag{15}$$

each of which presents a peak of amplitude A_k at $\tau = D_k$. Consequently, the ATs and SAs can be estimated by locating and measuring the maximum of each replica in (14) where, of course, the cyclic spectra are replaced by their estimates. This technique has been referred to as the BL-SPECCORR (Band-Limited SPECtral CORrelation Ratio) method [13] in accordance with [11] in the TDOA estimation context. Moreover, for $\alpha \neq 0$, it includes the BL-SPECCORR method of [11] (a variation of which was first proposed in [4, 14]) as the special case for which $N = 1$ and the estimate of the SA is ignored (in which case the scale factor outside the $\operatorname{Re}\{\cdot\}$ operation in (14) can be replaced with unity).

3.4 Cyclic Prony Method

The cyclic methods considered above are expected to exhibit an arrival-time resolution limited by the width of the cyclic autocorrelation function $R_s^\alpha(\tau)$ (see (7) and (10)).

To overcome this limitation, we present now a cyclic method [8, 10] based on the autoregressive spectral analysis technique referred to as the Prony algorithm. This cyclic method, however, exhibits a lower tolerance to noise and interference.

Accounting for (6) and the generalization to cyclostationary input of the well-known time-invariant system identification formula (see [4, 5]), the method can be introduced by recognizing that the samples of the multipath-channel transfer function $H(f)$ at M evenly spaced frequencies can be written as

$$H(f_n + \alpha/2) = \frac{S_{yx}^{\alpha}(f_n)}{S_x^{\alpha}(f_n)} = \sum_{k=1}^{N} b_k^{\alpha} z_k^n \triangleq H_n^{\alpha} \qquad n = 0, 1, \ldots, M-1 \quad (16)$$

where

$$f_n \triangleq f_c + n\Delta f_c \quad (17)$$

$$b_k^{\alpha} \triangleq A_k e^{-j2\pi(f_c+\alpha/2)D_k} \quad (18a)$$

$$z_k \triangleq e^{-j2\pi\Delta f_c D_k}. \quad (18b)$$

The samples have an autoregressive structure, i.e., each sample can be expressed as a fixed weighted sum of its neighbors:

$$H_n^{\alpha} = -\sum_{m=1}^{N} \lambda_m H_{n-m}^{\alpha} \qquad n = N, N+1, \ldots, M-1 \quad (19)$$

where the weights λ_m are such that the predictor polynomial can be written as

$$P_N(z) \triangleq z^N + \sum_{m=1}^{N} \lambda_m z^{N-m} = \prod_{k=1}^{N}(z - z_k). \quad (20)$$

Then, the cyclic Prony procedure consists of the following steps:

1. Calculation of the weights λ_m by solving (19);
2. Evaluation of the z_k's by factoring the polynomial (20) and, hence, evaluation of the arrival times D_k by (18b);
3. Evaluation of the b_k^{α}'s by (16) and, hence, evaluation of the A_k's by (18a).

Note that this cyclic Prony method is identical to the usual Prony algorithm [3] where, however, the samples of the channel transfer function are obtained by cyclic spectrum measurements. In the following section, which deals with implementation problems, an improved cyclic Prony method is introduced.

4 IMPLEMENTATION OF THE METHODS

The cyclic correlation functions and the cyclic spectra involved in the methods presented in the previous section must be replaced by their estimates obtained from measurements.

The cyclic cross-correlation function $R^{\alpha}_{yx}(\tau)$ can be efficiently estimated by evaluating the cyclic cross-correlogram [4]

$$R^{\alpha}_{yx_T}(\tau) \triangleq \frac{1}{T}\int_{-(T-|\tau|)/2}^{(T-|\tau|)/2} y\left(u+\frac{\tau}{2}\right) x\left(u-\frac{\tau}{2}\right) e^{-j2\pi\alpha u} du \tag{21}$$

via inverse Fourier transforming the cyclic cross-periodogram

$$S^{\alpha}_{yx_T}(f) \triangleq \frac{1}{T} Y_T(f+\alpha/2) X^*_T(f-\alpha/2) \tag{22}$$

where T is the integration, or collection, time and the transforms

$$X_T(f) \triangleq \int_{-T/2}^{T/2} x(t) e^{-j2\pi f t} dt \tag{23a}$$

$$Y_T(f) \triangleq \int_{-T/2}^{T/2} y(t) e^{-j2\pi f t} dt \tag{23b}$$

are approximated by an FFT algorithm.

The cyclic spectra $S^{\alpha}_{yx}(f)$ and $S^{\alpha}_{x}(f)$ are efficiently estimated by using the frequency-smoothed cyclic periodograms [4]. For example, the estimation of $S^{\alpha}_{yx}(f)$ is performed by

$$S^{\alpha}_{yx_T}(f)_{\Delta f} \triangleq \frac{1}{\Delta f}\int_{f-\Delta f/2}^{f+\Delta f/2} S^{\alpha}_{yx_T}(\nu) d\nu \tag{24}$$

where Δf is the width of the frequency-smoothing window. Such a parameter Δf must be large enough to assure sufficient reliability of the cyclic spectrum estimates, that is, $\Delta f \gg 1/T$, and, moreover, small enough to resolve the oscillations in the cyclic spectra and to assure satisfactory operating range for the arrival times. Specifically, from (5b) and (21)–(24) it follows that

$$\begin{aligned} &\mathcal{F}^{-1}\left[S^{\alpha}_{yx_T}(f)_{\Delta f}\right] \\ &\quad = R^{\alpha}_{yx_T}(\tau)\frac{\sin(\pi\Delta f \tau)}{\pi \Delta f \tau} \simeq \sum_{k=1}^{N} A_k R^{\alpha}_{s_T}(\tau - D_k) e^{-j\pi\alpha D_k} \frac{\sin(\pi \Delta f \tau)}{\pi\Delta f \tau} \end{aligned} \tag{25}$$

where $\mathcal{F}^{-1}[\cdot]$ denotes the inverse Fourier transform operation and the approximation is close when $\max_k D_k \ll T$. Then, from (25) it follows that the condition $\Delta f < \lambda / \max_k D_k$ (where the chosen value of λ affects the estimation accuracy) must be satisfied.

From the previous discussion, it follows that for the CYCCOR method the condition $\max_k D_k \ll T$ must be satisfied, whereas for the SPECCOA, BL-SPECCORR, and Prony methods both $\max_k D_k < \lambda/\Delta f$ and $T\Delta f \gg 1$ must be satisfied. Therefore, the arrival-time operating range of the CYCCOR method is larger than that of the other methods which utilize the frequency-smoothed cyclic periodograms. Note that, when the two previously mentioned conditions cannot be simultaneously satisfied and CYCCOR is not expected to perform satisfactorily, one can obtain coarse estimates of the D_k's by the CYCCOR technique, thereby identifying the intervals $(D_{0i} - \lambda/\Delta f, D_{0i} + \lambda/\Delta f)$ where the D_k's lie. Then, one of the other methods can be applied by utilizing the smoothed cyclic periodogram $S^{\alpha}_{y_i x_T}(f)_{\Delta f}$ (with $y_i(t) \triangleq y(t + D_{0i})$) so that

$$R^{\alpha}_{y_i x_T}(\tau)\frac{\sin(\pi\Delta f\tau)}{\pi\Delta f\tau} \simeq e^{j\pi\alpha D_{0i}} R^{\alpha}_{yx_T}(\tau + D_{0i})\frac{\sin(\pi\Delta f\tau)}{\pi\Delta f\tau} \tag{26}$$

provided that the D_{0i}'s are sufficiently smaller than T [7, 9].

As regards the implementation of the cyclic Prony procedure presented in Section 3, since N is generally unknown and, moreover, the data are noisy, one uses a predictor of higher order, say $N_0 \gg N$, by assuming $N_0 = M/2$, as recommended by Kay and Marple in [15].

With reference to the implementation of step 1, to gain greater immunity against the added noise in the data, the weights λ_m are evaluated [8, 10] by solving (using the standard least-squares "pseudoinverse" method) the system constituted not only by the forward prediction equations (19) in their finite collection time versions

$$\widetilde{H}^{\alpha}_n \doteq -\sum_{m=1}^{N_0} \widetilde{\lambda}_m \widetilde{H}^{\alpha}_{n-m} \qquad n = N_0, N_0 + 1, \ldots, M - 1 \tag{27a}$$

but also by the backward prediction equations

$$\widetilde{H}^{\alpha *}_n \doteq -\sum_{m=1}^{N_0} \widetilde{\lambda}_m \widetilde{H}^{\alpha *}_{n+m} \qquad n = 0, 1, \ldots, M - N_0 - 1. \tag{27b}$$

Tilde identifies, here and in the following, quantities obtained by measurements over a finite collection time and $\doteq$ denotes approximation in the least-squares sense.

As regards step 3 of the procedure, a modification aimed at assuring more accurate estimates can be introduced [8, 10]. More specifically, the estimates of b^{α}_k are obtained by considering the system constituted not only by the finite collection time versions of the M equations (16)

$$\widetilde{H}^{\alpha}_n \doteq \sum_{k=1}^{N_0} \widetilde{b}^{\alpha}_k \widetilde{z}^{\,n}_k \qquad n = 0, 1, \ldots, M - 1 \tag{28a}$$

but also by the M equations

$$\widetilde{H}^{\alpha}_n \doteq \sum_{k=1}^{N_0} \widetilde{b}^{\alpha}_k (\widetilde{z}^{*}_k)^{-n} \qquad n = 0, 1, \ldots, M - 1. \tag{28b}$$

Note that, in the ideal case of infinite collection time T (i.e., noise-free samples H^{α}_n), the system (28b) is equivalent to (28a), whereas, for finite collection time, this is not true, in that the roots $\widetilde{z}_k$ do not necessarily lie on the unit circle.

The modified cyclic Prony technique considered here has been referred to as the improved cyclic Prony method [10].

5 SIMULATION RESULTS

The simulation results presented here are aimed at substantiating the tolerance of the cyclostationarity-exploiting methods to severely corruptive noise and interference and at quantitatively assessing their performances. The superiority with respect to the conventional techniques, which utilize measurements of the power spectra (i.e., $\alpha = 0$), is also corroborated.

The estimates of the conventional and cyclic spectra have been carried out by discretizing both time and frequency with sampling increments $T_s = T/(K-1)$ and $F_s = 1/T$ and by using $K = 2^{15}$. The ES is a binary phase-shift-keyed (BPSK) signal with carrier frequency $f_0 = 0.125/T_s$, baud rate $\alpha_0 = 0.03125/T_s$, and full-duty-cycle rectangular envelope. It exhibits cyclostationarity with cycle frequencies $\alpha = k\alpha_0$ and $\alpha = \pm 2f_0 + k\alpha_0$ for all integers k such that $\alpha \neq 0$. In all experiments, a two-path channel with $A_1 = A_2 = 1$ is considered.

The first experiment qualitatively shows that, unlike the corresponding conventional techniques, the CYCCOR, SPECCOA, and BL-SPECCORR methods exhibit a high tolerance to noise and correlated interference. The interference-plus-noise $n(t)$ and $m(t)$ contain additive white Gaussian noise (AWGN) components independent from each other, whose discretized versions have equal power. Moreover, to introduce correlation between $n(t)$ and $m(t)$, it is assumed that an interfering BPSK signal contaminates $s(t)$ and that the multipath signal is corrupted by a delayed version of the same interfering signal with a delay $D_I = (D_1 + D_2)/2$. This BPSK interfering signal has carrier frequency $f_I = 0.101/T_s$, baud rate $\alpha_I = 0.02525/T_s$, and full-duty-cycle rectangular envelope. The cyclic spectra utilized here are those with $\alpha = 2f_0$ and the ATs are fixed at $D_1 = 314T_s$ and $D_2 = 454T_s$, so that both (7) and (10) are satisfied and, then, nonoverlapping replicas are expected.

Figures 1–3 present results referring to a single trial of the experiment for a signal-to-noise ratio for $x(t)$ (SNR_x) and a signal-to-interference ratio for $x(t)$ (SIR_x) both fixed at 0 dB.[1] More specifically, Fig. 1a shows the magnitude of the normalized cyclic cross-correlogram $| R^{\alpha}_{yx_T}(\tau) | / | R^{\alpha}_{x_T}(0) |$ as a function of τ/T_s. The AT estimates $\widetilde{D}_1$ and $\widetilde{D}_2$ are essentially error free, whereas the SAs are estimated with an error of about 2% for A_1 and 1% for A_2. The graph of Fig. 1b shows the failure of the conventional ($\alpha = 0$) cross-correlation method owing to the presence of a replica centered at D_I. For the SPECCOA, BL-SPECCORR, and their corresponding conventional methods, which have been implemented (here and in the following) with a smoothing product $T\Delta f = 100$, analogous behavior is shown in Figs. 2 and 3, although less accurate amplitude estimates are obtained by the BL-SPECCORR method. Note that BL-SPECCORR and its corresponding conventional method have been implemented by assuming $B_\alpha = 2\alpha_0$ and $f_\alpha = 0$, for $\alpha = 2f_0$, and $f_\alpha = f_0$ for $\alpha = 0$.

[1] Once SNR_x and SIR_x are fixed, the assumptions made yield the values of SNR_y and SIR_y.

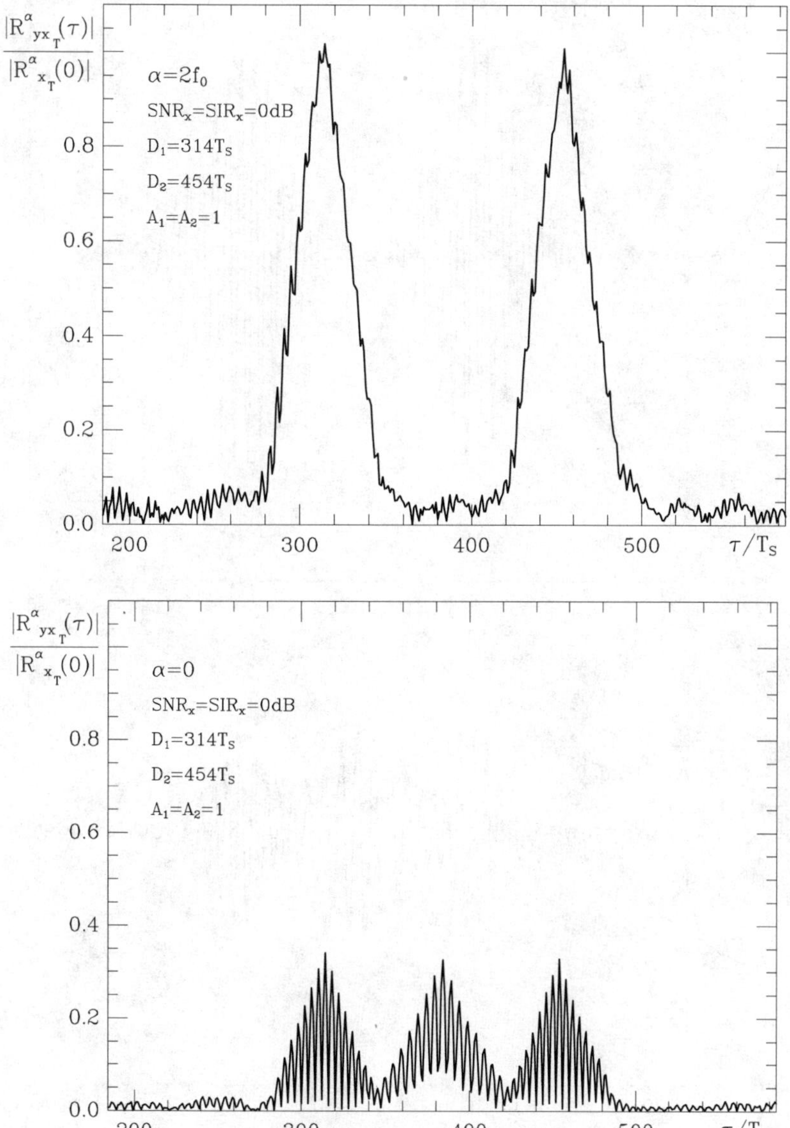

Figure 1: (a) Graph of magnitude of normalized cyclic cross-correlogram. (b) Graph of magnitude of normalized cross-correlogram.

To quantitatively analyze the performance degradation due to the presence of correlated interference, a second experiment has been considered where the environment is the same as the previous one, except for the absence of noise. Figures 4 and 5 present the sample root mean-square errors (RMSEs), computed on the basis of 400 trials, ϵ_{D_1} and ϵ_{A_1} on the estimates of D_1 and A_1 (respectively) as functions of SIR_x.

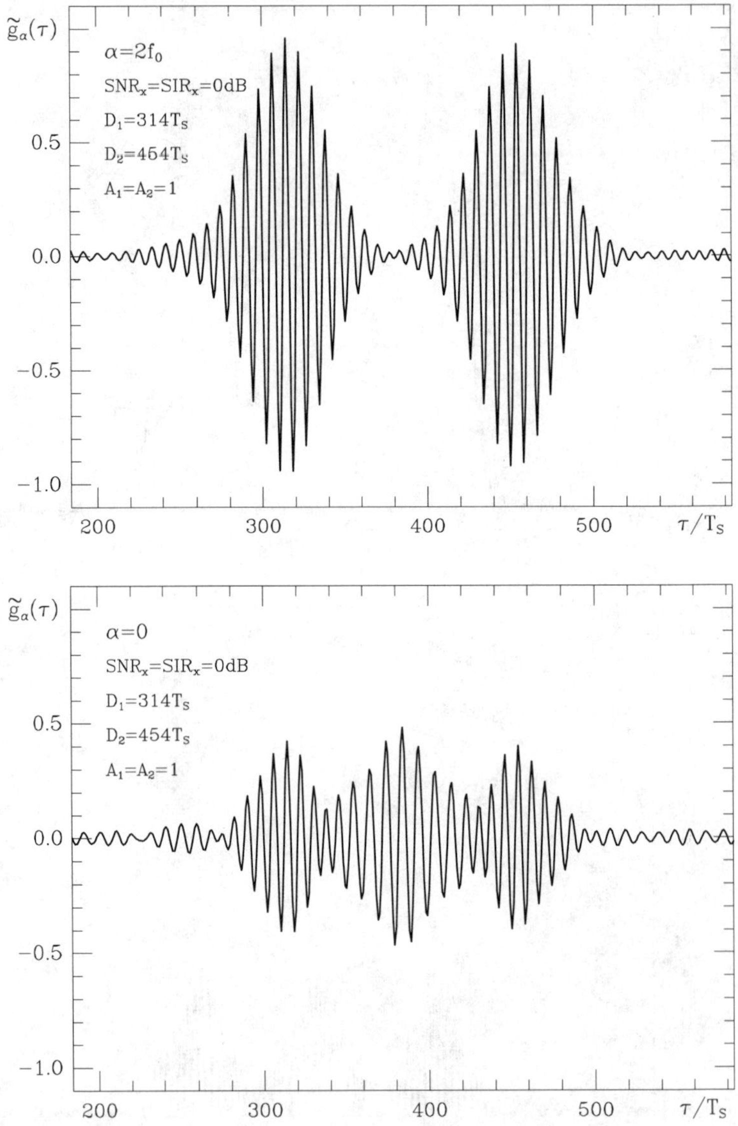

Figure 2: (a) Graph of estimate $\tilde{g}_\alpha(\tau)$ of SPECCOA function (11). (b) Graph of estimated SPECCOA function with $\alpha = 0$ (GCC method with a weighting function $S_x(f)$).

The results confirm the capability of the cyclic methods to provide, unlike the corresponding conventional methods, accurate estimates in severely corruptive correlated interference environments. In particular, SPECCOA outperforms CYCCOR for the AT estimate, but CYCCOR is superior to SPECCOA for the SA estimate. Moreover,

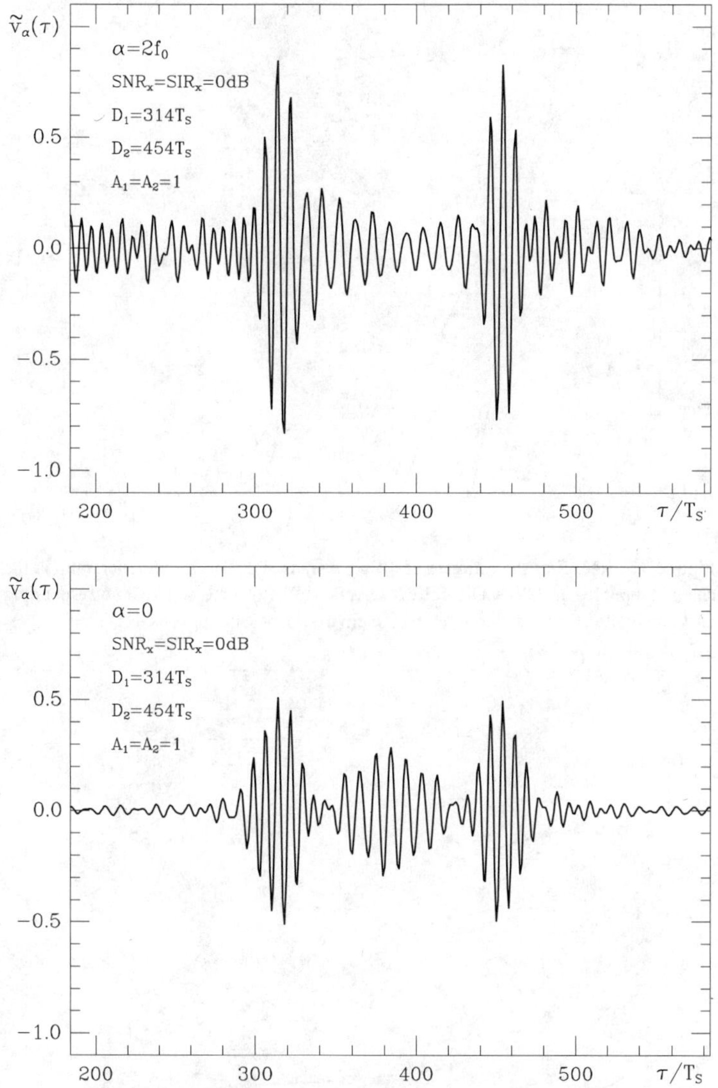

Figure 3: (a) Graph of estimate $\tilde{v}_\alpha(\tau)$ of BL-SPECCORR function (14). (b) Graph of estimated BL-SPECCORR function with $\alpha = 0$ (GCC method with a weighting function $1/S_x(f)$ in the support band for $S_s(f)$ and zero otherwise).

both methods outperform BL-SPECCORR in both AT and SA estimates. The peak at $\text{SIR}_x = 5$ dB in ϵ_{D_1} of BL-SPECCORR is caused by insufficient reliability, as was confirmed by results obtained on the basis of 600 trials. Finally, note that, here and in the following, the curves referring to ϵ_{D_2} and ϵ_{A_2} are not reported since they are very close to those concerning ϵ_{D_1} and ϵ_{A_1} (respectively).

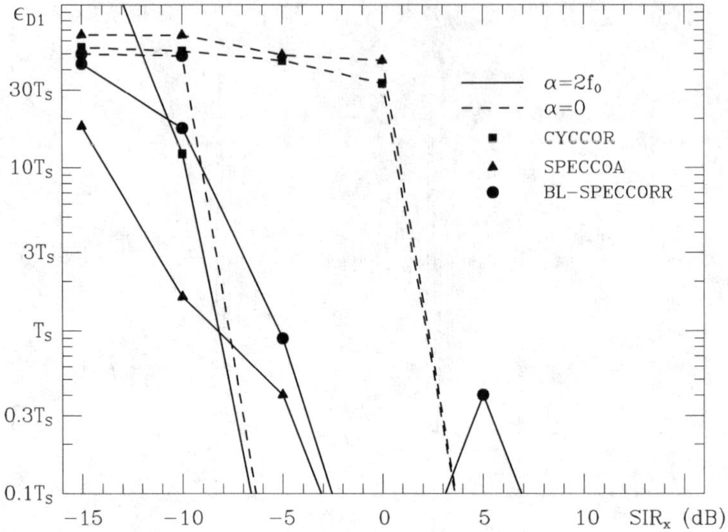

Figure 4: RMSE of the estimate of AT D_1 versus signal-to-interference ratio at the channel input for the CYCCOR, SPECCOA, BL-SPECCORR, and their corresponding GCC methods in a correlated interference environment without AWGNs.

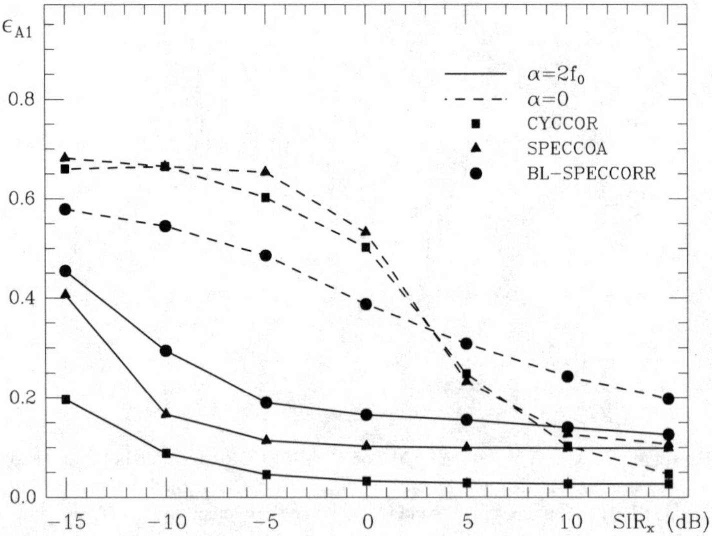

Figure 5: RMSE of the estimate of SA A_1 versus signal-to-interference ratio at the channel input for the CYCCOR, SPECCOA, BL-SPECCORR, and their corresponding GCC methods in a correlated interference environment without AWGNs.

The third experiment is aimed at assessing the performance of the estimation methods under consideration when the measurements at the input and the output of the channel are corrupted only by the independent AWGNs considered in the first experiment. Figures 6 and 7 present the RMSEs ϵ_{D_1} and ϵ_{A_1}, computed on the basis of 400 trials, versus SNR_x. As regards the AT estimate (Fig. 6), both cyclic and conventional methods work well for high SNR_xs; for low SNR_xs, vice versa, only SPECCOA is competitive with the corresponding conventional technique. With reference to the SA estimate (Fig. 7), the conventional methods are far inferior to the corresponding cyclic methods and, moreover, CYCCOR significantly outperforms both SPECCOA and BL-SPECCORR.

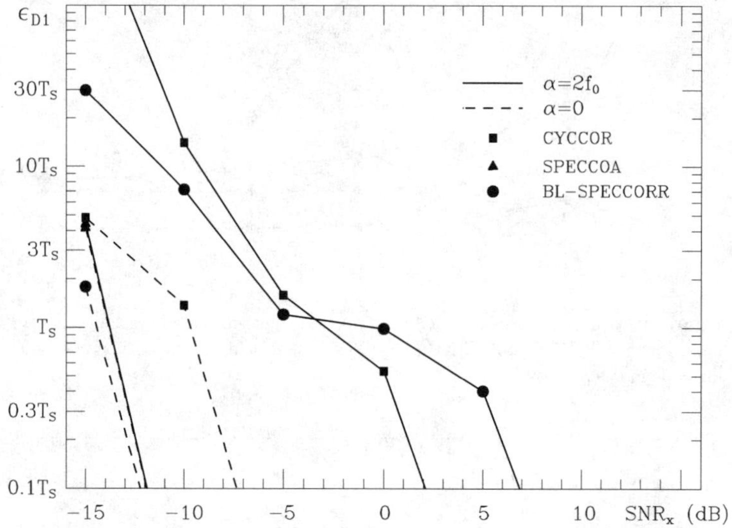

Figure 6: RMSE of the estimate of AT D_1 versus signal-to-noise ratio at the channel input for the CYCCOR, SPECCOA, BL-SPECCORR, and their corresponding GCC methods in independent AWGNs.

To test the behavior of the methods when both independent noises and correlated interferences are present, the first experiment has again been considered. The performances from 400 trials are shown in Figs. 8 and 9, which present ϵ_{D_1} and ϵ_{A_1} (respectively) as functions of SIR_x for $\text{SNR}_x = 0$ dB. The expected superiority of the cyclic methods with respect to the conventional ones is confirmed. Moreover, the SPECCOA technique largely outperforms both CYCCOR and BL-SPECCORR methods in the AT estimation, whereas the CYCCOR method provides the most accurate SA estimates.

To analyze the influence of the smoothing window width Δf on the performance of both SPECCOA and BL-SPECCORR methods, 400 trials of the first experiment were made for various values of Δf for $\text{SNR}_x = -8$ dB and $\text{SIR}_x = -5$ dB. The results showed that very small and very large values of Δf must be avoided. The former lead to inadequate smoothing (see (24)); the latter provide a correlation tapering window width too small to assure (see (25)) a satisfactory estimation accuracy. The optimum value of Δf for $\alpha = 2f_0$ was approximately $\alpha_0/10$.

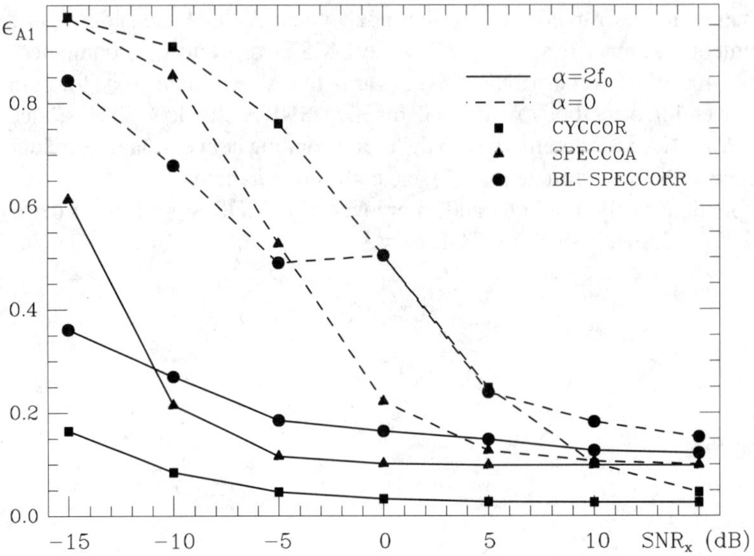

Figure 7: RMSE of the estimate of SA A_1 versus signal-to-noise ratio at the channel input for the CYCCOR, SPECCOA, BL-SPECCORR, and their corresponding GCC methods in independent AWGNs.

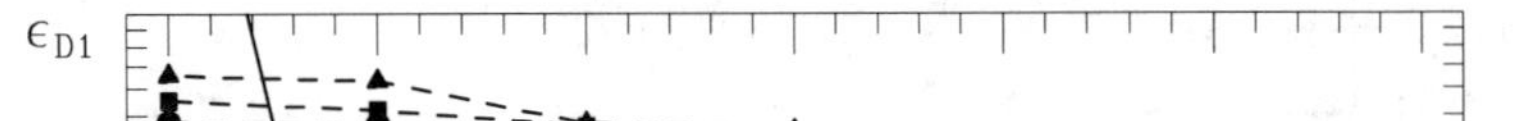

Figure 8: RMSE of the estimate of AT D_1 versus signal-to-interference ratio at the channel input for the CYCCOR, SPECCOA, BL-SPECCORR, and their corresponding GCC methods in the presence of both independent AWGNs and correlated interferences.

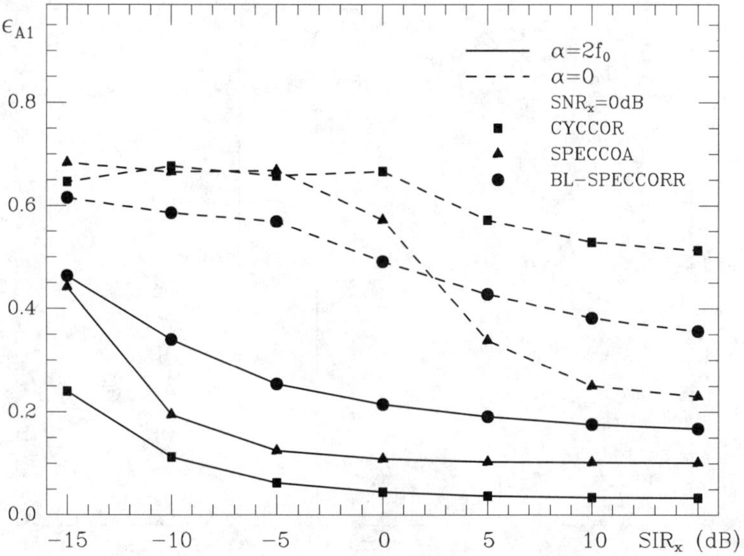

Figure 9: RMSE of the estimate of SA A_1 versus signal-to-interference ratio at the channel input for the CYCCOR, SPECCOA, BL-SPECCORR, and their corresponding GCC methods in the presence of both independent AWGNs and correlated interferences.

As regards the Prony techniques presented in the previous sections, to qualitatively compare their performances, the first experiment has again been considered. In this case, however, accounting for the fact that the Prony methods are more sensitive to noisy measurements than the other methods, a smoothing product $T\Delta f = 1000$ rather than 100 has been utilized and higher SNR$_x$ and SIR$_x$ have been used. The true ATs can be chosen closer than those considered in the previous methods since in this case no condition on the AT separation must be satisfied. Therefore, in the experiment they have been fixed at $D_1 = 3T_s$ and $D_2 = 10T_s$. Moreover, accounting for the regions of support of the BPSK cyclic spectrum in the bifrequency (f, α) plane [4], the equispaced samples $\widetilde{H}_n^\alpha$ of the channel transfer function have been selected in the spectral band $|f| \leq 1/2T_s - |\alpha|/2$. Finally, since the value of Δf_c and, hence, the number of samples M must be chosen according to the constraint $2\Delta f_c < 1/\max_k D_k$ to avoid phase ambiguity (see (18b)), in the experiment $M = 32$ and, hence, $N_0 = 16$ have been used.

Figure 10 presents results for a single trial with SNR$_x$ = 3 dB and SIR$_x$ = 5 dB. In particular, Fig. 10a shows that the improved cyclic Prony (ICP) method, utilizing cyclic spectra with $\alpha = 2f_0$, provides fully satisfactory estimates, in that two values of amplitudes ($\widetilde{A}_1 = 0.96$ and $\widetilde{A}_2 = 1.08$) are significantly greater than all others and close to the actual values and the corresponding arrival times ($\widetilde{D}_1 = 3.24T_s$ and $\widetilde{D}_2 = 10.05T_s$) are very close to the true values. Figure 10b presents the results of the cyclic Prony method, which evaluates the b_k^α's (and, hence, the A_k's)

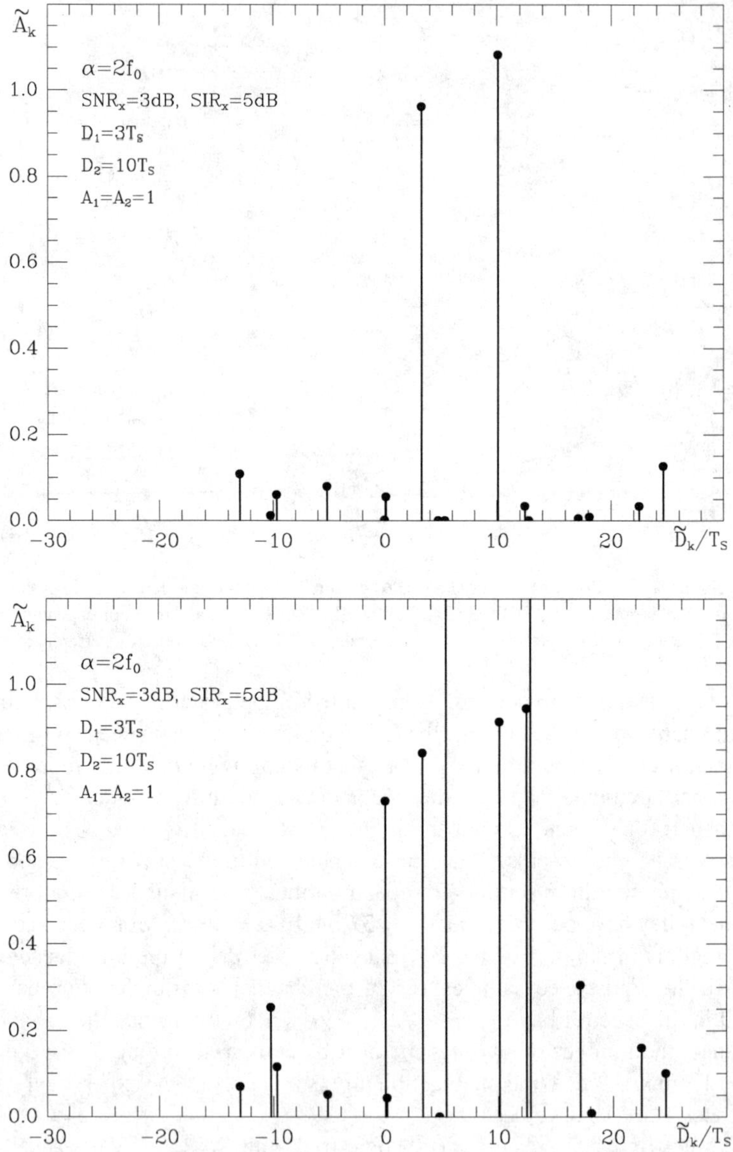

Figure 10: (a) SA and AT estimates by the improved cyclic Prony method working at $\alpha = 2f_0$. (b) SA and AT estimates by the cyclic Prony method working at $\alpha = 2f_0$. (c) SA and AT estimates by the conventional Prony method.

by solving the system constituted by (28a) only. The failure of the method, and, consequently, the effectiveness of the modification proposed in step 3 of the Prony algorithm, is evident. Finally, Fig. 10c shows the ineffectiveness of the conventional

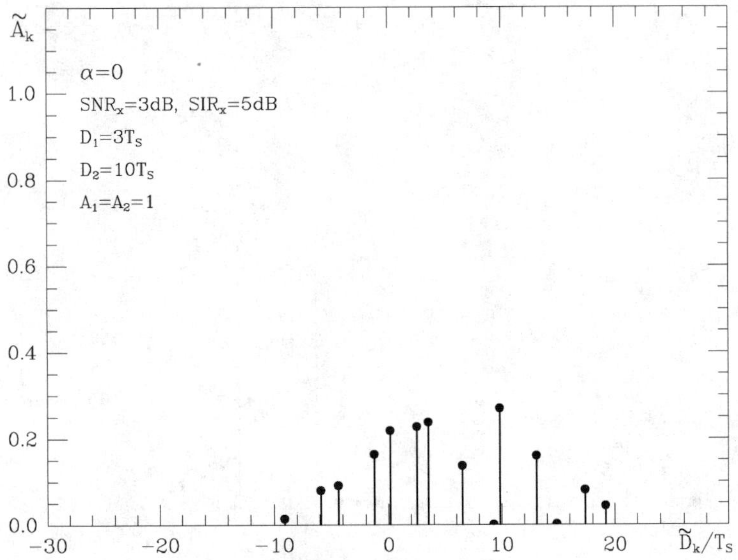

Figure 10: (Continued)

Prony technique (i.e., the one utilizing measurements based on the power spectra), even though the modification of step 3 has been used here.

An experiment, which differs from the previous one in that the interfering BPSK signal and the ES have the same carrier frequency ($f_I = f_0 = 0.125/T_s$) but different baud rates ($\alpha_I = 0.0625/T_s$ and $\alpha_0 = 0.03125/T_s$), has also been considered. In such a case, since (6a) and (6b) do not hold for $\alpha = 2f_0$, a cycle frequency related to the baud rate of the ES must be used and, therefore, $\alpha = \alpha_0$ has been chosen. Figure 11 presents the results (referring again to a single trial) obtained by the ICP method with reference to $\text{SNR}_x = 7$ dB and $\text{SIR}_x = 0$ dB. The results are fully satisfactory, unlike those (not reported here for the sake of brevity) obtained by the (not improved) cyclic Prony method and the conventional Prony technique.

To carry out a statistical performance analysis of the Prony methods, we still refer to the last two experiments. The performances are evaluated in terms of the sample mean normalized integrated square error ϵ_H (computed on the basis of 400 trials) on the reconstruction of the multipath-channel transfer function:

$$\epsilon_H \triangleq \frac{E\left\{\int_0^{1/2T_S} \left|H(f) - \widetilde{H}(f)\right|^2 df\right\}}{\int_0^{1/2T_S} |H(f)|^2\, df} \tag{29}$$

where $\widetilde{H}(f)$ is the reconstructed transfer function, obtained from the estimates $\widetilde{A}_k$ and $\widetilde{D}_k$ ($k = 1, 2, \ldots, N_0$), and E denotes sample mean.

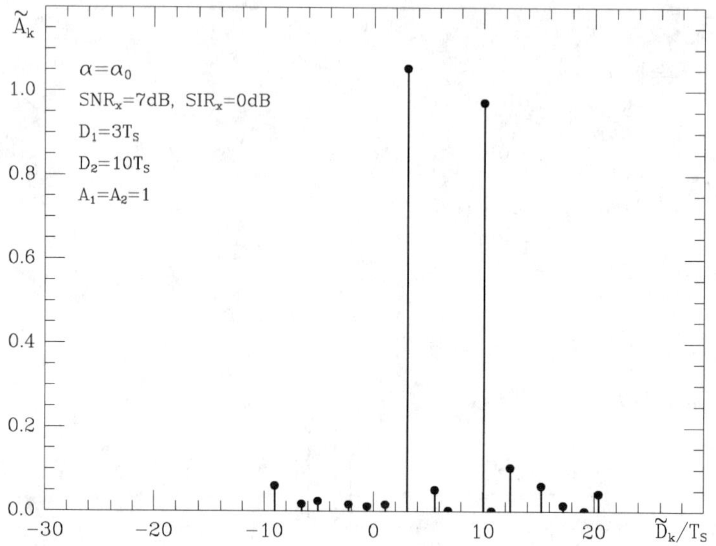

Figure 11: SA and AT estimates by the improved cyclic Prony method working at $\alpha = \alpha_0$.

With reference to the first of the two experiments, Fig. 12 shows the behavior of ϵ_H versus SIR_x for all considered Prony techniques. The ICP method is the only one that can assure satisfactory performance. Such a result is confirmed (Fig. 13) with reference to the second experiment where the ES and the interfering signal have the same carrier frequency.

Finally, let us consider an experiment aimed at comparing all cyclostationarity-based methods in terms of arrival-time resolution and operating range. The ES and the noise and interference environment are those utilized to obtain the results shown in Figs. 1–3. As in the previous experiments, the smoothing product $T\Delta f$ has been fixed at 100 for the SPECCOA and BL-SPECCORR techniques, and at 1000 for the ICP method. With reference to $\text{SNR}_x = \text{SIR}_x = 5$ dB, Figs. 14 and 15 present the RMSEs ϵ_{D_1} and ϵ_{A_1} (respectively), computed on the basis of 400 trials, as functions of the normalized arrival-time separation $\Delta D/T_s$ between the true values of the ATs. The ICP method exhibits a resolution largely greater than those (comparable) of the CYC-COR and SPECCOA methods, which are superior to that of the BL-SPECCORR technique. Moreover, the results show that CYCCORR and, especially, SPECCOA can assure satisfactory performances also in the presence of partially overlapped replicas, i.e., when the conditions (7) and (10) are not satisfied. As regards the arrival-time operating range, the CYCCOR method, as expected, largely outperforms the other methods, which utilize frequency-smoothed cyclic periodograms. In particular, the ICP method exhibits the smallest arrival-time operating range since a very large smoothing product $T\Delta f$ has been adopted to assure adequately accurate spectrum estimates.

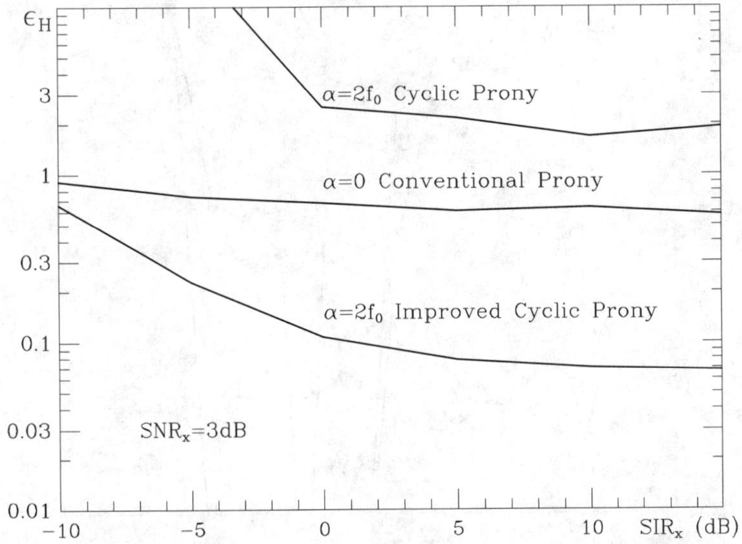

Figure 12: Sample-mean-normalized integrated squared error of the reconstruction of the multipath-channel transfer function versus signal-to-interference ratio at the channel input (the cyclic methods work at $\alpha = 2f_0$).

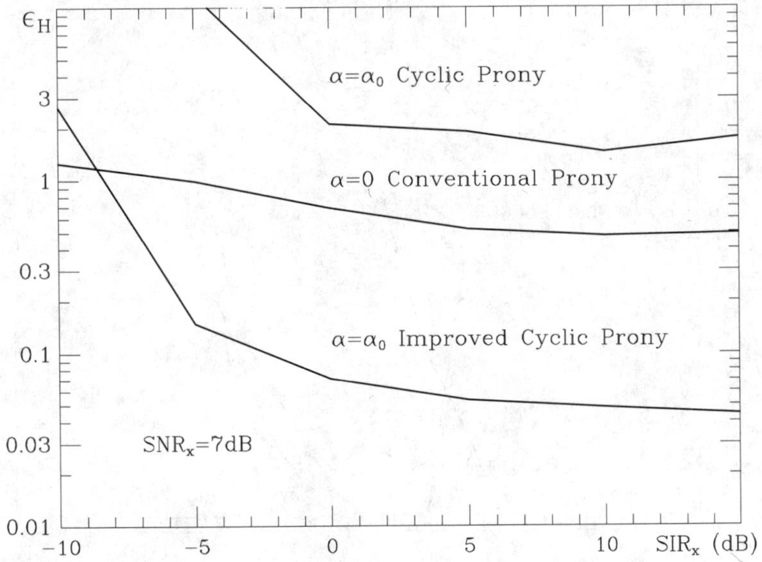

Figure 13: Sample-mean-normalized integrated squared error of the reconstruction of the multipath-channel transfer function versus signal-to-interference ratio at the channel input (the cyclic methods work at $\alpha = \alpha_0$).

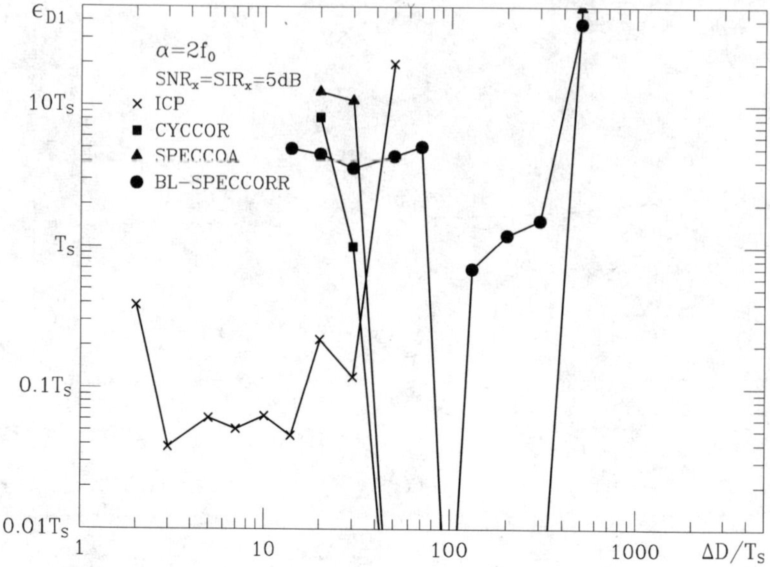

Figure 14: RMSE of the estimate of AT D_1 versus normalized AT separation for the ICP, CYCCOR, SPECCOA, and BL-SPECCORR methods.

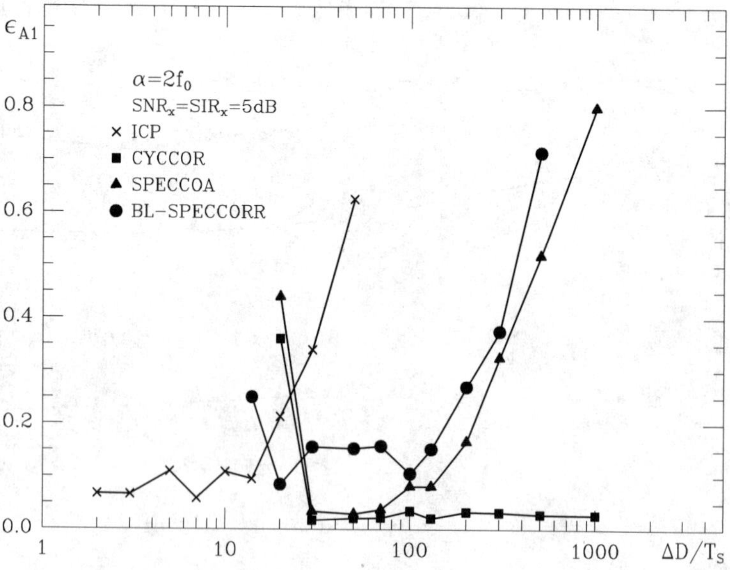

Figure 15: RMSE of the estimate of SA A_1 versus normalized AT separation for the ICP, CYCCOR, SPECCOA, and BL-SPECCORR methods.

6 CONCLUSIONS

This article provides a tutorial review of recently proposed methods for multipath-channel identification. By exploiting the cyclostationarity property of the exciting signal, these methods ideally (for infinite collection time) remove the effects of noise and interfering signals present at the input and the output of the channel. Simulation results have shown that, unlike the conventional techniques, the cyclostationarity-exploiting methods can assure satisfactory performances in severely corruptive noise and interference environments, such as when the interfering signals are comparable to, or greater than, the exciting signal in spectral density level and bandwidth and/or when they exhibit an arbitrarily high degree of temporal correlation. As regards the comparison among the cyclic methods in terms of multipath-parameter estimation accuracy, the SPECCOA and improved cyclic Prony methods are the least and the most (respectively) sensitive to noise and interference effects. Moreover, both the CYCCOR and BL-SPECCORR techniques significantly outperform the improved cyclic Prony method. This last technique, however, exhibits an arrival-time resolution capability that is much greater than those (comparable) of the CYCCOR and SPECCOA methods, which are significantly superior to that of the BL-SPECCORR method. Finally, as regards the arrival-time operating range, the CYCCOR method is strongly superior to the SPECCOA and BL-SPECCORR techniques, and, even more, to the improved cyclic Prony method.

APPENDIX

The optimization problems stated in (9) and (13) can both be formulated as the minimization, with respect to $\gamma_1, \gamma_2, \ldots, \gamma_N$ and $\beta_1, \beta_2, \ldots, \beta_N$, of

$$\begin{aligned}\int_{-\infty}^{+\infty} \left| W_\alpha(f) \left[\frac{S_{yx}^\alpha(f)}{S_x^\alpha(f)} - \sum_{h=1}^{N} \gamma_h e^{-j2\pi(f+\alpha/2)\beta_h} \right] \right|^2 df \\ = \int_{-\infty}^{+\infty} |W_\alpha(f)|^2 \left| \frac{S_{yx}^\alpha(f)}{S_x^\alpha(f)} \right|^2 df \\ + \sum_{l=1}^{N} \sum_{h=1}^{N} \gamma_l \gamma_h \int_{-\infty}^{+\infty} |W_\alpha(f)|^2 e^{-j2\pi(f+\alpha/2)(\beta_l - \beta_h)} df \\ -2 \sum_{h=1}^{N} \gamma_h \mathrm{Re} \left\{ \int_{-\infty}^{+\infty} |W_\alpha(f)|^2 \frac{S_{yx}^\alpha(f)}{S_x^\alpha(f)} e^{j2\pi(f+\alpha/2)\beta_h} df \right\}.\end{aligned} \tag{A-1}$$

In (A-1), the weighting function $W_\alpha(f)$ is equal to $S_x^\alpha(f)$ when SPECCOA is considered, whereas, for BL-SPECCORR, it assumes a value of unity in the support band for $S_s^\alpha(f)$ and a value of zero otherwise.

By equating to zero the derivatives of (A-1) with respect to γ_k and β_k ($k = 1, 2, \ldots, N$), one obtains the system of $2N$ equations in $2N$ unknowns

$$\sum_{h=1}^{N} \gamma_h \mathrm{Re}\left\{\left[\int_{-\infty}^{+\infty} |W_\alpha(f)|^2 e^{-j2\pi(f+\alpha/2)\tau} df\right]_{\tau=\beta_k-\beta_h}\right\} - \mathrm{Re}\left\{\left[\int_{-\infty}^{+\infty} |W_\alpha(f)|^2 \frac{S_{yx}^\alpha(f)}{S_x^\alpha(f)} e^{j2\pi(f+\alpha/2)\tau} df\right]_{\tau=\beta_k}\right\} = 0 \quad \text{(A-2)}$$

$$\sum_{h=1}^{N} \gamma_h \mathrm{Re}\left\{\left[\frac{\partial}{\partial\tau}\int_{-\infty}^{+\infty} |W_\alpha(f)|^2 e^{-j2\pi(f+\alpha/2)\tau} df\right]_{\tau=\beta_k-\beta_h}\right\} - \mathrm{Re}\left\{\left[\frac{\partial}{\partial\tau}\int_{-\infty}^{+\infty} |W_\alpha(f)|^2 \frac{S_{yx}^\alpha(f)}{S_x^\alpha(f)} e^{j2\pi(f+\alpha/2)\tau} df\right]_{\tau=\beta_k}\right\} = 0. \quad \text{(A-3)}$$

The solution of such a system is very difficult. However, the problem is greatly simplified if one assumes that the true values D_k satisfy the resolution relations

$$|D_h - D_k| \geq d'_\alpha \qquad h \neq k; \quad h, k = 1, 2, \ldots, N$$

where d'_α is the width of the autocorrelation function

$$r_{W_\alpha}(\tau) \triangleq \int_{-\infty}^{+\infty} |W_\alpha(f)|^2 e^{j2\pi f\tau} df. \quad \text{(A-5)}$$

In fact, on this assumption, the optimization is performed under the constraint

$$|\beta_h - \beta_k| \geq d'_\alpha \qquad h \neq k; \quad h, k = 1, 2, \ldots, N \quad \text{(A-6)}$$

which leads to

$$\left[\int_{-\infty}^{+\infty} |W_\alpha(f)|^2 e^{-j2\pi(f+\alpha/2)\tau} df\right]_{\tau=\beta_k-\beta_h} = \left[e^{-j\pi\alpha\tau} r_{W_\alpha}(-\tau)\right]_{\tau=\beta_k-\beta_h} = 0 \qquad h \neq k. \quad \text{(A-7)}$$

Consequently, from (A-2) and (A-3) it follows that

$$\gamma_k = \frac{1}{\int_{-\infty}^{+\infty} |W_\alpha(f)|^2 df} \mathrm{Re}\left\{\int_{-\infty}^{+\infty} |W_\alpha(f)|^2 \frac{S_{yx}^\alpha(f)}{S_x^\alpha(f)} e^{j2\pi(f+\alpha/2)\beta_k} df\right\} \quad \text{(A-8)}$$

$$k = 1, 2, \ldots, N$$

$$\frac{\partial}{\partial \beta_k}\mathrm{Re}\left\{\int_{-\infty}^{+\infty} |W_\alpha(f)|^2 \frac{S_{yx}^\alpha(f)}{S_x^\alpha(f)} e^{j2\pi(f+\alpha/2)\beta_k}\,df\right\} = 0 \qquad k = 1, 2, \ldots, N. \quad \text{(A-9)}$$

Equations (A-8) show that the γ_k's can be determined once the β_k's have been evaluated. Moreover, accounting for (6a) and (A-5), one has

$$\frac{1}{\int_{-\infty}^{+\infty} |W_\alpha(f)|^2\,df}\mathrm{Re}\left\{\int_{-\infty}^{+\infty} |W_\alpha(f)|^2 \frac{S_{yx}^\alpha(f)}{S_x^\alpha(f)} e^{j2\pi(f+\alpha/2)\tau}\,df\right\}$$

$$= \frac{1}{r_{W_\alpha}(0)} \sum_{k=1}^{N} A_k \mathrm{Re}\left\{e^{j\pi\alpha(\tau - D_k)} r_{W_\alpha}(\tau - D_k)\right\}. \quad \text{(A-10)}$$

In other words, in (A-10) each of the N nonoverlapping replicas of $\mathrm{Re}\left\{r_{W_\alpha}(\tau - D_k)\right\}$ presents a peak at $\tau = D_k$. Then, accounting for (A-8) and (A-9), the ATs and SAs can be estimated by locating and measuring the maximum of each replica in (A-10) where, of course, the cyclic spectra are replaced by their estimates.

Finally, we note that, for the SPECCOA method, the width d'_α of the autocorrelation function $r_{W_\alpha}(\tau)$ is twice the width d_α of $R_s^\alpha(\tau)$. For the BL-SPECCORR method, d'_α is not finite, but one can obtain useful approximations by assuming $d'_\alpha = 2d_\alpha$.

REFERENCES

[1] C. H. Knapp and G. C. Carter, "The generalized correlation method for estimation of time delay," *IEEE Trans. Acoust., Speech, and Signal Process.*, vol. ASSP-24, pp. 320–327, August 1976.

[2] G. C. Carter, "Coherence and time delay estimation," *Proc. IEEE*, vol. 75, pp. 236–255, February 1991.

[3] A. Hewitt, W. H. Lau, J. Austin, and E. Vilar, "An autoregressive approach to the identification of multipath ray parameters from field measurements," *IEEE Trans. Commun.*, vol. 37, pp. 1136–1143, November 1989.

[4] W. A. Gardner, *Statistical Spectral Analysis: A Nonprobabilistic Theory*, Englewood Cliffs, NJ: Prentice-Hall, 1988.

[5] W. A. Gardner, "Identification of systems with cyclostationary input and correlated input/output measurement noise," *IEEE Trans. Automat. Contr.*, vol. 35, pp. 449–452, April 1990.

[6] L. Izzo, A. Napolitano, and L. Paura, " Interference-tolerant estimation of amplitude and time-delay parameters of a composite signal," in *Proc. of Fifth European Signal Processing Conference (EUSIPCO'90)*, Barcelona, Spain, September 1990.

[7] L. Izzo, A. Napolitano, L. Paura, and M. Tanda, "Estimation of multipath-signal parameters by exploitation of spectral coherence," in *Proc. of 2nd International Conference on Communication Systems (ICCS'90)*, Singapore, November 1990.

[8] G. Gelli, L. Izzo, A. Napolitano, and L. Paura, "Spectral-coherence based Prony algorithm for multipath channel identification," in *Proc. of International Conference on Digital Signal Processing '91*, Florence, Italy, September 1991.

[9] L. Izzo, A. Napolitano, and M. Tanda, "Spectral-correlation based methods for multipath-channel identification," *European Transactions on Telecommunications and Related Technologies*, vol. III, pp. 341–348, July–August 1992.

[10] G. Gelli, L. Izzo, A. Napolitano, and L. Paura, "Multipath-channel identification by an improved Prony algorithm based on spectral correlation measurements," *Signal Processing*, vol. 31, pp. 17–29, March 1993.

[11] W. A. Gardner and C.-K. Chen, "Signal-selective time-difference-of-arrival estimation for passive location of man-made signal sources in highly corruptive environments, Part I: Theory and method," *IEEE Trans. Signal Processing*, vol. 40, pp. 1168–1184, May 1992.

[12] W. A. Gardner and C.-K. Chen, "Selective source location by exploitation of spectral coherence," in *Proc. of IEEE ASSP 4th Workshop Spectrum Estimat. and Model.*, Minneapolis, MN, August 1988, pp. 271–276.

[13] L. Izzo and A. Napolitano, "A cyclic-feature based method for multipath-channel identification," in *Proc. of International Conference on DSP Applications and Technology (ICSPAT'92)*, Cambridge, MA, November 1992.

[14] W. A. Gardner and C.-K. Chen, "Interference-tolerant time-difference-of-arrival estimation for modulated signals," *IEEE Trans. Acoust., Speech, and Signal Process.*, vol. 36, pp. 1385–1395, September 1988.

[15] S. M. Kay and S. L. Marple, "Spectrum analysis—A modern perspective," *Proc. IEEE*, vol. 69, pp. 1380–1419, November 1981.

Blind Channel Identification and Equalization Using Spectral Correlation Measurements, Part I: Frequency-Domain Analysis

Zhi Ding
Department of Electrical Engineering
Auburn University
Auburn, AL 36849-5201

Abstract

Recent work by Gardner [1] and Tong, Xu, and Kailath [2] presents novel techniques that exploit cyclostationarity for channel identification in QAM data communication systems. In this article, we investigate the identifiability of linear time-invariant (LTI) channels based on the use of second-order cyclic statistics. Using a frequency-domain approach, we show that channel identification is achievable for a class of linear channels without the need for pilot signals or other special assistance from the transmitter. We also study the identification of linear rationally parameterized channels from cyclic spectral measurements. Since the estimation of second-order statistics requires less data compared to the estimation of higher-order statistics, faster algorithms can be realized by exploiting the cyclostationarity of the channel output. Moreover, channel identification based on cyclic statistics does not preclude Gaussian or near-Gaussian inputs.

Work supported by the National Science Foundation under Grant MIP-92-10100.

1 INTRODUCTION

Blind adaptive channel equalizers eliminate the need for training signals in digital communication systems when the transmission of a training sequence is impractical or very costly [3, 4, 5]. This ability of blind startup enables a blind equalizer to start its self-adaptation or to self-recover from system breakdowns, during which the equalizer may have lost track of the desired parameter settings, without special assistance from the transmitter.

This article summarizes the work to date on blind equalization and channel estimation and studies possible alternatives to traditional solutions by utilizing the cyclostationarity of communication signals. This presentation describes various approaches and some shortcomings in the traditional solution to the blind equalization problem. Through the investigation of channel identifiability based on cyclostationary statistics of the channel output, the possible future development of some simpler and faster blind channel estimation methods is illustrated.

Most blind equalization schemes begin by sampling the channel output at the baud rate [3, 4, 6] to produce a stationary channel output sequence for processing. Consequently, blind channel identification based on input/output statistics (without direct access to the channel input) must require the use of higher than second-order statistics [3] since second-order statistics are sufficient only to recover the magnitude but not the phase of the channel transfer function. Some well-known blind equalization schemes based on explicit or implicit use of higher-order statistics can be found in [3, 4, 7, 8, 9, 10, 11, 12, 13]. Because of the large amount of data necessary to estimate higher-order statistics, algorithms based on higher-order statistics tend to be rather slow.

It is important to note, however, that the actual analog channel outputs of pulse-amplitude-modulated (PAM) and quadrature-amplitude-modulated (QAM) systems are in fact cyclostationary processes instead of stationary. Thus a very important question arises as to whether it is possible to identify the unknown linear channel based solely on the second-order (cyclic) statistics of the cyclostationary channel output. Gardner [1] recently investigated the use of the second-order cyclic spectrum in channel identification and proposed a scheme requiring the use of a training period during which the unknown training sequence is transmitted at a very low rate. Based on a certain rank condition of the channel convolution matrix, Tong, Xu, and Kailath [2] also proposed an effective finite impulse response (FIR) channel identification scheme by using the second-order cyclostationary autocorrelation of the channel output sampled at a rate higher than the baud rate.

While the work of Gardner [1] and Tong, Xu, and Kailath [2] opened up possibilities for a class of new blind equalization schemes, the key issue that still needs to be addressed is whether, in principle, second-order statistics alone can be sufficient for blind channel identification *without training*. Motivated by [1] and [2], this article, along with Article 5, is aimed at establishing channel identifiability based on second-order cyclic spectra and at deriving various conditions under which blind channel identification can be accomplished. It is shown that for a class of channels with limited phase variations, such as those with rational transfer functions, the channel dynamics can be identified through the use of cyclic spectra of the cyclostationary channel output signal. Thus, in contrast to blind equalizers for baud-spaced time samples, which require the use of higher-order statistics, the channel output signal sampled at higher than the baud rate can contain important second-order statistical information for channel identification. Since second-order (cyclic) statistics require less data for estimation, faster algorithms can be devised by exploiting the cyclostationarity of the channel output.

This article is organized as follows: In Section 2 the problem of blind equalization and channel estimation is outlined. The best-known and most popular approaches using higher-order statistics are briefly described and some key weaknesses associated with various approaches are pointed out. In Section 3 the cyclostationarity of the channel output signal to be processed is clarified, and the possible use of the cyclostationary statistics for blind channel estimation is addressed. It is shown that while, in general, an arbitrary channel phase cannot be identified based on cyclostationary statistics alone, the identifiability of a class of practical channels can be established. In Section 4, a new parametric channel identification scheme based on the use of cyclostationary statistics is presented. The proof of its feasibility corroborates the results in [2] and shows a potentially fruitful new research direction in the development of future blind equalization strategies. Simulation results are given in Section 5. Although most of our analysis is carried out in the frequency domain, a time-domain approach is presented in Article 5.

2 FUNDAMENTALS OF BLIND CHANNEL EQUALIZATION AND IDENTIFICATION

2.1 Channel Equalization in QAM Systems

The complex baseband model for a typical QAM data communication system, as shown in Fig. 1, simply consists of an unknown linear time-invariant (LTI) channel, which represents all the interconnections between the transmitter and the receiver. The baseband-equivalent transmitter generates a sequence of complex-valued random input data $\{s_n\}$, each element of which comes from a finite complex alphabet $\mathcal{A}$ (or constellation) of QAM symbols. The data sequence $\{s_n\}$ is sent through a baseband-equivalent complex LTI channel whose output $x(t)$ is observed by the receiver. The function of the receiver is to restore the original data $\{s_n\}$ from the observation $x(t)$ by outputting a sequence of estimates for $\{s_n\}$.

The complex LTI communication channel is assumed to be causal with impulse response $h(t)$. The input/output relation for the QAM system can be written as

$$x(t) = \sum_{k=-\infty}^{\infty} s_k h(t - kT - t_0) + w(t), \quad s_k \in \mathcal{A}, \tag{1}$$

where T is the symbol baud period. The noise $w(t)$ is stationary, white, and independent of the channel input s_n, but not necessarily Gaussian. When the distortion caused

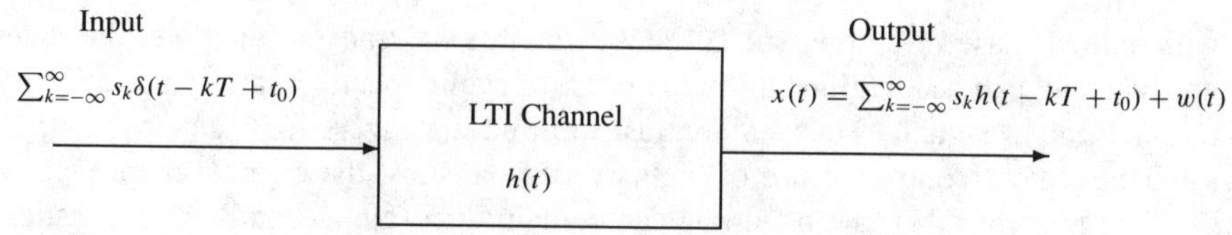

Figure 1: Baseband representation of QAM data communication system.

by a nonideal LTI channel is significant, as is often the case in practice, equalization is needed to remove the intersymbol interference (ISI) at the sampling instants. Due to the presence of ISI, the recovery of the input signal sequence s_n requires that the channel impulse response $h(t-t_0)$ be identified. The channel identification process is explicit in nonlinear channel equalization schemes such as the decision feedback equalizer (DFE) and the maximum-likelihood sequence estimator (MLSE) and implicit in linear equalizers where the channel inverse is identified.

2.2 Blind Channel Equalization/Identification Based on Stationary Statistics

In traditional blind equalization systems, the channel output is sampled at the known baud rate $1/T$. The sampled channel output

$$x(nT)=\sum_{k=-\infty}^{\infty} s_k h(nT-kT-t_0)+w(nT)=s_n \circledast h(nT-t_0)+w(nT) \quad (2)$$

is a stationary process. Using the notation $x_n \triangleq x(nT)$, $w_n \triangleq w(nT)$, and $h_n \triangleq h(nT-t_0)$, the relationship (2) can be written as a noisy discrete convolution

$$x_n=\sum_{k=-\infty}^{\infty} s_k h_{n-k}+w_n. \quad (3)$$

The function of the receiver is to restore the original data $\{s_n\}$ from the observation $\{x_n\}$.

The discrete channel transfer function is defined by

$$H(q^{-1})=\sum_{n=0}^{\infty} h_n q^{-n}, \quad (4)$$

where $\{h_n\}$ is the discrete impulse response of the channel. When the channel is noiseless and ideal with no ISI such that only one nonzero element exists in the sequence $\{h_n\}$ (Nyquist I Criterion), the channel output becomes

$$x_n=h_m s_{n-m}, \quad h_m \neq 0,$$

which is simply a scaled version of the input with finite integer delay m. The difficulty arises when the channel is imperfect such that more than one nonzero element exists in the channel impulse response $\{h_n\}$. In this case, undesirable ISI is introduced at the channel output x_n from which a simple memoryless decision device may not be able to recover the original data sequence.

In blind equalization the original sequence is unknown to the receiver except for its probabilistic or statistical properties over the known alphabet $\mathcal{A}$. Since both the channel input s_n and the channel impulse response h_n are unknown, traditional minimum-mean-squared-error approaches are not applicable due to the lack of a reference or training signal. Blind equalization or channel estimation is necessary to restore s_n or to identify h_n from the channel output without direct measurement of the channel input data. The following summary of facts about stationary blind channel equalization is taken from [3]:

1. A mixed-phase linear dynamical channel is identifiable from its output (and knowledge of the distribution of its input) only when the input is not Gaussian.
2. Second-order statistics alone are insufficient for blind equalization or estimation of channels with mixed phase.
3. Most practical QAM systems employ a complex-valued constellation $\mathcal{A}$ which has symmetrical properties such that the constellation is $k\pi/2$-rotation-invariant. Thus, blind channel equalization or estimation will be subjected to a phase ambiguity of $k\pi/2$.

Blind channel equalization and estimation methods based on higher-order statistics (HOS) generally belong to two distinct classes, which shall be referred to as explicit HOS methods and implicit HOS methods. Because stationary blind deconvolution of nonminimum phase systems must rely on the use of HOS, all existing blind equalization and estimation algorithms based on baud-rate data samples can be thus classified depending on their direct or indirect computation of higher-order statistics.

2.3 Blind Adaptive Equalization Algorithms: Implicit HOS Methods

A linear channel equalizer is a linear filter $G(q^{-1})$ that is applied to the channel output in order to eliminate its ISI by essentially cancelling the channel dynamics. All the ISI is removed if

$$H(q^{-1})G(q^{-1}) = cq^{-m}, \quad c \neq 0 \tag{5}$$

for some integer m. The desired response of (5) can be rewritten as

$$G(q^{-1}) = cq^{-m}H^{-1}(q^{-1}), \tag{6}$$

which means that the equalizer attempts to achieve the inverse of the channel transfer function with a possible gain difference and/or a constant time delay.

The objective of the blind equalizer is to adjust its parameter vector θ such that the desired response (6) can be achieved. The key in the development of a blind equalizer is therefore to design the rule for self-adjustment or adaptation. Clearly, a blind equalizer adaptation scheme needs to minimize some special type of cost function, other than mean-squared error (MSE), which implicitly exploits higher-order statistics of the channel output. Consequently, almost all existing blind adaptive equalization algorithms are designed as stochastic gradient descent (SGD) schemes to update the parameter vector by seeking to minimize some special (non-MSE) mean cost functions that do not involve the use of the original input s_n but still reflect the current level of ISI in the equalizer output. The *mean cost function* is defined by

$$J(\theta) \triangleq E\{\Psi(z_n)\}, \tag{7}$$

where Ψ is a real-valued scalar cost function of z_n. The mean cost function $J(\theta)$ should be at its minimum when θ is such that $z_n = e^{j\phi}a_{n-m}$.

With mean cost functions defined, stochastic gradient descent algorithms can be used to adjust the equalizer parameters [3]. A blind equalizer can be defined either by the cost function Ψ or, equivalently, by its derivative ψ which is called the *error function* since it resembles the role of a prediction error in the least mean square (LMS) algorithm. The design of blind equalization now translates to the selection of the function Ψ (or ψ) such that local minima of $J(\theta)$ correspond to significant removal of ISI.

Although adaptive algorithms have been developed for channel equalization with training since the early work of Lucky [14], the first well-known implicit HOS method for *blind* adaptive equalization was presented by Sato [7] in 1975. This rather heuristic method was later extended and analyzed by Benveniste, Goursat, and Ruget [3] in their seminal work of 1980. In 1982, Godard presented his well-known algorithm [4], which also generalized Sato's work for QAM systems by providing a class of cost-functions to be minimized. A similar scheme, known as the "constant modulus algorithm" (CMA), was also developed separately by Treichler and coworkers [15, 16] through a property restoral philosophy. Since then, there have been a plethora of algorithms proposed that are based on various approaches [8, 9, 13, 17, 18, 19, 20, 21, 22, 23, 24, 25]. The recent tutorial paper [6] contains many useful references, and a special section devoted to blind equalization has also appeared recently in *Optical Engineering* [26].

Nevertheless, almost all the existing on-line blind equalization algorithms can be viewed as SGD minimization algorithms differing only in the specification of the error function $\psi(\cdot)$. The special error function implicitly utilizes higher-order statistics of the channel output in order to obtain phase information. With few exceptions, existing implicit HOS algorithms are various forms of SGD algorithms.

2.4 Explicit HOS Methods for Channel Identification and Equalization

The use of explicit HOS for blind system equalization originated in geophysics where blind deconvolution was needed for non-Gaussian inputs. In explicit HOS methods, the second-order statistics, namely, estimates of the power spectral density, are used to determine the magnitude of the channel transfer function. The phase of the channel is determined from the higher-order statistics of the channel output. Due to the symmetry of QAM constellations, the lowest order of the cumulants of the channel output signal that can be used for phase identification is the fourth order. For a simple illustration, the relationship between the phase $\Psi(\omega_1, \omega_2, \omega_3)$ of the tri-spectrum (fourth-order spectral cumulant) and the phase $\Phi(\omega)$ of the channel for zero-mean, independent identically distributed (i.i.d.) and non-Gaussian input can be written as

$$\Psi(\omega_1, \omega_2, \omega_3) = \Phi(\omega_1) + \Phi(\omega_2) + \Phi(\omega_3) - \Phi(\omega_1 + \omega_2 + \omega_3). \tag{8}$$

This relationship can be exploited to extract the phase of the channel $\Phi(\omega)$.

The first detailed study of explicit HOS methods was presented by Donoho in [27] in which theoretical analysis and extension were provided for the rather ad hoc minimum entropy deconvolution approach of Wiggins [28]. Based on the relationship between the cumulants of input and output signals of a linear time-invariant system given by Brillinger and Rosenblatt [29] and by Kuznetsov, Stratonovich, and Tikhonov [30, 31], various identification methods were proposed to identify the impulse response of linear systems with autoregressive moving average (ARMA) models. Some well-known studies on HOS blind deconvolution include those by Lii and Rosenblatt [32], Tugnait [11, 33], Giannakis and Mendel [34], Nikias [10], Giannakis and Swami [35], Mendel and Swami [36], Friedlander and Porat [12], and Hatzinakos and Nikias [13]. In addition to two tutorial papers given by Nikias and Raghuveer [38] and Mendel [37], as well as the references therein, other related work can be found in the two special issues on higher-order statistics [39, 40].

2.5 Limitations of HOS-Based Methods

The implicit and explicit HOS methods based on the baud-rate sampled channel output can be generally effective, judging from successes reported by various authors. Nevertheless, they also exhibit some critical weaknesses such as the possible convergence of some algorithms to local minima [41] and the sensitivity to timing jitter [42].

One particularly important limitation of HOS-based methods is the slow rate of convergence due to the need for accurate time-average approximations of higher-order statistics. Time-average approximation of higher-order statistics demands a much larger number of data samples than that required for second-order statistics. In order to maintain similarly low estimation bias and variance for n-th order statistics, the required amount of data increases almost exponentially with the order n [43]. Since channel identification requires at least fourth-order statistics, slow rate of convergence is typical of HOS-based blind equalization and channel estimation algorithms.

Typically, several thousands of data samples are needed for the success of existing blind equalization and channel estimation schemes. The resulting delay in equalization cannot be accommodated in many practical communication systems. In land mobile communications, for example, such a delay can make it impossible to track the time-varying channel, or at least represents a loss of several seconds of useful information, which could potentially obliterate other advantages of the no-training, blind equalization system. This problem can be further exasperated when certain coding schemes utilized by the channel input may cause the channel output to have near zero fourth-order cumulants [44], resulting in the need for even higher order statistics whose approximations are even harder to obtain. By contrast, adaptive equalizers with training take only a few hundred data samples to converge. Clearly, the slow convergence exhibited by the existing blind equalization schemes is detrimental to their wide application in many data communication areas such as mobile communications where rapid adaptivity is essential.

3 CYCLOSTATIONARITY OF THE CHANNEL OUTPUT SIGNAL

3.1 Second-Order Statistics of the Cyclostationary Channel Output

Although the use of higher-order statistics is necessary for the identification of non-minimum phase channels based on channel outputs sampled at the baud rate, the actual channel output $x(t)$ as in (1) is cyclostationary instead of stationary. It can be verified that for a stationary channel input s_n and noise $w(t)$ with autocorrelation functions

$$R_s[k-l] = E\{s_k s_l^*\}, \quad R_w(t_2 - t_1) = E\{w(t_1) w^*(t_2)\},$$

we have

$$\begin{aligned} R_x(t_1, t_2) &\triangleq E\{x(t_1) x^*(t_2)\} \\ &= \sum_{k=-\infty}^{\infty} \sum_{l=-\infty}^{\infty} R_s[k-l] h(t_1 - kT - t_0) h^*(t_2 - lT - t_0) + R_w(t_2 - t_1) \quad (9) \\ &= R_x(t_1 + T, t_2 + T), \end{aligned}$$

where the symbol duration T is the fundamental period of the cyclostationary process. Hence, failure to identify the channel phase using second-order statistics of the stationary channel-output baud-rate samples does not necessarily imply failure for second-order cyclic-statistics of the analog cyclostationary channel output signal or the channel output sampled at a rate higher than the baud rate.

The notation for second-order statistics of cyclostationary processes used here generally follows that in the pioneering work of Gardner [45, 46]. Specifically, for cyclostationary process $x(t)$ with fundamental period T, the *cyclic autocorrelation function* is defined by

$$R_x^\alpha(\tau) \triangleq \frac{1}{T} \int_{-T/2}^{T/2} R_x\left(t + \frac{\tau}{2}, t - \frac{\tau}{2}\right) e^{-j2\pi\alpha t} dt, \quad \text{where } \alpha = k/T. \quad (10)$$

Correspondingly, the *spectral-correlation density* (SCD) is given by

$$S_x^\alpha(j\omega) = \int_{-\infty}^{\infty} R_x^\alpha(\tau) e^{-j\omega\tau} d\tau. \quad (11)$$

For the QAM signal $x(t)$ of (1), the SCD can be shown to relate to the input *power spectral density* (PSD) and the frequency response of the channel through the expression [45]

$$S_x^\alpha(j\omega) = \tfrac{1}{T} H(j\omega + j\pi\alpha) H^*(j\omega - j\pi\alpha) S_s(j\omega + j\pi\alpha) e^{-j2\pi\alpha t_0} + S_w^\alpha(j\omega), \tag{12}$$

in which

$$H(j\omega) = \int_{-\infty}^{\infty} h(t) \exp(-j\omega t) dt$$

is the channel frequency response and

$$S_s(j\omega) = \sum_{k=-\infty}^{\infty} R_s[k] \exp(-j\omega k T) \tag{13}$$

is the PSD of the stationary channel input sequence s_n. Note that for white stationary noise $w(t)$, we have

$$S_w^{k/T}(j\omega) \triangleq \int_{-\infty}^{\infty} R_w^{k/T}(\tau) \exp(-j\omega\tau) d\tau = N_0 \delta[k]. \tag{14}$$

If the channel input $\{s_n\}$ is i.i.d. with zero mean, its autocorrelation function is simply

$$R_s[k] = E\{s_{n+k} s_n^*\} = E\{|s_n|^2\} \delta[k], \tag{15}$$

and its PSD is a constant $S_s(j\omega) = R_s[0] = E\{|s_n|^2\}$.

3.2 Channel Identifiability Based on the Channel Output SCD

The objective here is to study the possibility of identifying the channel frequency response $H(j\omega)$ from the SCD of the channel output $x(t)$. Without loss of generality, the discussion is limited to the case of a zero-mean and i.i.d. input. Thus, since the magnitude of $H(j\omega)$ can be identified from the PSD of the channel output, our objective is to identify the phase of $H(\omega)$ from

$$S_x^{k/T}(j\omega) = \frac{R_s[0]}{T} H\left[j\left(\omega + \frac{\pi k}{T}\right)\right] H^*\left[j\left(\omega - \frac{\pi k}{T}\right)\right] \exp\left(-j\frac{2\pi k t_0}{T}\right), \quad k = \pm 1, \pm 2, \ldots. \tag{16}$$

It is apparent that our discussion can be applied also to correlated channel inputs if the correlation is known.

Let $\Psi_k(j\omega)$ be the phase of SCD $S_x^{k/T}(\omega)$ and $\Phi(\omega)$ be the phase of the frequency response $H(j\omega)$. For $k \neq 0$, the following relationship between the two phases can be expressed according to (16) by

$$\Psi_k(\omega) = \Phi\left(\omega + \frac{\pi k}{T}\right) - \Phi\left(\omega - \frac{\pi k}{T}\right) - \frac{2\pi k t_0}{T}, \quad k = \pm 1, \pm 2, \ldots. \tag{17}$$

For an arbitrary channel, the above equations indexed by k, in which phase unwrapping is necessary to obtain $\Psi_k(\omega)$, are the only sources of phase information. Clearly, t_0 can at best be identified as modulo T. For analytical simplicity, it is temporarily assumed that t_0 is known. The identifiability of $\Phi(\omega)$ from $\Psi_k(\omega)$ can be better illustrated through a cepstral approach. Let

$$\psi_k(\tau) \triangleq \frac{1}{2\pi}\int_{-\infty}^{\infty}\left[\Psi_k(\omega) + \frac{2\pi k t_0}{T}\right] e^{j\omega\tau} d\omega, \tag{18}$$

$$\phi(\tau) \triangleq \frac{1}{2\pi}\int_{-\infty}^{\infty} \Phi(\omega) e^{j\omega\tau} d\omega. \tag{19}$$

We have from (17)

$$\begin{aligned}\psi_k(\tau) &= \phi(\tau)\exp(-j\pi k\tau/T) - \phi(\tau)\exp(j\pi k\tau/T)\\ &= -2j\phi(\tau)\sin(\pi k\tau/T).\end{aligned} \tag{20}$$

From this particular relationship we can make the following observations:

- No cepstral channel phase information can be extracted when

$$\sin(\pi k\tau/T) = 0, \quad \text{or when} \quad \tau = mT/k, \quad m \in \mathbb{Z}, \tag{21}$$

  where $\mathbb{Z}$ is the set of integers. Therefore, an arbitrary channel transfer function cannot be completely identified from the cyclostationary statistics alone.
- It is clear from (20) that no additional phase information can be extracted from the SCD by setting $k = \pm 2, \pm 3, \ldots$. Consequently, we simply consider the use of the SCD with $k = 1$ which has the largest support region.

Theoretically, most of the imaginary cepstrum of $H(j\omega)$ can be identified through

$$\phi(\tau) = \frac{j\psi_1(\tau)}{2\sin(\pi\tau/T)}, \tag{22}$$

except for $\tau = mT$. If $\phi(\tau)$ is bounded at $\tau = mT$, the L_2-normed error between the true phase and the phase estimate from (22) will be zero since the set $\{\tau : \tau = mT\}$ has zero measure. However, when $\phi(\tau)$ is not bounded at $\tau = mT$, the channel phase cannot be determined even theoretically. For example, if $\phi(\tau)$ has an impulse at $\tau = mT$, then correspondingly $\Phi(\omega)$ will contain a sinusoidal part with frequency mT (e.g., $\cos(mT\omega)$) and the L_2-normed phase error can be infinite. It is in fact clear from (17) that the SCD contains no information about any oscillatory phase content with frequency mT. This fact is reinforced in Article 5 where it is shown that discrete channels with zeros uniformly spaced on a circle (resulting in oscillatory phase content) cannot be identified. In a recent private communication, the author learned that similar results were also independently obtained earlier by Gardner.

In practical communication systems, channels are effectively bandlimited. It is therefore reasonable to consider the special case where

$$\Phi(\omega) = \tilde{\Phi}(\omega) + c_1\,\mathrm{sgn}(\omega) + c_2, \quad c_1, c_2 \in \mathbb{R}, \tag{23}$$

in which $\tilde{\Phi}$ is bounded and has finite support (ω_1, ω_2) and $\mathbb{R}$ is the set of real numbers. Based on (17) and (23), t_0 can be identified as modulo T from

$$\Psi_1\left(\omega_2 + \frac{\pi}{T}\right) = \Phi\left(\omega_2 + \frac{2\pi}{T}\right) - \Phi(\omega_2) - \frac{2\pi t_0}{T} = -\frac{2\pi t_0}{T}. \tag{24}$$

Hence $\psi_1(\tau)$ can be determined and it satisfies the following relationship

$$\psi_1(\tau) = -2j\sin(\pi\tau/T)\left[\tilde{\phi}(\tau) - \frac{c_1}{j\pi\tau} + c_2\delta(\tau)\right], \tag{25}$$

where $\tilde{\phi}$ is the inverse Fourier transform of $\tilde{\Phi}$. Both $\tilde{\phi}$ and $\tilde{\Phi}$ have finite energy according to Parseval's Theorem. Therefore the only source of estimation error from using (22) lies in our inability to determine c_2, which is merely a constant phase shift. Thus in theory, the phase of bandlimited channels satisfying the assumption (23) can be identified from cyclostationary statistics except for a possible constant phase ambiguity.

The preceding argument establishes the theoretical identifiability of bandlimited channels. In practice, however, the estimate $\phi(\tau)$ will be highly inaccurate whenever $\tau \approx mT$. Moreover, due to the bandwidth limitation of the channel, the support of the SCD is finite and is typically small. It becomes necessary, therefore, to estimate the missing information through interpolation assuming the continuity of $\phi(\tau)$ or through a parametric approach in which we define a parametric model for the channel and identify the unknown parameters from cepstral phase information obtained for values of τ sufficiently far removed from those points where $\sin(2\pi\tau/T) = 0$. In the following section, we describe one such parametric method.

4 PARAMETRIC CHANNEL IDENTIFICATION

4.1 Basic Assumptions and the Uniqueness of the SCD Model

It is assumed that the channel transfer function is given or can be closely approximated by the following rational transfer function

$$H(j\omega) = Ae^{-j\omega t_d}\frac{\Pi_{i=1}^{M}(j\omega - z_i)}{\Pi_{i=1}^{N}(j\omega - p_i)}, \quad M \le N. \tag{26}$$

The orders of the channel model (M, N) should be no less than those of the actual rational channel. It is required that the channel be strictly stable such that all of its poles are inside the stability region $\{Re(p_i) < 0\}$. It is also required that the channel transfer function have no allpass factor. The channel bandwidth B_w is required to be above the minimum bandwidth $B_w > 1/2T$ needed for zero ISI.

The identification of the channel transfer function makes use of the following resultant rational function model for the SCD

$$S_x^{k/T}(j\omega) = A^2 \frac{R_s[0]}{T} \exp[-j2\pi k(t_0 + t_d)/T] \cdot \frac{\Pi_{i=1}^{M}[j(\omega + \pi k/T) - z_i][-j(\omega - \pi k/T) - z_i^*]}{\Pi_{i=1}^{N}[j(\omega + \pi k/T) - p_i][-j(\omega - \pi k/T) - p_i^*]}. \tag{27}$$

It has $2N$ poles that are not on the imaginary axis in the complex plane $\mathbb{C}$ because of the stability condition $Re(p_i) < 0$. Thus the complex rational function $S_x^{k/T}(j\omega)$ is analytic on a complex open set Ω, which includes the entire real axis. The following well-known lemma [47] can therefore be invoked:

Lemma 4.1.1 *If $S_1(z)$ and $S_2(z)$ are two analytic functions in a region Ω and if $S_1(z) = S_2(z)$ for all z in some set that has a limit point in Ω, then $S_1(z) = S_2(z)$ $\forall z \in \Omega$.*

$S_x^{k/T}(j\omega)$ is therefore uniquely determined by its values given over any frequency (open or closed) interval. In other words, the complex function $S_x^{k/T}(j\omega)$ is uniquely specified by

$$S_x^{k/T}(j\omega), \quad \omega \in [\omega_1, \omega_2] \subset \mathbb{R}. \tag{28}$$

This property is the key to extracting phase information from the limited support of the SCD. In practice, however, the smaller the length $\omega_2 - \omega_1$ (of the interval) is, the more accurate the estimate of $S_x^{k/T}(j\omega)$ on this interval must be.

4.2 Identifiability

It is clear that the channel poles can be identified from the PSD

$$S_x^0(j\omega) = A^2 \frac{R_s[0]}{T} \frac{\Pi_{i=1}^{M}|j\omega - z_i|^2}{\Pi_{i=1}^{N}|j\omega - p_i|^2} + N_0 \tag{29}$$

by finding the roots of its denominator except for those allpass factors that have $p_i = -z_i^*$. Since it is required that the channel have no such pole/zero relationship, the approach to channel identification proposed here is to first identify the poles from the PSD and then to identify all the zeros based on the cyclostationary information contained in the SCD $S_x^{k/T}(j\omega)$.

Given that all N poles have been identified, the function defined by

$$D_x^{k/T}(j\omega) \triangleq S_x^{hs1k/T}(j\omega)\left(\Pi_{i=1}^{N}[j(\omega + \pi k/T) - p_i][-j(\omega - \pi k/T) - p_i^*]\right) \tag{30}$$

can be calculated. It follows from (27) that the zeros of $D_x^{k/T}(j\omega)$ are shifted (parallel to the imaginary axis) versions of the zeros of the transfer function and their reflections about the imaginary axis; that is, their locations are given by

$$\pm Re(z_i) + jIm(z_i) \mp j\pi k/T, \quad i = 1, 2, \ldots, M.$$

The fact that $D_x^{k/T}(j\omega)$ is a complex polynomial function of the complex variable $j\omega$ is essential to the identification procedure presented here.

Without loss of generality, $k > 0$ is chosen. Then the following procedure can be adopted:

1. Find the set of all zeros of $D_x^{k/T}(j\omega)$ and label it U_M;
2. Find the element a_M in U_M that has the maximum imaginary part—if it is not unique, arbitrarily choose one—to obtain

$$Im(z_M) = Im(a_M) - \pi k/T \quad \text{and} \quad Re(z_M) = -Re(a_M);$$

3. Form a new set U_{M-1} by removing two zeros,

$$U_{M-1} = U_M - \{\pm Re(z_M) + jIm(z_M) \mp j\pi k/T\};$$

4. Repeat steps 2 and 3 for $U_i, \quad i = M-1, M-2, \cdots, 1$ to identify all M zeros of the channel transfer function $H(j\omega)$,

$$a_i = \text{Arg} \max_{x \in U_i} Im(x),$$

$$Im(z_i) = Im(a_i) - \pi k/T \quad \text{and} \quad Re(z_i) = -Re(a_i)$$
$$U_{i-1} = U_i - \{\pm Re(z_i) + jIm(z_i) \mp j\pi k/T\}.$$

This procedure identifies all the zeros of the channel transfer function. Once the poles and zeros of $S_x^{k/T}(j\omega)$ are identified, the combined attenuation factor $|A|R_s[0]$ and the combined delay $t_0 + t_d$ modulo T can also be identified. Hence, from the SCD $S_x^{k/T}(j\omega)$ of the received channel output signal, the zeros of the channel transfer function can be uniquely identified, and the attenuation factor $|A|$ can also be identified if $R_s[0]$ is known (from the knowledge of the symbol constellation). But the exact delay t_d cannot be identified. Although k can be any positive integer, it is best to rely on $S_x^{1/T}(j\omega)$ since it has the largest support. The ambiguity of a possible linear phase term ωt_d is not crucial to the objective of recovering the input sequence $\{s_n\}$ since we have identified the delayed channel impulse response $h(t - t_0)$. On the other hand, the sign ambiguity in $|A|$ is inherent in the blind identification/equalization problem when all odd-order cumulants of the channel output are zero [11].

4.3 Remarks

The major cause of ISI is the limited bandwidth of the channel. From the channel output SCD given in (16), the bandwidth limitation of $H(j\omega)$ can also cause the SCD $S_x^{1/T}(j\omega)$ to be negligible over most of the Nyquist band $[-1/2T,\ 1/2T]$. However, since it is assumed that the unknown channel does have bandwidth higher than the minimum Nyquist bandwidth, i.e., $B > 1/2T$, and since the support band of the SCD is given (from (16)) by

$$S_x^{1/T}(j\omega) \neq 0, \quad |\omega| < 2\pi(B - 1/2T), \tag{31}$$

then the SCD of the channel output is, in general, nonzero over the frequency interval $(1/2T - B, B - 1/2T)$ and can be used to parametrically determine the function $S_x^{1/T}(j\omega)$ which can then be used to identify the linear channel.

The constructive proof of rational parametric channel identifiability based on second-order cyclic statistics outlines a procedure for channel identification based on the SCD. Essentially, $S_x^0(j\omega)$ and $S_x^{1/T}(j\omega)$ are first parametrically determined from data via SCD estimation. The channel transfer function $H(j\omega)$ is then identified from the poles, zeros, and attenuation factor of the SCD models. This method is used only to illustrate the concept and should not be taken as an effective approach.

Although it was assumed that the channel has no allpass factor, this condition can be relaxed if the allpass poles do not cause any pole-zero cancellation in $H(j\omega + j\pi\alpha)H^*(j\omega - j\pi\alpha)$. Under this relaxed condition, the poles of $D_x^\alpha(j\omega)$ can also be identified using the same procedure as that used for identifying its zeros.

For stationary signals and systems, if the input is Gaussian or nearly Gaussian, it then becomes impossible to identify the unknown channel and to deconvolve the unknown input. However, if a cyclostationary output is available, the result obtained herein shows that even when the input is Gaussian, channel identification and blind deconvolution can still be accomplished based on second-order cyclic statistics. This feature has practical importance since recent work [44] has shown that certain constellation shaping tends to make the channel input nearly Gaussian.

5 SCD ESTIMATION AND SIMULATION

5.1 Computing the SCD

The computation of the SCD can be conveniently accomplished from discrete-time samples of the channel output $x(t)$,

$$x(n\Delta) = \sum_{k=-\infty}^{\infty} s_k h(n\Delta - kT + t_0) + w(n\Delta), \qquad s_k \in \mathcal{A}, \tag{32}$$

where Δ is the sampling increment. This is a discrete-time cyclostationary process when T/Δ is an integer greater than unity. Its cyclic autocorrelation function and SCD can be estimated using the estimators

$$\widehat{R}_x^\alpha(l\Delta) = \frac{1}{2N+1} \sum_{n=-N}^{N} x(n\Delta + l\Delta)x^*(n\Delta)\exp(-j2\pi\alpha(n + l/2)\Delta) \tag{33}$$

$$\widehat{S}_x^\alpha(j\omega) = \sum_{l=-M}^{M} \widehat{R}_x^\alpha(l\Delta)e^{-j\omega l\Delta} \tag{34}$$

where $N >> T/\Delta$ and $N >> M$ must be satisfied sufficiently strongly for reliability (low variance) and M must be sufficiently large for spectral resolution (low bias) [48]. The sampling interval Δ must be sufficiently small to avoid spectral aliasing (see [46]). Once $\widehat{S_x^\alpha}(j\omega)$ is obtained, a parametric model $S_x^\alpha(j\omega|\,\theta)$ with known orders can be fitted to $\widehat{S_x^\alpha}(j\omega)$ using the method of least squares (or some other method) for $\alpha = 0$ and $\alpha = 1/T$. The resulting parameter vector θ can then be used for pole/zero channel identification.

Several FFT-based smoothing methods for computing the SCD that are computationally more efficient than the direct method used here (34) are available. The details of these algorithms are given in [48, 49].

5.2 Simulation Example

We simulated a noiseless binary phase-shift-keyed (BPSK) ($s_k = \pm 1$) signal in which the data rate was 1 Kbits/s. The linear channel was a mixed phase channel with four poles and two zeros, one of which was in the right-half plane. The channel impulse response is given by

$$\begin{aligned} h(t) = {} & 10e^{-900t}\sin(1200t + 2.1475) \\ & -12.8151e^{-1200t}\sin(900t + 2.4286), \quad t \geq 0. \end{aligned} \tag{35}$$

The channel output was sampled at 16 kHz. Stationary white Gaussian noise was also added to the channel output. Channel identification results were obtained for 1024, 256, and 64 input symbols, respectively, corresponding to the use of $N = 8192, 2048$, and 512 in the estimator (33). $M = 16$ was adopted in the SCD estimator (34) for all three cases. The nonlinear least square fit was accomplished through gradient descent search initialized with a linear least square fit.

In Fig. 2, the actual channel frequency response $H(j\omega)$ is shown indicating its lowpass feature. The true channel impulse response $h(t)$ is shown in Fig. 3 along with the channel estimates based on 256 symbols (4096 data samples) and channel output signal-to-noise ratio (SNR) of 20 dB. The comparison of identification based on different numbers of data samples and SNRs is made by computing the normalized mean squared error (NMSE) of the channel response estimate $\widehat{h}(t)$,

$$\text{NMSE} = \frac{E\{||\widehat{h}(t) - h(t)||^2\}}{||h(t)||^2}, \tag{36}$$

where the average E is over 50 simulation runs. The comparison of NMSE is given in Fig. 4. Although asymptotically the stationary noise should not affect channel identification, the effect of strong noise (SNR = 10 dB) is evident for finite data length. Overall, good identification results are obtained from cyclostationary statistics even with only 256 input symbols and an SNR of 20 dB. Such a short data set would be insufficient for most identification methods based on higher-order statistics.

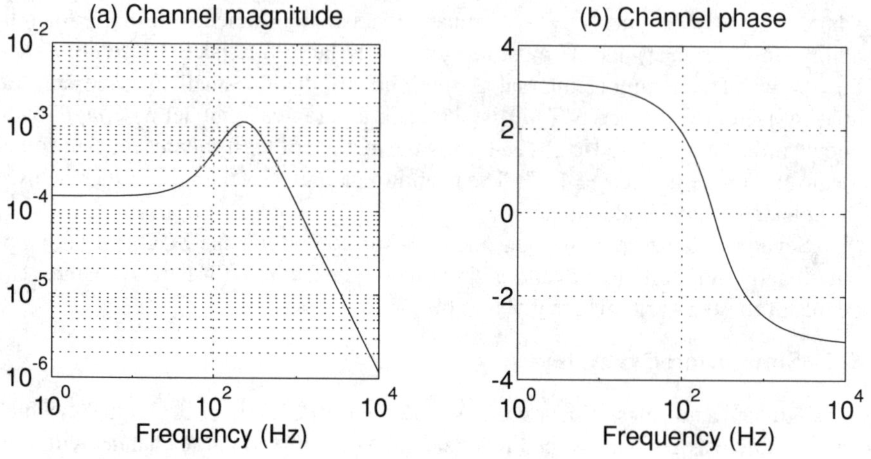

Figure 2: True frequency response of the channel used in the simulation.

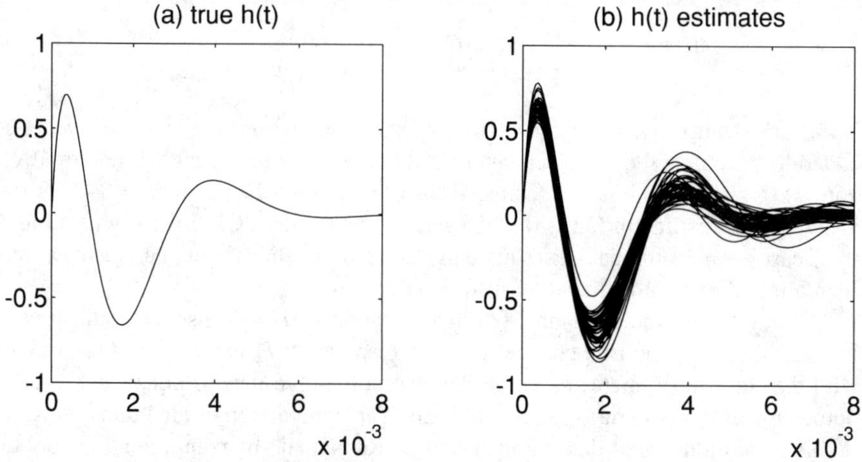

Figure 3: True channel impulse response and estimated channel impulse response based on 256 input symbols with 20 dB SNR.

6 CONCLUSIONS

The cyclostationarity of the QAM channel output can be used for blind identification of unknown channels. By merely using the second-order cyclic statistics or cyclic spectra of the channel output, the phase as well as the magnitude of a class of transfer functions for lowpass rational channels can be identified. The potential advantage of using second-order statistics for blind equalization is the significant improvement in convergence rate compared to algorithms based on higher-order statistics. Although the identification procedure presented here is used only to illustrate the identifiability of rationally parameterized channels and may not be efficient, the identifiability result

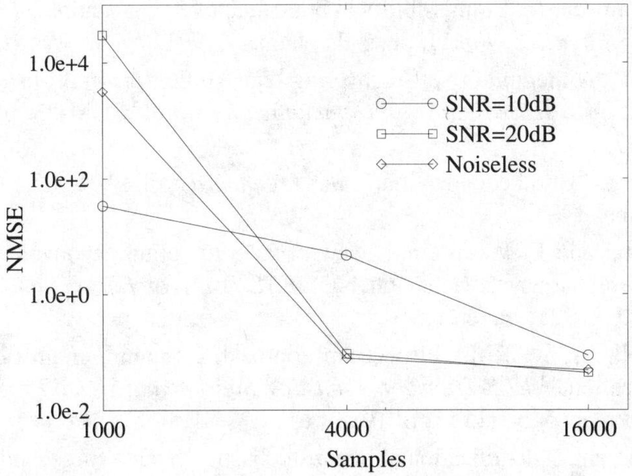

Figure 4: Normalized mean squared error between the true channel and the identified channel responses with i.i.d. binary input and stationary white Gaussian noise.

may lead to future development of highly effective and fast algorithms based on second-order statistics only. To this end, the recent results of Tong, Xu, and Kailath [2] are very encouraging. More work is needed to develop effective channel identification and equalization systems utilizing cyclostationarity, and to extensively show actual advantages for algorithms based on second-order cyclic statistics in achieving fast and global convergence of blind adaptive equalizers. Finally, it is expected that schemes combining cyclostationary higher-order statistics might have more potential for improving the speed and accuracy of blind channel equalizers.

REFERENCES

[1] W. A. Gardner, "A new method of channel identification," *IEEE Trans. on Communications*, vol. COM-39, pp. 813–817, 1991.

[2] L. Tong, G. Xu, and T. Kailath, "A new approach to blind identification and equalization of multipath channels," in *Proc. 25th Asilomar Conf. Signals, Systems, and Computers*, pp. 856–860, November 1991.

[3] A. Benveniste, M. Goursat, and G. Ruget, "Robust identification of a nonminimum phase system: Blind adjustment of a linear equalizer in data communications," *IEEE Trans. on Automatic Control*, vol. AC-25, pp. 385–399, June 1980.

[4] D. N. Godard, "Self-recovering equalization and carrier tracking in two-dimensional data communication systems," *IEEE Trans. on Communications*, vol. COM-28, pp. 1867–1875, November 1980.

[5] A. Benveniste and M. Goursat, "Blind equalizers," *IEEE Trans. on Communications*, vol. COM-32, pp. 871–882, August 1982.

[6] C. R. Johnson, Jr, "Admissibility in blind adaptive equalization," *IEEE Control Systems Magazine*, vol. 11, pp. 3–15, January 1991.

[7] Y. Sato, "A method of self-recovering equalization for multi-level amplitude modulation," *IEEE Trans. on Communications*, vol. COM-23, pp. 679–682, June 1975.

[8] S. Bellini, "Blind equalization," *Alta Frequenza*, vol. LVII-N.7, pp. 445–450, September 1988.

[9] O. Shalvi and E. Weinstein, "New criteria for blind deconvolution of non-minimum phase systems (channels)," *IEEE Trans. on Information Theory*, vol. 36, pp. 312–321, March 1990.

[10] C. L. Nikias, "ARMA bispectrum approach to nonminimum phase system identification," *IEEE Trans. on Acoustics, Speech, and Signal Processing*, vol. ASSP-36, pp. 513–525, April 1988.

[11] J. K. Tugnait, "Identification of linear stochastic systems via second and fourth-order cumulant matching," *IEEE Trans. Information Theory*, vol. IT-33, pp. 393–407, May 1987.

[12] B. Friedlander and B. Porat, "Asymptotically optimal estimation of MA and ARMA parameters of non-Gaussian process from high-order statistics," *IEEE Trans. on Automatic Control*, vol. COM-35, pp. 27–36, January 1990.

[13] D. Hatzinakos and C. L. Nikias, "Blind equalization using a tricepstrum based algorithm," *IEEE Trans. on Communications*, vol. 39, pp. 669–682, May 1991.

[14] R. W. Lucky, "Automatic equalization for digital communication," *Bell Syst. Tech. Journal*, vol. 44, pp. 547–588, April 1965.

[15] J. R. Treichler and M. G. Larimore, "New processing techniques based on the constant modulus adaptive algorithm," *IEEE Trans. on Acoustics, Speech, and Signal Processing*, vol. ASSP-33, pp. 420–431, April 1985.

[16] J. R. Treichler and B. G. Agee, "A new approach to multipath correction of constant modulus signals," *IEEE Trans. on Acoustics, Speech, and Signal Processing*, vol. ASSP-31, pp. 349–472, April 1983.

[17] G. Picchi and G. Prati, "Blind equalization and carrier recovery using a 'stop-and-go' decision-directed algorithm," *IEEE Trans. on Communications*, vol. COM-35, pp. 877–887, September 1987.

[18] R. Godfrey and F. Rocca, "Zero memory non-linear deconvolution," *Geophysical Prospecting*, vol. 29, pp. 189–228, 1981.

[19] W. T. Rupprecht, "Analysis of the mean-absolute-deviation-error for adaptive equalizers in data transmission systems," in *Proc. IASTED Int. Symp. Applied Signal Processing and Signal Filtering*, pp. 99–102, Paris, 1985.

[20] R. A. Kennedy and Z. Ding, "Blind adaptive equalizers for QAM communication systems based on convex cost functions," *Optical Engineering*, vol. 31, pp. 1189–1199, June 1992.

[21] V. Weerackody, S. A. Kassam, and K. R. Laker, "Convergence analysis of an algorithm for blind equalization," *IEEE Trans. on Communications*, vol. COM-39, pp. 856–865, June 1991.

[22] K. Giridhar, J. J. Shynk, and R. A. Iltis, "Bayesian/decision-feedback algorithm for blind adaptive equalization," *Optical Engineering*, vol. 31, pp. 1211–1223, June 1992.

[23] N. Seshadri, "Joint data and channel estimation using fast blind trellis search techniques," in *Proc. IEEE Global Telecom. Conf.*, pp. 1659–1663, San Diego, CA, 1990.

[24] M. H. Meyers, "Robust control of decision directed loops," in *Proc. IEEE Int. Conf. on Communications*, pp. 1030–1036, Boston, MA, May 1989.

[25] E. Zervas, J. Proakis, and V. Eyuboglu, "A quantized channel approach to blind equalization," in *Proc. 1992 Int. Conf. on Communications*, pp. 1539–1543, Chicago, June 1992.

[26] Section on Adaptive Signal Processing, *SPIE Optical Engineering*, June 1992.

[27] D. Donoho, "On minimum entropy deconvolution," in *Applied Time Series Analysis II*, pp. 565–608. Academic Press, 1981.

[28] R. A. Wiggins, "Minimum entropy deconvolution," *Geoexploration*, vol. 16, pp. 21–35, 1978.

[29] D. R. Brillinger and M. Rosenblatt, "Computation and interpretation of k-th order spectra," in B. Harris, ed., *Spectral Analysis of Time Series*, New York: Wiley, 1967.

[30] P. I. Kuznetsov, R. L. Stratonovich, and V. I. Tikhonov, "The transmission of certain random functions through linear systems," in P. I. Kuznetsov, ed., *Non-linear Transformations of Stochastic Processes*. Oxford: Pergamon Press, 1965.

[31] P. I. Kuznetsov, R. L. Stratonovich, and V. I. Tikhonov, "The transmission of random functions through non-Linear systems." in P. I. Kuznetsov, ed., *Non-linear Transformations of Stochastic Processes*. Oxford: Pergamon Press, 1965.

[32] K. S. Lii and M. Rosenblatt, "Deconvolution and estimation of transfer function phase and coefficients for non-Gaussian linear processes," *Annals of Statistics*, vol. 10, pp. 1195–1208, 1982.

[33] J. K. Tugnait, "Identification of non-minimum phase linear stochastic systems," *Automatica*, vol. 22, pp. 454–464, 1986.

[34] G. B. Giannakis and J. M. Mendel, "Identification of nonminimum phase systems using higher order statistics," *IEEE Trans. on Acoustics, Speech, and Signal Processing*, vol. 37, pp. 360–377, 1989.

[35] G. B. Giannakis and A. Swami, "On estimating non-causal non-minimum phase ARMA models of non-Gaussian processes," *IEEE Trans. on Acoustics, Speech, and Signal Processing*, vol. 38, pp. 478–495, 1990.

[36] J. M. Mendel and A. Swami, "ARMA parameter estimation using only output cumulants," *IEEE Trans. on Acoustics, Speech, and Signal Processing*, vol. 38, pp. 1257–1265, July 1990.

[37] J. M. Mendel, "Tutorial on higher-order statistics (spectra) in signal processing and system theory: Theoretical results and some applications," *Proc. IEEE*, vol. 79, pp. 277–305, March 1991.

[38] C. L. Nikias and M. R. Raghuveer, "Bispectrum estimation: A digital signal processing framework," *Proc. IEEE*, vol. 75, pp. 869–891, July 1987.

[39] Special issue on higher order statistics in system theory and signal processing. *IEEE Trans. on Automatic Control*, vol. AC-35, January 1990.

[40] Special section on higher order spectral analysis, *IEEE Trans. on Acoustics, Speech, and Signal Processing*, vol. ASSP-38, pp. 1236–1317, July 1990.

[41] Z. Ding, R. A. Kennedy, B. D. O. Anderson, and C. R. Johnson, Jr., "Ill-convergence of Godard blind equalizers in data communications," *IEEE Trans. on Communications*, vol. 39, pp. 1313–1328, September 1991.

[42] S. U. H. Qureshi, "Adaptive equalization," *Proc. IEEE*, vol. 73, pp. 1349–1387, September 1985.

[43] D. R. Brillinger, *Times Series Data Analysis and Theory*, New York: Holt, Rinehart and Winston, Inc., 1975.

[44] E. Zervas, J. Proakis, and V. Eyuboglu, "Effect of constellation shaping on blind equalization," in Simon Haykin, ed., *Adaptive Signal Processing, Proc. SPIE 1565*, pp. 178–187, 1991.

[45] W. A. Gardner, "Exploitation of spectral redundancy in cyclostationary signals," *IEEE Signal Processing Magazine*, vol. 8, pp. 14–36, April 1991.

[46] W. A. Gardner, *Introduction to Random Processes with Applications to Signals and Systems*, 2nd ed., New York: McGraw-Hill, 1990.

[47] W. Rudin, *Real and Complex Analysis*, 3rd ed., New York: McGraw-Hill, 1987.

[48] W. A. Gardner, *Statistical Spectral Analysis: A Nonprobabilistic Theory*, Englewood Cliffs, NJ: Prentice-Hall, 1988.

[49] R. S. Roberts, W. A. Brown, and H. H. Loomis, Jr, "Computationally efficient algorithms for cyclic spectral analysis," *IEEE Signal Processing Magazine*, vol. 8, pp. 38–49, April 1991.

Blind Channel Identification and Equalization Using Spectral Correlation Measurements, Part II: A Time-Domain Approach

Lang Tong
Department of Electrical and Systems Engineering
University of Connecticut
U-157, Storrs, CT 06269-3157

Guanghan Xu
Department of Electrical and Computer Engineering
The University of Texas at Austin
Austin, TX 78712

Thomas Kailath
Hitachi America Professor of Engineering
Stanford University
Stanford, CA 94305

This research was supported in part by the SDIO/IST Program managed by the Army Research Office under Grant DAAH04-93-G-0029, in part by the Advanced Research Projects Agency of the Department of Defense monitored by the Air Force Office of Scientific Research under Contract F49620-93-1-0085, and in part by Joint Service Program at Stanford University (US Army, US Navy, US Air Force) under contract DAAL03-91-C-0011.

Abstract

A time-domain approach to blind channel identification and equalization is described in this article. Based on the second-order cyclostationary statistics, the time-domain approach provides a faster convergence rate than most techniques based on higher-order statistics. An explicit formula for the identification of possibly nonminimum phase FIR channels is given. Such an approach can be applied to the identification and equalization of multipath channels using an array of receivers. A connection between the time-domain approach and the frequency-domain approach presented in Article 4 is also explained.

1 INTRODUCTION

The frequency-domain approach of Article 4 provides insight into some new blind channel identification and equalization methods. However, a time-domain approach [10] is perhaps simpler as far as the implementation is concerned. Here we present an overview of this method and explore its relationship to frequency-domain concepts.

The main features of the time-domain approach can be summarized as follows:

1. The algorithm provides an explicit identification of possibly *nonminimum phase* finite impulse response (FIR) channels.
2. The algorithm is insensitive to the uncertainties associated with timing recovery. This is achieved by exploiting the signal subspace structure and by sampling the received signal at a rate higher than the baud rate, an idea perhaps first used in fractionally spaced equalization [14].
3. The algorithm can be used to initialize various adaptive schemes, and the equalized output can be used to facilitate decision feedback adaptation. The identified channel can be used to implement a maximum-likelihood sequence estimator [2] to further reduce intersymbol interference (ISI).
4. The algorithm relies on only the second-order statistics of the received signal. Therefore, it requires fewer symbols than most other schemes suggested to date which rely on higher-order statistics. This property also implies that the algorithm can be used to estimate rapidly varying channels.
5. There is no restriction imposed on the probability distribution of the source symbols. The random source can be real or complex, continuous or discrete, or even Gaussian.
6. The algorithm can be extended easily to incorporate multiple receivers. This feature is most attractive for wireless communication where spatial diversity is particularly important.

2 EXPLOITATION OF CYCLOSTATIONARITY IN THE TIME DOMAIN

2.1 A Time-Domain Formulation

For time-invariant communication channels, a discrete-time baseband model of a digital QAM (quadrature-amplitude-modulation) system is given by

$$r(t) = \sum_{k=-\infty}^{\infty} s_k h(t - kT), \quad x(t) = r(t) + w(t), \tag{1}$$

where $x(t)$ is the received signal; $\{s_k\}$ is the sequence of information symbols; $h(\cdot)$ is the discrete-time "composite" channel impulse response including shaping filters at the transmitters, channel dispersion, and receiving filters; T is the symbol interval, and $w(\cdot)$ is an additive noise process.

The objective of blind channel identification is to identify $h(\cdot)$ given only the received signal $x(\cdot)$ and some a priori information about the transmitted signal. Once such identification is achieved, the estimation of the information symbols $\{s_k\}$ becomes more or less a solved problem, and various channel equalization and symbol-sequence estimation techniques can be applied.

The following is assumed throughout the presentation:

A1: The symbol interval T is a known integer and $T > 1$.

A2: The channel has a finite impulse response (FIR).

A3: $E(s_k s_l^*) = \delta(k - l)$.

A4: $w(\cdot)$ is zero mean, uncorrelated with $\{s_k\}$, and
$E(w(t_1)w^*(t_2)) = \sigma^2 \delta(t_1 - t_2)$.

As a general notational convention, symbols for matrices (in capital letters) and vectors are in boldface; the symbols $(\cdot)^H$, $(\cdot)^t$, $(\cdot)^*$, and $(\cdot)^\dagger$ stand for hermitian transpose, transpose, complex conjugate, and the Moore-Penrose generalized inverse, respectively; the symbols $\mathbf{I}$ and $\mathbf{0}$ stand for the identity and zero matrices with appropriate dimensions, respectively; and $E(\cdot)$ denotes the probabilistic expectation.

2.2 A Stationary Representation of Cyclostationary Processes

One of the important properties of a cyclostationary process is that it can be represented by a set of jointly stationary processes. Several stationary representations were derived by Gardner and Franks [4], the simplest of which is the special case of the translation series representation (TSR) that is obtained by sampling the received signal. Such a representation leads naturally to a time-invariant vector linear regression model of the channel, as we shall now explain.

As illustrated by Fig. 1, an important observation is that, under assumption A2, the signal space of the observation restricted to any finite time interval I (referred to as an *observation interval*) is a finite dimensional linear space spanned by a finite number of time-shifted copies of $h(\cdot)$. As shown in Fig. 1, consider the received (noise-free) signal $r(\cdot)$ in an observation interval $I_0 = (t_0, t_0 + L)$ for some finite L. The signal space is spanned by $h(t - K_0T), \ldots, h(t - (K_0 + (d - 1))T)$ with the $h(t - iT)$ being restricted to $(t_0, t_0 + L)$. It is then not surprising that this observation leads to a linear regression model of the channel.

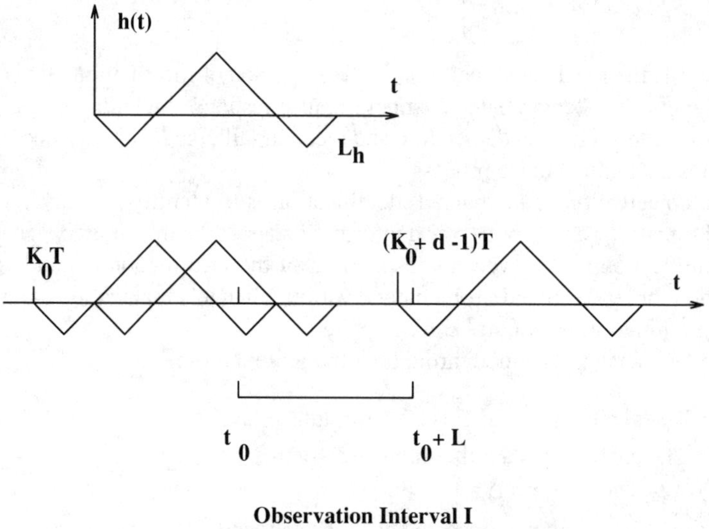

Figure 1: Basis functions in a sampling interval.

If an $m =$ vector $\mathbf{x}(k)$ is formed by

$$\mathbf{x}(k) = [x(kT), x(kT - 1), \cdots, x(kT - m + 1)]^t, \tag{2}$$

then the vector $\mathbf{x}(k)$ is related to a vector of input symbols $\mathbf{s}(k)$ by a memoryless time-invariant matrix filter $\mathbf{H}$ of dimension $m \times d$ (plus noise):

$$\mathbf{x}(k) = \mathbf{H}\,\mathbf{s}(k) + \mathbf{w}(k), \tag{3}$$

where

$$\mathbf{s}(k) = [s_k, s_{k-1}, \cdots, s_{k-d+1}]^t, \tag{4}$$

$$\mathbf{w}(k) = [w(kT), w(kT - 1), \cdots, w(kT - m + 1)]^t. \tag{5}$$

Each row of $\mathbf{H}$ corresponds to a baud-rate sampled channel impulse response, and the columns of $\mathbf{H}$ are related directly to the basis in Fig. 1. The identification of $\mathbf{H}$ leads to the identification of the impulse response $h(\cdot)$.

3 CHANNEL IDENTIFICATION

To simplify the presentation of the key ideas for channel identifiability, we first ignore the noise in Section 3.1. Similar results can be obtained when the noise statistics (second-order) are known by using the techniques to be described in Section 3.2.

3.1 Channel Identifiability

Following the development in the previous section, the *blind* channel identification problem can be restated as follows:

Consider a vector process $\mathbf{r}(i)$ *obtained from a linear model*

$$\mathbf{r}(i) = \mathbf{H}\mathbf{s}(i), i = 0, 1, \ldots \tag{6}$$

along with the following constraints:

1. $\mathbf{H}$ *is an* $m \times d$ *complex matrix of full column rank.*
2. $\mathbf{s}(i)$ *is a zero-mean stationary vector process with the matrix of autocorrelation functions*
$$\mathbf{R}_{\mathbf{s}}(k) = E\{\mathbf{s}(i)\mathbf{s}^H(i-k)\} \tag{7}$$
of the form
$$\mathbf{R}_{\mathbf{s}}(k) = \begin{cases} \mathbf{J}^k & k \geq 0 \\ (\mathbf{J}^H)^{|k|} & k < 0, \end{cases} \tag{8}$$
where $\mathbf{J}$ *is the* $d \times d$ *shifting matrix*
$$\mathbf{J} = \begin{pmatrix} 0 & 0 \ldots & 0 & 0 \\ 1 & 0 \ldots & 0 & 0 \\ \vdots & \ddots & \vdots & \vdots \\ 0 & 0 \ldots & 1 & 0 \end{pmatrix}. \tag{9}$$

The objectives of blind channel identification and equalization are to identify $\mathbf{H}$ *(channel identification) and to estimate* $\mathbf{s}(i)$ *from* $\mathbf{x}(i)$ *(channel equalization).*

Remarks

1. In most practical cases, the rank constraint on $\mathbf{H}$ can be satisfied by sufficiently fast sampling combined with choosing a sufficiently large m, or by designing the pulse shaping filter in the transmitter properly. As we shall see in Section 3.3, unless the channel is unidentifiable from second-order cyclostationary statistics, the full-rank condition can always be satisfied.
2. The idea of sampling the received signal at a rate higher than the symbol rate has been used in fractionally spaced equalization (FSE), and is known to provide robustness to timing recovery uncertainties [14].
3. Equations (8–9) are a direct consequence of A3 and (4).

In what follows, an algebraic approach is used to establish channel identifiability and an identification algorithm. To provide some insight into the derivation, it is noted that because of the special forward-shift structure of the correlation matrices of the source, the rank of $E\{\mathbf{r}(t)\mathbf{r}^H(t-k)\} = \mathbf{H}\,\mathbf{R_s}(k)\mathbf{H}^H$ decreases as k increases. On the other hand, the range space of $\mathbf{R_r}(k) = E\{\mathbf{r}(t)\mathbf{r}^H(t-k)\}$ is spanned by the columns of the channel parameter matrix $\mathbf{H}$. It is the change in the rank of $\mathbf{R_r}(k)$ that provides the information for the identification of the column vectors of $\mathbf{H}$ and, consequently, the identification of the channel.

The following theorem given in [10] provides an identifiability result and suggests an explicit identification algorithm.

Theorem 1 *Let $\mathbf{R_r}(0)$ have the following singular value decomposition (SVD),*

$$\mathbf{U}^H\mathbf{R_r}(0)\mathbf{U} = \operatorname{diag}(\sigma_1^2, \ldots, \sigma_d^2, 0, \ldots, 0). \tag{10}$$

Let $\mathbf{u}_i$ denote the ith column of $\mathbf{U}$, and let

$$\mathbf{U_s} = [\mathbf{u}_1, \ldots, \mathbf{u}_d],\ \ \mathbf{\Sigma} = \operatorname{diag}(\sigma_1, \ldots, \sigma_d),\ \ \mathbf{F} = \mathbf{\Sigma}^{-1}\mathbf{U}_\mathbf{s}^H. \tag{11}$$

Let $\mathbf{R} = \mathbf{F}\,\mathbf{R_r}(1)\mathbf{F}^H$ have an SVD of the form

$$[\,\mathbf{y}_1, \ldots, \mathbf{y}_d]^H\mathbf{R}[\mathbf{z}_1, \ldots, \mathbf{z}_d] = \operatorname{diag}(\gamma_1^2, \ldots, \gamma_d^2). \tag{12}$$

Then there exists a real phase ϕ such that

$$\mathbf{H} = \mathbf{U_s}\mathbf{\Sigma}\mathbf{Q}e^{j\phi}, \tag{13}$$

where

$$\mathbf{Q} = [\,\mathbf{y}_d, \mathbf{R}\mathbf{y}_d, \ldots, \mathbf{R}^{(\mathbf{d-1})}\mathbf{y}_d], \tag{14}$$

or, equivalently,

$$\mathbf{Q} = [(\mathbf{R}^\dagger)^{(d-1)}\mathbf{z}_d, (\mathbf{R}^\dagger)^{(d-2)}\mathbf{z}_d, \ldots, \mathbf{z}_d]. \tag{15}$$

The proof of this theorem, given in [10], may be obtained by first assuming that $\mathbf{H}$ is orthogonal. We then have

$$\mathbf{R_r}(1) = \mathbf{H}\,\mathbf{J}\,\mathbf{H}^H. \tag{16}$$

When $\mathbf{H}$ is orthogonal, the right-hand side of the above equation is in fact a Jordan decomposition of $\mathbf{R_r}(1)$. This leads to a recursive relation among the column vectors of $\mathbf{H}$. The first column vector of $\mathbf{H}$ turns out to be the left singular vector associated with the (unique) zero singular value, which results in the identification of $\mathbf{H}$. The matrix $\mathbf{H}$ is of course not orthogonal in general. However, a commonly used transform defined by $\mathbf{F}$ in (11) can be applied to orthogonalize $\mathbf{H}$.

3.2 Algorithm Implementation with Noisy Data

Theorem 1 provides the essential parts for proposing a blind channel identification algorithm. Although the previous development was based on a noise-free model, it is not hard to think of ways of handling additive white noise. In particular, the ideas now widely used in sensor array processing (see e.g., [15]) can be used. For the vectorized process $\mathbf{x}(i)$ satisfying

$$\mathbf{x}(i) = \mathbf{H}\mathbf{s}(i) + \mathbf{w}(i), \tag{17}$$

the correlation matrix $\mathbf{R}_{\mathbf{x}}(k)$ is given by

$$\mathbf{R}_{\mathbf{x}}(k) = \mathbf{H}\mathbf{R}_{\mathbf{s}}(k)\mathbf{H}^H + \mathbf{R}_{\mathbf{w}}(k). \tag{18}$$

Under the assumption of white noise, the noise correlation matrix $\mathbf{R}_{\mathbf{w}}(k)$, for $k > 0$, has the form

$$\mathbf{R}_{\mathbf{w}}(k) = E\{\mathbf{w}(i)\mathbf{w}^H(i-k)\} \tag{19}$$

$$= \sigma^2 \mathbf{J}^{kT}. \tag{20}$$

Note that the superscript T in (20) denotes the baud period. Although neither the noise covariance, nor the signal space dimension d is known a priori, they can be obtained from the data covariance matrix $\mathbf{R}_{\mathbf{x}}(0)$. That is, it can be shown that the SVD of $\mathbf{R}_{\mathbf{x}}(0)$ has the form

$$\mathbf{U}^H\mathbf{R}_{\mathbf{x}}(0)\mathbf{U} = \text{diag}(\lambda_1 + \sigma^2, \cdots, \lambda_d + \sigma^2, \sigma^2, \cdots, \sigma^2), \tag{21}$$

where $\lambda_1 \geq \lambda_2 \geq \cdots \geq \lambda_d > 0$. Therefore, both σ^2 and d can in principle be obtained by determining the most significant singular values of $\mathbf{R}_{\mathbf{x}}(0)$. In practice, a threshold test can be employed to determine d, and then to estimate σ^2 from the singular values of the estimated data covariance matrix. Readers are referred to [5] and [16] for suitable methods. Once the noise covariance σ^2 is determined, the identification procedure suggested in Theorem 1 can be easily extended to handle noisy data by subtracting the corresponding noise correlation matrices from the observation correlation matrices.

The identification algorithm is first outlined and some important technical points of the algorithm are addressed.

3.2.1 A Blind Channel Identification and Equalization Algorithm

The proposed algorithm consists of the following seven steps:

1. Select a sufficiently large m and form the observation vector $\mathbf{x}(i) = [x(iT), x(iT-1), \cdots, x(iT-m+1)]^t$.
2. Estimate $\mathbf{R}_{\mathbf{x}}(0)$ and $\mathbf{R}_{\mathbf{x}}(1)$ from $\mathbf{x}(i)$ by time averaging,

$$\hat{\mathbf{R}}_{\mathbf{x}}(0) = \frac{1}{N}\sum_{i=1}^{N}\mathbf{x}(i)\mathbf{x}^H(i), \tag{22}$$

$$\hat{\mathbf{R}}_{\mathbf{x}}(1) = \frac{1}{N}\sum_{i=1}^{N}\mathbf{x}(i)\mathbf{x}^H(i-1). \tag{23}$$

3. From $\hat{\mathbf{R}}_{\mathbf{x}}(0)$, use the SVD to estimate the noise covariance σ^2 and the dimension d of the signal space.
4. Compute the SVD of $\mathbf{R}_0 = \hat{\mathbf{R}}_{\mathbf{x}}(0) - \hat{\sigma}^2\mathbf{I}$ and form $\mathbf{U}_s$ which consists of the singular vectors associated with the d largest singular values, $\mathbf{\Sigma}$ which consists of the positive square roots of the d largest singular values, and then form $\mathbf{F} = \mathbf{\Sigma}^{-1}\mathbf{U}_s^H$.
5. Compute the SVD of $\mathbf{R} \triangleq \mathbf{F}(\hat{\mathbf{R}}_{\mathbf{x}}(1) - \hat{\mathbf{R}}_{\mathbf{w}}(1))\mathbf{F}^H$, where
$$\hat{\mathbf{R}}_{\mathbf{w}}(1) = \hat{\sigma}^2\mathbf{J}^T. \tag{24}$$
Let $\mathbf{y}_d$ and $\mathbf{z}_d$ denote the left and right singular vectors corresponding to the smallest singular value of $\mathbf{R}$.
6. Form an estimate of $\mathbf{H}$ (and consequently $h(\cdot)$ if necessary) using the formula
$$\hat{\mathbf{H}} = \mathbf{U}_s\mathbf{\Sigma}\mathbf{Q}, \tag{25}$$
where
$$\mathbf{Q} = [\mathbf{y}_d, \mathbf{R}\mathbf{y}_d, \ldots, \mathbf{R}^{(d-1)}\mathbf{y}_d] \tag{26}$$
or, equivalently,
$$\mathbf{Q} = [(\mathbf{R}^\dagger)^{(d-1)}\mathbf{z}_d, (\mathbf{R}^\dagger)^{(d-2)}\mathbf{z}_d, \ldots, \mathbf{z}_d] \tag{27}$$
or a certain combination of the above. For example, one can reduce the error propagation caused by $\mathbf{R}^k$ and $(\mathbf{R}^\dagger)^k$ by using (26) for the first $d/2$ columns of $\mathbf{Q}$ and using (27) for the last $d/2$ columns of $\mathbf{Q}$.
7. Extract the information symbols using the estimator
$$\hat{\mathbf{s}}(i) = \hat{\mathbf{H}}^\dagger\mathbf{x}(i) \tag{28}$$
$$= \mathbf{Q}^H\mathbf{F}\mathbf{x}(i), \tag{29}$$
or by implementing any equalizer or maximum-likelihood estimation scheme that is based on the estimated channel.

3.3 Connections Between the Time-Domain and Frequency-Domain Approaches

Insight can be gained by examining the connection between the time-domain identification results presented here and a frequency-domain solution. Recall from Article 4 that the cyclic spectrum[1] (or spectral correlation density) $S_x^k(j\omega)$ of the observation is related to the channel frequency response $H(j\omega)$ by
$$S_x^{k/T}(j\omega) = H(j\omega)H^*(j\omega - 2\pi k/T) + \sigma^2\delta(k). \tag{30}$$

---

[1]Note that the definition of the cyclic spectrum used here differs from the original definition in [3] which is used in Article 4 by a shift in the frequency in the amount of $\frac{\pi}{T}$ radian/sec. Also, the cycle frequency parameter in [3] and Article 4 has units of Hertz.

For notational convenience, the product in (30) is denoted by

$$\Gamma^k(\omega) = H(j\omega)H^*(j\omega - 2\pi k/T). \tag{31}$$

These functions can be estimated by estimating the spectral correlation densities ($k \neq 0$) and power spectral density ($k = 0$). In the z-domain, this leads to the following identification equation

$$\Gamma^k(z) = H(z)H^*(e^{jk2\pi/T}\frac{1}{z^*}), \quad k = 0, \cdots, T-1. \tag{32}$$

The problem of channel identification is then equivalent to identifying $H(z)$ using $\Gamma^k(z)$, which can be approached by identifying the zeros of $H(z)$ from those of $\Gamma^k(z)$. The following theorem was given in [11].

Theorem 2 *The channel transfer function $H(z)$ is uniquely determined (identified) by $\{\Gamma^k(z)\}$ up to a multiplicative constant if and only if $H(z)$ does not have zeros uniformly spaced around a circle with separation of $2\pi/T$ radians. Moreover, if the channel is identifiable,*

$$\mathcal{Z}(H(z)) = \bigcap_k \mathcal{Z}(\Gamma^k(z)) \tag{33}$$

where $\mathcal{Z}(H(z))$ stands for the set of zeros of $H(z)$.

The above theorem suggests an identification algorithm in the frequency domain. By estimating the spectral correlation density functions (SCD), the zeros of the channel (assumed to be FIR) can be identified by using (33). However, in comparison to the time-domain approach, such an implementation appears to be more complicated and perhaps more sensitive to noise and estimation error.

Recall that the time-domain result was presented under the assumption that the channel impulse response parameter matrix **H** has full column rank. In light of the above frequency-domain result, the full-rank condition turns out to be a necessary and sufficient condition. The following theorem from [6, 13] provides an important link between the time-domain and frequency-domain approaches.

Theorem 3 *$H(z)$ has zeros uniformly spaced around a circle with separation of $2\pi/T$ radians if and only if the channel parameter matrix* **H** *has linearly dependent columns. In other words, the channel is identifiable if and only if the matrix* **H** *has full column rank.*

Two approaches can be used to prove this result. The approach by Li and Ding [6] exploits the root structure of the observation process whereas the one of Tong, Xu, and Kailath exploits the fact that **H** is a Sylvester matrix [9, 13].

3.4 A Simulation Example

In this example, the impulse response of a three-ray multipath channel is a weighted sum of delayed raised cosine pulses. A single such pulse is denoted by $c(t, \alpha)$ where

α is the roll-off factor. In particular,

$$h(t) = [c(t, 0.11) - 0.8c(t - 0.5, 0.11) + 0.4c(t - 3, 0.11)]W_{6T}(t), \quad (34)$$

where W_{6T} is a square window of duration 6 symbol intervals. The resultant impulse response is shown in Fig. 2(a). Among the 23 zeros of the system, there are 21 nonminimum phase zeros. The source symbols were drawn from a 16-QAM signal constellation with a uniform distribution on a rectangular grid.

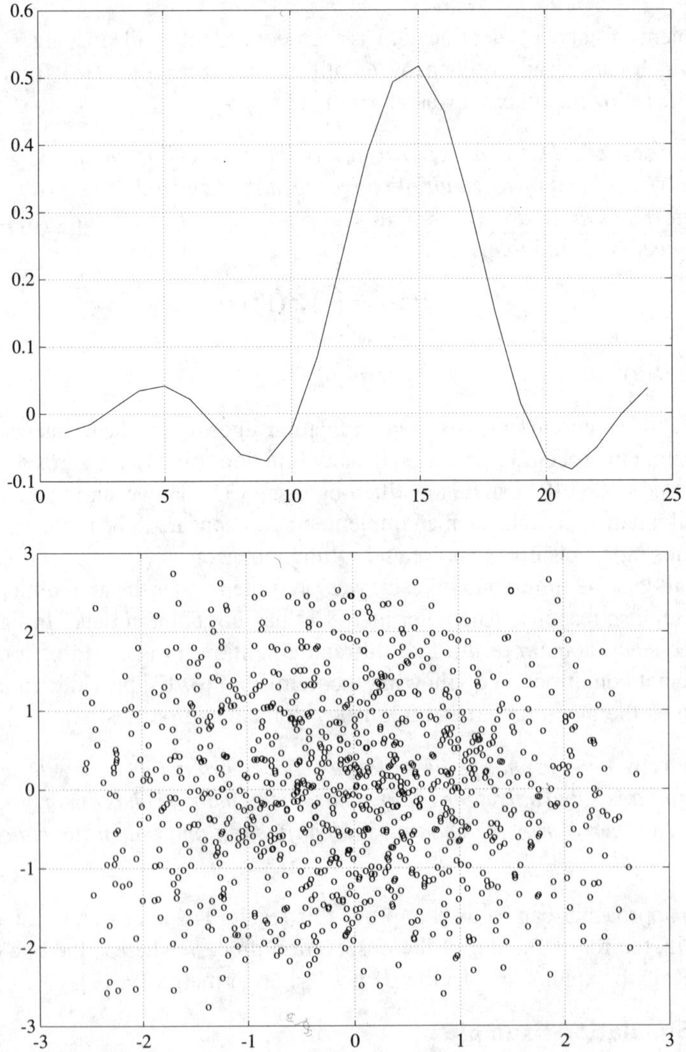

Figure 2: (a) The impulse response of a three-ray multipath channel. (b) Constellation at the output of the unequalized channel at SNR $= 30dB$.

The signal-to-noise ratio (SNR) is defined by

$$\text{SNR} = 20 \log \frac{\| r(\cdot) \|_2}{\| w(\cdot) \|_2} (\text{dB}), \tag{35}$$

where $\| r(\cdot) \|_2$ stands for the l_2 norm of $r(\cdot)$. Figure 2(b) is a plot of 1000 output symbols of the unequalized channel at SNR $= 30$ dB (obtained by sampling the received signal at kT, $k = 1, 2, \ldots$). Clearly, the intersymbol interference is severe and a high error rate is expected.

To evaluate the time-domain approach, a Monte Carlo simulation of 100 independent trials was conducted. The sampling frequency was chosen to be 4 times faster than the baud rate, i.e., $T = 4$, and five bauds of data were used to form the 20×1 complex vector $\mathbf{x}(i)$. The signal subspace dimension was assumed to be known in the simulation ($d = 10$). At each Monte Carlo run, 100 symbols were used to estimate $\mathbf{R}_\mathbf{x}(0)$ and $\mathbf{R}_\mathbf{x}(1)$, and the identification algorithm of Section 3.2 was applied. Figure 3 shows the sample mean of 100 estimates and a scatter plot of the channel estimates. For demonstration purposes, the unknown multiplicative constant factor was removed by scaling the estimated channel. Such scaling is of course not possible in practice, nor is it necessary when differential encoding schemes are used.

To measure the performance of the channel estimator, the normalized root-mean-square error (NRMSE) of the estimator

$$\text{NRMSE} = \frac{1}{\|\mathbf{h}\|} \sqrt{\frac{1}{M} \sum_{i=1}^{M} \|\hat{\mathbf{h}}_i - \mathbf{h}\|^2} \tag{36}$$

is used, where M is the number of Monte Carlo trials (100 in our case), $\mathbf{h}$ is the vector of the channel impulse response, $\hat{\mathbf{h}}_i$ is the estimated channel impulse response vector at the ith trial, and $\|\mathbf{h}\|$ stands for the 2-norm of the vector $\mathbf{h}$. Figure 4(a) shows the NRMSE versus SNR.

The Bit-Error-Rate (BER) was evaluated as a function of SNR. In this case, a binary phase-shift-keyed (BPSK) source was used to estimate the channel. At each Monte Carlo run, the channel parameter matrix $\mathbf{H}$ was first estimated by the proposed algorithm, and the source symbol vector $\mathbf{s}(i)$ was estimated by a (suboptimal) minimum-variance equalizer given by

$$\hat{\mathbf{s}}(i) = \hat{\mathbf{H}}^H (\hat{\mathbf{H}}\hat{\mathbf{H}}^H + \hat{\sigma}^2 \mathbf{I})^{-1} \mathbf{x}(i) = \mathbf{T}\mathbf{x}(i). \tag{37}$$

The fractionally spaced equalizer was defined by the row of $\mathbf{T}$ with minimum norm in order to reduce noise enhancement. The output of this equalizer was followed by a baud-rate sampler and a slicer.

At each Monte Carlo run, a fractionally spaced equalizer was obtained, and the probability of bit-error was computed (analytically). The BER is defined here as the probability of error averaged over 100 Monte Carlo runs. Figure 4(b) shows the effect of noise on the BER. Whereas the performance is excellent at high SNRs, the error rate is high when SNR is below 25 dB. To achieve better performance, a larger sample size will be necessary.

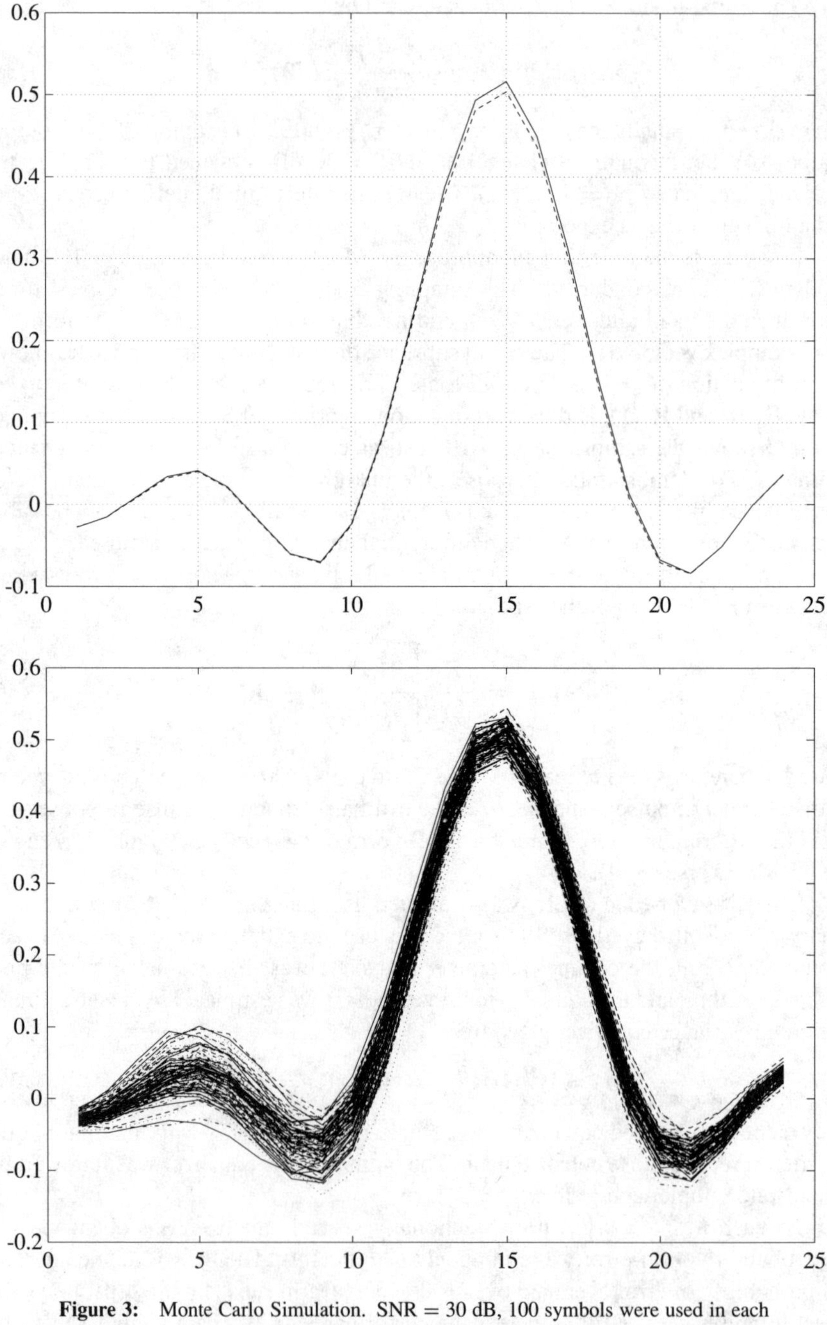

Figure 3: Monte Carlo Simulation. SNR = 30 dB, 100 symbols were used in each estimation and 100 runs were made. (a) The actual channel (solid) and the sample mean (dashed) of 100 channel estimates. (b) Scatter plot of 100 estimates.

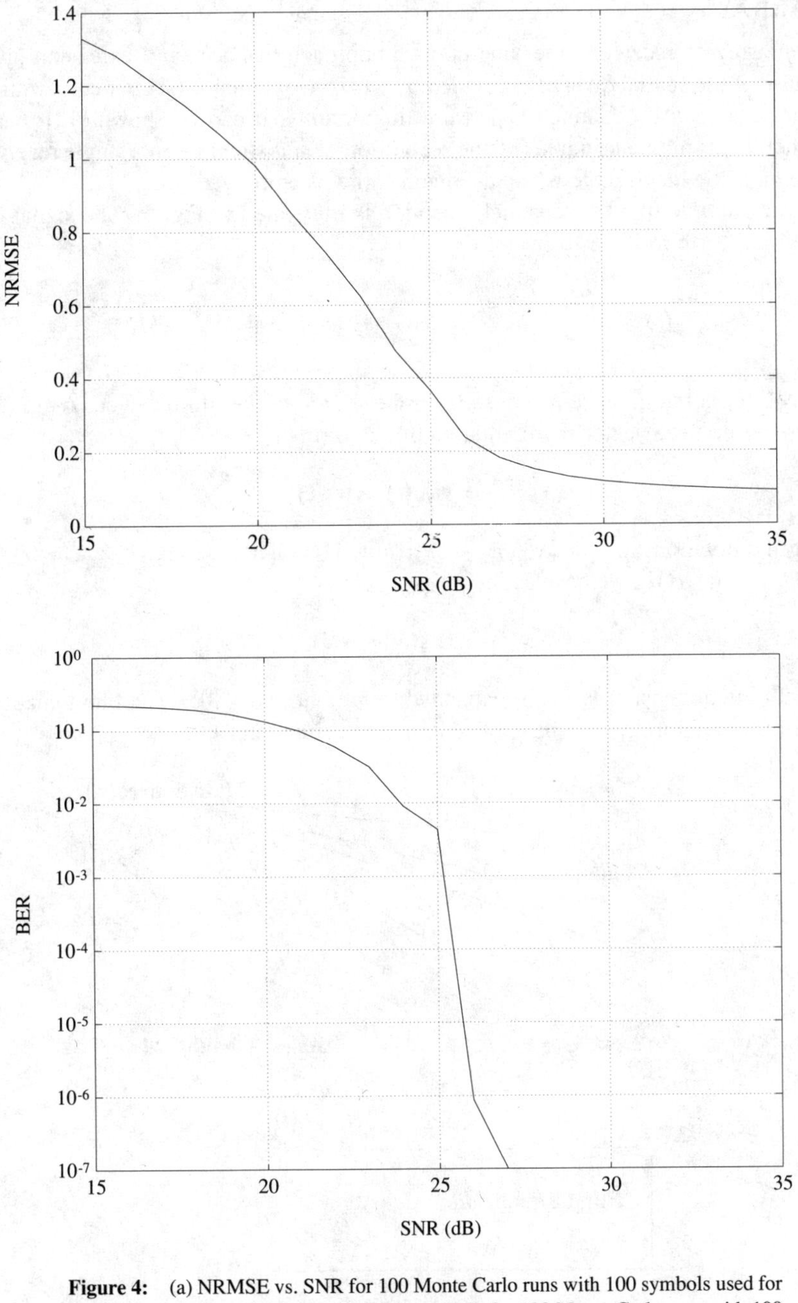

Figure 4: (a) NRMSE vs. SNR for 100 Monte Carlo runs with 100 symbols used for channel estimation in each run. (b) BER vs. SNR for 100 Monte Carlo runs with 100 symbols used for channel estimation in each run.

4 BLIND EQUALIZATION WITH ANTENNA ARRAYS

For an array of receivers, the time-domain approach has a natural extension [12]. Because of the spatial diversity provided by an array of antennas, better equalization performance can be obtained [1]. More importantly, it can be shown [12] that a channel that is not identifiable (by the second-order statistics) when a single receiver is used may be identifiable when an antenna array is employed.

An example of an M-channel reception is illustrated in Fig. 5. The signal received at the jth receiver is given by

$$x_j(t) = \sum_{k=-\infty}^{\infty} s_k h_j(t - kT) + w_j(t), \quad j = 1, \cdots, M, \tag{38}$$

where $h_j(t)$ is the impulse response from the source to the jth receiver. As in (3), each receiver has a vector representation of the form

$$\mathbf{x}_j(i) = \mathbf{H}_j \mathbf{s}(i) + \mathbf{w}_j(i). \tag{39}$$

Using the notation $\mathbf{x}(i) = [\mathbf{x}_1^t(i), \cdots, \mathbf{x}_M^t(i)]^t$, $\mathbf{H} = [\mathbf{H}_1^t, \cdots, \mathbf{H}_M^t]^t$, and $\mathbf{w}(i) = [\mathbf{w}_1^t(i), \cdots, \mathbf{w}_M^t(i)]^t$, we have again

$$\mathbf{x}(i) = \mathbf{H}\,\mathbf{s}(i) + \mathbf{w}(i), \tag{40}$$

but with the dimension m in (3) expanded to mM here in (40). The blind channel identification algorithm can be applied directly to equation (40).

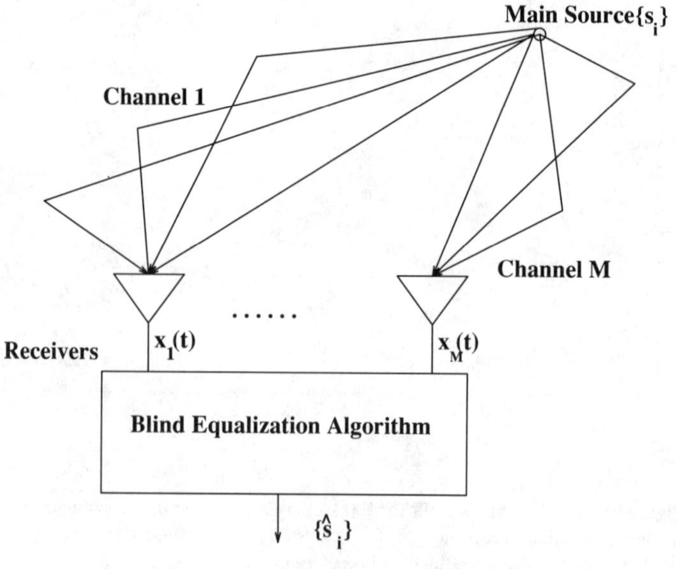

Figure 5: An antenna-array-based blind equalizer.

To illustrate the effect of spatial diversity, we consider a three-antenna three-ray specular multipath channel [1]

$$h_j(t) = \alpha_{1j} g(t) + \alpha_{2j} g(t - \tau_1) + \alpha_{3j} g(t - \tau_2), \ j = 1, 2, 3, \tag{41}$$

where $\tau_1 = 0.75T$, $\tau_2 = 0.875T$, and

$$[\alpha_{ij}] = \begin{pmatrix} 0.3310 - 0.6545i & -1.2125 - 1.7851i & 0.0655 - 0.1734i \\ -0.3081 - 1.1443i & -1.4500 - 0.2900i & -0.1519 - 0.0175i \\ -0.2541 + 0.0529i & -0.2198 - 1.2166i & 0.1321 - 0.2432i \end{pmatrix}. \tag{42}$$

Quadrature-phase-shift-keyed (QPSK) modulation is used in this simulation. Again, $g(\cdot)$ is a raised-cosine pulse with 11% roll-off. The impulse response is truncated after 6 symbol intervals. Figures 6(a,b) show the constellations of the outputs of the unequalized channels associated with the first two receivers (the output of the third is similar). The SNR is 20 dB.

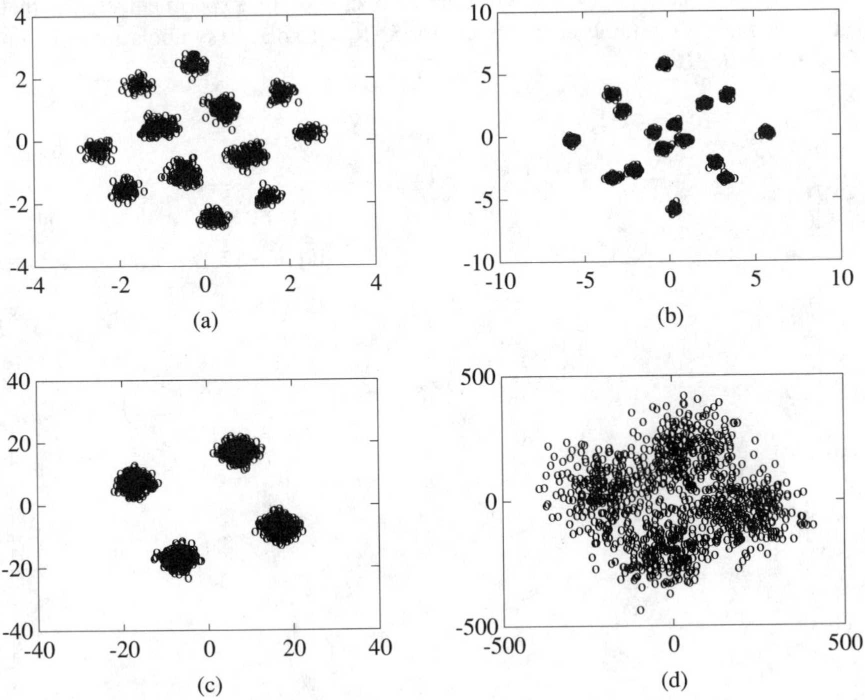

Figure 6: Constellations for SNR = 20 dB with 20 symbols used for channel estimation. (a) Unequalized output of receiver no. 1. (b) Unequalized output of receiver no. 2. (c) Equalized output using three receivers. (d) Equalized output using receiver no. 1.

In applying the blind identification algorithm, we experimented with a simple order determination scheme. The dimension of the signal subspace d was chosen so that the singular value spread is less than 10. In this simulation, the dimension was determined to be $d = 3$ (the actual dimension is 6). Figure 6(c) shows the constellation of the equalized channel output when the data from all three receivers were used. Figure 6(d) shows the constellation of the equalized channel output when only the first receiver is active. The improvement obtained by using three receivers results from the spatial diversity.

A Monte Carlo simulation was conducted to evaluate the performance of the three-channel equalizer. The ISI performance measure used is (from [8])

$$\text{ISI} = \frac{\sum_t |h_e(t)|^2 - \max_t |h_e(t)|^2}{\max_t |h_e(t)|^2}, \tag{43}$$

where $h_e(t)$ is the single composite equalized channel impulse response. The number N of symbols used in the identification process and the SNR were varied. For each N and SNR, 500 Monte Carlo runs were conducted. Figure 7 is a plot of the averaged ISI (43) vs. SNR. Since blind equalization can be switched to decision-directed equalization when ISI is below 0.1 [7], the simulation results of this experiment suggest that, on the average, 20 symbols are sufficient for SNR > 15 dB, 40 symbols are sufficient for SNR > 10 dB.

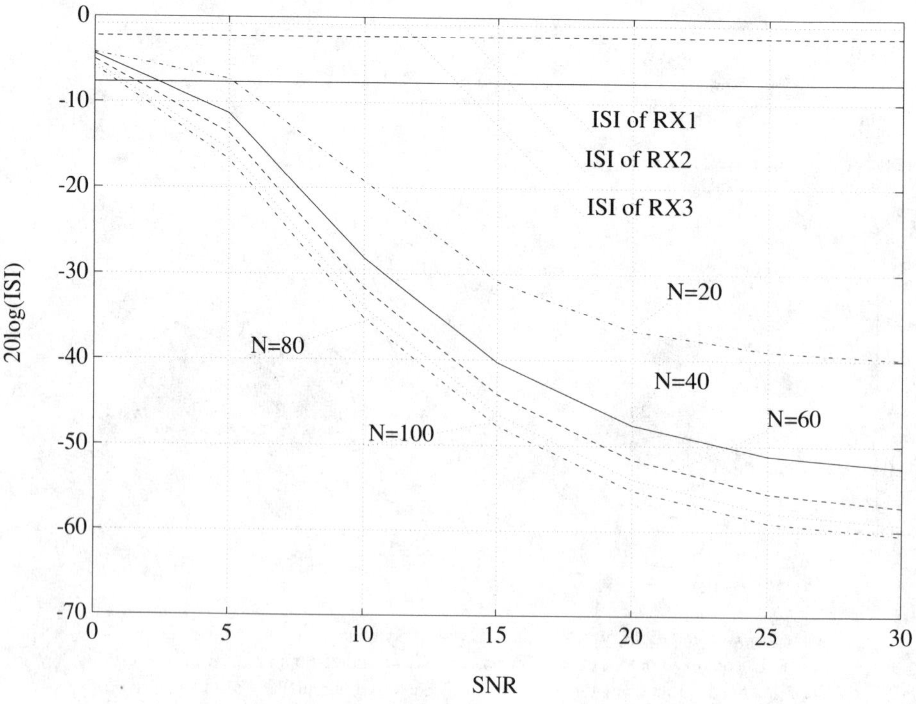

Figure 7: Average ISI vs. SNR for 100 Monte Carlo runs.

5 CONCLUSIONS

Blind equalization is of significant value in many communication problems, and cyclostationarity turns out to hold a key to rapidly converging blind equalization methods. The cyclostationarity properties of communication signals can be exploited in either the frequency domain or the time domain, and this article describes and compares some of the relationships between these two approaches. An explicit algorithm for identification of channels resulting from using either a single-receiver or multiple-receivers is described and the results of initial simulations are very encouraging (see also Chapter 3 in this volume). The multiple-receiver scheme exploits the spatial diversity of an array to enhance the identification performance.

ACKNOWLEDGMENTS

The authors wish to thank Dr. W. A. Gardner for many helpful suggestions on the content and the presentation of both Articles 4 and 5.

REFERENCES

[1] P. Balaban and J. Salz, "Optimum diversity combining and equalization in digital data transmission with applications to cellular mobile radio—Part I: Theoretical considerations," *IEEE Trans. Communications*, vol. 40, pp. 885–894, May 1992.

[2] G. D. Forney, "Maximum-likelihood sequence estimation of digital sequences in the presence of intersymbol interference," *IEEE Trans. Inform. Theory*, vol. IT-18, pp. 363–378, May 1972.

[3] W. A. Gardner, *Statistical Spectral Analysis: A Nonprobabilistic Theory*, Englewood Cliffs, NJ: Prentice-Hall, 1987.

[4] W.A. Gardner and L.E. Franks, "Characterization of cyclostationary random signal processes," *IEEE Trans. Inform. Theory*, vol. IT-21, pp. 4–14, Jan. 1975.

[5] K. Konstantinides and K. Yao, "Statistical analysis of effective singular values in matrix rank determination," *IEEE Trans. ASSP*, vol. 36, pp. 757–763, May 1988.

[6] Y. Li and Z. Ding, "Blind channel identification based on second order cyclostationary statistics," in *Proc. ICASSP93 Conf.*, pp. IV81–IV83, Minneapolis, MN, April 1993.

[7] J. E. Mazo, "Analysis of decision directed equalizer convergence," *Bell Syst. Tech. Jour.*, vol. 59, pp. 1857–1876, Dec 1980.

[8] O. Shalvi and E. Weinstein, "New criteria for blind deconvolution of nonminimum phase systems (channels)," *IEEE Trans. Inform. Theory*, vol. 36, pp. 312–320, March 1990.

[9] L. Tong, G. Xu, B. Hassibi, and T. Kailath, "Blind identification and equalization of multipath channels: A frequency domain approach," Submitted to *IEEE Trans. Inform. Theory*.

[10] L. Tong, G. Xu, and T. Kailath, "Blind identification and equalization based on second-order statistics: A time domain approach," *IEEE Trans. Inform. Theory*. Accepted for publication.

[11] L. Tong, G. Xu, and T. Kailath, "Blind identification and equalization of multipath channels," in *Proc. 1992 Intel. Conf. on Comm.*, pp. 1513–1517, Chicago, IL, June 1992.

[12] L. Tong, G. Xu, and T. Kailath, "Fast blind equalization of multipath channels via antenna arrays," in *Proc. ICASSP'92*, pp. IV272–IV275, Minneapolis, MN, April 1993.

[13] L. Tong, G. Xu, and T. Kailath, "Necessary and sufficient conditions for channel identification based on second-order statistics," in *Proc. 1993 Intl. Symp. Inform. Theory*, p. 188, San Antonio, TX, Jan. 1993.

[14] G. Ungerboeck, "Fractional tap-spacing equalizer and consequences for clock recovery in data modems," *IEEE Trans. Commun.*, vol. COM-24, pp. 856–864, August 1976.

[15] M. Viberg, B. Ottersten, and T. Kailath, "Detection and estimation in sensor arrays using weighted subspace fitting," *IEEE Trans. Signal Processing*, vol. 39, pp. 2436–2449, November 1991.

[16] M. Wax and T. Kailath, "Detection of signals by information theoretic criteria," *IEEE Trans. ASSP*, vol. ASSP-33, pp. 387–392, April 1985.

A Review of Digital Spectral Correlation Analysis: Theory and Implementation

Randy S. Roberts
Los Alamos National Laboratory
Los Alamos, NM 87545

William A. Brown
Mission Research Corporation
2300 Garden Road, Suite 2
Monterey, CA 93940

Herschel H. Loomis, Jr.
Department of Electrical and Computer Engineering
Naval Postgraduate School
Monterey, CA 93943

1 INTRODUCTION

Estimation of the spectral correlation function, also called the cyclic spectrum, is important in a number of signal processing applications. For example, time-difference-of-arrival, signal detection, and modulation recognition techniques based on spectral correlation theory require estimates of the spectral correlation function over some

This work was supported in part by a grant from ESL Inc. with partial matching support from the California State MICRO Program (PI: W. A. Gardner) and was also supported in part by the U. S. Department of the Air Force (PI: H. H. Loomis, Jr.).

region of the bifrequency plane. Some applications demand rapid computation of these quantities, and thus require rapid estimation of the spectral correlation function. If the problem has a priori knowledge of the cycle frequencies of interest, and the number of cycle frequencies is not large, estimation of the spectral correlation function has a computational cost comparable to ordinary spectral analysis. On the other hand, if the cycle frequencies of interest are unknown, then estimation of the spectral correlation function over the entire bifrequency plane can be required. In this case the computational burden can be much greater than it is for ordinary spectral analysis.

The theory and implementation of digital spectral correlation analysis algorithms has been covered in a number of publications. The basic time and frequency smoothing methods of spectral correlation analysis were introduced in [1] and proof of their equivalence was given in [2 and 3]. The additional dimension of cycle frequency, which is absent in ordinary spectral analysis, presents issues peculiar to spectral correlation analysis. In particular, because of the large computational burden in some applications, the computational efficiency of cyclic spectrum estimation algorithms is a central issue and has been the focus of research in this area. The computationally efficient algorithms reviewed in this paper were introduced in [4] and discussed in [5, 6, 7]. They exploit the computational efficiency of the Fast Fourier Transform (FFT) to produce cyclic spectrum estimates efficiently when they are needed over a large portion of the bifrequency plane. The algorithms are generalizations of techniques used previously for measuring the correlation frequency density between sonar signals with unknown relative doppler shift [8, p. 198].

This article reviews the theory behind digital spectral correlation analysis algorithms and describes various aspects of implementing the algorithms. The discussion begins with time smoothing algorithms. In general, FFT based time smoothing algorithms are considered most attractive for computing estimates of the spectral correlation function over the entire bifrequency plane. Two computationally efficient algorithms, the FFT Accumulation Method and the Strip Spectral Correlation Algorithm, are described here. Frequency smoothing methods are generally considered best for computing estimates of the spectral correlation function along lines of constant cycle frequency for moderate numbers of cycle frequencies. In particular, a frequency smoothing algorithm called the Digital Frequency Smoothing Method is useful for this type of estimation problem. Although spectral correlation algorithms are generally classified as time or frequency smoothing, hybrid algorithms (i.e., algorithms that smooth in both time and in frequency) can be advantageous in certain applications [9].

2 TIME SMOOTHING ALGORITHMS

We begin our study of time smoothing algorithms with the time smoothed cyclic (cross) periodogram. By itself, the time smoothed cyclic periodogram is not computationally efficient for computing estimates of the spectral correlation function over large regions of the bifrequency plane. However, modification of the time smoothed cyclic periodogram leads to several computationally efficient algorithms. In particular, the FFT Accumulation Method (FAM), and the Strip Spectral Correlation Algorithm (SSCA) can be derived by modifying the time smoothed cyclic periodogram. Kernel representations of the algorithms are also developed. These representations are useful in quantifying the performance of the algorithms. References [6 and 7] present detailed studies of the time smoothing algorithms.

The time smoothed cyclic cross periodogram is the basis for all time smoothing algorithms discussed in this article. Consider a pair of continuous-time waveforms $\tilde{x}(t)$ and $\tilde{y}(t)$ with sampled versions $x(n) = \tilde{x}(nT_s)$ and $y(n) = \tilde{y}(nT_s)$, where T_s is the sample period. The time smoothed cyclic cross periodogram is defined by

$$S_{xy_T}^{\alpha_0}(n, f_0)_{\Delta t} = \left\langle X_T(n, f_0 + \alpha_0/2) Y_T^*(n, f_0 - \alpha_0/2) \right\rangle_{\Delta t} \tag{1}$$

where the quantities $X_T(n, \cdot)$ and $Y_T(n, \cdot)$ are the complex demodulates of $x(n)$ and $y(n)$, respectively. A complex demodulate such as $X_T(n, f)$ is obtained by passing $x(n)$ through a one-sided bandpass filter centered at frequency f with bandwidth $\Delta f = 1/T$, and frequency shifting the output to baseband. It is a complex-valued lowpass waveform and represents the spectral components of $x(n)$ within a band of width Δf centered at frequency f. Hence, (1) can be interpreted as correlating spectral components of $x(n)$ with those of $y(n)$ over a time span of Δt seconds. The time averaging operation in (1) employs a unity area weighting function $g(n)$ of duration $\Delta t = NT_s$ seconds,

$$S_{xy_T}^{\alpha_0}(n, f_0)_{\Delta t} = \sum_r X_T(r, f_1) Y_T^*(r, f_2) g(n - r). \tag{2}$$

In (2) the complex demodulate frequencies are related to the frequency and cycle frequency of the point estimate by $f_0 = (f_1 + f_2)/2$ and $\alpha_0 = f_1 - f_2$. The complex demodulate $X_T(n, \cdot)$ can be computed from the formula

$$X_T(n, f) = \sum_{r=-N'/2}^{N'/2-1} a(r) x(n - r) e^{-i2\pi f(n-r)T_s} \tag{3}$$

where $a(r)$ is a data tapering window of duration $T = N'T_s$ seconds. Complex demodulates computed by (3) have bandwidths on the order of $\Delta f = \Delta a = f_s/N'$ where $f_s = 1/T_s$. Figure 1 illustrates the parameters of the time smoothed cyclic periodogram.

Reliable estimation of the spectral correlation function requires that $\Delta t \gg T$, or more succinctly, that $\Delta t \Delta f \gg 1$. If the data tapering window is normalized according to $\sum_r |a(r)|^2 = 1$, then in the limit as $\Delta t \to \infty$ and $\Delta f \to 0$, the time smoothed cyclic periodogram approaches the limit cyclic cross spectrum [3]

$$\lim_{\Delta f \to 0} \lim_{\Delta t \to \infty} S_{xy_T}^{\alpha_0}(n, f_0)_{\Delta t} = S_{xy}^{\alpha_0}(f_0) \tag{4}$$

$$= f_s S_{\tilde{x}\tilde{y}}^{\alpha_0}(f_0) \tag{5}$$

where the last equation is valid for $|f_0| < f_s/2 - |\alpha_0|/2$ and $f_s > 2B$, where B is the bandwidth of the waveforms $\tilde{x}(t)$ and $\tilde{y}(t)$. A basic time smoothing algorithm can be formed by evaluating the time smoothed cyclic periodogram at frequency/cycle-frequency pairs (f_0, α_0) of interest. In many applications, computation of the entire cyclic cross spectrum estimate (i.e., all possible point estimates) using this basic algorithm would require an enormous number of computations. However, the basic algorithm can be used to derive other, more computationally efficient, algorithms.

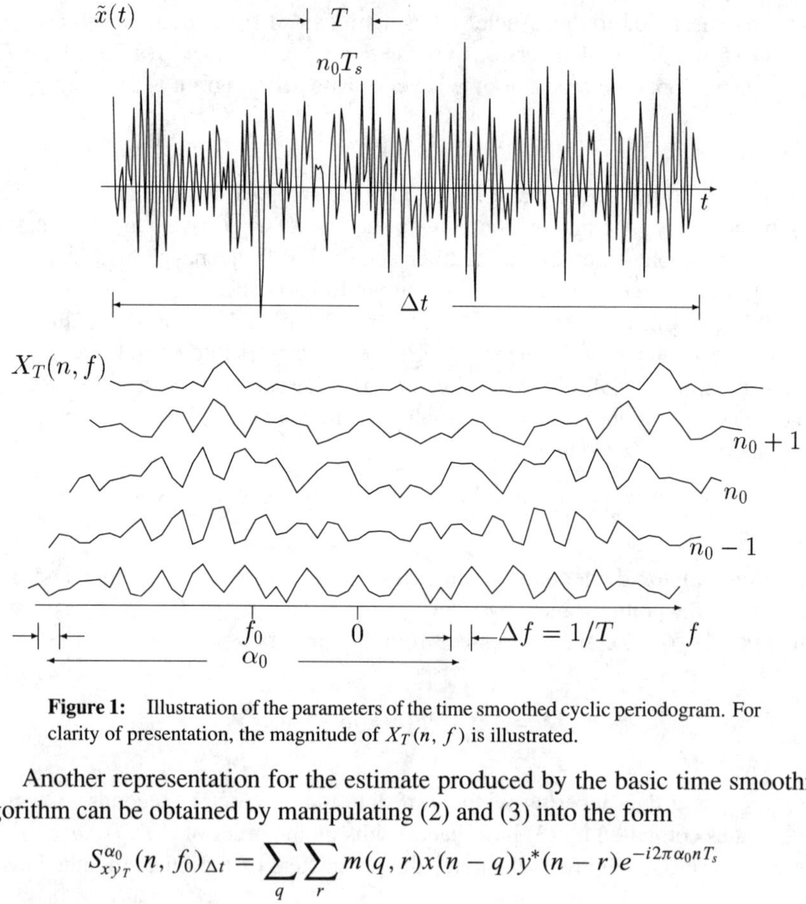

Figure 1: Illustration of the parameters of the time smoothed cyclic periodogram. For clarity of presentation, the magnitude of $X_T(n, f)$ is illustrated.

Another representation for the estimate produced by the basic time smoothing algorithm can be obtained by manipulating (2) and (3) into the form

$$S_{xy_T}^{\alpha_0}(n, f_0)_{\Delta t} = \sum_q \sum_r m(q, r) x(n-q) y^*(n-r) e^{-i2\pi\alpha_0 n T_s} \tag{6}$$

where the kernel $m(q, r)$ is

$$m(q, r) = \sum_p g(p) a(q-p) a(r-p) e^{i2\pi f_0 (q-r) T_s} e^{i\pi\alpha_0 (q+r) T_s}. \tag{7}$$

Equation (6) is a quadratic transformation that represents the estimate in terms of the parameters f_0, α_0, Δt, and Δf. Note that all of these system parameters are contained within the kernel $m(q, r)$. Of greater use in describing algorithms is the rotated Fourier Series Transform (FST) of the kernel

$$M(\alpha, f) = \sum_q \sum_r m(q, r) e^{-i2\pi(f+\alpha/2) q T_s} e^{i2\pi(f-\alpha/2) r T_s}. \tag{8}$$

In terms of the rotated kernel $M(\alpha, f)$, the estimate is given by

$$S_{xy_T}^{\alpha_0}(n, f_0)_{\Delta t} = T_s \sum_{\beta \in \mathcal{R}} \int_{-f_s/2}^{f_s/2} M(\beta, \nu) S_{xy}^{\beta}(\nu)\, d\nu\, e^{i2\pi(\beta-\alpha_0)nT_s} + R(n) \qquad (9)$$

where the first term on the right side is the mean of the left side, the second term is a randomly fluctuating residual, and where $\mathcal{R}$ is the region $-f_s/2 < \beta - \alpha_0 \leq f_s/2$. Equation (9) indicates that cycle features of the underlying cyclic cross spectrum $S_{xy}^{\alpha}(f)$ that are within the domain of $M(\alpha, f)$ are summed in cycle frequency and integrated in frequency to produce the value of the estimate. Cycle features outside of $M(\alpha, f)$ do not appear in the estimate.

The relationship between the amplitude scaling of the estimate and the form of the kernel transform is important in connection with the FAM algorithm discussed in the next section. The amplitude scaling of the estimate is determined as follows. In the limit as $\Delta t \to \infty$ and as $\Delta f \to 0$ the random residual in (9) disappears, and (9) becomes

$$\lim_{\Delta f \to 0} \lim_{\Delta t \to \infty} S_{xy_T}^{\alpha_0}(n, f_0)_{\Delta t} = \left[\lim_{\Delta f \to 0} \lim_{\Delta t \to \infty} \int_{-f_s/2}^{f_s/2} T_s M(\alpha_0, f) df \right] S_{xy}^{\alpha_0}(f_0). \qquad (10)$$

The area under the kernel transform evaluated at $\alpha = \alpha_0$ scales the amplitude of the estimate. With the weighting functions $a(n)$ and $g(n)$ scaled as described earlier, this scaling factor is unity for the unmodified time smoothing algorithm.

Under the condition $\Delta t \Delta f \gg 1$, the rotated kernel can be approximated by the separable form

$$M(\alpha, f) \approx G_{1/\Delta t}(\alpha - \alpha_0) H_{1/T}(f - f_0). \qquad (11)$$

The functions $G_{1/\Delta t}(\alpha)$ and $H_{1/T}(f)$ are frequency windows and are related in a simple manner to the FSTs of the data tapering windows used to compute the estimate. For example, the basic time smoothing algorithm described by (2) and (3) constitutes a quadratic system and has a kernel transform given by

$$M(\alpha, f) = G(\alpha - \alpha_0) A\left(f - f_0 + \frac{\alpha - \alpha_0}{2}\right) A^*\left(f - f_0 - \frac{\alpha - \alpha_0}{2}\right) \qquad (12)$$

$$\approx G(\alpha - \alpha_0) |A(f - f_0)|^2 \text{ for } \Delta t \Delta f \gg 1 \qquad (13)$$

where $A(f)$ and $G(\alpha)$ are the FSTs of $a(n)$ and $g(n)$. Equation (13) indicates that $M(\alpha, f)$ is a two dimensional pulse that has most of its area concentrated in the region defined by $|f - f_0| \leq \Delta a/2$ and $|\alpha - \alpha_0| \leq 1/2\Delta t$. Since cycle features within this region contribute to the estimate (cf. (9)) the region determines the frequency resolution Δf and the cycle frequency resolution $\Delta\alpha$ of the estimate. A troublesome type of spectral leakage, called cycle leakage, can occur if the kernel transform $M(\alpha, f)$ is incorrectly designed. If cycle features are present in the sidelobes of $M(\alpha, f)$ they will contribute to the estimate. Cycle leakage can be minimized by properly designing the kernel transform so that it has low sidelobes and low skirts.

Two modifications to the basic time smoothing algorithm can be made to improve computational efficiency: use a computationally efficient Fourier Transform to perform time smoothing, and use a computationally efficient algorithm to compute the complex demodulates. An algorithm based on the first modification can be derived by introducing a frequency shift ϵ into the complex demodulate product sequence before smoothing. Consider the expression

$$S_{xy_T}^{\alpha_0}(n, f_0)_{\Delta t} = \sum_r X_T(r, f_1)Y_T^*(r, f_2)g(n-r)e^{-i2\pi\epsilon r T_s} \tag{14}$$

where $|\epsilon| < \Delta a$. Note that the cycle frequency parameter has been redefined to $\alpha_0 = f_1 - f_2 + \epsilon$. Since the product sequence has a bandwidth on the order of $2\Delta a$, and the averaging operation is a lowpass filter with bandwidth on the order of $\Delta\alpha = 1/\Delta t$, frequency shifting the product sequence moves different bands of the product sequence into the lowpass filter. The kernel transform associated with $|q| < \frac{\Delta a}{\Delta\alpha}$, (14) is

$$M(\alpha, f) = G(\alpha - f_1 + f_2 - \epsilon)A(f - f_1 + \alpha/2)A^*(f - f_2 - \alpha/2) \tag{15}$$

and is a pulse centered at $(f_0 = (f_1 + f_2)/2, \alpha_0 = f_1 - f_2 + \epsilon)$ with a width of $\Delta\alpha = 1/\Delta t$ and a length of $\Delta f = \Delta a - |\epsilon|$. By discretizing the frequency shift to $\epsilon = q\Delta\alpha$, (14) becomes

$$S_{xy_T}^{\alpha_0}(n, f_0)_{\Delta t} = \sum_r X_T(r, f_1)Y_T^*(r, f_2)g(n-r)e^{-i2\pi rq/N} \tag{16}$$

and it follows that the sum in (16) can be efficiently evaluated by an N-point FFT. Note that (16) computes a column of N point estimates with the column centered at $(f_0, f_1 - f_2)$, and that the qth point estimate has a variable frequency resolution of $\Delta f = \Delta a - |q| f_s/N$, and a constant cycle frequency resolution of $\Delta\alpha = f_s/N$.

The second modification improves the computational efficiency of the complex demodulate calculation. For now, modify (3) so that L samples are skipped between computations,

$$X_T(pL, f) = \sum_{r=-N'/2}^{N'/2-1} a(r)x(pL - r)e^{-i2\pi f(pL-r)T_s}. \tag{17}$$

By hopping the channelizer L samples between computations the number of samples in each complex demodulate sequence shortens from N to $P = N/L$. As described in the next section, an FFT can be used to efficiently compute this sum for a set of discrete frequencies. (See [10] for a general discussion on efficient algorithms for computing complex demodulates.) If $g(\cdot)$ in (2) is replaced by its comb-filter version

$$g_c(n) = \sum_m \delta(n - mL)g(n)/\sum_k g(KL)$$

and the output is sampled at $n = PL$, the estimation equation (2) can be written in terms of the decimated complex demodulates,

$$S_{xy_T}^{\alpha_0}(pL, f_0)_{\Delta t} = \sum_r X_T(rL, f_1)Y_T^*(rL, f_2)g_d(p - r) \tag{18}$$

where $g_d(r) = g(rL)$. To determine the kernel transform for this system, $G(\cdot)$ in (12) is replaced by the FST of the comb filter $g_c(\cdot)$,

$$G_c(\alpha) = \sum_r g_c(r)e^{-i2\pi\alpha r T_s} \tag{19}$$

$$= \sum_n G_d\left(\alpha + \frac{nf_s}{L}\right), \tag{20}$$

where $G_d(\alpha)$ consists of one period of $G_c(\alpha)$,

$$G_d(\alpha) = \begin{cases} \sum_m g_d(m)e^{-i2\pi\alpha m L T_s} & \text{for } |\alpha| < f/2L \\ 0 & \text{otherwise.} \end{cases}$$

The kernel transform for this algorithm is thus

$$M(\alpha, f) = \sum_n G_d\left(\alpha - \alpha_0 + \frac{nf_s}{L}\right) A\left(f - f_1 + \frac{\alpha}{2}\right) A^*\left(f - f_2 - \frac{\alpha}{2}\right) \tag{21}$$

$$= \sum_n G_d\left(\alpha - \alpha_0 + \frac{nf_s}{L}\right) M_1(\alpha, f). \tag{22}$$

The teeth of the comb filter have bandwidths on the order of $1/\Delta t$, and are spaced f_s/L Hz apart. Cycle leakage can result if the teeth of the comb overlap the region of support of $M_1(\alpha, f)$. Proper values of decimation factor L are discussed in conjunction with the FFT Accumulation Method.

2.1 The FFT Accumulation Method

The two modifications to the basic time smoothing algorithm can be combined to form the FFT Accumulation Method (FAM). But first, note that an N'-point FFT can be used to efficiently compute the complex demodulates in (17). Equation (17) can be manipulated into the form

$$X_T(pL, f_m) = \left[\sum_{k=0}^{N'-1} a(d-k)x(pL-d+k)e^{-i2\pi mk/N'}\right] e^{-i2\pi m(pL-d)/N'} \tag{23}$$

where $d = N'/2 - 1$. In this form, it is immediately apparent that the summation in the brackets can be efficiently computed using an FFT. Complex demodulates computed in this manner have a discrete set of center frequencies $f_m = mf_s/N'$, $-N'/2 < m < N'/2 - 1$. With this set of complex demodulates, and combining the results of (16) and (18), a point estimate can be expressed as

$$S_{xy_T}^{\alpha_0}(pL, f_{kl})_{\Delta t} = \sum_r X_T(rL, f_k)Y_T^*(rL, f_l)g_d(p-r)e^{-i2\pi rq/P} \tag{24}$$

where $\alpha_{kl} = f_k - f_l$, $\alpha_0 = \alpha_{kl} + q\Delta\alpha$, and $f_0 = f_{kl} = (f_k + f_l)/2$. Equation (24) can be efficiently evaluated by a P-point FFT, thereby producing a block of point estimates. The kernel transform of the FAM is given by

$$M(\alpha, f; k, l, q) = G_c(\alpha - \alpha_0)A\left(f - f_k + \frac{\alpha}{2}\right) A^*\left(f - f_l - \frac{\alpha}{2}\right). \tag{25}$$

The idealized region of support (no sidelobes or skirts) of $A(f-f_k+\alpha/2)A^*(f-f_l-\alpha/2)$ is called a channel-pair region. As is evident from Fig. 2, computation of (24) for all combinations of channel pairs (k,l) and FFT bins $|q| < LP/N'$ completely covers the bifrequency plane.

The cycle frequency resolution of the FAM is $\Delta\alpha = f_s/PL$ and the frequency resolution, which is variable, is $\Delta f = \Delta a - |q|\Delta\alpha$. Likewise, the time-frequency resolution product is variable, and is $\Delta t\Delta f = \Delta t\Delta a - |q|$. Typically, the time-frequency resolution product is referenced to the point estimate at $q = 0$. The amplitude scaling of the estimate, determined according to (10), also varies with q:

$$T_s \int_{-f_s/2}^{f_s/2} M(\alpha_0, f)df = T_s G_c(0) \int_{-f_s/2}^{f_s/2} A(f+q\Delta\alpha/2)A^*(f-q\Delta\alpha/2)df \quad (26)$$

$$\approx 1 - |q|\frac{\Delta\alpha}{\Delta a}. \quad (27)$$

The design of the channelizer is particularly important in the FAM approach. The possibility of severe cycle frequency leakage, which can occur if the replica teeth in the comb filter G_c intersect the channel-pair region, requires careful design of the channelizer passband shape $A(f)$ and selection of the decimation factor L. From Fig. 3, it can be seen that the extent in cycle frequency of the nominal channel-pair region is $2\Delta a$ and that the teeth of the comb filter are separated by f_s/L. Therefore, for all values of α_0 spanned by the channel-pair region, the replica teeth will not intersect the channel-pair region provided that $f_s/L > 2\Delta a$ or, equivalently, $L < N'/2$. However, any actual tapering function $a(n)$ of length N' will have significant response for frequencies greater than $\Delta a/2 = f_s/2N'$. Therefore, we recommend designing $a(n)$ to produce a sharp transition band and using

$$L = N'/4. \quad (28)$$

In applications where computation speed is paramount, there can be advantage in selecting $L = N'/2$ in order to minimize the FFT size $P = N/L$. This can be accomplished without cycle frequency leakage by increasing the length of $a(n)$ to achieve a more rectangular passband shape. For example, replace N' in (23) by $2N'$ and design $a(n)$ to produce a nearly rectangular passband of width $\Delta a = f_s/N'$. The increased cost of channelization is offset by the decreased cost of performing the P-point FFT for all channel-pair regions when P is reduced from $4N/N'$ to $2N/N'$.

The variable time-frequency resolution product is troublesome near the extremes of channel-pair regions since point estimates there have a high degree of variability. The trouble can be eliminated by simply discarding the offending point estimates; for example, retain the point estimates whose index parameter lies within the range

$$-\frac{\Delta a}{2} \leq q\Delta\alpha < \frac{\Delta a}{2} \quad (29)$$

or

$$-\frac{PL}{2N'} \leq q \leq \frac{PL}{2N'} - 1. \quad (30)$$

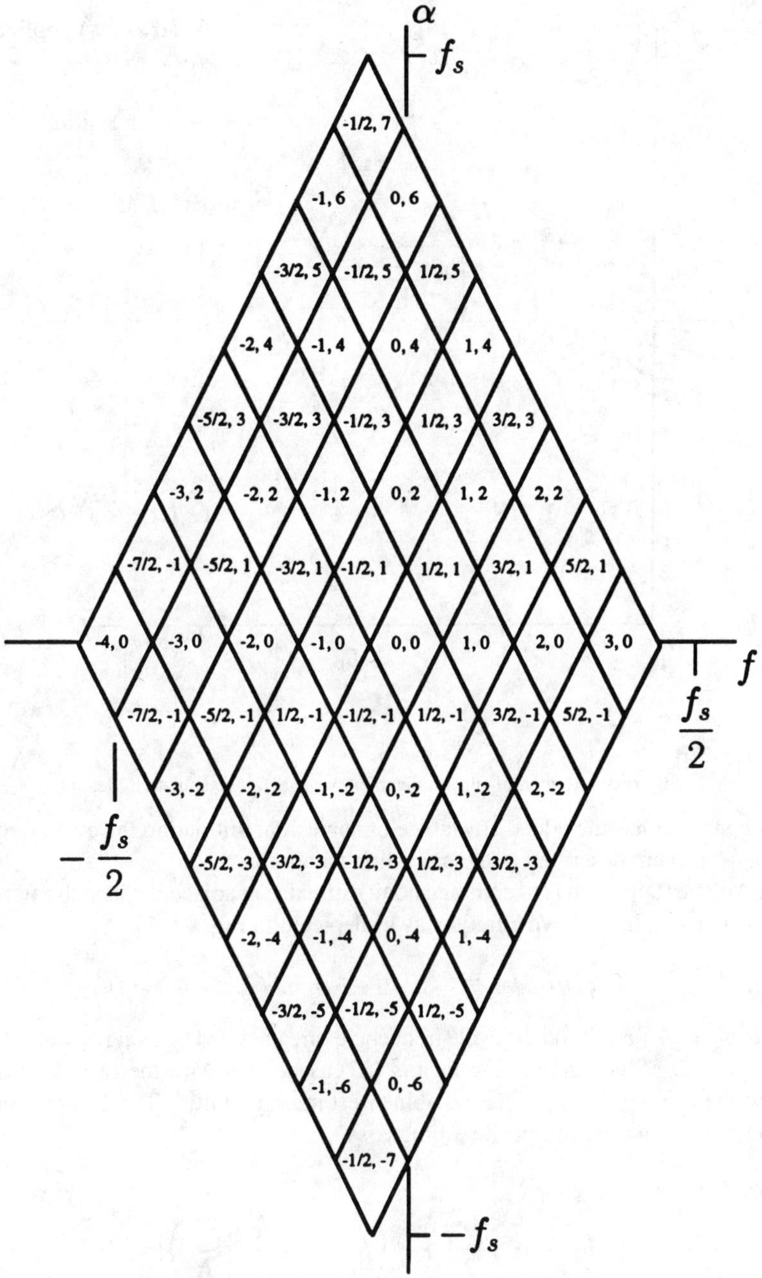

Figure 2: Tiling the bifrequency plane with channel-pair regions for $N' = 8$. The channel-pair region index for channels f_k and f_l is $((k+l)/2, k-l)$.

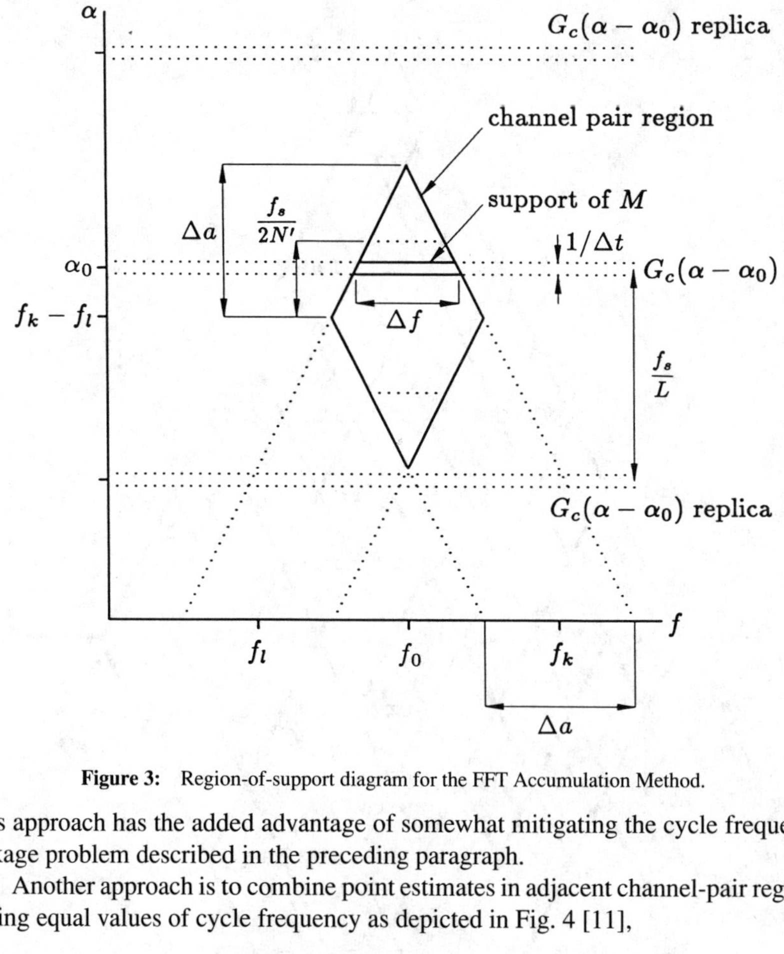

Figure 3: Region-of-support diagram for the FFT Accumulation Method.

This approach has the added advantage of somewhat mitigating the cycle frequency leakage problem described in the preceding paragraph.

Another approach is to combine point estimates in adjacent channel-pair regions having equal values of cycle frequency as depicted in Fig. 4 [11],

$$S_{xy_T}^{\alpha_0}(pL, f_0) = S_{xy_T}^{\alpha_0}(pL, f_{kl}) + S_{xy_T}^{\alpha_0}(pL, f_{k+1,l}) \tag{31}$$

where $f_0 = f_l + \alpha_0/2$ and the cycle frequencies are matched by using $q_2 = q_1 - N/N'$ with $q_1 > 0$. Since the scaling factors (27) associated with the two terms sum to unity, the scaling factor of the combined estimate is unity. The kernel transform corresponding to this composite estimate is

$$\begin{aligned} &M(\alpha, f; k, k+1, l, q_1, q_2) \\ &= G_c(\alpha - \alpha_0)\left[A\left(f - f_k + \frac{\alpha}{2}\right) + A\left(f - f_k + \frac{\alpha}{2} - \frac{f_s}{N'}\right)\right] A^*\left(f - f_l - \frac{\alpha}{2}\right). \end{aligned} \tag{32}$$

If $A(f)$ is rectangular and of width f_s/N' then the frequency resolution associated with this kernel is indeed independent of α_0 and equal to f_s/N'. There are two drawbacks to this approach. First, the need for rectangular channelizer passbands increases the computational cost of channelization, and second, combining estimates from adjacent channel-pair regions complicates the partitioning of the algorithm onto multiple processors operating in parallel.

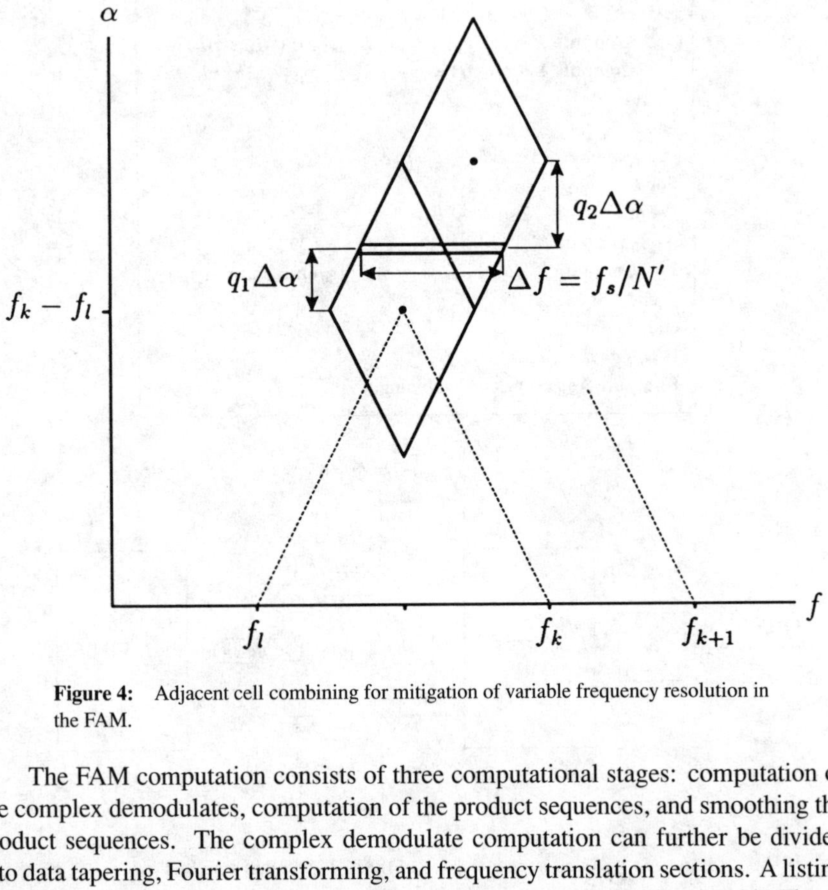

Figure 4: Adjacent cell combining for mitigation of variable frequency resolution in the FAM.

The FAM computation consists of three computational stages: computation of the complex demodulates, computation of the product sequences, and smoothing the product sequences. The complex demodulate computation can further be divided into data tapering, Fourier transforming, and frequency translation sections. A listing of the algorithm is given in Table 1, and a digital realization is illustrated in Fig. 5. The listing is an implementation of the FAM on a uniprocessor (i.e., single CPU) computer and computes estimates of the cyclic cross spectrum. An optional step of adjacent cell combining is included for completeness. The algorithm is easily modified to computing estimates of the cyclic spectrum of a real-valued signal by replacing references to y with x and adjusting the indices on the product computation loop to be $0 \leq k \leq N'/2 - 1$ and $0 \leq l \leq N'/2 - 1$. It is important to realize that other operations are needed for postprocessing such as data thinning, graphics computations, etc. Although these computations are important in the design and implementation of a digital spectral correlation analyzer, they are not discussed here.

Table 1: A Uniprocessor Implementation of the FFT Accumulation Method

```
/* Compute Complex Demodulates of the Input Sequences */
Do p := 0 to P − 1
   Compute x_T(pL, f_k) = FFT{a(r)x(pL + r)}
   Compute y_T(pL, f_k) = FFT{a(r)y(pL + r)}
   Do k := −N'/2 to N'/2 − 1
      Compute X_T(pL, f_k) = x_T(pL, f_k) exp{−i2πkpL/N'}
      Compute Y_T(pL, f_k) = y_T(pL, f_k) exp{−i2πkpL/N'}
   end
end
/* Compute and Smooth Product Sequences */
For k := −N'/2 to N'/2 − 1
   For l := −N'/2 to N'/2 − 1
      Compute S_xyT^{α_kl}(pL, f_kl) = X_T(pL, f_k) Y_T^*(pL, f_l)
      Compute S_xyT^{α_0}(pL, f_kl)_Δt = FFT{g_d(p) S_xyT^{α_kl}(pL, f_kl)}
   end
end
/* Optional */
Compute Adjacent Cell Combining.
```

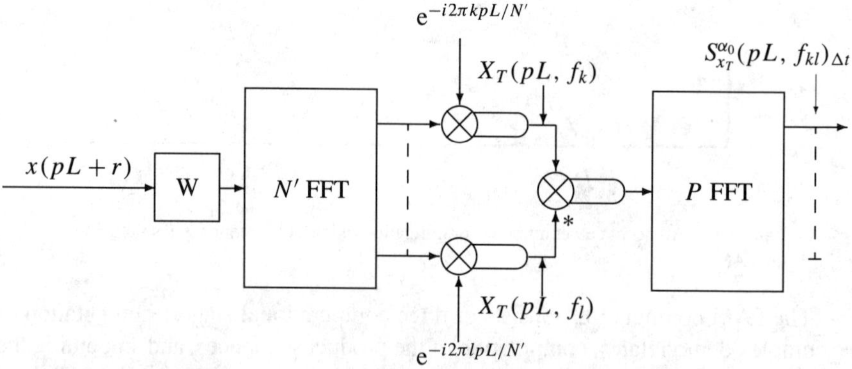

Figure 5: Digital implementation of the FFT Accumulation Method.

The computational complexity of the FAM is given in Table 2, where complex multiplications are used as a metric. From that table, we can develop an expression for the number of real multiplications required by the FAM to compute an estimate of the cyclic spectrum of a real-valued signal. That expression, for $L = N'/4$, is

$$C_{rm} = 2NN'\log_2\left(\frac{4N}{N'}\right) + 8N\log_2 N' + 4NN' + 20N. \tag{33}$$

Relating the parameters of our digital implementation, N, P, L, and N', to the fundamental signal processing parameters Δf and Δt, we note that $\Delta f = f_s/N'$, $\Delta t = NT_s$ and $P = N/L = 4N/N'$ (for $L = N'/4$). Thus the number of real multiply operations required in the FAM is

$$C_{rm} = \frac{2\,(\Delta t \Delta f)}{(\Delta f/f_s)^2}\log_2[4(\Delta t \Delta f)] + \frac{8(\Delta t \Delta f)}{(\Delta f/f_s)}\log_2(f_s/\Delta f) + \frac{4(\Delta t \Delta f)}{(\Delta f/f_s)^2} + \frac{20(\Delta t \Delta f)}{(\Delta f/f_s)}. \tag{34}$$

The dominant term in this expression is

$$o\left[\frac{(\Delta t \Delta f)}{(\Delta f/f_s)^2}\log(\Delta t \Delta f)\right] \tag{35}$$

which represents computation of the smoothing FFTs. In order to give the reader a feel for the magnitude of the computations involved in estimating the cyclic spectrum, a numerical example is given.

Table 2: Computational Complexity of the FAM in Terms of the Number of Complex Multiplications

Computation Section	Number of Complex Multiplications	
	Cyclic Cross Spectrum of Two Complex Signals	Cyclic Spectrum of a Single Real Signal
Channelizer		
Data Tapering	$2N'P$	[a]
N'-point FFT	$PN'\log_2 N'$	$P(N'/2)\log_2 N'$
Frequency Shift	$2N'P$	$N'P$
Cross Multiply	$P(N')^2$	$P[(N')^2]/4$
FFT Product Sequences	$(N')^2(P/2)\log_2 P$	$[(N')^2/4](P/2)\log_2 P$

[a] $N'P$ real multiplications are required for data tapering.

Example: Consider computing an estimate of the cyclic spectrum of a real-valued bandpass signal, whose bandwidth is 20 MHz, having been sampled at $f_s =$ 50 MHz, where the frequency resolution is $\Delta f = 2^{-4} \cdot f_s$ and the time-frequency resolution product is $\Delta t \Delta f = 2^8$. (The requirement on the time-frequency resolution product is to ensure detection with a sufficient SNR. Determination of the time-frequency resolution product required to achieve a desired output SNR as a function of input SNR is treated in [7].)

In this example, the number of real multiplies required by FAM to estimate the cyclic spectrum over one quarter of the bifrequency plane is 1.6×10^6, for a collect of $N = 2^{16}$ samples, collected in $\Delta t = 82 \times 10^{-6}$ seconds. Using a 33 MFLOP processor efficiently, we would be able to estimate the cyclic spectrum in about 48×10^{-3} seconds, or 591 times real-time. (A computation is said to be real-time if the amount of time required to compute the estimate is equal to the amount of time required to collect the data for the estimate.) To achieve additional computation speed, we could employ parallelism by assigning computation of different correlation products and P-point FFTs to different processors, operating in parallel.

While the uniprocessor implementation of the algorithm might be satisfactory for some problems, applications demanding rapid estimation of the cyclic (cross) spectrum require a multiprocessor approach. Parallel FAM algorithms can be developed by observing several aspects of the uniprocessor algorithm. First, note that several operations in the uniprocessor algorithm can be vectorized. In particular, windowing the input data and frequency translation in the complex demodulate computation can be vectorized. Similarly, computation of the product sequences can be vectorized. However, the greatest gains in improving the computation rate are obtained by parallelizing the computation and smoothing of the product sequences. Since the product sequences are independent of one another, (if adjacent cell combining is not used) parallelization is relatively easy. A parallel algorithm based on these observations is described in [6]. The tricky part in exploiting this natural parallelism is in the efficient distribution of the complex demodulate sequences to the processors.

2.2 The Strip Spectral Correlation Algorithm

The Strip Spectral Correlation Algorithm (SSCA) is another efficient algorithm for computing estimates of the spectral correlation function [6]. The primary difference between the FAM and SSCA is that in the SSCA the complex demodulates $X_T(n, f_k)$ are multiplied by $y^*(n)$ instead of by $Y_T^*(n, f_l)$. Point estimates are computed by

$$S_{xy_T}^{\alpha_0}(n, f_0)_{\Delta t} = \sum_r X_T(r, f_k) y^*(r) g(n-r) e^{-i2\pi qr/N} \tag{36}$$

where $\alpha_0 = f_k + q\Delta\alpha$ and $f_0 = f_k/2 - q\Delta\alpha/2$. The sum in (36) can be efficiently evaluated with an N-point FFT. The resulting point estimates lie along the line $\alpha = 2f_k - 2f$, which is a strip in the bifrequency plane. For this algorithm it appears that the channelizer decimation factor must be $L = 1$ since $X_T(r, f_k)$ and $y^*(r)$ must be at the same sampling rate to be properly multiplied. However, a decimation factor of $L = N'/4$, like that of the FAM, can be used if the demodulate sequence is interpolated to match the sampling rate of $y(n)$. It has been found that adequate interpolation is obtained by holding the value of each complex demodulate sample for L samples. Thus the interpolated sequence $\tilde{X}_T(n, f_k) = X_T(pL, f_k)$ for $pL \leq n < (p+1)L$ can be used in place of $X_T(n, f_k)$ in (36).

The kernel transform for the SSCA is

$$M(\alpha, f) = G(\alpha - \alpha_0) A\left(f - f_0 + \frac{\alpha - \alpha_0}{2}\right) \tag{37}$$

where

$$f_0 = \frac{f_k}{2} - q\frac{\Delta\alpha}{2} \tag{38}$$

$$\alpha_0 = f_k + q\Delta\alpha \tag{39}$$

and $G(\alpha)$ and $A(f)$ are the FSTs of $g(n)$ and $a(n)$. The frequency resolution of the SSCA is $\Delta f = f_s/N'$, and the time-frequency resolution product is $\Delta t \Delta f = N/N'$. The uniprocessor description of the SSCA is similar to that of the FAM and consists of three steps: computing the complex demodulates, computing the product sequences, and smoothing the product sequences. A uniprocessor algorithm for the SSCA is given in Table 3, and a digital realization of the algorithm is illustrated in Fig. 6.

Table 3: Uniprocessor Implementation of the Strip Spectral Correlation Algorithm

```
/* Compute Complex Demodulates of x */
Do p := 0 to P − 1
   Compute x_T(pL, f_k) = FFT{a(r)x(pL + r)}
   Do k := −N'/2 to N'/2 − 1
      Compute X_T(pL, f_k) = x_T(pL, f_k) exp{−i2πpLk/N'}
   end
end
/* Interpolate X_T(pL, f_k) */
   Compute X_T(pL, f_k) → X̃_T(n, f_k)
/* Compute and Smooth Product Waveforms */
Do k := −N'/2 to N'/2 − 1
   Compute S_{xy_T}^{f_k}(n, f_k) = y*(n) X̃_T(n, f_k)
   Compute S_{xy_T}^{α_0}(n, f_0)_{Δt} = FFT{g(n) S_{xy_T}^{f_k}(n, f_k)}
end.
```

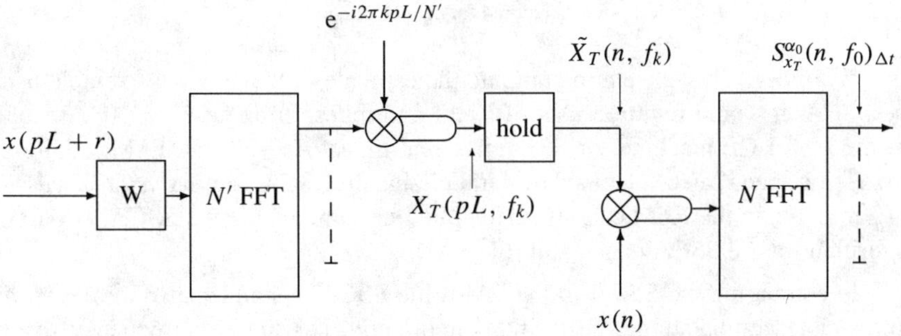

Figure 6: Digital implementation of the Strip Spectral Correlation Algorithm.

The computational complexity of the SSCA, for two typical estimation problems, is given in Table 4. In a similar manner to that of the FAM, using the information in Table 4, we can derive an expression for the number of real multiplication operations required by SSCA for computing an estimate of the cyclic spectrum of a real-valued signal. That expression, for $L = N'/4$, is

$$C_{rm} = NN'\log_2 N + 2NN' + 8N\log_2(N') + 12N. \tag{40}$$

Making the same substitutions as in FAM for the signal processing parameters in terms of Δt, Δf, and f_s yields

$$C_{rm} = \frac{(\Delta t \Delta f)}{(\Delta f/f_s)^2}\log_2\left[\frac{(\Delta t \Delta f)}{(\Delta f/f_s)}\right] + \frac{8\Delta t \Delta f}{(\Delta f/f_s)}\log_2(f_s/\Delta f) + \frac{2\Delta t \Delta f}{(\Delta f/f_s)^2} + \frac{12\Delta t \Delta f}{(\Delta f/f_s)}. \tag{41}$$

Note that the dominant term in this expression is the same as in the FAM,

$$o\left[\frac{(\Delta t \Delta f)}{(\Delta f/f_s)^2}\log(\Delta t \Delta f)\right]. \tag{42}$$

Table 4: Computational Complexity of the SSCA in Terms of the Number of Complex Multiplications

Computation Section	Number of Complex Multiplications	
	Cyclic Cross Spectrum of Two Complex Signals	Cyclic Spectrum of a Single Signal
Channelizer		
Data Tapering	$4N$	a
N'-point FFT	$N(N'/2)\log_2 N'$	$P(N'/2)\log_2 N'$
Frequency Shift	NN'	$2N$
Compute Product Sequences	NN'	$N(N'/2)$
FFTs Product Sequences	$N'(N/2)\log_2 N$	$(N'/2)(N/2)\log_2 N$

[a] $4N$ real multiplications are required for data tapering.

Example: If we were to compute the estimate of the previous example using the SSCA, it would require 9.2×10^5 real multiplies, or about 27×10^{-3} seconds on a 33 MFLOP machine, or 338 times real-time. As with the FAM, additional processing speed can be obtained by parallelizing the SSCA. Similarly, the discussion on parallelizing the FAM can also be applied to parallelizing the SSCA. A parallel algorithm for the SSCA is given in [6].

In comparing the SSCA to the FAM, the SSCA is simpler than the FAM, has comparable computational cost, and, importantly, has uniform frequency resolution. The disadvantages of the SSCA relative to the FAM are that larger size FFTs are required (large FFTs can be inefficient if the data array size exceeds the local memory capacity of the FFT processor) and there is less flexibility to partition the algorithm for parallel processing.

3 FREQUENCY SMOOTHING ALGORITHMS

Frequency smoothing algorithms can be classified as direct or indirect algorithms. Direct frequency smoothing algorithms first compute the spectral components of the data and then perform spectral correlation operations directly on the spectral components. Indirect algorithms compute a related quantity, such as the cyclic cross correlation function or the Wigner-Ville distribution, and Fourier transform one variable of the bivariate function to get the frequency smoothed cyclic cross periodogram. In general, direct frequency smoothing algorithms are computationally superior to indirect frequency smoothing algorithms [3]. This article focuses attention on a direct algorithm called the Digital Frequency Smoothing Method (DFSM).

The basis for the DFSM is the frequency smoothed cyclic cross periodogram represented by

$$S_{xy_{\Delta t}}^{\alpha_0}(n, f_0)_{\Delta f} = \left\langle X_{\Delta t}(n, f_0 + \alpha_0/2) Y_{\Delta t}^*(n, f_0 - \alpha_0/2) \right\rangle_{\Delta f}. \tag{43}$$

An explicit formula suitable for digital implementation is obtained by employing discrete frequency averaging,

$$S_{xy_{\Delta t}}^{\alpha_i}(n, f_j)_{\Delta f} = \sum_r W(j-r) \tilde{X}_{\Delta t}(n, r + \lfloor i/2 \rfloor) \tilde{Y}_{\Delta t}^*(n, r - \lceil i/2 \rceil) \tag{44}$$

where $\alpha_i = i f_s/N$ and $f_j = j f_s/N$ and the spectral components are computed by

$$\tilde{X}_{\Delta t}(n, k) = \sum_r b(r) x(n-r) e^{-i2\pi k(n-r)/N} \tag{45}$$

in which $b(r)$ is a data-tapering window of length $N = \Delta t/T_s$. The shape of data tapering window $b(r)$ is typically rectangular. The frequency smoothing function $W(r)$ ranges over $|r| \leq M/2$, which corresponds to $\Delta f \approx M f_s/N$. A kernel scaling factor of unity results if $\sum_r |b(r)|^2 = T_s$ and $\sum_m W(m) = 1$. Figure 7 illustrates the parameters of the frequency smoothed cyclic periodogram (and of the DFSM). The kernel transform for the DFSM is

$$M(\alpha, f) = \sum_m B\left(f - f_0 + \frac{\alpha - \alpha_0}{2} + \frac{m f_s}{N}\right) B^*\left(f - f_0 - \frac{\alpha - \alpha_0}{2} + \frac{m f_s}{N}\right) W(m). \tag{46}$$

For $\Delta t \Delta f \gg 1$, the frequency smoothing kernel and the time smoothing kernel in (13) closely approximate one another [3].

A uniprocessor algorithm for the DFSM consists of computing (44) for all desired values of α_i and f_j. There are several approaches to organizing the computations, each of which might be better suited for a particular analysis task. One approach is to compute spectral products along a line of constant cycle frequency and then convolve the products with a smoothing window. This approach yields point estimates spaced f_s/N Hz apart in frequency and can ease interpretation of the overall estimate. Another approach is to compute (44) along a line of constant cycle frequency, but space the point estimates by $f_s M/N$ Hz in frequency. Such an arrangement provides full coverage at minimal computational expense (particularly

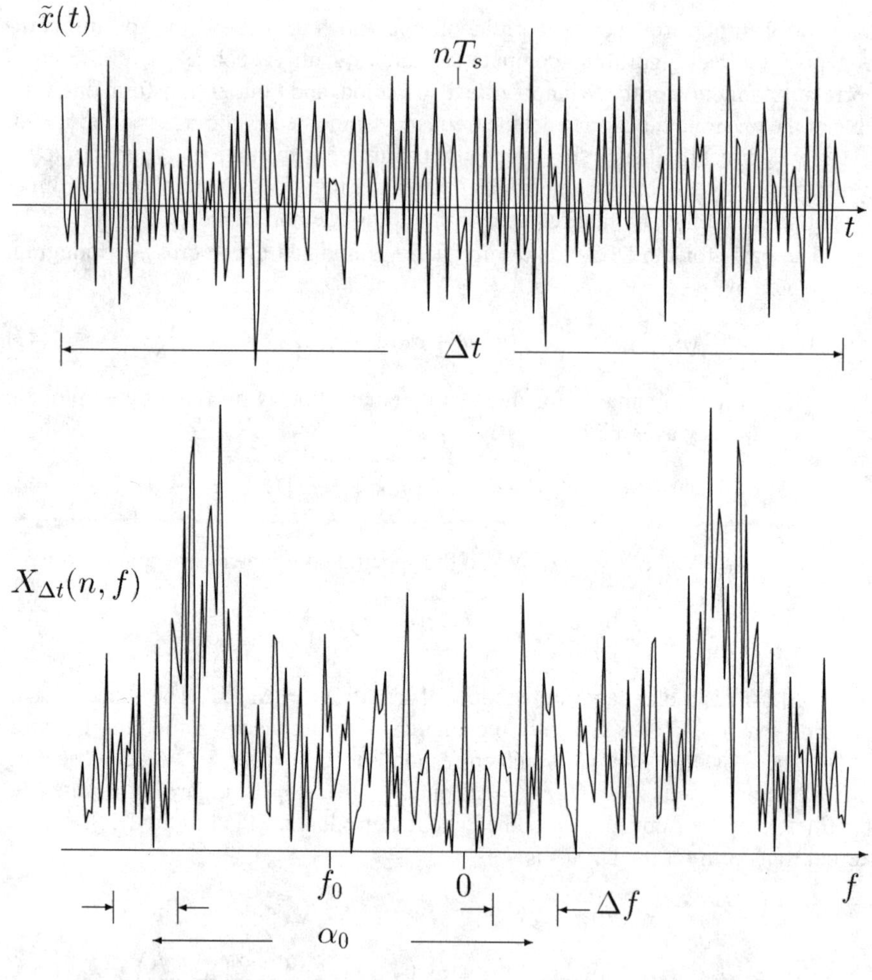

Figure 7: Illustration of the parameters of the frequency smoothed cyclic periodogram. For clarity of presentation, the magnitude of $X_{\Delta t}(n, f)$ is illustrated.

if a rectangular smoothing window is used), but may make interpretation of the overall estimate more difficult. A digital realization of the DFSM is illustrated in Fig. 8.

The computational complexity of the DFSM, in terms of complex multiplications, is easily estimated. For the problem of estimating the cyclic cross spectrum of two complex signals (each consisting of N samples), two N-point FFTs (each with $(N/2)\log_2 N$ complex multiplications) and N^2 cross multiplications are required for a total of $N^2 + N\log_2 N$ complex multiplications. For the problem of estimating the cyclic spectrum of a real signal, one N-point FFT and $N^2/4$ cross multiplications are required for a total of $N^2/4 + N/2\log_2 N$ complex multiplications. Note that the contributions to computational complexity by smoothing is not considered. (If a rectangular window is used to smooth the estimates, its contribution to the number of complex multiplications is nil.) The number of real multiplications required by the DFSM to compute an estimate of the cyclic spectrum of a real signal is

$$C_{rm} = N^2 + 2N\log_2 N. \tag{47}$$

Expressing C_{rm} as a function of Δf, Δt, and f_s yields

$$C_{rm} = \frac{(\Delta t\Delta f)^2}{(\Delta f/f_s)^2} + \frac{2(\Delta t\Delta f)}{(\Delta f/f_s)}\log_2\left[\frac{\Delta t\Delta f}{(\Delta f/f_s)}\right]. \tag{48}$$

Note that the dominant term for estimating the cyclic spectrum is $o\left[\frac{(\Delta t\Delta f)^2}{(\Delta f/f_s)^2}\right]$, as opposed to $o\left[\frac{(\Delta t\Delta f)}{(\Delta f/f_s)^2}\log(\Delta t\Delta f)\right]$ for the same estimation problem using the FAM or SSCA.

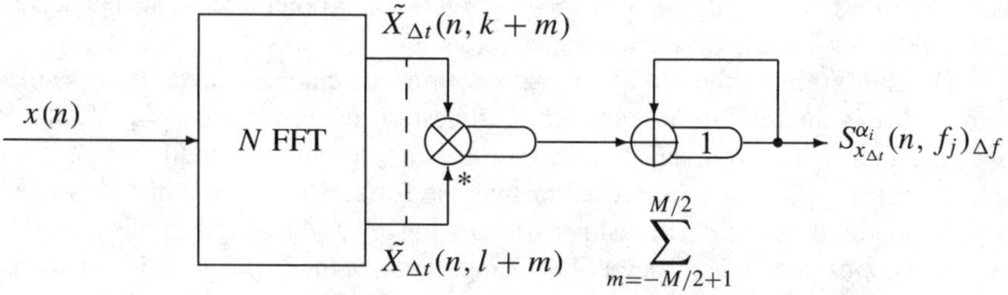

Figure 8: Digital implementation of the Direct Frequency Smoothing Method.

Example: If we wished to use the DFSM to estimate the cyclic spectrum in the previous example, we would need to compute 16×10^6 real multiplies, requiring on the order of 0.5 seconds, for 6×10^3 times real-time.

From the preceding discussion and example it appears that the DFSM is not well suited for computing estimates over the entire bifrequency plane. In passing, it is noted that the DFSM is a highly parallel algorithm [5]. However, in order to exploit this parallelism for large regions of the bifrequency plane, a massively parallel computer is required. For now, we content ourselves with employing this algorithm for small estimation tasks, such as estimating the spectral correlation function over a small number of cycle frequencies. But even for restricted problems the computational burden can be substantial. Fortunately, the algorithm is relatively easy to parallelize on even quite small problems. A simple way to partition the DFSM is into two sections: computation of input FFTs, and correlation and smoothing of the spectral data. Further partitioning can occur by breaking the spectral correlation and smoothing operations along bands of cycle frequency.

Another approach to improving the computational efficiency of the DFSM is to use the One Bit Spectral Correlation Algorithm (OBSCA). The OBSCA technique, which applies to both time and frequency smoothing, converts the complex multiplications in the spectral correlation operation into sign change and data multiplexing operations. Thus, the complex multiplication operations indicated in (44) can be performed with greatly simplified hardware. Details of the OBSCA technique are given in [12], and digital implementations of the OBSCA, along with parallel computation structures for the DFSM, are given in [13].

4 ON THE DESIGN OF DIGITAL SPECTRAL CORRELATION ANALYZERS

To illustrate some aspects of the design of digital spectral correlation analyzers, two implementation examples are presented. The examples are not meant to be exhaustive, or case studies of existing systems, but rather to illuminate various design considerations that apply to implementing the algorithms discussed in this article. As a first example, consider the design of an instrument for testing communications signals. For this type of instrument, the cycle frequencies of interest are known a priori, so estimation can be limited to a restricted number of cycle frequencies. For specificity, assume that a DSP board capable of 33 million floating point operations per second (MFLOPS) is available for the implementation.

For this problem the DFSM is the algorithm of choice. Given the specified hardware it is appropriate to consider the question of how long it would take to compute a particular estimate. Such a timing estimate can be found by considering the number of operations the processor must perform as a function of the problem size. To begin, the processor must perform $5N \log_2 N$ real multiplications and additions to compute the input N-point FFT. Computation of the cyclic periodogram along N_α lines of constant cycle frequency requires at most $4NN_\alpha$ real multiplications and $2NN_\alpha$ real additions. (Note that this is the precise operation count for $\alpha = 0$. As $|\alpha|$ increases toward f_s, the number of computations in the cyclic periodogram computation decreases.) In order to increase computational efficiency, frequency smoothing is accomplished by the simple hopped summing operations previously described. This operation requires at most $2NN_\alpha$ real additions. Finally, say that a floating point operation, either multiplication or addition, can be performed in T_{fp} seconds. Thus, the cyclic spectrum of a real valued signal can be estimated along N_α lines of constant cycle frequency in approximately

$$T_c \approx (5N \log_2 N + 8NN_\alpha)T_{fp} \tag{49}$$

seconds. Estimates of the computation time with a 33 MFLOPS processor are listed in Table 5 for a variety of problem sizes.

Table 5 indicates that the processing time for some problems might be unacceptable. One approach to decreasing the computation time is to obtain a faster processor. For purposes of processor sizing, say that a fixed amount of time is available for the computation. For this type of instrument a reasonable delay time, i.e., the time from when the data is available for processing to when the results are displayed, is 2 seconds. The processor speeds required to perform the computations in 2 seconds are listed in Table 6. Note that other computational facets of the system, such as graphics computations for display of the results, might require a significant number of computations, but are ignored here.

Table 5: Table for Estimating Time Required to Compute Estimates along N_α Lines of Constant Cycle Frequency with the DFSM. (Times are listed in seconds for a 33 MFLOPS processor.)

		N									
		2^{11}	2^{12}	2^{13}	2^{14}	2^{15}	2^{16}	2^{17}	2^{18}	2^{19}	2^{20}
N_α	4	0.01	0.01	0.02	0.05	0.11	0.22	0.46	0.97	2.02	4.19
	8	0.01	0.02	0.03	0.07	0.14	0.29	0.59	1.22	2.53	5.21
	16	0.01	0.02	0.05	0.10	0.20	0.41	0.85	1.73	3.54	7.24
	32	0.02	0.04	0.08	0.16	0.33	0.67	1.35	2.75	5.58	11.31
	64	0.04	0.07	0.14	0.29	0.58	1.18	2.37	4.78	9.64	19.45
	128	0.07	0.13	0.27	0.54	1.09	2.19	4.40	8.85	17.78	35.72
	256	0.13	0.26	0.52	1.05	2.11	4.23	8.47	16.98	34.05	68.25

Table 6: Table for Estimating MFLOPS Required to Compute Estimates along N_α Lines of Constant Cycle Frequency with the DFSM (computation time is two seconds.)

		N									
		2^{11}	2^{12}	2^{13}	2^{14}	2^{15}	2^{16}	2^{17}	2^{18}	2^{19}	2^{20}
N_α	4	0.2	0.4	0.8	1.7	3.5	7.3	15.3	32.0	66.6	138.4
	8	0.2	0.5	1.1	2.2	4.6	9.4	19.5	40.4	83.4	172.0
	16	0.4	0.8	1.6	3.2	6.7	13.6	27.9	57.1	116.9	239.1
	32	0.6	1.3	2.6	5.3	10.8	22.0	44.7	90.7	184.0	373.3
	64	1.2	2.3	4.7	9.5	19.2	38.8	78.2	157.8	318.2	641.7
	128	2.2	4.4	8.9	17.9	36.0	72.4	145.4	292.0	586.7	1178.6
	256	4.3	8.6	17.3	34.7	69.6	139.5	279.6	560.5	1123.5	2252.3

A second alternative to obtaining a processor with the speeds indicated in Table 6 is to parallelize the computations as previously described. A natural partition is to compute the input FFT in one second and then compute and smooth the cyclic periodograms in the remaining second. For the problems indicated in the tables, the FFT processor would have to perform on the order of 105 MFLOPS for the largest computation. The correlation processors would have to perform on the order of 2150 MFLOPS. Both the FFT computation and the cyclic periodogram computation can be further parallelized. Computation of the cyclic periodogram calculation is achieved by assigning sets of cycle frequencies to individual processors. For example, the design goals are obtained if $R = 32$ processors each with 67 MFLOPS processing speed are used.

As a second example, consider the design of a system to estimate the cyclic spectrum of a real-valued signal over the entire bifrequency plane and to perform the computation in real-time. Recall that a system is considered to operate in real-time if the amount of time required to compute an estimate is equal to the amount of time required to collect the data for the computation. From the preceding discussions, the need for parallel processing is evident. The high degree of parallelism inherent in the FAM makes that algorithm a good choice for this problem. The system consists of a channelizer to compute the complex demodulates and R processors to compute and smooth the product sequences. As with the previous problem, the characteristics of the system can be inferred from estimates of the number of computations required for a variety of values of Δf and $\Delta t \Delta f$.

The complex demodulates of the input sequence are computed by the channelizer. In order to improve computational efficiency, the decimation factor is set to $L = N'/2$ instead of the typical value of $L = N'/4$. Each of the R correlation processors is assigned to compute and smooth the product sequences associated with a subset of the $(N')^2/4$ channel-pair regions. The processors are assigned channel-pair regions in such a way as to form regions where adjacent-cell combining can be performed with a minimum of data transfer between processors. After the correlation processors have been loaded with appropriate data, they begin to cross multiply complex demodulate streams, and Fourier transform (smooth) product sequences. The FAM generates a large amount of data for many problems of interest. For example, if the input stream consists of $N = 2^{23}$ samples and $N' = 256$, then approximately 2^{29} point estimates are generated by the FAM. Most of these point estimates are irrelevant. It is therefore appropriate to "thin" the data by applying a threshold test. Point estimates that fall below the threshold can be discarded with little effect on the overall estimate.

The processing speed required by the correlation processors is estimated by the following simplified analysis. The analysis uses the following simplifications:

1. Each correlation processor can compute a P-point FFT in $T_{FFT} = \mu P \log_2 P$ seconds where μ^{-1} is on the order of the correlation processor FLOP rate.
2. The channelizer has sufficient resources to channelize the data in less time than it takes R correlation processors to process an epoch.
3. The time to compute the product sequences, perform adjacent-cell combining, and thin the data is negligible compared to T_{FFT}.
4. Data transfer times are negligible.
5. The time required to compute the display is ignored.

With these assumptions, the average time required to process an epoch is the number of channel-pair regions times T_{FFT} divided by the number of correlation processors,

$$T_E \approx \frac{\mu (N')^2 P \log_2 P}{4R}. \tag{50}$$

Substituting $P = N/L$ and $L = N'/2$, approximation (50) becomes

$$T_E \approx \left(\frac{\mu}{R}\right)\left[\frac{NN'}{2} \log_2 \frac{2N}{N'}\right]. \tag{51}$$

The quantity in brackets is roughly the number of operations required to perform the estimation. Table 7 lists the quantity in brackets for a number of values of N and N'.

Table 7: Table for Estimating the Number of Operations Required to Compute an Epoch with the FAM. (Table entries are scaled by 10^9. The decimation factor for this table is set to $L = N'/2$. For reference, $\Delta t \Delta f = N/N'$ and $\Delta f = f_s/N'$.)

		N						
		2^{17}	2^{18}	2^{19}	2^{20}	2^{21}	2^{22}	2^{23}
	8	0.0079	0.0168	0.0357	0.0755	0.1594	0.3355	0.7046
	16	0.0147	0.0315	0.0671	0.1426	0.3020	0.6375	1.3422
N'	32	0.0273	0.0587	0.1258	0.2684	0.5704	1.2080	2.5501
	64	0.0503	0.1091	0.2349	0.5033	1.0737	2.2817	4.8318
	128	0.0923	0.2013	0.4362	0.9395	2.0133	4.2950	9.1268
	256	0.1678	0.3691	0.8053	1.7448	3.7581	8.0531	17.1799

Consider a system that meets the previous simplifications and has $R = 32$ processors, where each processor has $\mu = 20 \times 10^{-9}$ seconds. Say that an estimate is performed with a frequency resolution of $\Delta f = f_s/256$ and a time-frequency resolution product of $\Delta t \Delta f = 2048$. Using the relationships $\Delta f = f_s/N'$ and $\Delta t \Delta f = N/N'$, it is determined that $N' = 256$ and $N = 2^{19}$. From Table 7 and approximation (51), the approximate time to compute an epoch using the FAM is $T_E \approx 0.5033$ seconds. The real-time bandwidth of the system is determined by equating the collection time T to the time required to compute an epoch, T_E. Say that the system sampling frequency is $f_s = 2.5B$, where B is the signal bandwidth. Then, since $N = Tf_s = 2.5TB$, equating $T = T_E$ and solving for B, it is found that the system's real-time bandwidth is $B = 416.7$ kHz.

5 CONCLUSIONS

The objective of this article is to describe the theory and implementation of digital spectral correlation algorithms. In particular, the FFT Accumulation Method (FAM), the Strip Spectral Correlation Algorithm (SSCA), and the Digital Frequency Smoothing Method (DFSM) are described. All three of the algorithms are based on either the time or frequency smoothed cyclic periodogram. Each is derived by manipulating the smoothed cyclic periodogram equations. Each algorithm can be represented as a quadratic system characterized by its kernel function. With the kernel representation, attributes such as frequency resolution, time-frequency resolution product, cycle leakage, amplitude scaling, and computational complexity are easily determined. Regarding computational complexity, the FAM and SSCA are well suited for computing estimates of the cyclic (cross) spectrum over the entire bifrequency plane. The DFSM is best suited for computing estimates of the cyclic (cross) spectrum along a moderate number of lines of constant cycle frequency. Parallelization of the algorithms is briefly mentioned, and several examples illustrate some implementation issues.

It is appropriate at this point to list some relevant topics that are not covered. With regard to algorithms, the FAM and SSCA produce a great deal of data, much of which is irrelevant because most signals have spectral correlation features at relatively few cycle frequencies. Thinning algorithms are desirable to reduce the amount of data transmitted to postprocessing systems using the spectral correlation estimates. It is very important to have algorithms that convert raw estimates into a more intelligible form. Such algorithms might be graphical, if a human provides the interpretation, or perhaps might be automatic pattern recognition classification algorithms, if a machine performs the interpretation. Algorithms for computing estimates over the entire bifrequency plane and along lines of constant cycle frequency are discussed. Algorithms for computing estimates along lines of constant frequency are not discussed. For completeness, it is noted that either time smoothing approaches, such as those discussed in [6] or frequency smoothing approaches, such as those discussed in [13], are available for this purpose.

With regard to implementation, we hope it is apparent that parallel processing is the key to real-time spectral correlation analysis. This article only touches on parallelization of the FAM and SSCA. Aspects such as network topology, load balancing, and interprocessor communications (particularly for adjacent cell combining in the FAM) are not discussed. Finally, only one type of channelization is mentioned. Other types of channelizers are available and indeed might be better suited for a particular analysis task.

REFERENCES

[1] W. A. Gardner, "The spectral correlation theory of cyclostationary time-series," *Signal Processing*, vol. 11, pp. 13–36, July 1986.

[2] W. A. Gardner, "Measurement of spectral correlation," *IEEE Trans. Acoust., Speech, Signal Processing*, vol. ASSP-34, pp. 1111–1123, October 1986.

[3] W. A. Gardner, *Statistical Spectral Analysis: A Nonprobabilistic Theory*, Englewood Cliffs, NJ: Prentice-Hall, 1987.

[4] W. A. Brown, "On the theory of cyclostationary signals," Ph.D. dissertation, Department of Electrical Engineering and Computer Science, University of California, Davis, September 1987.

[5] R. S. Roberts, "Architectures for digital cyclic spectral analysis," Ph.D. dissertation, Department of Electrical Engineering and Computer Science, University of California, Davis, September 1989.

[6] R. S. Roberts, W. A. Brown, and H. H. Loomis, Jr., "Computationally efficient algorithms for cyclic spectral analysis," *IEEE Signal Processing Magazine*, vol. 8, pp. 38–49, April 1991.

[7] W. A. Brown and H. H. Loomis, Jr., "Digital implementations of spectral correlation analyzers," *IEEE Trans. Signal Processing*, vol. 41, no. 2, pp. 703–720, February, 1993.

[8] A. A. Gerlock, "High-speed coherence processing using the sectionalized Fourier transform," *IEEE Trans. Acoust., Speech, and Signal Processing*, vol. ASSP-30, pp. 189–205, April 1982.

[9] R. S. Roberts, "A time-frequency smoothing algorithm for cyclic spectral analysis," Workshop on Cyclostationary Signals, Yountville, CA, August 16–18, 1992, Los Alamos National Laboratory document LA–UR–93–1984.

[10] R. Crochiere and L. Rabiner, *Multirate Digital Signal Processing*. Englewood Cliffs, NJ: Prentice-Hall, 1983.

[11] B. G. Agee and W. A. Gardner, "Cyclic spectrum analysis study: Executive summary and final briefing," ARGOSystems Technical Report No. B83-0014, June 1984.

[12] W. A. Gardner and R. S. Roberts, "One-bit spectral-correlation algorithms," *IEEE Trans. Signal Processing*, vol. 41, pp. 423–427, January 1993.

[13] R. S. Roberts and H. H. Loomis, Jr., "Parallel computation structures for a class of cyclic spectral analysis algorithms," to appear in the *Journal of VLSI Signal Processing*.

Article 7

On Recent Developments in Prediction Theory for Cyclostationary Processes

Abol G. Miamee
Department of Mathematics
Hampton University
Hampton, VA 23668

1 INTRODUCTION

Interest in prediction theory of cyclostationary (also called periodically correlated) processes is growing rapidly. This is because of its potential applications in economics and various fields in science and engineering, including climatology, meteorology, hydrology, oceanology, medicine, and biology (cf. [1,2,3,4,5] and Chapter 1 in this volume and references therein). The foundations for mathematical treatment of (discrete-time) cyclostationary sequences were laid down in [6]. Gladyshev showed that every cyclostationary sequence $x = \{x(n) : n \in \mathbf{Z}\}$ is strongly harmonizable and obtained a complete description for their spectrum. He also proved that any cyclostationary sequence can be represented as a linear combination of components of a certain multivariate stationary sequence which has some common approximation properties with the sequence x itself. Since then prediction theory of cyclostationary sequences has been studied by various authors including [7,8,9,10,11,12]. These mathematical foundations have been, to some degree, also extended to (continuous-time) cyclostationary processes in [3,13,14,15,16,17,18]. It seems that prediction of continuous-time cyclostationary processes inevitably leads to consideration of infinite dimensional stationary processes. The fact that a continuous-time cyclostationary

Supported by Office of Naval Research Grant N00014-89-J-1824 and U.S. Army Research Office Grant DAAL 03-91-G-0238.

process need not be harmonizable [13,15] has obscured its spectral analysis. Under the assumption of strong harmonizability, the description of the spectrum and a certain series representation of a continuous-time cyclostationary process were given in [15,19]; also [16] obtained a characterization of the spectrum and the random spectrum of bounded cyclostationary processes.

In this article we briefly review these and some other recent developments of prediction theory for cyclostationary processes. Section 2 is devoted to notation and preliminaries. Discrete-time cyclostationary processes are studied in Section 3, and continuous-time cyclostationary processes are studied in Section 4.

2 PRELIMINARIES

Let β be a σ-field of subsets of the sample space Ω and P be a probability measure on β. The triple (Ω, β, P) is called a probability space. Throughout this article, the term *random variable* means a measurable function on the probability space (Ω, β, P) with zero mean and finite variance. So a random variable x is an element of the Hilbert space $L^2(\Omega, \beta, P)$, and for most of our purposes can be considered to be an element in any complex Hilbert space $\mathbf{H}$.

Definition: Let $\mathbf{G}$ represent the set of real numbers $\mathbf{R}$ or integers $\mathbf{Z}$. A bounded measurable $\mathbf{H}$-valued function $x = \{x(t) : t \in \mathbf{G}\}$ is said to be a ($\mathbf{H}$-valued) stochastic process. To distinguish the discrete-time case $\mathbf{G} = \mathbf{Z}$ from the continuous-time case $\mathbf{G} = \mathbf{R}$, we usually refer to the former one as a sequence and to the latter one as a process.

Definition: Let $\mathbf{F}$ be a locally convex topological vector space and $B(\mathbf{F}, \mathbf{H})$ denote the space of all bounded linear transformations from $\mathbf{F}$ into $\mathbf{H}$. For $\mathbf{G} = \mathbf{R}$ or $\mathbf{G} = \mathbf{Z}$, a $B(\mathbf{F}, \mathbf{H})$-valued function $\mathbf{x} = \{\mathbf{x}(t) : t \in \mathbf{G}\}$ is said to be stationary if $\mathbf{x}(t) f$ is continuous for every $f \in \mathbf{F}$ and

$$(\mathbf{x}(t) f, \mathbf{x}(s) g) = (\mathbf{x}(t+h) f, \mathbf{x}(s+h) g), \qquad \text{for all } f, g \in \mathbf{F} \text{ and } t, s, h \in \mathbf{G}.$$

Here $(\cdot, \cdot)$ denotes the inner product of two elements in the Hilbert space $\mathbf{H}$. If $\dim \mathbf{F} = 1$ the process is simply a (univariate) stationary process and if $\dim\ \mathbf{F} = \infty$ the process is called infinite dimensional. If $\dim \mathbf{F} = q < \infty$, then $\mathbf{x}$ will be referred to as a q-variate or a multivariate stationary process and is usually denoted by $\mathbf{x}(t) = (x^1(t), x^2(t), \ldots, x^q(t))'$, with $x^i(t) = \mathbf{x}(t) e_i$, and $\{e_j\}$ being the standard basis of $\mathbf{F}$. We use $(\ldots)'$ to denote the transpose of the vector $(\ldots)$.

Definition: An ($\mathbf{H}$-valued) stochastic process $x = \{x(t) : t \in \mathbf{G}\}$ is said to be cyclostationary (CS) or periodically correlated if there exists some $T \in \mathbf{G}$ such that

$$(x(s), x(t)) = (x(s+T), x(t+T)), \qquad \text{for all } s, t \in \mathbf{G}.$$

In this case the smallest such T is called the period of the process $\mathbf{x}$. The following subspaces play a crucial role in prediction theory of an $\mathbf{H}$-valued stochastic process

$x = \{x(t) : t \in \mathbf{G}\}$:

$$\begin{aligned} H(x) &= \text{sp}\{x(t) : t \in \mathbf{G}\}, && \text{time domain of } x \\ H(x; s) &= \text{sp}\{x(t) : t \leq s\}, && \text{past of } x \text{ at time } s \\ H(x; -\infty) &= \cap H(x; s), && \text{remote past of } x, \end{aligned}$$

where sp{ } stands for the closed linear span of all elements inside { } in the **H**-topology.

Definition: An **H**-valued stochastic sequence $x = \{x(t) : t \in \mathbf{G}\}$ is called deterministic if $H(x; -\infty) = H(x)$, and it is called nondeterministic otherwise. It is called purely nondeterministic if $H(x; -\infty) = \{0\}$.

Definition: The time domain, past, and remote past of a $B(\mathbf{F}, \mathbf{H})$-valued process $\mathbf{x} = \{\mathbf{x}(t) : t \in \mathbf{G}\}$ are similarly defined to be the following subspaces of **H**:

$$\begin{aligned} H(\mathbf{x}) &= \text{sp}\{\mathbf{x}(t)f : f \in \mathbf{F}, t \in \mathbf{G}\}, && \text{time domain of } \mathbf{x} \\ H(\mathbf{x}; s) &= \text{sp}\{\mathbf{x}(t)f : f \in \mathbf{F}, t \leq s\}, && \text{past of } \mathbf{x} \text{ at time } s \\ H(\mathbf{x}; -\infty) &= \cap H(\mathbf{x}; s), && \text{remote past of } \mathbf{x}. \end{aligned}$$

For a $B(\mathbf{F}, \mathbf{H})$-valued process **x**, the notions of determinism, and pure nondeterminism can be similarly defined.

3 DISCRETE-TIME CYCLOSTATIONARY PROCESSES

In this section we discuss some prediction problems for CS sequences. It is well known that CS sequences are closely related to multivariate (MV) stationary sequences. In fact CS sequences are connected to MV stationary sequences in a number of different ways. These connections have been exploited by several authors in their investigations of different prediction questions concerning CS sequences. Here these ideas are discussed and their applications to some prediction problems for cyclostationary sequences (CSSs) are given.

It is well known that any CSS is harmonizable and hence has a spectrum. This is important because it allows for the use of the powerful tools of harmonic analysis in developing a prediction theory for CSSs similar to that of stationary sequences. However since the spectrum of a CSS is given through a bimeasure rather than a measure, this approach is not as fruitful as might be expected. For instance, in order to use the spectral approach for developing a successful prediction theory similar to that for stationary sequences, the time domain must be complete. But this does not seem to be the case. Hence, the time-domain approach seems more appropriate for studying CSSs and, in this section, time-domain techniques are adopted. We first investigate their moving average and autoregressive properties and then exploit the close tie between CSSs and multivariate stationary sequences (MVSS) to study some prediction problems concerning CSSs.

3.1 Autoregressive and Moving Average Representations

One way of predicting a seasonal time-series is that of fitting it to an autoregressive (AR) or moving average (MA) model. Here we consider some properties of these models. For a CSS, its AR and MA coefficients can vary with time; however, these coefficients must vary in a periodic manner.

An **H**-valued stochastic process $x = \{x(t) : t \in \mathbf{Z}\}$ is said to be MA of order q if

$$x(t) = \psi(t) + b_1(t)\psi(t-1) + \ldots + b_q(t)\psi(t-q), \quad t \in \mathbf{Z},$$

and it is said to be AR of order q if

$$x(t) + a_1(t)x(t-1) + \ldots + a_q(t)x(t-q) = \psi(t), \quad t \in \mathbf{Z}, \tag{1}$$

where $\psi(t)$ is a white noise sequence. For each integer k, $b_k(t)$ and $a_k(t)$ are constant if x is stationary but change with t, in a periodic fashion, if x is CS.

If $x = \{x(t) : t \in \mathbf{Z}\}$ is a CSS of period T, then its AR coefficients $a_k(t)$ are periodic with period T and hence can be expanded in a Fourier series:

$$a_k(t) = \sum_{j=0}^{T-1} \left[c_{kj} \cos \frac{2\pi jt}{T} + d_{kj} \sin \frac{2\pi jt}{T} \right].$$

Substituting these series representations into the AR representation (1) yields the following AR representation for x

$$x(t) = \psi(t) - \sum_{k=1}^{q} c_{k0}x(t-k) - \sum_{k=1}^{q}\sum_{j=1}^{T-1} [c_{kj}x(t-k) \cos \frac{2\pi jt}{T}$$
$$+ d_{kj}x(t-k) \sin \frac{2\pi jt}{T}]$$

This shows that the best linear predictor of a CSS in terms of its past can be expressed in terms of its past data and the past data multiplied by sines and cosines of the fundamental and harmonic frequencies of the period [8,20].

We now consider MA representations of a CSS $x = \{x(t) : t \in \mathbf{Z}\}$. Let

$$\hat{x}(t : t-1) = P_{H(x;t-1)}x(t)$$

denote the best (in the least squares sense) linear predictor of $x(t)$ based on its past (P_M denotes the orthogonal projection from the Hilbert space **H** onto its subspace M). When $x(t) = \hat{x}(t; t-1)$, we define $\psi(t) = 0$ otherwise we take $\psi(t)$ to be the normalized innovation

$$\psi(t) = \frac{x(t) - \hat{x}(t; t-1)}{\|x(t) - \hat{x}(t; t-1)\|}, \qquad t \in \mathbf{Z}.$$

One can then express $H(x; t)$ as the direct sum $H(x; t) = sp\{\psi(t)\} \oplus H(x; t-1)$. Continuing this process q more times

$$H(x; t) = \text{sp}\{\psi(t), \psi(t-1), \ldots, \psi(t-q)\} \oplus H(x; t-q-1).$$

Since $x(t) \in H(x; t)$, one can see that

$$x(t) = \sum_{k=0}^{q} b_k(t)\psi(t-k) + v^q(t), \quad \text{with } v^q(t) = P_{H(x;t-q-1)}x(t). \tag{2}$$

This general decomposition of x is not as useful as one might like it to be because, in contrast to the stationary case, the coefficients $b_k(t)$ here vary with time t. However in the case of a CSS x, these coefficients vary in a periodic manner. That is, for each k the coefficient $b_k(t)$ is periodic in t with period T and hence it can be expanded as

$$b_k(t) = \sum_{j=0}^{T-1} e_{kj} \exp\left(\frac{2\pi ijt}{T}\right).$$

Substituting these representations for the b_k's into (2), one obtains

$$x(t) = \sum_{j=0}^{T-1} \left(\exp\left(\frac{2\pi ijt}{T}\right) \sum_{k=0}^{q} e_{kj}\Psi(t-k) \right) + v^q(t). \tag{3}$$

This representation can be used to prove the following decomposition of the MA part of CSS in the sum of finitely many MA stationary sequences.

Proposition *Suppose $x = \{x(t) : t \in \mathbf{Z})$ is a MA CSS with period T. Then x can be decomposed into the sum of T moving average stationary sequences $y^j = \{y^j(t) : t \in \mathbf{Z}\}$, $j = 0, 1, \ldots, T-1$, each having the same order as that of the original sequence x.*

Proof: If the CSS x has an MA model of order q, then all the $v^q(t)$'s in (3) vanish,

$$x(t) = \sum_{j=0}^{T-1} \left(\exp\left(\frac{2\pi ijt}{T}\right) \sum_{k=0}^{q} e_{kj}\Psi(t-k) \right) = \sum_{j=0}^{T-1} y_j. \tag{4}$$

One can then check that each MA sequence $y^j = \{y^j(t) : t \in \mathbf{Z}\}$, $j = 0, 1, \ldots, T-1$ is stationary.

Remarks: We can rewrite (4) as

$$\begin{aligned} x(t) &= \sum_{j=0}^{T-1} \left(\sum_{k=0}^{q} e_{kj} \exp\left(\frac{2\pi ijk}{T}\right) \exp\left(\frac{2\pi ij(t-k)}{T}\right) \Psi(t-k) \right) \\ &= \sum_{j=0}^{T-1} \sum_{k=0}^{q} e_{kj} \exp\left(\frac{2\pi ijk}{T}\right) \Psi^j(t-k), \end{aligned}$$

where each $\psi^j(t) = \exp(2\pi ijt/T)\,\psi(t)$, $j = 0, 1, \ldots, T-1$, is a white noise sequence. This time-independent MA representation for the CSS x shows that in order to complete our modeling one needs to estimate only Tq parameters and use them with the innovations data as well as their products with some appropriate harmonic functions.

Using the above decompositions one can immediately prove the following Wold decompositions with time-independent coefficients.

Theorem *A CSS $x = \{x(t) : t \in \mathbf{Z}\}$ with period T can be decomposed as $x(t) = u(t) + v(t)$ with*

$$u(t) = z^0(t) + \exp\left(\frac{2\pi it}{T}\right) z^1(t) + \cdots + \exp\left(\frac{2\pi it(T-1)}{T}\right) z^{T-1}(t),$$

where

i. *the sequences $z^j = \{z^j(t) : t \in \mathbf{Z}\}$, $j = 0, 1, \ldots, T-1$, are purely nondeterministic and jointly stationary,*

ii. *the sequence $v = \{v(t) : t \in \mathbf{Z}\}$ is CS and deterministic,*

iii. *the sequence $v = \{v(t) : t \in \mathbf{Z}\}$ is uncorrelated with each of the sequences $z^j = \{z^j(t) : t \in \mathbf{Z}\}$, $j = 0, 1, \ldots, T-1$.*

Proof: It is sufficient in (3) to take

$$z^j(t) = \sum_{k=0}^{q} e_{kj}\Psi(t-k).$$

Theorem *Let $x = \{x(t) : t \in \mathbf{Z}\}$ be a CSS with period T. Then it can be decomposed as $x(t) = u(t) + v(t)$ with*

$$u(t) = y^0(t) + y^1(t) + \cdots + y^{T-1}(t)$$

such that

i. *each sequence $y^j = \{y^j(t) : t \in \mathbf{Z}\}$ is stationary and purely nondeterministic,*

ii. *$v(t)$ is CS, deterministic, and uncorrelated to each $y^j(t)$.*

Proof: It is sufficient in (3) to take

$$y^j(t) = \exp\left(\frac{2\pi ijt}{T}\right) z^j(t) = \exp\left(\frac{2\pi ijt}{T}\right) \sum_{k=0}^{q} e_{kj}\Psi(t-k).$$

3.2 Prediction of Discrete-Time Cyclostationary Sequences

There is a well known connection between any CSS of period T and the MVSS formed from its consecutive blocks of length T. This connection can be used to study some prediction properties of CSSs such as their Wold decomposition and Wold-Cramer concordance.

The correlation function $R(t,\tau) = (x_{t+\tau}, x_t)$ of a CSS $x = \{x(t) : t \in \mathbf{Z}\}$ of period T is periodic in t with period T. Hence it can be decomposed as

$$R(t,\tau) = \sum_{k=0}^{T-1} R_k(\tau) \exp\left(\frac{2\pi i j t}{T}\right). \tag{5}$$

The functions $R_k(\tau), k = 0, 1, \ldots, T-1$ are extended to all integers by $R_k(\tau) = R_{k+T}(\tau)$. There exist measures $F_k, k = 0, 1, \ldots, T-1$ and bimeasure

$$F(A, B) = \sum_{k=0}^{T-1} F_k(A \cap (B - 2\pi k/T))$$

such that

$$R_k(\tau) = \int_0^{2\pi} e^{-i\tau\lambda} dF_k(\lambda) \quad \text{and} \quad R(t,\tau) = \int_0^{2\pi}\int_0^{2\pi} e^{-i(t+\tau)\lambda + it\theta} dF(\lambda, \theta).$$

For any subset B of real numbers and any real number λ, $B - \lambda$ denotes the set $\{b - \lambda : b \in B\}$. This shows that the bimeasure F of a CSS is concentrated on $2T-1$ line segments $\lambda - \theta = 2\pi k/T, k = 1-T, \ldots, T-1$ in the square $[0, 2\pi)^2$, with F_k being the mass of F on each of these lines. When we talk about the absolute continuity of this spectral bimeasure F, its derivative F', its density f, or its Cramer components F^a and F^s, we mean with respect to the measure whose restriction to each of these $2T-1$ segments is Lebesgue measure.

For a fixed integer T, to each univariate sequence $x = \{x(t) : t \in \mathbf{Z})$ we correspond the MV sequence $y = \{y(t) : t \in \mathbf{Z}\}$ defined by $y(t) = (y^0(t), y^1(t), \ldots, y^{T-1}(t))' = (x_{tT}, x_{tT+1}, \ldots, x_{tT+T-1})'$. This correspondence is clearly one to one. It is well known that x is CS with period T iff its corresponding MV sequence y is stationary. If we decompose y (according to the Wold-decomposition for MVS sequences) as the sum of two uncorrelated MVSSs $y(t) = y^r(t) + y^s(t)$ with y^r being purely nondeterministic and y^s deterministic, then we can go back and consider their corresponding CSSs x^r and x^s to get the Wold-decomposition $x(t) = x^r(t) + x^s(t)$ of x as the sum of two uncorrelated CSSs x^r and x^s, which are, respectively, purely nondeterministic and deterministic. We define the rank of a CSS $x = \{x(t) : t \in \mathbf{Z}\}$ to be the number of $x(t)$'s inside one period with $x(t) \notin H(x; t-1)$. It is not hard to see that the rank of a CSS x is equal to the rank of its corresponding MVSS y as given in [21,22]. Using this result in conjunction with the available Wold-Cramer concordance results for MVSSs in [21,22,23] one can similarly prove that the Wold-decomposition $x(t) = x^r(t) + x^s(t)$ of a CSS x is in concordance with the Cramer-decomposition

$F = F^a + F^s$ of its spectral bimeasure F. That is, one can prove that the spectral bimeasures of the CS components x^s and x^r are identical to F^s and F^a, respectively, iff the rank of x equals the rank of its spectral density F'.

Since prediction of the deterministic component is usually not of interest, throughout the remainder of this section the CSSs are assumed to be purely nondeterministic with an absolutely continuous (a.c.) spectral bimeasure F that has a spectral density f. Let $x = \{x(t) : t \in \mathbf{Z}\}$ be CSS and y be its MVSS in the preceding. Under appropriate assumptions, one can continue to use this linkage and obtain an algorithm for finding the best linear predictor $\hat{x}(t; t-1)$ of $x(t)$ based on its past $H(x; t-1)$. Because of the periodicity, our attention can be restricted to just one period, namely to $t = 0, 1, \ldots, T-1$. For $t = 0$ the coefficients a_k of our predictor

$$\hat{x}(0, -1) = \sum_{k<0} a_k x_{-k}$$

are exactly the coefficients of the 0-th component of the predictor $\hat{y}(0; -1)$ of the corresponding MVSS y and hence can be obtained using prediction theory of MVSSs [21,22,24]. To obtain the predictor $\hat{x}(t, t-1)$, for $0 < t < T$ we first note the following orthogonal decomposition of $H(x, t-1)$

$$H(x, t-1) = H(x, -1) \ominus \text{sp}\,\{x(i) - \hat{x}(i, -1) : 0 \le i \le t-1\}$$

and then we use this decomposition to write

$$\begin{aligned}\hat{x}(t : t-1) &= P_{H(x;t-1)}x(t) = P_{H(x;-1)}x(t) + \sum_{i=0}^{t-1} c_i(x(i) - \hat{x}(i, -1)) \\ &= \sum_{k<0} b_k(t)x_k + \sum_{i=0}^{t-1} c_i(x(i) - \hat{x}(i, -1)).\end{aligned}$$

The $b^k(t)$'s are exactly the coefficients of the t-th component of the predictor $\hat{y}(0; -1)$ of the corresponding MVSS y and hence can, again, be obtained using prediction theory of MVSSs. The other coefficients, namely c_j's can be obtained from a system of t linear equations resulting from the following orthogonality requirements

$$x(t) - \sum_{k<0} a_k x_k - \sum_{i=0}^{t-1} c_i(x(i) - \hat{x}(i, -1)) \perp x(s), \quad s = 0, \ldots, t-1.$$

The prediction problem for CSSs can be handled more efficiently using another related MVSS z appearing in the following theorem. As we shall see, this new technique results in an explicit formula for the predictor of CSSs which can be calculated in a single step.

Theorem (Gladyshev [6]) *The function $R(t, \tau)$ in (5) is the correlation function of a CSS iff the matrix-valued function*

$$\mathbf{R}(\tau) = (R_{kk'}(\tau))_{k,k'=0}^{T-1}, \quad \textit{with } R_{kk'}(\tau) = R_{k'-k}(\tau) \exp\left(\frac{2\pi i k}{T}\right)$$

is the correlation function of some MVSS $\mathbf{z} = \{\mathbf{z}^k(t) : t \in \mathbf{Z}\}$.

In order to use this new MVSS $\mathbf{z}$ for prediction purposes it is crucial to have an explicit representation of this $\mathbf{z}$ sequence in terms of the original CSS x, and vice versa. The following theorem which was first proved in [12] and is given as Theorem 4 in [25] gives such a representation.

Theorem (Miamee [12]) *Let $R(t, \tau)$ and $R(\tau) = (R_{kk'})$ be the correlation function of an* $\mathbf{H}$*-valued CSS $x = \{x(t) : t \in \mathbf{Z}\}$ and its MVSS $\mathbf{z} = \{\mathbf{z}^k(t), k = 1, \ldots, T, t \in \mathbf{Z}\}$, respectively, related as in (5). Then*

a) components $\mathbf{z}^k(t)$ of $\mathbf{z}$ can be expressed as elements in the direct sum $\mathbf{H} \oplus \ldots \oplus \mathbf{H}$ by

$$z^k(t) = \oplus \sum_{j=0}^{T-1} \frac{1}{\sqrt{T}} x(t+j) \exp\left(\frac{2\pi i k(t+j)}{T}\right),$$

b) Identifying $x(t)$ in H as $x(t) \oplus 0 \oplus 0 \oplus \ldots \oplus 0$ in $H \oplus H \oplus H \oplus \ldots \oplus H$ one has

$$x(t) = \frac{1}{\sqrt{T}} \sum_{j=0}^{T-1} z^k(t) \exp\left(\frac{-2\pi i k t}{T}\right).$$

This new MVSS captures more prediction properties of the original CSS than the widely used y sequence mentioned before. This closer relation gives the following explicit formula for the predictors of CSSs, which naturally results in a more efficient algorithm for finding the predictors. Details omitted here are given in [12].

Theorem *A CSS x has an autoregressive representation of the form*

$$\hat{x}(v; 0) = P_{H(x;0)} x(v) = \frac{1}{\sqrt{t}} \sum_{k=0}^{\infty} \left(\sum_{j=0}^{T-1} a_{k0j}^{v} \exp\left(\frac{-2\pi i k}{T}\right)\right) x(-k), \tag{6}$$

whenever its corresponding MVSS $\mathbf{z}$ has an autoregressive representation

$$\hat{\mathbf{z}}(v; 0) = P_{H(z;0)} \mathbf{z}(v) = \sum_{k=0}^{\infty} \mathbf{A}_k^v \mathbf{z}(-k). \tag{7}$$

Matrix coefficients in the representation (7) for the predictor at lag v of the MVSS $\mathbf{z}$ can be obtained using the available multivariate prediction techniques (cf. for example, [21,22,24]). Hence (6) furnishes an explicit formula for evaluating the predictor of CSSs under some appropriate assumptions.

4 CONTINUOUS-TIME CYCLOSTATIONARY PROCESSES

This section is devoted to prediction problems for continuous-time cyclostationary processes (CSP). The spectrum of these processes has been studied in [3,15,16,17]. In general, the spectrum of a CSP is a distribution rather than a measure. Hence spectral approaches here are even less promising than those in the discrete-time case and are, therefore, not discussed here. Our investigation of prediction properties of CSPs is limited to time-domain techniques and is again based on their close tie with stationary processes.

Two infinite dimensional stationary processes are associated with a bounded CSP $x = \{x(t) : t \in \mathbf{R}\}$. These processes can be considered as counterparts of the two processes we have already discussed in the discrete-time case. As we will see they capture some approximation properties of the original CSP, a matter that is crucial for developing a Wold-decomposition and obtaining some regularity conditions for CSPs. These two associated stationary processes have appeared in some earlier work [5,19]. They have recently been reconsidered in [16,17,18] after their discrete counterparts were proved to be very useful in [11,12].

Definition: Let $x = \{x(t) : t \in \mathbf{R}\}$ be an $\mathbf{H}$-valued CSP with period T. For each $k \in \mathbf{Z}$ and $t \in \mathbf{R}$ let $z_k(t)$ be the $\mathbf{H}$-valued function on $[0, T)$ defined by $z_k(t)(s) = \exp(2\pi ik(t+s)/T)x(t+s)$, $s \in [0, T)$.

The process $\mathbf{z} = \{z_k(t) : k \in \mathbf{Z}, t \in \mathbf{R}\}$, regarded as a $B(\mathbf{s_0}, L^2([0, T), \mathbf{H}))$-valued (continuous-time) process, turns out to be stationary. Here $L^2([0, T), \mathbf{H})$ is the space of all $\mathbf{H}$-valued functions on $[0, T)$ with a square integrable norm, and $\mathbf{s}_0$ is the space of all sequences indexed by $\mathbf{Z}$ with finitely many nonzero entries. Proof of stationarity of $\mathbf{z}(t) = \{z_k(t) : k \in \mathbf{Z}, t \in \mathbf{R}\}$ is easy and is therefore omitted.

Definition: Let $x = \{x(t) : t \in \mathbf{R}\}$ be any $\mathbf{H}$-valued stochastic process and $T > 0$. For every $f \in L^2[0, T)$ and $k \in \mathbf{Z}$ define

$$y_k f = \int_0^T x(kT + t) f(t) dt.$$

Clearly $y_k \in B(L^2([0, T), \mathbf{H}))$ for $k \in \mathbf{Z}$. The $B(L^2[0, T), \mathbf{H}))$-valued sequence $\mathbf{y} = \{y_k : k \in \mathbf{Z}\}$ will be referred to as the y-sequence associated with $x = \{x(t) : t \in \mathbf{R}\}$.

Proposition *$x = \{x(t) : t \in \mathbf{Z}\}$ is a CSP if and only if its associated infinite dimensional sequence y is stationary.*

Proof: If x is a CSP then for any two functions $f, g \in L^2[0, T)$ we have

$$(y_{m+1} f, y_{n+1} g) = \iint (x(mT+T+t), x(nT+T+s)) f(t) g(s) dt ds = (y_m f, y_n g),$$

which means $\mathbf{y}$ is stationary. Conversely, if $\mathbf{y}$ is stationary, that is, if

$$(y_{m+1} f, y_{n+1} g) = (y_m f, y_n g), \qquad \text{for all } f, g \in L^2[0, T),$$

then one can prove

$$\iint \varphi(s)\psi(t)(x(s+T), x(t+T)) ds dt = \iint \varphi(s)\psi(t)(x(s), x(t)) ds dt$$

for all rapidly decreasing functions ψ, φ on $\mathbf{R}$, which completes the proof.

Remarks: The z-process seems more promising using spectral-domain techniques, whereas the $\mathbf{y}$-sequence seems better using time-domain techniques. This follows because prediction theory of infinite dimensional stationary sequences is more developed than that of infinite dimensional stationary processes.

Now we use these ideas to study some prediction properties of CSPs. We start by showing that a CSP x and its associated $\mathbf{y}$-sequence and $\mathbf{z}$-process share the same prediction properties. In particular, we show that the CSP x and its associated $\mathbf{z}$-process and $\mathbf{y}$-sequence are deterministic (purely nondeterministic) simultaneously.

To proceed we need the following lemma whose proof is given in [16,17].

Lemma *Let $M(t)$ be a measurable subspace-valued function in a separable Hilbert space $\mathbf{H}$. Then the orthogonal complement $L^2([0, T), \mathbf{H}) \ominus \{f \in L^2([0, T), \mathbf{H}) : f(t) \in M(t)$ a.e.$\}$ of the subspace $\{f \in L^2([0, T), \mathbf{H}) : f(t) \in M(t)$ a.e.$\}$ in $L^2([0, T), \mathbf{H})$ is given by*

$$L^2([0, T), \mathbf{H}) \ominus \{f \in L^2([0, T), \mathbf{H}) : f(t) \in M(t) \text{ a.e.}\}$$
$$= \{f \in L^2([0, T), \mathbf{H}) : f(t) \in M(t)^{\perp} \text{ a.e.}\}.$$

Using this lemma one can prove the following two propositions.

Proposition *Let $x = \{x(t) : t \in \mathbf{R}\}$ be a CSP and let $\mathbf{y} = \{y_k : k \in \mathbf{Z}\}$ be its associated $\mathbf{y}$-sequence. Then $H(x; (n + 1)T) = H(y; n)$ for every $n \in \mathbf{Z}$.*

Proposition *Let $x = \{x(t) : t \in \mathbf{R}\}$ be a CSP with period T and let $\mathbf{z}(t) = \{z_k(t) : k \in \mathbf{Z}\}$, $t \in \mathbf{R}$, be its associated $\mathbf{z}$-process. Then*

i. $H(\mathbf{z}; \infty) = \{f \in L^2([0, T), \mathbf{H}) : f(t) \in H(x; \infty) \text{ a.e.}\}$,
ii. $H(\mathbf{z}; s) = \{f \in L^2([0, T), \mathbf{H}) : f(t) \in H(x; s + t) \text{ a.e.}\}$,
iii. $H(\mathbf{z}; -\infty) = \{f \in L^2([0, T), \mathbf{H}) : f(t) \in H(x; -\infty) \text{ a.e.}\}$.

The next theorem is an immediate consequence of the last two propositions.

Theorem *Consider a CSP $x = \{x(t) : t \in \mathbf{R}\}$ with period T and its associated $\mathbf{z}$-process and $\mathbf{y}$-sequence. The following conditions are equivalent:*

a. x is purely nondeterministic (respectively, deterministic),
b. z is purely nondeterministic (respectively, deterministic),
c. y is purely nondeterministic (respectively, deterministic).

This theorem makes it possible to utilize the prediction theory of infinite dimensional stationary processes for studying CSPs. As an example, we employ this idea to obtain a Wold-decomposition for CSPs.

Theorem (Wold-Decomposition) *Let $x = \{x(t) : t \in \mathbf{R}\}$ be a CSP with period T. Define $x^s(t) = P_{H(x;-\infty)}x(t)$ and $x^r(t) = x(t) - x^s(t)$ for $t \in \mathbf{R}$. Then*

i. $x^s = \{x^s(t) : t \in \mathbf{R}\}$ and $x^r = \{x^r(t) : t \in \mathbf{R}\}$ are two jointly CSPs with period T;

ii. $x^s = \{x^s(t) : t \in \mathbf{R}\}$ is deterministic;

iii. $x^r = \{x^r(t) : t \in \mathbf{R}\}$ is purely nondeterministic;

iv. $H(x^s; t) \oplus H(x^r; t) = H(x; t)$, for every $t \in \mathbf{R}$.

Furthermore, this decomposition is unique. That is if $x^s = \{x^s(t) : t \in \mathbf{R}\}$ *and* $x^r = \{x^r(t) : t \in \mathbf{R}\}$ *are any two CSPs satisfying* (**i**)–(**iv**) *as well as* $x(t) = x^s(t) + x^r(t)$, *for all* $t \in \mathbf{R}$, *then* $x^s(t)$ *must be equal to* $P_{H(x;-\infty)}x(t)$ *for all* $t \in \mathbf{R}$.

The following is a sketch of a proof of this theorem

Proof: Let $\mathbf{z}(t) = \{z_k(t) : k \in \mathbf{Z}, t \in \mathbf{R}\}$ be the $\mathbf{z}$-process associated with our CSP x. The Wold-decomposition theorem for infinite dimensional stationary processes given for example in [26] decomposes the $\mathbf{z}$-process uniquely into the sum of two $B(s_0, L^2([0, T), \mathbf{H}))$-valued processes $z^s(t) = \{z_k^s(t) : k \in \mathbf{Z}, t \in \mathbf{R}\}$ and $\mathbf{z}^r(t) = \{z_k^r(t) : k \in \mathbf{Z}, t \in \mathbf{R}\}$, such that

a. $\mathbf{z}(t) = \mathbf{z}^s(t) + \mathbf{z}^r(t), \quad$ for all $t \in \mathbf{R}$,

b. $H(\mathbf{z}; t) = H(\mathbf{z}^s; t) \oplus H(\mathbf{z}^r; t)$, for all $t \in \mathbf{R}$,

c. $H(\mathbf{z}^s; t) = H(\mathbf{z}^s; \infty), \quad$ for all $t \in \mathbf{R}$,

d. $H(\mathbf{z}^r; -\infty) = \{0\}$.

Now let $x^s(t) = P_{H(z;-\infty)}x(t)$ and $x^r(t) = x(t) - x^s(t)$, $t \in \mathbf{R}$. Then $H(x^s; t) \oplus H(x^r; t) = H(x; t)$ for $t \in \mathbf{R}$. If U is the unitary operator defined by $U(\sum a_k x(t_k)) = \sum a_k x(t_k + T)$, then $P_{H(x;-\infty)}U = UP_{H(x;-\infty)}$, because $UH(x; -\infty) \subseteq H(x; -\infty)$ and $U^{-1}H(x; -\infty) = H(x; -\infty)$. Therefore $x^s(t+T) = Ux^s(t)$ and $x^r(t+T) = Ux^r(t)$, for all $t \in \mathbf{R}$. This implies that the processes x^s and x^r are both CS with period T. The proof that $x^r = \{x^r(t) : t \in \mathbf{R}\}$ is purely nondeterministic and $x^s = \{x^s(t) : t \in \mathbf{R}\}$ is deterministic follows from (**c**) and (**d**).

Remarks: One can continue this line of work in conjunction with the known results on the spectrum of CSS processes to obtain some spectral characterizations for a CSP to be purely nondeterministic. However, because of their limited applications, this is not pursued here.

REFERENCES

[1] W. R. Bennett, "Statistics of regenerative digital transmission," *Bell System Tech. J.*, vol. 37, pp. 1501–1542, 1958.

[2] L. I. Gudzenko, "On periodic nonstationary processes," *Radio Eng. Electron. Phys.* (Russian), vol. 4, pp. 220–224, 1959.

[3] H. Ogura, "Spectral representation of periodically nonstationary random processes," *IEEE Trans. Inform. Theory*, vol. IT-17, pp. 143–149, 1971.

[4] H. Sakai, "Spectral analysis and lattice filter for periodic autoregressive processes," *Electronics and Communications in Japan*, Part 3, vol. 73, pp. 9–15, 1990.

[5] W. A. Gardner, *Introduction to Random Processes with Applications to Signals and Systems*, New York: Macmillan, 1985 (2nd ed., 1989, McGraw-Hill).

[6] E. G. Gladyshev, "Periodically correlated random sequences," *Soviet Math. Dokl.*, vol. 2, pp. 385–388, 1961.

[7] H. L. Hurd, "An investigation of periodically correlated stochastic processes," Ph.D. Dissertation, Duke University, Durham, North Carolina, 1969.

[8] M. Pagano, "On periodic and multiple autoregressions," *Ann Statist.*, vol. 6, pp. 1310–1317, 1978.

[9] A. G. Miamee and H. Salehi, "On the prediction of periodically correlated stochastic processes," *Multivariate Analysis-V*, (Ed., P. R. Krishnaiah), Amsterdam: North Holland Publishing Company, pp. 167–179, 1980.

[10] M. Pourahmadi and H. Salehi, "On subordination and linear transformation of harmonizable and periodically correlated processes," in *Probability Theory on Vector Spaces, III* (Lublin), Berlin-New York: Springer, pp. 195–213, 1983.

[11] A. G. Miamee, "Periodically correlated processes and their stationary dilations," *SIAM J. Appl. Math.* vol. 50, pp. 1194–1199, 1990.

[12] A. G. Miamee, "Explicit formulas for the best linear predictor of periodically correlated sequences," *SIAM J. Math. Anal.* vol. 24, pp. 703–711, 1993.

[13] E. G. Gladyshev, "Periodically and almost periodically correlated random processes with continuous time parameter," *Theory Prob. Appl.* vol. 8, pp. 173–177, 1963.

[14] H. L. Hurd, "Stationarizing properties of random shifts," *SIAM J. Appl. Math.*, vol. 26, pp. 203–212, 1974.

[15] H. L. Hurd, "Representation of strongly harmonizable periodically correlated processes and their covariances," *J. Multivariate Anal.*, vol. 29, pp. 53–67, 1989.

[16] A. Makagon, A. G. Miamee, and H. Salehi, "Periodically correlated processes and their spectrum," in *Nonstationary Stochastic Processes and Their Application* (Ed., A. G. Miamee), Singapore: World Scientific, pp. 147–164, 1992.

[17] A. Makagon, A. G. Miamee, and H. Salehi, "Continuous time periodically correlated processes: spectrum and predictions" *Stochastic Processes and their Applications*, 1994.

[18] H. L. Hurd and G. Kallianpur, "Periodically correlated processes and their relationship to $L_2[0, T)$-valued stationary sequence," in *Nonstationary Stochastic Processes and Their Application* (Ed., A. G. Miamee) Singapore: World Scientific, pp. 256–284, 1992.

[19] W. A. Gardner and L. E. Franks, "Characterization of cyclostationary random signal processes," *IEEE Trans. Inform. Theory*, vol. IT-21, pp. 4–14, 1975.

[20] W. M. Brelsford and R. H. Jones, "Time series with periodic structure," *Biometrica*, vol. 54, pp. 403–407, 1967.

[21] N. Wiener and P. Masani, "The prediction theory of multivariate stochastic processes I," *Acta. Math.*, vol. 98, pp. 111–150, 1957.

[22] N. Wiener and P. Masani, "The prediction theory of multivariate stochastic processes II," *Acta Math.*, vol. 99, pp. 93–137, 1958.

[23] J. B. Robertson and M. Rosenberg, "The decomposition of matrix-valued measures", *Michigan Math. J.*, vol. 15, pp. 353–368, 1968.

[24] P. Masani, "The prediction theory of multivariate stochastic processes III," *Acta Math.*, vol. 104, pp. 141–162, 1960.

[25] A. Makagon and H. Salehi, "Structure of periodically distributed stochastic sequences," in *Stochastic Processes, A Festschrift in Honor of Gopinath Kallianpur*, (Eds., S. Cambanis et al.), New York: Springer Verlag, pp. 246–251, 1993.

[26] S. A. Chobanyan and A. Weron, "Banach space valued stationary processes and their linear prediction," *Dissertationes Math.*, vol. 125, pp. 1–45, 1975.

INDEX

A

Acoustic noise analysis for rotating machinery, 9
ACS (*see* Almost cyclostationary)
Adapting a spatial filter
conventional methods for, 182–83
cyclostationarity-exploiting methods for, 183–92
Algorithms
for blind channel identification/equalization, 8, 79–80, 204–12, 417–33, 437–53
for blind adaptive spatial filtering, 71, 182–95
conjugate cross-SCORE, 185, 198, 199
constant modulus, 422
cross-SCORE, 184–86
cyclic least squares, 218–19
least-squares SCORE, 183–84, 186
PCCA, 191–92
phase-SCORE, 194, 215
SCORE, 170, 183–95
wideband-SCORE, 195–96, 202–3
for cyclic spectral analysis, 455–78
for direction finding, 72, 213–19
cyclic least squares, 218–19
cylic MUSIC, 214–15
MUSIC, 213–14
XK cyclic MUSIC 217–18
for estimating number of signals 222–26
CCST, 223
CCT, 224
DBM, 224–25
MDL, 222–23
VCBM, 225
for estimating time-difference of arrival, 70–71, 155–56, 394–97
cyclic phase difference, 156
CYCCOR methods, 394–97
SPECCOA, 71, 156, 395–96
SPECCORR, 71, 396
for multipath channel identification, 394–97
cyclic Prony, 396–97
CYCCOR methods, 394–97
SPECCOA, 395–96
SPECCORR, 396
for nonlinear system identification, 80
for prediction, 77–78
for signal detection/classification, 69, 151–54
for signal extraction, 73–77, 247–64
for spectral correlation analysis, 455–78
by time-smoothing, 456–70
FAM, 457, 461–68
SSCA, 457, 468–70
by frequency smoothing, 470–74
DFSM, 471–73
OBSCA, 473–74
Aliasing effects
avoiding spectral, 431
due to sampling, 135, 244
Almost cyclostationary (ACS) processes, 296, (*see also* Polycyclostationary processes)
Almost periodically correlated (APC) processes, 295, 199–301, 310, 316–20
Almost periodically unitary (APU) processes, 310
AM (*see* Amplitude modulated signals)
Amplitude modulated (AM) signals, 8, 28, 43, 50, 73, 75
Amplitude-shift-keyed (ASK) signals, 8, 28, 62–64, 73, 75, 363
Antenna arrays
blind adaptive spatial filtering with, 182–95
blind channel equalization with, 204–12
estimating number of signals with, 222–26
direction finding with, 213–19
Antenna pattern
of array at base station, 197
of a spatial filter, 179
Aperture vector, 171
APU (*see* Almost periodically unitary processes)
Array manifold, 171
Array response vector, 171, 172
Arrival times (ATs), estimation of, 392
ASK (*see* Amplitude-shift-keyed signals)
ATs (*see* Arrival times)
Autocorrelation function,
conjugate cyclic, 41, 45
cyclic, 34, 39–46
Markovian representation of, 276–78

probability-space, 32
time-space, 33
(*see also* Cyclic autocorrelation)
Autoregressive (AR) model, 483–85

B

Backward prediction equations, 399
Band-limited spectral correlation ratio (BL-SPECCORR) method, 396, 398
Baseband representation of QAM data communication system, 135–40, 420
Binary phase-shift-keyed (BPSK) signals
and CCA beamformer results, 192–93
compared to ASK, QPSK, and SQPSK, 62–64
in Cross-SCORE signal separation, 185
in multiple AM interference and noise, 67–69
and polyperiodic filtering, 263
and signal extraction, 73
and spread spectrum system, 202
used to estimate channel, 447
Blind-adaptive spatial filtering, 71–72, 182–95
application examples, 195–04
summary, 203–4
Blind adaptive channel identification/equalization
CS-based, 79–80, 204–12, 437–53
with antenna arrays, 204–12, 450–52
frequency-domain analysis of, 423–29
HOS-based, 421–24
time-domain, analysis of 441–42
BL-SPECCORR (*see* Band-limited spectral correlation ratio)
BPSK (*see* Binary phase-shift-keyed signals)

C

Canonical configuration for *M*th-order PTV system, 247
Canonical correlation analysis (CCA)
spatial filter derived as solution to, 189
programmable, 191–92
simulation results, 192–93
Canonical correlation significance test (CCST), 223–24
Canonical correlation test (CCT), 224
Canonical innovations, 268, 281, 282–83
Canonical representation of PARMA models, 286–89
Causality between time-series, 28, 77–78
CCA (*see* Canonical correlation analysis)
CCML (*see* Constrained conditional maximum likelihood)
CCST (*see* Canonical correlation significance test)
CCT (*see* Canonical correlation test)
CDMA (*see* Code division multiple-access)
CE (*see* Cycloergodic processes)
Cellular communication system, design of, 169
Cepstral approach to channel identification, 426
CF (*see* Cumulative function)
Channel equalization (*see* Blind adaptive channel identification/equalization)
Channel identification (*see* Blind adaptive channel identification/equalization)
Circulant matrix of channel responses, 342
CL (*see* Conjugate linear transformation)
Classification of signals, 9, 69, 151–54
Climatology/meteorology, use of cyclostationary models in, 7
CLS (*see* Cyclic least squares algorithm)
CMA (*see* Constant modulus algorithm)
Cochannel interference
elimination of, 73–77, 258–64
improved system design in the presence of, 329
Code division multiple-access (CDMA), 194, 196, 329
Computational complexity
of FAM, 467
of SSCA, 470
Conjugate cyclic autocorrelation, 41, 45
Conjugate linear (CL) transformation, 258
Constant-component extractor, 20
Constant modulus algorithm (CMA), 422
Constrained conditional maximum likelihood (CCML), 186–89, 190–91
Control problems in presence of CS disturbances, 268
CP (*see* Cyclic polyspectrum)
Cramer-Rao lower bound (CRLB), 221
Cross-coupled transmit matrix, two-user case, 332
Cross-SCORE
algorithm, 184–86, 195–98
processor, block diagram of, 191
crosstalk suppression, 330
CS (*see* Cyclostationary)
CSD (*see* Cyclic spectral density function)
CTCF (*see* Cyclic temporal cumulant function)
CTMF (*see* Cyclic temporal moment function)
Cumulants, 110–15, 122–24
addition rule for, 113
applications of, 151–59
cyclic, 125
and cyclic polyspectra, 123
derivation in terms of sine-wave generation, 31, 92, 108–14
effects of signal processing operations on, 131–35
formulas for QAM, 135–37
history of, 98–103
and interpretation of sine-wave generation, 96
measurement of, 140–51

Cumulants (*cont.*)
of polycyclostationary signals, structure of, 91
and polyspectra, 98–103
relationship between moments and, 112–13
of a set of random variables, 111–12
signal selectivity property of, 115
simple, 112
of a single random variable, 110–11
special cases of 125–27
spectral, 363
statistics, 99
superiority of over moments, 100
of a time-series, 114
utility of theory of, 125
Cumulative function (CF), 99, 102
Cumulative moment functions, 99
CYCCOR (*see* Cyclic cross-correlation method)
Cycle frequency, 43, 92, 300, 456
Cycle leakage, 316, 459, 462
Cycle spectrum, 43, 260
Cyclic autocorrelation
for an AM signal, 43–46
estimation of, 229–31, 312–14, 316–19
function, 34, 39–46, 126, 260, 424
for a PAM signal, 61–62
Cyclic Correlation Significance Test, 190
Cyclic cross-correlation (CYCCOR) method, 394–95
Cyclic cumulants, 125
Cyclic DF algorithms, 214–22
performance, 219–22
Cyclic MUSIC algorithms, 214–16
generalizations of, 215–16
Cyclic least squares (CLS) algorithm, 218–19
Cyclic periodogram
frequency-smoothed, 471–74
time-smoothed, 456–70
Cyclic polyspectrum (CP)
definition of, 123
effects of signal processing operations on, 131–35
and HOS, 127
measurement of, 141–42, 144–51
of digital QAM signals, 136–37
and relation to RT-CTCF, 123, 129
signal selectivity of, 125
Cyclic Prony method, 396–97, 399–400
Cyclic spectral density (CSD) function, 49
(*see also* Spectral correlation density)
Cyclic spectrum (*see* Cyclic spectral density function)
Cyclic temporal cumulant function (CTCF)
applications of, 151–56
definition of, 114
of digital QAM signals, 136–37
effects of signal processing operations on, 131–35
reduced-dimension version of, 116
and its relation to CTMF, 114, 129
signal selectivity of, 115
Cyclic temporal moment function (CTMF)
definition of, 105
and its relation to SMF, 121, 129
effect of signal processing operations on, 131–35
Cyclic Wiener filter, 77, 179, 250–64, 331
Cyclic Wiener relation, 49, 119, 127, 129
Cycloergodic (CE) processes
terminology, 7
definition, 24
Cycloergodicity, 13
and HOS, 128
and measurements of cumulants, 147–51
and pitfalls of stochastic processes, 26–29
property, 317–18
and refined stochastic-process definitions, 24–25
Cyclospectral factorization (*see* Periodic spectral factorization theorem)
Cyclostationarity
background, 2–10
detection of, 321
to enhance reliability of information, 3
exploitation of, 65–80, 125, 151–56, 168–227, 250–65, 327–59, 362–88, 391–413, 424–31, 435–52
fundamental concepts, philosophy, and definitions, 10–32
hidden, 24–25
higher-order, introduction to, 91–104
histogram of papers on, 5
mathematical motives for studying, 9–10
measurement of, 455–78
practical motivations for studying, 7–9
*n*th-order, 94
of order 2 (in the wide sense), 34
research interest in, 4–7
second-order, 32–65
seminal contribution to study of, 6–7
and sensor array processing problem, 174–78
signal selectivity associated with, 67, 115
usefulness of, 3
viability of exploitation of, 4
what is, 2
why publish a book on, 4
Cyclostationary autocovariance function, Markovian
representation of, 276–78
Cyclostationary models
advantages of, 7
of communication signals (*see* AM, ASK, BPSK, QPSK, SQPSK)

periodic autoregressive (PAR), 7, 284, 289
periodic autoregressive-moving average (PARMA), 7, 268, 283, 284–91
periodic moving average (PMA), 7
Cyclostationary noise sources, 9
Cyclostationary (CS) processes
definiton of, 21, 25, 94
PARMA representation of, 284–91
periodic innovations representation of, 281–82
prediction theory for, 480–91
representation of and estimation for, 295–321
representation, prediction, and identification of, 267–92
state-space representation of, 274–84
stationary representations of, 299, 303–10, 482–91
as subclass of nonstationary processes, 9
synonymous with periodically correlated, 296
(*see also* Cyclostationary signals; Cyclostationary time-series)
Cyclostationary signals
definition of, 2, 23
introduction to, 1–80
two approaches to, 1–2, 10–32
of second order, 32–80
(*see also* Cyclostationary time-series)
Cyclostationary time-series
as conceptual aid for solving practical problems, 32
definition, 23, 94
nonstochastic theory of, 7
(*see also* Cyclostationary signals)

D

DBM (*see* Determinant-based method)
Detection of signals, 9, 69, 151–54
general search problem, 152–53
known-cycle frequency problem, 153–54
known-modulation-problem, 154
using the SCD, 69
Determinant-based method (DBM), 224–25
DF (*see* Direction finding)
Digital cellular radio, 196–98
Digital frequency smoothing method (DFSM), 456, 471–73
Digital implementation
of FAM, 466
of SSCA, 469
Digital quadrature-amplitude modulation, 135–40
Digital spectral correlation
analysis, review of, 455–78
analyzers, design of, 474–77
Direction finding (DF)
conventional algorithms for 213–14
cyclostationarity-exploiting algorithms for, 214–19
exploiting signal selectivity in, 72–73
Direction vector, 171
Direct-sequence spread spectrum signal, 202, 254–58
Double-sideband (DSB) AM signals, 36–38, 43–45, 50–51, 67, 73, 364, 370
Double-sideband suppressed carrier (DSBSC) signals (*see* Double-sideband AM signals)
DSB (*see* Double sideband AM signals)
DSBSC (*see* Double-sideband suppressed-carrier signals)
Duality between probability-space theory and time-space theory, 20, 32
Dual theoretical frameworks, 17–20

E

Economics, use of cyclostationary models in, 8
Equalization (*see* Blind adaptive channel identification/equalization; Fractionally-spaced equalizer)
Equalizers, fractionally spaced, 26–29, 97
Ergodicity (E)
absence of, 26–29, 97
relevant types of, 23–24
(*see also* Cycloergodicity)
Ergodic theorem, need for fundamental, 29
Estimation
of channels, 8, 79–80, 204–17, 319–413, 417–31, 437–53
of CTMF and CTCF, 140–44
of cyclic polyspectrum, 140–42, 144–50
of signal parameters, 8, 69–71, 72–73, 362–86
of signals, 8, 73–77, 329–59
of spectral correlation densities, 64, 314–16, 319–20, 453–78
of TDOA, 70–71, 155–56
of spectral correlation function, 455
Excess bandwidth, optimization of, 340–42

F

Fast Fourier transform accumulation method (FAM), 456, 457, 461–68
Filter
design, theory of, 241
jointly optimum transmit and receive, 329–59
for optimum reception of baseband PAM, 253
Filtering
blind-adaptive spatial, 71–72, 178–204
effects of on spectral correlation density (SCD) function, 51–53

Filtering (*cont.*)
frequency-shift (FRESH), 73–77, 263–64, 331
input-output CTMF, CTCF, and CP relations for, 133–34
input-output SCD relation for, 51–52
periodically, 58–64, 240–58
polyperiodic, 240–64
structures for spatial, 178–81
FOT (*see* Fraction-of-time probability)
Fourier series transform (FST), 458
Fractionally-spaced equalizer (FSE)
blindly adapting a, 204–11
matched filter/tapped-delay-line filter corresponding to, 253–54
as special case of LCL-PTV filter, 180
Fraction-of-time (FOT) probability, 19–20, 22–23, 93–95
concept, 15
operational origin of, 13
and development of HOCS, 96, 103
Frequency-shift (FRESH) filtering (*see* Filtering)
Frequency smoothed cyclic periodograms, 398, 470–74
FSE (*see* Fractionally spaced-equalizer)
FST (*see* Fourier series transform)
Fundamental concepts of cyclostationarity, 10–32
Fundamental theorem
of expectation, 18
of polyperiodic component extraction, 23
of time-averaging, 19–21

G

GCC (*see* Generalized cross-correlation methods)
General search problem, 151–53
Generalization of the second-order parameters, 127
Generalized cross-correlation (GCC) methods, 97, 155, 392

H

Half invariants, cumulants defined as, 98
Harmonic series representation, 247–51, 308, 310
Harmonizable processes, 301–5
Hidden cyclostationarity, 24–25, 27
Hidden polycyclostationarity, 25, 27
Hidden statistical dependence, 26
Higher-order cyclostationarity (HOCS)
applications, 151–57
and cyclic spectrum, 125–27
development of the theory, 104–24, 128–35
introduction to, 91–104
and higher-order stationarity, 96–97, 127–28
importance of theory of, 125, 150
measurement of the parameters of, 140–51
motivations for developing theory of, 97–98
and power spectrum, 125–27
relationship between parameters of, 125, 129
signal processing operations in, 131–35
spectral parameters of 119–24
study of, defined, 92
temporal parameters of, 104–19
utility of the theory, 125
Higher-order statistics (HOS)
explicit methods for channel identification and equalization, 422
implicit methods for blind adaptive equalization algorithms, 421
-based methods of channel identification/ equalization, limitations of, 423–24
limitations of theory of, 96–97, 128
for nonlinear processing of cyclostationary signals, 91–159
and relation to HOCS, 96–97, 127–28
relevant work in the areas of, 98–103
Higher-order stationarity, 127–28
Historical perspective on stochastic processes in communications, 16–17
HOCS (*see* Higher-order cyclostationarity)
HOS (*see* Higher-order statistics)
HSR (*see* Harmonic series representation)
Hybrid (time and frequency) smoothing algorithms, 456
Hydrology, use of cyclostationary models in, 7

I

Identification
of PARMA models, 289–90
of multipath channels, 391–413
(*see also* Blind adaptive channel identification/ equalization)
IL (*see* Innovationslike representations)
Impure cycle frequency of order *n*, 105
Impure *n*th-order sine waves, 106–8
Impure third-order cycle frequencies, 119
Inference and decision, statistical, 9, 15–16, 27
Innovationslike (IL) representations, 268, 281–83
In-phase components
statistics, for squared PAM, 379–81
variances and covariances of, for PAM, 367–73
Interference-tolerance (*see* Signals)
Intersymbol interference (ISI)
removal of, 8
(*see also* Blind adaptive channel identification/equalization)
ISI (*see* Intersymbol interference)

J

Joint optimization of MIMO systems, 331
coordinated users, 332–42
uncoordinated users, 342–53
numerical examples, 354–59
Joint transmitter/receiver optimization for multiuser communications, 329–59

K

Kalman filter/predictor, 267, 278–80, 287, 291
K-statistics, 99

L

Laws of presumptive errors, 98
Leakage problem, cycle, 145, 146, 316
Least-squares SCORE (LS-SCORE), 183–84, 186
LCL-PTV (*see* Linear-conjugate-linear polyperiodic time varying)
Lifted representation, 273
Linear-conjugate-linear (LCL) transformation, 259
Linear-conjugate-linear polyperiodic time varying (LCL-PTV), 175, 179–81, 187, 189, 190
Linear periodic systems, basics of, 242–47, 270–74
Linear prediction (*see* Prediction)
Linear quadratic Gaussian (LQG) methods, 268
Linearly constrained minimum variance beamforming, 182
LQG (*see* Linear quadratic Gaussian methods)
LS-SCORE (*see* Least-squares SCORE)

M

Markovian model, 267
Markovian representation of a CS autocovariance function, 276–78
Matched filter (MF), 253–54
Maximum likelihood timing-phase estimator, 363
MDL (*see* Minimum description length principle)
Measurement of the parameters of HOCs, 140–51
Measurement of the parameters of SOCS, 311–20, 455–78
Medicine/biology, use of cyclostationary models in, 8
MF (*see* Matched filter)
MIMO (*see* Multiinput, multioutput models)
Minimum description length (MDL) principle, 222–27
Minimum variance distortionless response (MVDR) beamforming, 182
Mitigation of multipath fading effects, 8
Mixing conditions, 32
Mobile communication system, 169, 332
Moments
of cyclic correlation estimates, 229–31
operational origin of temporal, 13
properties of temporal, 115–19
relationship between cumulants and, 112–13
spectral, 120–22
temporal, 104–8
Monodromy matrix and stability, 270
Multiinput, multioutput (MIMO) models, 330
optimization of transmitter and receiver for fully coordinated, 332–42
optimization of transmitter and receiver for uncoordinated, 342–53
numerical examples of optimized transmitters and receivers, 354–59
Multipath-channel identification 391–415
cyclostationary model for, 393–94
Multiple Signal Classification (MUSIC) algorithm, 213–14
Multiple-sine-wave-component extractor, 22
Multiplication, signal, 54–57, 131–33
Multiple incommensurate periodicities, 58
Multiuser communication, 329–59
Multivariate moment and cumulant relations, 112–13
MUSIC (*see* Multiple Signal Classification)
MVDR (*see* Minimum variance distortionless response beamforming)

N

Nonlinear system identification, 79–80
Nonlinear system models, 92–93
Nonlinear transformations
characterization of, 11
and essence of cyclostationarity, 30
sine-wave components present in, 104, 125
Nonstochastic operational model, stochastic vs., 13–15
Nonstochastic statistical inference and decision, 15
Nonstochastic time-series approach, limited exposure of, 14

O

OBSCA (*see* One bit spectral correlation algorithm)
Oceanology, use of cyclostationary models in, 8

One bit spectral correlation algorithm (OBSCA), 473
Operational definition of polycyclostationarity, 11
Optimal transmitter and receiver (*see* Multiinput, multioutput models)
Orthogonality condition in polyperiodic linear filtering, 250, 156, 160, 264

P

PAM (*see* Pulse-amplitude modulated signals)
PAR (*see* Periodic autoregressive cyclostationarity models)
Parameter estimation (*see* Estimation)
Parameters
  frequency-domain of HOCS, 119–24, 144–46
  time-domain of HOCS, 104–18, 143–44
PARMA (*see* Periodic autoregressive-moving average cyclostationarity models)
PC (*see* Periodically correlated processes)
PCCA (*see* Programmable canonical correlation analyzer)
PCE (*see* Polycycloergodic process definition)
PCS (*see* Polycyclostationary)
Pd (*see* Probability density)
PD (*see* Probability distribution)
Periodic and polyperiodic linear systems, 240–65
Periodic autoregressive (PAR) cyclostationary models, 7, 284, 289, 483
Periodic autoregressive moving average (PARMA) cyclostationary models, 7, 268, 283, 284–91
Periodically correlated (PC) processes, 295–96 (*see also* Cyclostationary processes)
Periodically time-variant filtering, 8, 58–64, 247–58
Periodically-time-varying (PTV) linear systems, 240–65
Periodicity
  first-order, 35
  hidden, 35
  *n*th-order, 106
  second-order, 40
Periodic moving average (PMA) cyclostationary models, 7, 483–85
Periodic prediction of cyclostationary processes, 8, 77–78
Periodic pulse train, random-amplitude modulated, 34 (*see also* Pulse-amplitude modulated signals)
Periodic Riccati equation, 279–83
Periodic spectral factorization theorem, 287, 289
Periodic time-sampling, 55–58, 134–35
Philosophy of cyclostationarity, 10–32
Phase randomization, 25–29, 128
Phase-SCORE algorithm, 194, 215–16
Phase-shift-keyed (PSK) signals, 8, 62–64, 135–40
PMA (*see* Periodic moving average cyclostationarity models)
Polycycloergodicity, 3, 6, 10
  nonexistent theory of, 27
  see also cycloergodicity)
Polycycloergodic (PCE) process definition, 24, 29
Polycycloergodic theorem, need for fundamental, 29
Polycyclostationarity, operational definition of, 11 (*see also* Cyclostationarity)
Polycyclostationary (PCS) signals, 3, 11
  applicability of mathematics of, 14
  of second order, 32–80
  (*see also* Cyclostationary signals; Polycyclostationary time-series)
Polycyclostationary processes definition, 21 (*see also* Cyclostationary processes)
Polycyclostationary time-series, definition, 23 (*see also* Cyclostationary time-series; Polycyclostationary signals)
Polyperiodic component
  extraction, fundamental theorem of, 23
  extractor, equivalence with probabilistic expectation, 13, 22–23
  linear operation for extraction of, 12
  potential generation of, 11
Polyperiodic filtering (*see* Filtering)
Polyperiodic FOT Pd, 22
Polyperiodic FOT PD, 22
Polyperiodic nonlinear systems, identification of, 80
Polyperiodic time-variation
  mathematical characterization of, 13
  physical evidence of, 11, 13
Polyspectrum, 100, 127–28
  measurement of, 141
Power spectral density (PSD), 46
Prediction
  and causality, 28, 77–78
  of discrete-time cyclostationary sequences, 27, 77–78, 486–88
  error (PE) representation of CS processes, 268, 280
  for PARMA models, 286–89
  theory for cyclostationary processes, recent developments in 480–91
Probabilistic analysis using time-space theory, 31
Probability density function, 18, 98
Probability distribution function, 17
Probabilistic models
  for nonstochastic statistical inference and decision, 15–16

operational origin of, 11
stochastic vs. nonstochastic, 13–15, 321
Probability-space
approach, abstractions introduced by, 30
definitions of S, CS, PCS, 21, 25
theory, duality with time-space theory, 20
Programmable canonical correlation analyzer (PCCA), 191–92
Prony algorithm, 392, 397
PSD (*see* Power spectral density)
PSK (*see* Phase-shift-keyed signals)
PTV (*see* Periodically-time-varying linear filters)
Pulse-amplitude modulated signals (PAM)
analysis of, 362–88
cyclic autocorrelation of, 61–62
cyclic polyspectrum of, 135–40
cyclic temporal cumulant, 135–40
deterministic relationship between pairs of spectral components of, 362
measured cumulants of, 147–51
power spectral density of, 38–40
separated using periodic filters, 8, 28, 73–77
sine-wave generation from, 38
spectral correlation density of, 59–62
temporal and spectral cumulants of, 135–40
Pulse-shaping at the transmitter, 331
Pure and impure components of an *n*th-order sine wave, identifying, 107
Pure *n*th-order sine waves, 103, 108–10
Pure second-order sine waves, 108
Pure sine waves and temporal cumulants, 114
Pure stationarity/cyclostationarity, 23

Q

QAM (*see* Quadrature-amplitude modulation signals, digital)
QPSK (*see* Quaternary phase-shift keyed signals)
Quadrature-amplitude modulation (QAM) signals, digital
cyclic polyspectra of, 135–40
cyclic temporal cumulants of, 135–40
equalization for, 204–12, 415–33, 437–53
optimization of transmitters and receivers for, 329–59
separated by frequency-shift filters, 8, 28, 73–77
Quadrature components
statistics of, for squared PAM, 379–81
variances and covariances of, for PAM, 367–73
Quadrature self-noise, vanishing of, 370, 381–82, 387
Quaternary phase-shift keyed (QPSK) signals
and blind equalization with antenna arrays, 451
complex envelope of, 61
and frequency shift filtering, 73–77
measuring spectral correlation of, 64–65

R

Radio-signal analysis, 9
Random processes (*see* Cyclostationary processes; Polycyclostationary processes)
Rational CS process, 277
RD (*see* Reduced dimension functions)
Realization problem, 277
Reduced-dimension (RD) functions, 116–17, 122–24, 129
Representation
and estimation for periodically and almost periodically correlated random processes, 295–321
prediction, and identification of cyclostationary processes, 267–92
Riccati equation, periodic, 268, 278–81, 283

S

S (*see* Stationary time series definition)
Sampling effects on spectral moments and cumulants, 55–58, 134–35
SAs (*see* Scaling amplitudes)
Scaling Amplitudes, 392
SCD (*see* Spectral correlation density)
SCORE (*see* Spectral-coherence-restoral algorithms)
SCORE-STFDMA (*see* Spectral-coherence-restoral-space-time-frequency-division multiple access)
SDMA (*see* Space division multiple access)
Second-order cyclic spectra, establishing channel identifiability based on, 418
Second-order polycyclostationarity (SOCS), 92
Semi-invariants, 100
Sensor array processing
conclusions about, 227
for cyclostationary signals, 168–230
implications of cyclostationarity for, 177–78
Signals
of interest (SOI), 174
not of interest (SNOI), 174–221
introduction to cyclostationary, 1–80
selectivity, 66–67, 115–16, 125, 169
stochastic vs. nonstochastic operational models, 13–15
types of interest to study of cyclostationarity, 11
(*see also* Cyclostationary signals; Polycyclostationary signals)
Sine-wave
component extractor, 22
generation, 2, 11–13, 34–39
impure *n*th-order, 106–8
pure *n*th-order, 108–10

Single-input, single-output (SISO) system, 329
SISO (*see* Single-input, single-output)
SMF (*see* Spectral moment function)
SNOI (*see* Signal not of interest)
SOCS (*see* Second-order cyclostationary)
SOI (*see* Signal of interest)
Space division multiple access (SDMA), 196
Space-time-frequency division multiple access (STFDMA), 197–202
Spatial and fractionally spaced temporal equalization, 204
Spatial filtering
  conventional methods for adapting, 182–83
  cyclostationarity-exploiting methods for adapting, 183–92
  (*see also* Filtering)
Spatial resolution, superior, 72
Spatio-temporal equalization, 204–11
Spatio-temporal filter (STF), 179–81
Spectral analyzer for measuring PSD, 46
Spectral-coherence-alignment (SPECCOA) 71, 156, 395–96
Spectral-coherence-restoral (SCORE) algorithms, 72, 170, 183–95
Spectral-coherence-restoral-space-time-frequency division multiple access (SCORE-STFDMA),199–202
Spectral correlation
  analysis, 46–48, 64–65, 314–16, 319–20, 455–78
  function (*see* Spectral correlation density)
Spectral correlation density (SCD)
  conjugate, 47
  effects of signal processing operations on, 51–62
    for an AM signal, 50–51
    for digital signals, 62–64
    for a PAM signal, 59–60
  function, 34, 46–61
  and HOCS, 125–27
Spectral cumulant function (SCF), 122–24
Spectral estimation (*see* Spectral correlation analysis)
Spectral factorization, periodic 287–89
Spectral leakage, 146 (*see also* Cycle leakage)
Spectral line generation (*see* Sine-wave generation)
Spectral measure, time dependent, 309
Spectral moment function (SMF), 120–22
Spectral parameters of HOCS, 119–24
Spectral redundancy, 66–69
  exploitation of, 2, 8, 66–80
  measure of the degree of, 49
Spectral theory
  for the covariance, 297–301
  for harmonizable case, 301–3
  of PC and APC processes, 295
Spectrum analyzer, for measuring PSD, 46
Spread-spectrum signal, optimum demodulation filter for, 257
Spread spectrum and spatial filtering, 202–3
SSPS (*see* Symmetric periodic positive semidefinite)
SQPSK (*see* Staggered quaternary phase-shift-keying)
SSCA (*see* Strip spectral correlation algorithm)
Staggered quaternary phase-shift keying signals (SQPSK), 63–64
State-space approach to cyclostationary processes, 267–92
State-space representations
  equivalence of, 272
  properties of, 269, 274–84
Stationary linear time-varying systems, similarity to polyperiodic systems, 243
Stationary (S) time-series definition, 23
Statistical analysis of time-series, other theories for, 95–97
Statistical inference and decision
  in communications engineering, 16
  and HOCS, 157
  nonstochastic, 15–16
  problems, 9
STF (*see* Spatio-temporal filter)
STFDMA (*see* Space-time-frequency division multiple access)
Stochastic periodic systems, 274
  evolution of moments of, 275–76
Stochastic-process
  approach, advantages of, 29–30
  definitions, 21–22, 24–25
  framework, pitfalls of, 26–29
Stochastic processes
  texts in, 102
  Venn diagram of classes of, 21, 25
  (*see also* Cyclostationary processes; Polycyclostationary processes)
Stochastic vs. nonstochastic operational models, 13–15
Strip spectral correlation algorithm (SSCA) 457, 468–70
Submanifolds for CP, 142
Symmetric periodic positive semidefinite (SSPS) matrices, 277
Synchronization, 362–88

T

Tapped-delay-line (TDL) filter, 253
TCF (*see* Temporal cumulant function)
TDL (*see* Tapped-delay-line filter)
TDMA (*see* Time division multiple access)

TDOA (*see* Time-difference-of-arrival estimation)
Telemetry signals, 8
Temporal correlation coefficient, 42
Temporal crosscovariance, 42
Temporal cumulant function (TCF), 103–14
Temporal-expectation operation, 19
Temporal moment function (TMF), 13, 105–6, 129, 133
Temporal parameters of HOCS, 104–19
Temporal probability
  approach, 31
  distribution functions, 13
Terminology, 7, 43
Theoretical frameworks, dual, 17–20
Time division multiple access (TDMA), 196
Time-dependent spectral representation, 310
Time-difference-of-arrival (TDOA) estimation, 70–71, 155–56, 391–413
Time-invariant representation (TIR), 272
Time-invariant system identification formula, 397
Time sampling; (*see* Sampling effects on spectral moments and cumulants)
Time-series
  approach to cyclostationarity, 1
  models, construction of, 30
  vs. stochastic processes, 13–15, 95
  (*see also* Cyclostationary time-series; Polycyclostationary time-series)
Time-smoothed cyclic (cross) periodogram, 456–58
Time-smoothing algorithms for spectral correlation analysis FFT-based, 456–70
Time-space
  approach, need to adopt, 30
  theory, 20, 31, 33
TIR (*see* Time-invariant representation)
TMF (*see* Temporal moment function)
Transmitter and receiver optimization (*see* Multiinput, multioutput models)
Translation series representation (TSR), 248, 309, 310
TSR (*see* Translation series representation)

U

ULA (*see* Uniform linear array)
Uniform linear array (ULA), 172
Uniformly APC (UAPC) processes, 300
Uniprocessor implementation
  of FAM, 466
  of SSCA, 469
Unitary operators
  family of, 306
  of PC processes, 305–8
  representations based on, 308
Unitary transformation, 337

V

VCBM (*see* Variable coefficient based method)
Variable coefficient based method (VCBM), 225
Volterra series
  model identification, 80, 93
  in origin of temporal probability model, 11–13, 104

W

Weakly harmonizable processes, 302
Weakly stationary processes, 296
Weakly stationary vector sequences, 307
Weak-signal detection, 69, 97, 98, 151–54
Whitening filter, 282
Wideband cyclic MUSIC, 215
Wideband SCORE algorithm, 195–96, 202–3
Wide-sense cyclostationary (WSCS) processes, 32–65
Wiener filter
  cyclic, 77, 179, 150–64, 331
  multiinput-multioutput, 251
Wiener relation, 32, 49, 119
  cyclic, 49, 119, 127, 129
Wiener's generalized harmonic analysis, 296
Wold decompositions, 485, 489–91
WSCS (*see* Wide-sense cyclostationary processes)

Editor's Biography

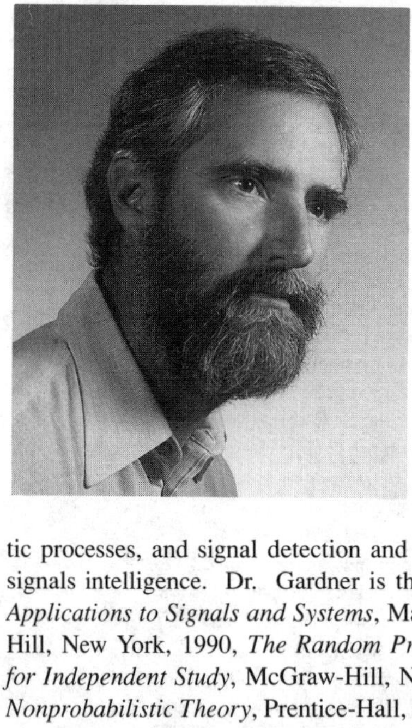

William A. Gardner (S'64–M'67–SM'84–F'91) was born in Palo Alto, CA, on November 4, 1942. He received the M.S. degree from Stanford University, in 1967, and the Ph.D. degree from the University of Massachusetts, Amherst, in 1972, both in electrical engineering.

He was a Member of the Technical Staff at Bell Laboratories in MA, from 1967 to 1969. He has been a faculty member at the University of California Davis, since 1972, where he is Professor of Electrical Engineering and Computer Science. Since 1982, he has also been President of the engineering consulting firm Statistical Signal Processing, Inc., Yountville, CA. His research interests are in the general area of statistical signal processing, with primary emphasis on the theories of time-series analysis, stochastic processes, and signal detection and estimation and applications to communications and signals intelligence. Dr. Gardner is the author of *Introduction to Random Processes with Applications to Signals and Systems*, Macmillan, New York, 1985, second edition, McGraw-Hill, New York, 1990, *The Random Processes Tutor: A Comprehensive Solutions Manual for Independent Study*, McGraw-Hill, New York, 1990, and *Statistical Spectral Analysis: A Nonprobabilistic Theory*, Prentice-Hall, Englewood Cliffs, NJ, 1987. He holds several patents and is the author of numerous research-journal papers. He received the Best Paper of the Year Award from the European Association for Signal Processing in 1986 for the paper entitled "The spectral correlation theory of cyclostationary signals," the 1987 Distinguished Engineering Alumnus Award from the University of Massachusetts, and the Stephen O. Rice Prize Paper Award in the Field of Communication Theory from the IEEE Communications Society in 1988 for the paper entitled "Signal interception: A unifying theoretical framework for feature detection." He organized and chaired the NSF/ONR/ARO/AFOSR-sponsored workshop on Cyclostationary Signals, 1992. He is a member of the American Association for the Advancement of Science, the European Association for Signal Processing, and a member of the honor societies Sigma Xi, Tau Beta Pi, Eta Kappa Nu, and Alpha Gamma Sigma.